Fundamentals of
Matrix Computations

Fundamentals of Matrix Computations

Third Edition

DAVID S. WATKINS

Washington State University
Department of Mathematics
Pullman, WA

A JOHN WILEY & SONS, INC., PUBLICATION

Library of Congress Cataloging-in-Publication Data:

Watkins, David S.
   Fundamentals of matrix computations / David S. Watkins. — 3rd ed.
      p. cm.
   Includes bibliographical references and index.
   ISBN 978-0-470-52833-4 (cloth)
   1. Matrices. I. Title.
   QA188.W38 2010
   512.9'434—dc22   2010001933

10 9 8 7 6 5 4

CONTENTS

Preface

This book was written for advanced undergraduates, graduate students, and mature scientists in mathematics, computer science, engineering, and all disciplines in which numerical methods are used. At the heart of most scientific computer codes lie matrix computations, so it is important to understand how to perform such computations efficiently and accurately. This book meets that need by providing a detailed introduction to the fundamental ideas of numerical linear algebra.

The prerequisites are a first course in linear algebra and some experience with computer programming. For understanding of some of the examples, especially in the second half of the book, the student will find it helpful to have had a first course in differential equations.

The book has found use mostly as a graduate text, but it is also suitable for advanced undergraduates. It was written to be readable, and much of it is quite elementary. It is meant to be easier to read than the other excellent text and reference books that cover this material, for example the books by Trefethen and Bau [91], Demmel [15], Golub and Van Loan [43] (of which a new edition is in the works), and Stewart [85, 87]. I have been gratified by the strong positive student response to the first two editions, and I am now pleased to offer this third edition.

The book contains many exercises, ranging from easy to moderately difficult. Some are interspersed with the textual material and others are collected at the end of each section. Those that are interspersed with the text are meant to be worked immediately by the reader. This is my way of getting students actively involved in the learning process. In order to get something out, you have to put something in. Many of the exercises at the ends of sections are lengthy and may appear intimidating at first. However, the persistent student will find that s/he can make it through them with the help of the ample hints and advice that are given. I encourage every student to work as many of the exercises as possible.

Use of MATLAB

By now MATLAB[1] is firmly established as the most widely used vehicle for teaching matrix computations. MATLAB is an easy to use, very high-level language that allows the student to perform much more elaborate computational experiments than before. MATLAB is also widely used in industry. I have therefore included many examples and exercises that make use of MATLAB. This book is not, however, an introduction to MATLAB, nor is it a MATLAB manual. For those purposes there are other books available, for example, the *MATLAB Guide* by Higham and Higham [51]. However, MATLAB's extensive help facilities are good enough that the reader may feel no need for a supplementary text. In an effort to make it easier for the student to use MATLAB with this book, I have included an index of MATLAB terms, separate from the ordinary index.

MATLAB is proprietary software. The instructor or student who would prefer to use free open-source software can find what s/he needs in Sage[2], for example. There are also several packages with functionality similar to that of MATLAB that can be downloaded for free from the internet.

I used to make my students write and debug their own Fortran programs. I have left the Fortran exercises from the first edition largely intact. I hope a few students will choose to work through some of these worthwhile projects.

Applications

In order to help the student better understand the importance of the subject matter of this book, I have included many examples and exercises on applications (solved using MATLAB), mostly at the beginnings of chapters. I have chosen very simple applications: electrical circuits, mass-spring systems, simple partial differential equations. In my opinion the simplest examples are the ones from which we can learn the most.

Numbering Scheme

Nearly all numbered items in this book, including theorems, lemmas, numbered equations, examples, and exercises, share a single numbering scheme. For example, the first numbered item in Section 1.3 is Theorem 1.3.1. The next two numbered

---

[1]MATLAB is a registered trademark of The MathWorks, Inc. (http://www.mathworks.com/)
[2]http://www.sagemath.org/

items are displayed equations, which are numbered (1.3.2) and (1.3.3), respectively. These are followed by the first exercise of the section, which bears the number 1.3.4. Thus each item has a unique number: the only item in the book that has the number 1.3.4 is Exercise 1.3.4. Although this scheme is unusual, I believe that most readers will find it perfectly natural, once they have gotten used to it. Its big advantage is that it makes things easy to find: The reader who has located Exercises 1.4.15 and 1.4.25 but is looking for Example 1.4.20, knows for sure that this example lies somewhere between the two exercises.

There are a couple of exceptions to the scheme. For technical reasons related to the type setting, tables and figures (the so-called *floating bodies*) are numbered separately by chapter. For example, the third figure of Chapter 1 is Figure 1.3.

New Features of the Third Edition

The first four chapters have been changed very little, and the same can be said for the first five sections of Chapter 5. The major changes are in the treatment of the eigenvalue problem, starting in the middle of Chapter 5.

Francis's Algorithm

The most important method for computing eigenvalues/vectors is Francis's algorithm, which is more commonly known as the implicitly-shifted QR algorithm. The traditional (and universally followed) approach to presenting this algorithm is to begin with a basic QR iteration that involves repeated QR decompositions. Then refinements such as origin shifts and upper Hessenberg form are introduced and shown to be helpful. Finally, the implicitly-shifted QR algorithm is introduced. This is a completely different procedure that involves repeatedly disturbing the Hessenberg form and then restoring it by a bulge-chasing process. This new algorithm is then shown to be equivalent to the shifted QR algorithm by the so called implicit-Q theorem.

This approach is unnecessarily cumbersome. A simpler approach is to introduce Francis's algorithm right away and explain its properties without reference to the basic QR algorithm. Once I made the decision to do this, I found that I was able to eliminate an entire section from Chapter 5. For the instructor who wants to expose his/her students to this important material, the path is now much shorter. No time is spent playing around with QR decompositions. As soon as the reduction to Hessenberg form has been explained, Francis's algorithm (which is the method that is actually used in practice) can be introduced. The students can write a quick-and-dirty MATLAB code to execute Francis's algorithm, and they can observe its excellent performance for themselves.

Along with this change in approach, I am advocating a change of name. I am emphasizing the name *Francis's algorithm* and de-emphasizing the name *implicitly shifted QR algorithm*. There are two reasons for this. First of all, the execution of Francis's bulge-chasing algorithm does not involve any QR decompositions, so why call it the QR algorithm? Secondly, beginning students often confuse the QR

decomposition with the QR algorithm, not realizing that they are different things. If we start referring to the latter as *Francis's algorithm*, the confusion will go away.

Two eminent friends, Nick Trefethen and Michael Overton, have told me not to do this. Trefethen says, "Why change the name of our best product?" Overton says that name-change attempts generally turn out badly. Well, then, I will not tell you to *stop* calling it the QR algorithm. But I will suggest that you *start* calling it Francis's algorithm. The influential tome of Golub and Van Loan [43, p. 358] refers to a "Francis QR step".

Explanation of Francis's Algorithm

The explanation of why Francis's algorithm works is largely postponed to Chapter 6. It was also the case in the previous editions that some explanations were postponed for one chapter, but now the explanations have been changed. The notion of *Krylov subspace* is introduced earlier (§ 6.3) than it was in the second edition. Each Francis iteration $\hat{A} = Q^*AQ$ is a change of coordinate system. There are Krylov subspaces lurking in the columns of the transforming matrix Q, and the properties of these Krylov subspaces can be used to demonstrate directly that each Francis iteration is a nested subspace iteration (simultaneous iteration) with a change of coordinate system. There is no need to make reference to the basic QR algorithm, nor is the implicit-Q theorem needed.

There is, of course, a connection between Francis's algorithm and the QR decomposition. I have chosen to de-emphasize that connection, but ultimately it is worth knowing. We make the connection at the end of Section 6.3.

Other Changes

- The discussion of roundoff errors in the Gram-Schmidt process (Section 3.4) has been streamlined. I am now advocating the use of classical Gram-Schmidt with reorthogonalization, which has been shown in both theory and practice to be an effective means of obtaining vectors that are orthogonal to working precision. The material on modified Gram-Schmidt is still there, but it has been relegated to exercises.

- The explanation of the Golub-Reinsch (QR-SVD) algorithm for computing the SVD (§ 5.8) has been changed. Previously I used the implicit-Q theorem. Now I am embedding the bidiagonal matrix B in a larger matrix, performing a perfect shuffle, and deriving the Golub-Reinsch algorithm as a special case of Francis's algorithm.

- In Section 6.1 I have added a discussion of the computation of invariant subspaces by eigenvalue swapping in the Schur form.

- In Section 6.5 on the implicitly-restarted Arnoldi process, I have added a description of the Krylov-Schur method of restarting.

- The Jacobi-Davidson method has been given a section of its own (§ 6.6), and the description of the method has been expanded.

- There is a new section (§ 7.3) on product eigenvalue problems.

- In Section 7.4 the justification of the Moler-Stewart (QZ) algorithm for solving the generalized eigenvalue problem has been changed. Previously the implicit-Q theorem was used. Now I am justifying it as an instance of Francis's algorithm, using the newly added material on product eigenvalue problems (§ 7.3).

- The material on eigenvalue problems has now been split over three chapters. This is more a matter of reorganization than expansion. In the second edition there were two chapters totaling 16 sections; now there are three chapters totaling 18 sections.

- Chapter 8 contains a new section (§ 8.5) on stopping criteria for iterative methods for solving linear systems. A rational stopping criterion based on backward error is recommended.

In addition I have made numerous other small changes and corrections. I sincerely hope that I have not introduced too many new errors. I wanted to write a new edition that was not too much longer that the previous one, and I am pleased that I have succeeded in doing so.

DAVID S. WATKINS

Pullman, Washington
June, 2010

Acknowledgments

I am greatly indebted to the authors of some of the early works in this field. These include A. S. Householder [54], J. H. Wilkinson [103], G. E. Forsythe and C. B. Moler [29], G. W. Stewart [84], C. L. Lawson and R. J. Hanson [63], B. N. Parlett [69], A. George and J. W. Liu [37], as well as the authors of the Handbook [105], the EISPACK Guide [81], and the LINPACK Users' Guide [21]. Every one of these books is still worth reading today. Special thanks go to Cleve Moler for inventing MATLAB, which has changed everything.

Most of the first edition was written while I was on leave, at the University of Bielefeld, Germany. I am pleased to thank once again my host and longtime friend, Ludwig Elsner. During that stay I received financial support from the Fulbright commission. A big chunk of the second edition was also written in Germany, at the Technical University of Chemnitz. I thank my host (and another longtime friend), Volker Mehrmann. On that visit I received financial support from Sonderforschungsbereich 393, TU Chemnitz. The third edition was mostly written at home in Pullman. I am indebted to my home institution, Washington State University, for its support of my work on all three editions.

Numerous people have sent me corrections, feedback, and comments on the first and second editions. These include P. Abeles, J. Burman, C. Cacho, A. Cline, I. Dhillon, L. Dieci, J. Edwards, I. Gladwell, J. Haag, K. Hegewisch, D. Holland, E. Jessup, Y.-F. Jing, D. Koya, A. Loveless, D. D. Olesky, B. N. Parlett, A. C. Raines,

A. Witt, K. Wright, J. Wu, K. Yates, and T. J. Ypma. In preparing this new edition I also benefited from input from several anonymous referees. Finally, I thank the many students who have helped me learn how to teach this material over the years.

D. S. W.

CHAPTER 1

GAUSSIAN ELIMINATION AND ITS VARIANTS

One of the most frequently occurring problems in all areas of scientific endeavor is that of solving a system of n linear equations in n unknowns. For example, in Section 1.2 we will see how to compute the voltages and currents in electrical circuits, analyze simple elastic systems, and solve differential equations numerically, all by solving systems of linear equations. The main business of this chapter is to study the use of Gaussian elimination to solve such systems. We will see that there are many ways to organize this fundamental algorithm.

1.1 MATRIX MULTIPLICATION

Before we begin to study the solution of linear systems, let us take a look at some simpler matrix computations. In the process we will introduce some of the basic themes that will appear throughout the book. These include the use of operation counts (flop counts) to measure the complexity of an algorithm, the use of partitioned matrices and block matrix operations, and an illustration of the wide variety of ways in which a simple matrix computation can be organized.

Fundamentals of Matrix Computations, Third Edition. By David S. Watkins
Copyright © 2010 John Wiley & Sons, Inc.

Multiplying a Matrix by a Vector

Of the various matrix operations, the most fundamental one that cannot be termed entirely trivial is that of multiplying a matrix by a vector. Consider an $n \times m$ matrix, that is, a matrix with n rows and m columns:

$$A = \begin{bmatrix} a_{11} & a_{12} & \cdots & a_{1m} \\ a_{21} & a_{22} & \cdots & a_{2m} \\ \vdots & \vdots & & \vdots \\ a_{n1} & a_{n2} & \cdots & a_{nm} \end{bmatrix}.$$

The entries of A might be real or complex numbers. Let us assume that they are real for now. Given an m-tuple x of real numbers:

$$x = \begin{bmatrix} x_1 \\ x_2 \\ \vdots \\ x_m \end{bmatrix},$$

we can multiply A by x to get a product $b = Ax$, where b is an n-tuple. Its ith component is given by

$$b_i = a_{i1}x_1 + a_{i2}x_2 + \cdots + a_{im}x_m = \sum_{j=1}^{m} a_{ij}x_j. \tag{1.1.1}$$

In words, the ith component of b is obtained by taking the inner product (dot product) of ith row of A with x.

Example 1.1.2 An example of matrix-vector multiplication with $n = 2$ and $m = 3$ is

$$\begin{bmatrix} 1 & 2 & 3 \\ 4 & 5 & 6 \end{bmatrix} \begin{bmatrix} 7 \\ 8 \\ 9 \end{bmatrix} = \begin{bmatrix} 50 \\ 122 \end{bmatrix},$$

since $1 \times 7 + 2 \times 8 + 3 \times 9 = 50$ and $4 \times 7 + 5 \times 8 + 6 \times 9 = 122$. □

A computer code to perform matrix-vector multiplication might look something like this:

$$
\begin{aligned}
&b \leftarrow 0 \\
&\text{for } i = 1, \ldots, n \\
&\quad \lfloor \text{ for } j = 1, \ldots, m \\
&\quad\quad \lfloor \; b_i \leftarrow b_i + a_{ij}x_j
\end{aligned}
\tag{1.1.3}
$$

The j-loop accumulates the inner product b_i.

There is another way to view matrix-vector multiplication that turns out to be quite useful. Take another look at (1.1.1), but now view it as a formula for the entire vector b rather than just a single component. In other words, take the equation (1.1.1), which is really n equations for $b_1, b_2, \ldots b_n$, and stack the n equations into a single vector equation.

$$\begin{bmatrix} b_1 \\ b_2 \\ \vdots \\ b_n \end{bmatrix} = \begin{bmatrix} a_{11} \\ a_{21} \\ \vdots \\ a_{n1} \end{bmatrix} x_1 + \begin{bmatrix} a_{12} \\ a_{22} \\ \vdots \\ a_{n2} \end{bmatrix} x_2 + \cdots + \begin{bmatrix} a_{1m} \\ a_{2m} \\ \vdots \\ a_{nm} \end{bmatrix} x_m. \qquad (1.1.4)$$

This shows that b is a linear combination of the columns of A.

Example 1.1.5 Referring to Example 1.1.2, we have

$$\begin{bmatrix} 50 \\ 122 \end{bmatrix} = \begin{bmatrix} 1 \\ 4 \end{bmatrix} 7 + \begin{bmatrix} 2 \\ 5 \end{bmatrix} 8 + \begin{bmatrix} 3 \\ 6 \end{bmatrix} 9.$$

$\square$

Proposition 1.1.6 *If $b = Ax$, then b is a linear combination of the columns of A.*

If we let A_j denote the jth column of A, we have

$$b = \sum_{j=1}^{m} A_j x_j.$$

Expressing these operations as computer pseudocode, we have

$$b \leftarrow 0$$
$$\text{for } j = 1, \ldots, m$$
$$\left[\; b \leftarrow b + A_j x_j \right.$$

If we use a loop to perform each vector operation, the code becomes

$$b \leftarrow 0$$
$$\text{for } j = 1, \ldots, m$$
$$\left[\begin{array}{l} \text{for } i = 1, \ldots, n \\ \left[\; b_i \leftarrow b_i + a_{ij} x_j \right. \end{array} \right. \qquad (1.1.7)$$

Notice that (1.1.7) is identical to (1.1.3), except that the loops are interchanged. The two algorithms perform exactly the same operations but not in the same order. We call (1.1.3) a *row-oriented* matrix-vector multiply, because it accesses A by rows. In contrast, (1.1.7) is a *column-oriented* matrix-vector multiply.

Flop Counts

Real numbers are normally stored in computers in a floating-point format. The arithmetic operations that a computer performs on these numbers are called floating-point operations or *flops*, for short. The update $b_i \leftarrow b_i + a_{ij} x_j$ involves two flops, one floating-point multiply and one floating-point add.[3]

---

[3] We discuss floating-point arithmetic in Section 2.5.

Any time we run a computer program, we want to know how long it will take to complete its task. If the job involves large matrices, say 1000×1000, it may take a while. The traditional way to estimate running time is to count how many flops the computer must perform. Let us therefore count the flops in a matrix-vector multiply. Looking at (1.1.7), we see that if A is $n \times m$, the outer loop will be executed m times. On each of these passes, the inner loop is executed n times. Each execution of the inner loop involves two flops. Therefore the total flop count for the algorithm is $2nm$. This is a particularly easy flop count. On more complex algorithms it will prove useful to use the following device. Replace each loop by a summation sign Σ. Since the inner loop is executed for $i = 1, \ldots, n$, and there are two flops per pass, the total number of flops performed on each execution of the inner loop is $\sum_{i=1}^{n} 2$. Since the outer loop runs from $j = 1 \ldots m$, the total number of flops is

$$\sum_{j=1}^{m} \sum_{i=1}^{n} 2 = \sum_{j=1}^{m} 2n = 2nm.$$

The flop count gives a rough idea of how long it will take the algorithm to run. Suppose, for example, we have run the algorithm on a 300×400 matrix and observed how long it takes. If we now want to run the same algorithm on a matrix of size 600×400, we expect it to take about twice as long, since we are doubling n while holding m fixed and thereby doubling the flop count $2nm$.

Square matrices arise frequently in applications. If A is $n \times n$, the flop count for a matrix-vector multiply is $2n^2$. If we have performed a matrix-vector multiply on, say, a 500×500 matrix, we can expect the same operation on a 1000×1000 matrix to take about four times as long, since doubling n quadruples the flop count in this case.

An $n \times n$ matrix-vector multiply is an example of what we call an $O(n^2)$ process, or a process of *order* n^2. This just means that the amount of work is proportional to n^2. The notation is used as a way of emphasizing the dependence on n and de-emphasizing the proportionality constant, which is 2 in this case. Any $O(n^2)$ process has the property that doubling the problem size quadruples the amount of work.

It is important to realize that the flop count is only a crude indicator of the amount of work that an algorithm entails. It ignores many other tasks that the computer must perform in the course of executing the algorithm. The most important of these are fetching data from memory to be operated on and returning data to memory once the operations are done. On many computers the memory access operations are slower than the floating point operations, so it makes more sense to count memory accesses than flops. The flop count is useful, nevertheless, because it also gives us a rough count of the memory traffic. For each operation we must fetch operands from memory, and after each operation we must return the result to memory. This is a gross oversimplification of what actually goes on in modern computers. The execution speed of the algorithm can be affected drastically by how the memory traffic is organized. We will have more to say about this at the end of the section. Nevertheless, the flop count gives us a useful first indication of an algorithm's operation time, and we shall count flops as a matter of course.

rcise 1.1.8 Begin to familiarize yourself with MATLAB. Log on to a machine that has MATLAB installed, and start MATLAB. From MATLAB's command line type `A = randn(3,4)` to generate a 3 × 4 matrix with random entries. To learn more about the `randn` command, type `help randn`. Now type `x = randn(4,1)` to get a vector (a 4 × 1 matrix) of random numbers. To multiply A by x and store the result in a new vector b, type `b = A*x`.

To get MATLAB to save a transcript of your session, type `diary on`. This will cause a file named *diary*, containing a record of your MATLAB session, to be saved. Later on you can edit this file, print it out, turn it in to your instructor, or whatever. To learn more about the `diary` command, type `help diary`.

Other useful commands are `help` and `help help`. In addition to getting help from the command line, you can open MATLAB's help browser and search or browse for a large number of topics. Explore MATLAB to find out what other features it has. There are demonstrations and help for beginners that you might find useful. MATLAB also has a built in editor and a debugger.   □

rcise 1.1.9 Consider the following simple MATLAB program.

```
n = 500;
for jay = 1:4
  if jay > 1
    oldtime = time;
  end
  A = randn(n);
  x = randn(n,1);
  t = cputime;
  for rep = 1:100 % compute the product 100 times
    b = A*x;
  end
  matrixsize = n
  time = cputime - t
  if jay > 1
    ratio = time/oldtime
  end
  n = 2*n;
end
```

The syntax is simple enough that you can readily figure out what the program does. The commands `randn` and `b = A*x` are familiar from the previous exercise. We perform each matrix-vector multiplication 100 times in order to build up a significant amount of computing time. The function `cputime` tells how much computer (central processing unit) time the current MATLAB session has used. This program times the execution of 100 matrix-vector multiplications for square matrices A of dimension 500, 1000, 2000, and 4000.

Enter this program into a file called matvectime.m. Actually, you can call it whatever you please, but you must use the .m extension. (MATLAB programs are called m-files.) Now start MATLAB and type `matvectime` (without the .m) to execute the program. Depending on how fast your computer is, you may like to change the size of the matrix or the number of times the `jay` loop or the `rep` loop is executed.

MATLAB normally prints the output of all of its operations to the screen. You can suppress printing by terminating the operation with a semicolon. Looking at the program, we see that A, x, t, and b will not be printed, but `matrixsize`, `time`, and `ratio` will.

Look at the values of `ratio`. Are they close to what you would expect based on the flop count? □

Exercise 1.1.10 Write a MATLAB program that performs matrix-vector multiplication two different ways: (a) using the built-in MATLAB command `b = A*x`, and (b) using loops, as follows.

```
b = zeros(n,1);
for j = 1:n
  for i = 1:n
    b(i) = b(i) + A(i,j)*x(j);
  end
end
```

Time the two different methods on matrices of various sizes. Which method is faster? (You may want to use some of the code from Exercise 1.1.9.) □

Exercise 1.1.11 Write a Fortran or C program that performs matrix-vector multiplication using loops. How does its speed compare with that of MATLAB? □

Multiplying a Matrix by a Matrix

If A is an $n \times m$ matrix, and X is $m \times p$, we can form the product $B = AX$, which is $n \times p$. The (i, j) entry of B is

$$b_{ij} = \sum_{k=1}^{m} a_{ik} x_{kj}. \tag{1.1.12}$$

In words, the (i, j) entry of B is the dot product of the ith row of A with the jth column of X. If $p = 1$, this operation reduces to matrix-vector multiplication. If $p > 1$, the matrix-matrix multiply amounts to p matrix-vector multiplies: the jth column of B is just A times the jth column of X.

A computer program to multiply A by X might look something like this:

$$
\begin{aligned}
&B \leftarrow 0 \\
&\text{for } i = 1, \ldots, n \\
&\quad \left[\begin{array}{l} \text{for } j = 1, \ldots, p \\ \quad \left[\begin{array}{l} \text{for } k = 1, \ldots, m \\ \quad \left[\begin{array}{l} b_{ij} \leftarrow b_{ij} + a_{ik} x_{kj} \end{array} \right. \end{array} \right. \end{array} \right.
\end{aligned}
\tag{1.1.13}
$$

The decision to put the i-loop outside the j-loop was perfectly arbitrary. In fact, the order in which the updates $b_{ij} \leftarrow b_{ij} + a_{ik} x_{kj}$ are made is irrelevant, so the three loops can be nested in any way. Thus there are six basic variants of the matrix-matrix multiplication algorithm. These are all equivalent in principle. In practice, some versions may run faster than others on a given computer because of the order in which the data are accessed.

It is a simple matter to count the flops in matrix-matrix multiplication. Since there are two flops in the innermost loop of (1.1.13), the total flop count is

$$
\sum_{i=1}^{n} \sum_{j=1}^{p} \sum_{k=1}^{m} 2 = 2nmp.
$$

In the important case when all of the matrices are square of dimension $n \times n$, the flop count is $2n^3$. Thus square matrix multiplication is an $O(n^3)$ operation. This function grows rather quickly with n: each time n is doubled, the flop count is multiplied by eight. (However, this is not the whole story. See the remarks on fast matrix multiplication at the end of this section.)

rcise 1.1.14 Modify the MATLAB code from Exercise 1.1.9 by changing the matrix-vector multiply b = A*x to a matrix-matrix multiply B = A*X, where X is $n \times n$. You may also want to decrease the initial matrix dimension n and the number of times the rep loop is executed. Run the code and check the ratios. Are they close to what you would expect them to be, based on the flop count? □

Block Matrices and Block Matrix Operations

The idea of partitioning matrices into blocks is simple but powerful. It is a useful tool for proving theorems, constructing algorithms, and developing faster variants of algorithms. We will use block matrices again and again throughout the book.

Consider the matrix product $AX = B$, where the matrices have dimensions $n \times m$, $m \times p$, and $n \times p$, respectively. Suppose we partition A into blocks:

$$
A = \begin{array}{c} \\ n_1 \\ n_2 \end{array} \begin{array}{c} m_1 \quad m_2 \\ \left[\begin{array}{cc} A_{11} & A_{12} \\ A_{21} & A_{22} \end{array} \right] \end{array} \qquad \left\{ \begin{array}{l} n_1 + n_2 = n \\ m_1 + m_2 = m. \end{array} \right.
\tag{1.1.15}
$$

The labels n_1, n_2, m_1, and m_2 indicate that the block A_{ij} has dimensions $n_i \times m_j$.

We can partition X similarly.

$$X = \begin{array}{c} \\ m_1 \\ m_2 \end{array} \begin{array}{cc} p_1 & p_2 \\ \left[\begin{array}{cc} X_{11} & X_{12} \\ X_{21} & X_{22} \end{array} \right] \end{array} \qquad \left\{ \begin{array}{l} m_1 + m_2 = m \\ p_1 + p_2 = p. \end{array} \right. \tag{1.1.16}$$

The numbers m_1 and m_2 are the same as in (1.1.15). Thus, for example, the number of rows in X_{12} is the same as the number of columns in A_{11} and A_{21}. Continuing in the same spirit, we partition B as follows:

$$B = \begin{array}{c} \\ n_1 \\ n_2 \end{array} \begin{array}{cc} p_1 & p_2 \\ \left[\begin{array}{cc} B_{11} & B_{12} \\ B_{21} & B_{22} \end{array} \right] \end{array} \qquad \left\{ \begin{array}{l} n_1 + n_2 = n \\ p_1 + p_2 = p. \end{array} \right. \tag{1.1.17}$$

The row partition of B is the same as that of A, and the column partition is the same as that of X. The product $AX = B$ can now be written as

$$\left[\begin{array}{cc} A_{11} & A_{12} \\ A_{21} & A_{22} \end{array} \right] \left[\begin{array}{cc} X_{11} & X_{12} \\ X_{21} & X_{22} \end{array} \right] = \left[\begin{array}{cc} B_{11} & B_{12} \\ B_{21} & B_{22} \end{array} \right]. \tag{1.1.18}$$

We know that B is related to A and X by the equations (1.1.12), but how are the blocks of B related to the blocks of A and X? We would hope to be able to multiply the blocks as if they were numbers. For example, we hope that $A_{11}X_{11} + A_{12}X_{21} = B_{11}$. Theorem 1.1.19 states that this is indeed the case.

Theorem 1.1.19 *Let A, X, and B be partitioned as in (1.1.15), (1.1.16), and (1.1.17), respectively. Then $AX = B$ if and only if*

$$A_{i1}X_{1j} + A_{i2}X_{2j} = B_{ij}, \qquad i, j = 1, 2.$$

You can easily convince yourself that Theorem 1.1.19 is true. It follows more or less immediately from the definition of matrix multiplication. We will skip the tedious but routine exercise of writing out a detailed proof. You might find the following exercise useful.

Exercise 1.1.20 Consider matrices A, X, and B, partitioned as indicated.

$$A = \left[\begin{array}{cc|cc} 1 & 3 & 2 \\ 2 & 1 & 1 \\ \hline -1 & 0 & 1 \end{array} \right] \qquad X = \left[\begin{array}{cc|cc} 1 & 0 & 1 \\ \hline 2 & 1 & 1 \\ -1 & 2 & 0 \end{array} \right] \qquad B = \left[\begin{array}{cc|cc} 5 & 7 & 4 \\ 3 & 3 & 3 \\ \hline -2 & 2 & -1 \end{array} \right]$$

Thus, for example, $A_{12} = \left[\begin{array}{cc} 3 & 2 \\ 1 & 1 \end{array} \right]$ and $A_{21} = \left[\begin{array}{c} -1 \end{array} \right]$. Show that $AX = B$ and $A_{i1}X_{1j} + A_{i2}X_{2j} = B_{ij}$ for $i, j = 1, 2$. □

Once you believe Theorem 1.1.19, you should have no difficulty with the following generalization. Make a finer partition of A into r block rows and s block columns.

$$
A = \begin{array}{c} \\ n_1 \\ \vdots \\ n_r \end{array} \begin{array}{ccc} m_1 & \cdots & m_s \\ \left[\begin{array}{ccc} A_{11} & \cdots & A_{1s} \\ \vdots & & \vdots \\ A_{r1} & \cdots & A_{rs} \end{array} \right] \end{array} \qquad \left\{ \begin{array}{l} n_1 + \cdots + n_r = n \\ m_1 + \cdots + m_s = m. \end{array} \right. \qquad (1.1.21)
$$

Then partition X *conformably* with A; that is, make the block row structure of X identical to the block column structure of A.

$$
X = \begin{array}{c} \\ m_1 \\ \vdots \\ m_s \end{array} \begin{array}{ccc} p_1 & \cdots & p_t \\ \left[\begin{array}{ccc} X_{11} & \cdots & X_{1t} \\ \vdots & & \vdots \\ X_{s1} & \cdots & X_{st} \end{array} \right] \end{array} \qquad \left\{ \begin{array}{l} m_1 + \cdots + m_s = m \\ p_1 + \cdots + p_t = p. \end{array} \right. \qquad (1.1.22)
$$

Finally, partition the product B conformably with both A and X.

$$
B = \begin{array}{c} \\ n_1 \\ \vdots \\ n_r \end{array} \begin{array}{ccc} p_1 & \cdots & p_t \\ \left[\begin{array}{ccc} B_{11} & \cdots & B_{1t} \\ \vdots & & \vdots \\ B_{r1} & \cdots & B_{rt} \end{array} \right] \end{array} \qquad \left\{ \begin{array}{l} n_1 + \cdots + n_r = n \\ p_1 + \cdots + p_t = p. \end{array} \right. \qquad (1.1.23)
$$

Theorem 1.1.24 *Let A, X, and B be partitioned as in (1.1.21), (1.1.22), and (1.1.23), respectively. Then $B = AX$ if and only if*

$$
B_{ij} = \sum_{k=1}^{s} A_{ik} X_{kj} \qquad i = 1, \ldots, r, \ j = 1, \ldots, t.
$$

rcise 1.1.25 Make a partition of the matrix-vector product $Ax = b$ that demonstrates that b is a linear combination of the columns of A. □

Use of Block Matrix Operations to Decrease Data Movement Delays

Suppose we wish to multiply A by X to obtain B. For simplicity, let us assume that the matrices are square, $n \times n$, although the idea to be discussed in this section can be applied to non-square matrices as well. Assume further that A can be partitioned into s block rows and s block columns, where each block is $r \times r$.

$$
A = \left[\begin{array}{ccc} A_{11} & \cdots & A_{1s} \\ \vdots & & \vdots \\ A_{s1} & \cdots & A_{ss} \end{array} \right].
$$

Thus $n = rs$. We partition X and B in the same way. The assumption that the blocks are all the same size is again just for simplicity. (In practice we want nearly all of the blocks to be approximately square and nearly all of approximately the same size.) Writing a block version of (1.1.13), we have

$$
\begin{aligned}
&B \leftarrow 0 \\
&\text{for } i = 1, \ldots, s \\
&\quad \left[\begin{aligned}
&\text{for } j = 1, \ldots, s \\
&\quad \left[\begin{aligned}
&\text{for } k = 1, \ldots, s \\
&\quad \left[\; B_{ij} \leftarrow B_{ij} + A_{ik} X_{kj} \right.
\end{aligned} \right.
\end{aligned} \right.
\end{aligned}
\tag{1.1.26}
$$

A computer program based on this layout would perform the following operations repeatedly: grab the blocks A_{ik}, X_{kj}, and B_{ij}, multiply A_{ik} by X_{kj} and add the result to B_{ij}, using something like (1.1.13), then set B_{ij} aside.

Varying the block size will not affect the total flop count, which will always be $2n^3$, but it can affect the performance of the algorithm dramatically nevertheless, because of the way the data are handled. Every computer has a memory hierarchy. This has always been the case, although the details have changed with time. Nowadays a typical computer has a small number of registers, a smallish, fast cache, a much larger, slower, main memory, and even larger, slower, bulk storage areas (e.g. disk drives). Data stored in the main memory must first be moved to cache and then to registers before it can be operated on. The transfer from main memory to cache is much slower than that from cache to registers, and it is also slower than the rate at which the computer can perform arithmetic. If we can move the entire arrays A, X, and B into cache, then perform all of the operations, then move the result, B, back to main memory, we expect the job to be done much more quickly than if we must repeatedly move data back and forth. Indeed, if we can fit all of the arrays into cache, the total number of data transfers between slow and fast memory will be about $4n^2$, whereas the total number of flops is $2n^3$. Thus the ratio of arithmetic to memory transfers is $\frac{1}{2}n$ flops per data item, which implies that the relative importance of the data transfers decreases as n increases. Unfortunately, unless n is fairly small, the cache will not be large enough to hold the entire matrices. Then it becomes beneficial to perform the matrix-matrix multiply by blocks.

Before we consider blocking, let us see what happens if we do not use blocks. Let us suppose the cache is big enough to hold two matrix columns or rows. Computation of entry b_{ij} requires the ith row of A and the jth column of X. The time required to move these into cache is proportional to $2n$, the number of data items. Once these are in fast memory, we can quickly perform the $2n$ flops. If we now want to calculate $b_{i,j+1}$, we can keep the ith row of A in cache, but we need to bring in column $j+1$ of X, which means we have a time delay proportional to n. We can then perform the $2n$ flops to get $b_{i,j+1}$. The ratio of arithmetic to data transfers is about 2 flops per data item. In other words, the number of transfers between main memory and cache is proportional to the amount of arithmetic done. This severely limits performance. There are many ways to reorganize the matrix-matrix multiplication algorithm (1.1.13) without blocking, but they all suffer from this same limitation.

Now let us see how we can improve the situation by using blocks. Suppose we perform algorithm (1.1.26) using a block size r that is small enough that the three blocks A_{ik}, X_{kj}, and B_{ij} can all fit into cache at once. Since the number of entries in these blocks is $O(r^2)$, the time to move them from main memory to cache is $O(r^2)$. Once they are in the fast cache, A_{ik} is multiplied by X_{kj} and the result is added to B_{ij} in $O(r^3)$ flops. Once this operation is done, the blocks can be moved back to main memory in time proportional to $O(r^2)$. The ratio of arithmetic time to data transfer time is clearly $O(r^3/r^2) = O(r)$. Thus the data transfer times, and the associated delays, can be made relatively insignificant by making r large. On the other hand r is limited by the size of the cache; we need to be able to fit several blocks at once into the cache.

Further efficiency can be obtained by performing the arithmetic and the data transfers at the same time. Suppose we allow enough room to fit five blocks into cache at once, the five blocks in cache are $A_{i,k-1}$, $X_{k-1,j}$, A_{ik}, X_{kj}, and B_{ij}, we have just multiplied $A_{i,k-1}$ by $X_{k-1,j}$ and added the product to B_{ij}, and we are now going to multiply A_{ik} by X_{kj} and add that result to B_{ij}. We can begin this multiplication at once. At the same time we can be transferring the used blocks $A_{i,k-1}$ and $X_{k-1,j}$ back to main memory. Once that transfer is complete, we can begin to transfer the next blocks we are going to use, probably $A_{i,k+1}$ and $X_{k+1,j}$, from main memory into cache. Since the arithmetic time is $O(r^3)$ and the data transfer time is only $O(r^2)$, we can have the new blocks in cache and ready to go by the time the operations on the current blocks are finished, if r is large enough. Thus, if we can make r large enough, we can eliminate almost all data transfer delays. Thus we should make r as large as possible, subject to the constraint that we must be able to fit five blocks into cache simultaneously.

We also remark that if the computer happens to have multiple processors, it can operate on several blocks in parallel. The use of blocks simplifies the organization of parallel computing. It also helps to minimize the bottlenecks associated with communication between processors. This latter benefit also stems from the fact that $O(r^3)$ flops are performed on $O(r^2)$ data items.

Many of the algorithms that will be considered in this book can be organized into blocks, as we shall see. The public-domain linear algebra subroutine library LAPACK [1] uses block algorithms wherever it can.

Fast Matrix Multiplication

If we multiply two $n \times n$ matrices together using the definition (1.1.12), $2n^3$ flops are required. In 1969 V. Strassen [88] amazed the computing world by presenting a method that can do the job in $O(n^s)$ flops, where $s = \log_2 7 \approx 2.81$. Since $2.81 < 3$, Strassen's method will be faster than conventional algorithms if n is sufficiently large. Tests have shown that even for n as small as 100 or so, Strassen's method can be faster. However, since 2.81 is only slightly less than 3, n has to be made quite large before Strassen's method wins by a large margin. Accuracy is also an issue. Strassen's method has not made a great impact so far, but that could change in the future.

Even "faster" methods have been found. The current record holder, due to Copper-smith and Winograd, can multiply two $n \times n$ matrices in about $O(n^{2.376})$ flops. But there is a catch. When we write $O(n^{2.376})$, we mean that there is a constant C such that the algorithm takes no more than $Cn^{2.376}$ flops. For this algorithm the constant C is so large that it does not beat Strassen's method until n is really enormous.

A good overview of fast matrix multiplication methods is given by Higham [52].

1.2 SYSTEMS OF LINEAR EQUATIONS

In the previous section we discussed the problem of multiplying a matrix A times a vector x to obtain a vector b. In scientific computations one is more likely to have to solve the inverse problem: Given A (an $n \times n$ matrix) and b, solve for x. That is, find x such that $Ax = b$. This is the problem of solving a system of n linear equations in n unknowns.

You have undoubtedly already had some experience solving systems of linear equations. We will begin this section by reminding you briefly of some of the basic theoretical facts. We will then look at several simple examples to remind you of how linear systems can arise in scientific problems.

Nonsingularity and Uniqueness of Solutions

Consider a system of n linear equations in n unknowns

$$
\begin{aligned}
a_{11}x_1 + a_{12}x_2 + \cdots + a_{1n}x_n &= b_1 \\
a_{21}x_1 + a_{22}x_2 + \cdots + a_{2n}x_n &= b_2 \\
&\vdots \\
a_{n1}x_1 + a_{n2}x_2 + \cdots + a_{nn}x_n &= b_n.
\end{aligned}
\tag{1.2.1}
$$

The coefficients a_{ij} and b_i are given, and we wish to find $x_1, \ldots, x_n$ that satisfy the equations. In most applications the coefficients are real numbers, and we seek a real solution. Therefore we will confine our attention to real systems. However, everything we will do can be carried over to the complex number field. (In some situations minor modifications are required. These will be covered in the exercises.)

Since it is tedious to write out (1.2.1) again and again, we generally prefer to write it as a single matrix equation

$$
Ax = b,
\tag{1.2.2}
$$

where

$$
A = \begin{bmatrix} a_{11} & a_{12} & \cdots & a_{1n} \\ a_{21} & a_{22} & \cdots & a_{2n} \\ \vdots & \vdots & & \vdots \\ a_{n1} & a_{n2} & \cdots & a_{nn} \end{bmatrix} \qquad x = \begin{bmatrix} x_1 \\ x_2 \\ \vdots \\ x_n \end{bmatrix} \qquad b = \begin{bmatrix} b_1 \\ b_2 \\ \vdots \\ b_n \end{bmatrix}.
$$

A and b are given, and we must solve for x. A is a square matrix; it has n rows and n columns.

Equation (1.2.2) has a unique solution if and only if the matrix A is nonsingular. Theorem 1.2.3 summarizes some of the simple characterizations of nonsingularity that we will use in this book.

First we recall some standard terminology. The $n \times n$ *identity matrix* is denoted by I. It is the unique matrix such that $AI = IA = A$ for all $A \in \mathbb{R}^{n \times n}$. The identity matrix has 1's on its main diagonal and 0's elsewhere. For example, the 3×3 identity matrix has the form

$$I = \begin{bmatrix} 1 & 0 & 0 \\ 0 & 1 & 0 \\ 0 & 0 & 1 \end{bmatrix}.$$

Given a matrix A, if there is a matrix B such that $AB = BA = I$, then B is called the *inverse* of A and denoted A^{-1}. Not every matrix has an inverse.

Theorem 1.2.3 *Let A be a square matrix. The following six conditions are equivalent; that is, if any one holds, they all hold.*

(a) A^{-1} *exists.*

(b) *There is no nonzero y such that $Ay = 0$.*

(c) *The columns of A are linearly independent.*

(d) *The rows of A are linearly independent.*

(e) $\det(A) \neq 0$.

(f) *Given any vector b, there is exactly one vector x such that $Ax = b$.*

(In condition (b), the symbol 0 stands for the vector whose entries are all zero. In condition (e), the symbol 0 stands for the real number 0. $\det(A)$ denotes the determinant of A.)

For a proof of Theorem 1.2.3 see any elementary linear algebra text. If the conditions of Theorem 1.2.3 hold, A is said to be *nonsingular* or *invertible*. If the conditions do not hold, A is said to be *singular* or *noninvertible*. In this case (1.2.2) has either no solution or infinitely many solutions. In this chapter we will focus on the nonsingular case.

If A is nonsingular, the unique solution of (1.2.2) can be obtained in principle by multiplying both sides by A^{-1}: From $Ax = b$ we obtain $A^{-1}Ax = A^{-1}b$, and since $A^{-1}A = I$, the identity matrix, we obtain $x = A^{-1}b$. This equation solves the problem completely in theory, but the method of solution that it suggests—first calculate A^{-1}, then multiply A^{-1} by b to obtain x—is usually a bad idea. As we shall see, it is generally more efficient to solve $Ax = b$ directly, without calculating A^{-1}. On most large problems the savings in computation and storage achieved by avoiding the use of A^{-1} are truly spectacular.

Situations in which the inverse really needs to be calculated are quite rare. This does not imply that the inverse is unimportant; it is an extremely useful theoretical tool.

Exercise 1.2.4 Prove that if A^{-1} exists, then there can be no nonzero y for which $Ay = 0$. □

Exercise 1.2.5 Prove that if A^{-1} exists, then $\det(A) \neq 0$.   □

We now move on to some examples.

Electrical Circuits

Example 1.2.6 Consider the electrical circuit shown in Figure 1.1. Suppose the circuit is in an equilibrium state; all of the voltages and currents are constant. The four unknown nodal voltages $x_1, \ldots, x_4$ can be determined as follows. At each of the four nodes, the sum of the currents away from the node must be zero (Kirchhoff's current law). This gives us an equation for each node. In each of these equations the currents can be expressed in terms of voltages using Ohm's law, which states that the voltage drop (in volts) is equal to the current (in amperes) times the resistance (in ohms). For example, suppose the current from node 3 to node 4 through the 5 Ω resistor is I. Then by Ohm's law, $x_3 - x_4 = 5I$, so $I = .2(x_3 - x_4)$. Treating the other two currents flowing from node 3 in the same way, and applying Kirchhoff's current law, we get the equation

$$.2(x_3 - x_4) + 1(x_3 - x_1) + .5(x_3 - 6) = 0$$

or

$$-x_1 + 1.7x_3 - .2x_4 = 3.$$

Applying the same procedure to nodes 1, 2, and 4, we obtain a system of four linear equations in four unknowns:

$$
\begin{array}{rrrrrrr}
2x_1 & - & x_2 & - & x_3 & & & = & 0 \\
- x_1 & + & 1.5x_2 & & & - & .5x_4 & = & 0 \\
- x_1 & & & + & 1.7x_3 & - & .2x_4 & = & 3 \\
& - & .5x_2 & - & .2x_3 & + & 1.7x_4 & = & 0
\end{array}
$$

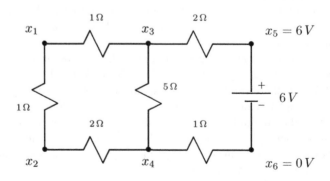

Figure 1.1   Solve for the nodal voltages.

which can be written as a single matrix equation

$$
\begin{bmatrix}
2 & -1 & -1 & 0 \\
-1 & 1.5 & 0 & -.5 \\
-1 & 0 & 1.7 & -.2 \\
0 & -.5 & -.2 & 1.7
\end{bmatrix}
\begin{bmatrix}
x_1 \\ x_2 \\ x_3 \\ x_4
\end{bmatrix}
=
\begin{bmatrix}
0 \\ 0 \\ 3 \\ 0
\end{bmatrix}.
$$

Though we will not prove it here, the coefficient matrix is nonsingular, so the system has exactly one solution. Solving it using MATLAB, we find that

$$
x =
\begin{bmatrix}
3.0638 \\ 2.4255 \\ 3.7021 \\ 1.1489
\end{bmatrix}.
$$

Thus, for example, the voltage at node 3 is 3.7021 volts. These are not the exact answers; they are rounded off to four decimal places. Given the nodal voltages, we can easily calculate the current through any of the resistors by Ohm's law. For example, the current flowing from node 3 to node 4 is $.2(x_3 - x_4) = 0.5106$ amperes.

□

rcise 1.2.7 Verify the correctness of the equations in Example 1.2.6. Use MATLAB (or some other means) to compute the solution. If you are unfamiliar with MATLAB, you can find out how to enter matrices by searching for the topic "entering matrices" in MATLAB's help browser or by exploring MATLAB's "demos" or "getting started". Once you have entered the matrix A and the vector b, you can type x = A\b to solve for x. A transcript of your whole session (which you can later edit, print out, and turn in to your instructor) can be made by using the command diary on. For more information about the diary command type help diary at the command line or search for "diary" in the help browser.   □

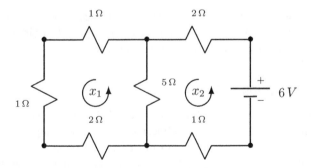

Figure 1.2   Solve for the loop currents.

Example 1.2.8 Another way to analyze a planar electrical circuit is to solve for loop currents instead of nodal voltages. Figure 1.2 shows the same circuit as before, but now we have associated currents x_1 and x_2 with the two loops. (These are clearly not the same x_i as in the previous figure.) For the resistors that lie on the boundary of the circuit, the loop current is the actual current flowing through the resistor, but the current flowing through the 5 Ω resistor in the middle is the difference $x_2 - x_1$. An equation for each loop can be obtained by applying the principle that the sum of the voltage drops around the loop must be zero (Kirchhoff's voltage law). The voltage drop across each resistor can be expressed in terms of the loop currents by applying Ohm's law. For example, the voltage drop across the 5 Ω resistor, from top to bottom, is $5(x_2 - x_1)$ volts. Summing the voltage drops across the four resistors in loop 1, we obtain

$$5(x_1 - x_2) + 1x_1 + 1x_1 + 2x_1 = 0.$$

Similarly, in loop 2,

$$5(x_2 - x_1) + 1x_2 - 6 + 2x_2 = 0.$$

Rearranging these equations, we obtain the 2×2 system

$$\begin{bmatrix} 9 & -5 \\ -5 & 8 \end{bmatrix} \begin{bmatrix} x_1 \\ x_2 \end{bmatrix} = \begin{bmatrix} 0 \\ 6 \end{bmatrix}.$$

Solving these equations by hand, we find that $x_1 = 30/47 = 0.6383$ amperes and $x_2 = 54/47 = 1.1489$ amperes. Thus, for example, the current through the 5 Ω resistor, from top to bottom, is $x_2 - x_1 = .5106$ amperes, and the voltage drop is 2.5532 volts. These results are in agreement with those of Example 1.2.6.   □

Exercise 1.2.9   Check that the equations in Example 1.2.8 are correct. Check that the coefficient matrix is nonsingular. Solve the system by hand, by MATLAB, or by some other means.   □

It is easy to imagine much larger circuits with many loops. See, for example, Exercise 1.2.19. Then imagine something much larger. If a circuit has, say, 100 loops, then it will have 100 equations in 100 unknowns.

Simple Mass-Spring Systems

In Figure 1.3 a steady force of 2 newtons is applied to a cart, pushing it to the right and stretching the spring, which is a linear spring with a spring constant (stiffness) 4 newtons/meter. How far will the cart move before stopping at a new equilibrium position? Here we are not studying the dynamics, that is, how the cart gets to its new equilibrium. For that we would need to know the mass of the cart and the frictional forces in the system. Since we are asking only for the new equilibrium position, it suffices to know the stiffness of the spring.

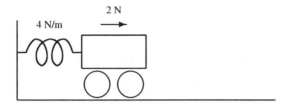

Figure 1.3   Single cart and spring

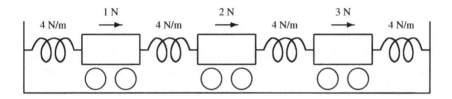

Figure 1.4   System of three carts

The new equilibrium will be at the point at which the rightward force of 2 newtons is exactly balanced by the leftward force applied by the spring. In other words, the equilibrium position is the one at which the sum of the forces on the cart is zero. Let x denote the (yet unknown) amount by which the cart moves to the right. Then the restoring force of the spring is -4 newtons/meter $\times$ x meters $= -4x$ newtons. It is negative because it pulls the cart leftward. The equilibrium occurs when $-4x+2 = 0$. Solving this system of one equation in one unknown, we find that $x = 0.5$ meter.

Example 1.2.10 Now suppose we have three masses attached by springs as shown in Figure 1.4. Let x_1, x_2, and x_3 denote the amount by which carts 1, 2, and 3, respectively, move when the forces are applied. For each cart the new equilibrium position is that point at which the sum of the forces on the cart is zero. Consider the second cart, for example. An external force of two newtons is applied, and there is the leftward force of the spring to the left, and the rightward force of the spring to the right. The amount by which the spring on the left is stretched is $x_2 - x_1$ meters. It therefore exerts a force -4 newtons/meter $\times$ $(x_2 - x_1)$ meters $= -4(x_2 - x_1)$ newtons on the second cart. Similarly the spring on the right applies a force of $+4(x_3 - x_2)$ newtons. Thus the equilibrium equation for the second cart is

$$-4(x_2 - x_1) + 4(x_3 - x_2) + 2 = 0$$

or
$$-4x_1 + 8x_2 - 4x_3 = 2.$$

Similar equations apply to carts 1 and 3. Thus we obtain a system of three linear equations in three unknowns, which we can write as a matrix equation

$$\begin{bmatrix} 8 & -4 & 0 \\ -4 & 8 & -4 \\ 0 & -4 & 8 \end{bmatrix} \begin{bmatrix} x_1 \\ x_2 \\ x_3 \end{bmatrix} = \begin{bmatrix} 1 \\ 2 \\ 3 \end{bmatrix}.$$

Entering the matrix A and vector b into MATLAB, and using the command $x = A\backslash b$ (or simply solving the system by hand) we find that

$$x = \begin{bmatrix} 0.625 \\ 1.000 \\ 0.875 \end{bmatrix}.$$

Thus the first cart is displaced to the right by a distance of 0.625 meters, for example.

The coefficient matrix A is called a *stiffness matrix*, because the values of its nonzero entries are determined by the stiffnesses of the springs.   □

Exercise 1.2.11   Check that the equations in Example 1.2.10 are correct. Check that the coefficient matrix of the system is nonsingular. Solve the system by hand, by MATLAB, or by some other means.   □

It is easy to imagine more complex examples. If we have n carts in a line, we get a system of n equations in n unknowns. See Exercise 1.2.20. We can also consider problems in which masses are free to move in two or three dimensions and are connected by a network of springs.

Numerical (Approximate) Solution of Differential Equations

Many physical phenomena can be modeled by differential equations. We shall consider one example without going into too many details.

Example 1.2.12 Consider a differential equation

$$-u''(x) + c\,u'(x) + d\,u(x) = f(x) \qquad 0 < x < 1 \tag{1.2.13}$$

with boundary conditions
$$u(0) = 0, \quad u(1) = 0. \tag{1.2.14}$$

The problem is to solve for the function u, given the constants c and d and the function f. For example, $u(x)$ could represent the unknown concentration of some chemical pollutant at distance x from the end of a pipe.[4] Depending on the function f, it may

---

[4]The term $-u''(x)$ is a diffusion term, $c\,u'(x)$ is a convection term, $d\,u(x)$ is a decay term, and $f(x)$ is a source term.

or may not be within our power to solve this boundary value problem exactly, but we can always solve it approximately by the finite difference method, as follows.

Pick a (possibly large) integer m, and subdivide the interval $[0, 1]$ into m equal subintervals of length $h = 1/m$. The subdivision points of the intervals are $x_j = jh$, $j = 0, \ldots, m$. These points constitute our computational grid. The finite difference technique will produce approximations to $u(x_1), \ldots, u(x_{m-1})$. Since (1.2.13) holds at each grid point, we have

$$-u''(x_i) + c\,u'(x_i) + d\,u(x_i) = f(x_i) \qquad i = 1, \ldots, m - 1.$$

If h is small, good approximations for the first and second derivatives are

$$u'(x_i) \approx \frac{u(x_i + h) - u(x_i - h)}{2h} = \frac{u(x_{i+1}) - u(x_{i-1})}{2h}$$

and

$$u''(x_i) \approx \frac{u(x_{i+1}) - 2u(x_i) + u(x_{i-1})}{h^2}.$$

(See Exercise 1.2.21.) Substituting these approximations for the derivatives into the differential equation, we obtain, for $i = 1, \ldots, m - 1$,

$$\frac{-u(x_{i+1}) + 2u(x_i) - u(x_{i-1})}{h^2} + c\left(\frac{u(x_{i+1}) - u(x_{i-1})}{2h}\right) + d\,u(x_i) \approx f(x_i).$$

We now approximate this by a system of difference equations

$$\frac{-u_{i+1} + 2u_i - u_{i-1}}{h^2} + c\left(\frac{u_{i+1} - u_{i-1}}{2h}\right) + d\,u_i = f_i, \tag{1.2.15}$$

$i = 1, \ldots, m - 1$. Here we have replaced the approximation symbol by an equal sign and $u(x_i)$ by the symbol u_i, which (hopefully) is an approximation of $u(x_i)$. We have also introduced the symbol f_i as an abbreviation for $f(x_i)$. This is a system of $m - 1$ linear equations in the unknowns $u_0, u_1, \ldots, u_m$. Applying the boundary conditions (1.2.14), we can take $u_0 = 0$ and $u_m = 0$, leaving only $m - 1$ unknowns $u_1, \ldots, u_{m-1}$.

Suppose, for example, $m = 6$ and $h = 1/6$. Then (1.2.15) is a system of five equations in five unknowns, which can be written as the single matrix equation

$$
\begin{bmatrix}
72 + d & -36 + 3c & 0 & 0 & 0 \\
-36 - 3c & 72 + d & -36 + 3c & 0 & 0 \\
0 & -36 - 3c & 72 + d & -36 + 3c & 0 \\
0 & 0 & -36 - 3c & 72 + d & -36 + 3c \\
0 & 0 & 0 & -36 - 3c & 72 + d
\end{bmatrix}
\begin{bmatrix}
u_1 \\ u_2 \\ u_3 \\ u_4 \\ u_5
\end{bmatrix}
=
\begin{bmatrix}
f_1 \\ f_2 \\ f_3 \\ f_4 \\ f_5
\end{bmatrix}.
$$

Given specific c, d, and f, we can solve this system of equations for $u_1, \ldots, u_5$. Since the difference equations mimic the differential equation, we expect that $u_1, \ldots, u_5$ will approximate the true solution of the boundary value problem at the points $x_1, \ldots x_5$.

Of course, we do not expect a very good approximation when we take only $m = 6$. To get a good approximation, we should take m much larger, which results in a much larger system of equations to solve.   □

Exercise 1.2.16 Write the system of equations from Example 1.2.12 as a matrix equation for (a) $m = 8$, (b) $m = 20$.   □

More complicated systems of difference equations arising from partial differential equations are discussed in Section 8.1.

Additional Exercises

Exercise 1.2.17 Consider the electrical circuit in Figure 1.5.

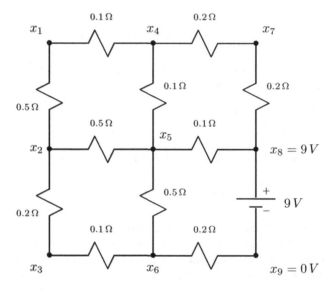

Figure 1.5   Electric circuit with nodal voltages

(a) Write down a linear system $Ax = b$ with seven equations for the seven unknown nodal voltages.

(b) Using MATLAB, for example, solve the system to find the nodal voltages. Calculate the residual $r = b - A\hat{x}$, where $\hat{x}$ denotes your computed solution. In theory r should be zero. In practice you will get a tiny but nonzero residual because of roundoff errors in your computation. Use the `diary` command to make a transcript of your session that you can turn in to your instructor.

□

rcise 1.2.18   The circuit in Figure 1.6 is the same as for the previous problem, but now let us focus on the loop currents.

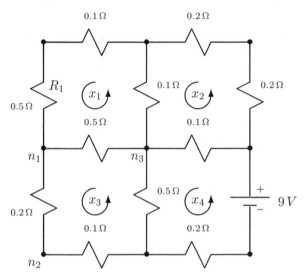

Figure 1.6   Electrical circuit with loop currents

(a) Write down a linear system $Ax = b$ of four equations for the four unknown loop currents.

(b) Solve the system for x. Calculate the residual $r = b - A\hat{x}$, where $\hat{x}$ denotes your computed solution.

(c) Using Ohm's law and the loop currents that you just calculated, find the voltage drops from the node labeled n_1 to the nodes labeled n_2 and n_3. Are these in agreement with your solution to Exercise 1.2.17?

(d) Using Ohm's law and the voltages calculated in Exercise 1.2.17, find the current through the resistor labeled R_1. Is this in agreement with your loop current calculation?

☐

rcise 1.2.19   In the circuit in Figure 1.7 all of the resistances are $1\,\Omega$.

(a) Write down a linear system $Ax = b$ that you can solve for the loop currents.

(b) Solve the system for x.

☐

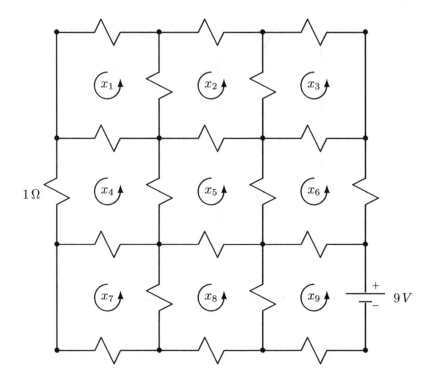

Figure 1.7 Calculate the loop currents.

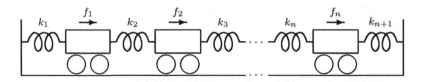

Figure 1.8 System of n masses

rcise 1.2.20 Consider a system of n carts connected by springs, as shown in Figure 1.8. The ith spring has a stiffness of k_i newtons/meter. Suppose that the carts are subjected to steady forces of $f_1, f_2, \ldots, f_n$ newtons, respectively, causing displacements of $x_1, x_2, \ldots, x_n$ meters, respectively.

(a) Write down a system of n linear equations $Ax = b$ that could be solved for $x_1, \ldots, x_n$. Notice that if n is at all large, the vast majority of the entries of A will be zeros. Matrices with this property are called *sparse*. Since all of the nonzeros are confined to a narrow band around the main diagonal, A is also called *banded*. In particular, the nonzeros are confined to three diagonals, so A is *tridiagonal*.

(b) Compute the solution in the case $n = 20$, $k_i = 1$ newton/meter for all i, and $f_i = 0$ except for $f_5 = 1$ newton and $f_{16} = -1$ newton. (Type help toeplitz to learn an easy way to enter the coefficient matrix in MATLAB.)

□

rcise 1.2.21 Recall the definition of the derivative from elementary calculus:

$$u'(x) = \lim_{h \to 0} \frac{u(x+h) - u(x)}{h}.$$

(a) Show that if h is a sufficiently small positive number, then both

$$\frac{u(x+h) - u(x)}{h} \quad \text{and} \quad \frac{u(x) - u(x-h)}{h}$$

are good approximations of $u'(x)$.

(b) Take the average of the two estimates from part (a) to obtain the estimate

$$u'(x) \approx \frac{u(x+h) - u(x-h)}{2h}.$$

Draw a picture (the graph of u and a few straight lines) that shows that this estimate is likely to be a better estimate of $u'(x)$ than either of the estimates from part (a) are.

(c) Apply the estimate from part (b) to $u''(x)$ with h replaced by $h/2$ to obtain

$$u''(x) \approx \frac{u'(x+h/2) - u'(x-h/2)}{h}.$$

Now approximate $u'(x+h/2)$ and $u'(x-h/2)$ using the estimate from part (b), again with h replaced by $h/2$, to obtain

$$u''(x) \approx \frac{u(x+h) - 2u(x) + u(x-h)}{h^2}.$$

□

1.3   TRIANGULAR SYSTEMS

A linear system whose coefficient matrix is triangular is particularly easy to solve. It is a common practice to reduce general systems to a triangular form, which can then be solved inexpensively. For this reason we will study triangular systems in detail.

A matrix $G = (g_{ij})$ is *lower triangular* if $g_{ij} = 0$ whenever $i < j$. Thus a lower-triangular matrix has the form

$$
G = \begin{bmatrix}
g_{11} & 0 & 0 & \cdots & 0 \\
g_{21} & g_{22} & 0 & \cdots & 0 \\
g_{31} & g_{32} & g_{33} & \ddots & \vdots \\
\vdots & \vdots & \vdots & \ddots & 0 \\
g_{n1} & g_{n2} & g_{n3} & \cdots & g_{nn}
\end{bmatrix}.
$$

Similarly, an *upper triangular* matrix is one for which $g_{ij} = 0$ whenever $i > j$. A *triangular* matrix is one that is either upper or lower triangular.

Theorem 1.3.1 *Let G be a triangular matrix. Then G is nonsingular if and only if $g_{ii} \neq 0$ for $i = 1, \ldots, n$.*

Proof. Recall from elementary linear algebra that if G is triangular, then $\det(G) = g_{11}g_{22} \cdots g_{nn}$. Thus $\det(G) \neq 0$ if and only if $g_{ii} \neq 0$ for $i = 1, \ldots, n$. See Exercises 1.3.23 and 1.3.24.   □

Lower-Triangular Systems

Consider the system

$$
Gy = b, \tag{1.3.2}
$$

where G is a nonsingular, lower-triangular matrix. It is easy to see how to solve this system if we write it out in detail:

$$
\begin{array}{rcl}
g_{11}y_1 & = & b_1 \\
g_{21}y_1 + g_{22}y_2 & = & b_2 \\
g_{31}y_1 + g_{32}y_2 + g_{33}y_3 & = & b_3 \\
\vdots & & \vdots \\
g_{n1}y_1 + g_{n2}y_2 + g_{n3}y_3 + \cdots + g_{nn}y_n & = & b_n.
\end{array}
$$

The first equation involves only the unknown y_1, the second involves only y_1 and y_2, and so on. We can solve the first equation for y_1:

$$
y_1 = b_1/g_{11}.
$$

The assumption that G is nonsingular ensures that $g_{11} \neq 0$. Now that we have y_1, we can substitute its value into the second equation and solve that equation for y_2:

$$
y_2 = (b_2 - g_{21}y_1)/g_{22}.
$$

Since G is nonsingular, $g_{22} \neq 0$. Now that we know y_2, we can use the third equation to solve for y_3, and so on. In general, once we have $y_1, y_2, \ldots, y_{i-1}$, we can solve for y_i, using the ith equation:

$$y_i = \frac{b_i - g_{i1}y_1 - g_{i2}y_2 - \cdots - g_{i,i-1}y_{i-1}}{g_{ii}},$$

which can be expressed more succinctly using sigma notation:

$$y_i = g_{ii}^{-1}\left(b_i - \sum_{j=1}^{i-1} g_{ij}y_j\right). \tag{1.3.3}$$

Since G is nonsingular, $g_{ii} \neq 0$.

This algorithm for solving a lower-triangular system is called *forward substitution* or *forward elimination*. This is the first of two versions we will consider. It is called *row-oriented* forward substitution because it accesses G by rows; the ith row is used at the ith step. It is also called the *inner-product* form of forward substitution because the sum $\sum_{j=1}^{i-1} g_{ij}y_j$ can be regarded as an inner or dot product.

Equation (1.3.3) describes the algorithm completely; it even describes the first step ($y_1 = b_1/g_{11}$), if we agree, as we shall throughout the book, that whenever the lower limit of a sum is greater than the upper limit, the sum is zero. Thus

$$y_1 = \frac{b_1 - \sum_{j=1}^{0} g_{1j}y_j}{g_{11}} = \frac{b_1 - 0}{g_{11}} = \frac{b_1}{g_{11}}.$$

rcise 1.3.4 Use pencil and paper to solve the system

$$\begin{bmatrix} 2 & 0 & 0 & 0 \\ -1 & 2 & 0 & 0 \\ 3 & 1 & -1 & 0 \\ 4 & 1 & -3 & 3 \end{bmatrix} \begin{bmatrix} y_1 \\ y_2 \\ y_3 \\ y_4 \end{bmatrix} = \begin{bmatrix} 2 \\ 3 \\ 2 \\ 9 \end{bmatrix}$$

by forward substitution. You can easily check your answer by substituting it back into the equations. This is a simple means of checking your work that you will be able to use on many of the hand computation exercises that you will be asked to perform in this chapter. □

It would be easy to write a computer program to do forward elimination. Before we write down the algorithm, notice that b_1 is only used in calculating y_1, b_2 is only used in calculating y_2, and so on. In general, once we have calculated y_i, we no longer need b_i. It is therefore usual for a computer program to store y over b. Thus we have a single array that contains b before the program is executed and y afterward.

The algorithm looks like this:

$$
\begin{aligned}
&\text{for } i = 1, \ldots, n \\
&\left[\begin{array}{l}
\text{for } j = 1, \ldots, i - 1 \text{ (not executed when } i = 1) \\
\quad \left[\; b_i \leftarrow b_i - g_{ij} b_j \right. \\
\text{if } g_{ii} = 0, \text{ set error flag, exit} \\
b_i \leftarrow b_i / g_{ii}
\end{array}\right.
\end{aligned}
\tag{1.3.5}
$$

There are no references to y, since it is stored in the array named b. The check of g_{ii} is included to make the program foolproof. There is nothing to guarantee that the program will not at some time be given (accidentally or otherwise) a singular coefficient matrix. The program needs to be able to respond appropriately to this situation. It is a good practice to check before each division that the divisor is not zero. In most linear algebra algorithms these checks do not contribute significantly to the time it takes to run the program, because the division operation is executed relatively infrequently.

To get an idea of the execution time of forward substitution, let us count the floating-point operations (flops). In the inner loop of (1.3.5), two flops are executed. These flops are performed $i - 1$ times on the ith time through the outer loop. The outer loop is performed n times, so the total number of flops performed in the j loop is $2 \times (0 + 1 + 2 + \cdots + n - 1) = 2 \sum_{i=1}^{n} (i - 1)$. Calculating this sum by a well-known trick (see Exercises 1.3.25, 1.3.26, and 1.3.28), we get $n(n - 1)$, which is approximated well by the simpler expression n^2. These considerations are summarized by the equations

$$
\sum_{i=1}^{n} \sum_{j=1}^{i-1} 2 = 2 \sum_{i=1}^{n} (i - 1) = n(n - 1) \approx n^2.
$$

Looking at the operations that are performed outside the j loop, we see that g_{ii} is compared with zero n times, and there are n divisions. Regardless of what each of these operations costs, the total cost of doing all of them is proportional to n, not n^2, and will therefore be insignificant if n is at all large. Making this assumption, we ignore the lesser costs and state simply that the cost of doing forward substitution is n^2 flops.

This figure gives us valuable qualitative information. We can expect that each time we double n, the execution time of forward substitution will be multiplied by approximately four.

Exploiting Leading Zeros in Forward Substitution

Significant savings can be achieved if some of the leading entries of b are zero. This observation will prove important when we study banded matrix computations in Section 1.5. First suppose $b_1 = 0$. Then obviously $y_1 = 0$ as well, and we do not need to make the computer do the computation $y_1 = b_1 / g_{11}$. In addition, all subsequent computations involving y_1 can be skipped. Now suppose that $b_2 = 0$ also. Then

$y_2 = b_2/g_{22} = 0$. There is no need for the computer to carry out this computation or any subsequent computation involving y_2. In general, if $b_1 = b_2 = \cdots b_k = 0$, then $y_1 = y_2 = \cdots = y_k = 0$, and we can skip all of the computations involving y_1, $\ldots y_k$. Thus (1.3.3) becomes

$$y_i = \frac{b_i - \displaystyle\sum_{j=k+1}^{n} g_{ij} y_j}{g_{ii}} \qquad i = k+1, \ldots, n. \tag{1.3.6}$$

Notice that the sum begins at $j = k + 1$.

It is enlightening to look at this from the point of view of partitioned matrices. If $b_1 = b_2 = \cdots = b_k = 0$, we can write

$$b = \begin{bmatrix} 0 \\ \hat{b}_2 \end{bmatrix} \begin{matrix} k \\ j \end{matrix} ,$$

where $j = n - k$. Partitioning G and y also, we have

$$G = \begin{matrix} k \\ j \end{matrix} \begin{matrix} k \quad\quad j \\ \begin{bmatrix} G_{11} & 0 \\ G_{21} & G_{22} \end{bmatrix} \end{matrix} \qquad y = \begin{bmatrix} \hat{y}_1 \\ \hat{y}_2 \end{bmatrix} \begin{matrix} k \\ j \end{matrix} ,$$

where G_{11} and G_{22} are lower triangular. The equation $Gy = b$ becomes

$$\begin{bmatrix} G_{11} & 0 \\ G_{21} & G_{22} \end{bmatrix} \begin{bmatrix} \hat{y}_1 \\ \hat{y}_2 \end{bmatrix} = \begin{bmatrix} 0 \\ \hat{b}_2 \end{bmatrix}$$

or

$$\begin{aligned} G_{11}\hat{y}_1 &= 0 \\ G_{21}\hat{y}_1 + G_{22}\hat{y}_2 &= \hat{b}_2. \end{aligned}$$

Since G_{11} is nonsingular (why?), the first equation implies $\hat{y}_1 = 0$. The second equation then reduces to

$$G_{22}\hat{y}_2 = \hat{b}_2.$$

Thus we only have to solve this $(n - k) \times (n - k)$ lower-triangular system, which is exactly what (1.3.6) does. G_{11} and G_{21} are not used because they interact only with the unknowns $y_1, \ldots, y_k$, (i.e. $\hat{y}_1$). Since the system now being solved is of order $n - k$, the cost is now $(n - k)^2$ flops.

rcise 1.3.7 Write a modified version of Algorithm (1.3.5) that checks for leading zeros in b and takes appropriate action. □

Column-Oriented Forward Substitution

We now derive a column-oriented version of forward substitution. Partition the system $Gy = b$ as follows:

$$\begin{bmatrix} g_{11} & 0 \\ \hat{h} & \hat{G} \end{bmatrix} \begin{bmatrix} y_1 \\ \hat{y} \end{bmatrix} = \begin{bmatrix} b_1 \\ \hat{b} \end{bmatrix}. \tag{1.3.8}$$

$\hat{h}$, $\hat{y}$, and $\hat{b}$ are vectors of length $n-1$, and $\hat{G}$ is an $(n-1) \times (n-1)$ lower-triangular matrix. The partitioned system can be written as

$$\begin{aligned} g_{11}y_1 \qquad\qquad &= b_1 \\ \hat{h}y_1 \quad + \quad \hat{G}\hat{y} &= \hat{b}. \end{aligned}$$

This leads to the following algorithm:

$$\begin{aligned} y_1 &= b_1/g_{11} \\ \tilde{b} &= \hat{b} - \hat{h}y_1 \\ &\text{solve } \hat{G}\hat{y} = \tilde{b} \text{ for } \hat{y}. \end{aligned} \qquad (1.3.9)$$

This algorithm reduces the problem of solving an $n \times n$ triangular system to that of solving the $(n-1) \times (n-1)$ system $\hat{G}\hat{y} = \tilde{b}$. This smaller problem can be reduced (by the same algorithm) to a problem of order $n-2$, which can in turn be reduced to a problem of order $n-3$, and so on. Eventually we get to the 1×1 problem $g_{nn}y_n = \check{b}_n$, which has the solution $y_n = \check{b}_n/g_{nn}$.

If you are a student of mathematics, this algorithm should remind you of proof by induction. If, on the other hand, you are a student of computer science, you might think of recursion, which is the computer science analogue of mathematical induction. Recall that a *recursive* procedure is one that calls itself. If you like to program in a computer language that supports recursion (and most modern languages do), you might enjoy writing a recursive procedure that implements (1.3.9). The procedure would perform steps one and two of (1.3.9) and then call itself to solve the reduced problem. All variables named b, $\hat{b}$, $\tilde{b}$, y, and so on, can be stored in a single array, which will contain b before execution and y afterward.

Although it is fun to write recursive programs, this algorithm can also be coded nonrecursively without difficulty. We will write down a nonrecursive formulation of the algorithm, but first it is worthwhile to work through one or two examples by hand.

Example 1.3.10 Let us use the column-oriented version of forward substitution to solve the lower-triangular system

$$\begin{bmatrix} 5 & 0 & 0 \\ 2 & -4 & 0 \\ 1 & 2 & 3 \end{bmatrix} \begin{bmatrix} y_1 \\ y_2 \\ y_3 \end{bmatrix} = \begin{bmatrix} 15 \\ -2 \\ 10 \end{bmatrix}.$$

First we calculate $y_1 = b_1/g_{11} = 15/5 = 3$. Then

$$\tilde{b} = \hat{b} - \hat{h}y_1 = \begin{bmatrix} -2 \\ 10 \end{bmatrix} - \begin{bmatrix} 2 \\ 1 \end{bmatrix} 3 = \begin{bmatrix} -8 \\ 7 \end{bmatrix}.$$

Now we have to solve $\hat{G}\hat{y} = \tilde{b}$:

$$\begin{bmatrix} -4 & 0 \\ 2 & 3 \end{bmatrix} \begin{bmatrix} y_2 \\ y_3 \end{bmatrix} = \begin{bmatrix} -8 \\ 7 \end{bmatrix}.$$

We achieve this by repeating the algorithm: $y_2 = -8/-4 = 2$,

$$\tilde{b} = \hat{b} - \hat{h}y_2 = \begin{bmatrix} 7 \end{bmatrix} - \begin{bmatrix} 2 \end{bmatrix} 2 = \begin{bmatrix} 3 \end{bmatrix},$$

$\begin{bmatrix} 3 \end{bmatrix} y_3 = \begin{bmatrix} 3 \end{bmatrix}$, and $y_3 = 3/3 = 1$. Thus

$$y = \begin{bmatrix} 3 \\ 2 \\ 1 \end{bmatrix}.$$

You can check that if you multiply G by y, you get the correct right hand side b.   □

rcise 1.3.11   Use column-oriented forward substitution to solve the system from Exercise 1.3.4.   □

rcise 1.3.12   Write a nonrecursive algorithm in the spirit of (1.3.5) that implements column-oriented forward substitution. Use a single array that contains b initially, stores intermediate results (e.g. $\hat{b}$, $\tilde{b}$) during the computation, and contains y at the end.   □

Your solution to Exercise 1.3.12 should look something like this:

$$\begin{array}{ll} \text{for } j = 1, \ldots, n \\ \quad \left[\begin{array}{l} \text{if } g_{jj} = 0, \text{ set error flag } (G \text{ is singular}), \text{ exit} \\ b_j \leftarrow b_j/g_{jj} \quad (\text{this is } y_j) \\ \text{for } i = j+1, \ldots, n \quad (\text{not executed when } j = n) \\ \quad \left[\begin{array}{l} b_i \leftarrow b_i - g_{ij}b_j \end{array} \right. \end{array} \right. \end{array} \qquad (1.3.13)$$

Notice that (1.3.13) accesses G by columns, as anticipated: on the jth step, the jth column is used. Each time through the outer loop, the size of the problem is reduced by one. On the last time through, the computation $b_n \leftarrow b_n/g_{nn}$ (giving y_n) is performed, and the inner loop is skipped.

rcise 1.3.14

(a) Count the operations in (1.3.13). Notice that the flop count is identical to that of the row-oriented algorithm (1.3.5).

(b) Convince yourself that the row- and column-oriented versions of forward substitution carry out exactly the same operations but not in the same order.

□

Like the row-oriented version, the column-oriented version can be modified to take advantage of any leading zeros that may occur in the vector b. On the jth time through the outer loop in (1.3.13), if $b_j = 0$, then no changes are made in b. Thus the jth step can be skipped whenever $b_j = 0$. (This is true regardless of whether or not b_j is a leading zero. However, once a nonzero b_j has been encountered, all subsequent b_j's will not be the originals; they will have been altered on a previous step. Therefore they are not likely to be zero.)

Which of the two versions of forward substitution is superior? The answer depends on how G is stored and accessed. This, in turn, depends on the programmer's choice of data structures and programming language and on the architecture of the computer.

Upper-Triangular Systems

As you might expect, upper-triangular systems can be solved in much the same way as lower-triangular systems. Consider the system $Ux = y$, where U is upper triangular. Writing out the system in detail we get

$$
\begin{array}{ccccccccc}
u_{11}x_1 & + & u_{12}x_2 & + & \cdots & + & u_{1,n-1}x_{n-1} & + & u_{1,n}x_n & = & y_1 \\
 & & u_{22}x_2 & + & \cdots & + & u_{2,n-1}x_{n-1} & + & u_{2n}x_n & = & y_2 \\
 & & & & & & \vdots & & & & \vdots \\
 & & & & & & u_{n-1,n-1}x_{n-1} & + & u_{n-1,n}x_n & = & y_{n-1} \\
 & & & & & & & & u_{nn}x_n & = & y_n.
\end{array}
$$

It is clear that we should solve the system from bottom to top. The nth equation can be solved for x_n, then the $(n-1)$st equation can be solved for x_{n-1}, and so on. The process is called *back substitution*, and it has row- and column-oriented versions. The cost of back substitution is obviously the same as that of forward substitution, about n^2 flops.

Exercise 1.3.15 Develop the row-oriented version of back substitution. Write pseudocode in the spirit of (1.3.5) and (1.3.13).   □

Exercise 1.3.16 Develop the column-oriented version of back substitution Write pseudocode in the spirit of (1.3.5) and (1.3.13).   □

Exercise 1.3.17 Solve the upper-triangular system

$$
\begin{bmatrix}
3 & 2 & 1 & 0 \\
0 & 1 & 2 & 3 \\
0 & 0 & -2 & 1 \\
0 & 0 & 0 & 4
\end{bmatrix}
\begin{bmatrix}
x_1 \\
x_2 \\
x_3 \\
x_4
\end{bmatrix}
=
\begin{bmatrix}
-10 \\
10 \\
1 \\
12
\end{bmatrix}
$$

(a) by row-oriented back substitution, (b) by column-oriented back substitution.   □

Block Algorithms

It is easy to develop block variants of both forward and back substitution. We will focus on forward substitution. Suppose the lower triangular matrix G has been partitioned into blocks as follows:

$$
G = \begin{array}{c} \\ r_1 \\ r_2 \\ \vdots \\ r_s \end{array}
\begin{array}{cccc} r_1 & r_2 & \cdots & r_s \\ \end{array}
\begin{bmatrix}
G_{11} & 0 & \cdots & 0 \\
G_{21} & G_{22} & & 0 \\
\vdots & \vdots & \ddots & \vdots \\
G_{s1} & G_{s2} & \cdots & G_{ss}
\end{bmatrix}.
$$

Each G_{ii} is square and lower triangular. Then the equation $Gy = b$ can be written in the partitioned form

$$
\begin{bmatrix}
G_{11} & 0 & \cdots & 0 \\
G_{21} & G_{22} & & 0 \\
\vdots & \vdots & \ddots & \vdots \\
G_{s1} & G_{s2} & \cdots & G_{ss}
\end{bmatrix}
\begin{bmatrix}
y_1 \\ y_2 \\ \vdots \\ y_s
\end{bmatrix}
=
\begin{bmatrix}
b_1 \\ b_2 \\ \vdots \\ b_s
\end{bmatrix}.
\tag{1.3.18}
$$

In this equation the entries b_i and y_i are not scalars; they are vectors with r_i components each. (The partition (1.3.8) is a special case of (1.3.18), and so is the partition used in Exercise 1.3.29 below.) Equation (1.3.18) suggests that we find y_1 by solving the system $G_{11}y_1 = b_1$. Once we have y_1, we can solve the equation $G_{21}y_1 + G_{22}y_2 = b_2$ for y_2, and so on. This leads to the block version of row-oriented forward substitution:

$$
\begin{array}{l}
\text{for } i = 1, \ldots, s \\
\quad\left[
\begin{array}{l}
\text{for } j = 1, \ldots, i - 1 \text{ (not executed when } i = 1) \\
\quad\left[\, b_i \leftarrow b_i - G_{ij}b_j \right. \\
\text{if } G_{ii} \text{ is singular, set error flag, exit} \\
b_i \leftarrow G_{ii}^{-1}b_i
\end{array}
\right.
\end{array}
\tag{1.3.19}
$$

This is nearly identical to (1.3.5). The operation $b_i \leftarrow G_{ii}^{-1}b_i$ does not require explicit computation of G_{ii}^{-1}. It can be effected by solving the lower-triangular system $G_{ii}x = b_i$ by either row- or column-oriented forward substitution.

Exercise 1.3.20 Write the block variant of the column-oriented forward substitution algorithm (1.3.13). □

Exercise 1.3.21 Convince yourself that the block versions of forward substitution perform exactly the same arithmetic as the scalar algorithms (1.3.5) and (1.3.13), but not in the same order. □

Additional Exercises

Exercise 1.3.22 Write Fortran subroutines to do each of the following: (a) row-oriented forward substitution, (b) column-oriented forward substitution, (c) row-oriented back substitution, (d) column-oriented back substitution. Invent some problems on which to test your programs.

An easy way to devise a problem $Ax = b$ with a known solution is to specify the matrix A and the solution x, then multiply A by x to get b. Give A and b to your program, and see if it can calculate x.   □

Exercise 1.3.23 Prove that if G is triangular, then $\det(G) = g_{11}g_{22}\cdots g_{nn}$.   □

Exercise 1.3.24 Devise a proof of Theorem 1.3.1 that does not use determinants. For example, use condition (c) or (d) of Theorem 1.2.3 instead.   □

Exercise 1.3.25 Write down the numbers $1, 2, 3, \ldots, n-1$ on a line. Immediately below these, write down the numbers $n-1, n-2, n-3, \ldots, 1$. Add these numbers up by summing column-wise first. Conclude that

$$2 \times (1 + 2 + 3 + \cdots + n - 1) = n(n-1).$$

□

Exercise 1.3.26   In this exercise you will use mathematical induction to draw the same conclusion as in the previous exercise. If you are weak on mathematical induction, you should definitely work this exercise. We wish to prove that

$$\sum_{i=1}^{n-1} i = \frac{n(n-1)}{2} \tag{1.3.27}$$

for all positive integers n. Begin by showing that (1.3.27) holds when $n = 1$. The sum on the left-hand side is empty in this case. If you feel nervous about this, you can check the case $n = 2$ as well. Next show that if (1.3.27) holds for $n = k$, then it holds also holds for $n = k + 1$. That is, show that

$$\sum_{i=1}^{k} i = \frac{(k+1)k}{2}$$

is true, assuming that

$$\sum_{i=1}^{k-1} i = \frac{k(k-1)}{2}$$

is true. This is just a matter of simple algebra. Once you have done this, you will have proved by induction that (1.3.27) holds for all positive integers n.   □

Exercise 1.3.28 This exercise introduces a useful approximation technique. Draw pictures that demonstrate the inequalities

$$\int_0^{n-1} x\,dx \le \sum_{i=1}^{n-1} i \le \int_0^n x\,dx.$$

Evaluate the integrals and deduce that

$$\sum_{i=1}^{n-1} i = \frac{n^2}{2} + O(n) \approx \frac{n^2}{2}.$$

Show that the same result holds for $\sum_{i=1}^{n} i$.   □

rcise 1.3.29 We derived the column-oriented version of forward substitution by partitioning the system $Gy = b$. Different partitions lead to different versions. For example, consider the partition

$$
\begin{bmatrix} \hat{G} & 0 \\ h^T & g_{nn} \end{bmatrix} \begin{bmatrix} \hat{y} \\ y_n \end{bmatrix} = \begin{bmatrix} \hat{b} \\ b_n \end{bmatrix},
$$

where $\hat{G}$ is $(n-1) \times (n-1)$.

(a) Derive a recursive algorithm based on this partition.

(b) Write a nonrecursive version of the algorithm. (*Hint:* Think about how your recursive algorithm would calculate y_i, given $y_1, \ldots, y_{i-1}$.)

Observe that your nonrecursive algorithm is nothing but row-oriented forward substitution.   □

1.4 POSITIVE DEFINITE SYSTEMS; CHOLESKY DECOMPOSITION

In this section we discuss the problem of solving systems of linear equations for which the coefficient matrix is of a special form, namely, positive definite. If you would prefer to read about general systems first, you can skip ahead to Sections 1.7 and 1.8.

Recall that the *transpose* of an $n \times m$ matrix $A = (a_{ij})$, denoted A^T, is the $m \times n$ matrix whose (i, j) entry is a_{ji}. Thus the rows of A^T are the columns of A. A square matrix A is *symmetric* if $A = A^T$, that is, $a_{ij} = a_{ji}$ for all i and j. The matrices in the electrical circuit examples of Section 1.2 are all symmetric, as are those in the examples of mass-spring systems. The matrix in Example 1.2.12 (approximation of a differential equation) is not symmetric unless c (convection coefficient) is zero.

Since every vector is also a matrix, every vector has a transpose: A column vector x is a matrix with one column. Its transpose x^T is a matrix with one row. The set of column n-vectors with real entries will be denoted $\mathbb{R}^n$. That is, $\mathbb{R}^n$ is just the set of real $n \times 1$ matrices.

If A is $n \times n$, real, symmetric, and also satisfies the property

$$
x^T A x > 0 \tag{1.4.1}
$$

for all nonzero $x \in \mathbb{R}^n$, then A is said to be *positive definite*.[5] The left-hand side of (1.4.1) is a matrix product. Examining the dimensions of x^T, A, and x, we find that $x^T A x$ is a 1×1 matrix, that is, a real number. Thus (1.4.1) is just an inequality between real numbers. It is also possible to define complex positive definite matrices. See Exercises 1.4.63 through 1.4.65.

---

[5] Some books, notably [43], do not include symmetry as part of the definition.

Positive definite matrices arise frequently in applications. Typically the expression $x^T Ax$ has physical significance. For example, the matrices in the electrical circuit problems of Section 1.2 are all positive definite. In the examples in which the entries of x are loop currents, $x^T Ax$ turns out to be the total power drawn by the resistors in the circuit (Exercise 1.4.66). This is clearly a positive quantity. In the examples in which the entries of x are nodal voltages, $x^T Ax$ is closely related to (and slightly exceeds) the power drawn by the circuit (Exercise 1.4.67).

The matrices of the mass-spring systems in Section 1.2 are also positive definite. In those systems $\frac{1}{2} x^T Ax$ is the *strain energy* of the system, the energy stored in the springs due to compression or stretching (Exercise 1.4.68). This is a positive quantity.

Other areas in which positive definite matrices arise are least-squares problems (Chapter 3), statistics (Exercise 1.4.69), and the numerical solution of partial differential equations (Chapter 8).

Theorem 1.4.2 *If A is positive definite, then A is nonsingular.*

Proof. We will prove the contrapositive form of the theorem: If A is singular, then A is not positive definite. Assume A is singular. Then by Theorem 1.2.3, part (b), there is a nonzero $y \in \mathbb{R}^n$ such that $Ay = 0$. But then $y^T Ay = 0$, so A is not positive definite. $\square$

Corollary 1.4.3 *If A is positive definite, the linear system $Ax = b$ has exactly one solution.*

Theorem 1.4.4 *Let M be any $n \times n$ nonsingular matrix, and let $A = M^T M$. Then A is positive definite.*

Proof. First we must show that A is symmetric. Recalling the elementary formulas $(BC)^T = C^T B^T$ and $B^{TT} = B$, we find that $A^T = (M^T M)^T = M^T M^{TT} = M^T M = A$. Next we must show that $x^T Ax > 0$ for all nonzero x. For any such x, we have $x^T Ax = x^T M^T Mx$. Let $y = Mx$, so that $y^T = x^T M^T$. Then $x^T Ax = y^T y = \sum_{i=1}^n y_i^2$. This sum of squares is certainly nonnegative, and it is strictly positive if $y \neq 0$. But clearly $y = Mx \neq 0$, because M is nonsingular, and x is nonzero. Thus $x^T Ax > 0$ for all nonzero x, and A is positive definite. $\square$

Theorem 1.4.4 provides an easy means of constructing positive definite matrices: Just multiply any nonsingular matrix by its transpose.

Example 1.4.5 Let $M = \begin{bmatrix} 1 & 2 \\ 3 & 4 \end{bmatrix}$. M is nonsingular, since $\det(M) = -2$. Therefore $A = M^T M = \begin{bmatrix} 10 & 14 \\ 14 & 20 \end{bmatrix}$ is positive definite. $\square$

Example 1.4.6 Let

$$M = \begin{bmatrix} 1 & 1 & 1 \\ 0 & 1 & 1 \\ 0 & 0 & 1 \end{bmatrix}.$$

M is nonsingular, since $\det(M) = 1$. Therefore

$$A = M^T M = \begin{bmatrix} 1 & 1 & 1 \\ 1 & 2 & 2 \\ 1 & 2 & 3 \end{bmatrix}$$

is positive definite.   □

The next theorem, the Cholesky Decomposition Theorem, is the most important result of this section. It states that every positive definite matrix is of the form $M^T M$ for some M. Thus the recipe given by Theorem 1.4.4 generates all positive definite matrices. Furthermore M can be chosen to have a special form.

Theorem 1.4.7 *(Cholesky Decomposition Theorem) Let A be positive definite. Then A can be decomposed in exactly one way into a product*

$$A = R^T R \qquad \text{(Cholesky Decomposition)}$$

such that R is upper triangular and has all main diagonal entries r_{ii} positive. R is called the Cholesky factor *of A.*

The theorem will be proved later in the section. Right now it is more important to discuss how the Cholesky decomposition can be used and figure out how to compute the Cholesky factor.

Example 1.4.8 Let

$$A = \begin{bmatrix} 1 & 1 & 1 \\ 1 & 2 & 2 \\ 1 & 2 & 3 \end{bmatrix} \quad \text{and} \quad R = \begin{bmatrix} 1 & 1 & 1 \\ 0 & 1 & 1 \\ 0 & 0 & 1 \end{bmatrix}.$$

R is upper triangular and has positive main-diagonal entries. In Example 1.4.6 we observed that $A = R^T R$. Therefore R is the Cholesky factor of A.   □

The Cholesky decomposition is useful because R and R^T are triangular. Suppose we wish to solve the system $Ax = b$, where A is positive definite. If we know the Cholesky factor R, we can write the system as $R^T R x = b$. Let $y = Rx$. We do not know x, so we do not know y either. However, y clearly satisfies $R^T y = b$. Since R^T is lower triangular, we can solve for y by forward substitution. Once we have y, we can solve the upper-triangular system $Rx = y$ for x by back substitution. The total flop count is a mere $2n^2$, if we know the Cholesky factor R.

If the Cholesky decomposition is to be a useful tool, we must find a practical method for calculating it. One of the easiest ways to do this is to write out the decomposition $A = R^T R$ in detail and study it:

$$
\begin{bmatrix}
a_{11} & a_{12} & a_{13} & \cdots & a_{1n} \\
a_{21} & a_{22} & a_{23} & \cdots & a_{2n} \\
a_{31} & a_{32} & a_{33} & \cdots & a_{3n} \\
\vdots & \vdots & \vdots & \ddots & \vdots \\
a_{n1} & a_{n2} & a_{n3} & \cdots & a_{nn}
\end{bmatrix}
$$
$$
=
\begin{bmatrix}
r_{11} & 0 & 0 & \cdots & 0 \\
r_{12} & r_{22} & 0 & \cdots & 0 \\
r_{13} & r_{23} & r_{33} & & 0 \\
\vdots & \vdots & \vdots & \ddots & \\
r_{1n} & r_{2n} & r_{3n} & \cdots & r_{nn}
\end{bmatrix}
\begin{bmatrix}
r_{11} & r_{12} & r_{13} & \cdots & r_{1n} \\
0 & r_{22} & r_{23} & \cdots & r_{2n} \\
0 & 0 & r_{33} & \cdots & r_{3n} \\
\vdots & \vdots & & \ddots & \vdots \\
0 & 0 & 0 & & r_{nn}
\end{bmatrix}.
$$

The element a_{ij} is the (inner) product of the ith row of R^T with the jth column of R. Noting that the first row of R^T has only one nonzero entry, we focus on this row:

$$a_{1j} = r_{11}r_{1j} + 0r_{2j} + 0r_{3j} + \cdots + 0r_{nj} = r_{11}r_{1j}.$$

In particular, when $j = 1$ we have $a_{11} = r_{11}^2$, which tells us that

$$r_{11} = +\sqrt{a_{11}}. \tag{1.4.9}$$

We know that the positive square root is the right one, because the main-diagonal entries of R are positive. Now that we know r_{11}, we can use the equation $a_{1j} = r_{11}r_{1j}$ to calculate the rest of the first row of R:

$$r_{1j} = a_{1j}/r_{11}, \quad j = 2, \ldots, n. \tag{1.4.10}$$

This is also the first column of R^T. We next focus on the second row, because the second row of R^T has only two nonzero entries. We have

$$a_{2j} = r_{12}r_{1j} + r_{22}r_{2j}. \tag{1.4.11}$$

Only elements from the first two rows of R appear in this equation. In particular, when $j = 2$ we have $a_{22} = r_{12}^2 + r_{22}^2$. Since r_{12} is already known, we can use this equation to calculate r_{22}:

$$r_{22} = +\sqrt{a_{22} - r_{12}^2}.$$

Once r_{22} is known, the only unknown left in (1.4.11) is r_{2j}. Thus we can use (1.4.11) to compute the rest of the second row of R:

$$r_{2j} = (a_{2j} - r_{12}r_{1j})/r_{22}, \quad j = 3, \ldots, n.$$

There is no need to calculate r_{21} because $r_{21} = 0$. We now know the first two rows of R (and the first two columns of R^T). Now, as an exercise, you can show how to calculate the third row of R.

Now let us see how to calculate the ith row of R, assuming that we already have the first $i - 1$ rows. Since only the first i entries in the ith row of R^T are nonzero,

$$a_{ij} = r_{1i}r_{1j} + r_{2i}r_{2j} + \cdots + r_{i-1,i}r_{i-1,j} + r_{ii}r_{ij}. \tag{1.4.12}$$

All entries of R appearing in (1.4.12) lie in the first i rows. Since the first $i - 1$ rows are known, the only unknowns in (1.4.12) are r_{ii} and r_{ij}. Taking $i = j$ in (1.4.12), we have

$$a_{ii} = r_{1i}^2 + r_{2i}^2 + \cdots + r_{i-1,i}^2 + r_{ii}^2,$$

which we can solve for r_{ii}:

$$r_{ii} = +\sqrt{a_{ii} - \sum_{k=1}^{i-1} r_{ki}^2}. \tag{1.4.13}$$

Now that we have r_{ii}, we can use (1.4.12) to solve for r_{ij}:

$$r_{ij} = \left(a_{ij} - \sum_{k=1}^{i-1} r_{ki}r_{kj} \right) \Big/ r_{ii} \qquad j = i + 1, \ldots, n. \tag{1.4.14}$$

We do not have to calculate r_{ij} for $j < i$ because those entries are all zero.

Equations (1.4.13) and (1.4.14) give a complete recipe for calculating R. They even hold for the first row of R ($i = 1$) if we make use of our convention that the sums $\sum_{k=1}^{0}$ are zero. Notice also that when $i = n$, nothing is done in (1.4.14); the only nonzero entry in the nth row of R is r_{nn}.

The algorithm we have just developed is called *Cholesky's method*. This, the first of several formulations that we will derive, is called the *inner-product formulation* because the sums in (1.4.13) and (1.4.14) can be regarded as inner products. Cholesky's method turns out to be closely related to the familiar Gaussian elimination method. The connection between them is established in Section 1.7.

A number of important observations can now be made. First, recall that the Cholesky decomposition theorem (which we haven't proved yet) makes two assertions: (i) R exists, and (ii) R is unique. In the process of developing the inner-product form of Cholesky's method, we have proved that R is unique: The equation $A = R^T R$ and the stipulation that R is upper triangular with $r_{11} > 0$ imply (1.4.9). Thus this value of r_{11} is the only one that will work; r_{11} is uniquely determined. Similarly, r_{1j} is uniquely determined by (1.4.10) for $j = 2, \ldots, n$. Thus the first row of R is uniquely determined. Now suppose the first $i - 1$ rows are uniquely determined. Then so is the ith row, for r_{ii} is uniquely specified by (1.4.13), and r_{ij} is uniquely determined by (1.4.14) for $j = i + 1, \ldots, n$. Notice the importance of the stipulation $r_{ii} > 0$ in determining which square root to choose. Without this stipulation we would not have uniqueness.

rcise 1.4.15 Let $A = \begin{bmatrix} 4 & 0 \\ 0 & 9 \end{bmatrix}$. (a) Prove that A is positive definite. (b) Calculate the Cholesky factor of A. (c) Find three other upper triangular matrices R such that $A = R^T R$. (d) Let A be any $n \times n$ positive definite matrix. How many upper-triangular matrices R such that $A = R^T R$ are there?   □

The next important observation is that Cholesky's method serves as a test of positive definiteness. By Theorems 1.4.4 and 1.4.7, A is positive definite if and only if there exists an upper triangular matrix R with positive main diagonal entries such that $A = R^T R$. Given any symmetric matrix A, we can attempt to calculate R by Cholesky's method. If A is not positive definite, the algorithm must fail, because any R that satisfies (1.4.13) and (1.4.14) must also satisfy $A = R^T R$. The algorithm fails if and only if at some step the number under the square root sign in (1.4.13) is negative or zero. In the first case there is no real square root; in the second case $r_{ii} = 0$. Thus, if A is not positive definite, there must come a step at which the algorithm attempts to take the square root of a number that is not positive. Conversely, if A is positive definite, the algorithm cannot fail. The equation $A = R^T R$ guarantees that the number under the square root sign in (1.4.13) is positive at every step. (After all, it equals r_{ii}^2.) Thus Cholesky's method succeeds if and only if A is positive definite. This is the best general test of positive definiteness known.

The next thing to notice is that (1.4.13) and (1.4.14) use only those a_{ij} for which $i \leq j$. This is not surprising, since in a symmetric matrix the entries above the main diagonal are identical to those below the main diagonal. This underscores the fact that in a computer program we do not need to store all of A; there is no point in duplicating information. If space is at a premium, the programmer may choose to store A in a long one-dimensional array with $a_{11}, a_{12}, \ldots, a_{1n}$ immediately followed by $a_{22}, a_{23}, \ldots, a_{2n}$, immediately followed by $a_{33}, \ldots, a_{3n}$, and so on. This compact storage scheme makes the programming more difficult but is worth using if space is scarce.

Finally we note that each element a_{ij} is used only to compute r_{ij}, as is clear from (1.4.13) and (1.4.14). It follows that in a computer program, r_{ij} can be stored over a_{ij}. This saves additional space by eliminating the need for separate arrays to store R and A.

Exercise 1.4.16 Write an algorithm based on (1.4.13) and (1.4.14) that checks a matrix for positive definiteness and calculates R, storing R over A.   □

Your solution to Exercise 1.4.16 should look something like this:

Cholesky's Algorithm (inner-product form)
for $i = 1, \ldots, n$
⎡ for $k = 1, \ldots, i-1$ (not executed when $i = 1$)
⎢ ⎡ $a_{ii} \leftarrow a_{ii} - a_{ki}^2$
⎢ if $a_{ii} \leq 0$, set error flag (A is not positive definite), exit
⎢ $a_{ii} \leftarrow \sqrt{a_{ii}}$ (this is r_{ii})   (1.4.17)
⎢ for $j = i+1, \ldots, n$ (not executed when $i = n$)
⎢ ⎡ for $k = 1, \ldots, i-1$ (not executed when $i = 1$)
⎢ ⎢ ⎡ $a_{ij} \leftarrow a_{ij} - a_{ki} a_{kj}$
⎣ ⎣ $a_{ij} \leftarrow a_{ij}/a_{ii}$ (this is r_{ij})

The upper part of R is stored over the upper part of A. There is no need to store the lower part of R because it consists entirely of zeros.

Example 1.4.18 Let

$$A = \begin{bmatrix} 4 & -2 & 4 & 2 \\ -2 & 10 & -2 & -7 \\ 4 & -2 & 8 & 4 \\ 2 & -7 & 4 & 7 \end{bmatrix} \quad \text{and} \quad b = \begin{bmatrix} 8 \\ 2 \\ 16 \\ 6 \end{bmatrix}.$$

Notice that A is symmetric. We will use Cholesky's method to show that A is positive definite and calculate the Cholesky factor R. We will then use the Cholesky factor to solve the system $Ax = b$ by forward and back substitution.

$$
\begin{aligned}
r_{11} &= \sqrt{a_{11}} = \sqrt{4} = 2 \\
r_{12} &= a_{12}/r_{11} = -2/2 = -1 \\
r_{13} &= 2 \\
r_{14} &= 1 \\
r_{22} &= \sqrt{a_{22} - r_{12}^2} = \sqrt{10 - (-1)^2} = 3 \\
r_{23} &= \frac{a_{23} - r_{13}r_{12}}{r_{22}} = \frac{-2 - (-1)2}{3} = 0 \\
r_{24} &= -2 \\
r_{33} &= \sqrt{a_{33} - r_{13}^2 - r_{23}^2} = \sqrt{8 - 2^2 - 0^2} = 2 \\
r_{34} &= \frac{a_{34} - r_{14}r_{13} - r_{24}r_{23}}{r_{33}} = \frac{4 - (2)(1) - (0)(-2)}{2} = 1 \\
r_{44} &= \sqrt{a_{44} - r_{14}^2 - r_{24}^2 - r_{34}^2} = \sqrt{7 - 1^2 - (-2)^2 - 1^2} = 1.
\end{aligned}
$$

Thus

$$R = \begin{bmatrix} 2 & -1 & 2 & 1 \\ 0 & 3 & 0 & -2 \\ 0 & 0 & 2 & 1 \\ 0 & 0 & 0 & 1 \end{bmatrix}.$$

Since we were able to calculate R, A is positive definite. We can check our work by multiplying R^T by R and seeing that the product is A.

To solve $Ax = b$, we first solve $R^T y = b$ by forward substitution and obtain $y = \begin{bmatrix} 4 & 2 & 4 & 2 \end{bmatrix}^T$. We then solve $Rx = y$ by back substitution to obtain $x = \begin{bmatrix} 1 & 2 & 1 & 2 \end{bmatrix}^T$. Finally, we check our work by multiplying A by x and seeing that we get b. □

Exercise 1.4.19 Calculate R (of Example 1.4.18) by the *erasure method*. Start with an array that has A penciled in initially (main diagonal and upper triangle only). As you calculate each entry r_{ij}, erase a_{ij} and replace it by r_{ij}. Do all of the operations in your head, using only the single array for reference. This procedure is surprisingly easy, once you get the hang of it.   □

Example 1.4.20 Let

$$A = \begin{bmatrix} 1 & 2 & 3 \\ 2 & 5 & 10 \\ 3 & 10 & 16 \end{bmatrix}.$$

We will use Cholesky's method to determine whether or not A is positive definite. Proceeding as in Example 1.4.18, we find that $r_{11} = 1$, $r_{12} = 2$, $r_{13} = 3$, $r_{22} = 1$, $r_{23} = 4$, and finally $r_{33} = \sqrt{a_{33} - r_{13}^2 - r_{23}^2} = \sqrt{-9}$. In attempting to calculate r_{33}, we encounter a negative number under the square root sign. Thus A is not positive definite.   □

Exercise 1.4.21 Let

$$A = \begin{bmatrix} 16 & 4 & 8 & 4 \\ 4 & 10 & 8 & 4 \\ 8 & 8 & 12 & 10 \\ 4 & 4 & 10 & 12 \end{bmatrix} \quad \text{and} \quad b = \begin{bmatrix} 32 \\ 26 \\ 38 \\ 30 \end{bmatrix}.$$

Notice that A is symmetric. (a) Use the inner-product formulation of Cholesky's method to show that A is positive definite and compute its Cholesky factor. (b) Use forward and back substitution to solve the linear system $Ax = b$.   □

Exercise 1.4.22 Determine whether or not each of the following matrices is positive definite.

$$A = \begin{bmatrix} 9 & 3 & 3 \\ 3 & 10 & 7 \\ 3 & 5 & 9 \end{bmatrix} \quad B = \begin{bmatrix} 4 & 2 & 6 \\ 2 & 2 & 5 \\ 6 & 5 & 29 \end{bmatrix}$$

$$C = \begin{bmatrix} 4 & 4 & 8 \\ 4 & -4 & 1 \\ 8 & 1 & 6 \end{bmatrix} \quad D = \begin{bmatrix} 1 & 1 & 1 \\ 1 & 2 & 2 \\ 1 & 2 & 1 \end{bmatrix}$$

□

Exercise 1.4.23 Rework Exercise 1.4.22 with the help of MATLAB. The MATLAB command R = chol(A) computes the Cholesky factor of A and stores it in R. The upper half of A is used in the computation. (MATLAB does not check whether or not A is symmetric. For more details about chol, type help chol or read about chol in MATLAB's help browser.)   □

Although Cholesky's method generally works well, a word of caution is appropriate here. Unlike the small hand computations that are scattered throughout the book, most matrix computations are performed by computer, in which case the arithmetic operations are subject to roundoff errors. In Chapter 2 we will see that the performance of Cholesky's method in the face of roundoff errors is as good as we could hope for. However, there are linear systems, called ill-conditioned systems, that simply cannot be solved accurately in the presence of errors. Naturally we cannot expect Cholesky's method (performed with roundoff errors) to solve ill-conditioned systems accurately. For more on ill-conditioned systems and roundoff errors, see Chapter 2.

Flop Count

To count the flops in Cholesky's algorithm (1.4.17), we need to know that

$$\sum_{i=1}^{n} i^2 = \frac{n^3}{3} + O(n^2).$$

The easiest way to obtain this is to approximate the sum by an integral:

$$\sum_{i=1}^{n} i^2 \approx \int_0^n i^2 \, di = \frac{n^3}{3}.$$

The details are discussed in Exercises 1.4.70 and 1.4.71.

Proposition 1.4.24 *Cholesky's algorithm (1.4.17) applied to an $n \times n$ matrix performs about $n^3/3$ flops.*

rcise 1.4.25 Prove Proposition 1.4.24 □

Proof. Examining (1.4.17), we see that in each of the two k loops, two flops are performed. To see how many times each loop is executed, we look at the limits on the loop indices. We conclude that the number of flops attributable to the first of the k loops is

$$\sum_{i=1}^{n} \sum_{k=1}^{i-1} 2 = 2 \sum_{i=1}^{n} (i-1) = n(n-1) \approx n^2$$

by Exercise 1.3.25 or 1.3.26. Applying the same procedure to the second of the k loops, we get a flop count of

$$\sum_{i=1}^{n} \sum_{j=i+1}^{n} \sum_{k=1}^{i-1} 2.$$

We have a triple sum this time, because the loops are nested three deep.

$$
\begin{aligned}
\sum_{i=1}^{n} \sum_{j=i+1}^{n} \sum_{k=1}^{i-1} 2 &= 2 \sum_{i=1}^{n} \sum_{j=i+1}^{n} (i-1) \\
&= 2 \sum_{i=1}^{n} (n-i)(i-1) \\
&= 2n \sum_{i=1}^{n} (i-1) - 2 \sum_{i=1}^{n} i^2 + 2 \sum_{i=1}^{n} i
\end{aligned}
$$

$$= n^3 - 2\frac{n^3}{3} + O(n^2)$$

$$\approx \frac{n^3}{3}.$$

Here we have used the estimates $2\sum_{i=1}^{n}(i-1) = n^2 + O(n)$ and $2\sum_{i=1}^{n} i = n^2 + O(n)$. In the end we discard the $O(n^2)$ term, because it is small in comparison with the term $n^3/3$, once n is sufficiently large. Thus about $n^3/3$ flops are performed in the second k loop. Notice that the number of flops performed in the first k loop is negligible by comparison.

In addition to the flops in the k loops, there are some divisions. The exact number is

$$\sum_{i=1}^{n}\sum_{j=i+1}^{n} 1 = \sum_{i=1}^{n}(n-i) = \frac{n(n-1)}{2},$$

which is also negligible. Finally, $\sum_{i=1}^{n} 1 = n$ error checks and square roots are done. We conclude that the flop count for (1.4.17) is $n^3/3 + O(n^2)$.   □

Since the flop count is $O(n^3)$, we expect that each time we double the matrix dimension, the time it takes to compute the Cholesky factor will be multiplied by about eight. See Exercise 1.4.72.

If we wish to solve a system $Ax = b$ by Cholesky's method, we must first compute the Cholesky decomposition at a cost of about $n^3/3$ flops. Then we must perform forward and back substitution using the Cholesky factor and its transpose at a total cost of about $2n^2$ flops. We conclude that the bulk of the time is spent computing the Cholesky factor; the forward and backward substitution times are negligible. Thus the cost of solving a large system using Cholesky's method can be reckoned to be $n^3/3$ flops. Each time we double the dimension of the system, we can expect the time it takes to solve $Ax = b$ by Cholesky's method to be multiplied by about eight.

Outer-Product Form of Cholesky's Method

The *outer-product* form of Cholesky's method is derived by partitioning the equation $A = R^T R$ in the form

$$\begin{bmatrix} a_{11} & b^T \\ b & \hat{A} \end{bmatrix} = \begin{bmatrix} r_{11} & 0 \\ s & \hat{R}^T \end{bmatrix} \begin{bmatrix} r_{11} & s^T \\ 0 & \hat{R} \end{bmatrix}. \tag{1.4.26}$$

Equating the blocks, we obtain

$$a_{11} = r_{11}^2, \qquad b^T = r_{11}s^T, \qquad \hat{A} = ss^T + \hat{R}^T\hat{R}. \tag{1.4.27}$$

The fourth equation, $b = sr_{11}$, is redundant. Equations (1.4.27) suggest the following procedure for calculating r_{11}, s^T, and $\hat{R}$ (and hence R):

$$\begin{aligned} r_{11} &= \sqrt{a_{11}} \\ s^T &= r_{11}^{-1}b^T \\ \tilde{A} &= \hat{A} - ss^T \\ \text{Solve} \quad \tilde{A} &= \hat{R}^T\hat{R} \quad \text{for } \hat{R}. \end{aligned} \tag{1.4.28}$$

This procedure reduces the $n \times n$ problem to that of finding the Cholesky factor of the $(n-1) \times (n-1)$ matrix $\tilde{A}$. This problem can be reduced to an $(n-2) \times (n-2)$ problem by the same algorithm, and so on. Eventually the problem is reduced to the trivial 1×1 case. This is called the *outer-product* formulation because at each step an outer product ss^T is subtracted from the remaining submatrix. It can be implemented recursively or nonrecursively with no difficulty.

rcise 1.4.29 Use the outer-product formulation of Cholesky's method to calculate the Cholesky factor of the matrix of Example 1.4.18. $\qquad\square$

rcise 1.4.30 Use the outer-product formulation of Cholesky's method to work Example 1.4.20. $\qquad\square$

rcise 1.4.31 Use the outer-product form to work part (a) of Exercise 1.4.21. $\qquad\square$

rcise 1.4.32 Use the outer-product form to work Exercise 1.4.22. $\qquad\square$

rcise 1.4.33 Write a nonrecursive algorithm that implements the outer-product formulation of Cholesky's algorithm (1.4.28). Your algorithm should exploit the symmetry of A by referencing only the main diagonal and upper part of A, and it should store R over A. Be sure to put in the necessary check before taking the square root. $\qquad\square$

rcise 1.4.34 (a) Do a flop count for the outer-product formulation of Cholesky's method. You will find that approximately $n^3/3$ flops are performed, the same number as for the inner-product formulation. (If you do an exact flop count, you will find that the counts are exactly equal.) (b) Convince yourself that the outer-product and inner-product formulations of the Cholesky algorithm perform exactly the same operations, but not in the same order. $\qquad\square$

Bordered Form of Cholesky's Method

The bordered form will prove useful in the next section, where we develop a version of Cholesky's method for banded matrices. We start by introducing some new notation and terminology. For $j = 1, \ldots, n$ let A_j be the $j \times j$ submatrix of A consisting of the intersection of the first j rows and columns of A. A_j is called the jth *leading principal submatrix* of A. In Exercise 1.4.54 you will show that if A is positive definite, then all of its leading principal submatrices are positive definite. Suppose A is positive definite, and let R be the Cholesky factor of A. Then R has leading principal submatrices R_j, $j = 1, \ldots, n$, which are upper triangular and have positive entries on the main diagonal.

rcise 1.4.35 By partitioning the equation $A = R^T R$ appropriately, show that R_j is the Cholesky factor of A_j for $j = 1, \ldots, n$. $\qquad\square$

It is easy to construct $R_1 = [r_{11}]$, since $a_{11} = r_{11}^2$. Thinking inductively, if we can figure out how to construct R_j, given R_{j-1}, then we will be able to construct

$R_n = R$ in n steps. Suppose, therefore, that we have calculated R_{j-1} and wish to find R_j. Partitioning the equation $A_j = R_j^T R_j$ so that A_{j-1} and R_{j-1} appear as blocks:

$$\begin{bmatrix} A_{j-1} & c \\ c^T & a_{jj} \end{bmatrix} = \begin{bmatrix} R_{j-1}^T & 0 \\ h^T & r_{jj} \end{bmatrix} \begin{bmatrix} R_{j-1} & h \\ 0 & r_{jj} \end{bmatrix}, \tag{1.4.36}$$

we get the equations

$$A_{j-1} = R_{j-1}^T R_{j-1}, \qquad c = R_{j-1}^T h, \qquad a_{jj} = h^T h + r_{jj}^2. \tag{1.4.37}$$

Since we already have R_{j-1}, we have only to calculate h and r_{jj} to get R_j. Equations (1.4.37) show how this can be done. First solve the equation $c = R_{j-1}^T h$ for h by forward substitution. Then calculate $r_{jj} = \sqrt{a_{jj} - h^T h}$. The algorithm that can be built along these lines is called the *bordered form* of Cholesky' method.

Exercise 1.4.38 Use the bordered form of Cholesky's method to calculate the Cholesky factor of the matrix of Example 1.4.18.   □

Exercise 1.4.39 Use the bordered form of Cholesky's method to work Example 1.4.20.   □

Exercise 1.4.40 Use the bordered form to work part (a) of Exercise 1.4.21.   □

Exercise 1.4.41 Use the bordered form to work Exercise 1.4.22.   □

Exercise 1.4.42 (a) Do a flop count for the bordered form of Cholesky's method. Again you will find that approximately $n^3/3$ flops are done. (b) Convince yourself that this algorithm performs exactly the same arithmetic operations as the other two formulations of Cholesky's method.   □

We have now introduced three different versions of Cholesky's method. We have observed that the inner-product formulation (1.4.17) has triply nested loops; so do the others. If one examines all of the possibilities, one finds that there are six distinct basic variants, associated with the six (= 3!) different ways of nesting three loops. Exercise 1.7.55 discusses the six variants.

Computing the Cholesky Decomposition by Blocks

All formulations of the Cholesky decomposition algorithm have block versions. As we have shown in Section 1.1, block implementations can have superior performance on large matrices because of more efficient use of cache and greater ease of parallelization.

Let us develop a block version of the outer-product form. Generalizing (1.4.26), we write the equation $A = R^T R$ by blocks:

$$\begin{bmatrix} A_{11} & B \\ B^T & \hat{A} \end{bmatrix} = \begin{bmatrix} R_{11}^T & 0 \\ S^T & \hat{R}^T \end{bmatrix} \begin{bmatrix} R_{11} & S \\ 0 & \hat{R} \end{bmatrix}.$$

A_{11} and R_{11} are square matrices of dimension $d_1 \times d_1$, say. A_{11} is symmetric and (by Exercise 1.4.54) positive definite. By the way, this equation also generalizes (1.4.36). Equating the blocks, we have the equations

$$A_{11} = R_{11}^T R_{11}, \quad B = R_{11}^T S, \quad \hat{A} = S^T S + \hat{R}^T \hat{R},$$

which suggest the following procedure for calculating R.

$$
\begin{aligned}
R_{11} &= \text{cholesky}(A_{11}) \\
S &= R_{11}^{-T} B \\
\tilde{A} &= \hat{A} - S^T S \\
\hat{R} &= \text{cholesky}(\tilde{A}).
\end{aligned}
\tag{1.4.43}
$$

The symbol R_{11}^{-T} means $(R_{11}^{-1})^T$, which is the same as $(R_{11}^T)^{-1}$. The operation $S = R_{11}^{-T} B$ does not require the explicit computation of R_{11}^{-T}. Instead we can refer back to the equation $R_{11}^T S = B$. Letting s and b denote the ith columns of S and B, respectively, we see that $R_{11}^T s = b$. Since R_{11}^T is lower triangular, we can obtain s from b by forward substitution. (Just such an operation is performed in the bordered form of Cholesky's method.) Performing this operation for each column of S, we obtain S from B. Notice that these forward substitutions need not be done one after the other; they can all be done at once.

cise 1.4.44 Let C be any nonsingular matrix. Show that $(C^{-1})^T = (C^T)^{-1}$. $\qquad \square$

The matrices $\hat{A}$, B, $\hat{R}$, and S may themselves be partitioned into blocks. Consider a finer partition of A:

$$
A = \begin{array}{c}
 \\ d_1 \\ d_2 \\ \vdots \\ d_s
\end{array}
\begin{array}{cccc}
d_1 & d_2 & \cdots & d_s \\
\left[\begin{array}{cccc}
A_{11} & A_{12} & \cdots & A_{1s} \\
 & A_{22} & \cdots & A_{2s} \\
 & & \ddots & \vdots \\
 & & & A_{ss}
\end{array}\right]
\end{array}.
$$

We have shown only the upper half because of symmetry. Then

$$B = \begin{bmatrix} A_{12} & \cdots & A_{1s} \end{bmatrix},$$

and the operation $S = R_{11}^{-T} B$ becomes

$$S_{1j} = R_{11}^{-T} A_{1j}, \quad j = 2, \ldots, s,$$

where we partition R conformably with A, and the operation $\tilde{A} = \hat{A} - S^T S$ becomes

$$\tilde{A}_{ij} = \hat{A}_{ij} - R_{1i}^T R_{1j}, \quad i, j = 1, \ldots, s.$$

Once we have $\tilde{A}$, we can calculate its Cholesky factor by applying (1.4.43) to it.

Exercise 1.4.45 Write a nonrecursive algorithm that implements the algorithm that we have just sketched. Your algorithm should exploit the symmetry of A by referencing only the main diagonal and upper part of A, and it should store R over A.   □

Your solution to Exercise 1.4.45 should look something like this:

Block Cholesky Algorithm (outer-product form)

for $k = 1, \ldots, s$
$$\left[\begin{array}{l} A_{kk} \leftarrow \text{cholesky}(A_{kk}) \quad \text{(if cholesky fails, set flag and exit)} \\ \text{for } j = k+1, \ldots, s \\ \quad \left[\; A_{kj} \leftarrow A_{kk}^{-T} A_{kj} \right. \\ \text{for } i = k+1, \ldots, s \\ \quad \left[\begin{array}{l} \text{for } j = i, \ldots, s \\ \quad \left[\; A_{ij} \leftarrow A_{ij} - A_{ki}^{T} A_{kj} \right. \end{array} \right. \end{array} \right. \tag{1.4.46}$$

In order to implement this algorithm, we need a standard Cholesky decomposition code (based on (1.4.17), for example) to perform the small Cholesky decompositions $A_{kk} \leftarrow \text{cholesky}(A_{kk})$. In the operation $A_{kj} \leftarrow A_{kk}^{-T} A_{kj}$, the block A_{kk} holds the triangular matrix R_{kk} at this point. Thus the operation can be effected by forward substitutions, as already explained; there is no need to calculate an inverse.

Exercise 1.4.47 Write a block version of the inner-product form of Cholesky's method.   □

Exercise 1.4.48 Convince yourself that the block versions of Cholesky's method perform exactly the same arithmetic as the standard versions, but not in the same order.   □

The benefits of organizing the Cholesky decomposition by blocks are exactly the same as those of performing matrix multiplication by blocks, as discussed in Section 1.1. To keep the discussion simple, let us speak as if all of the blocks were square and of the same size, $d \times d$. The bulk of the work is concentrated in the operation

$$A_{ij} \leftarrow A_{ij} - A_{ki}^{T} A_{kj},$$

which is triply nested in the algorithm. This is essentially a matrix-matrix multiply, which takes $2d^3$ flops. Since we are performing $O(d^3)$ flops on $O(d^2)$ data items, we can minimize the delays associated with swapping data in and out of cache by making d as large as possible. The only constraint on d is that we need to be able to fit several (actually 5) blocks into cache at once. Similar benefits are obtained for the operations $A_{kk} \leftarrow \text{cholesky}(A_{kk})$ and $A_{kj} \leftarrow A_{kk}^{-T} A_{kj}$, as these also require $O(d^3)$ flops on $O(d^2)$ data items.

Exercise 1.4.49 Show that the implementation of $A_{kj} \leftarrow A_{kk}^{-T} A_{kj}$ via forward substitutions requires $d^3 + O(d^2)$ flops if all three blocks are $d \times d$.   □

Let us also consider briefly the computation of the Cholesky factor on a parallel computer using (1.4.46). This may be worthwhile if the matrix is really huge. We continue to make the simplifying assumption that all of the blocks are $d \times d$. This implies that $n = ds$. If we increase d, we must decrease s, and vice versa. Let us focus on the part of the algorithm where most of the work is done. For each k, the operations

$$A_{ij} \leftarrow A_{ij} - A_{ki}^T A_{kj}$$

can be done simultaneously for $i = k + 1, \ldots, s$, for $j = i, \ldots, s$. Thus as many as about $s^2/2$ processors can be kept busy at once. This suggests that s should be made large (and, consequently, d small). On the other hand, a processor cannot perform the operation until it has received the data. Since $2d^3$ flops are performed on $3d^2$ data items, data movement is minimized by making d large. This suggests that d should be made large (and, consequently, s small). Thus we see that there are trade offs that keep us from making either d or s too large. Which choice of d and s is best will depend on the details of the parallel computer that is being used.

rcise 1.4.50 Assuming $d \times d$ blocks in (1.4.46), calculate the fraction of the total work (reckoned in flops) that is spent (a) calculating Cholesky decompositions of the main-diagonal blocks, and (b) executing the instruction $A_{kj} \leftarrow A_{kk}^{-T} A_{kj}$. In each case your answer should be a function of s. What are the respective fractions when $s = 10$? when $s = 20$? $\qquad\square$

Proof of the Cholesky Decomposition Theorem

Having discussed the use of the Cholesky decomposition, having determined how to calculate the Cholesky factor, and having noted the numerous ways the computation can be organized, we now take time to prove the Cholesky Decomposition Theorem. We begin by recalling that a symmetric matrix A is positive definite if $x^T A x > 0$ for all nonzero $x \in \mathbb{R}^n$. We can use this property to prove a few simple propositions.

Proposition 1.4.51 *If A is positive definite, then $a_{ii} > 0$ for $i = 1, \ldots, n$.*

rcise 1.4.52 Prove Proposition 1.4.51. Do not use the Cholesky decomposition in your proof; we want to use this result to prove that the Cholesky decomposition exists. (*Hint:* Find a nonzero vector x such that $x^T A x = a_{ii}$.) $\qquad\square$

Proposition 1.4.53 *Let A be positive definite, and consider a partition*

$$A = \left[\begin{array}{cc} A_{11} & A_{12} \\ A_{21} & A_{22} \end{array} \right],$$

in which A_{11} and A_{22} are square. Then A_{11} and A_{22} are positive definite.

rcise 1.4.54 Prove Proposition 1.4.53. As in the previous exercise, do not use the Cholesky decomposition in your proof; use the fact that $x^T A x > 0$ for all nonzero x. $\qquad\square$

Propositions 1.4.51 and 1.4.53 are clearly closely related; 1.4.53 is (essentially) a generalization of 1.4.51. Can you think of a generalization that encompasses both of these results? What is the most general result you can think of? After you give this some thought, take a look at Exercise 1.4.61.

Proposition 1.4.55 *If A and X are $n \times n$, A is positive definite, and X is nonsingular then the matrix $B = X^T A X$ is also positive definite.*

Considering the special case $A = I$ (which is clearly positive definite), we see that this proposition is a generalization of Theorem 1.4.4.

Exercise 1.4.56   Prove Proposition 1.4.55.   □

The Cholesky Decomposition Theorem states that if A is positive definite, then there is a unique R such that R is upper triangular, the main diagonal entries of R are positive, and $A = R^T R$. We have already demonstrated uniqueness, so we only have to prove existence. We do so by induction on n. When $n = 1$, we have $A = [a_{11}]$. By Proposition 1.4.51, $a_{11} > 0$. Let $r_{11} = +\sqrt{a_{11}}$ and $R = [r_{11}]$. Then R has the desired properties. Moving on to the interesting part of the proof, we will show that every $n \times n$ positive definite matrix has a Cholesky factor, given that every $(n - 1) \times (n - 1)$ positive definite matrix does. Given an $n \times n$ positive definite A, partition A as we did in the development of the outer-product formulation of Cholesky's method:

$$A = \begin{bmatrix} a_{11} & b^T \\ b & \hat{A} \end{bmatrix}.$$

Proposition 1.4.51 guarantees that $a_{11} > 0$. Using (1.4.27) and (1.4.28) as a guide, define

$$
\begin{aligned}
r_{11} &= +\sqrt{a_{11}} \\
s &= r_{11}^{-1} b \\
\tilde{A} &= \hat{A} - ss^T.
\end{aligned}
$$

Then, as one easily checks,

$$A = \begin{bmatrix} r_{11} & 0 \\ s & I \end{bmatrix} \begin{bmatrix} 1 & 0 \\ 0 & \tilde{A} \end{bmatrix} \begin{bmatrix} r_{11} & s^T \\ 0 & I \end{bmatrix}, \tag{1.4.57}$$

where I denotes the $(n - 1) \times (n - 1)$ identity matrix. The matrix

$$\begin{bmatrix} r_{11} & s^T \\ 0 & I \end{bmatrix}$$

is upper triangular and its main diagonal entries are nonzero, so it is nonsingular. Let us call its inverse X. Then if we let

$$B = \begin{bmatrix} 1 & 0 \\ 0 & \tilde{A} \end{bmatrix},$$

we have $B = X^T A X$, so B is positive definite, by Proposition 1.4.55. If we now apply Proposition 1.4.53 to B, we find that $\tilde{A}$ is positive definite. Since $\tilde{A}$ is

$(n-1) \times (n-1)$, there is an upper triangular matrix $\tilde{R}$ whose main-diagonal entries are positive, such that $\tilde{A} = \tilde{R}^T \tilde{R}$, by the induction hypothesis. Therefore, using (1.4.57),

$$A = \begin{bmatrix} r_{11} & 0 \\ s & I \end{bmatrix} \begin{bmatrix} 1 & 0 \\ 0 & \tilde{R}^T \end{bmatrix} \begin{bmatrix} 1 & 0 \\ 0 & \tilde{R} \end{bmatrix} \begin{bmatrix} r_{11} & s^T \\ 0 & I \end{bmatrix}$$

$$= \begin{bmatrix} r_{11} & 0 \\ s & \tilde{R}^T \end{bmatrix} \begin{bmatrix} r_{11} & s^T \\ 0 & \tilde{R} \end{bmatrix} = R^T R,$$

where R is upper triangular and has positive entries on the main diagonal. This completes the proof.

The matrix $\tilde{A} = \hat{A} - ss^T = \hat{A} - a_{11}^{-1} bb^T$, is called the *Schur complement* of a_{11} in A. The main business of the proof of the Cholesky Decomposition Theorem is to show that positive definiteness is inherited by the Schur complement. The argument is generalized in the following exercise.

Exercise 1.4.58 Let $A = \begin{bmatrix} A_{11} & A_{12} \\ A_{21} & A_{22} \end{bmatrix}$ be positive definite, and suppose A_{11} is $j \times j$ and A_{22} is $k \times k$. By Proposition 1.4.53, A_{11} is positive definite. Let R_{11} be the Cholesky factor of A_{11}, let $R_{12} = R_{11}^{-T} A_{12}$, and let $\tilde{A}_{22} = A_{22} - R_{12}^T R_{12}$. The matrix $\tilde{A}_{22}$ is called the *Schur complement* of A_{11} in A.

(a) Show that $\tilde{A}_{22} = A_{22} - A_{21} A_{11}^{-1} A_{12}$.

(b) Establish a decomposition of A that is similar to (1.4.57) and involves $\tilde{A}_{22}$.

(c) Prove that $\tilde{A}_{22}$ is positive definite.

$\square$

Exercise 1.4.59 Write down a second proof of the existence of the Cholesky factor based on the decomposition established in the previous exercise. $\square$

Exercise 1.4.60 Carefully prove by induction that the Cholesky decomposition is unique: Suppose $A = R^T R = S^T S$, where R and S are both upper-triangular matrices with positive main-diagonal entries. Partition A, R, and S conformably and prove that the parts of S must equal the corresponding parts of R. $\square$

Exercise 1.4.61 This Exercise generalizes Propositions 1.4.51 and 1.4.53. Let A be an $n \times n$ positive definite matrix, let $j_1, j_2, \ldots, j_k$ be integers such that $1 \le j_1 < j_2 < \cdots < j_k \le n$, and let $\hat{A}$ be the $k \times k$ matrix obtained by intersecting rows $j_1, \ldots, j_k$ with columns $j_1, \ldots, j_k$. Prove that $\hat{A}$ is positive definite. $\square$

Exercise 1.4.62 Prove that if A is positive definite, then $\det(A) > 0$. $\square$

Complex Positive Definite Matrices

The next three exercises show how the results of this section can be extended to complex matrices. The set of complex n-vectors will be denoted $\mathbb{C}^n$. The *conjugate transpose* A^* of a complex $m \times n$ matrix A is the $n \times m$ matrix whose (i, j) entry is $\bar{a}_{ji}$. The bar denotes the complex conjugate. A is *hermitian* if $A = A^*$, that is, $a_{ij} = \bar{a}_{ji}$ for all i and j.

Exercise 1.4.63 (a) Prove that if A and B are complex $m \times n$ and $n \times p$ matrices, then $(AB)^* = B^* A^*$. (b) Prove that if A is hermitian, then $x^* A x$ is real for every $x \in \mathbb{C}^n$. (*Hint:* Let $\alpha = x^* A x$. Then α is real if and only if $\alpha = \bar{\alpha}$. Think of α as a 1×1 matrix, and consider of the matrix α^*.) ☐

If A is hermitian and satisfies $x^* A x > 0$ for all nonzero $x \in \mathbb{C}^n$, then A is said to be *positive definite*.

Exercise 1.4.64 Prove that if M is any $n \times n$ nonsingular matrix, then $M^* M$ is positive definite. ☐

Exercise 1.4.65 Prove that if A is positive definite, then there exists a unique matrix R such that R is upper triangular and has positive (real!) main diagonal entries, and $A = R^* R$. This is the *Cholesky decomposition* of A. ☐

All of the algorithms that we have developed in this section can easily be adapted to the complex case.

Additional Exercises

Exercise 1.4.66

(a) The power drawn by an appliance (in watts) is equal to the product of the electromotive force (in volts) and the current (in amperes). Briefly, watts = volts × amps. Suppose a current I amperes passes through a resistor with resistance R ohms, causing a voltage drop of E volts. Show that the power dissipated by the resistor is (i) RI^2 watts, (ii) E^2/R watts.

(b) Figure 1.9 illustrates loop currents passing through resistors with resistances R and S. Obviously the S ohm resistor draws power Sx_i^2, which is positive unless $x_i = 0$. Notice that

$$Sx_i^2 = \begin{bmatrix} x_i & x_j \end{bmatrix} \begin{bmatrix} S & 0 \\ 0 & 0 \end{bmatrix} \begin{bmatrix} x_i \\ x_j \end{bmatrix}.$$

Show that the power drawn by the R ohm resistor is $R(x_i - x_j)^2$. Deduce that this resistor draws positive power unless $x_i = x_j$, in which case it draws zero power. Show that the power drawn by this resistor can also be expressed as

$$\begin{bmatrix} x_i & x_j \end{bmatrix} \begin{bmatrix} R & -R \\ -R & R \end{bmatrix} \begin{bmatrix} x_i \\ x_j \end{bmatrix}.$$

(c) Let

$$A = \begin{bmatrix} 9 & -5 \\ -5 & 8 \end{bmatrix},$$

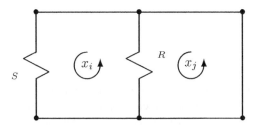

Figure 1.9 Two loop currents

the matrix of Example 1.2.8. Show that if $x = \begin{bmatrix} x_1 \\ x_2 \end{bmatrix}$ is the vector of loop currents, then the total power drawn by (all of the resistors in) the circuit is $x^T A x$. Deduce that A is positive definite.

(d) Show that the 9×9 coefficient matrix of Exercise 1.2.19 is positive definite. (*Hint:* Show that the power drawn by the circuit is a sum of terms of the form $R x_k^2$ and $R(x_i - x_j)^2$, and that this sum is positive unless all loop currents are zero. Then show that the power drawn by the circuit can also be expressed as $x^T A x$, where A is the coefficient matrix of the system.)

□

rcise 1.4.67

(a) The *conductance* C of a resistor (in siemens) is equal to the reciprocal of the resistance: $C = 1/R$. Show that if two nodes with voltages x_i and x_j are connected by a resistor with conductance C, the power drawn by the resistor is $C(x_i - x_j)^2$, which can also be expressed as

$$\begin{bmatrix} x_i & x_j \end{bmatrix} \begin{bmatrix} C & -C \\ -C & C \end{bmatrix} \begin{bmatrix} x_i \\ x_j \end{bmatrix}.$$

This quantity is positive, unless $x_i = x_j$.

(b) Consider the circuit in Figure 1.1, Example 1.2.6. Show that the power drawn by this circuit is a sum of terms of the form $C(x_i - x_j)^2$ and is positive unless all of the nodal voltages $x_1, \ldots, x_6$ are equal.

(c) Show that the power can be expressed as $x^T H x$, where H is a 6×6 symmetric matrix. This matrix is *positive semi-definite*, as $x^T H x \geq 0$ for all x.

(d) Let A denote the coefficient matrix of the system in Example 1.2.6. Show that A is a submatrix of the matrix H from part (c). Deduce that A is positive definite. (*Hint:* Set $x_5 = x_6 = 0$ in the expression $x^T H x$.)

(e) Show that the coefficient matrix of the circuit in Exercise 1.2.17 is positive definite.

□

Exercise 1.4.68

(a) When a spring is stretched (or compressed) from its equilibrium position (0 meters) to some new position (x meters), the energy stored in the spring (strain energy) is equal to the work required to move the spring to its new position. This is $\int_0^x f(s)\,ds$ joules, the integral of force with respect to distance, where $f(s)$ is the force (in newtons) exerted against the spring that has been stretched to s meters. Show that for a linear spring with stiffness k newtons/meter, the strain energy is $\frac{1}{2}kx^2$ joules.

(b) Show that if a spring is stretched or compressed by displacing its right and left endpoints by x_2 and x_1 meters, respectively, the strain energy of the spring is $\frac{1}{2}k(x_1 - x_2)^2$ joules, which is positive unless $x_1 = x_2$. Show that the strain energy can also be written in the form

$$\frac{1}{2}\begin{bmatrix} x_1 & x_2 \end{bmatrix}\begin{bmatrix} k & -k \\ -k & k \end{bmatrix}\begin{bmatrix} x_1 \\ x_2 \end{bmatrix}.$$

(c) Show that the total strain energy of the mass-spring system in Example 1.2.10 is a sum of terms of the form $\frac{1}{2}kx_i^2$ and $\frac{1}{2}k(x_i - x_{i+1})^2$ and is positive unless all of the displacements are zero.

(d) Let A be the coefficient matrix of the mass-spring system of Example 1.2.10. Show that the strain energy of the system is $\frac{1}{2}x^T Ax$. Deduce that A is positive definite.

(e) Show that the coefficient matrix of the mass-spring system of Exercise 1.2.20 is positive definite.

□

Exercise 1.4.69 Let v be a vector whose entries represent some statistical data. For example, v could be a vector with 365 entries representing the high temperature in Seattle on 365 consecutive days. We can normalize this vector by computing its mean and subtracting the mean from each of the entries to obtain a new vector with mean zero. Suppose now we have such a normalized vector. Then the *variance* of v is $\sum_{i=1}^{n} v_i^2$. This nonnegative number gives a measure of the variation in the data. Notice that if we think of v as a column vector, the variance can be expressed as $v^T v$. Now let v and w be two vectors with mean zero and variance (normalized to be) one. Then the *correlation* of v and w is defined to be $\sum_{i=1}^{n} v_i w_i = w^T v = v^T w$. This number, which can be positive or negative, measures whether the data in v and w vary with or against one another. For example, the temperatures in Seattle and Vancouver should have a positive correlation, while the temperatures in Seattle and Hobart, Tasmania, should have a negative correlation. Now consider k vectors $v_1, \ldots, v_k$ with mean zero and variance one. The *correlation matrix* C of the data $v_1, \ldots, v_k$ is the $k \times k$ matrix whose (i, j) entry is $c_{ij} = v_j^T v_i$, the correlation of v_i with v_j. Show that C is a symmetric matrix whose main-diagonal entries are all ones. Show that $C = V^T V$ for some appropriately constructed (nonsquare) matrix V. Show that C is positive definite if the vectors $v_1, \ldots, v_k$ are linearly independent. Show that if $v_1, \ldots, v_k$ are linearly dependent, then C is not positive definite, but it is positive semidefinite, i.e. $x^T C x \geq 0$ for all x. $\square$

rcise 1.4.70 Draw pictures that demonstrate that

$$\int_0^n i^2 \, di \leq \sum_{i=0}^n i^2 \leq \int_0^{n+1} i^2 \, di.$$

Deduce that

$$\sum_{i=0}^n i^2 = \frac{n^3}{3} + O(n^2).$$

$\square$

rcise 1.4.71 Prove by induction on n that

$$\sum_{i=0}^n i^2 = \frac{n(n + \frac{1}{2})(n + 1)}{3}.$$

$\square$

rcise 1.4.72 Figure out what the following MATLAB code does.

```
n = 500;
for jay = 1:4
   if jay > 1; oldtime = time ; end
   M = randn(n);
   A = M'*M;
   t = cputime;
   for rep = 1:5
      R = chol(A);
   end
   matrixsize = n
   time = cputime - t
   if jay > 1; ratio = time/oldtime, end
   n = 2*n;
end
```

If you put these instructions into a file named, say, zap.m, you can run the program by typing zap from the MATLAB command line. The functions randn, cputime, and chol are built-in MATLAB functions, so you can learn about them by typing help randn, etc. You might find it useful to type more on before typing help {*topic*}. Alternatively you can read about them in MATLAB's help browser.

Does the code produce reasonable values of `ratio` when you run it? What value would you expect in theory? Depending on the speed of the machine on which you are running MATLAB, you may want to adjust the initial value of `n` or the number of times the `rep` loop is executed. □

Exercise 1.4.73   Write Fortran subroutines to implement Cholesky's method in (a) inner-product form, (b) outer-product form, (c) bordered form. If you have already written a forward-substitution routine, you can use it in part (c). Your subroutines should operate only on the main diagonal and upper-triangular part of A, and they should overwrite A with R. They should either return R or set a warning flag indicating that A is not positive definite. Try out your routines on the following examples.

$$(i)\ A\ =\ \begin{bmatrix} 36 & 30 & 24 \\ 30 & 34 & 26 \\ 24 & 26 & 21 \end{bmatrix}$$

$$(ii)\ A\ =\ \begin{bmatrix} 1 & -1 & 1 & -1 & 1 \\ -1 & 2 & -2 & 2 & -2 \\ 1 & -2 & 3 & -3 & 3 \\ -1 & 2 & -3 & 4 & -4 \\ 1 & -2 & 3 & -4 & 5 \end{bmatrix}$$

$$(iii)\ A\ =\ \begin{bmatrix} 1 & 2 & 3 \\ 2 & 5 & 10 \\ 3 & 10 & 16 \end{bmatrix} \qquad \text{cf. Example 1.4.20.}$$

You might like to devise some additional examples. The easy way to do this is to write down R first and then multiply R^T by R to get A. With the help of MATLAB you can generate larger matrices. Use the MATLAB `save` command to export a matrix to an ASCII file. Type `help save` for details. □

Exercise 1.4.74   Write a Fortran program that solves positive definite systems $Ax = b$ by calling subroutines to (a) calculate the Cholesky factor, (b) perform forward substitution, and (c) perform back substitution. Try out your program on the following problems.

$$\textbf{(i)}\ \begin{bmatrix} 36 & -30 & 24 \\ -30 & 34 & -26 \\ 24 & -26 & 21 \end{bmatrix} \begin{bmatrix} x_1 \\ x_2 \\ x_3 \end{bmatrix} = \begin{bmatrix} 0 \\ 12 \\ -7 \end{bmatrix}$$

$$\textbf{(ii)}\ \begin{bmatrix} 1 & 1 & 1 & 1 & 1 \\ 1 & 2 & 2 & 2 & 2 \\ 1 & 2 & 3 & 3 & 3 \\ 1 & 2 & 3 & 4 & 4 \\ 1 & 2 & 3 & 4 & 5 \end{bmatrix} \begin{bmatrix} x_1 \\ x_2 \\ x_3 \\ x_4 \\ x_5 \end{bmatrix} = \begin{bmatrix} 5 \\ 9 \\ 12 \\ 14 \\ 15 \end{bmatrix}$$

You might like to make some additional examples. You can use MATLAB to help you build larger examples, as suggested in the previous exercise. □

1.5   BANDED POSITIVE DEFINITE SYSTEMS

Large systems of equations occur frequently in applications, and large systems are usually sparse. In this section we will study a simple yet very effective scheme for applying Cholesky's method to large, positive definite systems of equations that are banded or have an envelope structure. This method is less popular than it once was, as more efficient methods have become widely available [14] (and even implemented in MATLAB [38]). It is worth studying nevertheless, as it shows via very simple arguments that enormous savings in computer time and storage space can be achieved by exploiting sparseness. More sophisticated sparse matrix methods are discussed in Section 1.6. For extremely large systems, iterative methods are preferred. These are discussed in Chapter 8.

A matrix A is *banded* if there is a narrow band around the main diagonal such that all of the entries of A outside of the band are zero, as shown in Figure 1.10. More precisely, if A is $n \times n$, and there is an $s \ll n$ such that $a_{ij} = 0$ whenever

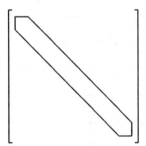

Figure 1.10   Banded matrix: All entries outside of the band are zero.

$|i - j| > s$, then all of the nonzero entries of A are confined to a band of $2s + 1$ diagonals centered on the main diagonal. We say that A is *banded* with *band width* $2s + 1$. Since we are concerned with symmetric matrices in this section, we only need half of the band. Since $a_{ij} = 0$ whenever $i - j > s$, there is a band of s diagonals above the main diagonal that, together with the main diagonal, contains all of the nonzero entries of A. We say that A has *semiband width* s.

Example 1.5.1 Consider the mass-spring system depicted in Figure 1.11. This is

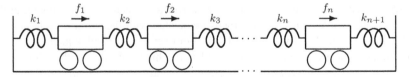

Figure 1.11   System of n masses

exactly the system that we discussed in Exercise 1.2.20. There are n carts attached by springs. If forces are applied to the carts, we can calculate their displacements x_i by solving a system $Ax = b$ of n equations in n unknowns. Since the ith cart is directly attached only to the two adjacent carts, the ith equation involves only the unknowns x_{i-1}, x_i, and x_{i+1}. Thus its form is

$$a_{i,i-1}x_{i-1} + a_{i,i}x_i + a_{i,i+1}x_{i+1} = b_i,$$

and $a_{ij} = 0$ whenever $|i - j| > 1$. This is an extreme example of a banded coefficient matrix. The band width is 3 and the semiband width is 1. Such matrices are called *tridiagonal*. ☐

Example 1.5.2 Consider a 100×100 system of equations $Ax = b$ associated with the grid depicted in Figure 1.12. The ith grid point has one equation and one

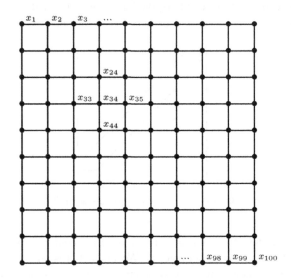

Figure 1.12   "Large" grid

unknown associated with it. For example, the unknown could be a nodal voltage (or a displacement, a pressure, a temperature, a hydraulic head, ...). Assume that the ith equation involves only the unknowns associated with the ith grid point and the grid points that are directly connected to it. For example, the 34th equation involves only the unknowns x_{24}, x_{33}, x_{34}, x_{35}, and x_{44}. This means that in the 34th row of A, only the entries $a_{34,24}$, $a_{34,33}$, $a_{34,34}$, $a_{34,35}$, and $a_{34,44}$ are nonzero. The other 95 are zero. Clearly the same is true for every other equation; no row of A contains more than five nonzero entries. Thus the matrix is very sparse. It is also banded, if the equations and unknowns are numbered as shown in the figure. Clearly $a_{ij} = 0$ if $|i - j| > 10$. Thus the system is 100×100 with a semiband width of 10. ☐

rcise 1.5.3 Make a rough sketch of the matrix A of Example 1.5.2, noting where the zeros and nonzeros lie. Notice that even within the band, most of the entries are zero. Most large application problems have this feature. □

rcise 1.5.4 Modern applications usually involve matrices that are much larger than the one discussed in Example 1.5.2. Figure 1.12 depicts a 10×10 network of nodes. Imagine an $m \times m$ network with $m \gg 10$. How many equations does the resulting system have? How many nonzeros does each row of the matrix have? What is the bandwidth of the system, assuming that the equations and unknowns are numbered as depicted in Figure 1.12? Answer these questions in general and also in the specific cases (a) $m = 100$, (b) $m = 1000$. □

Notice that the bandedness depends on how the nodes are numbered. If, for example, nodes 2 and 100 are interchanged in Figure 1.12, the resulting matrix is not banded, since $a_{100,1}$ and $a_{1,100}$ are nonzero. However it is still sparse; the number of nonzero entries in the matrix does not depend on how the nodes are ordered.

If a network is regular, it is easy to see how to number the nodes to obtain a narrow band. Irregular networks can also lead to banded systems, but it will usually be more difficult to decide how the nodes should be numbered.

Banded positive definite systems can be solved economically because it is possible to ignore the entries that lie outside of the band. For this it is crucial that the Cholesky factor inherits the band structure of the original matrix. Thus we can save storage space by using a data structure that stores only the semiband of A. R can be stored over A. Just as importantly, computer time is saved because all operations involving entries outside of the band can be skipped. As we shall soon see, these savings are substantial.

Instead of analyzing banded systems, we will introduce a more general idea, that of the envelope of a matrix. This will increase the generality of the discussion while simplifying the analysis. The *envelope* of a symmetric or upper-triangular matrix A is a set of ordered pairs (i, j), $i < j$, representing element locations in the upper triangle of A, defined as follows: (i, j) is in the envelope of A if and only if $a_{kj} \neq 0$ for some $k \leq i$. Thus if the first nonzero entry of the jth column is a_{mj} and $m < j$, then (m, j), $(m + 1, j)$, ..., $(j - 1, j)$ are the members of the envelope of A from the jth column.

The crucial theorem about envelopes (Theorem 1.5.7) states that if R is the Cholesky factor of A, then R has the same envelope as A. Thus A can be stored in a data structure that stores only its main diagonal and the entries in its envelope, and R can be stored over A. All operations involving the off-diagonal entries lying outside of the envelope can be skipped. If the envelope is small, substantial savings in computer time and storage space are realized. Banded matrices have small envelopes.

A simple example of an unbanded matrix with a small envelope is

$$
\begin{bmatrix}
2 & -1 & 0 & \cdots & 0 & 0 & -1 \\
-1 & 2 & -1 & \ddots & & 0 & 0 \\
0 & -1 & 2 & \ddots & & & 0 \\
\vdots & \ddots & \ddots & \ddots & & & \vdots \\
0 & & & & 2 & -1 & 0 \\
0 & 0 & & & -1 & 2 & -1 \\
-1 & 0 & 0 & \cdots & 0 & -1 & 2
\end{bmatrix},
\tag{1.5.5}
$$

which was obtained from discretization of an ordinary differential equation with a periodic boundary condition.

Exercise 1.5.6   Identify the envelope of the matrix in (1.5.5). Assuming the matrix is $n \times n$, approximately what fraction of the upper triangle of the matrix lies within the envelope?   □

Like the band width, the envelope of a matrix depends on the order in which the equations and unknowns are numbered. Often it is easy to see how to number the nodes to obtain a reasonably small envelope. For those cases in which it is hard to tell how the nodes should be ordered, there exist algorithms that attempt to minimize the envelope in some sense. For example, see the discussion of the reverse Cuthill-McKee algorithm in [37].

Theorem 1.5.7 *Let A be positive definite, and let R be the Cholesky factor of A. Then R and A have the same envelope.*

Proof. Consider the bordered form of Cholesky's method. At the jth step we solve a system

$$
R_{j-1}^T h = c,
$$

where $c \in \mathbb{R}^{j-1}$ is the portion of the jth column of A lying above the main diagonal, and h is the corresponding portion of R. (See equations (1.4.36) and (1.4.37).) Let $\hat{c} \in \mathbb{R}^{s_j}$ be the portion of c that lies in the envelope of A. Then

$$
c = \begin{bmatrix} 0 \\ \hat{c} \end{bmatrix}.
$$

In Section 1.3 we observed that if c has leading zeros, then so does h:

$$
h = \begin{bmatrix} 0 \\ \hat{h} \end{bmatrix},
$$

where $\hat{h} \in \mathbb{R}^{s_j}$. See (1.3.6) and the accompanying discussion. It follows immediately that the envelope of R is contained in the envelope of A. Furthermore it is not hard to show that the first entry of $\hat{h}$ is nonzero. Thus the envelope of R is exactly the envelope of A. □

Corollary 1.5.8 *Let A be a banded, positive definite matrix with semiband width s. Then its Cholesky factor R also has semiband width s.*

Referring to the notation in the proof of Theorem 1.5.7, if $\hat{c} \in \mathbb{R}^{s_j}$, then the cost of the arithmetic in the jth step of Cholesky's method is essentially equal to that of solving an $s_j \times s_j$ lower-triangular system, that is, s_j^2 flops. (See the discussion following (1.3.6).) If the envelope is not exploited, the cost of the jth step is j^2 flops. To get an idea of the savings that can be realized by exploiting the envelope structure of a matrix, consider the banded case. If A has semiband width s, then the portion of the jth row that lies in the envelope has at most s entries, so the flop count for the jth step is about s^2. Since there are n steps in the algorithm, the total flop count is about ns^2.

rcise 1.5.9 Let R be an $n \times n$ upper-triangular matrix with semiband width s. Show that the system $Rx = y$ can be solved by back substitution in about $2ns$ flops. An analogous result holds for lower-triangular systems. □

Example 1.5.10 The matrix of Example 1.5.2 has $n = 100$ and $s = 10$. If we perform a Cholesky decomposition using a program that does not exploit the band structure of the matrix, the cost of the arithmetic is about $\frac{1}{3}n^3 \approx 3.3 \times 10^5$ flops. In contrast, if we do exploit the band structure, the cost is about $ns^2 = 10^4$ flops, which is about 3% of the previous figure. In the forward and back substitution steps, substantial but less spectacular savings are achieved. The combined arithmetic cost of forward and back substitution without exploiting the band structure is about $2n^2 = 2 \times 10^4$ flops. If the band structure is exploited, the flop count is about $4ns = 4 \times 10^3$, which is 20% of the previous figure.

If the matrix is stored naively, space for $n^2 = 10,000$ numbers is needed. If only the semiband is stored, space for not more than $n(s+1) = 1100$ numbers is required. □

The results of Example 1.5.10, especially the savings in flops in the Cholesky decomposition, are already impressive, even though the matrix is not particularly large. Much more impressive results are obtained if larger matrices are considered, as the following exercise shows.

rcise 1.5.11 As in Exercise 1.5.4, consider the banded system of equations arising from an $m \times m$ network of nodes like Figure 1.12 but larger, with the nodes numbered by rows, as in Figure 1.12.

(a) For the case $m = 100$ (for which $n = 10^4$) calculate the cost of solving the system $Ax = b$ (Cholesky decomposition plus forward and back substitution) with and without exploiting the band structure. Show that exploiting the band structure cuts the flop count by a factor of several thousand. Show that if only the semiband is stored, the storage space required is only about 1% of what would be required to store the matrix naively.

(b) Repeat part (a) with $m = 1000$.

□

Savings such as these can make the difference between being able to solve a large problem and not being able to solve it.

The following exercises illustrate another important feature of banded and envelope matrices: The envelope structure of A is *not* inherited by A^{-1}. In fact, it is typical of sparse matrices that the inverse is not sparse. Thus it is highly uneconomical to solve a sparse system $Ax = b$ by finding the inverse and computing $x = A^{-1}b$.

Exercise 1.5.12 Consider a mass-spring system as in Figure 1.11 with six carts. Suppose each spring has a stiffness $k_i = 1$ newton/meter.

(a) Set up the tridiagonal, positive definite, coefficient matrix A associated with this problem.

(b) Use the MATLAB `chol` command to calculate the Cholesky factor R. Notice that R inherits the band structure of A. (To learn an easy way to enter this particular A, type `help toeplitz`.)

(c) Use the MATLAB `inv` command to calculate A^{-1}. Notice that A^{-1} does not inherit the band structure of A.

□

Exercise 1.5.13 In the previous exercise, the matrix A^{-1} is full; none of its entries are zero.

(a) What is the physical significance of this fact? (Think of the equation $x = A^{-1}b$, especially in the case where only one entry, say the jth, of b is nonzero. If the (i, j) entry of A^{-1} were zero, what would this imply? Does this make physical sense?)

(b) The entries of A^{-1} decrease in magnitude as we move away from the main diagonal? What does this mean physically?

□

Exercise 1.5.14 Consider the linear system $Ax = b$ from Exercise 1.2.19. The matrix A is banded and positive definite.

(a) Use MATLAB to compute the Cholesky factor R. Observe that the envelope is preserved.

(b) Use MATLAB to calculate A^{-1}. Observe that A^{-1} is full. (What is the physical significance of this?)

□

Envelope Storage Scheme

A fairly simple data structure can be used to store the envelope of a coefficient matrix. We will describe the scheme from [37]. A one-dimensional real array DIAG of length n is used to store the main diagonal of the matrix. A second one-dimensional real array ENV is used to store the envelope by columns, one after the other. A third array IENV, an integer array of length $n + 1$, is used to store pointers to ENV. Usually IENV(J) names the position in ENV of the first (nonzero) entry of column J of the matrix. However, if column J contains no nonzero entries above the main diagonal, then IENV(J) points to column $J + 1$ instead. Thus the absence of nonzero entries in column J above the main diagonal is signaled by IENV(J) = IENV($J + 1$). IENV($n + 1$) points to the first storage location after the envelope. These rules can be expressed more succinctly (and more accurately) as follows: IENV(1) = 1 and IENV($J + 1$) - IENV(J) equals the number of elements from column J of the matrix that lie in the envelope.

Example 1.5.15 The matrix

$$
\begin{bmatrix}
a_{11} & a_{12} & 0 & a_{14} & 0 & 0 \\
& a_{22} & a_{23} & a_{24} & 0 & 0 \\
& & a_{33} & a_{34} & 0 & 0 \\
& & & a_{44} & 0 & a_{46} \\
& & & & a_{55} & a_{56} \\
& & & & & a_{66}
\end{bmatrix}
$$

is stored as follows using the envelope scheme:

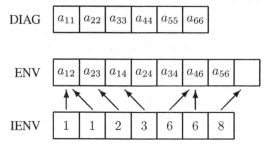

When the envelope storage scheme is used, certain formulations of the Cholesky decomposition, forward-substitution, and back-substitution algorithms are much more appropriate than others. For example, we would not want to use the outer-product formulation of Cholesky's method, because that algorithm operates on (the upper triangle of) A by rows. In the envelope storage scheme A is stored by columns; rows are hard to access. The inner-product formulation is inappropriate for the same reason.[6] From the proof of Theorem 1.5.7 it is clear that the bordered form of

---

[6]The inner-product formulation accesses A by both rows and columns.

Cholesky's method is appropriate. At each step virtually all of the work goes into solving a lower-triangular system $R_{i-1}^T h = c$, where

$$c = \begin{bmatrix} 0 \\ \hat{c} \end{bmatrix} \quad \text{and} \quad h = \begin{bmatrix} 0 \\ \hat{h} \end{bmatrix}.$$

If we partition R_{i-1}^T conformably,

$$R_{i-1}^T = \begin{bmatrix} H_{11} & 0 \\ H_{21} & H_{22} \end{bmatrix},$$

the equation $R_{i-1}^T h = c$ reduces to $H_{22}\hat{h} = \hat{c}$. H_{22} is a lower-triangular matrix consisting of rows and columns $i - s_i$ through $i - 1$ of R^T. A subroutine can be used to solve $H_{22}\hat{h} = \hat{c}$. What is needed is a forward-substitution routine that solves systems of the form $\hat{R}^T \hat{h} = \hat{c}$, where $\hat{R}$ is a submatrix of R consisting of rows and columns j through k, where j and k can be any integers satisfying $1 \le j \le k \le n$. Since R^T, hence $\hat{R}^T$, is stored by rows, the appropriate formulation of forward substitution is the row-oriented version. This subroutine can also be used with $j = 1$ and $k = n$ to perform the forward-substitution step ($R^T y = b$) after the Cholesky decomposition has been completed. Finally, a back-substitution routine is needed to solve $Rx = y$. Since R is stored by columns, we use the column-oriented version.

Exercise 1.5.16   Write a set of three Fortran subroutines to solve positive definite systems, using the envelope storage scheme:

(a) Row-oriented forward-substitution routine, capable of solving systems $\hat{R}^T \hat{h} = \hat{c}$, where $\hat{R}$ is a submatrix of R consisting of rows and columns j through k, $1 \le j \le k \le n$.

(b) Cholesky decomposition routine, bordered form, which calls the forward-substitution routine to do most of the work.

(c) column-oriented back-substitution routine.

Write a main program that allocates storage, handles input and output, and calls the subroutines to solve positive definite systems $Ax = b$. Test your programs using the test problems given below.

For storage you will need the arrays DIAG, ENV, and IENV discussed above and one additional real array of length n that holds b initially, gets changed to y during the forward-substitution step, and finally gets changed to x during the back-substitution step. The arrays DIAG and ENV contain A initially and get changed to R as the Cholesky decomposition is computed. This is all the storage space that is needed (except, of course, for the space occupied by the program itself). Test problems:

(a) $A = \begin{bmatrix} 4 & 0 & 6 & 0 & 2 \\ 0 & 1 & 3 & 0 & 2 \\ 6 & 3 & 19 & 2 & 6 \\ 0 & 0 & 2 & 5 & -5 \\ 2 & 2 & 6 & -5 & 16 \end{bmatrix}$ (i) $b = \begin{bmatrix} 12 \\ 6 \\ 36 \\ 2 \\ 21 \end{bmatrix}$ (ii) $b = \begin{bmatrix} 12 \\ 4 \\ 26 \\ -8 \\ 27 \end{bmatrix}$

The Cholesky decomposition only has to be done once. *Solution:*

$R = \begin{bmatrix} 2 & 0 & 3 & 0 & 1 \\ & 1 & 3 & 0 & 2 \\ & & 1 & 2 & -3 \\ & & & 1 & 1 \\ & & & & 1 \end{bmatrix}$

(i) $y = \begin{bmatrix} 6 \\ 6 \\ 0 \\ 2 \\ 1 \end{bmatrix}$ $x = \begin{bmatrix} 1 \\ 1 \\ 1 \\ 1 \\ 1 \end{bmatrix}$ (ii) $y = \begin{bmatrix} 6 \\ 4 \\ -4 \\ 0 \\ 1 \end{bmatrix}$ $x = \begin{bmatrix} 1 \\ -1 \\ 1 \\ -1 \\ 1 \end{bmatrix}$

(b) $A = \begin{bmatrix} 16 & 4 & 4 & 0 & 0 & 0 & 0 & 0 & 0 & 0 \\ 4 & 5 & -3 & 6 & 0 & 0 & 0 & 10 & 0 & 0 \\ 4 & -3 & 6 & -8 & 0 & 2 & 0 & -9 & 0 & 0 \\ 0 & 6 & -8 & 17 & 8 & -2 & 0 & 13 & 0 & 0 \\ 0 & 0 & 0 & 8 & 17 & 3 & 2 & -1 & 0 & 0 \\ 0 & 0 & 2 & -2 & 3 & 15 & 1 & -3 & 3 & 0 \\ 0 & 0 & 0 & 0 & 2 & 1 & 21 & -4 & -7 & 0 \\ 0 & 10 & -9 & 13 & -1 & -3 & -4 & 32 & -1 & 4 \\ 0 & 0 & 0 & 0 & 0 & 3 & -7 & -1 & 10 & 0 \\ 0 & 0 & 0 & 0 & 0 & 0 & 0 & 4 & 0 & 24 \end{bmatrix}$ $b = \begin{bmatrix} 36 \\ 109 \\ -76 \\ 188 \\ 141 \\ 113 \\ 68 \\ 281 \\ 51 \\ 272 \end{bmatrix}$

(c) Use your program to show that

$$A = \begin{bmatrix} 1 & 2 & 4 \\ 2 & 13 & 16 \\ 4 & 16 & 10 \end{bmatrix}$$

is not positive definite.

(d) Think about how your subroutines could be used to calculate the inverse of a positive definite matrix. Calculate A^{-1}, where

$$A = \begin{bmatrix} 1 & 1/2 & 1/3 \\ 1/2 & 1/3 & 1/4 \\ 1/3 & 1/4 & 1/5 \end{bmatrix}.$$

It turns out that the entries of A^{-1} are all integers. Notice that your computed solution suffers from significant roundoff errors. This is because A is (mildly) ill conditioned. This is the 3×3 member of a famous family of ill-conditioned matrices called *Hilbert matrices*; the condition gets rapidly worse as the size of the matrix increases. We will discuss ill-conditioned matrices in Chapter 2.

□

1.6 SPARSE POSITIVE DEFINITE SYSTEMS

If one compares a sparse matrix A with its Cholesky factor R, one normally finds that R has many more nonzero entries than the upper half of A does. The "new" nonzero entries are called *fill* or *fill-in*. How much fill one gets depends on how the equations are ordered.

Example 1.6.1 An *arrowhead matrix* like

$$A = \begin{bmatrix} 7 & 0 & 0 & 1 \\ 0 & 3 & 0 & 1 \\ 0 & 0 & 2 & 1 \\ 1 & 1 & 1 & 1 \end{bmatrix}$$

suffers no fill-in during the Cholesky decomposition. Its Cholesky factor is

$$R = \begin{bmatrix} \sqrt{7} & 0 & 0 & 1/\sqrt{7} \\ & \sqrt{3} & 0 & 1/\sqrt{3} \\ & & \sqrt{2} & 1/\sqrt{2} \\ & & & 1/\sqrt{42} \end{bmatrix},$$

which has as many zeros above the main diagonal as A has. Now consider the matrix

$$\hat{A} = \begin{bmatrix} 1 & 1 & 1 & 1 \\ 1 & 2 & 0 & 0 \\ 1 & 0 & 3 & 0 \\ 1 & 0 & 0 & 7 \end{bmatrix},$$

obtained from A by reversing the order of the rows and columns. This reversed arrowhead matrix has the Cholesky factor

$$\hat{R} = \begin{bmatrix} 1 & 1 & 1 & 1 \\ & 1 & -1 & -1 \\ & & 1 & -2 \\ & & & 1 \end{bmatrix},$$

which is completely filled in. Obviously the same will happen to any arrowhead matrix; the pattern of nonzeros is what matters, not their exact values. Furthermore, there is nothing special about the case $n = 4$; we can build arbitrarily large arrowhead matrices. Notice that A has a small envelope that contains no nonzero entries, whereas $\hat{A}$ has a large envelope that includes many zeros. Theorem 1.5.7 guarantees that the envelope of R will be no bigger than the envelope of A, but it says nothing about the fate of zeros within the envelope. Usually zeros within the envelope of A will turn into nonzeros in R.　　　　　　　　　　　　　　　　　　　　　　　　　　□

The key to making the sparse Cholesky decomposition economical is to keep the fill under control. Ideally one would like to find the reordering that minimizes fill. As it turns out, this is a difficult problem (there are $n!$ orderings to consider). The solution seems to be beyond reach. Fortunately there are several practical algorithms that do a reasonable job of keeping the fill-in under control [14, 37]. Two very effective methods are implemented in MATLAB. The *reverse Cuthill-McKee* algorithm attempts to make the envelope small. The *approximate minimum-degree* algorithm tries to minimize fill-in by analyzing the sparsity pattern of the matrix using graph-theoretic methods. Bandwidth is ignored. For descriptions of these two methods see [14]. We will explore their performance by means of some MATLAB examples.

MATLAB has considerable support for sparse matrix computations. There is an easy-to-use sparse matrix data structure. Most of the operations that are available for full (i.e. non-sparse) matrices can be applied to sparse matrices as well. For example, if A is a positive-definite matrix stored in the sparse format, the command R = chol(A) gives the Cholesky factor of A, also stored in the sparse format. There are also numerous commands that apply strictly to sparse matrices. Type help sparfun in MATLAB for a list of these, or look in MATLAB's help browser.

Example 1.6.2 A nice example of a sparse matrix that's not too big is bucky, which is the incidence matrix of the Bucky Ball (soccer ball). The Bucky Ball is a polyhedron with 60 vertices, 32 faces (12 pentagons and 20 hexagons), and 90 edges. An incidence matrix is obtained by numbering the vertices and building a 60×60 matrix whose (i, j) entry is 1 if vertices i and j are connected by an edge and 0 otherwise. Since each vertex is connected to exactly three others, each row of the Bucky Ball matrix has exactly three nonzero entries. This matrix is not positive definite; its main diagonal entries are all zero. Figure 1.13 shows the pattern of nonzeros in the Bucky Ball matrix in the "original" ordering specified by MATLAB and in three reorderings. Plots of this type are called *spy plots* in MATLAB and are generated by the command spy(A). Each of the four plots in Figure 1.13 has 180 dots, corresponding to the 180 nonzero entries of the matrix. We note that the reverse Cuthill-McKee ordering gathers the nonzeros into a band, whereas the minimum-degree ordering does not.　　　　　　　　　　　　　　　　□

Example 1.6.3 We now move on to a positive-definite example. We used the following commands to generate a very small scale 3d discrete Laplacian matrix A.

```
m = 5;
r = zeros(1,m); r(1) = 2; r(2) = -1;
B = sparse(toeplitz(r));
C = speye(m);
A = kron(kron(B,C),C) + ...
    kron(kron(C,B),C) + ...
    kron(kron(C,C),B);
```

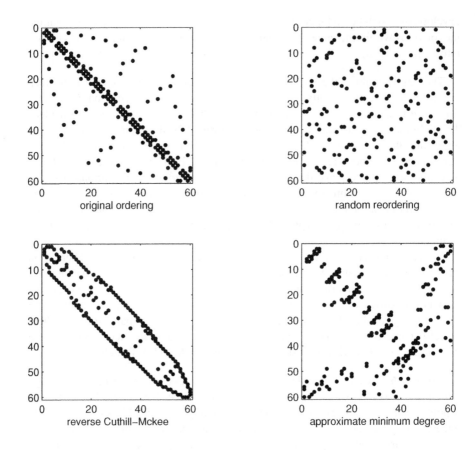

Figure 1.13 Spy plots of several orderings of the Bucky Ball matrix

This is a 125×125 positive-definite matrix. Figure 1.14 shows the spy plot of A and three of its reorderings. Notice that in its original ordering A is already banded. The reverse Cuthill-McKee ordering makes the envelope a bit narrower, but the approximate minimum-degree ordering destroys the band structure entirely.

Figure 1.15 shows spy plots of the Cholesky factors of A and each of the reorderings. In each case the number of nonzero entries is shown. Notice that the random ordering has much worse fill-in than the others. The reverse Cuthill-McKee ordering has slightly less fill than the original banded ordering, but the approximate minimum-degree ordering is much better. □

One small example proves nothing, but extensive tests on larger matrices have confirmed that the approximate minimum-degree algorithm does significantly better than reverse Cuthill-McKee on a wide variety of problems. However, effective

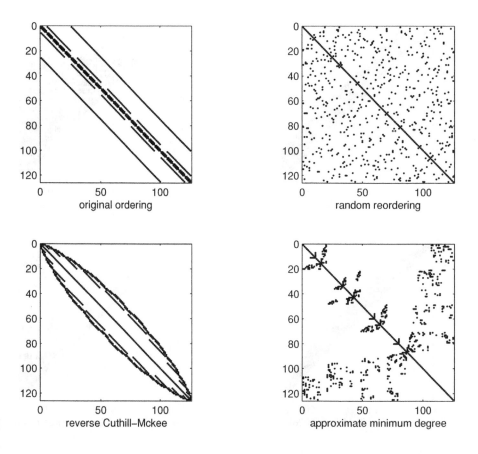

Figure 1.14   Spy plots of several orderings of a discrete Laplacian matrix

exploitation of the good fill properties of the approximate minimum-degree algorithm requires use of a more flexible data structure for sparse matrices since the fill is not restricted to a narrow band. In contrast, if we use the reverse Cuthill-McKee algorithm, we can use a simple band or envelope scheme that accommodates the fill automatically.

MATLAB's sparse matrix data structure is quite simple. Three numbers, m, n, and nz specify the number of rows, columns, and nonzero entries of the array. The information about the matrix entries and their locations is stored in three lists (one-dimensional arrays), i, j, and s, of length (at least) nz. Arrays i and j have integer entries, and s has (double precision) real entries. For each $k \leq nz$, $s(k)$ is the

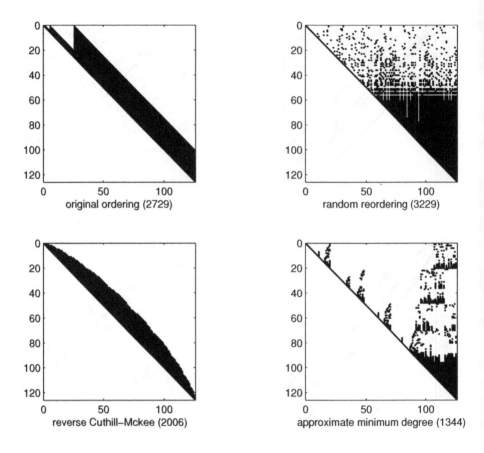

Figure 1.15 Spy plots of Cholesky factors of reorderings of a discrete Laplacian matrix. For each ordering, the number of nonzeros is given in parentheses.

value of a nonzero entry of A. Its position in the matrix is $(i(k), j(k))$.[7] When the MATLAB command R = chol(A) is used to compute the Cholesky factor of the sparse matrix A, MATLAB has to allocate space for the sparse matrix R. In doing so it must take into account not only the amount of space that A occupies but also the fill that occurs as R is computed. This all looks easy to the user because MATLAB does it all automatically.

---

[7]If more than one entry is assigned to a given (i, j) location, entries assigned to the same location are added together. For example, if sparse matrix A has $(i(3), j(3)) = (i(7), j(7)) = (i(50), j(50)) = (21, 36)$ (and all other (i, j) pairs differ from $(21,36)$), then $a_{21,36} = s(3) + s(7) + s(50)$.

rcise 1.6.4 Investigate the MATLAB commands in Example 1.6.3, and figure out how the 3D discrete Laplacian matrix was constructed.   □

rcise 1.6.5 Build a larger version of the 3D discrete Laplacian matrix of Example 1.6.3 by increasing the value of m in the code shown there. The dimension of the matrix is m^3, so if you take, for example, $m = 20$, you will get an 8000×8000 matrix. Once you have generated your matrix, type `size(A)` to confirm the dimensions and `issparse(A)` to find out whether A is stored as a sparse or a full matrix. The answer 1 indicates sparse and 0 indicates full. Type `spy(A)` to get a picture of the nonzero structure of A. (Type simply A to get a long list of the nonzero entries of A.) The label on the spy plot tells you how many nonzero entries A has. (In the case $m = 20$ it is 53,600.) This information can also be obtained by typing `nnz(A)`.

Calculate the Cholesky factor of A and several reorderings of A:

(a) First consider A in its original ordering. Calculate the Cholesky factor of A, keeping track of how long it takes to do so. For example, you can use the commands

```
tic, R = chol(A); toc
```

or

```
t = cputime; R = chol(A); time = cputime - t
```

How many nonzeros does R have? Take a look at the spy plot of R.

(b) Now permute the rows/columns of A randomly and repeat part (a). This is achieved as follows, for example:

```
p = randperm(size(A,1));
arnd = A(p,p);
spy(arnd)
tic, rrnd = chol(arnd); toc
nz = nnz(rrnd)
spy(rrnd)
```

The first command returns a random permutation of the integers 1 through k, where k is the dimension of A. The second produces a new matrix `arnd` whose rows and columns have been permuted according to the random permutation.

(c) Repeat part (b) using the reverse Cuthill-McKee ordering instead of a random reordering. The correct permutation is obtained by replacing the random permutation by `p = symrcm(A);`. Thus

```
p = symrcm(A);
arcm = A(p,p);
spy(arcm)
tic, rrcm = chol(arcm); toc
```

```
nz = nnz(rrcm)
spy(rrcm)
```

The command `symrcm` is a mnemonic for "SYMmetric Reverse Cuthill-McKee."

(d) Repeat part (c) using the approximate minimum-degree ordering instead of reverse Cuthill-McKee. The approximate minimum-degree permutation is obtained by `p = symamd(A);` `symamd` is a mnemonic for "SYMmetric Approximate Minimum Degree."

(e) Comment on your results.

$\square$

Exercise 1.6.6 Repeat Exercise 1.6.5 with a much larger 3D discrete Laplacian matrix, leaving out part (b). $\square$

Exercise 1.6.7 MATLAB has a command `delsq`, which generates 2D discrete Laplacian ("del squared") matrices associated with various regions. For example, try

```
m = 92
A = delsq(numgrid('S',m));
issparse(A)
size(A)
```

for a discrete Laplacian on a square. This produces an 8100×8100 matrix. If this matrix is too small or too big, a larger or smaller version can be obtained by increasing or decreasing m. In general the matrix A has dimension $(m-2)^2$. Its structure is the same as that of the matrices discussed in Example 1.5.2 and Exercise 1.5.4. For more information on `delsq` type `help delsq` and `help numgrid` in MATLAB, or search in the help browser. Numerous variations can be obtained by replacing the `'S'` by other letters in the `numgrid` command.

Using the matrix A generated as shown above (using a larger m if your computer allows it), calculate the Cholesky factor of A and several reorderings of A. Use the MATLAB commands that you learned in Exercise 1.6.5.

(a) Make a spy plot of A. Notice that the original ordering already gives a narrow bandwidth. Calculate the Cholesky factor of A, noting the CPU time. How many nonzeros does the Cholesky factor have? Take a look at its spy plot.

(b) Repeat part (a) using a random reordering of the rows/columns of A. (`p = randperm((m-2)^2); arnd = A(p,p);`).

(c) Repeat part (a) using the reverse Cuthill-McKee ordering.

(d) Repeat part (a) using the approximate minimum-degree ordering.

(e) Another ordering that is available for this particular example is the *nested-dissection* ordering. Type `Anest = delsq(numgrid('N',m));`. This gives the same matrix as before, except that the rows/columns are numbered according to a nested-dissection ordering [14, 37]. Repeat part (a) using the nested-dissection ordering.

(f) Discuss the results from parts (a)–(e).

□

rcise 1.6.8 Another interesting example that is supplied with MATLAB is the "NASA airfoil." Run the NASA Airfoil command-line Demo by typing `airfoil` in MAT-LAB. This begins by plotting a grid of triangles that has been used for computation of the flow around an airplane wing by the finite element method. The grid has 4253 points or nodes, each having from three to nine neighbors. The Demo also builds a positive definite matrix A related to the adjacency matrix of the grid. This is a 4253×4253 matrix with a -1 in the (i, j) position if $i \neq j$ and node i is adjacent to node j (the nodes having been numbered beforehand). The main diagonal entries are made large enough that the matrix is positive definite. The exact recipe for A is given in the Demo. The Demo then proceeds to produce spy plots of A and several reorderings of A. Once you have run the Demo, the matrix A remains in memory for your use.

(a) Calculate the Cholesky factor of A, noting the CPU time. How many nonzeros does the Cholesky factor have? Take a look at its spy plot.

(b) Repeat part (a) using the reverse Cuthill-McKee ordering.

(c) Repeat part (a) using the approximate minimum-degree ordering.

(d) Discuss the results from parts (a)–(c).

□

MATLAB comes equipped with many other test matrices, both sparse and full. Type `help elmat` for a partial list. In particular, the `gallery` collection of N. J. Higham contains many good specimens. Type `help gallery` and follow the instructions.

rcise 1.6.9 Repeat Exercise 1.6.7 (skipping part (e)) using a Wathen matrix from Higham's gallery. Type A = `gallery('wathen',70,60)`, for example. For a bigger matrix replace the 70 and 60 by larger numbers. Type `help private/wathen` for information about this matrix family. □

1.7 GAUSSIAN ELIMINATION AND THE LU DECOMPOSITION

In this and the next section, we will consider the problem of solving a system of n linear equations in n unknowns $Ax = b$ by Gaussian elimination. The algorithms developed here produce (in the absence of rounding errors) the unique solution whenever A is nonsingular. A is not assumed to have any special properties such as symmetry or positive definiteness.

Our strategy will be to transform the system $Ax = b$ to an equivalent system $Ux = y$ whose coefficient matrix is upper triangular. The system $Ux = y$ can then be solved easily by back substitution if U is nonsingular. To say that two systems are *equivalent* is to say that they have the same solutions.

We will transform the system by means of *elementary operations* of three types:

1. Add a multiple of one equation to another equation.

2. Interchange two equations.

3. Multiply an equation by a nonzero constant.

Proposition 1.7.1 *If $\hat{A}x = \hat{b}$ is obtained from $Ax = b$ by an elementary operation of type 1, 2, or 3, then the systems $Ax = b$ and $\hat{A}x = \hat{b}$ are equivalent.*

Exercise 1.7.2   Prove Proposition 1.7.1. *Discussion:* Suppose the system $Ax = b$ is transformed to $\hat{A}x = \hat{b}$ by an operation of type 1. You must show that (a) every solution of $Ax = b$ is a solution of $\hat{A}x = \hat{b}$ and (b) every solution of $\hat{A}x = \hat{b}$ is a solution of $Ax = b$. Part (a) should be easy. Part (b) becomes easy when you realize that $Ax = b$ can be recovered from $\hat{A}x = \hat{b}$ by an operation of type 1: If $\hat{A}x = \hat{b}$ was obtained from $Ax = b$ by adding m times the jth row to the ith row, then $Ax = b$ can be recovered from $\hat{A}x = \hat{b}$ by adding $-m$ times the jth row to the ith row. Analogous remarks apply to operations of types 2 and 3.   □

It is convenient to represent the system $Ax = b$ by an augmented matrix $[A \mid b]$. Each equation in the system $Ax = b$ corresponds to a row of the matrix $[A \mid b]$. The elementary operations on the equations amount to the following *elementary row operations* on $[A \mid b]$:

1. Add a multiple of one row to another row.

2. Interchange two rows.

3. Multiply a row by a nonzero constant.

One can also apply elementary row operations to A alone, leaving off the vector b.

Proposition 1.7.3 *Suppose $\hat{A}$ is obtained from A by an elementary row operation of type 1, 2, or 3. Then $\hat{A}$ is nonsingular if and only if A is.*

Proof. By Theorem 1.2.3, A is nonsingular if and only if the system $Ay = 0$ has no solutions other than $y = 0$. The same holds for $\hat{A}$. If $\hat{A}$ is obtained from A by some elementary row operation, then the same row operation will transform the augmented matrix $[A \mid 0]$ to $[\hat{A} \mid 0]$, since the row operation (regardless of type) cannot create nonzero entries from the zeros in the last column. Thus, by Proposition 1.7.1, the systems $Ay = 0$ and $\hat{A}y = 0$ are equivalent; that is, they have the same solutions. Therefore $\hat{A}$ is nonsingular if and only if A is.   □

Elementary row operations are explored in more depth in Exercises 1.7.34 through 1.7.36. These exercises relate row operations to multiplication by *elementary matrices* of types 1, 2, and 3. They also contain an alternate proof of Proposition 1.7.3 that uses determinants.

In this section our focus will be on operations of type 1. In the next section, type 2 operations will come into play. We will not use type 3 operations, except in the discussion of scaling in Section 2.8.

Gaussian Elimination without Row Interchanges

We will approach the problem of solving $Ax = b$ in two stages. In the first stage, which will occupy us for the rest of this section, we will assume that A satisfies a special property that makes it possible to transform A to upper triangular form using elementary row operations of type 1 only. The algorithm that we will derive for carrying out this transformation is called *Gaussian elimination without row interchanges* or *Gaussian elimination without pivoting*.

Recall that for $k = 1, \ldots, n$, the kth leading principal submatrix of A is the matrix A_k obtained by intersecting the first k rows and columns of A. For the rest of this section, the following assumption will hold:

$$A_k \text{ is nonsingular} \qquad k = 1, \ldots, n. \tag{1.7.4}$$

In particular, $A \; (= A_n)$ is itself nonsingular, so the system $Ax = b$ has a unique solution.

The reduction to triangular form is carried out in $n - 1$ steps. In the first step appropriate multiples of the first row are subtracted from each of the other rows to create zeros in positions $(2, 1), (3, 1), \ldots, (n, 1)$. It is clear that in order to do this we must have $a_{11} \neq 0$. This is guaranteed by the assumption that $A_1 = [a_{11}]$ is nonsingular. The appropriate multiplier for the ith row is

$$m_{i1} = a_{i1}/a_{11} \qquad i = 2, \ldots, n. \tag{1.7.5}$$

rcise 1.7.6 Verify that if m_{i1} times the first row is subtracted from the ith row of $[A \mid b]$, the resulting array has a zero in the $(i, 1)$ position.   □

Thus we carry out the operations

$$\begin{aligned} a_{ij}^{(1)} &= a_{ij} - m_{i1}a_{1j} & j = 2, \ldots, n, \; i = 2, \ldots, n \\ b_i^{(1)} &= b_i - m_{i1}b_1 & i = 2, \ldots, n \end{aligned} \tag{1.7.7}$$

to reduce $[A \mid b]$ to the form

$$
\begin{bmatrix}
a_{11} & a_{12} & \cdots & a_{1n} & b_1 \\
0 & a_{22}^{(1)} & \cdots & a_{2n}^{(1)} & b_2^{(1)} \\
\vdots & \vdots & & \vdots & \vdots \\
0 & a_{n2}^{(1)} & \cdots & a_{nn}^{(1)} & b_n^{(1)}
\end{bmatrix}.
$$

It is not necessary to calculate the entries $a_{i1}^{(1)}$, $i = 2, \ldots, n$, explicitly, because we know in advance that they are all zero.

In a computer implementation of this step, it is not necessary to store the zeros in the first column. Those storage spaces can be used for something else. As we shall soon see, it turns out to be a good idea to store the multipliers $m_{21}, m_{31}, \ldots, m_{n1}$ there. Thus the array that initially contained A and b would look as follows after the first step:

$$
\begin{bmatrix}
a_{11} & a_{12} & \cdots & a_{1n} & b_1 \\
m_{21} & a_{22}^{(1)} & \cdots & a_{2n}^{(1)} & b_2^{(1)} \\
\vdots & \vdots & & \vdots & \vdots \\
m_{n1} & a_{n2}^{(1)} & \cdots & a_{nn}^{(1)} & b_n^{(1)}
\end{bmatrix}.
$$

The cost of the arithmetic in the first step is easily determined by examining (1.7.5) and (1.7.7). There are $n-1$ divisions in (1.7.5) and $(n-1)^2 + (n-1)$ multiplications and an equal number of additions in (1.7.7). The total flop count is $(2n+1)(n-1)$, which is approximately $2n^2$.

The second step operates on rows $2, 3, \ldots, n$. Any operations that are performed on these rows will leave the zeros in column 1 undisturbed, because subtraction of a multiple of zero from zero leaves zero. Thus the second step ignores both the first row and the first column. Appropriate multiples of the second row are subtracted from rows $3, 4, \ldots, n$ to create zeros in positions $(3, 2), (4, 2), \ldots, (n, 2)$. Thus the second step is identical to the first, except that it operates on the submatrix

$$
\begin{bmatrix}
a_{22}^{(1)} & \cdots & a_{2n}^{(1)} & b_2^{(1)} \\
\vdots & & \vdots & \vdots \\
a_{n2}^{(1)} & \cdots & a_{nn}^{(1)} & b_n^{(1)}
\end{bmatrix}.
$$

The operations are

$$
m_{i2} = a_{i2}^{(1)} / a_{22}^{(1)} \qquad i = 3, \ldots, n
$$

and

$$
\begin{aligned}
a_{ij}^{(2)} &= a_{ij}^{(1)} - m_{i2} a_{2j}^{(1)} & j = 3, \ldots, n,\ i = 3, \ldots, n \\
b_i^{(2)} &= b_i^{(1)} - m_{i2} b_2^{(1)} & i = 3, \ldots, n.
\end{aligned}
$$

As in the first step, there is no need to calculate $a_{i2}^{(2)}$ explicitly for $i = 3, \ldots, n$, because the multipliers m_{i2} were chosen so that $a_{i2}^{(2)} = 0$.

In order to carry out this step, we need $a_{22}^{(1)} \neq 0$. That this is so follows from the assumption that

$$A_2 = \begin{bmatrix} a_{11} & a_{12} \\ a_{21} & a_{22} \end{bmatrix}$$

is nonsingular. After the first step, A_2 has been transformed to

$$\hat{A}_2 = \begin{bmatrix} a_{11} & a_{12} \\ 0 & a_{22}^{(1)} \end{bmatrix}.$$

$\hat{A}_2$ was obtained from A_2 by subtracting m_{21} times the first row from the second, a type 1 row operation. Therefore, by Proposition 1.7.3, the nonsingularity of A_2 implies nonsingularity of $\hat{A}_2$. Since the $\hat{A}_2$ is upper triangular and has $a_{22}^{(1)}$ on its main diagonal, we conclude that $a_{22}^{(1)} \neq 0$, as claimed.

After the second step the augmented matrix will have been transformed to

$$\left[\begin{array}{cccc|c} a_{11} & a_{12} & a_{13} & \cdots & a_{1n} & b_1 \\ 0 & a_{22}^{(1)} & a_{23}^{(1)} & \cdots & a_{2n}^{(1)} & b_2^{(1)} \\ 0 & 0 & a_{33}^{(2)} & \cdots & a_{3n}^{(2)} & b_3^{(2)} \\ \vdots & \vdots & \vdots & & \vdots & \vdots \\ 0 & 0 & a_{n3}^{(2)} & \cdots & a_{nn}^{(2)} & b_n^{(2)} \end{array} \right].$$

In a computer implementation the zeros will be replaced by the multipliers $m_{21}, \ldots,$ m_{n1} and $m_{32}, \ldots, m_{n2}$. Since the second step is the same as the first, but on a matrix with one less row and one less column, the flop count for the second step is about $2(n-1)^2$.

The third step is identical to the previous two, except that it operates on the smaller matrix

$$\left[\begin{array}{ccc|c} a_{33}^{(2)} & \cdots & a_{3n}^{(2)} & b_3^{(2)} \\ \vdots & & \vdots & \vdots \\ a_{n3}^{(2)} & \cdots & a_{nn}^{(2)} & b_n^{(2)} \end{array} \right].$$

In order to carry out the step, we need to have $a_{33}^{(2)} \neq 0$. This is guaranteed by the assumption that A_3 is nonsingular. After the first two steps, A_3 will have been transformed to

$$\hat{A}_3 = \begin{bmatrix} a_{11} & a_{12} & a_{13} \\ 0 & a_{22}^{(1)} & a_{23}^{(1)} \\ 0 & 0 & a_{33}^{(2)} \end{bmatrix}$$

via two elementary row operations of type 1. Therefore, by Proposition 1.7.3, the nonsingularity of A_3 implies the nonsingularity of $\hat{A}_3$. This implies, in turn, that $a_{33}^{(2)} \neq 0$. The arithmetic for the third step amounts to about $2(n-2)^2$ flops.

After $n-1$ steps the system will be reduced to $[U \mid y]$, where U is upper triangular. For each k, the possibility of carrying out step k is guaranteed by the assumption that

A_k is nonsingular. In the end we know that U is nonsingular because A is. Thus the system $Ux = y$ can be solved by back substitution to yield x, the unique solution of $Ax = b$.

The total flop count for the reduction to triangular form is approximately

$$2n^2 + 2(n-1)^2 + 2(n-2)^2 + \ldots = 2\sum_{k=1}^{n} k^2.$$

As in Section 1.4 (cf. Exercises 1.4.70 and 1.4.71) we can approximate this sum by an integral:

$$2\sum_{k=1}^{n} k^2 \approx 2\int_0^n k^2 \, dk = \tfrac{2}{3}n^3.$$

The additional cost of the back substitution is n^2 flops, which is relatively insignificant for large n. Thus the total cost of solving $Ax = b$ by this method is about $\tfrac{2}{3}n^3$ flops. This is about twice the cost of solving a positive definite system by Cholesky's method, which saves a factor of 2 by exploiting the symmetry of the coefficient matrix.

Example 1.7.8 Let

$$A = \begin{bmatrix} 2 & 1 & 1 \\ 2 & 2 & -1 \\ 4 & -1 & 6 \end{bmatrix} \quad \text{and} \quad b = \begin{bmatrix} 9 \\ 9 \\ 16 \end{bmatrix}.$$

Notice that $\det(A_1) = 2$, $\det(A_2) = 2$, and $\det(A_3) = -4$, so the leading principal submatrices are all nonsingular. This guarantees that we will be able to transform A to upper triangular form by row operations of type 1 only. It also guarantees that the system $Ax = b$ has a unique solution. In order to solve this system, we form the augmented matrix

$$[A \mid b] = \begin{bmatrix} 2 & 1 & 1 & 9 \\ 2 & 2 & -1 & 9 \\ 4 & -1 & 6 & 16 \end{bmatrix}.$$

The multipliers for the first step are $m_{21} = a_{21}/a_{11} = 1$ and $m_{31} = a_{31}/a_{11} = 2$. Thus we subtract 1 times the first row from the second row and 2 times the first row from the third row to create zeros in the $(2, 1)$ and $(3, 1)$ positions. Performing these row operations, we obtain

$$\begin{bmatrix} 2 & 1 & 1 & 9 \\ 0 & 1 & -2 & 0 \\ 0 & -3 & 4 & -2 \end{bmatrix}.$$

The multiplier for the second step is $m_{32} = a_{32}^{(1)}/a_{22}^{(1)} = -3$. Thus we subtract -3 times the second row from the third row to obtain

$$\begin{bmatrix} 2 & 1 & 1 & 9 \\ 0 & 1 & -2 & 0 \\ 0 & 0 & -2 & -2 \end{bmatrix}.$$

After two steps the reduction is complete. If we save the multipliers in place of the zeros, the array looks like this:

$$\left[\begin{array}{ccc|c} 2 & 1 & 1 & 9 \\ 1 & 1 & -2 & 0 \\ 2 & -3 & -2 & -2 \end{array}\right]. \tag{1.7.9}$$

We can now solve the system by solving

$$\left[\begin{array}{ccc} 2 & 1 & 1 \\ 0 & 1 & -2 \\ 0 & 0 & -2 \end{array}\right] \left[\begin{array}{c} x_1 \\ x_2 \\ x_3 \end{array}\right] = \left[\begin{array}{c} 9 \\ 0 \\ -2 \end{array}\right]$$

by back substitution. Doing so, we find that $x_3 = 1$, $x_2 = 2$, and $x_1 = 3$. $\quad\square$

rcise 1.7.10 Let

$$A = \left[\begin{array}{cccc} 2 & 1 & -1 & 3 \\ -2 & 0 & 0 & 0 \\ 4 & 1 & -2 & 6 \\ -6 & -1 & 2 & -3 \end{array}\right] \quad\text{and}\quad b = \left[\begin{array}{c} 13 \\ -2 \\ 24 \\ -14 \end{array}\right].$$

(a) Calculate the appropriate (four) determinants to show that A can be transformed to (nonsingular) upper-triangular form by operations of type 1 only. (By the way, this is strictly an academic exercise. In practice one never calculates these determinants in advance.)

(b) Carry out the row operations of type 1 to transform the system $Ax = b$ to an equivalent system $Ux = y$, where U is upper triangular. Save the multipliers for use in Exercise 1.7.18.

(c) Carry out the back substitution on the system $Ux = y$ to obtain the solution of $Ax = b$. Don't forget to check your work.

$\quad\square$

In real-life problems it will not be practical to verify in advance that the conditions (1.7.4) are satisfied. This does not of itself stop us from proceeding with the algorithm. In the course of performing the eliminations, if some A_k turns out to be singular (and all previous A_j were nonsingular), this fact will be signaled by $a_{kk}^{(k-1)} = 0$. This gives us, in principle, a test that we can apply in the course of transforming A to triangular form, to determine whether or not the conditions (1.7.4) are satisfied.

Although this test is perfectly satisfying in theory, it fails badly in practice. In real problems the entries are not all integers, the computations are done on a computer, and there are roundoff errors. Consequently an $a_{kk}^{(k-1)}$ that should be exactly zero in theory will typically turn out to be nonzero (albeit tiny) in practice. The test then fails to stop the computation, we divide by this erroneous number to obtain the multipliers for the kth step, and we end up with a disastrously inaccurate result. We will see how to deal with this problem in the next section.

Interpretation of the Multipliers

Now let us see why it is a good idea to save the multipliers m_{ij}. Suppose we have solved the system $Ax = b$, and we now wish to solve another system $Ax = \hat{b}$, which has the same coefficient matrix but a different right-hand side. We could form the augmented matrix $[A \mid \hat{b}]$ and solve the system from scratch, but this would be inefficient. Since the coefficient matrix is the same as before, all of the multipliers and row operations will be the same. If the multipliers have been saved, we can perform the row operations on the $\hat{b}$ column only and save a good many computations. Let's see how this works. The operations on b were as follows:

$$
\begin{aligned}
b_i^{(1)} &= b_i - m_{i1}b_1 & i &= 2, 3, \ldots, n \\
b_i^{(2)} &= b_i^{(1)} - m_{i2}b_2^{(1)} & i &= 3, 4, \ldots, n \\
&\vdots & &\vdots \\
b_i^{(j)} &= b_i^{(j-1)} - m_{ij}b_j^{(j-1)} & i &= j+1 \ldots, n \\
&\vdots & &\vdots \\
b_i^{(n-1)} &= b_i^{(n-2)} - m_{i,n-1}b_{n-1}^{(n-2)} & i &= n.
\end{aligned}
\tag{1.7.11}
$$

At the end of the operations b has been transformed to

$$
\begin{bmatrix} b_1 \\ b_2^{(1)} \\ b_3^{(2)} \\ \vdots \\ b_n^{(n-1)} \end{bmatrix} = \begin{bmatrix} y_1 \\ y_2 \\ y_3 \\ \vdots \\ y_n \end{bmatrix} = y.
\tag{1.7.12}
$$

The same operations applied to $\hat{b}$ would yield a vector $\hat{y}$. The system $Ax = \hat{b}$ could then be solved by solving the equivalent upper-triangular system $Ux = \hat{y}$, where U is the upper-triangular matrix that was obtained in the original reduction of A.

Example 1.7.13 Suppose we wish to solve the system $Ax = \hat{b}$, where

$$
A = \begin{bmatrix} 2 & 1 & 1 \\ 2 & 2 & -1 \\ 4 & -1 & 6 \end{bmatrix} \quad \text{and} \quad \hat{b} = \begin{bmatrix} 7 \\ 3 \\ 20 \end{bmatrix}.
$$

The coefficient matrix is the same as in Example 1.7.8, so the multipliers are as shown in (1.7.9), that is, $m_{21} = 1$, $m_{31} = 2$, and $m_{32} = -3$. This means that A can be transformed to upper-triangular form by subtracting 1 times the first row from the third row, 2 times the first row from the third row, and -3 times the (new) second row from the third row. Rather than performing these operations on the augmented matrix $[A \mid \hat{b}]$, we apply them to the column $\hat{b}$ only and get

$$
\hat{b}_2^{(1)} = \hat{b}_2 - m_{21}\hat{b}_1 = 3 - 1 \cdot 7 = -4
$$

$$\hat{b}_3^{(1)} = \hat{b}_3 - m_{31}\hat{b}_1 = 20 - 2 \cdot 7 = 6$$
$$\hat{b}_3^{(2)} = \hat{b}_3^{(1)} - m_{32}\hat{b}_2^{(1)} = 6 - (-3)(-4) = -6.$$

After these transformations, the new right-hand side is

$$\hat{y} = \begin{bmatrix} b_1 \\ b_2^{(1)} \\ b_3^{(2)} \end{bmatrix} = \begin{bmatrix} 7 \\ -4 \\ -6 \end{bmatrix}.$$

Now we can get the solution by solving $Ux = y$ by back substitution, where

$$U = \begin{bmatrix} 2 & 1 & 1 \\ 0 & 1 & -2 \\ 0 & 0 & -2 \end{bmatrix},$$

as in Example 1.7.8. Doing so, we find that $x_3 = 3$, $x_2 = 2$, and $x_1 = 1$.   □

A quick count shows that (1.7.11) takes about n^2 flops. Adding this to the n^2 flops for back substitution, we find that the total arithmetic cost of solving $Ax = \hat{b}$ is about $2n^2$ flops. When we compare this with the $\frac{2}{3}n^3$ flops that were needed to transform A to upper triangular form initially, we conclude that once we have solved $Ax = b$, we can solve additional systems with the same coefficient matrix at little additional cost.

A closer look at the transformation of b to y yields an important interpretation of Gaussian elimination. By (1.7.12), we can rewrite (1.7.11) in terms of the components of y as follows:

$$
\begin{array}{llll}
b_i^{(1)} &= b_i - m_{i1}y_1 & i = 2, 3, \ldots, n \\
b_i^{(2)} &= b_i^{(1)} - m_{i2}y_2 & i = 3, 4, \ldots, n \\
\vdots & & \vdots \\
b_i^{(n-1)} &= b_i^{(n-2)} - m_{i,n-1}y_{n-1} & i = n.
\end{array}
\tag{1.7.14}
$$

Together with (1.7.12), these equations can be used to derive expressions for y_1, y_2, $\ldots$, y_n. By the first equation of (1.7.14), $y_2 = b_2^{(1)} = b_2 - m_{21}y_1$. By the first two equations of (1.7.14), $y_3 = b_3^{(2)} = b_3^{(1)} - m_{32}y_2 = b_3 - m_{31}y_1 - m_{32}y_2$. Similarly $y_4 = b_4 - m_{41}y_1 - m_{42}y_2 - m_{43}y_3$, and in general

$$y_i = b_i - \sum_{j=1}^{i-1} m_{ij}y_j \qquad i = 1, 2, \ldots, n. \tag{1.7.15}$$

We can use (1.7.15) to calculate $y_1, y_2, \ldots y_n$. Clearly the operations are the same as those of (1.7.11), but in a different order. The equations (1.7.15) can be interpreted as a matrix operation if we rewrite them as

$$\sum_{j=1}^{i-1} m_{ij}y_j + y_i = b_i, \qquad i = 1, \ldots, n.$$

This can be expressed as the matrix equation

$$
\begin{bmatrix}
1 & 0 & 0 & \cdots & 0 \\
m_{21} & 1 & 0 & \ddots & \vdots \\
m_{31} & m_{32} & 1 & \ddots & \\
\vdots & & & \ddots & 0 \\
m_{n1} & m_{n2} & m_{n3} & \cdots & 1
\end{bmatrix}
\begin{bmatrix}
y_1 \\ y_2 \\ y_3 \\ \vdots \\ y_n
\end{bmatrix}
=
\begin{bmatrix}
b_1 \\ b_2 \\ b_3 \\ \vdots \\ b_n
\end{bmatrix}. \tag{1.7.16}
$$

Thus we see that y is just the solution of a linear system $Ly = b$, where L is lower triangular. In fact L is *unit lower triangular*, which means that its main diagonal entries are all ones. We learned in Section 1.3 that any lower-triangular system can be solved by forward substitution. In fact (1.7.15) is just row-oriented forward substitution. The divisions that are generally required (as in (1.3.3)) are absent from (1.7.15) because the main-diagonal entries of (1.7.16) are ones. You can easily check that (1.7.11) is nothing but column-oriented forward substitution.

A brief summary of what we have done so far will lead to an interesting and important conclusion: L and U can be interpreted as factors of A. In order to solve the system

$$Ax = b,$$

we reduced it to the form

$$Ux = y,$$

where U is upper triangular, and y is the solution of a unit lower-triangular system

$$Ly = b.$$

Combining these last two equations, we find that $LUx = b$. Thus $LUx = b = Ax$. These equations hold for any choice of b, and hence for any choice of x. (For a given x, the appropriate b is obtained by the calculation $b = Ax$.) Since the equation $LUx = Ax$ holds for all $x \in \mathbb{R}^n$, it must be that

$$A = LU.$$

We conclude that Gaussian elimination without row interchanges (saving the multipliers) can be viewed as a process of decomposing A into a product, $A = LU$, where L is lower triangular and U is upper triangular. In fact the usual procedure is not to form an augmented matrix $[A \mid b]$ but to do row operations on A alone. A is reduced, saving multipliers, to the form

$$
\begin{bmatrix}
u_{11} & u_{12} & u_{13} & \cdots & u_{1n} \\
m_{21} & u_{22} & u_{23} & \cdots & u_{2n} \\
m_{31} & m_{32} & u_{33} & \cdots & u_{3n} \\
\vdots & \vdots & \vdots & & \vdots \\
m_{n1} & m_{n2} & m_{n3} & & u_{nn}
\end{bmatrix},
$$

which contains all information about L and U. The system $LUx = b$ is then solved by first solving $Ly = b$ for y by forward substitution and then solving $Ux = y$ by back substitution.

Example 1.7.17 Solve the system $Ax = b$, where

$$A = \begin{bmatrix} 2 & 1 & 1 \\ 2 & 2 & -1 \\ 4 & -1 & 6 \end{bmatrix} \quad \text{and} \quad b = \begin{bmatrix} 3 \\ 0 \\ 11 \end{bmatrix}.$$

The coefficient matrix is the same as in Example 1.7.8. From (1.7.9) we know that $A = LU$, where

$$L = \begin{bmatrix} 1 & 0 & 0 \\ 1 & 1 & 0 \\ 2 & -3 & 1 \end{bmatrix} \quad \text{and} \quad U = \begin{bmatrix} 2 & 1 & 1 \\ 0 & 1 & -2 \\ 0 & 0 & -2 \end{bmatrix}.$$

Solving $Ly = b$ by forward substitution, we get $y = [3, -3, -4]^T$. Solving $Ux = y$ by back substitution, we get $x = [0, 1, 2]^T$.   □

rcise 1.7.18 Solve the linear system $Ax = \hat{b}$, where A is as in Exercise 1.7.10 and $\hat{b} = [12, -8, 21, -26]^T$. Use the L and U that you calculated in Exercise 1.7.10. □

We have already proved most of the following theorem.

Theorem 1.7.19 *(LU Decomposition Theorem) Let A be an $n \times n$ matrix whose leading principal submatrices are all nonsingular. Then A can be decomposed in exactly one way into a product*

$$A = LU,$$

such that L is unit lower triangular and U is upper triangular.

Proof. We have already shown that L and U exist.[8] It remains only to show that they are unique. Our uniqueness proof will yield a second algorithm for calculating the LU decomposition. Look at the equation $A = LU$ in detail.

$$\begin{bmatrix} a_{11} & a_{12} & a_{13} & \cdots & a_{1n} \\ a_{21} & a_{22} & a_{23} & \cdots & a_{2n} \\ a_{31} & a_{32} & a_{33} & \cdots & a_{3n} \\ \vdots & \vdots & \vdots & & \vdots \\ a_{n1} & a_{n2} & a_{n3} & \cdots & a_{nn} \end{bmatrix}$$

$$= \begin{bmatrix} 1 & 0 & 0 & \cdots & 0 \\ l_{21} & 1 & 0 & \cdots & 0 \\ l_{31} & l_{32} & 1 & \cdots & 0 \\ \vdots & \vdots & \vdots & & \vdots \\ l_{n1} & l_{n2} & l_{n3} & \cdots & 1 \end{bmatrix} \begin{bmatrix} u_{11} & u_{12} & u_{13} & \cdots & u_{1n} \\ 0 & u_{22} & u_{23} & \cdots & u_{2n} \\ 0 & 0 & u_{33} & \cdots & u_{3n} \\ \vdots & \vdots & \vdots & & \vdots \\ 0 & 0 & 0 & \cdots & u_{nn} \end{bmatrix}.$$

---

[8]Exercise 1.7.47 outlines a second existence proof.

The first row of L is known completely, and it has only one nonzero entry. Multiplying the first row of L by the jth column of U, we find that

$$a_{1j} = 1u_{1j} + 0u_{2j} + 0u_{3j} + \cdots + 0u_{nj}.$$

That is, $u_{1j} = a_{1j}$. Thus the first row of U is uniquely determined. Now that we know the first row of U, we see that the first column of U is also known, since its only nonzero entry is u_{11}. Multiplying the ith row of L by the first column of U, we find that

$$a_{i1} = l_{i1}u_{11}, \qquad i = 2, \ldots, n. \qquad (1.7.20)$$

The assumption that A is nonsingular implies that U is also nonsingular. (Why?) Hence $u_{kk} \neq 0$ for $k = 1, \ldots, n$, and, in particular, $u_{11} \neq 0$. Therefore (1.7.20) determines the first column of L uniquely:

$$l_{i1} = \frac{a_{i1}}{u_{11}}, \qquad i = 2, \ldots, n.$$

Thus the first column of L is uniquely determined. Now that the first row of U and first column of L have been determined, it is not hard to show that the second row of U is also uniquely determined. As an exercise, determine a formula for u_{2j} ($j \geq 2$) in terms of a_{2j} and entries of the first row of U and column of L. Once u_{22} is known, it is possible to determine the second column of L. Do this also as an exercise.

Now suppose the first $k - 1$ rows of U and columns of L have been shown to be uniquely determined. We will show that the kth row of U and column of L are uniquely determined; this will prove uniqueness by induction. The kth row of L is $[l_{k1} \, l_{k2} \, \cdots \, l_{k,k-1} \, 1 \, 0 \, \cdots \, 0]$. Since $l_{k1}, \ldots, l_{k,k-1}$ are all in the first $k - 1$ columns of L, they are uniquely determined. Multiplying the kth row of L by the jth column of U ($j \geq k$), we find that

$$a_{kj} = \sum_{m=1}^{k-1} l_{km}u_{mj} + u_{kj}. \qquad (1.7.21)$$

All of the u_{mj} (aside from u_{kj}) lie in the first $k - 1$ rows of U and are therefore known (i.e., uniquely determined). Therefore u_{kj} is uniquely determined by (1.7.21):

$$u_{kj} = a_{kj} - \sum_{m=1}^{k-1} l_{km}u_{mj}, \qquad j = k, k+1, \ldots, n. \qquad (1.7.22)$$

This proves that the kth row of U is uniquely determined and provides a way of computing it. Now that u_{kk} is known, the entire kth column of U is determined. Multiplying the ith row of L by the kth column of U, we find that

$$a_{ik} = \sum_{m=1}^{k-1} l_{im}u_{mk} + l_{ik}u_{kk}. \qquad (1.7.23)$$

All of the l_{im} (aside from l_{ik}) lie in the first $k-1$ columns of L and are therefore uniquely determined. Furthermore $u_{kk} \neq 0$. Thus (1.7.23) determines l_{ik} uniquely:

$$l_{ik} = u_{kk}^{-1} \left(a_{ik} - \sum_{m=1}^{k-1} l_{im} u_{mk} \right), \qquad i = k+1, \ldots, n. \qquad (1.7.24)$$

This proves that the kth column of L is uniquely determined and provides a way of calculating it. The proof that L and U are unique is now complete. $\qquad \square$

Equations (1.7.22) and (1.7.24), applied in the correct order, provide a means of calculating L and U. Because both (1.7.22) and (1.7.24) require inner-product computations, we will call this algorithm the *inner-product formulation* of Gaussian elimination.[9] Historically it has been known as the *Doolittle reduction*.[10]

In the calculation of L and U by row operations, the entries of A are gradually replaced by entries of L and U. The same can be done if the inner-product formulation is used: a_{kj} ($k \leq j$) is used only to compute u_{kj}, and a_{ik} ($i > k$) is used only to compute l_{ik}. Therefore each entry of L or U can be stored in place of the corresponding entry of A as soon as it is computed. You should convince yourself that the two methods perform exactly the same operations, although not in the same order. In Gaussian elimination by row operations, a typical entry is modified numerous times before the final result is obtained. In the inner-product formulation the same modifications are made, but all of the modifications of each entry are made at once.

Example 1.7.25 Let

$$A = \begin{bmatrix} 2 & 4 & 2 & 3 \\ -2 & -5 & -3 & -2 \\ 4 & 7 & 6 & 8 \\ 6 & 10 & 1 & 12 \end{bmatrix} \qquad \text{and} \qquad b = \begin{bmatrix} -3 \\ 3 \\ -1 \\ -16 \end{bmatrix}.$$

We will calculate L and U such that $A = LU$ by two different methods. First let's do Gaussian elimination by row operations.

Step 1:

$$\begin{bmatrix} 2 & 4 & 2 & 3 \\ \boxed{-1} & -1 & -1 & 1 \\ 2 & -1 & 2 & 2 \\ 3 & -2 & -5 & 3 \end{bmatrix}$$

Step 2:

$$\begin{bmatrix} 2 & 4 & 2 & 3 \\ \boxed{-1} & \boxed{-1} & -1 & 1 \\ 2 & \boxed{1} & 3 & 1 \\ 3 & \boxed{2} & -3 & 1 \end{bmatrix}$$

---

[9] In contrast, the calculation of L and U by row operations of type 1 is an *outer product* formulation. See Exercise 1.7.27.

[10] A well-known variant is the *Crout reduction*, which is very similar but produces a decomposition in which U, instead of L, has ones on the main diagonal.

Step 3:

$$\begin{bmatrix} 2 & 4 & 2 & 3 \\ -1 & -1 & -1 & 1 \\ 2 & 1 & 3 & 1 \\ 3 & 2 & -1 & 2 \end{bmatrix}$$

Now let's try the inner-product formulation.

Step 1:

$$\begin{bmatrix} 2 & 4 & 2 & 3 \\ -1 & -5 & -3 & -2 \\ 2 & 7 & 6 & 8 \\ 3 & 10 & 1 & 12 \end{bmatrix}$$

The first row of U and column of L have been calculated. The rest of the matrix remains untouched.

Step 2:

$$\begin{bmatrix} 2 & 4 & 2 & 3 \\ -1 & -1 & -1 & 1 \\ 2 & 1 & 6 & 8 \\ 3 & 2 & 1 & 12 \end{bmatrix}$$

Step 3:

$$\begin{bmatrix} 2 & 4 & 2 & 3 \\ -1 & -1 & -1 & 1 \\ 2 & 1 & 3 & 1 \\ 3 & 2 & -1 & 12 \end{bmatrix}$$

Now only u_{44} remains to be calculated.

Step 4:

$$\begin{bmatrix} 2 & 4 & 2 & 3 \\ -1 & -1 & -1 & 1 \\ 2 & 1 & 3 & 1 \\ 3 & 2 & -1 & 2 \end{bmatrix}$$

Both reductions yield the same result. You might find it instructive to try the inner-product reduction by the erasure method. Begin with the entries of A entered in pencil. As you calculate each entry of L or U, erase the corresponding entry of A and replace it with the new result. Do the arithmetic in your head.

Now that we have the LU decomposition of A, we perform forward substitution on

$$\begin{bmatrix} 1 & 0 & 0 & 0 \\ -1 & 1 & 0 & 0 \\ 2 & 1 & 1 & 0 \\ 3 & 2 & -1 & 1 \end{bmatrix} \begin{bmatrix} y_1 \\ y_2 \\ y_3 \\ y_4 \end{bmatrix} = \begin{bmatrix} -3 \\ 3 \\ -1 \\ 16 \end{bmatrix}$$

to get $y = [-3, 0, 5, -2]^T$. We then perform back substitution on

$$
\begin{bmatrix}
2 & 4 & 2 & 3 \\
0 & -1 & -1 & 1 \\
0 & 0 & 3 & 1 \\
0 & 0 & 0 & 2
\end{bmatrix}
\begin{bmatrix}
x_1 \\
x_2 \\
x_3 \\
x_4
\end{bmatrix}
=
\begin{bmatrix}
-3 \\
0 \\
5 \\
-2
\end{bmatrix}
$$

to get $x = [4, -3, 2, -1]^T$. $\quad\square$

rcise 1.7.26 Use the inner-product formulation to calculate the LU decomposition of the matrix A in Exercise 1.7.10 $\quad\square$

rcise 1.7.27 Develop an outer-product formulation of the LU decomposition algorithm in the spirit of the outer-product formulation of the Cholesky decomposition algorithm. Show that this algorithm is identical to Gaussian elimination by row operations of type 1. $\quad\square$

Variants of the LU Decomposition

An important variant of the LU decomposition is the LDV decomposition, which has a diagonal matrix sandwiched between two unit-triangular matrices. A matrix B is a *diagonal* matrix if $b_{ij} = 0$ whenever $i \neq j$. Thus all of the entries off of the main diagonal are zero.

Theorem 1.7.28 *(LDV Decomposition Theorem) Let A be an $n \times n$ matrix whose leading principal submatrices are all nonsingular. Then A can be decomposed in exactly one way as a product*

$$A = LDV,$$

such that L is unit lower triangular, D is diagonal, and V is unit upper triangular.

Proof. By Theorem 1.7.19 there exist unit lower-triangular L and upper-triangular U such that $A = LU$. Since U is nonsingular, $u_{kk} \neq 0$ for $k = 1 \ldots, n$. Let D be the diagonal matrix whose main-diagonal entries are $u_{11}, \ldots, u_{nn}$. Then D is nonsingular; D^{-1} is the diagonal matrix whose main-diagonal entries are $u_{11}^{-1}, \ldots, u_{nn}^{-1}$. Let $V = D^{-1}U$. You can easily check that V is unit upper triangular, and $A = LDV$.

To complete the proof, we must show that the decomposition is unique. Suppose $A = L_1 D_1 V_1 = L_2 D_2 V_2$. Let $U_1 = D_1 V_1$ and $U_2 = D_2 V_2$. Then obviously U_1 and U_2 are upper triangular, and $A = L_1 U_1 = L_2 U_2$. By the uniqueness of LU decompositions, $L_1 = L_2$ and $U_1 = U_2$. The latter equation implies $D_1 V_1 = D_2 V_2$; therefore

$$D_2^{-1}D_1 = V_2 V_1^{-1}. \qquad (1.7.29)$$

Since V_1 is unit upper triangular, so is V_1^{-1}. Since V_2 and V_1^{-1} are unit upper triangular, so is $V_2 V_1^{-1}$. (See Exercises 1.7.44, 1.7.45, and 1.7.46.) On the other hand, $D_2^{-1}D_1$ is clearly diagonal. Thus by (1.7.29) $V_2 V_1^{-1}$ is both unit upper triangular and diagonal; that is, $V_2 V_1^{-1} = I$. Therefore $V_2 = V_1$ and $D_2 = D_1$. $\quad\square$

Because of the symmetric roles played by L and V, the LDV decomposition is of special interest when A is symmetric.

Theorem 1.7.30 *Let A be a symmetric matrix whose leading principal submatrices are nonsingular. Then A can be expressed in exactly one way as a product $A = LDL^T$, such that L is unit lower triangular and D is diagonal.*

Proof. A has an LDV decomposition: $A = LDV$. We need only show that $V = L^T$. Now $A = A^T = (LDV)^T = V^T D^T L^T$. V^T is unit lower triangular, D^T is diagonal, and L^T is unit upper triangular, so $V^T D^T L^T$ is an LDV decomposition of A. By uniqueness of the LDV decomposition, $V = L^T$. □

In Section 1.4 we proved that if A is positive definite, then each of the leading principal submatrices A_k is positive definite and hence nonsingular. Therefore every positive definite matrix satisfies the hypotheses of Theorem 1.7.30.

Theorem 1.7.31 *Let A be positive definite. Then A can be expressed in exactly one way as a product $A = LDL^T$, such that L is unit lower triangular, and D is a diagonal matrix whose main-diagonal entries are positive.*

Exercise 1.7.32 Prove Theorem 1.7.31. To this end it suffices to show that A is positive definite if and only if the main-diagonal entries of D are positive. See Proposition 1.4.55. □

Theorem 1.7.31 leads to a second proof of the existence of the Cholesky decomposition: Since the main-diagonal entries of D are positive, we can take their square roots. Let $E = D^{1/2}$. This is the diagonal matrix whose main diagonal entries are $+\sqrt{d_{11}}, \ldots, +\sqrt{d_{nn}}$. Then $A = LDL^T = LE^2 L^T = LEE^T L^T = (LE)(LE)^T = R^T R$, where $R = (LE)^T$. Then R is upper triangular with positive main-diagonal entries, and $A = R^T R$.

The LDL^T decomposition is sometimes used in place of the Cholesky decomposition to solve positive definite systems. Algorithms analogous to the various formulations of Cholesky's method exist (Exercise 1.7.40). Each of these algorithms has block versions. The LDL^T decomposition is sometimes preferred because it does not require the extraction of square roots.

The next theorem presents a third decomposition for positive definite matrices that is preferred in some situations.

Theorem 1.7.33 *Let A be positive definite. Then A can be expressed in exactly one way as a product $A = MD^{-1}M^T$, such that M is lower triangular, D is a diagonal matrix whose main-diagonal entries are positive, and the main diagonal entries of M are the same as those of D.*

Proof. Starting with the decomposition $A = LDL^T$, let $M = LD$. Then clearly $A = MD^{-1}M^T$, and the main-diagonal entries of M are the same as those of D. The proof of uniqueness is left as an exercise. □

Algorithms for calculating the $MD^{-1}M^T$ decomposition can be derived by making slight modifications to algorithms for the Cholesky or LDL^T decomposition.

Like the LDL^T algorithms, the $MD^{-1}M^T$ algorithms do not require the computation of square roots. In a sense the $MD^{-1}M^T$ decomposition is more fundamental than the others; see the discussion of generic Gaussian elimination in Exercise 1.7.55.

The LDL^T and $MD^{-1}M^T$ algorithms can be applied to symmetric matrices that are not positive definite, provided the leading principal submatrices are not singular. However, it is difficult to check in advance whether these conditions are satisfied, and even if they are, the computation may be spoiled by roundoff errors. Stable, efficient algorithms for symmetric, indefinite linear systems do exist, but we will not discuss them here. See the discussion in [43, § 4.4].

Additional Exercises

rcise 1.7.34 In this exercise you will show that performing an elementary row operation of type 1 is equivalent to left multiplication by a matrix of a special type. Suppose $\hat{A}$ is obtained from A by adding m times the jth row to the ith row.

(a) Show that $\hat{A} = MA$, where M is the triangular matrix obtained from the identity matrix by replacing the zero by an m in the (i, j) position. For example, when $i > j$, M has the form

Notice that this is the matrix obtained by applying the type 1 row operation directly to the identity matrix. We call M an *elementary matrix* of type 1.

(b) Show that $\det(M) = 1$ and $\det(\hat{A}) = \det(A)$. Thus we see (again) that $\hat{A}$ is nonsingular if and only if A is.

(c) Show that M^{-1} differs from M only in that it has $-m$ instead of m in the (i, j) position. M^{-1} is also an elementary matrix of type 1. To which elementary operation does it correspond?

□

rcise 1.7.35 Suppose $\hat{A}$ is obtained from A by interchanging the ith and jth rows. This is an operation of type 2.

(a) Show that $\hat{A} = MA$, where M is the matrix obtained from the identity matrix by interchanging the ith and jth rows. We call this M an *elementary matrix* of type 2.

(b) Show that $\det(M) = -1$ and $\det(\hat{A}) = -\det(A)$. Thus we see (again) that $\hat{A}$ is nonsingular if and only if A is.

(c) Show that $M^{-1} = M$. Explain why this result would have been expected, in view of the action of M as the effector of an elementary operation.

$\square$

Exercise 1.7.36 Suppose $\hat{A}$ is obtained from A by multiplying the ith row by the nonzero constant c.

(a) Find the form of the matrix M (an *elementary matrix* of type 3) such that $\hat{A} = MA$.

(b) Find M^{-1} and state its function as an elementary matrix.

(c) Find $\det(M)$ and determine the relationship between $\det(\hat{A})$ and $\det(A)$. Deduce that $\hat{A}$ is nonsingular if and only if A is.

$\square$

Exercise 1.7.37 Let A be a symmetric matrix with $a_{11} \neq 0$. Suppose A has been reduced to the form

$$
\left[
\begin{array}{c|ccc}
a_{11} & a_{12} & \cdots & a_{1n} \\
\hline
0 & & & \\
\vdots & & A^{(1)} & \\
0 & & &
\end{array}
\right]
$$

by row operations of type 1 only. Prove that $A^{(1)}$ is symmetric. (It follows that by exploiting symmetry we can cut the arithmetic almost in half, provided A_k is nonsingular for $k = 1, \ldots, n$. This is the case, e.g., if A is positive definite.) $\square$

Exercise 1.7.38 (a) Develop a bordered form of Gaussian elimination analogous to the bordered form of the Cholesky decomposition algorithm. (b) Suppose A is sparse, its lower part is stored in a row-oriented envelope, and its upper part is stored in a column-oriented envelope. Prove that the envelope of L (by rows!) equals the envelope of the lower part of A, and the envelope of U (by columns) equals the envelope of the upper part of A. $\square$

Exercise 1.7.39 We have seen that if the leading principal submatrices A_k are nonsingular for $k = 1, \ldots, n$, then A has an LU decomposition. Prove the following converse. If A is nonsingular and has an LU decomposition, then A_k is nonsingular for $k = 1, \ldots, n$. (*Hint:* Partition the matrices in the decomposition $A = LU$.) $\square$

Exercise 1.7.40 Develop algorithms to calculate the LDL^T decomposition (Theorem 1.7.31) of a positive definite matrix: (a) inner-product formulation, (b) outer-product formulation, (c) bordered formulation. (d) Count the operations for each algorithm. You may find that $\frac{2}{3}n^3$ multiplications are required, twice as many for Cholesky's method.

In this case, show how half of the multiplications can be moved out of the inner loop to cut the flop count to $n^3/3$. A small amount of extra storage space is needed to store the intermediate results. (e) Which of the three formulations is the same as the one suggested by Exercise 1.7.37? □

rcise 1.7.41 Develop algorithms to calculate the $MD^{-1}M^T$ decomposition of a positive definite matrix (Theorem 1.7.33): (a) inner-product formulation, (b) outer-product formulation, (c) bordered formulation. Again the flop count is about $n^3/3$ if the algorithms are written carefully. □

rcise 1.7.42 Prove the uniqueness part of Theorem 1.7.33. □

rcise 1.7.43 Let A be a nonsymmetric matrix. (a) Prove that if the leading principal submatrices of A are all nonsingular, then there exist unique matrices M, D, and U, such that M is lower triangular, D is diagonal, U is upper triangular, M, D, and U all have the same entries on the main diagonal, and $A = MD^{-1}U$. (b) Derive an algorithm that computes M, D, and U and stores them over A. (Notice that the array that holds A *does* have enough room to store M, D, and U.) □

rcise 1.7.44 Let L be a nonsingular, lower-triangular matrix.

(a) Prove that L^{-1} is lower triangular. (*Outline:* Use induction on n, the dimension of L. Let $M = L^{-1}$. Partition L and M in the same way:

$$L = \begin{bmatrix} L_{11} & 0 \\ L_{21} & L_{22} \end{bmatrix}, \qquad M = \begin{bmatrix} M_{11} & M_{12} \\ M_{21} & M_{22} \end{bmatrix},$$

where L_{11} and M_{11} are square, and L_{11} and L_{22} are lower triangular and nonsingular. Since $M = L^{-1}$, we have $LM = I$. Partition this equation, and use it to deduce that $M_{12} = 0$, $M_{11} = L_{11}^{-1}$, and $M_{22} = L_{22}^{-1}$. Use the induction hypothesis to conclude that M_{11} and M_{22} are lower triangular. This is just one of numerous ways to organize this proof.)

(b) Prove that the entries on the main diagonal of L^{-1} are $l_{11}^{-1}, l_{22}^{-1}, \ldots, l_{nn}^{-1}$. Thus L^{-1} is unit lower triangular if L is.

□

rcise 1.7.45 Let L and M be lower-triangular matrices.

(a) Prove that LM is lower triangular.

(b) Prove that the entries on the main diagonal of LM are $l_{11}m_{11}, \ldots, l_{nn}m_{nn}$. Thus the product of two unit lower-triangular matrices is unit lower triangular.

□

rcise 1.7.46 Prove the upper-triangular analogues of the results in Exercises 1.7.44 and 1.7.45. (The easy way to do this is to take transposes and invoke the lower-triangular results.) □

Exercise 1.7.47 In this exercise you will extend Exercise 1.7.34, expressing the entire reduction to triangular form by type 1 operations in the language of matrices. Let A be a (nonsingular) matrix whose leading principal submatrices are all nonsingular.

(a) Define a matrix M_1 by

$$M_1 = \begin{bmatrix} 1 & 0 \\ m_1 & I_{n-1} \end{bmatrix},$$

where I_{n-1} denotes the $(n-1) \times (n-1)$ identity matrix, and $m_1 = [m_{21}, \ldots, m_{n1}]^T$. Show that

$$M_1^{-1} = \begin{bmatrix} 1 & 0 \\ -m_1 & I_{n-1} \end{bmatrix}.$$

(b) The first step of Gaussian elimination without row interchanges transforms A to

$$A^{(1)} = \begin{bmatrix} a_{11} & a_{12} & \cdots & a_{1n} \\ \hline 0 & a_{22}^{(1)} & \cdots & a_{2n}^{(1)} \\ \vdots & \vdots & & \vdots \\ 0 & a_{n2}^{(1)} & \cdots & a_{nn}^{(1)} \end{bmatrix}.$$

Prove that $A^{(1)} = M_1^{-1} A$, where M_1 is as in part (a), and $m_{21}, \ldots, m_{n1}$ are the multipliers for step 1, i.e. $m_{i1} = a_{i1}/a_{11}$.

(c) Let $A^{(k)}$ denote the matrix to which A has been transformed after k steps of Gaussian elimination. Show that $A^{(k)} = M_k^{-1} A^{(k-1)}$, where

$$M_k = \begin{bmatrix} I_{k-1} & & \\ & 1 & 0 \\ & m_k & I_{n-k} \end{bmatrix}.$$

The blank areas in the array are zeros, and $m_k = [m_{k+1,k}, \ldots, m_{nk}]^T$, where $m_{k+1,k}, \ldots, m_{nk}$ are the multipliers for the kth step.

Matrices of the form of M_k and M_k^{-1} are sometimes called *Gauss transforms*, because they effect the transformations of Gaussian elimination.

(d) After $n-1$ steps of Gaussian elimination, A has been transformed to the upper-triangular matrix $A^{(n-1)} = U$. Prove that $A = LU$, where $L = M_1 M_2 \cdots M_{n-1}$. By Exercise 1.7.45 L is unit lower triangular.

(e) Prove that the matrix $L = M_1 \cdots M_{n-1}$ from part (d) has the form

$$L = \begin{bmatrix} 1 & 0 & 0 & \cdots & 0 \\ m_{21} & 1 & 0 & \cdots & 0 \\ m_{31} & m_{32} & 1 & & 0 \\ \vdots & \vdots & & \ddots & \vdots \\ m_{n1} & m_{n2} & m_{n3} & \cdots & 1 \end{bmatrix}.$$

For example, verify that

$$\begin{bmatrix} 1 & 0 \\ m & I_j \end{bmatrix} \begin{bmatrix} 1 & 0 \\ 0 & \hat{M} \end{bmatrix} = \begin{bmatrix} 1 & 0 \\ m & \hat{M} \end{bmatrix},$$

then use this as the basis of an induction proof.

$\square$

rcise 1.7.48 Write down a careful proof of the existence of the LU decomposition, based on parts (a) through (d) of Exercise 1.7.47. $\square$

rcise 1.7.49 Consider the Gauss transforms M_i introduced in Exercise 1.7.47. Show that each such matrix has the form $M = I - vw^T$, where v and w are appropriately chosen vectors. Since vw^T is a matrix of rank 1, matrices of the form $I - vw^T$ are called *rank-one updates of the identity matrix*. $\square$

rcise 1.7.50 This exercise introduces the idea of *block* Gaussian elimination. Let A be a (nonsingular) matrix whose leading principal submatrices are all nonsingular. Partition A as

$$A = \begin{bmatrix} A_{11} & A_{12} \\ A_{21} & A_{22} \end{bmatrix},$$

where A_{11} is, say, $k \times k$. Since A_{11} is a leading principal submatrix, it is nonsingular.

(a) Show that there is exactly one matrix M such that

$$\begin{bmatrix} I_k & 0 \\ -M & I_{n-k} \end{bmatrix} \begin{bmatrix} A_{11} & A_{12} \\ A_{21} & A_{22} \end{bmatrix} = \begin{bmatrix} A_{11} & A_{12} \\ 0 & \tilde{A}_{22} \end{bmatrix}.$$

In this equation we place no restriction on the form of $\tilde{A}_{22}$. The point is that we seek a transformation that makes the $(2,1)$ block zero. This is a block Gaussian elimination operation; M is a block multiplier.

Show that the unique M that works is given by $M = A_{21} A_{11}^{-1}$, and this implies that

$$\tilde{A}_{22} = A_{22} - A_{21} A_{11}^{-1} A_{12}. \tag{1.7.51}$$

The matrix $\tilde{A}_{22}$ is called the *Schur complement* of A_{11} in A.

(b) Show that

$$\begin{bmatrix} A_{11} & A_{12} \\ A_{21} & A_{22} \end{bmatrix} = \begin{bmatrix} I & 0 \\ M & I \end{bmatrix} \begin{bmatrix} A_{11} & A_{12} \\ 0 & \tilde{A}_{22} \end{bmatrix}. \tag{1.7.52}$$

This is a *block LU decomposition*.

(c) The leading principal submatrices of A_{11} are, of course, all nonsingular. Prove that $\tilde{A}_{22}$ is nonsingular. More generally, prove that all of the leading principal submatrices of $\tilde{A}_{22}$ are nonsingular.

(d) By part (c), both A_{11} and $\tilde{A}_{22}$ have LU decompositions, say $A_{11} = L_1 U_1$ and $\tilde{A}_{22} = L_2 U_2$. Show that

$$A = \begin{bmatrix} L_1 & 0 \\ ML_1 & L_2 \end{bmatrix} \begin{bmatrix} U_1 & L_1^{-1} A_{12} \\ 0 & U_2 \end{bmatrix}.$$

This is the LU decomposition of A.

(e) After k steps of Gaussian elimination without row interchanges, as discussed in this section and in Exercise 1.7.47, A has been transformed to

$$A^{(k)} = \begin{bmatrix} A_{11}^{(k)} & A_{12}^{(k)} \\ 0 & A_{22}^{(k)} \end{bmatrix},$$

where $A_{11}^{(k)}$ is upper triangular. The subsequent Gaussian elimination operations act on $A_{22}^{(k)}$ only, ignoring the rest of the matrix. Prove that

   (i) $A_{11}^{(k)} = U_1$, the upper-triangular factor of A_{11},

   (ii) $A_{22}^{(k)} = \tilde{A}_{22}$, the Schur complement of A_{11} in A.

$\square$

Exercise 1.7.53 Write a careful induction proof of existence and uniqueness of the LU decomposition, based on the developments in parts (a) through (d) of the previous exercise. $\square$

Exercise 1.7.54 (a) Prove that if H is nonsingular and symmetric, then H^{-1} is symmetric. (b) Prove that the Schur complement $\hat{A}_{22}$ (1.7.51) is symmetric if A is. This result generalizes Exercise 1.7.37. $\square$

Exercise 1.7.55 Dongarra, Gustavson, and Karp [22] introduced the interesting concept of a *generic* Gaussian elimination algorithm. In this exercise we will develop the idea in the context of positive definite matrices. Let A be positive definite. Then by Theorem 1.7.33, A can be written as $A = MD^{-1}M^T$, where M is lower triangular, D is diagonal and positive definite, and M and D have the same main-diagonal entries.

(a) Rewrite the equation $A = MD^{-1}M^T$ in block form, where

$$A = \begin{bmatrix} a_{11} & b^T \\ b & \hat{A} \end{bmatrix}, \quad M = \begin{bmatrix} m_{11} & 0 \\ p & \tilde{M} \end{bmatrix}, \quad \text{and} \quad D = \begin{bmatrix} m_{11} & 0 \\ 0 & \tilde{D} \end{bmatrix}.$$

Show that $m_{11} = a_{11}$, $p = b$, and $\tilde{A} = \tilde{M}\tilde{D}^{-1}\tilde{M}^T$, if we define $\tilde{A} = \hat{A} - b a_{11}^{-1} b^T$. ($\tilde{A}$ is the Schur complement of a_{11} in A.)

(b) Use the result of part (a) to derive an outer-product formulation of the $MD^{-1}M^T$ decomposition. Write a pseudocode algorithm that accesses only the *lower* half of A and stores M over A.

Your solution should look something like this:

$$
\begin{array}{l}
\text{for } k = 1, \dots, n \\
\quad \left[\begin{array}{l}
\text{for } j = k+1, \dots, n \\
\quad \left[\begin{array}{l}
\text{for } i = j, \dots, n \\
\quad \left[\begin{array}{l}
a_{ij} \leftarrow a_{ij} - \dfrac{a_{ik} a_{jk}}{a_{kk}}
\end{array} \right.
\end{array} \right.
\end{array} \right.
\end{array}
\tag{1.7.56}
$$

In practice there should be a check for positivity of a_{kk} at an appropriate point, but we have left it out to avoid clutter. The version shown here is the one obtained if the Schur complement calculation $\tilde{A} = \hat{A} - b a_{11}^{-1} b^T$ is done by columns. If it is done by rows, the i and j loops should be interchanged:

$$
\begin{array}{l}
\left[\begin{array}{l}
\text{for } i = k+1, \dots, n \\
\quad \left[\begin{array}{l}
\text{for } j = k+1, \dots, i \\
\quad \left[\begin{array}{l}
a_{ij} \leftarrow a_{ij} - \dfrac{a_{ik} a_{jk}}{a_{kk}}
\end{array} \right.
\end{array} \right.
\end{array} \right.
\end{array}
\tag{1.7.57}
$$

(c) Count the flops in (1.7.56), and note that there are $n^3/6$ divisions. Show how to cut the number of divisions down to about $n^2/2$ by introducing one extra temporary storage array of length n and moving the divisions out of the innermost loop. Once this has been done, the total flop count is about $n^3/3$, as one would expect.

(d) Develop an inner-product formulation of the $MD^{-1}M^T$ algorithm.

Your solution should look something like this:

$$
\begin{array}{l}
\text{for } j = 1, \dots, n \\
\quad \left[\begin{array}{l}
\text{for } i = j, \dots, n \\
\quad \left[\begin{array}{l}
\text{for } k = 1, \dots, j-1 \\
\quad \left[\begin{array}{l}
a_{ij} \leftarrow a_{ij} - \dfrac{a_{ik} a_{jk}}{a_{kk}}
\end{array} \right.
\end{array} \right.
\end{array} \right.
\end{array}
\tag{1.7.58}
$$

Again we have left out the positivity check for simplicity. Again one could move the divisions out of the inner loop. We have left them in the inner loop to keep the algorithm as uncluttered as possible. Notice that (1.7.56) and (1.7.58) are identical, except that the loops are nested differently. It is natural to ask whether one can obtain a valid $MD^{-1}M^T$ decomposition algorithm with the loop indices in any desired order. The answer turns out to be yes; one need only choose the ranges of the loop indices carefully. Thus we can write down a generic $MD^{-1}M^T$ algorithm

$$
\begin{array}{l}
\text{for } \underline{\qquad} \\
\quad \left[\begin{array}{l}
\text{for } \underline{\qquad} \\
\quad \left[\begin{array}{l}
\text{for } \underline{\qquad} \\
\quad \left[\begin{array}{l}
a_{ij} \leftarrow a_{ij} - \dfrac{a_{ik} a_{jk}}{a_{kk}}
\end{array} \right.
\end{array} \right.
\end{array} \right.
\end{array}
\tag{1.7.59}
$$

The blanks can be filled with i, j, and k in any order. We can refer to the outer-product formulation (1.7.56) as the kji-algorithm, because kji is the order of the loops. If the j and i loops of (1.7.56) are replaced by (1.7.57), we get the kij-algorithm. The inner-product formulation (1.7.58) is the jik-algorithm. Since i, j, and k can be placed in any order, there are six basic algorithms. This is true not only for the $MD^{-1}M^T$ decomposition algorithms, but also for the LDL^T and Cholesky decomposition algorithms, since these can be obtained from the $MD^{-1}M^T$ algorithms by a few simple modifications.

(e) Develop the bordered form of the $MD^{-1}M^T$ decomposition. Show that this is either the ijk-algorithm or the ikj-algorithm, depending whether the forward substitutions are done in a row-oriented or column-oriented manner.

(f) We may fill in the loop indices i, j, and k in (1.7.59) in any order, provided that certain relationships between the variables i, j, k, and n are maintained. Determine these relationships, assuming (as above) that only the lower half of the array is used.

Solution: $1 \leq k < j \leq i \leq n$.

(g) Now we have derived all of the variants except jki. Write down the jki-algorithm, starting from (1.7.59), being careful to get the ranges of the loop indices right.

(h) Modify your algorithm from part (g) so that it computes the Cholesky decomposition instead of the $MD^{-1}M^T$ decomposition. Instead of computing M, compute G such that $A = GG^T$ (i.e. $G = R^T$). (*Outline:* You will need to take the square root of a_{jj} at the appropriate point to get g_{jj}. [Put in a positivity check before the square-root operation.] Then you will also need to put in a loop that divides a column by g_{jj}. Notice that this eliminates the need for the divisions in the inner-most loop.)

□

1.8 GAUSSIAN ELIMINATION WITH PIVOTING

We now begin the second phase of our study of the solution of $Ax = b$ by Gaussian elimination, in which we drop the assumption that the leading principal submatrices are nonsingular. We will develop an algorithm that uses elementary row operations of types 1 and 2 either to solve the system $Ax = b$ or (in theory, at least) to determine that it is singular. The algorithm is identical to the algorithm that we developed in Section 1.7, except that at each step a row interchange can be made. Thus we are now considering *Gaussian elimination with row interchanges*, which is also known as *Gaussian elimination with pivoting*. Let us consider the kth step of the algorithm. After $k - 1$ steps, the array that originally contained A has been transformed to the form

$$
\begin{bmatrix}
u_{11} & u_{12} & \cdots & u_{1,k-1} & u_{1k} & & \cdots & u_{1n} \\
m_{21} & u_{22} & \cdots & u_{2,k-1} & u_{2k} & & \cdots & u_{2n} \\
\vdots & & \ddots & \ddots & \vdots & & & \vdots \\
m_{k-1,1} & & & u_{k-1,k-1} & u_{k-1,k} & & \cdots & u_{k-1,n} \\
m_{k1} & \cdots & & m_{k,k-1} & a_{k,k}^{(k-1)} & a_{k,k+1}^{(k-1)} & \cdots & a_{k,n}^{(k-1)} \\
& & & & a_{k+1,k}^{(k-1)} & a_{k+1,k+1}^{(k-1)} & \cdots & a_{k+1,n}^{(k-1)} \\
\vdots & & & \vdots & \vdots & \vdots & & \vdots \\
m_{n1} & \cdots & & m_{n,k-1} & a_{n,k}^{(k-1)} & a_{n,k+1}^{(k-1)} & \cdots & a_{n,n}^{(k-1)}
\end{bmatrix} .
$$

The u_{ij} are entries that will not undergo any further changes (eventual entries of the matrix U). The m_{ij} are stored multipliers (eventual entries of the matrix L.) The $a_{ij}^{(k-1)}$ are the entries that are still active.

To calculate the multipliers for the kth step, we should divide by $a_{kk}^{(k-1)}$. If $a_{kk}^{(k-1)} = 0$, we will have to use a type 2 row operation (row interchange) to get a nonzero entry into the (k, k) position. In fact, even if $a_{kk}^{(k-1)} \neq 0$, we may still choose to do a row interchange.

Consider the following possibility. If $|a_{kk}^{(k-1)}|$ is very small, it may be that $a_{kk}^{(k-1)}$ should be exactly zero and is nonzero only because of roundoff errors made on previous steps. If we now calculate multipliers by dividing by this number, we will surely get erroneous results. For this and other reasons, we will always carry out row interchanges in such a way as to avoid having a small entry in the (k, k) position. The effects of roundoff errors will be studied in detail in Chapter 2.

Returning to the description of our algorithm, we examine the entries $a_{kk}^{(k-1)}$, $a_{k+1,k}^{(k-1)}, \ldots, a_{nk}^{(k-1)}$. If all are zero, then A is singular (Exercise 1.8.1). Set a flag to warn that this is the case. At this point we can either stop or go on to step $k + 1$. If not all of $a_{kk}^{(k-1)}, \ldots, a_{nk}^{(k-1)}$ are zero, let $a_{mk}^{(k-1)}$ be the one whose absolute value is greatest. Interchange rows m and k, including the stored multipliers. Keep a record of the row interchange. This is easily done in an integer array of length n. Store the number m in position k of the array to indicate that at step k, rows k and m were interchanged. Now subtract the appropriate multiples of the new kth row from rows $k + 1, \ldots, n$ to produce zeros in positions $(k + 1, k), \ldots, (n, k)$. Of course, we actually store the multipliers $m_{k+1,k}, \ldots, m_{nk}$ in those positions instead of zeros. This concludes the description of step k.

rcise 1.8.1 After $k - 1$ steps of the Gaussian elimination process, the coefficient matrix has been transformed to the form

$$
B = \begin{bmatrix} B_{11} & B_{12} \\ 0 & B_{22} \end{bmatrix},
$$

where B_{11} is $(k - 1) \times (k - 1)$ and upper triangular. Prove that B is singular if the first column of B_{22} is zero.   □

The eventual (k, k) entry, by which we divide to form the multipliers, is called the *pivot* for step k. The kth row, multiples of which are subtracted from each of the remaining rows at step k, is called the *pivotal row* for step k. Our strategy, which is to make the pivot at step k (i.e., at *each* step) as far from zero as possible in order to protect against disasters due to roundoff errors, is called *partial pivoting*. (Later on we will discuss *complete pivoting*, in which both rows and columns are interchanged.) Notice that the pivots end up on the main diagonal of the matrix U of the LU decomposition. Also, the choice of pivots implies that all of the multipliers will satisfy $|m_{ij}| \leq 1$. Thus in the matrix L of the LU decomposition, all entries will have absolute values less than or equal to 1.

After $n - 1$ steps, the decomposition is complete. One final check must be made: If $a_{nn}^{(n-1)} = 0$, A is singular; set a flag. This is the last pivot. It is not used to create zeros in other rows, but, being the (n, n) entry of U, it is used as a divisor in the back-substitution process.

Each time we did a row interchange, we interchanged the stored multipliers associated with those rows as well. Because of this, the effect of the row interchanges is the same as if the rows of A had been interchanged initially. In other words, suppose we took a copy of the original matrix A and, without doing any elimination operations, performed the same sequence of row interchanges as we did during the Gaussian elimination process. This would yield a matrix $\hat{A}$ that is just A with the rows scrambled. If we now carried out Gaussian elimination without row interchanges on $\hat{A}$, we would get exactly the same result as we got when we made the interchanges during the elimination process. Gaussian elimination without row interchanges yields an LU decomposition. Thus the result of our new algorithm is an LU decomposition of $\hat{A}$, not of A.

Solving the system $Ax = b$ is the same as solving a system $\hat{A}x = \hat{b}$, obtained by interchanging the equations. Since we have saved a record of the row interchanges, it is easy to permute the entries of b to obtain $\hat{b}$. We then solve $Ly = \hat{b}$ by forward substitution and $Ux = y$ by back substitution to get the solution vector x.

Example 1.8.2 We will solve the system

$$\begin{bmatrix} 0 & 4 & 1 \\ 1 & 1 & 3 \\ 2 & -2 & 1 \end{bmatrix} \begin{bmatrix} x_1 \\ x_2 \\ x_3 \end{bmatrix} = \begin{bmatrix} 9 \\ 6 \\ -1 \end{bmatrix}$$

by Gaussian elimination with partial pivoting.

In step 1 the pivotal position is $(1, 1)$. Since there is a zero there, a row interchange is absolutely necessary. Since the largest potential pivot is the 2 in the $(3, 1)$ position, we interchange rows 1 and 3 to get

$$\begin{bmatrix} 2 & -2 & 1 \\ 1 & 1 & 3 \\ 0 & 4 & 1 \end{bmatrix}.$$

The multipliers for the first step are $1/2$ and 0. Subtracting $1/2$ the first row from the second row and storing the multipliers, we have

$$\begin{bmatrix} 2 & -2 & 1 \\ \boxed{1/2} & 2 & 5/2 \\ 0 & 4 & 1 \end{bmatrix} .$$

The pivotal position for the second step is $(2, 2)$. Since the 4 in the $(3, 2)$ position is larger than the 2 in the $(2, 2)$ position, we interchange rows 2 and 3 (including the multipliers) to get the 4 into the pivotal position:

$$\begin{bmatrix} 2 & -2 & 1 \\ \boxed{0} & 4 & 1 \\ \boxed{1/2} & 2 & 5/2 \end{bmatrix} .$$

The multiplier for the third step is $1/2$. Subtracting $1/2$ the second row from the third row and storing the multiplier, we have

$$\begin{bmatrix} 2 & -2 & 1 \\ \boxed{0} & 4 & 1 \\ 1/2 & \boxed{1/2} & 2 \end{bmatrix} . \tag{1.8.3}$$

This completes the Gaussian elimination process. Noting that the pivot in the $(3, 3)$ position is nonzero, we conclude that A is nonsingular, and the system has a unique solution. Encoded in (1.8.3) is the LU decomposition of

$$\hat{A} = \begin{bmatrix} 2 & -2 & 1 \\ 0 & 4 & 1 \\ 1 & 1 & 3 \end{bmatrix} ,$$

which was obtained from A by making appropriate row interchanges. You can check that $\hat{A} = LU$, where

$$L = \begin{bmatrix} 1 & 0 & 0 \\ 0 & 1 & 0 \\ 1/2 & 1/2 & 1 \end{bmatrix} \quad \text{and} \quad U = \begin{bmatrix} 2 & -2 & 1 \\ 0 & 4 & 1 \\ 0 & 0 & 2 \end{bmatrix} .$$

To solve $Ax = b$, we first transform b to $\hat{b}$. Since we interchanged rows 1 and 3 at step 1, we must first interchange components 1 and 3 of b to obtain $[-1, 6, 9]^T$. At step 2 we interchanged rows 2 and 3, so we must interchange components 2 and 3 to get $\hat{b} = [-1, 9, 6]^T$. We now solve

$$\begin{bmatrix} 1 & 0 & 0 \\ 0 & 1 & 0 \\ 1/2 & 1/2 & 1 \end{bmatrix} \begin{bmatrix} y_1 \\ y_2 \\ y_3 \end{bmatrix} = \begin{bmatrix} -1 \\ 9 \\ 6 \end{bmatrix}$$

to get $y = [-1, 9, 2]^T$. Finally we solve

$$\begin{bmatrix} 2 & -2 & 1 \\ 0 & 4 & 1 \\ 0 & 0 & 2 \end{bmatrix} \begin{bmatrix} x_1 \\ x_2 \\ x_3 \end{bmatrix} = \begin{bmatrix} -1 \\ 9 \\ 2 \end{bmatrix}$$

to get $x = [1, 2, 1]^T$. You can easily check that this is correct by substituting it back into the original equation. $\qquad\square$

Exercise 1.8.4 Let

$$A = \begin{bmatrix} 2 & 2 & -4 \\ 1 & 1 & 5 \\ 1 & 3 & 6 \end{bmatrix} \quad \text{and} \quad b = \begin{bmatrix} 10 \\ -2 \\ -5 \end{bmatrix}.$$

Use Gaussian elimination with partial pivoting (by hand) to find matrices L and U such that U is upper triangular, L is unit lower triangular with $|l_{ij}| \leq 1$ for all $i > j$, and $LU = \hat{A}$, where $\hat{A}$ can be obtained from A by making row interchanges. Use your LU decomposition to solve the system $Ax = b$. $\qquad\square$

Gaussian elimination with partial pivoting works very well in practice, but it is important to realize that an accurate answer is not absolutely guaranteed. There exist ill-conditioned systems (Chapter 2) that simply cannot be solved accurately in the presence of roundoff errors. The most extreme case of an ill-conditioned system is one whose coefficient matrix is singular; there is no unique solution to such a system. Our algorithm is supposed to detect this case, but unfortunately it usually will not. If the coefficient matrix is singular, there will be a step at which all potential pivots $a_{kk}^{(k-1)}, \ldots, a_{nk}^{(k-1)}$ are zero. However, due to roundoff errors on previous steps, the actual computed quantities will not be exact zeros. Consequently the singularity of the matrix will not be detected, and the algorithm will march ahead and produce a nonsensical "solution." An obvious precautionary measure would be to issue a warning whenever the algorithm is forced to use a very small pivot. Better ways to detect inaccuracy will be discussed in Chapter 2.

The additional costs associated with row interchanges in the partial-pivoting strategy are not great. There are two costs to consider, that of finding the largest pivot at each step, and that of physically interchanging the rows. At the kth step, $n - k + 1$ numbers must be compared to find the one that is largest in magnitude. This does not involve any arithmetic, but it requires the comparison of $n - k$ pairs of numbers. The total number of comparisons made in the $n - 1$ steps is therefore

$$\sum_{k=1}^{n-1} (n - k) = \frac{n(n-1)}{2} \approx \frac{n^2}{2}.$$

Regardless of the exact cost of making one comparison, the cost of making $n^2/2$ comparisons is small compared with the $O(n^3)$ arithmetic cost of the algorithm. The interchange of rows does not require any arithmetic either, but it takes time to fetch and store numbers. At each step, at most one row interchange is carried out. This involves fetching and storing $2n$ numbers. Since there are $n - 1$ steps, the total number of fetches and stores is not more than about $2n^2$. Thus the cost of interchanging rows is also negligible compared with the cost of the arithmetic of the algorithm. It is fair to say, as a rough approximation, that the cost of Gaussian elimination with partial pivoting is $\frac{2}{3}n^3$ flops.

Permuting the rows of a matrix is equivalent to multiplying the matrix by a permutation matrix.

Example 1.8.5 Let A and $\hat{A}$ be as in Example 1.8.2. Then one easily checks that $\hat{A} = PA$, where

$$P = \begin{bmatrix} 0 & 0 & 1 \\ 1 & 0 & 0 \\ 0 & 1 & 0 \end{bmatrix}.$$

$\square$

A *permutation matrix* is a matrix that has exactly one 1 in each row and in each column, all other entries being zero. For example, the matrix P in Example 1.8.5 is a permutation matrix.

rcise 1.8.6 Let A be an $n \times m$ matrix, and let $\hat{A}$ be a matrix obtained from A by scrambling the rows. Show that there is a unique $n \times n$ permutation matrix P such that $\hat{A} = PA$.

$\square$

rcise 1.8.7 Show that if P is a permutation matrix, then $P^T P = PP^T = I$. Thus P is nonsingular, and $P^{-1} = P^T$.

$\square$

We now easily obtain the following result.

Theorem 1.8.8 *Gaussian elimination with partial pivoting, applied to an $n \times n$ matrix A produces a unit lower-triangular matrix L such that $|l_{ij}| \leq 1$, an upper triangular matrix U, and a permutation matrix P such that*

$$A = P^T LU.$$

Proof. We have seen that Gaussian elimination with partial pivoting produces L and U of the required form such that $\hat{A} = LU$, where $\hat{A}$ was obtained from A by permuting the rows. By Exercise 1.8.6, there is a permutation matrix P such that $\hat{A} = PA$. Thus $PA = LU$ or $A = P^{-1}LU = P^T LU$.

$\square$

rcise 1.8.9 Let A be the matrix in Exercise 1.8.4. Determine matrices P, L, and U with the properties stated in Theorem 1.8.8, such that $A = P^T LU$.

$\square$

Given a decomposition $A = P^T LU$, we can use it to solve a system $Ax = b$, by writing $P^T LUx = b$, then solving, successively, $P^T \hat{b} = b$, $Ly = \hat{b}$, and $Ux = y$. The step $P^T \hat{b} = b$, or $\hat{b} = Pb$, just amounts to rearranging the entries of b in accordance with the row interchanges that were carried out during the elimination process. Thus this process is really no different from the process illustrated at the end of Example 1.8.2. In practice there is never really any need to build the permutation matrix P. Keeping a record of the row interchanges is just as good and much more compact. However, MATLAB will display P for you if you ask it to.

Earlier in the chapter there were quite a few exercises that used MATLAB to solve systems of linear equations. Whenever the command x=A\b is invoked to solve the system $Ax = b$ (where A is $n \times n$ and nonsingular), MATLAB uses Gaussian elimination with partial pivoting to solve the system. If you want to see the LU decomposition, use the MATLAB command `lu`.

Exercise 1.8.10 Use MATLAB to check the LU decomposition obtained in Example 1.8.2. Enter the matrix A, then type `[L,U,P] = lu(A)`. Once you have L, U, and P, you can put them back together by typing `P'*L*U`. (In MATLAB the prime symbol, applied to a real matrix, means "transpose.") You can also try `[K,V] = lu(A)`, and see what that gives you. Give a matrix equation that relates K to L. For a description of the `lu` command, type `help lu` or look in MATLAB's help browser.   □

Exercise 1.8.11 Repeat Exercise 1.8.10 using the matrix from Exercises 1.8.4 and 1.8.9.   □

Programming Gaussian Elimination

Exercise 1.8.12 Write an algorithm that implements Gaussian elimination with partial pivoting. Store L and U over A, and save a record of the row interchanges.   □

Your solution to Exercise 1.8.12 should look something like this:

Gauss $(A, n, intch, flag)$
clear $flag$
for $k = 1, 2, \ldots, n - 1$
$\quad \Big[\quad amax \leftarrow \max\{|a_{kk}|, |a_{k+1,k}|, \ldots, |a_{nk}|\}$
$\qquad$ if $(amax = 0)$ then
$\qquad \quad \Big[\begin{array}{l} \text{set } flag \\ intch(k) \leftarrow 0 \end{array} \Big\} \quad (A \text{ is singular})$
$\qquad$ else
$\qquad \quad \Big[\quad m \leftarrow \text{smallest integer} \geq k \text{ for which } |a_{mk}| = amax$
$\qquad \qquad intch(k) \leftarrow m \qquad \text{(record row interchange)}$
$\qquad \qquad$ if $(m \neq k)$ then
$\qquad \qquad \quad [\ \text{interchange rows } k \text{ and } m, \text{ including multipliers}$
$\qquad \qquad$ for $i = k + 1, \ldots, n$
$\qquad \qquad \quad [\ a_{ik} \leftarrow a_{ik}/a_{kk} \qquad \text{calculate multipliers}$
$\qquad \qquad$ for $i = k + 1, \ldots, n$
$\qquad \qquad \quad \Big[\begin{array}{l} \text{subtract } a_{ik} \text{ times } [a_{k,k+1}, \ldots, a_{k,n}] \\ \qquad \text{from } [a_{i,k+1}, \ldots, a_{i,n}] \end{array} \Big\} \quad \text{(row operations)}$
if $(a_{nn} = 0)$ then
$\quad \Big[\begin{array}{l} \text{set } flag \\ intch(n) \leftarrow 0 \end{array} \Big\} \quad (A \text{ is singular})$
else
$\quad [\ intch(n) \leftarrow n \ ]$

If A is found to be singular, the algorithm sets a flag but does not stop. It finishes the LU decomposition, but U is singular. The array A now contains the LU decomposition of $\hat{A} = PA$, and the array $intch$ contains a record of the row interchanges (i.e. an encoding of the permutation matrix P).

The bulk of the work is in the row operations. These can be organized several ways. If each row operation is executed as a loop, we have

$$\begin{array}{l} \text{for } i = k+1,\ldots,n \\ \left[\begin{array}{l} \text{for } j = k+1,\ldots,n \\ \left[\begin{array}{l} a_{ij} \leftarrow a_{ij} - a_{ik}a_{kj} \end{array} \right. \end{array} \right. \end{array} \qquad (1.8.13)$$

Each time through the outer loop, a complete row operation is performed. However there is no logical reason why the j loop should be on the inside. Clearly the code segment

$$\begin{array}{l} \text{for } j = k+1,\ldots,n \\ \left[\begin{array}{l} \text{for } i = k+1,\ldots,n \\ \left[\begin{array}{l} a_{ij} \leftarrow a_{ij} - a_{ik}a_{kj} \end{array} \right. \end{array} \right. \end{array} \qquad (1.8.14)$$

performs exactly the same operations, and there is no *a priori* reason to prefer (1.8.13) over (1.8.14). If (1.8.13) is used, the algorithm is said to be *row oriented*, because in the inner loop the row index i is held fixed. If (1.8.14) is used, the algorithm is said to be *column oriented*. In this case the elements a_{ij} and a_{ik} traverse the jth and kth columns, respectively, as the row index i is stepped forward.

Which orientation is best depends on the language in which the algorithm is implemented. If Fortran is used, for example, the column-oriented code is generally faster, because Fortran stores arrays by columns. As each column is traversed, elements are taken from consecutive locations in the computer memory.

In order to solve systems of linear equations, we need not only an LU decomposition routine, but also a routine to permute the b vector and perform forward and back substitution. Such a routine would look something like this:

Solve $(A, b, n, intch, flag)$
for $k = 1,\ldots,n-1$
$\left[\begin{array}{l} m \leftarrow intch(k) \\ \text{interchange } b_k \text{ and } b_m \end{array} \right.$
for $j = 1,\ldots,n-1$
$\left[\begin{array}{l} \text{for } i = j+1,\ldots,n \\ \left[\begin{array}{l} b_i \leftarrow b_i - a_{ij}b_j \end{array} \right. \end{array} \right\}$ (column-oriented forward substitution)
clear $flag$
for $j = n, n-1,\ldots,1$
$\left[\begin{array}{l} \text{if } (a_{jj} = 0) \text{ then} \\ \left[\begin{array}{l} \text{set } flag \\ exit \end{array} \right\} (U \text{ is singular}) \\ b_j \leftarrow b_j/a_{jj} \\ \text{for } i = 1,\ldots,j-1 \\ \left[\begin{array}{l} b_i \leftarrow b_i - a_{ij}b_j \end{array} \right. \end{array} \right\}$ (column-oriented back substitution)

This algorithm takes as inputs the outputs A and $intch$ from the algorithm Gauss. The array A contains an LU decomposition, and $intch$ contains a record of the row interchanges. The array b contains the vector b initially and the solution vector x afterward.

A main program to drive the subroutines Gauss and Solve might be organized something like this:

> read n and A
> call Gauss $(A, n, intch, flag)$
> if (flag is set) then
> $\quad\lceil$ print 'coefficient matrix is singular'
> $\quad\lfloor$ *stop*
> read num (number of right-hand sides)
> for $m = 1, \ldots, num$
> $\quad\lceil$ read b
> $\quad\mid$ call Solve $(A, b, n, intch, flag)$
> $\quad\lfloor$ print 'solution number' m 'is' b (prints solution)

This program allows for the possibility that the user would like to solve several problems $Ax = b^{(1)}$, $Ax = b^{(2)}$, $\ldots$, $Ax = b^{(k)}$ with the same coefficient matrix. The LU decomposition is only done once.

Exercise 1.8.15 Write a Fortran program and subroutines to solve linear systems $Ax = b$ by Gaussian elimination with partial pivoting. Your subroutines should be column oriented. Try out your program on the test problems $Ax = b$ and $Ax = c$, where

$$
A = \begin{bmatrix} 2 & 10 & 8 & 8 & 6 \\ 1 & 4 & -2 & 4 & -1 \\ 0 & 2 & 3 & 2 & 1 \\ 3 & 8 & 3 & 10 & 9 \\ 1 & 4 & 1 & 2 & 1 \end{bmatrix} \quad b = \begin{bmatrix} 52 \\ 14 \\ 12 \\ 51 \\ 15 \end{bmatrix} \quad c = \begin{bmatrix} 50 \\ 4 \\ 12 \\ 48 \\ 12 \end{bmatrix}.
$$

After the decomposition your transformed matrix should be

$$
\begin{bmatrix} 3 & 8 & 3 & 10 & 9 \\ 2/3 & 14/3 & 6 & 4/3 & 0 \\ 1/3 & 2/7 & -33/7 & 2/7 & -4 \\ 1/3 & 2/7 & 4/11 & -20/11 & -6/11 \\ 0 & 3/7 & -1/11 & -4/5 & 1/5 \end{bmatrix}.
$$

The row interchanges are given by $intch = [4, 4, 4, 5, 5]$. The solution vectors are $[1, 2, 1, 2, 1]^T$ and $[2, 1, 2, 1, 2]^T$. Once you have your program working this problem correctly, try it out on some other problems of your own devising.   □

Exercise 1.8.16 Write a program in your favorite computer language to solve systems of linear equations by Gaussian elimination with partial pivoting. Determine whether your language stores arrays by rows or by columns, and write your subroutines accordingly. Test your program on the problem from Exercise 1.8.15 and other problems of your own devising.   □

rcise 1.8.17 Write two versions of your Gaussian elimination program, one that is row oriented and one that is column oriented. Time them on some big problems (using randomly-generated coefficient matrices, for example.) Which is faster? Is this what you expected? □

Some modern languages, for example Fortran-90 and newer Fortran standards, allow the programmer to leave the choice of row-oriented or column-oriented code to the compiler. The row operations (1.8.13) or (1.8.14) amount to an outer-product update (This is the outer-product formulation!)

$$\hat{A} \leftarrow \hat{A} - vw^T, \tag{1.8.18}$$

where

$$\hat{A} = \begin{bmatrix} a_{k+1,k+1} & \cdots & a_{k+1,n} \\ \vdots & & \vdots \\ a_{n,k+1} & \cdots & a_{nn} \end{bmatrix}, \quad v = \begin{bmatrix} a_{k+1,k} \\ \vdots \\ a_{nk} \end{bmatrix}, \quad \text{and} \quad w = \begin{bmatrix} a_{k,k+1} \\ \vdots \\ a_{kn} \end{bmatrix}.$$

Fortran-90 allows matrix operations on arrays or sections of arrays. The submatrix $\hat{A}$ is denoted $a(k + 1 : n, k + 1 : n)$ in Fortran-90 (just as in MATLAB!), and the update (1.8.18) is given by

$$a(k + 1 : n, k + 1 : n) = a(k + 1 : n, k + 1 : n) \tag{1.8.19}$$
$$- \text{matmul}(a(k + 1 : n, k : k), a(k : k, k + 1 : n))$$

The `matmul` command performs matrix multiplication. This single instruction tells the computer to perform $2(n - k)^2$ flops. It is left to the Fortran-90 compiler to decide on the order in which the flops are performed. This is certainly more compact than either (1.8.13) or (1.8.14). It may be faster or slower, depending on how well the `matmul` command has been implemented in your compiler on your machine.

Our introduction to row interchanges has focused on the outer-product formulation of Gaussian elimination. However, row interchanges can easily be incorporated into other formulations of the LU decomposition algorithm (e.g. inner-product formulation, bordered form) as well. Thus there is a great variety of ways to implement Gaussian elimination with partial pivoting.

For larger matrices the fastest codes are those that make effective use of cache. To this end, we should perform Gaussian elimination by blocks. The advantages of blocking were discussed in the context of matrix multiplication in Section 1.1 and again in the context of the Cholesky decomposition in Section 1.4. A block Gaussian elimination algorithm can be built using the ideas discussed in Exercise 1.7.50 and in the discussion of block variants of Cholesky's method in Section 1.4. Every version of Gaussian elimination has a block variant. The need to perform row interchanges complicates the procedures but does not change the basic ideas.

Reliable Fortran routines that perform Gaussian elimination with partial pivoting can be obtained from the public-domain software packages LINPACK and LAPACK [1]. The LAPACK codes are organized by blocks. There is also a C wrapper for LAPACK called CLAPACK.

As we have seen, there are many ways to organize Gaussian elimination with partial pivoting. There are many more details than have been mentioned here, as one sees by examining some of the codes from LINPACK or LAPACK.

Exercise 1.8.20 Visit the NETLIB repository at http://www.netlib.org. Browse the repository and find the LINPACK and LAPACK libraries.

(a) Find the LINPACK subroutine dgefa, which does Gaussian elimination with partial pivoting in a column-oriented manner. Make a copy of the subroutine and annotate it, telling what each part does. Notice that most of the work is done by the subroutines idamax, dscal, and (especially) daxpy. These are Basic Linear Algebra Subroutines or BLAS. Find the BLAS library to figure out what these subroutines do.

(b) Find the LAPACK subroutine dgetrf. This does Gaussian elimination using scalar code (dgetf2) or by blocks, depending on the size of the matrix. If the elimination is done by blocks, the work is done by level-3 BLAS, subroutines that perform matrix operations. Which one does the bulk of the work?

$\square$

Calculating A^{-1}

Your program to solve $Ax = b$ can be used to calculate the inverse of a matrix. Letting $X = A^{-1}$, we have $AX = I$. Rewriting this equation in partitioned form as

$$A[x_1 \ x_2 \ \cdots \ x_n] = [e_1 \ e_2 \ \cdots \ e_n],$$

where $x_1, \ldots, x_n$ and $e_1, \ldots, e_n$ are the columns of X and I, respectively, we find that the equation $AX = I$ is equivalent to the n equations

$$Ax_i = e_i, \qquad i = 1, \ldots, n. \tag{1.8.21}$$

Solving these n systems by Gaussian elimination with partial pivoting, we obtain A^{-1}.

Exercise 1.8.22 Use your program to find A^{-1}, where A is the matrix in Exercise 1.8.15. $\square$

How much does it cost to calculate A^{-1}? The LU decomposition has to be done once, at a cost of $\frac{2}{3}n^3$ flops. Each of the n systems in (1.8.21) has to be solved by forward and back substitution at a cost of $2n^2$ flops. Thus the total flop count is $\frac{8}{3}n^3$. A more careful count shows that the job can be done for a bit less.

Exercise 1.8.23 The forward-substitution phase requires the solution of $Ly_i = e_i$, $i = 1, \ldots, n$. Some operations can be saved by exploiting the leading zeros in e_i (See Section 1.3). Do a flop count that takes these savings into account, and conclude that A^{-1} can be computed in about $2n^3$ flops. $\square$

rcise 1.8.24 Modify your Gaussian elimination program so that the forward substitution segment exploits leading zeros in the right-hand side. You now have a program that can calculate an inverse in $2n^3$ flops. □

You may be wondering how this method compares with other methods of calculating A^{-1} that you have seen. One method that is sometimes presented in elementary linear algebra classes starts with an augmented matrix $[A \mid I]$ and performs row operations that transform the augmented matrix to the form $[I \mid X]$. That algorithm also takes $2n^3$ flops, *if* it is organized efficiently. In fact it is basically the same algorithm as has been proposed here. Another method you may have seen is the cofactor method, which is summarized by the equation

$$A^{-1} = \frac{1}{\det(A)} \operatorname{adj}(A).$$

That method requires the computation of many determinants. If the determinants are calculated in the classical way (row or column expansion), the cost is more than $n!$ flops. Since $n!$ grows much more rapidly than n^3, the cofactor method is not competitive unless n is very small.

By the way, there are ways to compute the determinants in fewer than $n!$ flops. One very good way is to calculate an LU decomposition. Then $\det(A) = \pm \det(U) = \pm u_{11} u_{22} \cdots u_{nn}$, where the sign is minus (plus) if and only if an odd (even) number of row interchanges was made. This method obviously costs $\frac{2}{3} n^3$ flops.

rcise 1.8.25 Verify that this procedure does indeed yield the determinant of A. □

MATLAB has commands `inv` and `det` that will compute the inverse and determinant for you.

Complete Pivoting

A more cautious pivoting strategy known as *complete pivoting* deserves at least brief mention. Both row and column interchanges are allowed. At the first step the entire matrix is searched. The element of largest magnitude is found and moved to the $(1,1)$ position by a row interchange and a column interchange. (Note that the effect of an interchange of columns i and j is to interchange the unknowns x_i and x_j.) The maximal element is then used as a pivot to create zeros below it. The second step is the same is the first, but it operates on the $(n-1) \times (n-1)$ submatrix obtained by ignoring the first row and column, and so on. The complete pivoting strategy gives extra protection against the bad effects of roundoff errors and is quite satisfactory both in theory and in practice. Its disadvantage is that it is somewhat more expensive than partial pivoting. In the first step about n^2 pairs of numbers have to be compared in order to find the largest entry. In the second step, $(n-1)^2$ comparisons are required, in the third step $(n-2)^2$, and so on. Thus the total number of comparisons made during the pivot searches is approximately $\sum_{k=0}^{n-2}(n-k)^2 \approx \frac{1}{3} n^3$. Since the cost of making a comparison is not insignificant, this means that the cost of the pivot searches is of the same order of magnitude as the cost of the arithmetic. In contrast the total cost of the pivot searches in the partial-pivoting strategy is $O(n^2)$ and is therefore insignificant if n is large. The extra cost of complete pivoting would be worth paying if it gave significantly better results. However, it has been found that partial pivoting works very satisfactorily in practice, so partial pivoting is much more widely used.

Exercise 1.8.26   Let A be an $n \times m$ matrix, and let $\hat{A}$ be a matrix obtained by shuffling the *columns* of A. Express the relationship between A and $\hat{A}$ as a matrix equation involving a permutation matrix.   □

Exercise 1.8.27   (a) Show that Gaussian elimination with complete pivoting, applied to a nonsingular matrix A, produces unit lower-triangular L, upper-triangular U, and permutation matrices P and Q, such that

$$A = P^T L U Q^T.$$

(b) What additional properties do L and U satisfy? (c) In solving a system $Ax = b$ by Gaussian elimination with complete pivoting, an additional final step $x = Q\hat{x}$ must be carried out. How should this be done in practice?   □

On Competing Methods

A variant of Gaussian elimination that is taught in some elementary linear algebra textbooks is *Gauss-Jordan elimination*. In this variant the augmented matrix $[A \mid b]$ is converted to the form $[I \mid x]$ by elementary row operations. At the kth step the pivot is used to create zeros in column k both above and below the main diagonal. The disadvantage of this method is that it is more costly than the variants that we have discussed.

Exercise 1.8.28   (a) Do a flop count for Gauss-Jordan elimination and show that about n^3 flops are required, which is 50% more than ordinary Gaussian elimination needs. (b) Why is Gauss-Jordan elimination more expensive than transformation to the form $[U \mid y]$, followed by back substitution? (c) How could the operations in Gauss-Jordan elimination be reorganized so that only $\frac{2}{3}n^3$ flops are needed to get to the form $[I \mid x]$?   □

Another method for solving $Ax = b$ that was at one time covered in most elementary linear algebra texts is Cramer's rule, which is closely related to the cofactor method of calculating A^{-1}. Cramer's rule states that each entry x_i in the solution of $Ax = b$ is a quotient of two determinants:

$$x_i = \frac{\det(A^{(i)})}{\det(A)} \qquad i = 1, \dots, n,$$

where $A^{(i)}$ denotes the matrix obtained from A by replacing its ith column by b. This truly elegant formula is too expensive to be a practical computational tool, except when $n = 2$ or 3, because it requires the computation of determinants.

At the end of Section 1.1 we mentioned Strassen's method [88] and other methods that multiply two $n \times n$ matrices together in fewer than $O(n^3)$ flops. For example, Strassen's method takes $O(n^s)$ flops, where $s = \log_2 7 \approx 2.81$. These methods can be modified so that they compute A^{-1}. Therefore there exist methods that solve $Ax = b$ in fewer than $O(n^3)$ flops. Some of these methods could become important for large, dense (i.e. not sparse) matrix computations at some future date.

All of the methods that we have discussed so far are *direct methods*; if they were executed in exact arithmetic, they would produce the exact solution to $Ax = b$ after a finite, prespecified sequence of operations. A completely different type of method is the *iterative method*, which produces a sequence of successively better approximations to the solution. For extremely large, sparse matrices, iterative methods are the best choice. We discuss iterative methods in Chapter 8.

The Symmetric Case

Consider the problem of solving $Ax = b$ when A is symmetric. As in the positive-definite case, one would hope to halve the flop count by exploiting symmetry. If A is not positive definite, we can expect to have to do some pivoting, but now pivoting is complicated by the fact that every row interchange should be accompanied by a matching column interchange to preserve symmetry. Because of this constraint, it is not always possible to get a big enough entry into the pivotal position (e.g., consider a symmetric matrix whose main-diagonal entries are all zero). The solution is to perform block Gaussian elimination (cf. Exercise 1.7.50) using a 2×2 block pivot whenever a sufficiently large pivot is not readily available. The most popular algorithm of this type is the Bunch-Kaufman algorithm, which has a flop count of $n^3/3$ and spends only $O(n^2)$ effort on pivoting. It is discussed along with some of its competitors in [43, § 4.4], for example.

1.9 SPARSE GAUSSIAN ELIMINATION

If we wish to solve $Ax = b$, where A is large and sparse, we are well advised to exploit the structure of A. In Section 1.5 we observed that in the symmetric, positive definite case, huge savings can be realized by using a simple envelope scheme. More sophisticated methods can yield even better results, as we saw in Section 1.6. At several places in this chapter we have mentioned that the best alternative for really large systems is to use iterative methods (Chapter 8). Nevertheless, there remain many situations where the best alternative is to perform Gaussian elimination with pivoting, using some sort of sparse data structure. For example, if we want to solve $Ax = b^{(i)}$ accurately for $i = 1, 2, 3, \ldots$, with one coefficient matrix and many right-hand sides, often the best course is to compute a sparse LU decomposition once and use it over and over again. Situations like this arise in the solution of large, sparse eigenvalue problems, for example.

In Section 1.6 we observed that the amount of fill-in incurred during Cholesky's method depends on how the equations and unknowns are ordered. Reordering is a form of pivoting. In the positive definite case, there is no danger of hitting a zero pivot, but we may choose to pivot (i.e. reorder) to keep the Cholesky factor as sparse as possible. However, each row permutation has to be matched by the identical column permutation to preserve symmetry. In the nonsymmetric case, the row and column permutations can be different, because there is no symmetry to preserve. However, we must choose our permutations so that zero pivots and small pivots are avoided. Thus there are now two objectives, which are potentially in conflict with one another: (i) to avoid fill, and (ii) to avoid small pivots. For a discussion of these issues see [14, 26]. We shall content ourselves with some simple illustrations of fill-in using MATLAB.

Example 1.9.1 Let A denote the 81×81 sparse matrix generated by the command $A = \text{condif}(9,[0.2\ 5])$, using the m-file condif.m shown in Exercise 1.9.3. This is the matrix of a discretized convection-diffusion operator, with equations and unknowns numbered in a natural way. Its spy plot, shown in Figure 1.16 shows that it is a banded matrix. Its sparsity pattern is symmetric, but the matrix itself is

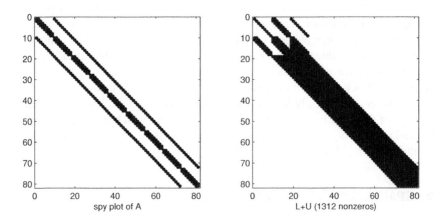

Figure 1.16 Small convection-diffusion matrix with original ordering

nonsymmetric. For example, $a_{10,1} = -6$ and $a_{1,10} = 4$.

MATLAB can calculate sparse LU decompositions. The command $[\text{L},\text{U},\text{P}] = \text{lu}(\text{A})$, applied to a sparse matrix A, produces an LU decomposition with partial pivoting. The output is exactly the same as if A were a full matrix, except that L, U, and P are stored as sparse matrices. Thus L is unit lower triangular with all entries satisfying $|l_{ij}| \leq 1$, U is upper triangular, P is a permutation matrix, and $PA = LU$.[11]

---

[11]MATLAB offers a *threshold pivoting* option that can be used instead of partial pivoting. Use of threshold pivoting sometimes reduces fill. Type `help lu` for details.

Applying this command to our 81×81 matrix, we obtain a sparse LU decomposition. To observe the fill in the L and U factors simultaneously, we type `spy(L+U)`. The result is shown in Figure 1.16. We observe that the band structure is disrupted slightly by pivoting in early steps. Aside from that, the band is almost completely filled in. Typing `nnz(L+U)`, we find that there are 1312 nonzero entries in the sum.

For comparison purposes we now apply a random permutation to the columns of A:

```
q = randperm(81);
Ar = A(:,q);   % Reorder the columns only.
spy(Ar)
```

The result is shown in Figure 1.17. Performing the LU decomposition and typing

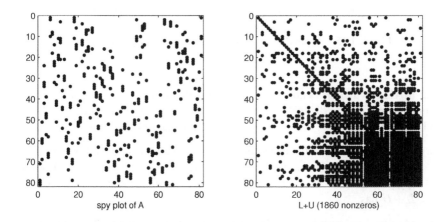

Figure 1.17   Convection-diffusion matrix with random reordering of columns

`spy(L+U)`, we find that this time there is substantially greater fill. Thus the natural ordering is significantly better than a random ordering. If you perform this experiment, you will get different results, because `randperm` will give you a different permutation.

For a second comparison, use the *column approximate minimum-degree ordering* supplied by MATLAB:

```
q = colamd(A);
Am = A(:,q);
spy(Am)
```

The result is shown in Figure 1.18. Performing the LU decomposition and typing `spy(L+U)`, we find that this column reordering results in significantly less fill than the original ordering. On larger versions of this matrix, the column minimum degree ordering yields even better results. See Exercise 1.9.3.

Since the LU decomposition produces L, U, and P for which $PA = LU$, you might find it useful to look at `spy(P*A)`. Compare this with `spy(L+U)` for all three orderings.   □

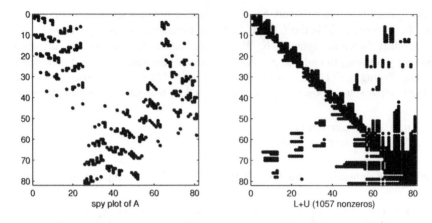

Figure 1.18   Approximate minimum-degree reordering of columns

Exercise 1.9.2 MATLAB includes a nonsymmetric, sparse demonstration matrix `west0479` of dimension 479×479, that can be accessed by typing `load west0479` in MATLAB. We type `A = west0479;` to give it an abbreviated name.

(a) Calculate the sparse LU decomposition, recording the time it takes to perform the operation; for example, `tic, [L,U,P] = lu(A); toc`. Look at spy plots of A, PA, and $L + U$, and make note of how many nonzero entries $L + U$ has.

(b) Modify A by applying a random permutation to the columns. Then repeat part (a).

(c) Modify A by applying the column minimum-degree ordering to the columns. Then repeat part (a).

(d) Compare your results from parts (a)–(c).

(e) If your goal is to solve a linear system $Ax = b$, you can use any of the LU decompositions computed in (a)–(c), together with the appropriate row and column permutations, to do the job. However, MATLAB does not really require that you go to all that trouble. The MATLAB command `x = A\b` works just as well for sparse matrices as for full matrices. This command causes MATLAB to compute a sparse LU decomposition of A and use it to solve the system $Ax = b$.

Generate a vector b at random (`b = randn(479,1);` for example), and have MATLAB solve the system $Ax = b$ by the command `x=A\b`. Time the operation, and compare it with the times for the LU decompositions in parts (a)–(c).

□

rcise 1.9.3 The MATLAB m-file shown below produces matrices of discretized convection-diffusion operators of various sizes.

```
function C = condif(n,c)
  %  C is a square convection-diffusion matrix
  %  of order n^2.  The user must supply n and
  %  the convection coefficients c(1:2).
I = speye(n);
N = sparse(1:n-1,2:n,1,n,n);
F1 = 2*I - (1-c(1))*N - (1+c(1))*N';
F2 = 2*I - (1-c(2))*N - (1+c(2))*N';
C = kron(I,F1) + kron(F2,I);
```

In Example 1.9.1 we used a small matrix of this type. To get some experience with a larger sparse matrix, generate a matrix A = condif(n,c) using, say, $n = 100$ (or larger). This gives a 10000×10000 matrix. You can use c = [0.2 5], as in Example 1.9.1. Repeat Exercise 1.9.2 using this matrix.   □

rcise 1.9.4 Repeat Exercise 1.9.3 with a much larger value of n, omitting the random permutation.   □

CHAPTER 2

SENSITIVITY OF LINEAR SYSTEMS; EFFECTS OF ROUNDOFF ERRORS

When we solve a system of linear equations, we seldom solve exactly the system we intended to solve; rather, we solve one that approximates it. In a system $Ax = b$, the coefficients in A and b will typically be known from some measurements and will therefore be subject to measurement error. For example, in the electrical circuit problems introduced in Section 1.2, the entries of the coefficient matrices depend on the values of the resistances, numbers that are known only approximately in practice. Thus the A and b with which we work are slightly different from the true A and b. Additional approximations must be made when the numbers are entered into a computer; the real or complex entries of A and b must be approximated by numbers from the computer's finite set of floating point numbers. (However, this error is usually much smaller than the measurement error.)

Because errors are inevitable, it is important to ask what effect small perturbations in the coefficients have on the solution of a system. When we change the system slightly, does this cause only a slight change in the solution or an enormous, unacceptable change? As we shall see, not all systems are alike in this regard; some are much more sensitive than others. The most important task of this chapter is to study the *sensitivity* of linear systems.

Fundamentals of Matrix Computations, Third Edition. By David S. Watkins
Copyright © 2010 John Wiley & Sons, Inc.

A second question is one that was already mentioned in Chapter 1. If we solve a system by Gaussian elimination on a computer, the result will be contaminated by the roundoff errors made during the computation. How do these errors affect the accuracy of the computed solution? We shall see that this question is closely related to the sensitivity issue.

The most comprehensive work on this subject is N. J. Higham's book, *Accuracy and Stability of Numerical Algorithms* [52].

2.1 VECTOR AND MATRIX NORMS

In order to study the effects of perturbations in vectors (such as b) and matrices (such as A), we need to be able to measure them. For this purpose we introduce vector and matrix norms.

Vector Norms

The vectors used in this book are generally n-tuples of real (or complex) numbers. Recall that the set of all real n-tuples is denoted $\mathbb{R}^n$. It is useful to visualize the members of $\mathbb{R}^2$ as points in a plane or as geometric vectors (arrows) in a plane with their tails at the origin. Likewise the elements of $\mathbb{R}^3$ can be viewed as points or vectors in space. Any two elements of $\mathbb{R}^n$ can be added in the obvious manner to yield an element of $\mathbb{R}^n$, and any element of $\mathbb{R}^n$ can be multiplied by any real number (scalar) to yield an element of $\mathbb{R}^n$. The vector whose components are all zero will be denoted 0. Thus the symbol 0 can stand for either a number or a vector. The careful reader will not be confused by this.

The set of all n-tuples of complex numbers is denoted $\mathbb{C}^n$. In this chapter, as in Chapter 1, we will restrict our attention to real numbers. However, everything that we do here can be carried over to complex numbers.

A *norm* (or *vector norm*) on $\mathbb{R}^n$ is a function that assigns to each $x \in \mathbb{R}^n$ a non-negative real number $\|x\|$, called the norm of x, such that the following three properties are satisfied for all $x, y \in \mathbb{R}^n$ and all $\alpha \in \mathbb{R}$:

$$\|x\| > 0 \text{ if } x \neq 0, \quad \text{and} \quad \|0\| = 0 \qquad \text{(positive definite property)} \quad (2.1.1)$$

$$\|\alpha x\| = |\alpha| \, \|x\| \qquad \text{(absolute homogeneity)} \qquad (2.1.2)$$

$$\|x + y\| \leq \|x\| + \|y\| \qquad \text{(triangle inequality)} \qquad (2.1.3)$$

Exercise 2.1.4

(a) In the equation $\|0\| = 0$, what it the nature of each zero (number or vector)?

(b) Show that the equation $\|0\| = 0$ actually follows from (2.1.2). Thus it need not have been stated explicitly in (2.1.1).

$\square$

Any norm can be used to measure the lengths or magnitudes (in a generalized sense) of vectors in $\mathbb{R}^n$. In other words, we think of $\| x \|$ as the (generalized) *length* of x. The (generalized) *distance* between two vectors x and y is $\| x - y \|$.

Example 2.1.5 The *Euclidean norm* is defined by

$$\| x \|_2 = \left(\sum_{i=1}^{n} |x_i|^2 \right)^{1/2}.$$

You can easily verify that this function satisfies (2.1.1) and (2.1.2). The triangle inequality (2.1.3) is not so easy. It follows from the Cauchy-Schwarz inequality, which we will prove shortly (Theorem 2.1.6). The Euclidean distance between two vectors x and y is given by

$$\| x - y \|_2 = \sqrt{\sum_{i=1}^{n} |x_i - y_i|^2}.$$

In the cases $n = 1, 2$, and 3, this measure coincides with our usual notion of distance between points in a line, in a plane, or in space, respectively.

Notice that the absolute value signs in the formula for $\| x \|_2$ are redundant, as $|x_i|^2 = x_i^2$ for any real number x_i. However, for complex numbers it is not generally true that $|x_i|^2 = x_i^2$, and the absolute value signs would be needed. Thus the inclusion of the absolute value signs gives a formula for $\| x \|_2$ that is correct for both real and complex vectors. $\square$

Theorem 2.1.6 *(Cauchy-Schwarz inequality) For all x, $y \in \mathbb{R}^n$*

$$\left| \sum_{i=1}^{n} x_i y_i \right| \le \left(\sum_{i=1}^{n} x_i^2 \right)^{1/2} \left(\sum_{i=1}^{n} y_i^2 \right)^{1/2}.$$

Proof. For every real number t we have

$$0 \le \sum_{i=1}^{n} (x_i + t y_i)^2 = \sum_{i=1}^{n} x_i^2 + 2t \sum_{i=1}^{n} x_i y_i + t^2 \sum_{i=1}^{n} y_i^2$$
$$= c + bt + at^2,$$

where $a = \sum_{i=1}^{n} y_i^2$, $b = 2 \sum_{i=1}^{n} x_i y_i$, and $c = \sum_{i=1}^{n} x_i^2$. Since $at^2 + bt + c \ge 0$ for all real t, the quadratic polynomial $at^2 + bt + c$ cannot have two distinct real zeros. Therefore the discriminant satisfies $b^2 - 4ac \le 0$. Rewriting this inequality as

$$(b/2)^2 \le ac$$

and taking square roots, we obtain the desired result. $\square$

Now we are ready to prove that the triangle inequality holds for the Euclidean norm. Thus the Euclidean norm is indeed a norm.

Theorem 2.1.7 *For all x, $y \in \mathbb{R}^n$, $\|x+y\|_2 \leq \|x\|_2 + \|y\|_2$.*

Proof. It suffices to show that $\|x+y\|_2^2 \leq (\|x\|_2 + \|y\|_2)^2$.

$$\|x+y\|_2^2 = \sum_{i=1}^n (x_i + y_i)^2 = \sum_{i=1}^n x_i^2 + 2\sum_{i=1}^n x_i y_i + \sum_{i=1}^n y_i^2.$$

Applying the Cauchy-Schwarz inequality to the middle term on the right-hand side, we find that

$$
\begin{aligned}
\|x+y\|_2^2 &\leq \sum_{i=1}^n x_i^2 + 2\left(\sum_{i=1}^n x_i^2\right)^{1/2}\left(\sum_{i=1}^n y_i^2\right)^{1/2} + \sum_{i=1}^n y_i^2 \\
&= \left[\left(\sum_{i=1}^n x_i^2\right)^{1/2} + \left(\sum_{i=1}^n y_i^2\right)^{1/2}\right]^2 \\
&= (\|x\|_2 + \|y\|_2)^2.
\end{aligned}
$$

$\square$

Example 2.1.8 Generalizing Example 2.1.5, we introduce the *p*-norms. For any real number $p \geq 1$, we define

$$\|x\|_p = \left(\sum_{i=1}^n |x_i|^p\right)^{1/p}.$$

Again it is easy to verify (2.1.1) and (2.1.2), but not (2.1.3). This is the *Minkowski Inequality*, which we will not prove because we are not going to use it. $\square$

The most important *p*-norm is the 2-norm, which is just the Euclidean norm.

Example 2.1.9 Another important *p*-norm is the 1-norm

$$\|x\|_1 = \sum_{i=1}^n |x_i|.$$

In this case (2.1.3) is not hard to prove; it follows directly from the triangle inequality for real numbers. $\square$

Exercise 2.1.10 Prove that the 1-norm is a norm. $\square$

Exercise 2.1.11 Let x, $y \in \mathbb{R}^2$. With respect to the 1-norm, the "distance" between x and y is $\|x-y\|_1 = |x_1 - y_1| + |x_2 - y_2|$. Explain why the 1-norm is sometimes called the *taxicab norm* (or *Manhattan metric*). $\square$

Example 2.1.12 The ∞-norm is defined by

$$\|x\|_\infty = \max_{1 \leq i \leq n} |x_i|.$$

$\square$

rcise 2.1.13 Prove that the ∞-norm is a norm. $\quad\square$

rcise 2.1.14 Given any norm on $\mathbb{R}^2$, the *unit circle* with respect to that norm is the set $\{x \in \mathbb{R}^2 \mid \|x\| = 1\}$. Thinking of the members of $\mathbb{R}^2$ as points in the plane, the unit circle is just the set of points whose distance from the origin is 1. On a single set of coordinate axes, sketch the unit circle with respect to the p-norm for $p = 1, 3/2, 2, 3, 10$, and ∞. $\quad\square$

The analytically inclined reader might like to prove that for all $x \in \mathbb{R}^n$, $\|x\|_\infty = \lim_{p \to \infty} \|x\|_p$.

Example 2.1.15 Given a positive definite matrix $A \in \mathbb{R}^{n \times n}$, define the A-norm on $\mathbb{R}^n$ by

$$\|x\|_A = (x^T A x)^{1/2}.$$

In Exercise 2.1.17 you will show that this is indeed a norm. $\quad\square$

rcise 2.1.16 Show that when $A = I$, the A-norm is just the Euclidean norm. $\quad\square$

rcise 2.1.17

(a) Let A be a positive definite matrix, and let R be its Cholesky factor, so that $A = R^T R$. Verify that for all $x \in \mathbb{R}^n$, $\|x\|_A = \|Rx\|_2$.

(b) Using the fact that the 2-norm is indeed a norm on $\mathbb{R}^n$, prove that the A-norm is a norm on $\mathbb{R}^n$.

$\quad\square$

Matrix Norms

The set of real $m \times n$ matrices is denoted $\mathbb{R}^{m \times n}$. Like vectors, the matrices in $\mathbb{R}^{m \times n}$ can be added and multiplied by scalars in the obvious manner. In fact the matrices in $\mathbb{R}^{m \times n}$ can be viewed simply as vectors in $\mathbb{R}^{mn}$ with the components arranged differently. In the case $m = n$, the theory becomes richer. Unlike ordinary vectors, two matrices in $\mathbb{R}^{n \times n}$ can be multiplied together (using the usual matrix multiplication) to yield a product in $\mathbb{R}^{n \times n}$. A *matrix norm* is a function that assigns to each $A \in \mathbb{R}^{n \times n}$ a real number $\|A\|$, called the norm of A, such that the three vector norm properties hold, as well as one additional property, *submultiplicativity*, which relates the norm function to the operation of matrix multiplication. Specifically, for all $A, B \in \mathbb{R}^{n \times n}$ and all $\alpha \in \mathbb{R}$,

$$\begin{align}
\|A\| &> 0 \text{ if } A \neq 0 & (2.1.18)\\
\|\alpha A\| &= |\alpha| \|A\| & (2.1.19)\\
\|A + B\| &\leq \|A\| + \|B\| & (2.1.20)\\
\|AB\| &\leq \|A\| \|B\| \quad \text{(submultiplicativity)} & (2.1.21)
\end{align}$$

Example 2.1.22 The *Frobenius norm* is defined by

$$\|A\|_F = \left(\sum_{i=1}^{n}\sum_{j=1}^{n}|a_{ij}|^2\right)^{1/2}.$$

Because it is the same as the Euclidean norm on vectors, we already know that it satisfies (2.1.18), (2.1.19), and (2.1.20). The submultiplicativity (2.1.21) can be deduced from the Cauchy-Schwarz inequality as follows. Let $C = AB$. Then $c_{ij} = \sum_{k=1}^{n}a_{ik}b_{kj}$. Thus

$$\|AB\|_F^2 = \|C\|_F^2 = \sum_{i=1}^{n}\sum_{j=1}^{n}|c_{ij}|^2 = \sum_{i=1}^{n}\sum_{j=1}^{n}\left|\sum_{k=1}^{n}a_{ik}b_{kj}\right|^2.$$

Applying the Cauchy-Schwarz inequality to the expression $\sum_{k=1}^{n}a_{ik}b_{kj}$, we find that

$$
\begin{aligned}
\|AB\|_F^2 &\leq \sum_{i=1}^{n}\sum_{j=1}^{n}\left(\sum_{k=1}^{n}|a_{ik}^2|\sum_{k=1}^{n}|b_{kj}^2|\right) \\
&= \left(\sum_{i=1}^{n}\sum_{k=1}^{n}|a_{ik}|^2\right)\left(\sum_{j=1}^{n}\sum_{k=1}^{n}|b_{kj}|^2\right) \\
&= \|A\|_F^2\|B\|_F^2.
\end{aligned}
$$

Thus the Frobenius norm is a matrix norm.   □

Exercise 2.1.23 Define $\|A\|_{\max} = \max_{1\leq i,j\leq n}|a_{ij}|$. Clearly this function satisfies (2.1.18), (2.1.19), and (2.1.20). Show by example that it violates (2.1.21) and is therefore not a matrix norm.   □

Every vector norm on $\mathbb{R}^n$ can be used to define a matrix norm on $\mathbb{R}^{n\times n}$ in a natural way. Given a vector norm $\|\cdot\|_v$, the matrix norm *induced* by $\|\cdot\|_v$ is defined by

$$\|A\|_M = \max_{x\neq 0}\frac{\|Ax\|_v}{\|x\|_v}.$$

Theorem 2.1.26 will show that the induced norm is indeed a matrix norm. Another name for the induced norm is the *operator norm*.

The induced norm has geometric significance that can be understood by viewing A as a linear operator that maps $\mathbb{R}^n$ into $\mathbb{R}^n$: Each $x \in \mathbb{R}^n$ is mapped by A to the vector $Ax \in \mathbb{R}^n$. The ratio $\|Ax\|_v/\|x\|_v$ is the magnification that takes place when x is transformed to Ax. The number $\|A\|_M$ is then the maximum magnification that A can cause.

It is a common practice not to use distinguishing suffixes v and M. One simply uses the same symbol for both the vector norm and the matrix norm and writes, for example,

$$\|A\| = \max_{x \neq 0} \frac{\|Ax\|}{\|x\|}.$$

We will adopt this practice. This need not lead to confusion because the meaning of the norm can always be deduced from the context.

Before proving that the induced norm is indeed a matrix norm, it is useful to make note of the following simple but important fact.

Theorem 2.1.24 *A vector norm and its induced matrix norm satisfy the inequality*

$$\|Ax\| \leq \|A\|\,\|x\| \tag{2.1.25}$$

for all $A \in \mathbb{R}^{n \times n}$ and $x \in \mathbb{R}^n$. This inequality is sharp in the following sense. For all $A \in \mathbb{R}^{n \times n}$ there exists a nonzero $x \in \mathbb{R}^n$ for which equality holds.

Proof. If $x = 0$, equality holds trivially. Otherwise

$$\frac{\|Ax\|}{\|x\|} \leq \max_{\hat{x} \neq 0} \frac{\|A\hat{x}\|}{\|\hat{x}\|} = \|A\|.$$

Thus $\|Ax\| \leq \|A\|\,\|x\|$. Equality holds if and only if x is a vector for which the maximum magnification is attained. (That such a vector exists is actually not obvious. It follows from a compactness argument that works because $\mathbb{R}^n$ is a finite-dimensional space. We omit the argument.) □

The fact that equality is attained in (2.1.25) is actually less important than the (obvious) fact that there exist vectors for which equality is approached as closely as one pleases.

Theorem 2.1.26 *The induced norm is a matrix norm.*

Proof. The proof is not particularly difficult. You are encouraged to provide one of your own before reading on. Only the submultiplicativity property (2.1.21) depends upon (2.1.25).

Each of the first three norm properties follows from the corresponding property of the vector norm. To prove (2.1.18), we must show that $\|A\| > 0$ if $A \neq 0$. If $A \neq 0$, there exists a (nonzero) vector $\hat{x} \in \mathbb{R}^n$ for which $A\hat{x} \neq 0$. Since the vector norm satisfies (2.1.1), we have $\|A\hat{x}\| > 0$ and $\|\hat{x}\| > 0$. Thus

$$\|A\| = \max_{x \neq 0} \frac{\|Ax\|}{\|x\|} \geq \frac{\|A\hat{x}\|}{\|\hat{x}\|} > 0.$$

To prove (2.1.19), we note that for every $x \in \mathbb{R}^n$ and every $\alpha \in \mathbb{R}$, $\|\alpha(Ax)\| = |\alpha|\,\|Ax\|$. This is because the vector norm satisfies (2.1.2). (*Remember:* Ax is a vector, not a matrix.) Thus

$$
\begin{aligned}
\|\alpha A\| &= \max_{x \neq 0} \frac{\|(\alpha A)x\|}{\|x\|} = \max_{x \neq 0} \frac{\|\alpha(Ax)\|}{\|x\|} = \max_{x \neq 0} \frac{|\alpha|\,\|Ax\|}{\|x\|} \\
&= |\alpha| \max_{x \neq 0} \frac{\|Ax\|}{\|x\|} = |\alpha|\,\|A\|.
\end{aligned}
$$

Applying similar ideas we prove (2.1.20).

$$
\begin{aligned}
\| A + B \| &= \max_{x \neq 0} \frac{\| (A + B)x \|}{\| x \|} = \max_{x \neq 0} \frac{\| Ax + Bx \|}{\| x \|} \\
&\leq \max_{x \neq 0} \frac{\| Ax \| + \| Bx \|}{\| x \|} \leq \max_{x \neq 0} \frac{\| Ax \|}{\| x \|} + \max_{x \neq 0} \frac{\| Bx \|}{\| x \|} \\
&= \| A \| + \| B \|.
\end{aligned}
$$

Finally we prove (2.1.21). Replacing x by Bx in (2.1.25), we have $\| ABx \| \leq \| A \| \| Bx \|$ for any x. Applying (2.1.25) again, we have $\| Bx \| \leq \| B \| \| x \|$. Thus

$$
\| ABx \| \leq \| A \| \| B \| \| x \|.
$$

For nonzero x we can divide both sides by the positive number $\| x \|$ and conclude that

$$
\| AB \| = \max_{x \neq 0} \frac{\| ABx \|}{\| x \|} \leq \| A \| \| B \|.
$$

$\square$

Exercise 2.1.27

(a) Show that for any nonzero vector x and scalar c, $\| A(cx) \| / \| cx \| = \| Ax \| / \| x \|$. Thus rescaling a vector does not change the amount by which it is magnified under multiplication by A. In geometric terms, the magnification undergone by x depends only on its direction, not on its length.

(b) Prove that the induced matrix norm satisfies

$$
\| A \| = \max_{\| x \| = 1} \| Ax \|.
$$

This alternative characterization is often useful.

$\square$

Some of the most important matrix norms are induced by p-norms. For $1 \leq p \leq \infty$, the norm induced by the p-norm is called the *matrix p-norm*:

$$
\| A \|_p = \max_{x \neq 0} \frac{\| Ax \|_p}{\| x \|_p}.
$$

The matrix 2-norm is also known as the *spectral norm*. As we shall see in subsequent chapters, this norm has great theoretical importance. Its drawback is that it is expensive to compute; it is *not* the Frobenius norm.

Exercise 2.1.28

(a) Calculate $\| I \|_F$ and $\| I \|_2$, where I is the $n \times n$ identity matrix, and notice that they are different.

(b) Use the Cauchy-Schwarz inequality (Theorem 2.1.6) to show that for all $A \in \mathbb{R}^{n \times n}$, $\| A \|_2 \leq \| A \|_F$.

$\square$

Other important cases are $p = 1$ and $p = \infty$. These norms can be computed easily.

Theorem 2.1.29   *(a)* $\| A \|_1 = \max\limits_{1 \leq j \leq n} \sum\limits_{i=1}^{n} |a_{ij}|$.   *(b)* $\| A \|_\infty = \max\limits_{1 \leq i \leq n} \sum\limits_{j=1}^{n} |a_{ij}|$.

Thus $\| A \|_1$ is found by summing the absolute values of the entries in each column of A and then taking the largest of these column sums. Therefore the matrix 1-norm is sometimes called the *column-sum norm*. Similarly, the matrix ∞-norm is called the *row-sum norm*.

Proof. We will prove part (a), leaving part (b) as an exercise. We first show that $\| A \|_1 \leq \max\limits_{j} \sum\limits_{i=1}^{n} |a_{ij}|$. For all $x \in \mathbb{R}^n$,

$$
\begin{aligned}
\| Ax \|_1 &= \sum_{i=1}^{n} |(Ax)_i| = \sum_{i=1}^{n} \left| \sum_{j=1}^{n} a_{ij} x_j \right| \leq \sum_{i=1}^{n} \sum_{j=1}^{n} |a_{ij}| |x_j| \\
&= \sum_{j=1}^{n} \sum_{i=1}^{n} |a_{ij}| |x_j| \leq \sum_{j=1}^{n} \left(\max_{k} \sum_{i=1}^{n} |a_{ik}| \right) |x_j| \\
&= \left(\max_{k} \sum_{i=1}^{n} |a_{ik}| \right) \left(\sum_{j=1}^{n} |x_j| \right) = \left(\max_{k} \sum_{i=1}^{n} |a_{ik}| \right) \| x \|_1.
\end{aligned}
$$

Therefore $\| Ax \|_1 / \| x \|_1 \leq \max\limits_{k} \sum\limits_{i=1}^{n} |a_{ik}|$ for all $x \neq 0$. From this $\| A \|_1 \leq \max\limits_{k} \left(\sum\limits_{i=1}^{n} |a_{ik}| \right)$. To prove equality, we must simply find an $\hat{x} \in \mathbb{R}^n$ for which

$$
\frac{\| A\hat{x} \|_1}{\| \hat{x} \|_1} = \max_{k} \left(\sum_{i=1}^{n} |a_{ik}| \right).
$$

Suppose that the maximum is attained in the mth column of A. Let $\hat{x}$ be the vector with a 1 in position m and zeros elsewhere. Then $\| \hat{x} \|_1 = 1$, $A\hat{x} = [a_{1m} \, a_{2m} \, \cdots \, a_{nm}]^T$, and $\| A\hat{x} \|_1 = \sum_{i=1}^{n} |a_{im}|$. Thus

$$
\frac{\| A\hat{x} \|_1}{\| \hat{x} \|_1} = \sum_{i=1}^{n} |a_{im}| = \max_{j} \left(\sum_{i=1}^{n} |a_{ij}| \right).
$$

☐

Exercise 2.1.30   Prove part (b) of Theorem 2.1.29. (The argument is generally similar to that of the proof of part (a), but your special vector $\hat{x}$ should be chosen to have components ± 1, with the sign of each component chosen carefully.)   ☐

Additional Exercises

Exercise 2.1.31   Let v, $w \in \mathbb{R}^n$, and let $A = vw^T$. This is a matrix of rank one. Show that

$$\|A\|_2 = \|v\|_2 \|w\|_2.$$

☐

Exercise 2.1.32   Show that for all $x \in \mathbb{R}^n$

$$\|x\|_\infty \leq \|x\|_2 \leq \|x\|_1 \leq \sqrt{n}\|x\|_2 \leq n\|x\|_\infty.$$

The two outer inequalities are fairly obvious. The inequality $\|x\|_2 \leq \|x\|_1$ becomes obvious on squaring both sides. the inequality $\|x\|_1 \leq \sqrt{n}\|x\|_2$ is obtained by applying the Cauchy-Schwarz inequality (Theorem 2.1.6) to the vectors x and $y = [1, 1, \cdots, 1]^T$.   ☐

Exercise 2.1.33   Make systematic use of the inequalities from Exercise 2.1.32 to prove that for all $A \in \mathbb{R}^{n \times n}$

$$\|A\|_1 \leq \sqrt{n}\|A\|_2 \leq n\|A\|_1$$

and

$$\|A\|_\infty \leq \sqrt{n}\|A\|_2 \leq n\|A\|_\infty.$$

☐

2.2   CONDITION NUMBERS

In this section we introduce and discuss the *condition number* of a matrix A. This is a simple but useful measure of the sensitivity of the linear system $Ax = b$.

Consider a linear system $Ax = b$, where A is nonsingular and b is nonzero. There is a unique solution x, which is nonzero. Now suppose we add a small vector δb to b and consider the perturbed system $A\hat{x} = b + \delta b$. This system also has a unique solution $\hat{x}$, which is hoped to be not too far from x. Let δx denote the difference between $\hat{x}$ and x, so that $\hat{x} = x + \delta x$. We would like to say that if δb is small, then δx is also small. A more precise statement would involve relative terms: when we say that δb is small, we really mean that it is small in comparison with b; when we say δx is small, we really mean small compared with x. In order to quantify the size of vectors, we introduce a vector norm $\|\cdot\|$. The size of δb relative to b is then given by $\|\delta b\|/\|b\|$, and the size of δx relative to x is given by $\|\delta x\|/\|x\|$. We would like to say that if $\|\delta b\|/\|b\|$ is small, then $\|\delta x\|/\|x\|$ is also small.

The equations $Ax = b$ and $A(x + \delta x) = b + \delta b$ imply that $A\delta x = \delta b$, that is, $\delta x = A^{-1}\delta b$. Whatever vector norm we have chosen, we will use the induced matrix norm to measure matrices. The equation $\delta x = A^{-1}\delta b$ and Theorem 2.1.24 imply that

$$\| \delta x \| \le \| A^{-1} \| \, \| \delta b \|. \tag{2.2.1}$$

Similarly the equation $b = Ax$ and Theorem 2.1.24 imply $\| b \| \le \| A \| \, \| x \|$, or equivalently

$$\frac{1}{\| x \|} \le \| A \| \frac{1}{\| b \|}. \tag{2.2.2}$$

Multiplying inequality (2.2.1) by (2.2.2), we arrive at

$$\frac{\| \delta x \|}{\| x \|} \le \| A \| \, \| A^{-1} \| \frac{\| \delta b \|}{\| b \|}, \tag{2.2.3}$$

which provides a bound for $\| \delta x \| / \| x \|$ in terms of $\| \delta b \| / \| b \|$. The factor $\| A \| \, \| A^{-1} \|$ is called the *condition number* of A and denoted $\kappa(A)$ [92]. That is,

$$\kappa(A) = \| A \| \, \| A^{-1} \|.$$

With this new notation, we rephrase (2.2.3) as the conclusion of a theorem.

Theorem 2.2.4 *Let A be nonsingular, and let x and $\hat{x} = x + \delta x$ be the solutions of $Ax = b$ and $A\hat{x} = b + \delta b$, respectively. Then*

$$\frac{\| \delta x \|}{\| x \|} \le \kappa(A) \frac{\| \delta b \|}{\| b \|}. \tag{2.2.5}$$

Since (2.2.1) and (2.2.2) are sharp, (2.2.5) is also sharp; that is, there exist b and δb (and associated x and δx) for which equality holds in (2.2.5).

rcise 2.2.6

(a) Show that $\kappa(A) = \kappa(A^{-1})$.

(b) Show that for any nonzero scalar c, $\kappa(cA) = \kappa(A)$.

□

From (2.2.5) we see that if $\kappa(A)$ is not too large, then small values of $\| \delta b \| / \| b \|$ imply small values of $\| \delta x \| / \| x \|$. That is, the system is not overly sensitive to perturbations in b. Thus if $\kappa(A)$ is not too large, we say that A is *well conditioned*. If, on the other hand, $\kappa(A)$ is large, a small value of $\| \delta b \| / \| b \|$ does not guarantee that $\| \delta x \| / \| x \|$ will be small. Since (2.2.5) is sharp, we know that in this case there definitely are choices of b and δb for which the resulting $\| \delta x \| / \| x \|$ is much larger than the resulting $\| \delta b \| / \| b \|$. In other words, the system is potentially very sensitive to perturbations in b. Thus if $\kappa(A)$ is large, we say that A is *ill conditioned*.

Proposition 2.2.7 *For any induced matrix norm, (a) $\| I \| = 1$ and (b) $\kappa(A) \ge 1$.*

Proof. Part (a) follows immediately from the definition of the induced matrix norm. To prove part (b), we note that $I = AA^{-1}$, so $1 = \|I\| = \|AA^{-1}\| \leq \|A\| \|A^{-1}\| = \kappa(A)$. $\qquad\qquad\square$

Thus the best possible condition number is 1. Of course the condition number depends on the choice of norm. We will use mainly the 1-, 2-, and ∞-norms, which give roughly comparable values for the condition numbers of matrices. We will use the notation

$$\kappa_p(A) = \|A\|_p \|A^{-1}\|_p \quad \text{for } 1 \leq p \leq \infty.$$

Example 2.2.8 Let $A = \begin{bmatrix} 1000 & 999 \\ 999 & 998 \end{bmatrix}$. You can easily verify that

$$A^{-1} = \begin{bmatrix} -998 & 999 \\ 999 & -1000 \end{bmatrix}.$$

Thus $\|A\|_\infty = \|A\|_1 = 1999 = \|A^{-1}\|_\infty = \|A^{-1}\|_1$, and

$$\kappa_\infty(A) = \kappa_1(A) = (1999)^2 = 3.996 \times 10^6.$$

The process of computing $\kappa_2(A)$ is more involved; we are not yet ready to describe it. However, MATLAB can easily do the job. Using the command cond(A) or cond(A, 2) , we find that $\kappa_2(A) \approx 3.992 \times 10^6$.

Now consider a linear system having A as its coefficient matrix:

$$\begin{bmatrix} 1000 & 999 \\ 999 & 998 \end{bmatrix} \begin{bmatrix} x_1 \\ x_2 \end{bmatrix} = \begin{bmatrix} b_1 \\ b_2 \end{bmatrix}. \tag{2.2.9}$$

This is a system of two linear equations,

$$\begin{aligned} 1000x_1 + 999x_2 &= b_1 \\ 999x_1 + 998x_2 &= b_2, \end{aligned}$$

each of which represents a line in the plane. The slopes of the lines are easily seen to be

$$m_1 = -1000/999 \approx -1.001001 \quad \text{and} \quad m_2 = -999/998 \approx -1.001002,$$

respectively. Thus the solution of the system is the intersection point of two nearly parallel lines. If either line is moved just slightly, the solution will be changed drastically. The situation is illustrated in Figure 2.1. The point labelled p is the solution of the system. A small perturbation in b_1 (for example) causes a small parallel shift in the first line. The perturbed line is represented by the dashed line in Figure 2.1. Since the two lines are nearly parallel, a small shift in one of them causes a large shift in the solution from point p to point q. In contrast, in the well-conditioned case, the rows are not nearly dependent, and the lines determined by the two equations are not nearly parallel. A small perturbation in one or both of the lines gives rise to a small perturbation in the solution.

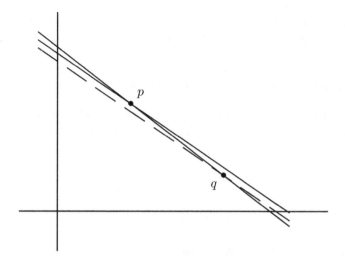

Figure 2.1   Perturbing an ill-conditioned system

The system depicted in Figure 2.1 is actually not very ill conditioned at all. It is not possible to draw a good picture of a truly ill-conditioned system such as (2.2.9); the lines would be so nearly parallel as to be indistinguishable in the vicinity of the solution.   □

So far we have said that a matrix that has a large condition number is ill conditioned, but we have not said anything about where the cutoff line between well-conditioned and ill-conditioned matrices lies. Of course there is no point in looking for a precise boundary. Furthermore the (fuzzy) boundary depends upon a number of factors, including the accuracy of the data being used, the precision of the floating point numbers, and the amount of error we are willing to tolerate in our computed solution. Suppose, for example, that the components of b are correct to about four decimal places. We do not know the exact value of b; in the computation we actually use $b + \delta b$, where $\| \delta b \| / \| b \| \approx 10^{-4}$. If we solve the problem accurately, we get not x but $x + \delta x$, where an upper bound for $\| \delta x \| / \| x \|$ is given by (2.2.5).

Now suppose $\kappa(A) \leq 10^2$. Then by (2.2.5) the worst that can happen is $\| \delta x \| / \| x \| \approx 10^{-2}$. That is, the error in x is not bigger than about one hundredth the size of x. In many problems, this much error in the solution is acceptable. By contrast, if $\kappa(A) \approx 10^4$, then (2.2.5) tells us that it can happen that $\| \delta x \| / \| x \| \approx 1$; that is, the error could be as big as the solution itself. In this case we would have to say that the condition number is unacceptably high. Thus it appears that in this problem the boundary between well-conditioned and ill-conditioned matrices lies somewhere in the range 10^2 to 10^4.

Sometimes the accuracy of the floating point arithmetic can be the deciding factor. It may be that we know b with extreme accuracy, but if numbers are only stored to about seven decimal places accuracy in the computer, then we will be forced to work with $b + \delta b$, where $\|\delta b\| / \|b\| \approx 10^{-7}$. Then if we have $\kappa(A) \approx 10^7$, we cannot be sure to get a reasonable answer, even if we solve the system very accurately. On the other hand a condition number of 10^3, 10^4, or even 10^5 may be small enough, depending on how accurate we require the solution to be.

Example 2.2.10 The the most famous examples of ill conditioned matrices are the *Hilbert matrices*, defined by $h_{ij} = 1/(i + j - 1)$. If we let H_n denote the $n \times n$ Hilbert matrix, then

$$H_4 = \begin{bmatrix} 1 & 1/2 & 1/3 & 1/4 \\ 1/2 & 1/3 & 1/4 & 1/5 \\ 1/3 & 1/4 & 1/5 & 1/6 \\ 1/4 & 1/5 & 1/6 & 1/7 \end{bmatrix},$$

for example. These matrices are symmetric, can be shown to be positive definite, and are increasingly ill conditioned as n is increased. For example, according to MATLAB, $\kappa_2(H_4) \approx 1.6 \times 10^4$ and $\kappa_2(H_8) \approx 1.5 \times 10^{10}$. □

Exercise 2.2.11 In MATLAB you can type A = `hilb(7)` to get the 7×7 Hilbert matrix, for example. Type `help cond` to find out how to use MATLAB's condition number function. Use it to calculate $\kappa_1(H_n)$, $\kappa_2(H_n)$, and $\kappa_\infty(H_n)$ for $n = 3, 6, 9$, and 12. □

The Condition Number and Magnification

We begin by introducing some new terms. The *maximum* and *minimum magnification* by A are defined by

$$\text{maxmag}(A) = \max_{x \neq 0} \frac{\|Ax\|}{\|x\|}$$

and

$$\text{minmag}(A) = \min_{x \neq 0} \frac{\|Ax\|}{\|x\|}.$$

Of course, maxmag(A) is nothing but the induced matrix norm $\|A\|$.

Exercise 2.2.12 Prove that if A is a nonsingular matrix, then

$$\text{maxmag}(A) = \frac{1}{\text{minmag}(A^{-1})} \quad \text{and} \quad \text{maxmag}(A^{-1}) = \frac{1}{\text{minmag}(A)}.$$

□

From this exercise it follows easily that $\kappa(A)$ is just the ratio of the maximum magnification to the minimum magnification.

Proposition 2.2.13 $\kappa(A) = \dfrac{\text{maxmag}(A)}{\text{minmag}(A)}$ *for all nonsingular* A.

rcise 2.2.14 Prove Proposition 2.2.13. □

An ill-conditioned matrix is one for which the maximum magnification is much larger than the minimum magnification.

If the matrix A is nonzero but singular, then there exists $x \neq 0$ such that $Ax = 0$. Thus minmag$(A) = 0$, and it is reasonable to say that $\kappa(A) = \infty$. That is, we view singularity as the extreme case of ill conditioning. Reversing the viewpoint, we can say that an ill-conditioned matrix is one that is "nearly" singular.

Since a matrix A is singular if and only if $\det(A) = 0$, it is natural to expect that the determinant is somehow connected to the condition number. This turns out to be wrong; there is no useful relationship between the condition number and the determinant. See Exercise 2.2.15. When it comes to assessing sensitivity of linear systems, the condition number is useful and the determinant is not.

rcise 2.2.15

(a) Let α be a positive number, and define

$$A_\alpha = \begin{bmatrix} \alpha & 0 \\ 0 & \alpha \end{bmatrix}.$$

Show that for any induced matrix norm we have $\| A_\alpha \| = \alpha$, $\| A_\alpha^{-1} \| = 1/\alpha$, and $\kappa(A_\alpha) = 1$. Thus A_α is well conditioned. On the other hand, $\det(A_\alpha) = \alpha^2$, so we can make it as large or small as we please by choosing α appropriately.

(b) More generally, given any nonsingular matrix A, discuss the condition number and determinant of αA, where α is any positive real number.

□

Example 2.2.16 Let us take another look at the ill-conditioned matrices

$$A = \begin{bmatrix} 1000 & 999 \\ 999 & 998 \end{bmatrix} \quad \text{and} \quad A^{-1} = \begin{bmatrix} -998 & 999 \\ 999 & -1000 \end{bmatrix}$$

from Example 2.2.8. Notice that

$$A \begin{bmatrix} 1 \\ 1 \end{bmatrix} = \begin{bmatrix} 1999 \\ 1997 \end{bmatrix}. \tag{2.2.17}$$

If we use the ∞-norm to measure lengths, the magnification factor $\| Ax \|_\infty / \| x \|_\infty$ is 1999, which equals $\| A \|_\infty$. Thus $\begin{bmatrix} 1 \\ 1 \end{bmatrix}$ is a vector that is magnified maximally by

A. Since the amount by which a vector is magnified depends only on its direction and not on its length, we say that $\begin{bmatrix} 1 \\ 1 \end{bmatrix}$ is in a *direction of maximum magnification* by A. Equivalently we can say that $\begin{bmatrix} 1999 \\ 1997 \end{bmatrix}$ lies in a *direction of minimum magnification* by A^{-1}. Looking now at A^{-1}, we note that

$$A^{-1} \begin{bmatrix} -1 \\ 1 \end{bmatrix} = \begin{bmatrix} 1997 \\ -1999 \end{bmatrix}.$$

The magnification factor $\| A^{-1}x \|_\infty / \| x \|_\infty$ is 1999, which equals $\| A^{-1} \|_\infty$, so $\begin{bmatrix} -1 \\ 1 \end{bmatrix}$ is in a direction of maximum magnification by A^{-1}. Equivalently

$$A \begin{bmatrix} 1997 \\ -1999 \end{bmatrix} = \begin{bmatrix} -1 \\ 1 \end{bmatrix}, \tag{2.2.18}$$

and $\begin{bmatrix} 1997 \\ -1999 \end{bmatrix}$ is in a direction of minimum magnification by A. We will use the vectors in (2.2.17) and (2.2.18) to construct a spectacular example.

Suppose we wish to solve the system

$$\begin{bmatrix} 1000 & 999 \\ 999 & 998 \end{bmatrix} \begin{bmatrix} x_1 \\ x_2 \end{bmatrix} = \begin{bmatrix} 1999 \\ 1997 \end{bmatrix}, \tag{2.2.19}$$

that is, $Ax = b$, where $b = \begin{bmatrix} 1999 \\ 1997 \end{bmatrix}$. Then by (2.2.17) the unique solution is $x = \begin{bmatrix} 1 \\ 1 \end{bmatrix}$.

Now suppose that we solve instead the slightly perturbed system

$$\begin{bmatrix} 1000 & 999 \\ 999 & 998 \end{bmatrix} \begin{bmatrix} \hat{x}_1 \\ \hat{x}_2 \end{bmatrix} = \begin{bmatrix} 1998.99 \\ 1997.01 \end{bmatrix}. \tag{2.2.20}$$

This is $A\hat{x} = b + \delta b$, where $\delta b = \begin{bmatrix} -.01 \\ .01 \end{bmatrix} = .01 \begin{bmatrix} -1 \\ 1 \end{bmatrix}$, which is in a direction of maximum magnification by A^{-1}. By (2.2.18), $A\delta x = \delta b$, where $\delta x = \begin{bmatrix} 19.97 \\ -19.99 \end{bmatrix}$. Therefore $\hat{x} = x + \delta x = \begin{bmatrix} 20.97 \\ -18.99 \end{bmatrix}$. Thus the nearly identical problems (2.2.19) and (2.2.20) have very different solutions. $\square$

It is important to recognize that this example was concocted in a special way. The vector b was chosen to be in a direction of minimum magnification by A^{-1}, so that the resulting x is in a direction of maximum magnification by A, and equality is attained in (2.2.2). The vector δb was chosen in a direction of maximum magnification by

A^{-1}, so that equality holds in (2.2.1). As a consequence, equality also holds in (2.2.5). Had we not made such special choices of b and δb, the result would have been less spectacular.

In some applications, for example, numerical solution of partial differential equations, if the solution is known to be a smooth function, it can be shown that b must necessarily lie in a direction of large magnification by A^{-1}. Under that restriction it is impossible to create an example that is anywhere near as spectacular as Example 2.2.16.

We have seen that if a system is ill conditioned, we can build examples where $\|\delta x\|/\|x\|$ is much larger than $\|\delta b\|/\|b\|$. In fact it can also happen that $\|\delta x\|/\|x\|$ is much smaller than $\|\delta b\|/\|b\|$. Inequality (2.2.5) has a companion inequality

$$\frac{\|\delta b\|}{\|b\|} \leq \kappa(A) \frac{\|\delta x\|}{\|x\|}, \tag{2.2.21}$$

which can be obtained by interchanging the roles of x and δx with b and δb, respectively, and which is also sharp.

rcise 2.2.22   Prove Inequality (2.2.21). Under what conditions does equality hold in (2.2.21)?   □

Example 2.2.23 We use the same ill-conditioned matrix A as in the two recent examples. By (2.2.18) the system

$$\begin{bmatrix} 1000 & 999 \\ 999 & 998 \end{bmatrix} \begin{bmatrix} x_1 \\ x_2 \end{bmatrix} = \begin{bmatrix} -1 \\ 1 \end{bmatrix}$$

has $x = \begin{bmatrix} 1997 \\ -1999 \end{bmatrix}$ as its unique solution. If we now perturb this solution by $\delta x = \begin{bmatrix} 0.01 \\ 0.01 \end{bmatrix}$, we obtain $x + \delta x = \begin{bmatrix} 1997.01 \\ -1998.99 \end{bmatrix}$, which hardly differs from x at all. However, using (2.2.17),

$$A(x + \delta x) = Ax + A\delta x = \begin{bmatrix} 18.99 \\ 20.97 \end{bmatrix},$$

which is nowhere near Ax.   □

Because of their great sensitivity it is generally futile, even meaningless, to try to solve ill-conditioned systems in the presence of uncertainty in the data.

rcise 2.2.24 Let $A = \begin{bmatrix} 375 & 374 \\ 752 & 750 \end{bmatrix}$.

(a) Calculate A^{-1} and $\kappa_\infty(A)$.

(b) Find b, δb, x, and δx such that $Ax = b$, $A(x + \delta x) = b + \delta b$, $\|\delta b\|_\infty / \|b\|_\infty$ is small, and $\|\delta x\|_\infty / \|x\|_\infty$ is large.

(c) Find b, δb, x, and δx such that $Ax = b$, $A(x + \delta x) = b + \delta b$, $\|\delta x\|_\infty / \|x\|_\infty$ is small, and $\|\delta b\|_\infty / \|b\|_\infty$ is large.

□

III Conditioning Caused by Poor Scaling

Some linear systems are ill conditioned simply because they are out of scale. Consider the following example.

Example 2.2.25 Let ϵ be a small positive number. The system

$$\begin{bmatrix} 1 & 0 \\ 0 & \epsilon \end{bmatrix} \begin{bmatrix} x_1 \\ x_2 \end{bmatrix} = \begin{bmatrix} 1 \\ \epsilon \end{bmatrix}$$

has the unique solution $x = \begin{bmatrix} 1 \\ 1 \end{bmatrix}$. You can easily check that if $\epsilon \ll 1$, then the coefficient matrix is ill conditioned with respect to the usual norms. In fact $\kappa_1(A) = \kappa_2(A) = \kappa_\infty(A) = 1/\epsilon$. This system is subject to everything we have said so far about ill-conditioned systems. For example, one can find a small perturbation in b that that causes a large perturbation in x: Just take $b + \delta b = \begin{bmatrix} 1 \\ 2\epsilon \end{bmatrix}$, for which $\|\delta b\|_\infty / \|b\|_\infty = \epsilon$, to get $x + \delta x = \begin{bmatrix} 1 \\ 2 \end{bmatrix}$, which is far from $\begin{bmatrix} 1 \\ 1 \end{bmatrix}$. Notice that this perturbation of b is small with respect to $\|b\|$ but not small compared to the component that was perturbed.

If we multiply the second equation of our system by $1/\epsilon$, we get a new system

$$\begin{bmatrix} 1 & 0 \\ 0 & 1 \end{bmatrix} \begin{bmatrix} x_1 \\ x_2 \end{bmatrix} = \begin{bmatrix} 1 \\ 1 \end{bmatrix},$$

which is perfectly well conditioned. Thus the ill conditioning was just a consequence of poor scaling.

□

Theorem 2.2.26 *Let A be any nonsingular matrix, and let $a_1, a_2, \ldots, a_n$ denote its columns. Then for any i and j,*

$$\kappa_p(A) \geq \frac{\|a_i\|_p}{\|a_j\|_p}, \quad 1 \leq p \leq \infty.$$

Proof. Clearly $a_i = Ae_i$, where e_i is the vector with a one in the ith position and zeros elsewhere. Thus

$$\text{maxmag}(A) = \max_{x \neq 0} \frac{\|Ax\|_p}{\|x\|_p} \geq \frac{\|Ae_i\|_p}{\|e_i\|_p} = \|a_i\|_p,$$

$$\text{minmag}(A) = \min_{x \neq 0} \frac{\|Ax\|_p}{\|x\|_p} \leq \frac{\|Ae_j\|_p}{\|e_j\|_p} = \|a_j\|_p,$$

and

$$\kappa_p(A) = \frac{\text{maxmag}(A)}{\text{minmag}(A)} \geq \frac{\|a_i\|_p}{\|a_j\|_p}.$$

□

Theorem 2.2.26 implies that any matrix that has columns that differ by several orders of magnitude is ill conditioned. The same can be said of rows, since A is ill conditioned if and only if A^T is. (You can easily verify that $\kappa_\infty(A) = \kappa_1(A^T)$. In Section 4.2 (Corollary 4.2.2) we will see that $\kappa_2(A) = \kappa_2(A^T)$.) Thus a necessary condition for a matrix to be well conditioned is that all of its rows and columns be of roughly the same magnitude. This condition is not sufficient, as the matrices in Example 2.2.8 and Exercise 2.2.24 show.

If a system is ill conditioned because its rows or columns are badly out of scale, one must refer back to the underlying physical problem in order to determine whether the ill conditioning is inherent in the problem or simply a consequence of poor choices of measurement units. The system in Example 2.2.25 was easy to handle only because it really consists of two independent problems

$$\begin{bmatrix} 1 \end{bmatrix} \begin{bmatrix} x_1 \end{bmatrix} = \begin{bmatrix} 1 \end{bmatrix} \quad \text{and} \quad \begin{bmatrix} \epsilon \end{bmatrix} \begin{bmatrix} x_2 \end{bmatrix} = \begin{bmatrix} \epsilon \end{bmatrix},$$

each of which is well conditioned. In general a more careful analysis is required. Although the rows and columns of any matrix can easily be rescaled so that all of the rows and columns have about the same norm, there is no unique way of doing it, nor is it guaranteed that the resulting matrix is well conditioned. Issues associated with scaling will be discussed in Section 2.8, but no definite advice will be given. Any decision about whether to rescale or not, and how to rescale, should be guided by the underlying physical problem.

For a summary of some of the most important known results on scaling to (nearly) minimize the condition number, see Higham [52]. A different kind of condition number that is invariant under row scaling is introduced in Section 2.9.

Estimating the Condition Number

The developments of this section have made clear the importance of the condition number of a matrix. Obviously we would like to be able to compute, or at least estimate, this quantity at will. In principle the condition number is not hard to calculate; one simply finds A^{-1} and then calculates $\|A\| \|A^{-1}\|$. (Or, if one is using MATLAB, one simply types cond(A).) This is fine if A is not too large. However, for really large A, we would prefer to save the considerable expense of computing A^{-1}. We do not need to know the condition number exactly; an order-of-magnitude estimate is good enough. What is needed is an inexpensive estimate of $\kappa(A)$.

Let us suppose that we have already solved the system $Ax = b$ by Gaussian elimination, and now we would like to estimate $\kappa(A)$ in order to help us assess the quality of our computed solution. Suppose we choose to estimate $\kappa_1(A) = $

$\|A\|_1\|A^{-1}\|_1$. From Theorem 2.1.29 we know that it is easy to compute $\|A\|_1$. What is more challenging is to get an estimate of $\|A^{-1}\|_1$. We begin by noting that for any nonzero $w \in \mathbb{R}^n$

$$\frac{\|A^{-1}w\|_1}{\|w\|_1} \leq \max_{y \neq 0} \frac{\|A^{-1}y\|_1}{\|y\|_1} = \|A^{-1}\|_1.$$

Thus, taking $w = b$, we have $A^{-1}w = x$,

$$\frac{\|x\|_1}{\|b\|_1} \leq \|A^{-1}\|_1 \quad \text{and} \quad \kappa_1(A) \geq \frac{\|A\|_1\|x\|_1}{\|b\|_1}.$$

This gives an inexpensive lower bound for $\kappa_1(A)$. More generally, for any nonzero $w \in \mathbb{R}^n$,

$$\kappa_1(A) \geq \frac{\|A\|_1\|A^{-1}w\|_1}{\|w\|_1}.$$

Since we already have an LU decomposition of A at hand, we can calculate $A^{-1}w$ by solving $Ac = w$ at a cost of only some $2n^2$ flops. If w is chosen in a direction of near maximum magnification by A^{-1}, the estimate

$$\kappa_1(A) \approx \frac{\|A\|_1\|A^{-1}w\|_1}{\|w\|_1} \tag{2.2.27}$$

will be quite good. Actually any w chosen at random is likely to have a significant component in the direction of maximum magnification by A^{-1} and therefore to give a reasonable estimate in (2.2.27). Since a random w will occasionally give a severe underestimate of $\kappa_1(A)$, the cautious operator might like to try several different choices of w.

More sophisticated approaches conduct systematic searches for a w that points nearly in the direction of maximum magnification. The most successful method to date has been the method of Hager, as modified by Higham (see [52]), which uses ideas from convex optimization to search for a w that maximizes $\|A^{-1}w\|_1/\|w\|_1$. This method, which usually gives an excellent estimate, is the basis of the condition number estimators in LAPACK [1] and MATLAB.

Exercise 2.2.28 Recall that in MATLAB you can type A = hilb(3) to get the 3×3 Hilbert matrix H_3, for example. Use MATLAB's condition number estimator condest to estimate $\kappa_1(H_n)$ for $n = 3, 6, 9$, and 12. Compare it with the true condition number, as computed by cond(A,1). Note the excellent agreement. □

Exercise 2.2.29 Try MATLAB's condition number estimator on a larger matrix. For example, try

```
m = 50;   % Make m larger or smaller, as needed.
A = delsq(numgrid('N',m));  % sparse matrix
size(A)                      % of dimension (m-2)^2.
issparse(A)
```

```
B = full(A);                      % nonsparse version of A
issparse(B)
tic; c1 = condest(A), toc
tic; c2 = cond(B,1), toc
```

Comment on the speed and accuracy of `condest`. You might also like to try `tic;` `c3 = cond(B,2), toc`, which computes $\kappa_2(B)$. However, you may find that this takes too long unless you decrease the size of the problem by decreasing m. This is a time-consuming calculation, because it requires the singular values of B (Section 4.2). $\qquad\square$

MATLAB's `condest` function has to compute the LU decomposition of the matrix, since it cannot assume that an LU decomposition is available. Thus `condest` is less efficient than it would be if the decomposition were assumed available. However, if the matrix under consideration is sparse, `condest` will do a sparse LU decomposition (Section 1.9), thereby saving a lot of work. This explains the good outcome in Exercise 2.2.29.

2.3 PERTURBING THE COEFFICIENT MATRIX

Up to this point we have considered only the effect of perturbing b in the system $Ax = b$. We must also consider perturbations of A, as A is also known and represented only approximately. Thus, let us compare two systems $Ax = b$ and $(A + \delta A)\hat{x} = b$, where $\|\delta A\|/\|A\|$ is small. Our first task is to establish a condition that guarantees that the system $(A + \delta A)\hat{x} = b$ has a unique solution, given that the system $Ax = b$ does. This is given by the following theorem, which, along with the subsequent theorems in this section, is valid for any vector norm and its induced matrix norm and condition number.

Theorem 2.3.1 *If A is nonsingular and*

$$\frac{\|\delta A\|}{\|A\|} < \frac{1}{\kappa(A)}, \tag{2.3.2}$$

then $A + \delta A$ is nonsingular.

Proof. The hypothesis $\|\delta A\|/\|A\| < 1/\kappa(A)$ can be rewritten in various ways, for example, $\|\delta A\| < 1/\|A^{-1}\|$ and $\|\delta A\| \, \|A^{-1}\| < 1$. We'll use this last form of the inequality, and we'll prove the contrapositive form of the theorem: If $A + \delta A$ is singular, then $\|\delta A\| \, \|A^{-1}\| \geq 1$.

Suppose $A + \delta A$ is singular. Then, by Theorem 1.2.3, there is a nonzero vector y such that $(A + \delta A)y = 0$. Reorganizing this equation, we obtain $y = -A^{-1}\delta Ay$, which implies $\|y\| = \|A^{-1}\delta Ay\| \leq \|A^{-1}\| \, \|\delta A\| \, \|y\|$. Since $\|y\| > 0$, we can divide both sides of the inequality by $\|y\|$ to obtain $1 \leq \|A^{-1}\| \, \|\delta A\|$, which is the desired result. $\qquad\square$

Theorem 2.3.1 demonstrates another important function of the condition number; it gives us an idea of the distance from A to the nearest singular matrix: If $A + \delta A$ is singular, then $\|\delta A\|/\|A\|$ must be at least $1/\kappa(A)$. It turns out that for the spectral norm this result is exact: If $A + \delta A$ is the singular matrix closest to A, in the sense that $\|\delta A\|_2$ is as small as possible, then $\|\delta A\|_2/\|A\|_2$ is exactly $1/\kappa_2(A)$. We will prove this in Corollary 4.2.22.

As long as (2.3.2) is satisfied, we are assured that the equation $(A + \delta A)\hat{x} = b$ has a unique solution. Notice that (2.3.2) is hard to satisfy if A is ill conditioned; that is, it is satisfied only for very small perturbations δA. If, on the other hand, A is well conditioned, (2.3.2) holds even for relatively large perturbations.

Now let us consider the relationship between the solutions of $Ax = b$ and $(A + \delta A)\hat{x} = b$. Let $\delta x = \hat{x} - x$, so that $\hat{x} = x + \delta x$. Under what conditions can we conclude that $\|\delta x\|/\|x\|$ is small? We would like an upper bound on $\|\delta x\|/\|x\|$ in the spirit of Theorem 2.2.4. We will obtain such a bound eventually, but it turns out to be easier to bound $\|\delta x\|/\|\hat{x}\|$. In most cases there will not be much difference between $\|x\|$ and $\|\hat{x}\|$, so it makes little difference which one we use in the denominator.

Theorem 2.3.3 *Let A be nonsingular, let $b \neq 0$, and let x and $\hat{x} = x + \delta x$ be solutions of $Ax = b$ and $(A + \delta A)\hat{x} = b$, respectively. Then,*

$$\frac{\|\delta x\|}{\|\hat{x}\|} \leq \kappa(A) \frac{\|\delta A\|}{\|A\|}. \tag{2.3.4}$$

Proof. Rewriting the equation $(A + \delta A)\hat{x} = b$ as $Ax + A\delta x + \delta A\hat{x} = b$, using the equation $Ax = b$, and reorganizing the resulting equation, we obtain $\delta x = -A^{-1}\delta A\hat{x}$. Thus

$$\|\delta x\| \leq \|A^{-1}\| \, \|\delta A\| \, \|\hat{x}\|. \tag{2.3.5}$$

Dividing through by $\|\hat{x}\|$ and using the definition $\kappa(A) = \|A\| \, \|A^{-1}\|$, we obtain the desired result. $\qquad\square$

Theorem 2.3.3 shows that once again the condition number of A plays the decisive role. If $\kappa(A)$ is not too large, then a small perturbation in A results in a small perturbation in x, in the sense that $\|\delta x\|/\|\hat{x}\|$ is small.

It is interesting to note that Theorem 2.3.3 does not rely on nonsingularity of $A + \delta A$, nor on any assumption to the effect that δA is small. In contrast, the next theorem, which provides a bound on $\|\delta x\|/\|x\|$, does make such an assumption.

Theorem 2.3.6 *If A is nonsingular, $\|\delta A\|/\|A\| < 1/\kappa(A)$, $b \neq 0$, $Ax = b$, and $(A + \delta A)(x + \delta x) = b$, then*

$$\frac{\|\delta x\|}{\|x\|} \leq \frac{\kappa(A)\dfrac{\|\delta A\|}{\|A\|}}{1 - \kappa(A)\dfrac{\|\delta A\|}{\|A\|}}. \tag{2.3.7}$$

If A is well conditioned and $\|\delta A\|/\|A\|$ is sufficiently small, then $\|\delta A\|/\|A\| \ll 1/\kappa(A)$. In this case the denominator on the right side of (2.3.7) is approximately 1.

Then (2.3.7) states *roughly* that

$$\frac{\|\delta x\|}{\|x\|} \le \kappa(A)\frac{\|\delta A\|}{\|A\|},$$

which is almost the same as (2.3.4). This shows that if A is well conditioned and $\|\delta A\|/\|A\|$ is small, then $\|\delta x\|/\|x\|$ is small.

If, on the other hand, A is ill conditioned, then (2.3.7) allows (but does not prove) that $\|\delta x\|/\|x\|$ could be large, even if $\|\delta A\|/\|A\|$ is small.

Proof. The proof of Theorem 2.3.6 is the same as that of Theorem 2.3.3, up to (2.3.5). Rewriting $\hat{x}$ as $x + \delta x$ in (2.3.5) and using the triangle inequality, we find that

$$
\begin{aligned}
\|\delta x\| &\le \|A^{-1}\|\,\|\delta A\|\,(\|x\| + \|\delta x\|)\\
&= \kappa(A)\frac{\|\delta A\|}{\|A\|}(\|x\| + \|\delta x\|).
\end{aligned}
$$

Now rewrite this inequality so that all of the terms involving $\|\delta x\|$ are on the left-hand side.

$$\left(1 - \kappa(A)\frac{\|\delta A\|}{\|A\|}\right)\|\delta x\| \le \kappa(A)\frac{\|\delta A\|}{\|A\|}\|x\|$$

The assumption $\|\delta A\|/\|A\| < 1/\kappa(A)$ guarantees that the factor that multiplies $\|\delta x\|$ is positive, so we can divide by it without reversing the inequality. If we also divide through by $\|x\|$, we obtain the desired result. $\square$

So far we have considered the effects of perturbing b and A separately. This was done not out of necessity but from a desire to keep the analysis simple. The combined effects of perturbations in A and b can be expressed in a single inequality, as the next two theorems show. The first is in the spirit of Theorem 2.3.3, and the second is in that of Theorem 2.3.6.

Theorem 2.3.8 *Let A be nonsingular, and suppose x and $\hat{x}$ satisfy $Ax = b$ and $\hat{A}\hat{x} = \hat{b}$, respectively, where $\hat{A} = A + \delta A$, $\hat{x} = x + \delta x \ne 0$, and $\hat{b} = b + \delta b \ne 0$. Then*

$$\frac{\|\delta x\|}{\|\hat{x}\|} \le \kappa(A)\left(\frac{\|\delta A\|}{\|A\|} + \frac{\|\delta b\|}{\|\hat{b}\|} + \frac{\|\delta A\|}{\|A\|}\frac{\|\delta b\|}{\|\hat{b}\|}\right).$$

Of the terms on the right-hand side, the product term is usually negligible. For example, if $\frac{\|\delta A\|}{\|A\|} = 10^{-5}$ and $\frac{\|\delta b\|}{\|\hat{b}\|} = 10^{-5}$, then $\frac{\|\delta A\|}{\|A\|}\frac{\|\delta b\|}{\|\hat{b}\|} = 10^{-10}$.

Theorem 2.3.9 *If A is nonsingular, $\|\delta A\|/\|A\| < 1/\kappa(A)$, $b \ne 0$, $Ax = b$, and $(A + \delta A)(x + \delta x) = b + \delta b$, then*

$$\frac{\|\delta x\|}{\|x\|} \le \frac{\kappa(A)\left(\dfrac{\|\delta A\|}{\|A\|} + \dfrac{\|\delta b\|}{\|b\|}\right)}{1 - \kappa(A)\dfrac{\|\delta A\|}{\|A\|}}. \tag{2.3.10}$$

Example 2.3.11 In Example 1.2.6 we considered an electrical circuit that leads to the linear system

$$
\begin{bmatrix}
2 & -1 & -1 & 0 \\
-1 & 1.5 & 0 & -.5 \\
-1 & 0 & 1.7 & -.2 \\
0 & -.5 & -.2 & 1.7
\end{bmatrix}
\begin{bmatrix}
x_1 \\
x_2 \\
x_3 \\
x_4
\end{bmatrix}
=
\begin{bmatrix}
0 \\
0 \\
3 \\
0
\end{bmatrix},
$$

which we solved to determine the voltages at the nodes of the circuit. If we solve the system using MATLAB or any other reliable software, we obtain an extremely accurate solution (See Example 2.4.2 below). It is the accurate solution of the given system, but what if the entries of A and b are incorrect? The entries of A depend on the resistances in the circuit, and the one nonzero entry of b depends also on the voltage of the battery. None of these quantities are known exactly. Theorem 2.3.9 gives us information about the effects of inaccuracies. Suppose, for example, the resistances and the voltage are known to be in error by less than one one hundredth of one percent. This means that the relative error is less than 10^{-4}, so, roughly speaking,

$$
\frac{\|\delta A\|_2}{\|A\|_2} \approx 10^{-4} \quad \text{and} \quad \frac{\|\delta b\|_2}{\|b\|_2} \approx 10^{-4}.
$$

Using MATLAB's `cond` function, we get $\kappa_2(A) = 12.7$. Substituting these values into (2.3.10), we find that

$$
\frac{\|\delta x\|_2}{\|x\|_2} \leq 2.5 \times 10^{-3}.
$$

Thus the computed nodal voltages are off by at most one quarter of one percent. It should be noted that the actual error is likely to be much less than this. Results obtained using an upper bound like the one in Theorem 2.3.9 tend to be quite pessimistic. □

Exercise 2.3.12 Prove Theorem 2.3.8. Do it your way, or use the following outline.

(a) Show that

$$
\delta x = A^{-1}\left(\delta b - \delta A \hat{x}\right),
$$

$$
\|\delta x\| \leq \|A^{-1}\| \left(\|\delta b\| + \|\delta A\| \|\hat{x}\|\right),
$$

and

$$
\frac{\|\delta x\|}{\|\hat{x}\|} \leq \kappa(A)\left(\frac{\|\delta A\|}{\|A\|} + \frac{\|\delta b\|}{\|A\| \|\hat{x}\|}\right).
$$

(b) Show that

$$
\|\hat{b}\| \leq \left(\|A\| + \|\delta A\|\right)\|\hat{x}\|,
$$

and therefore

$$
\frac{1}{\|\hat{x}\|} \leq \frac{\left(\|A\| + \|\delta A\|\right)}{\|\hat{b}\|}.
$$

(c) Combine the results of (a) and (b) to finish the proof.

□

rcise 2.3.13 Prove Theorem 2.3.9 by combining elements of the proofs of Theorems 2.2.4 and 2.3.6. □

2.4 A POSTERIORI ERROR ANALYSIS USING THE RESIDUAL

So far we have been studying the sensitivity of the solution of $Ax = b$ to perturbations in the data. Now we switch to a related question. If we solve the system $Ax = b$ using Gaussian elimination or some other method, how do the roundoff (and other) errors affect the accuracy of the computed solution? Before getting into the details of floating-point arithmetic and roundoff error analysis, let us pause to make note of a simple error test that uses the residual and the condition number.

Suppose we have computed a solution of the system $Ax = b$ by any method whatsoever. Call this computed solution $\hat{x}$. Regardless of how we obtained $\hat{x}$, we can easily compute the *residual* $\hat{r} = b - A\hat{x}$, which gives a measure of how well $\hat{x}$ fits the equations. The fit is good if $\hat{r}$ is small (more precisely, if $\|\hat{r}\|/\|b\|$ is small); $\hat{r} = 0$ if and only if $\hat{x}$ is the true solution of $Ax = b$.

A tiny residual is reassuring. It guarantees that $\hat{x}$ is the solution of a system that is close to $Ax = b$: If we define $\delta b = -\hat{r}$, then $\hat{x}$ is the exact solution of the system $A\hat{x} = b + \delta b$, which is just a slight perturbation of the system $Ax = b$. Unfortunately this does not guarantee that $\hat{x}$ is close to x; we have to take the condition number of A into account. Writing $\hat{x} = x + \delta x$, as in Section 2.2, we see that Theorem 2.2.4 gives us an upper bound on the relative error $\|\delta x\|/\|x\|$. Restating Theorem 2.2.4 as a statement about residuals, we have the following result.

Theorem 2.4.1 *Let A be nonsingular, let $b \neq 0$, and let $\hat{x}$ be an approximation to the solution of $Ax = b$. (In other words, let $\hat{x}$ be any vector whatsoever.) Let $\hat{r} = b - A\hat{x}$. Then*

$$\frac{\|x - \hat{x}\|}{\|x\|} \leq \kappa(A)\frac{\|\hat{r}\|}{\|b\|}.$$

From this simple theorem we see that if the residual is tiny and A is well conditioned, then $\hat{x}$ is an extremely accurate approximation to x. The cost of calculating $\hat{r}$ is only about $2n^2$ flops if A is full and even less if A is sparse. If we also have an efficient means of calculating or estimating the condition number (as discussed at the end of Section 2.2), then we may be able to use this theorem to guarantee the accuracy of our computed solution.

Theorem 2.4.1 is an example of an *a posteriori* error bound. It is a bound we obtain after we have solved the problem (i.e. computed $\hat{x}$). In contrast, an *a priori* error analysis attempts to determine, before solving a problem, whether the method is going to produce an accurate solution. *A posteriori* analyses are generally easier and more informative than *a priori* analyses because they can make use of the computed solution and any other information that was obtained in the course of the computation. In Section 2.7 we will develop an *a priori* analysis of the accuracy of the solution of a linear system by Gaussian elimination in the presence of roundoff errors.

Example 2.4.2 Consider the linear system

$$
\begin{bmatrix}
2 & -1 & -1 & 0 \\
-1 & 1.5 & 0 & -.5 \\
-1 & 0 & 1.7 & -.2 \\
0 & -.5 & -.2 & 1.7
\end{bmatrix}
\begin{bmatrix}
x_1 \\
x_2 \\
x_3 \\
x_4
\end{bmatrix}
=
\begin{bmatrix}
0 \\
0 \\
3 \\
0
\end{bmatrix}
$$

from Example 1.2.6. The components of the solution x are the nodal voltages of the circuit shown in that example. Entering the matrix A and vector b into MATLAB and using the commands

```
format long
xhat = A\b
```

we find that

$$
\hat{x} =
\begin{bmatrix}
3.06382978723404 \\
2.42553191489362 \\
3.70212765957447 \\
1.14893617021277
\end{bmatrix}.
$$

We write $\hat{x}$ to indicate that this result is not exactly the true solution, since roundoff errors occurred during the calculation. To check the accuracy of the result, we perform the additional operations

```
rhat = b - A*xhat;
nr = norm(rhat)
ca = cond(A)
errbound = ca*nr/norm(b)
```

to find that $\|\hat{r}\|_2 = 1.05 \times 10^{-15}$, $\kappa_2(A) = 12.7$, and

$$
\frac{\|x - \hat{x}\|_2}{\|x\|_2} \leq \frac{\kappa_2(A)\|\hat{r}\|_2}{\|b\|_2} = 4.45 \times 10^{-15}.
$$

This shows that our computed solution is very accurate. Roughly speaking, its entries agree with the correct nodal voltages to at least fourteen decimal places.

This assumes, of course, that the data (A and b) are correct. However, since A is well-conditioned, we know from results in Section 2.3 that slight errors in A and b will perturb the solution only slightly. See Example 2.3.11. □

Exercise 2.4.3 Rework Exercise 1.2.17. Compute the solution using MATLAB. Compute the residual and condition number, and conclude that the computed solution is extremely accurate. □

Exercise 2.4.4 If the matrix is large, we prefer to estimate the condition number instead of trying to compute it exactly. Using MATLAB, solve a system $Ax = b$, where A is a large discrete Laplacian operator:

```
m = 50;   % Make m larger or smaller, as needed.
A = delsq(numgrid('N',m));
n = size(A,1)
b = ones(n,1); % ``all ones'' right-hand side.
```

```
xhat = A\b;
ca = condest(A)
```

Compute the norm of the residual, and use `condest` and Theorem 2.4.1 to estimate the error. Since `condest` estimates the 1-condition number $\kappa_1(A)$, use 1-norms in your estimate. ☐

2.5 ROUNDOFF ERRORS; BACKWARD STABILITY

This section begins with a discussion of floating-point arithmetic and the effects of roundoff errors. The accuracy of arithmetic operations in the presence of errors is studied. It is found that a sudden loss of (relative) accuracy can occur when a cancellation occurs in the addition or subtraction of two numbers. Because of the threat of cancellation, it is impossible to analyze the errors in a complicated algorithm like Gaussian elimination in a forward or direct way. Therefore a more modest approach, backward error analysis, is introduced.

Floating-point Arithmetic

Most scientific computations are performed on computers using *floating-point arith-metic*, which is the computer version of scientific notation. We will not define the term but instead give some examples. The number 1.23456×10^7 is a six-digit decimal floating-point number. It has a *significand* or *mantissa* 1.23456 and an *exponent* 7. It is called a decimal number because the number base is 10 (and of course the significand is interpreted as a base-10 number). Because the number has an exponent, the decimal point can "float" rather than remaining in a fixed position. For example, $1.23456 \times 10^3 = 1234.56$, $1.23456 \times 10^8 = 123456000.$, and $1.23456 \times 10^{-3} = .00123456$. The advantage of the floating-point representation is that it allows very large and very small numbers to be represented accurately. Other examples of floating-point numbers are 6.542×10^{36}, a large four-digit decimal number, and -7.1236×10^{-42}, a small, five-digit decimal number. A floating-point number is said to be *normalized* if the first digit of its significand is nonzero. Thus the examples we have looked at so far are normalized, whereas 0.987×10^6 and $-0.0012346 \times 10^{-4}$ are not normalized. With few exceptions, nonzero floating-point numbers are stored in normalized form.

Each significand and each exponent takes up space in the computer's memory. In a floating-point number system a fixed amount of space is allocated for each number, so there are limitations to the precision with which numbers can be represented. For example, if we are using a decimal (base-10) machine that stores four decimal digits, we cannot represent the number 1.1112×10^5 exactly. We must approximate it by the nearest floating-point number, which is 1.111×10^5. Also, each arithmetic operation will be accompanied by a roundoff error. Suppose, for example, we multiply the floating point number 1.111×10^1 by 1.111×10^2. Multiplying the significands and adding the exponents, we obtain the exact result 1.234321×10^3. The significand has more digits than we can store, so we must approximate it by the nearest floating-point number 1.234×10^2. This is where the *roundoff error* occurs.

Since each number's exponent must also be stored within some prescribed amount of space, there is also a limit to the size of exponents that can be represented. Numbers that are past a certain threshold cannot be represented in the floating-point system. If a computation results in a number that is too big to be represented, an *overflow* is said to have occurred. For example, if our system can represent numbers up to about 10^{99}, an overflow will occur if we try to multiply, say, 10^{55} by 10^{50}. This is obviously something to be avoided. If a number that is nonzero but too small to be represented is computed, an *underflow* results. For example, if our number system allows storage of numbers as small as about 10^{-99} but no smaller, an underflow will occur if we multiply 10^{-55} by 10^{-50}. A common remedy for underflow is to set the result to zero. This action is usually, but not always, harmless.

Fortunately overflows and underflows can normally be avoided. One common computation that entails the risk of over/underflow is the computation of the Euclidean norm

$$\| x \|_2 = \left(\sum_{k=1}^{n} x_k^2 \right)^{1/2}.$$

First we will look at an example of harmful underflow; then we will consider a remedy.

Example 2.5.1 Consider the calculation of $\| x \|_2$, where

$$x = [10^{-49}, 10^{-50}, 10^{-50}, \ldots, 10^{-50}]^T \in \mathbb{R}^{101},$$

on a computer that underflows at 10^{-99}. Any number smaller than that is set to zero. Now $x_1^2 = 10^{-98}$, and $x_2^2 = \cdots = x_{101}^2 = 10^{-100}$, each of which is set to zero by the machine. Thus the computed norm is $\sqrt{x_1^2} = 10^{-49}$. This is in error by more than 40 percent, as the true value of the norm is $\sqrt{2 \times 10^{-98}} \approx 1.414 \times 10^{-49}$.   □

Problems with overflows and underflows in the calculation of $\| x \|_2$ can be avoided by the following simple procedure: Let $\beta = \| x \|_\infty = \max_{1 \le i \le n} | x_i |$. If $\beta = 0$, then $\| x \|_2 = 0$. Otherwise let $\hat{x} = (1/\beta)x$. Then $\| x \|_2 = \beta \| \hat{x} \|_2 = \beta \sqrt{\hat{x}_1^2 + \cdots + \hat{x}_n^2}$. This scaling procedure eliminates any practical possibility of overflow because $| \hat{x}_i | \le 1$ for all i. Underflows are still possible, but they occur only when some terms are so much smaller than the largest term $(= 1)$ that the relative error incurred by ignoring them is negligible. Thus these underflows are harmless and can safely be set to zero. We summarize the procedure as follows.

$$
\begin{aligned}
&\beta \leftarrow \max_{1 \le i \le n} | x_i | \\
&\text{if } \beta = 0 \\
&\quad \left[\ \| x \|_2 \leftarrow 0 \right. \\
&\text{else} \\
&\quad \left[\begin{array}{l} \hat{x} \leftarrow (1/\beta)x \\ \| x \|_2 \leftarrow \beta \sqrt{\hat{x}_1^2 + \cdots + \hat{x}_n^2} \end{array} \right.
\end{aligned}
\qquad (2.5.2)
$$

rcise 2.5.3 Use this scaling procedure to calculate the norm of the vector x of Example 2.5.1. □

All of our examples use base-10 arithmetic because that is what we humans are used to. Hand-held calculators aside, most computers do not use a base-10 representation; a power of two is more convenient architecturally. In the early days of computing, each manufacturer made its own decisions about the characteristics of its computers' floating point arithmetic (e.g. number base, how many digits are allocated to the significand and how many to the exponent). Thus there were many different floating point systems in use, some better than others. Since the adoption of the IEEE floating-point standard (ANSI/IEEE Standard 754-1985) in 1985, the situation has improved dramatically. Nowadays all of the inexpensive, widely-available microprocessors conform to this standard.

Here we will outline some of the basic properties of IEEE arithmetic. For more details see [15], [52], or [66]. The IEEE standard supports both single precision and double precision floating-point numbers. In both cases the number base is two. Single precision numbers are stored in 32-bit words, of which 24 bits are used for the significand and the other 8 for the exponent. Since $2^{24} \approx 10^7$, 24 bits are enough to store approximately seven decimal digits worth of information. In other words, single precision numbers can be accurate to about seven decimal places. With eight bits for the exponent it is possible to represent exponents from about -128 to $+128$. Since these are exponents of the base 2 (not 10), the range of numbers that can be represented is from about $2^{-128} \approx 10^{-38}$ to $2^{128} \approx 10^{38}$.

Double precision IEEE floating-point numbers are allocated 64 bits, of which 53 are used for the significand and 11 for the exponent. Since $2^{53} \approx 10^{16}$, double precision numbers can be accurate to about 16 decimal places. Eleven bits of exponent allow the representation of numbers from about $2^{-1024} \approx 10^{-308}$ to $2^{1024} \approx 10^{308}$, an extremely wide range.

The IEEE floating-point standard also includes useful features for automatic handling of exceptional events such as overflow, underflow, and division by zero. See [15], [52], or [66].[12]

Computing in the Presence of Errors

Our task is to assess the cumulative effects of roundoff errors on our calculations. To this end we introduce the notation $\mathrm{fl}(C)$ to denote the floating-point result of some computation C. For example, if we multiply x by y, the result calculated by the computer will be denoted $\mathrm{fl}(xy)$. We can apply this notation to more complicated expressions as well, as long as the order in which the computations are to be performed

---

[12] IEEE 754-1985 has been superseded by IEEE 754-2008. The single and double precision floating point numbers are the same as before, but they are now known as binary32 and binary64, respectively. A quadruple precision standard known as binary128 has been added. Standards for decimal floating point arithmetic have also been added.

is clear. For example, $\text{fl}(\sum_{i=1}^{n} x_i y_i)$ is a perfectly acceptable expression, as long as we have agreed on the order in which the terms are to be added.

Denoting the exact result of a computation also by the letter C, we have $\text{fl}(C) = C + e$, where e is the *absolute error* of the computation. A more useful measure is the *relative error* $\epsilon = e/C$, provided that $C \neq 0$. You can easily verify that the relative error ϵ satisfies

$$\text{fl}(C) = C(1 + \epsilon). \tag{2.5.4}$$

Example 2.5.5 One example suffices to show that the relative error is more meaningful than the absolute error. Suppose we perform a computation on a 7-digit decimal machine and get the result $\text{fl}(C) = 9.876572 \times 10^{17}$, whereas the correct value is $C = 9.8765432 \times 10^{17}$. The computed value is clearly a good approximation to the true value, but the absolute error is $e = C - \text{fl}(C) = 2.88 \times 10^{12}$, which looks large unless it is compared with C. In the relative error the magnitude of C is automatically taken into account: $\epsilon = e/C = 2.91 \times 10^{-6}$. Now consider a different computation in which $\text{fl}(C) = 9.876572 \times 10^{-15}$ and $C = 9.8765432 \times 10^{-15}$. Now $e = 2.88 \times 10^{-20}$, which appears extremely small until it is compared with C. In contrast $\epsilon = e/C = 2.91 \times 10^{-6}$, the same as before. That the relative error is less than 10^{-5} is reflected in the fact that C and $\text{fl}(C)$ agree in their first five decimal places. Because of this agreement, the difference between C and $\text{fl}(C)$ is about five powers of 10 smaller that C. That is, the relative error is approximately 10^{-5}. $\quad\square$

Another point worth mentioning is that the absolute error has the same units as C. If C is measured in volts (seconds, meters), then e is also measured in volts (seconds, meters). By contrast the relative error is a dimensionless number; it has no units.

In this text we will normally measure errors in relative terms. The relative error appears in various guises. For example, the expressions $\| \delta x \| / \| x \|$, $\| \delta b \| / \| b \|$, and $\| \delta A \| / \| A \|$, with which we worked in Section 2.2, are expressions of relative error. Also we have already observed that statements about the number of correct digits are actually vague statements about the relative error. Finally, statements about percent error are also statements about the relative error: percent error $= |$ relative error $| \times 100$.

In our analysis we will not assume that our machine satisfies the IEEE standard. Instead we will assume an ideal computer that performs each operation exactly and then rounds the result to the nearest floating point number. The IEEE standard satisfies this assumption, so long as no overflows or underflows occur.[13] Our analysis will ignore the possibility of overflow or underflow. Each individual computation produces a roundoff error that is tiny in the relative sense. We will define the *unit roundoff* u to be the largest relative error that can occur in a rounding operation. The value of u depends on the system. In a system with a significand of s decimal digits, the value of u will be around 10^{-s}, since the rounded value and the original value agree to s decimal places. A careful analysis (Exercise 2.5.12) shows that $u = \frac{1}{2} \times 10^{1-s}$. For IEEE single precision numbers $u = 2^{-24} \approx 6 \times 10^{-8}$, and

---

[13]This assumes that the default "round to nearest" rounding mode is being used. The IEEE standard also supports several other rounding modes. In all modes (2.5.6) continues to hold if we replace u by $2u$.

for double precision numbers $u = 2^{-53} \approx 10^{-16}$. Using the form (2.5.4), our ideal floating-point operations satisfy

$$
\begin{array}{rcll}
\text{fl}(x \pm y) & = & (x \pm y)(1 + \epsilon_1) & |\epsilon_1| \le u \\
\text{fl}(xy) & = & (xy)(1 + \epsilon_2) & |\epsilon_2| \le u \\
\text{fl}(x/y) & = & (x/y)(1 + \epsilon_3) & |\epsilon_3| \le u.
\end{array}
\qquad (2.5.6)
$$

Our analysis will take the results (2.5.6) as a starting point. These can give one a false sense of security. Since the error made in each individual operation is small, one might think that a great many operations would have to be made before significant errors could accumulate. Unfortunately this turns out to be false.

To get a realistic idea of what can happen, we need to take account of the fact that the operands x and y normally have some error in them already. Instead of the correct values, x and y, the computer works with polluted values $\hat{x} = x(1 + \epsilon_1)$ and $\hat{y} = y(1 + \epsilon_2)$. Instead of calculating xy or $\text{fl}(xy)$, the computer calculates $\text{fl}(\hat{x}\hat{y})$. We need to compare $\text{fl}(\hat{x}\hat{y})$ with xy. We would like to be able to say that if $|\epsilon_1| \ll 1$ and $|\epsilon_2| \ll 1$, then $\text{fl}(\hat{x}\hat{y}) = xy(1 + \epsilon)$, where $|\epsilon| \ll 1$. It turns out that such a result does hold for multiplication, and there is an analogous result for division. Unfortunately, addition and subtraction do not always behave so well.

Let us begin with the well behaved operations. The computer multiplies $\hat{x}$ by $\hat{y}$ to get $\text{fl}(\hat{x}\hat{y}) = \hat{x}\hat{y}(1 + \epsilon_3)$, where the roundoff error ϵ_3 satisfies $|\epsilon_3| \le u \ll 1$ by (2.5.6). Thus

$$
\begin{array}{rcl}
\text{fl}(\hat{x}\hat{y}) & = & x(1 + \epsilon_1)y(1 + \epsilon_2)(1 + \epsilon_3) \\
& = & xy(1 + \epsilon_1 + \epsilon_2 + \epsilon_3 + \epsilon_1\epsilon_2 + \epsilon_1\epsilon_3 + \epsilon_2\epsilon_3 + \epsilon_1\epsilon_2\epsilon_3) \\
& = & xy(1 + \epsilon),
\end{array}
$$

where $\epsilon = \epsilon_1 + \epsilon_2 + \epsilon_3 + \epsilon_1\epsilon_2 + \epsilon_1\epsilon_3 + \epsilon_2\epsilon_3 + \epsilon_1\epsilon_2\epsilon_3$. The terms involving products of two or more ϵ_i are negligible because all ϵ_i are small. Thus $\epsilon \approx \epsilon_1 + \epsilon_2 + \epsilon_3$. Since $|\epsilon_1| \ll 1$, $|\epsilon_2| \ll 1$, and $|\epsilon_3| \ll 1$, it also holds that $|\epsilon| \ll 1$. We conclude that multiplication is well behaved in the presence of errors in the operands.

In order to analyze division, we begin by recalling from the theory of geometric series that

$$
\frac{1}{1 + \epsilon_2} = 1 - \epsilon_2 + \epsilon_2^2 - \epsilon_2^3 + \cdots
$$

Since $|\epsilon_2| \ll 1$, the approximation $1/(1 + \epsilon_2) \approx 1 - \epsilon_2$, obtained by ignoring quadratic and higher terms, is good. Thus

$$
\begin{array}{rcl}
\text{fl}\left(\dfrac{\hat{x}}{\hat{y}}\right) & = & \dfrac{x(1 + \epsilon_1)}{y(1 + \epsilon_2)}(1 + \epsilon_3) \\[2ex]
& \approx & \dfrac{x}{y}(1 + \epsilon_1)(1 - \epsilon_2)(1 + \epsilon_3) \\[2ex]
& \approx & \dfrac{x}{y}(1 + \epsilon_1 - \epsilon_2 + \epsilon_3).
\end{array}
$$

Therefore $\text{fl}(\hat{x}/\hat{y}) = (x/y)(1 + \epsilon)$, where $\epsilon \approx \epsilon_1 - \epsilon_2 + \epsilon_3$. We conclude that division is well behaved in the presence of errors.

Our analysis of addition will be a little bit different. We know that the difference between $\mathrm{fl}(\hat{x} + \hat{y})$ and $\hat{x} + \hat{y}$ is relatively small, so we will simply compare $\hat{x} + \hat{y}$ with $x + y$. (We could have done the same in our analyses of multiplication and division.) This simplifies the analysis slightly and has the advantage of making it clear that any serious damage that is done is attributable not to the roundoff error from the current operation but to the errors that had been made previously.

$$\begin{aligned} \hat{x} + \hat{y} &= x(1 + \epsilon_1) + y(1 + \epsilon_2) \\ &= (x + y) + x\epsilon_1 + y\epsilon_2 \\ &= (x + y)\left(1 + \frac{x}{x + y}\epsilon_1 + \frac{y}{x + y}\epsilon_2\right). \end{aligned}$$

Thus $\hat{x} + \hat{y} = (x + y)(1 + \epsilon)$, where

$$\epsilon = \frac{x}{x + y}\epsilon_1 + \frac{y}{x + y}\epsilon_2.$$

Given that $|\epsilon_1| \ll 1$ and $|\epsilon_2| \ll 1$, we can say that $|\epsilon| \ll 1$ provided that neither x nor y is large compared with $x + y$. If, on the other hand, x or y is large compared with $x + y$, then ϵ can and probably will be large. That is, the computed result is probably inaccurate. This occurs when (and only when) x and y are almost exactly opposites of each other, so that they nearly cancel one another out when they are added.

An identical analysis holds for subtraction.

Exercise 2.5.7 Show that if $\hat{x} = x(1 + \epsilon_1)$ and $\hat{y} = y(1 + \epsilon_2)$, where $|\epsilon_1| \ll 1$ and $|\epsilon_2| \ll 1$, then $\hat{x} - \hat{y} = (x - y)(1 + \epsilon)$, where $|\epsilon| \ll 1$ unless x or y is much larger than $x - y$.   □

If x and y are nearly equal, so that $x - y$ is much smaller than both x and y, then the computed result $\hat{x} - \hat{y}$ can and probably will be inaccurate.

We conclude that both addition and subtraction are well behaved in the presence of errors, except when the operands nearly cancel one another out. Because cancellation generally signals a sudden loss of accuracy, it is sometimes called *catastrophic cancellation*.

Example 2.5.8 It is easy to see intuitively how cancellation leads to inaccurate results. Suppose an eight-digit decimal machine is to calculate $x - y$, where $x = 3.1415927\ldots \times 10^0$ and $y = 3.1415916\ldots \times 10^0$. Due to errors in the computation of x and y, the numbers that are actually stored in the computer's memory are $\hat{x} = 3.1415929\ldots \times 10^0$ and $\hat{y} = 3.1415914\ldots \times 10^0$. Clearly these numbers are excellent approximations to x and y; they are correct in the first seven decimal places. That is, the relative errors ϵ_1 and ϵ_2 are of magnitude about 10^{-7}. Since $\hat{x}$ and $\hat{y}$ are virtually equal, all but one of the seven accurate digits are canceled off when $\hat{x} - \hat{y}$ is formed: $\hat{x} - \hat{y} = 0.0000015 \times 10^0 = 1.5000000 \times 10^{-6}$. In the normalized result only the first digit is correct. The second digit is inaccurate as are all of the zeros that follow it. Thus the computed result is a poor approximation to the true result $x + y = 1.1 \ldots \times 10^{-6}$. The relative error is about 36 percent.   □

This example demonstrates a relatively severe case of cancellation. In fact a whole range of severities is possible. Suppose, for example, two numbers are accurate to seven decimal places and they agree with each other in the first three places. Then when their difference is taken, three accurate digits will be lost, and the result will be accurate to four decimal places.

We have demonstrated not only that cancellation can cause a sudden loss of relative accuracy, but also that it is the only mechanism by which a sudden loss of accuracy can occur. The only other way an inaccurate result can occur is by gradual accumulation of small errors. Although it is possible to concoct examples where this happens, it is seldom a problem in practice. The small errors that occur are just as likely to cancel each other out, at least in part, as they are to reinforce one another, so they tend to accumulate very slowly. Thus as a practical matter we can say that if a computation has gone bad, there must have been at least one cancellation at some crucial point. In other words, if no cancellations occur during a computation (and the original operands were accurate), the result will generally be accurate.

Unfortunately it is usually difficult to verify that no cancellation will occur in a given computation, and this makes it hard to prove that roundoff errors will not spoil the computation. The first attempts at error analysis took the *forward* or *direct* approach, in which one works through the algorithm, attempting to bound the error in each intermediate result. In the end one gets a bound for the error in the final result. This approach usually fails, because each time an addition or subtraction is performed, one must somehow prove either that cancellation cannot occur or that a cancellation at that point will not cause any damage in the subsequent computations. This is usually impossible.

Backward Error Analysis

Because of the threat of sudden loss of accuracy through cancellation, the pioneers of scientific computing were quite pessimistic about the possible effects of roundoff errors on their computations. It was feared that any attempt to solve, say, a system of 50 equations in 50 unknowns would yield an inaccurate result. The early attempts to solve systems of linear equations on computers turned out generally better than expected, although disasters did sometimes occur. The issues were not well understood until a new approach, called *backward error analysis*, was developed. The new approach does not attempt to bound the error in the result directly. Instead it pushes the effects of the errors back onto the operands.

Suppose, for example, we are given three floating-point numbers x, y, and z, and we wish to calculate $C = (x + y) + z$. The computer actually calculates $\hat{C} = \mathrm{fl}(\mathrm{fl}(x + y) + z)$. Even if the operands are exact, we cannot assert that the relative error in $\hat{C}$ is small: $\mathrm{fl}(x + y)$ is (probably) slightly in error, so there can be large relative error in $\hat{C}$ if cancellation takes place when $\mathrm{fl}(x + y)$ is added to z.

However, there is something else we can do. We have

$$\hat{C} = [(x + y)(1 + \epsilon_1) + z] (1 + \epsilon_2),$$

where $|\epsilon_1|$, $|\epsilon_2| \leq u \ll 1$. Define ϵ_3 by $(1 + \epsilon_3) = (1 + \epsilon_1)(1 + \epsilon_2)$, so that $|\epsilon_3| \approx |\epsilon_1 + \epsilon_2| \ll 1$. Then $\hat{C} = (x+y)(1+\epsilon_3)+z(1+\epsilon_2)$. Defining $\overline{x} = x(1+\epsilon_3)$, $\overline{y} = y(1 + \epsilon_3)$, and $\overline{z} = z(1 + \epsilon_2)$, we have

$$\hat{C} = (\overline{x} + \overline{y}) + \overline{z}.$$

Notice that $\overline{x}$, $\overline{y}$, and $\overline{z}$ are extremely close to x, y, and z, respectively. This shows that $\hat{C}$ is the exact result of performing the computation $(x + y) + z$ using the slightly perturbed data $\overline{x}$, $\overline{y}$, and $\overline{z}$. The errors have been shoved back onto the operands. The same can be done with subtraction, multiplication, division, and (with a bit of ingenuity) longer computations.

In general suppose we wish to analyze some long computation $C(z_1, \ldots, z_m)$ involving m operands or input data $z_1, \ldots, z_m$. Instead of trying to show directly that $\mathrm{fl}(C(z_1, \ldots, z_m))$ is close to $C(z_1, \ldots, z_m)$, a backward error analysis attempts to show that $\mathrm{fl}(C(z_1, \ldots, z_m))$ is the exact result of performing the computation with slightly perturbed input data; that is,

$$\mathrm{fl}(C(z_1, \ldots, z_m)) = C(\overline{z}_1, \ldots, \overline{z}_m),$$

where $\overline{z}_1, \ldots, \overline{z}_m$ are extremely close to $z_1, \ldots, z_m$. By *extremely close* we usually mean that the error is a modest multiple of the unit roundoff u. If such a result holds, the computation is said to be *backward stable*.

Of course the analysis does not end here. The backward error analysis has to be combined with a *sensitivity analysis* of the problem. If (1) the computation is backward stable, and (2) we can show that small perturbations in the operands lead to small perturbations in the results, then we can conclude that our computed result is accurate.

For example, if our problem is to solve a nonsingular linear system $Ax = b$, then the operands or inputs are the entries of A and b, and the output is x. We obtain x from A and b by some specified computation $x = C(A, b)$, for example, Gaussian elimination with complete pivoting. In this context we normally use the word *algorithm* instead of *computation*. If we perform the computations in exact arithmetic, we get x exactly, but if we use floating-point arithmetic, we obtain an approximate solution $\hat{x} = \mathrm{fl}(C(A, b))$. In this context the algorithm is backward stable if there exist $\hat{A}$ and $\hat{b}$ that are extremely close to A and b, respectively, such that $\hat{x} = C(\hat{A}, \hat{b})$, that is, $\hat{A}\hat{x} = \hat{b}$ exactly. Another way to say this is that $\hat{x}$ satisfies $(A+\delta A)\hat{x} = b+\delta b$ exactly for some δA and δb such that $\| \delta A \|/\| A \|$ and $\| \delta b \|/\| b \|$ are tiny, that is, a small multiple of u.

As we well know by now, backward stability does not imply that $\hat{x}$ is close to x. The sensitivity of the problem is given by the condition number of A. If the algorithm is backward stable, and the coefficient matrix is well conditioned, then the computed solution is accurate. If, on the other hand, the matrix is ill conditioned, the solution may well be inaccurate, even though the algorithm is backward stable.

The backward approach to error analysis separates clearly the properties of the problem (e.g. solve $Ax = b$) from the properties of the algorithm (e.g. Gaussian elimination with complete pivoting). The sensitivity analysis pertains to the problem, and the backward error analysis pertains to the algorithm. We say that the *problem* is *well conditioned* if small changes in the input lead to small changes in the results. Otherwise it is *ill conditioned*. An *algorithm* is backward stable if it returns answers that exactly solve some slightly perturbed problem. It follows that a backward stable algorithm will be able to solve well-conditioned problems accurately. We will adopt the attitude that it is unreasonable to expect any algorithm to solve ill-conditioned problems accurately. Therefore we will judge an algorithm to be satisfactory if it is backward stable.

The backward approach to error analysis succeeds because it is much less ambitious than the forward approach. The latter attempts to prove that a given algorithm always produces an accurate result, regardless of the sensitivity of the problem. This is usually impossible.

Small Residual Implies Backward Stability

If we say that an algorithm is backward stable (with no further qualification), we mean that it performs in a backward stable manner on all possible sets of input data. However, the term can also be used in a much broader sense. If we use an algorithm, say Gaussian elimination, to solve a particular problem $Ax = b$ (for some specific choice of A and b), we will say that the algorithm is backward stable *on that particular problem* if it produces an $\hat{x}$ that is the exact solution of a nearby problem $(A + \delta A)\hat{x} = b + \delta b$.

For the linear system problem (and many other problems) there is a simple *a posteriori* method checking the backward stability of a computation: check the residual. The problem is to solve $Ax = b$. Whatever method we use to get a solution $\hat{x}$, we can easily calculate the residual $\hat{r} = b - A\hat{x}$. As we already noted in Section 2.4, $\hat{x}$ is the exact solution of $A\hat{x} = b + \delta b$, where $\delta b = -\hat{r}$. If $\|\delta b\|/\|b\|$ is tiny, then $\hat{x}$ is indeed the solution of a nearby system. Thus the algorithm is backward stable on this problem. To summarize: a tiny residual implies backward stability.

The following exercise draws essentially the same conclusion by a different approach, in which the residual is associated with a perturbation in A instead of b.

rcise 2.5.9 Let $\hat{x}$ be an approximation to the solution of $Ax = b$, and let $\hat{r} = b - A\hat{x}$. Define $\delta A \in \mathbb{R}^{n \times n}$ by $\delta A = \alpha \hat{r} \hat{x}^T$, where $\alpha = \|\hat{x}\|_2^{-2}$.

(a) Show that $\hat{x}$ is the exact solution of $(A + \delta A)\hat{x} = b$.

(b) Show that $\|\delta A\|_2 = \|\hat{r}\|_2/\|\hat{x}\|_2$ and

$$\frac{\|\delta A\|_2}{\|A\|_2} = \frac{\|\hat{r}\|_2}{\|A\|_2 \|\hat{x}\|_2}.$$

$\square$

Thus if $\|\hat{r}\|_2$ is tiny relative to $\|A\|_2\|\hat{x}\|_2$, then the algorithm (whichever algorithm was used) is backward stable for this problem.

Additional Exercises

Exercise 2.5.10 Learn more about your computer's arithmetic by running the following MATLAB programs.

(a) What do you learn from running the following program?

```
a = 1;
u = 1;
b = a + u;
while b ~= a
    u = .5*u;
    b = a + u;
end
u
```

(b) What does this one tell you?

```
a = 1;
while a ~= Inf
    a= 10*a
end
```

To get a more precise result, replace the 10 by a 2. You might find `help inf` informative.

(c) What does this one tell you?

```
a = 1;
while a ~= 0
    a= .1*a
end
```

The outcome of this one is probably different from what you expected, based on the information given in the text. The IEEE standard allows *gradual underflow* through the use of subnormal numbers, once the minimum exponent is reached.

(d) What kind of arithmetic does MATLAB appear to be using? To learn more about your computer (as used by MATLAB) try `help computer`, `help inf`, and `help nan`, for example.

□

Exercise 2.5.11 Write Fortran or C programs that perform the same functions as the programs from Exercise 2.5.10. Make both single and double precision versions and try them out. □

rcise 2.5.12

(a) Show that in a base-β floating-point number system the largest relative gap between two consecutive normalized numbers occurs between

$$1.000\ldots000 \times \beta^t \quad \text{and} \quad 1.000\ldots001 \times \beta^t.$$

(The value of the exponent t is irrelevant.) Thus the largest relative error (the unit roundoff) occurs when one tries to represent the number that lies half way between these two.

(b) Show that the unit roundoff is $\frac{1}{2} \times \beta^{1-s}$, where β is the number base, and s is the number of base-β digits in the significand.

$\square$

2.6 PROPAGATION OF ROUNDOFF ERRORS IN GAUSSIAN ELIMINATION

Now that we know something about floating-point arithmetic and roundoff errors, we are ready to analyze the effects of roundoff errors on Gaussian elimination. Our formal tool will be backward error analysis, but we defer that to Section 2.7. In this section we will see that even without backward error analysis we can get some good insights through the study of well-chosen examples.

Gaussian Elimination with Ill-Conditioned Matrices

Our studies of the sensitivity of linear systems have given us ample reason to believe that there is no point in trying to solve severely ill-conditioned systems. Further insight can be gained by taking a heuristic look at what happens when one tries to solve an ill-conditioned system by Gaussian elimination with partial pivoting. We will assume that the rows and columns of the coefficient matrix are not out of scale.

When we do row operations, we take linear combinations of rows in such a way that zeros are deliberately created. Since the rows of an ill-conditioned matrix are nearly linearly dependent, there is the possibility of entire rows being made almost exactly zero by row operations. This possibility is encouraged by the progressive introduction of zeros into the array. Let us say that a row is *bad* if it is nearly a linear combination of previous rows. Suppose the kth row of A is the first bad row. After $k - 1$ steps of Gaussian elimination we will have subtracted multiples of the first $k - 1$ rows from the kth row in such a way that there are now zeros in the first $k - 1$ positions. If the kth row were exactly a linear combination of the previous rows (and exact arithmetic were used), the entire kth row would now be zero. (Why?) Since it is only approximately a linear combination of the previous rows, it will still contain nonzero entries, but these entries will typically be tiny. They are not only tiny but also inaccurate, because they became tiny through cancellation, as multiples of the earlier rows were subtracted from row k.

One of these tiny, inaccurate entries is the potential pivot in the (k, k) position. Because it is small, the kth row will be interchanged with a lower row that has a larger entry in its kth position, if such a row exists. In this way the bad rows get shifted downward. Eventually a step will be reached at which only bad rows remain. At this point all choices of pivot are tiny and inaccurate. Although the presence of small, inaccurate numbers is not necessarily disastrous to the computation, the use of one as a pivot must be avoided if possible. In the present scenario we are forced to use a tiny, inaccurate pivot. This is used as a divisor in the computation of not-so-small, inaccurate multipliers, whose error pollutes all subsequent rows. The pivots are also used as divisors in the last step of the back-substitution process. Each component of the computed solution is a quotient whose divisor is a pivot. We cannot expect these quotients to be accurate if the divisors are not.

The above analysis is obviously heuristic and is not claimed to be universally valid. A different flavor of ill conditioning is exhibited by the *Kahan matrix* [52].

Example 2.6.1 Consider the ill-conditioned matrix

$$A = \left[\begin{array}{cc} 1000 & 999 \\ 999 & 998 \end{array} \right],$$

which we have discussed previously. Since the rows are nearly linearly dependent, when a zero is created in the $(2, 1)$ position, the entry in the $(2, 2)$ position should become nearly zero as well. Indeed the multiplier is $l_{21} = .999$, and the $(2, 2)$ entry becomes

$$998 - (.999)(999) = 998 - 998.001 = -.001.$$

This is indeed small, and the result was obtained by severe cancellation. There is no error in the result, because it was computed by exact arithmetic. Consider what happens when five-digit decimal floating-point arithmetic is used. The computation yields

$$998.00 - (.99900)(999.00) = 998.00 - 998.00 = 0.$$

The matrix appears to be singular!   □

This example might remind you of a remark that was made in Chapter 1. Not only can a nonsingular matrix appear singular (as just happened); the reverse can occur as well and is actually a much more common occurrence. We remarked that if a Gaussian elimination program attempts to calculate the LU decomposition of a singular matrix, it almost certainly will not recognize that the matrix is singular: certain entries that should become zero in the course of the calculation will turn out to be nonzero because of roundoff errors. Thus in numerical practice it is impossible to distinguish between ill-conditioned matrices and singular matrices. In contrast to the theoretical situation, where there is a clear distinction between singular and nonsingular, we have instead a continuum of condition numbers, ranging from the well conditioned to the severely ill conditioned, with no clear dividing line in the middle. (Exceptions to this picture are certain matrices that are obviously singular, such as a matrix with a row or column of zeros or two equal rows.)

The next example shows that the distinction between good and bad rows is not always clear. It can happen that the accuracy of a computation deteriorates gradually over a number of steps.

Example 2.6.2 We introduced the ill-conditioned *Hilbert matrices*, defined by $h_{ij} = 1/(i + j - 1)$, in Example 2.2.10. For example,

$$H_4 = \begin{bmatrix} 1 & 1/2 & 1/3 & 1/4 \\ 1/2 & 1/3 & 1/4 & 1/5 \\ 1/3 & 1/4 & 1/5 & 1/6 \\ 1/4 & 1/5 & 1/6 & 1/7 \end{bmatrix}.$$

The rows look very much alike, which suggests ill conditioning. According to MATLAB, $\kappa_2(H_4) \approx 1.6 \times 10^4$.

Let us see how the ill conditioning manifests itself during Gaussian elimination. For the first step, there is no need to make a row interchange, since the largest entry is already in the pivotal position. You can easily check that after one step the transformed array is

$$\begin{bmatrix} 1 & 1/2 & 1/3 & 1/4 \\ \hline 1/2 & 1/12 & 1/12 & 3/40 \\ 1/3 & 1/12 & 4/45 & 1/12 \\ 1/4 & 3/40 & 1/12 & 9/112 \end{bmatrix}.$$

The second step operates on the submatrix

$$\begin{bmatrix} 1/12 & 1/12 & 3/40 \\ 1/12 & 4/45 & 1/12 \\ 3/40 & 1/12 & 9/112 \end{bmatrix},$$

all of whose entries are smaller than the original matrix entries. Thus each entry has undergone a small amount of cancellation. Of course these numbers are perfectly accurate because we calculated them by exact arithmetic. If we had used floating-point arithmetic, each of the entries would have suffered a slight loss of accuracy due to the cancellation. Notice that all of the entries of this submatrix are quite close to $1/12$; the rows are almost equal.

The potential pivots for the second step are smaller than those for the first step. Again there is no need for a row interchange, and after the step the transformed submatrix is

$$\begin{bmatrix} 1/12 & 1/12 & 3/40 \\ \hline 1 & 1/180 & 1/120 \\ 9/10 & 1/120 & 9/700 \end{bmatrix}.$$

The entries of the submatrix

$$\begin{bmatrix} 1/180 & 1/120 \\ 1/120 & 9/700 \end{bmatrix}$$

are even smaller than before; more cancellation has taken place. The potential pivots for the third step, 1/180 and 1/120, are both quite small. Since the latter is larger, we interchange the rows (although this has little effect on the outcome). After the third step, we have

$$\begin{bmatrix} 1/120 & 9/700 \\ 2/3 & -1/4200 \end{bmatrix}.$$

The final pivot is $-1/4200$, which is even smaller.

Now let us see what happens when the same operations are performed in three-digit decimal floating-point arithmetic. The original array is

$$\begin{bmatrix} 1.00 & .500 & .333 & .250 \\ .500 & .333 & .250 & .200 \\ .333 & .250 & .200 & .167 \\ .250 & .200 & .167 & .143 \end{bmatrix}.$$

Some of the entries are already slightly in error. On the first step the $(4, 3)$ entry (for example) is modified as follows:

$$.167 - (.250)(.333) = .167 - .833 \times 10^{-1} = .837 \times 10^{-1}.$$

Comparing it with the correct value $1/12 \approx .833 \times 10^{-1}$, we see that there is a substantial error in the third digit. The complete result of the first step is

$$\begin{bmatrix} 1.000 & .500 & .333 & .250 \\ .500 & .830 \times 10^{-1} & .830 \times 10^{-1} & .750 \times 10^{-1} \\ .333 & .830 \times 10^{-1} & .890 \times 10^{-1} & .837 \times 10^{-1} \\ .250 & .750 \times 10^{-1} & .837 \times 10^{-1} & .805 \times 10^{-1} \end{bmatrix}.$$

The result of the second step is (ignoring the first row and column)

$$\begin{bmatrix} .830 \times 10^{-1} & .830 \times 10^{-1} & .750 \times 10^{-1} \\ 1.00 & .600 \times 10^{-2} & .870 \times 10^{-2} \\ .904 & .870 \times 10^{-2} & .127 \times 10^{-1} \end{bmatrix}.$$

Significant cancellations have now taken place; most of these numbers have only one correct digit. For example, the following computation produced the $(4, 3)$ element:

$$.837 \times 10^{-1} - (.904)(.830 \times 10^{-1}) = (.837 - .750) \times 10^{-1} = .870 \times 10^{-2}.$$

Comparing this result with the correct value $1/120 \approx .833 \times 10^{-2}$, we see that it has only one correct digit. The result of the third and final step is

$$\begin{bmatrix} .870 \times 10^{-2} & .127 \times 10^{-1} \\ .690 & -.600 \times 10^{-4} \end{bmatrix}.$$

The $(4, 4)$ entry $-.600 \times 10^{-4}$ is not even close to the correct value $-1/4200 \approx -.238 \times 10^{-3}$. □

rcise 2.6.3 Work through the computations in Example 2.6.2, observing the cancellations and the accompanying loss of accuracy. Remember that if you wish to use your calculator to simulate three-digit decimal floating-point arithmetic, it does not suffice simply to set the display to show three digits. Although only three digits are displayed, nine or more digits are stored internally. Correct simulation of three-digit arithmetic requires that each intermediate result be rounded off before it is used in the next computation. A simple way to do this is to write down each intermediate result and then enter it back into the calculator when it is needed.   □

rcise 2.6.4 Work Example 2.6.2 using four-digit arithmetic instead of three. You will see that the outcome is not nearly so bad.   □

rcise 2.6.5 Use MATLAB to explore the extent of cancellation when Gaussian elimination is performed on larger Hilbert matrices.

(a) In MATLAB, type $A = \text{hilb}(7)$ to get H_7. To get the LU decomposition with partial pivoting, type $[L, U] = \text{lu}(A)$. Notice that the matrix L is not itself unit lower triangular, but it can be made unit lower triangular by permuting the rows. This is because MATLAB's lu command incorporates the row interchanges in the L matrix. Our real object of interest is the matrix U, which is the triangular matrix resulting from Gaussian elimination. Notice that the further down in U you go, the smaller the numbers become. The ones at the bottom appear to be zero. To get a more accurate picture, type format long and redisplay U.

(b) Generate H_{12} and its LU decomposition. Observe the U matrix using format long. It suffices to print out the last column of U, as this tells the story.

□

rcise 2.6.6 Let z denote the vector whose entries are all ones (Using MATLAB: $z = \text{ones}(n, 1)$), and let $b = H_n z$, where H_n is again the $n \times n$ Hilbert matrix. If we now solve the system $H_n x = b$ for x, we should get z as the solution in theory. Using MATLAB ($\text{xhat} = A \backslash b$), try solving $H_n x = b$ for $n = 4, 8, 12$, and 16, and see what you get. In each case compute the condition number $\kappa_2(H_n)$ and the norm of the difference: $\|\hat{x} - z\|_2$, where $\hat{x}$ is the computed solution. Calculate the residual norm $\|\hat{r}\|_2 = \|b - H_n \hat{x}\|_2$, too.   □

rcise 2.6.7 The previous exercise exaggerates somewhat the intractability of Hilbert matrices. There we saw that already for $n = 12$ we get a bad solution (assuming double-precision IEEE arithmetic). Actually, how bad the solution is depends not only on the matrix, but also on the vector b. Most choices of b will not give nearly such bad results as those we just saw. Carry out the following computations with $n = 12$. Use double precision (which you get automatically from MATLAB) throughout.

(a) Let z denote the vector of ones, as before, and solve the system $H_{12} y = z$ for y. Note that $\|y\|$ is huge.

(b) Define a vector b by $b = H_{12}y$. In principle b should be the same as z, but in the presence of roundoff errors it is a little different. Calculate b and $\|b - z\|_2$.

(c) Now consider the system $H_{12}x = b$. The way b was defined, the solution x ought to be the same as y. However, our experience from the previous problem suggests that the computed solution $\hat{x}$ may be far from y. Go ahead and solve $H_{12}x = b$ and compare the computed $\hat{x}$ with y. Notice that they agree to almost two decimal places. This is not nearly as bad as what we saw in the previous problem. Calculate the relative error $\|\hat{x} - y\|_2 / \|y\|_2$.

(d) Calculate the norm of the residual $\hat{r} = b - H_{12}\hat{x}$, and use it and the condition number to compute an upper bound on the relative error (Theorem 2.4.1). Note that this bound is very pessimistic compared to the actual relative error.

□

Another family of ill-conditioned matrices is the *Lotkin matrices*. The $n \times n$ Lotkin matrix is identical to the Hilbert matrix H_n, except that its first row consists entirely of ones. These nonsymmetric matrices are just as ill conditioned as the Hilbert matrices. To get the 6×6 Lotkin matrix in MATLAB type A = gallery('lotkin',6). For more information about the gallery of Higham test matrices, type help gallery. For more information about Lotkin matrices, type help private/lotkin.

Exercise 2.6.8 Rework each of the following exercises, using Lotkin matrices in place of Hilbert matrices: (a) Exercise 2.6.5, (b) Exercise 2.6.6, (c) Exercise 2.6.7. □

Why Small Pivots Should Be Avoided

In Section 1.8 we introduced the partial pivoting strategy, in which the pivot for the kth step is chosen to be the largest in magnitude of the potential pivots in column k. The justification given at that time was that we wanted to avoid using as a pivot a small number that would have been zero except for roundoff errors made in previous steps. Since then we have studied cancellation and ill-conditioned matrices and can make a more general statement: we wish to avoid using a small pivot because it may have become small as the result of cancellations in previous steps, in which case it could be very inaccurate. The dangers of using an inaccurate pivot were stated in the first part of this section, in connection with ill conditioning.

In the following example we consider a different scenario. We show what can go wrong if a small pivot is used even though large pivots are available. Here the pivot is not inaccurate; it ruins the computation simply by being small.

Example 2.6.9 Consider the linear system

$$\begin{bmatrix} .002 & 1.231 & 2.471 \\ 1.196 & 3.165 & 2.543 \\ 1.475 & 4.271 & 2.142 \end{bmatrix} \begin{bmatrix} x_1 \\ x_2 \\ x_3 \end{bmatrix} = \begin{bmatrix} 3.704 \\ 6.904 \\ 7.888 \end{bmatrix}. \tag{2.6.10}$$

This system is well conditioned ($\kappa_2(A) \approx 30$), and we will see that it can be solved accurately by Gaussian elimination with partial pivoting. But first let us observe what happens when we use Gaussian elimination without interchanges. We will see that the use of the exceptionally small $(1, 1)$ entry as a pivot destroys the computation. The computations will be done in four-digit decimal floating-point arithmetic. You should think of the numbers in (2.6.10) as being exact. You can easily check that the exact solution is $x = [1 \ 1 \ 1]^T$. The roundoff error effects you are about to observe are caused not because the small pivot is inaccurate (it is not!), but simply because it is small.

The multipliers for the first step are

$$l_{21} = \frac{1.196}{.002} = 598.0 \quad \text{and} \quad l_{31} = \frac{1.475}{.002} = 737.5. \tag{2.6.11}$$

These multiplied by the first row are subtracted from the second and third rows, respectively. For example, the $(2, 2)$ entry is altered as follows: 3.165 is replaced by

$$3.165 - (598.0)(1.231) = 3.165 - 736.1 = -732.9. \tag{2.6.12}$$

These equations are not exact; four-digit arithmetic was used. Notice that the resulting entry is much larger than the number it replaced and that the last two digits of 3.165 were lost (rounded off) when a large number was added to (i.e. subtracted from) it. This type of information loss is called *swamping*. The small number was *swamped* by the large one. You can check that swamping also occurs when the $(2, 3)$, $(3, 2)$, and $(3, 3)$ entries are modified. In fact three digits are swamped in the $(2, 3)$ and $(3, 3)$ positions. At the end of the first step the modified coefficient matrix looks like

$$\begin{bmatrix} .002 & 1.231 & 2.471 \\ \overline{598.0} & -732.9 & -1475. \\ 737.5 & -903.6 & -1820. \end{bmatrix}.$$

The second step of Gaussian elimination works with the submatrix

$$\tilde{A} = \begin{bmatrix} -732.9 & -1475. \\ -903.6 & -1820. \end{bmatrix}.$$

Since this matrix was obtained by subtracting very large multiples of the row [1.231 2.471] from the much smaller numbers that originally occupied these rows, the two rows of $\tilde{A}$ are almost exact multiples of [1.231 2.471]. Thus the rows of $\tilde{A}$ are nearly linearly dependent; that is, $\tilde{A}$ is ill conditioned. You can easily check that $\kappa_2(\tilde{A}) \approx 6400$, which is huge, considering that four-digit arithmetic is being used.

The multiplier for the second step is $l_{32} = (-903.6)/(-732.9) = 1.233$. It is used only to modify the $(3, 3)$ entry as follows:

$$-1820. - (1.233)(-1475.) = -1820. + 1819. = -1.000.$$

Severe cancellation occurs here. This is just an attempt to recover the information that was lost through swamping in the previous step. Unfortunately, that information is gone, and consequently the result -1.000 is inaccurate. At the end of the second step, the LU decomposition is complete:

$$
\begin{bmatrix}
.002 & 1.231 & 2.471 \\
598.0 & -732.9 & -1475. \\
737.5 & 1.233 & -1.000
\end{bmatrix}.
$$

The forward substitution step yields

$$
\begin{aligned}
y_1 &= 3.704 \\
y_2 &= 6.904 - (598.0)(3.704) = 6.904 - 2215. = -2208. \\
y_3 &= 7.888 - (737.5)(3.704) - (1.233)(-2208.) \\
&= 7.888 - 2732. + 2722. = -2724. + 2722. = -2.000
\end{aligned}
$$

Notice that the last three digits of 6.904 were swamped in the computation of y_2, and severe cancellation occurred in the calculation of y_3. Thus y_3 is inaccurate.

The first step of back substitution is

$$
x_3 = \frac{y_3}{u_{33}} = \frac{-2.000}{-1.000} = 2.000.
$$

Being the quotient of two inaccurate numbers, x_3 is also inaccurate. Recall that the correct value is 1.000. You can carry out the rest of the back substitution process and find that the computed solution is $[4.000, \; -1.012, \; 2.000]^T$, which is nothing like the true solution.

Let us summarize what went wrong, speaking in general terms and heuristically. When a pivot that is much smaller than the other potential pivots is used, large multipliers will result. Thus very large multiples of the pivotal row will be subtracted from the remaining rows. In the process the numbers that occupied those rows will be swamped. The resulting submatrix (the matrix that will be operated on in the next step) will be ill conditioned because each of its rows is almost exactly a multiple of the pivotal row. Because of the ill conditioning, there will be cancellations in later steps. These cancellations are actually just an attempt to uncover the information that was lost due to swamping, but that information is gone.

Now let us see what happens when we solve (2.6.10) using partial pivoting. Interchanging rows 1 and 3, we obtain the system

$$
\begin{bmatrix}
1.475 & 4.271 & 2.142 \\
1.196 & 3.165 & 2.543 \\
.002 & 1.231 & 2.471
\end{bmatrix}
\begin{bmatrix}
x_1 \\
x_2 \\
x_3
\end{bmatrix}
=
\begin{bmatrix}
7.888 \\
6.904 \\
3.704
\end{bmatrix}.
$$

After one step the partially reduced matrix has the form

$$
\begin{bmatrix}
1.475 & 4.271 & 2.142 \\
.8108 & -.2980 & .8060 \\
1.356 \times 10^{-3} & 1.225 & 2.468
\end{bmatrix}.
$$

When you carry out this computation, you can see that the information in the $(2, 2)$, $(2, 3)$, $(3, 2)$, and $(3, 3)$ positions is not swamped. There is, however, a slight cancellation in the $(2, 2)$ and $(2, 3)$ positions. The partial pivoting strategy dictates that we interchange rows 2 and 3. In this way we avoid using the slightly inaccurate number $-.2980$ as a pivot. After step 2 the LU decomposition is complete:

$$
\begin{bmatrix}
1.475 & 4.271 & 2.142 \\
1.356 \times 10^{-3} & 1.225 & 2.468 \\
.8108 & -.2433 & 1.407
\end{bmatrix}.
$$

Forward substitution yields $y = [7.888, \ 3.693, \ 1.407]^T$, and back substitution gives the computed result

$$
x = \begin{bmatrix}
1.000 \\
1.000 \\
1.000
\end{bmatrix}.
$$

It is a matter of luck that the computed solution agrees with the true solution exactly, but it is not luck that the computation yielded an accurate result. Accuracy is guaranteed by the well-conditioned coefficient matrix together with Theorem 2.7.14. □

rcise 2.6.13 Work through the details of the computations performed in Example 2.6.9.

□

2.7 BACKWARD ERROR ANALYSIS OF GAUSSIAN ELIMINATION

A major operation in many algorithms, including Gaussian elimination, is the accumulation of sums

$$
\sum_{j=1}^{n} w_j.
$$

There are many ways to add n numbers together. For example, if we have four numbers, we can add them in the "natural" way: first we add w_1 to w_2, then we add on w_3, then w_4, that is, $((w_1 + w_2) + w_3) + w_4$, or we can apply this process to any reordering of $w_1, \ldots, w_4$. Another possibility is to add w_1 to w_2, add w_3 to w_4, then add the two intermediate sums, that is, $(w_1 + w_2) + (w_3 + w_4)$.

If we accumulate a sum in different ways using floating-point arithmetic, we will get different results, because the different ways have different roundoff errors. The relative differences in the computed sums will usually be tiny, but they can be large if there is a cancellation in the end (the summands add "nearly" to zero).

Our first task is to show that if a sum is accumulated in floating-point arithmetic, the computation is always backward stable, regardless of the manner in which the sum was accumulated. As before, we will let u be the unit roundoff. We will use the notation $O(u^2)$ to denote terms of order u^2. These are tiny terms that can be neglected. For example, $(1 + \alpha_1)(1 + \alpha_2) = (1 + \beta)$, where $\beta = \alpha_1 + \alpha_2 + \alpha_1\alpha_2$. If $|\alpha_1| \approx u$ and $|\alpha_2| \approx u$, then $\alpha_1\alpha_2 = O(u^2)$, and we write $\beta = \alpha_1 + \alpha_2 + O(u^2)$.

Proposition 2.7.1 *Suppose we compute the sum* $\displaystyle\sum_{j=1}^{n} w_j$ *using floating-point arithmetic with unit roundoff* u. *Then*

$$\text{fl}\left(\sum_{j=1}^{n} w_j\right) = \sum_{j=1}^{n} w_j(1 + \gamma_j),$$

where $|\gamma_j| \leq (n-1)u + O(u^2)$, *regardless of the order in which the terms are accumulated.*

Proof. The proof is by induction on n. The proposition is trivially true when $n = 1$. Now let m be any positive integer. We shall show that the proposition holds for $n = m$, assuming that it holds for $n < m$, that is, for all sums of fewer than m terms.

Suppose we have reached the point in our computation of $\sum_{j=1}^{m} w_j$ at which we are within one addition of being done. At this point we will have accumulated two sums, one being the sum of, say, k of the w_j and the other being the sum of the other $m - k$. It might be that $k = 1$ or $k = m - 1$. In any event, $1 \leq k \leq m - 1$. For notational convenience relabel the terms so that the terms in the first sum are $w_1, \ldots, w_k$. Then the two sums that we have so far are

$$\sum_{j=1}^{k} w_j \quad \text{and} \quad \sum_{j=k+1}^{m} w_j.$$

Since each of these sums has fewer than m terms, we have, by the induction hypothesis,

$$\text{fl}\left(\sum_{j=1}^{k} w_j\right) = \sum_{j=1}^{k} w_j(1 + \alpha_j)$$

and

$$\text{fl}\left(\sum_{j=k+1}^{m} w_j\right) = \sum_{j=k+1}^{m} w_j(1 + \alpha_j),$$

where

$$|\alpha_j| \leq \begin{cases} (k-1)u + O(u^2) & \text{for } j = 1, \ldots, k, \\ (m-k-1)u + O(u^2) & \text{for } j = k+1, \ldots, m, \end{cases}$$

regardless of the order in which each of these sums was accumulated. Since $k - 1 \leq m - 2$ and $m - k - 1 \leq m - 2$, we have

$$|\alpha_j| \leq (m-2)u + O(u^2) \quad \text{for } j = 1, \ldots, m.$$

Adding the two sums together, we have

$$\text{fl}\left(\sum_{j=1}^{m} w_j\right) = \text{fl}\left(\text{fl}\left(\sum_{j=1}^{k} w_j\right) + \text{fl}\left(\sum_{j=k+1}^{m} w_j\right)\right)$$

$$= \left(\sum_{j=1}^{k} w_j(1 + \alpha_j) + \sum_{j=k+1}^{m} w_j(1 + \alpha_j) \right) (1 + \beta)$$

$$= \sum_{j=1}^{m} w_j(1 + \alpha_j)(1 + \beta).$$

β is the roundoff error of the current addition and satisfies $|\beta| \leq u$ by (2.5.6). Let $\gamma_j = \alpha_j + \beta + \alpha_j\beta$, so that $(1 + \gamma_j) = (1 + \alpha_j)(1 + \beta)$. Then

$$\begin{aligned}
|\gamma_j| &\leq |\alpha_j| + |\beta| + O(u^2) \\
&\leq (m - 2)u + O(u^2) + u + O(u^2) \\
&= (m - 1)u + O(u^2).
\end{aligned}$$

We have thus shown that

$$\mathrm{fl}\left(\sum_{j=1}^{m} w_j \right) = \sum_{j=1}^{m} w_j(1 + \gamma_j),$$

where $|\gamma_j| \leq (m - 1)u + O(u^2)$. □

The numbers γ_j in Proposition 2.7.1 can be termed *relative backward errors*. There are at least three reasons why the inequalities $|\gamma_j| \leq (n - 1)u + O(u^2)$ grossly overestimate $|\gamma_j|$. First of all, the factor $n - 1$ reflects the fact that each summand participates in at most $n - 1$ additions and is therefore subjected to at most $n - 1$ roundoff errors. γ_j is approximately the sum of all the roundoff errors that occur in sums involving w_j (recall $\gamma_j \approx \alpha_j + \beta$). The exact number of roundoff errors that each term suffers depends on the method of summation and is, on average, much less than $n - 1$. (You can clarify this for yourself by working Exercise 2.7.24.) Thus most γ_j are sums of many fewer than $n - 1$ roundoffs.

Secondly, u is an upper bound on each roundoff error. A typical round off will be significantly less than u. Finally, and most importantly, when roundoff errors are added together, they sometimes reinforce one another and they sometimes (partially) cancel each other out. Bounds like $|\gamma_j| \leq (n - 1)u + O(u^2)$ have to take into account the worst possible (and highly unlikely) case, where all roundoff errors are maximal and reinforce one another.

For these reasons, the γ_j are more likely to be much closer to u than $(n - 1)u$, so the factor $(n - 1)$ can be ignored in practice. Thus we consider Proposition 2.7.1 to be proof that the accumulation of sums is backward stable, regardless of how large n is.

Backward Stability of Forward and Back Substitution

Let G be a nonsingular, lower-triangular matrix, and let b be a nonzero vector. Then we can solve the system $Gy = b$ in about n^2 flops by forward substitution. Our next theorem shows that the forward substitution algorithm is backward stable. First we introduce some simplifying notation. Given an $n \times m$ matrix (or, in particular, a vector) C with (i, j) entry c_{ij}, we define $|C|$, the *absolute value* of C, to be the $n \times m$ matrix whose (i, j) entry is $|c_{ij}|$. Also, given two $n \times m$ matrices C and F, we will write $C \leq F$ if and only if $c_{ij} \leq f_{ij}$ for all i and j. With these new definitions, we can now make the following statement.

Theorem 2.7.2 *Let $G \in \mathbb{R}^{n \times n}$ be a nonsingular, lower-triangular matrix, and let $b \neq 0$. If the system $Gy = b$ is solved by any variant of forward substitution using floating-point arithmetic, then the computed solution $\hat{y}$ satisfies*

$$(G + \delta G)\hat{y} = b, \tag{2.7.3}$$

where δG satisfies

$$|\delta G| \leq 2nu|G| + O(u^2). \tag{2.7.4}$$

This inequality means that $|\delta g_{ij}| \leq 2nu|g_{ij}| + O(u^2)$ for all i and j. Thus the term $O(u^2)$ in (2.7.4) stands for an $n \times n$ matrix, each of whose entries is of order u^2.

This is not the tightest possible result. For a more careful argument that gets rid of the factor 2, see [52].

Proof. Once we have $y_1, \ldots, y_{i-1}$, we compute

$$y_i = \frac{b_i - \sum_{j=1}^{i-1} g_{ij} y_j}{g_{ii}},$$

in principle. In practice we use the computed quantities $\hat{y}_1, \ldots, \hat{y}_{i-1}$ and make further rounding errors, so

$$\hat{y}_i = \mathrm{fl}\left(\frac{b_i - \sum_{j=1}^{i-1} g_{ij} \hat{y}_j}{g_{ii}}\right).$$

The numerator is just a sum of i terms, but before we can do any additions, we have to do the multiplications. We have $\mathrm{fl}(g_{ij} \hat{y}_j) = g_{ij} \hat{y}_j (1 + \alpha_{ij})$, where $|\alpha_{ij}| \leq u$ by (2.5.6). Once we have the products, we can accumulate the numerator. Different variants of forward substitution will do this in different ways. Since it is a sum of i terms, Proposition 2.7.1 guarantees that no matter how it is done,

$$\mathrm{fl}\left(b_i - \sum_{j=1}^{i-1} g_{ij} \hat{y}_j (1 + \alpha_{ij})\right) = b_i(1 + \gamma_{ii}) - \sum_{j=1}^{i-1} g_{ij} \hat{y}_j (1 + \alpha_{ij})(1 + \gamma_{ij}),$$

where $|\gamma_{ij}| \leq (i-1)u + O(u^2)$ for $j = 1, \ldots, i$. Once this is done, we obtain $\hat{y}_i$ by a division, which introduces one more rounding error:

$$\hat{y}_i = \left(\frac{b_i(1 + \gamma_{ii}) - \sum_{j=1}^{i-1} g_{ij} \hat{y}_j (1 + \alpha_{ij})(1 + \gamma_{ij})}{g_{ii}}\right)(1 + \beta_i), \tag{2.7.5}$$

where $|\beta_i| \le u$.

We are aiming for the final result (2.7.3), in which all of the errors have been pushed back onto G and, in particular, not onto b. To obtain this effect we divide numerator and denominator in (2.7.5) by $(1 + \gamma_{ii})(1 + \beta_i)$ to get rid of the error terms that multiply b_i. If we define ϵ_{ij} through the equations

$$1 + \epsilon_{ij} = \begin{cases} \frac{(1+\alpha_{ij})(1+\gamma_{ij})}{1+\gamma_{ii}} & \text{if } j < i, \\ \frac{1}{(1+\gamma_{ii})(1+\beta_i)} & \text{if } j = i, \end{cases} \tag{2.7.6}$$

we then have

$$\hat{y}_i = \frac{b_i - \sum_{j=1}^{i-1} g_{ij}(1 + \epsilon_{ij})\hat{y}_j}{g_{ii}(1 + \epsilon_{ii})}. \tag{2.7.7}$$

This last equation can be rewritten as

$$\sum_{j=1}^{i} g_{ij}(1 + \epsilon_{ij})\hat{y}_j = b_i.$$

Since this holds for all i, we can write it as a single matrix equation

$$(G + \delta G)\hat{y} = b,$$

where δG is the lower-triangular matrix defined by $\delta g_{ij} = \epsilon_{ij} g_{ij}$ for $i \ge j$. To complete the proof we just have to show that $|\epsilon_{ij}| \le 2nu + O(u^2)$ for all i and j.

Referring back to (2.7.6), we see that we need to deal with the factor $1/(1 + \gamma_{ii})$. Since $\gamma_{ii} = O(u)$, we have

$$\frac{1}{1 + \gamma_{ii}} = 1 - \gamma_{ii} + \gamma_{ii}^2 - \gamma_{ii}^3 + \cdots = 1 - \gamma_{ii} + O(u^2).$$

Thus, when $j < i$,

$$\begin{aligned} 1 + \epsilon_{ij} &= (1 + \alpha_{ij})(1 + \gamma_{ij})(1 - \gamma_{ii} + O(u^2)) \\ &= 1 + \alpha_{ij} + \gamma_{ij} - \gamma_{ii} + O(u^2), \end{aligned}$$

and

$$\begin{aligned} |\epsilon_{ij}| &\le |\alpha_{ij}| + |\gamma_{ij}| + |\gamma_{ii}| + O(u^2) \\ &\le u + (i-1)u + (i-1)u + O(u^2) \\ &= (2i-1)u + O(u^2) \le 2nu + O(u^2). \end{aligned}$$

By a similar analysis we get that $|\epsilon_{ii}| \le iu + O(u^2) \le 2nu + O(u^2)$. $\square$

rcise 2.7.8 Check that (2.7.7) is valid for $i = 1$. $\square$

Corollary 2.7.9 *Under the conditions of Theorem 2.7.2 the computed solution* $\hat{y}$ *satisfies*

$$(G + \delta G)\hat{y} = b,$$

where

$$\|\delta G\|_\infty \le 2nu\|G\|_\infty + O(u^2). \tag{2.7.10}$$

Proof. The bound (2.7.10) follows directly from (2.7.4), using the properties proved in Exercise 2.7.11.   □

Exercise 2.7.11

   (a) Show that if $|C| \le |F|$ (elementwise), then $\|C\|_\infty \le \|F\|_\infty$.

   (b) Show that $\|C\|_\infty = \| \, |C| \, \|_\infty$.

                                                                        □

   The properties established in Exercise 2.7.11 hold also for the matrix 1-norm and the Frobenius norm. Corollary 2.7.9 holds for any norm that satisfies these properties.
   We have already noted that the factor 2 in the bounds (2.7.4) and (2.7.10)) can be eliminated. The factor n can also be ignored, just as in Proposition 2.7.1.
   Corollary 2.7.9 shows that forward substitution is *normwise backward stable*; that is, the computed solution $\hat{y}$ is the exact solution of a nearby problem $(G + \delta G)\hat{y} = b$, where $\|\delta G\|_\infty / \|G\|_\infty$ is tiny.
   Theorem 2.7.2 is actually a stronger result. It states not just that $\|\delta G\|$ is tiny relative to $\|G\|$, but each element perturbation δg_{ij} is tiny relative to g_{ij}, the element it is perturbing. This property is called *componentwise backward stability*.

Exercise 2.7.12 Construct an example of 2×2 matrices G and δG such that $\|\delta G\|_\infty / \|G\|_\infty$ is tiny but $|\delta g_{ij}|/|g_{ij}|$ is not tiny for at least one component (i, j).   □

   Theorems essentially identical to Theorem 2.7.2 and Corollary 2.7.9 hold for back substitution applied to upper-triangular systems. We see no need to state these results.

Backward Error of Gaussian Elimination

Gaussian elimination is sometimes backward stable, sometimes not, depending on circumstances. The basic results are the following two theorems, which are statements about Gaussian elimination without pivoting. However, both of these results are applicable to Gaussian elimination with row and column interchanges, since the latter is equivalent to Gaussian elimination without interchanges, applied to a matrix whose rows and columns were interchanged in advance. Thus these results can be applied to Gaussian elimination with partial pivoting, complete pivoting, or any other pivoting strategy.

Theorem 2.7.13 *Suppose the LU decomposition of A is computed by Gaussian elimination in floating-point arithmetic, and suppose no zero pivots are encountered in the process. Let $\hat{L}$ and $\hat{U}$ denote the computed factors. Then*

$$A + E = \hat{L}\hat{U},$$

where

$$|E| \le 2nu\,|\hat{L}|\,|\hat{U}| + O(u^2)$$

and

$$\|E\|_\infty \le 2nu\,\|\hat{L}\|_\infty\,\|\hat{U}\|_\infty + O(u^2).$$

Theorem 2.7.14 *Under the conditions of Theorem 2.7.13, suppose we solve $Ax = b$ numerically by performing forward substitution with $\hat{L}$ followed by back substitution with $\hat{U}$ in floating-point arithmetic. Then the computed solution $\hat{x}$ satisfies*

$$(A + \delta A)\hat{x} = b,$$

where

$$|\delta A| \le 6nu\,|\hat{L}|\,|\hat{U}| + O(u^2)$$

and

$$\|\delta A\|_\infty \le 6nu\,\|\hat{L}\|_\infty\,\|\hat{U}\|_\infty + O(u^2).$$

We will defer the proofs of these results until after we have discussed their implications.[14]

Looking at either of these Theorems, we see that the backward stability (or lack thereof) of Gaussian elimination depends upon how big $\hat{L}$ and $\hat{U}$ are. If $\|\hat{L}\|_\infty\|\hat{U}\|_\infty$ is only a modest multiple of $\|A\|_\infty$, then we can conclude that $\|\delta A\|_\infty/\|A\|_\infty$ is a modest multiple of the unit roundoff, and the operation is backward stable. If, on the other hand, $\|\hat{L}\|_\infty\|\hat{U}\|_\infty$ is much larger than $\|A\|_\infty$, then we can draw no such conclusion. In this case the computation is probably not backward stable.

Let us first consider Gaussian elimination without pivoting. The use of small pivots can result in large multipliers, which are entries of $\hat{L}$. Thus $\|\hat{L}\|_\infty$ can be arbitrarily large. The large multipliers cause large multiples of some rows to be added to other rows, with the effect that $\|\hat{U}\|_\infty$ is also large. These effects were seen in Example 2.6.9. There we showed how the unnecessary use of small pivots can destroy the accuracy of Gaussian elimination. See Exercise 2.7.25 as well. We conclude that Gaussian elimination without pivoting is unstable.

Now consider partial pivoting. This guarantees that all of the multipliers $\hat{l}_{ij}$ have modulus less than or equal to 1 and has the tendency of keeping the norm of $\hat{L}$ from being too large. In fact $\|\hat{L}\|_\infty \le n$. Thus, to guarantee backward stability, we need only show that $\|\hat{U}\|_\infty/\|A\|_\infty$ cannot be too large. Unfortunately there exist classes of matrices for which $\|\hat{U}\|_\infty/\|A\|_\infty \approx 2^{n-1}$. An example is given in

---

[14]These theorems have been stated in the form in which they will be proved. They are not the best possible results; a factor of 2 can be removed from each of the bounds.

Exercise 2.7.26. Because 2^{n-1} grows exponentially with n and is enormous even for modest n (e.g. $2^{n-1} > 10^{29}$ when $n = 100$), we cannot claim that Gaussian elimination with partial pivoting is unconditionally backward stable, except when n is quite small. The example of Exercise 2.7.26 is artificial, but naturally occurring classes of matrices that exhibit exponential growth of $\|\hat{U}\|_\infty$ have been reported in [30, 106].

Despite this bad news, partial pivoting is now and will continue to be widely used. For example, it is the main method for solving $Ax = b$ in MATLAB and LAPACK. Years of testing and experience have shown that the exponential element growth exhibited by the class of matrices in Exercise 2.7.26 is extremely rare in practice. Typically we see

$$\frac{\|\hat{U}\|_\infty}{\|A\|_\infty} \approx \sqrt{n}.$$

A statistical explanation of the good behavior of partial pivoting is given by Trefethen and Bau [91]. See also Exercise 2.7.27.

Since partial pivoting occasionally performs badly, one would like to have a way of checking whether one's results are good or not. Fortunately such tests exist. One such test would be simply to compute the ratio

$$\|\hat{L}\|_\infty \|\hat{U}\|_\infty / \|A\|_\infty.$$

If this is not too large, the computation was backward stable. An even simpler test, once $\hat{x}$ has been computed, is to calculate the residual $\hat{r} = b - A\hat{x}$. As we pointed out at the end of Section 2.5, if $\|\hat{r}\|/\|b\|$ is tiny, the computation was backward stable.

In cases where backward stability is not achieved, the result can often be improved by iterative refinement, as explained in Section 2.9.

Another test is simply to compute the backward error in the LU decomposition: $E = \hat{L}\hat{U} - A$. If $\|E\|$ is tiny, the computation was backward stable. However, this test is expensive. The computation of $\hat{L}\hat{U}$ costs $O(n^3)$ flops, even if the triangular form of the matrices is taken into account (Exercise 2.7.28).

Example 2.7.15 Using MATLAB we computed the LU decomposition (with partial pivoting) of the 12×12 Hilbert matrix (A = hilb(12); [L,U] = lu(A);) and then computed $E = LU - A$. We found that $\|E\|_\infty/\|A\|_\infty \approx 2.7 \times 10^{-17}$. Also, $\|L\|_\infty \approx 5.6$, $\|U\|_\infty \approx 3.1$, and $\|L\|_\infty \|U\|_\infty/\|A\|_\infty \approx 5.6$. All of these results signal stability. This shows that the poor results obtained in Exercises 2.6.7 and (especially) 2.6.6 are due entirely to the ill conditioning of the Hilbert matrix, not to instability of the algorithm. In those exercises you also computed residuals and found that they are tiny. This also demonstrates backward stability.   □

Exercise 2.7.16

(a) Perform the computations indicated in Example 2.7.15 for Hilbert matrices of dimension 12, 24, and 48. Notice that we have stability in every case.

(b) Repeat part (a) using Lotkin Matrices (A = gallery('lotkin',n)).

□

Complete pivoting is better behaved than partial pivoting. Element growth of the type exhibited by partial pivoting in Exercise 2.7.26 is impossible. In the worst known cases $\|\hat{U}\|/\|A\| = O(n)$. It is also true that $\|\hat{L}_\infty\| \leq n$, as for partial pivoting. Thus Gaussian elimination with complete pivoting is considered to be a backward stable algorithm.

In spite of the theoretical superiority of complete pivoting over partial pivoting, the latter is much more widely used. The reasons are simple: 1.) Partial pivoting works well in practice, and it is significantly less expensive. 2.) Inexpensive *a posteriori* stability tests exist.

Theorem 2.7.14 is also important for Gaussian elimination in sparse matrices. Here there are conflicting objectives: one wishes not only to perform the elimination in a stable manner, but also to keep fill-in as small as possible. One might therefore pursue a strategy that does not always select the largest possible pivots. The stability of the decomposition can be monitored by checking the size of the entries of $\hat{L}$ and $\hat{U}$ as they are produced.

For symmetric, positive definite systems a result like Theorem 2.7.14 holds for Cholesky's method with $\hat{L}$ and $\hat{U}$ replaced by $\hat{R}^T$ and $\hat{R}$, respectively. It can also be shown that $\|\hat{R}^T\|_F\|\hat{R}\|_F$ cannot be large relative to $\|A\|_F$. Therefore Cholesky's method is unconditionally backward stable. You can work out the details in Exercise 2.7.29.

In summary, Gaussian elimination with partial or complete pivoting and Cholesky's method for positive definite systems are stable in practice. For partial pivoting this assertion is based upon years of experience; for Cholesky's method it is an iron-clad fact. The factors of n that appear in the theorems are gross overestimates and can be ignored. In practice we get a computed solution $\hat{x}$ satisfying $(A + \delta A)\hat{x} = b$, where $\|\delta A\|/\|A\| \approx Cu$, with C a modest multiple of 1. Thus the total effect of roundoff errors is not much greater than that of the initial roundoff errors in the representation of A. These errors are usually much smaller than the original measurement errors in A and b. These considerations lead to a useful rule of thumb.

Rule of Thumb 2.7.17 *Suppose the linear system $Ax = b$ is solved by Gaussian elimination with partial or complete pivoting (or by Cholesky's method in the positive definite case). If the entries of A and b are accurate to about s decimal places and $\kappa(A) \approx 10^t$, where $t < s$, then the entries of the computed solution are accurate to about $s - t$ decimal places.*

"Proof." We intended to solve $Ax = b$, but our computed solution $\hat{x}$ satisfies a perturbed equation $(A + \delta A)\hat{x} = b + \delta b$, where δA is the sum of measurement error, initial rounding error, and the effect of roundoff errors made during the computation, and δb is the sum of measurement error and initial rounding error. We assume that measurement error dominates. Since the entries of A and b are accurate to about s decimal places,

$$\frac{\|\delta A\|}{\|A\|} \approx 10^{-s} \quad \text{and} \quad \frac{\|\delta b\|}{\|b\|} \approx 10^{-s}.$$

Preparing to apply Theorem 2.3.9, we note that

$$\kappa(A)\frac{\|\delta A\|}{\|A\|} \approx 10^{t-s} \ll 1 \quad \text{so} \quad 1 - \kappa(A)\frac{\|\delta A\|}{\|A\|} \approx 1.$$

Thus Theorem 2.3.9 gives roughly

$$\frac{\|\delta x\|}{\|x\|} \approx \kappa(A)\left(\frac{\|\delta A\|}{\|A\|} + \frac{\|\delta b\|}{\|b\|}\right) \approx 10^{t-s},$$

where $\hat{x} = x + \delta x$. That is, the entries of $\hat{x}$ are accurate to about $s - t$ decimal places.   □

Proofs of Theorems 2.7.13 and 2.7.14

Proof of Theorem 2.7.13. We begin by recalling (cf. (1.7.22) and (1.7.24)) that if $A = LU$, then L and U are given by the formulas

$$l_{ij} = \frac{a_{ij} - \sum_{k=1}^{j-1} l_{ik}u_{kj}}{u_{jj}} \quad \text{for } i > j \qquad (2.7.18)$$

and

$$u_{ij} = a_{ij} - \sum_{k=1}^{i-1} l_{ik}u_{kj} \quad \text{for } i \leq j. \qquad (2.7.19)$$

All versions of Gaussian elimination perform the computations indicated in (2.7.18) and (2.7.19), but different versions organize the computations differently.

Let us first consider the computation of l_{ij}. In practice the computed values $\hat{l}_{ik}$ and $\hat{u}_{kj}$ are used, and further roundoff errors occur. Thus the computed $\hat{l}_{ij}$ satisfies

$$\hat{l}_{ij} = \text{fl}\left(\frac{a_{ij} - \sum_{k=1}^{j-1} \hat{l}_{ik}\hat{u}_{kj}}{\hat{u}_{jj}}\right).$$

Different versions of Gaussian elimination accumulate the sum in the numerator in different ways. Proposition 2.7.1 shows that no matter how it is done,

$$\hat{l}_{ij} = \frac{a_{ij}(1 + \gamma_{ij}) - \sum_{k=1}^{j-1} \hat{l}_{ik}\hat{u}_{kj}(1 + \alpha_k)(1 + \gamma_{ik})}{\hat{u}_{jj}}(1 + \beta), \qquad (2.7.20)$$

where $|\gamma_{ij}| \leq (j-1)u + O(u^2)$. The quantities α_k and β are the roundoff errors associated with the multiplications and the division, respectively, and they satisfy $|\alpha_k| \leq u$ and $|\beta| \leq u$.

Proceeding just as in the proof of Theorem 2.7.2, we divide the numerator and denominator in (2.7.20) by $(1 + \gamma_{ij})(1 + \beta)$ to remove the error from the a_{ij} term. This move is not strictly necessary, but it yields a more elegant result. We then simplify the resulting equation by consolidating the errors. Define δ_{ik} by

$$1 + \delta_{ik} = \begin{cases} \frac{(1+\alpha_k)(1+\gamma_{ik})}{1+\gamma_{ij}} & \text{if } k < j, \\ \frac{1}{(1+\gamma_{ij})(1+\beta)} & \text{if } k = j. \end{cases}$$

Recalling that $1/(1 + \gamma_{ij}) = 1 - \gamma_{ij} + \gamma_{ij}^2 - \gamma_{ij}^3 + \cdots$, we have, for $k < j$, $\delta_{ik} = \alpha_k + \gamma_{ik} - \gamma_{ij} + O(u^2)$, and

$$
\begin{aligned}
|\delta_{ik}| &\leq |\alpha_k| + |\gamma_{ik}| + |\gamma_{ij}| + O(u^2) \\
&\leq u + (j-1)u + O(u^2) + (j-1)u + O(u^2) = (2j-1)u + O(u^2).
\end{aligned}
$$

Similarly, $\delta_{ij} = -\gamma_{ij} - \beta + O(u^2)$, and

$$
\begin{aligned}
|\delta_{ij}| &\leq |\gamma_{ij}| + |\beta| + O(u^2) \\
&\leq (j-1)u + O(u^2) + u + O(u^2) = ju + O(u^2).
\end{aligned}
$$

Thus

$$
|\delta_{ik}| \leq 2nu + O(u^2), \quad k = 1, \ldots, j.
$$

In terms of δ_{ik}, (2.7.20) becomes

$$
\hat{l}_{ij} = \frac{a_{ij} - \sum_{k=1}^{j-1} \hat{l}_{ik} \hat{u}_{kj} (1 + \delta_{ik})}{\hat{u}_{jj}(1 + \delta_{ij})}.
$$

Multiplying through by $\hat{u}_{jj}(1 + \delta_{ij})$ and rewriting the resulting expression so that all of the error terms are consolidated into a single term e_{ij}, we have

$$
a_{ij} + e_{ij} = \sum_{k=1}^{j} \hat{l}_{ik} \hat{u}_{kj} \tag{2.7.21}
$$

for $i > j$ (because we started from (2.7.18), which holds only for $i > j$), where

$$
e_{ij} = -\sum_{k=1}^{j} \hat{l}_{ik} \hat{u}_{kj} \delta_{ik}.
$$

Since $|\delta_{ik}| \leq 2nu + O(u^2)$,

$$
|e_{ij}| \leq 2nu \sum_{k=1}^{j} |\hat{l}_{ik}| |\hat{u}_{kj}| + O(u^2) \tag{2.7.22}
$$

for $i > j$.

We obtained (2.7.21) and (2.7.22) for $i > j$ by starting from (2.7.18). The same results can be obtained for $i \leq j$ by performing a similar analysis starting from (2.7.19). Thus (2.7.21) and (2.7.22) hold for all i and j. Writing (2.7.21) as a matrix equation, we have

$$
A + E = \hat{L}\hat{U}.
$$

Writing (2.7.22) as a matrix inequality, we have

$$
|E| \leq 2nu|\hat{L}| |\hat{U}| + O(u^2).
$$

The bound on $\| E \|_\infty$ follows immediately from the bound on $|E|$, using the properties of the matrix ∞-norm established in Exercise 2.7.11.   □

Exercise 2.7.23 Starting from (2.7.19), demonstrate that (2.7.21) and (2.7.22) hold for $i \le j$.   □

Proof of Theorem 2.7.14. In the forward substitution phase we compute $\hat{y}$ such that $(\hat{L} + \delta\hat{L})\hat{y} = b$, where

$$|\delta\hat{L}| \le 2nu |\hat{L}|$$

by Theorem 2.7.2. In the back substitution phase we compute $\hat{x}$ such that $(\hat{U} + \delta\hat{U})\hat{x} = \hat{y}$, where

$$|\delta\hat{U}| \le 2nu |\hat{U}|$$

by the upper-triangular analogue of Theorem 2.7.2. Thus $\hat{x}$ satisfies exactly

$$(\hat{L} + \delta\hat{L})(\hat{U} + \delta\hat{U})\hat{x} = b.$$

Multiplying out the product of matrices and using the fact (Theorem 2.7.13) that $\hat{L}\hat{U} = A + E$, where $|E| \le 2nu |\hat{L}| |\hat{U}| + O(u^2)$, we have

$$(A + \delta A)\hat{x} = b,$$

where

$$\delta A = E + (\delta\hat{L})\hat{U} + \hat{L}(\delta\hat{U}) + (\delta\hat{L})(\delta\hat{U}).$$

Applying the upper bounds that we have for E, $\delta\hat{L}$, and $\delta\hat{U}$, we obtain

$$|\delta A| \le 6nu |\hat{L}| |\hat{U}| + O(u^2).$$

The bound on $\| \delta A \|_\infty$ follows immediately from this bound.   □

Additional Exercises

Exercise 2.7.24 In Proposition 2.7.1 we showed that summation is backward stable, regardless of the order. In the notation of Proposition 2.7.1 the backward errors γ_j are bounded by $|\gamma_j| \le (n-1)u + O(u^2)$. In this exercise we obtain tighter bounds on $|\gamma_j|$ for two specific orders of summation of $\sum_{j=1}^{n} w_j$.

(a) First we consider the "obvious" order, reversed for notational convenience. Show that if we perform the computation in the order

$$(\cdots(((w_n + w_{n-1}) + w_{n-2})) + \cdots + w_2) + w_1$$

in floating-point arithmetic, then

$$|\gamma_j| \leq \begin{cases} ju + O(u^2) & \text{if } j < n, \\ (n-1)u + O(u^2) & \text{if } j = n. \end{cases}$$

(Look at small cases like $n = 3$ and $n = 4$, and observe the pattern.)

(b) Now consider summing by pairs. Calculate $w_1 + w_2$, $w_3 + w_4$, $w_5 + w_6$, and so on. If there is an odd term, just let it sit. Now you have a new list, which you can sum by pairs. Keep summing by pairs until you have a single sum. This is easiest to discuss when n is a power of 2, say $n = 2^k$. In this case, how many additions does each term participate in? Express your answer in terms of n. Show that if we carry out this process in floating-point arithmetic, we have

$$|\gamma_j| \leq (\log_2 n)u + O(u^2).$$

This is an excellent result, as $\log_2 n$ grows much more slowly than n.

$\square$

rcise 2.7.25 In this exercise you will assess the backward stability of Gaussian elimination by calculating backward error in the LU decomposition: $E = \hat{L}\hat{U} - A$. Write a MATLAB program that does Gaussian elimination without pivoting, for example

```
A = randn(n);
L = zeros(n); U = zeros(n);
for k = 1:n
   U(k,k:n)   = A(k,k:n)   - L(k,1:k-1)*U(1:k-1,k:n);
   L(k+1:n,k) = (A(k+1:n,k) - ...
                 L(k+1:n,1:k-1)*U(1:k-1,k))/U(k,k);
end
L = L + eye(n);
```

(a) Calculate the LU decomposition of random matrices (A = randn(n);) for several choices of n (e.g. $n = 40, 80, 160$), and note $\|L\|_\infty$, $\|U\|_\infty$, and the norm of the backward error: $\|E\|_\infty = \|LU - A\|_\infty$. On the same matrices do Gaussian elimination with partial pivoting ([L,U] = lu(A);) and calculate the same quantities. Notice that partial pivoting decreases the backward error and the norms of L and U, but the performance of Gaussian elimination without pivoting is usually not conspicuously bad. That is, usually Gaussian elimination without pivoting is able to calculate the LU decomposition (of a random matrix) more or less stably.

(b) To demonstrate the weakness of Gaussian elimination without pivoting, give it a matrix for which at least one of the pivots is guaranteed to be small. The easiest way to do this is to use matrices whose $(1,1)$ entry is tiny. Repeat the experiments from part (a) using matrices for which a_{11} is tiny. For example, take A = randn(n); A(1,1) = 50*eps*A(1,1); . MATLAB's *machine epsilon* eps equals $2u$, twice the unit roundoff. For these experiments n need not be large. Try several choices of n, but $n = 2$ is already big enough.

$\square$

Exercise 2.7.26 Let A_n denote the $n \times n$ matrix whose form is illustrated by

$$
A_5 = \begin{bmatrix}
1 & 0 & 0 & 0 & 1 \\
-1 & 1 & 0 & 0 & 1 \\
-1 & -1 & 1 & 0 & 1 \\
-1 & -1 & -1 & 1 & 1 \\
-1 & -1 & -1 & -1 & 1
\end{bmatrix}.
$$

Show that if Gaussian elimination with partial pivoting is used, then A_n can be reduced to upper-triangular form without row interchanges, and the resulting matrix U has $u_{nn} = 2^{n-1}$. Thus $\|U\|_\infty / \|A\|_\infty = 2^{n-1}/n$.   □

Exercise 2.7.27 It is pointed out in [91] that the LU decomposition of a random matrix is anything but random. If $A = LU$, then $U = L^{-1}A$. We have stability if $\|U\|$ is not too much larger than $\|A\|$, and this will be the case if $\|L^{-1}\|$ is not large. In other words, a necessary condition for instability is that $\|L^{-1}\|$ be large. In this exercise we will see by experiment that Gaussian elimination with partial pivoting tends to return matrices for which L^{-1} is not large.

(a) Write a MATLAB program that generates random unit lower-triangular matrices with entries between 1 and -1. For example, you can start with the identity matrix (L = eye(n)) and then fill in the lower triangular part. The command L(i,j) = 2*rand-1 gives a random number uniformly distributed in $[-1, 1]$. Calculate the norm of L^{-1} for several such matrices with $n = 40$, 80, and 160.

(b) Now generate the LU factors of random $n \times n$ matrices by

```
A = randn(n);
[L,U,P] = lu(A);
```

MATLAB's lu command does Gaussian elimination with partial pivoting. This way of using it produces a truly unit lower-triangular matrix L; the row interchanges are incorporated into the permutation matrix P. Calculate $\|L^{-1}\|$ for several matrices generated in this way with $n = 40$, 80, and 160. Contrast your results with those of part (a).

(c) Extend your code from part (b) so that it computes not only $\|L^{-1}\|$ but also $\|M^{-1}\|$, where M is generated from L by reversing the signs of the pivots: $M = 2I - L$.

    □

Exercise 2.7.28 Show that if $L \in \mathbb{R}^{n \times n}$ and $U \in \mathbb{R}^{n \times n}$ are full lower and upper triangular, respectively, the flop count for computing the product LU is $\frac{2}{3}n^3$.   □

Exercise 2.7.29 Here we prove the backward stability of Cholesky's method in detail. Let A be a positive definite matrix, and let $\hat{R}$ denote its Cholesky factor computed by some variant of Cholesky's method in floating-point arithmetic. Assume that square roots are calculated accurately: $\text{fl}(\sqrt{s}) = \sqrt{s}\,(1 + \epsilon)$, where $|\epsilon| \leq u$.

(a) Using the proof of Theorem 2.7.13 as a model, prove that $A + E = \hat{R}^T \hat{R}$, where

$$|E| \leq 2nu\,|\hat{R}|^T\,|\hat{R}| + O(u^2)$$

and

$$\|E\|_F \leq 2nu\|\hat{R}\|_F^2 + O(u^2).$$

(b) The *trace* of a matrix $B \in \mathbb{R}^{n \times n}$ is $\text{tr}(B) = \sum_{i=1}^n b_{ii}$. Use the Cauchy-Schwarz inequality to prove that $|\text{tr}(B)| \leq \sqrt{n}\,\|B\|_F$. (Notice that equality is attained when $B = I$. More commonly $|\text{tr}(B)| \approx \|B\|_F$.)

(c) Prove that if $A + E = \hat{R}^T \hat{R}$, then $\|\hat{R}\|_F^2 = \text{tr}(A + E) = \text{tr}(A) + \text{tr}(E)$. (This holds regardless of whether or not $\hat{R}$ is triangular.) Thus $\|\hat{R}\|_F^2 \leq \sqrt{n}\,(\|A\|_F + \|E\|_F)$.

(d) Substituting this last inequality into the result of part (a), show that

$$\|E\|_F \leq 2n^{3/2}u\,(\|A\|_F + \|E\|_F) + O(u^2)$$

and, if $2n^{3/2}u < 1$,

$$\|E\|_F \leq \frac{2n^{3/2}u}{1 - 2n^{3/2}u}\|A\|_F + O(u^2).$$

For realistic values of n and u we generally have $2n^{3/2}u \ll 1$, so the denominator in this expression is about 1. (Consider, e.g., the huge value $n = 10^6$ and IEEE double precision's $u \approx 10^{-16}$.) Furthermore, the factor $2n^{3/2}$ in the numerator is based on a pessimistic worst-case analysis and can be ignored. Thus we have, in practice,

$$\|E\|_F \approx Cu\|A\|_F,$$

where C is a modest multiple of 1, not $n^{3/2}$. Thus Cholesky's method for positive definite matrices is stable.

$\square$

2.8 SCALING

In a linear system $Ax = b$, any equation can be multiplied by any nonzero constant without changing the solution of the system. Such an operation is called a *row scaling* operation. A similar operation can be applied to a column of A. In contrast to row scaling operations, *column scaling* operations do change the solution.

Exercise 2.8.1 Show that if the nonsingular linear system $Ax = b$ is altered by multiplication of its jth column by $c \neq 0$, then the solution is altered only in the jth component, which is multiplied by $1/c$. $\square$

Scaling operations can be viewed as changes of measurement units. Suppose the entries of the jth column of A are masses expressed in grams, and x_j is an acceleration measured in meters/sec^2. Multiplication of the jth column by $1/1000$ is the same as changing the units of its entries from grams to kilograms. At the same time x_j is multiplied by 1000, which is the same as changing its units from meters/second^2 to millimeters/second^2.

A discussion of scaling operations is made necessary by the fact that these operations affect the numerical properties of the system. This discussion has been placed near the end of the chapter because in most cases rescaling is unnecessary and undesirable; usually an appropriate scaling is determined by the physical units of the problem. Consider for example the electrical circuit problem in Example 1.2.6. In the linear system derived there, all entries of the coefficient matrix have the same units (1/ohm), all components of the solution have the same units (volts), and all components of the right-hand side have the same units (amperes). One could rescale this system so that, for example, one of the unknowns is expressed in millivolts while the others remain in volts, but this should not be done without a good reason. In most cases it is best not to rescale.

Let us look at some examples that illustrate some of the effects of scaling.

Example 2.8.2 The first example shows that a small pivot cannot be "cured" by multiplying its row by a large number. Consider the system

$$\begin{bmatrix} 2.000 & 1231. & 2471. \\ 1.196 & 3.165 & 2.543 \\ 1.475 & 4.271 & 2.142 \end{bmatrix} \begin{bmatrix} x_1 \\ x_2 \\ x_3 \end{bmatrix} = \begin{bmatrix} 3704. \\ 6.904 \\ 7.888 \end{bmatrix},$$

which was obtained from the system (2.6.10) of Example 2.6.9 by multiplying the first row by 1000. We used (2.6.10) to illustrate the damaging effects of using a small number as a pivot. Now that the first row has been multiplied by 1000, the $(1,1)$ entry is no longer small. It is now the largest entry in the first column. Let us see what happens when it is used as a pivot.

Using four-digit decimal arithmetic, the multipliers for the first step are

$$l_{21} = \frac{1.196}{2.000} = .5980 \quad \text{and} \quad l_{31} = \frac{1.475}{2.000} = .7375.$$

Comparing these with (2.6.11), we see that they are 1000 times smaller than before. In step 1 the $(2,2)$ entry is altered as follows:

$$3.165 - (.5980)(1231.) = 3.165 - 736.1 = -732.9.$$

Comparing this with (2.6.12), we see that in spite of the smaller pivot, the outcome is the same as before. This time the number 3.165 is swamped by the large entry

1231. Notice that this computation is essentially identical with (2.6.12). The result is *exactly* the same, including the roundoff errors. You can check that when the $(2, 3)$, $(3, 2)$, and $(3, 3)$ entries are modified, swamping occurs just as before, and indeed the computations are essentially the same as before and yield exactly the same result. Thus, after the first step, the modified coefficient matrix is

$$\begin{bmatrix} 2.000 & 1231. & 2471. \\ \underline{.5980} & -732.9 & -1475. \\ .7375 & -903.6 & -1820. \end{bmatrix}.$$

The submatrix for the second step is

$$\begin{bmatrix} -732.9 & -1475. \\ -903.6 & -1820. \end{bmatrix},$$

which is exactly the same as before. If we continue the computation, we will have the same disastrous outcome. This time swamping occurred not because large multiples of the first row were subtracted from the other rows, but because the first row itself is large.

How could this disaster have been predicted? Looking at the coefficient matrix, we can see that it is ill conditioned: the rows (and the columns) are out of scale. It is interesting that we have two different explanations for the same disaster. With the original system we blamed a small pivot; with the rescaled system we blame ill conditioning. □

This example illustrated an interesting theorem of F. L. Bauer. Suppose we solve the system $Ax = b$ by Gaussian elimination, using some specified sequence of row and column interchanges, on a computer that uses base β floating-point arithmetic. If the system is then rescaled by multiplying the rows and columns by powers of β and solved again using the same sequence of row and column interchanges, the result will be exactly the same as before, including roundoff errors. All roundoff errors at all steps are the same as before. In our present example $\beta = 10$; multiplying the first row by 10^3 has no effect on the arithmetic. It is not hard to prove Bauer's theorem; you might like to do so as an exercise or at least convince yourself that it is true. The examples that follow should help.

Bauer's theorem has an interesting consequence. If the scaling factors are always chosen to be powers of β, then the only way rescaling affects the numerical properties of Gaussian elimination is by changing the choices of pivot. If scaling factors that are not powers of β are used, there will be additional roundoff errors associated with the rescaling, but it remains true that the principal effect of rescaling is to alter the pivot choices.

Example 2.8.3 Let us solve the system

$$\begin{bmatrix} .003 & .217 \\ .277 & .138 \end{bmatrix} \begin{bmatrix} x_1 \\ x_2 \end{bmatrix} = \begin{bmatrix} .437 \\ .553 \end{bmatrix} \tag{2.8.4}$$

using three-digit decimal floating-point arithmetic without row or column inter-
changes. The exact solution is $x = [1, \ 2]^T$. The multiplier is $l_{21} = .277/.003 = 92.3$, and $u_{22} = .138 - (92.3)(.217) = .138 - 20.0 = -19.9$, so the computed LU
decomposition is

$$\begin{bmatrix} 1 & 0 \\ 92.3 & 1 \end{bmatrix} \begin{bmatrix} .003 & .217 \\ 0 & -19.9 \end{bmatrix}.$$

The forward substitution gives $y_1 = .437$ and $y_2 = .553 - (92.3)(.437) = .553 - 40.3 = -39.7$. Finally the back substitution gives $x_2 = (-39.7)/(-19.9) = 1.99$
and $x_1 = .437 - (.217)(1.99)]/(.003) = (.437 - .432)/(.003) = (.005)/(.003) = 1.67$. Thus the computed solution is $\hat{x} = [1.67, \ 1.99]^T$, whose first component is
inaccurate.   □

Exercise 2.8.5

    (a) Calculate $\kappa_\infty(A)$, where A is the coefficient matrix of (2.8.4). Observe that A
        is well conditioned.

    (b) Perform Gaussian elimination on (2.8.4) with the rows interchanged, using
        three-digit decimal floating-point arithmetic, and note that an accurate solution
        is obtained. (Remember to round each intermediate result to three decimal
        places before using it in the next calculation.)

                                                               □

Example 2.8.6 Now let us solve

$$\begin{bmatrix} .300 & 21.7 \\ .277 & .138 \end{bmatrix} \begin{bmatrix} x_1 \\ x_2 \end{bmatrix} = \begin{bmatrix} 43.7 \\ .553 \end{bmatrix}, \qquad (2.8.7)$$

which was obtained by multiplying the first row of (2.8.4) by 10^2. Now the $(1, 1)$
entry is the largest entry in the first column. Again we use three-digit decimal
arithmetic and no row or column interchanges. By Bauer's theorem the outcome
should be the same as in Example 2.8.3. Let us check that it is. The multiplier is
$l_{21} = .277/.300 = .923$, and $u_{22} = .138 - (.923)(21.7) = .138 - 20.0 = -19.9$,
so the computed LU decomposition is

$$\begin{bmatrix} 1 & 0 \\ .923 & 1 \end{bmatrix} \begin{bmatrix} .300 & 21.7 \\ 0 & -19.9 \end{bmatrix}.$$

The forward substitution gives $y_1 = 43.7$ and $y_2 = .553 - (.923)(43.7) = .553 - 40.3 = -39.7$. Finally the back substitution yields $x_2 = (-39.7)/(-19.9) = 1.99$
and $x_1 = 43.7 - (21.7)(1.99)]/(.300) = (43.7 - 43.2)/(.300) = (.500)/(.300) = 1.67$. Thus the computed solution is again $\hat{x} = [1.67, \ 1.99]^T$. All intermediate
results are identical to those in Example 2.8.3, except for powers of 10.   □

rcise 2.8.8

    (a) Calculate $\kappa_\infty(A)$, where A is the coefficient matrix of (2.8.7). A is ill conditioned (relative to three-digit decimal arithmetic) because its rows (and columns) are out of scale.

    (b) Perform Gaussian elimination on (2.8.7) with the rows interchanged, using three-digit decimal arithmetic. Note that, as guaranteed by Bauer's theorem, the computations and outcome are identical to those in part (b) of Exercise 2.8.5. Thus an ill-conditioned coefficient matrix does not absolutely guarantee an inaccurate result. (However, if the partial-pivoting strategy had been used, the row interchange would not have been made, and the outcome would have been bad.)

$\square$

rcise 2.8.9   Solve (2.8.7) by Gaussian elimination with the columns interchanged, using three-digit decimal arithmetic. This is the complete pivoting strategy. Note that a good result is obtained. $\square$

Example 2.8.10 Now let us solve

$$\begin{bmatrix} .300 & .217 \\ .277 & .00138 \end{bmatrix} \begin{bmatrix} x_1 \\ x_2 \end{bmatrix} = \begin{bmatrix} 43.7 \\ .553 \end{bmatrix}, \tag{2.8.11}$$

which was obtained from (2.8.7) by multiplying the second column by $1/100$. The exact solution is therefore $x = [1, \, 200]^T$. The $(1,1)$ entry is now the largest entry in the matrix. Again we use three-digit decimal arithmetic and no row or column interchanges (which is the choice that both the partial and complete pivoting strategies would make). By Bauer's theorem the outcome should be the same as in Examples 2.8.3 and 2.8.6. The multiplier is $l_{21} = .277/.300 = .923$, and $u_{22} = .00138 - (.923)(.217) = .00138 - .200 = -.199$, so the computed LU decomposition is

$$\begin{bmatrix} 1 & 0 \\ .923 & 1 \end{bmatrix} \begin{bmatrix} .300 & .217 \\ 0 & -.199 \end{bmatrix}.$$

The forward substitution gives $y_1 = 43.7$ and $y_2 = .553 - (.923)(43.7) = .553 - 40.3 = -39.7$. Finally the back substitution yields $x_2 = (-39.7)/(-.199) = 199.$, and $x_1 = 43.7 - (.217)(199.)]/(.300) = (43.7 - 43.2)/(.300) = (.500)/(.300) = 1.67$. Thus the computed solution is $\hat{x} = [1.67, \, 199.]^T$. All computations were identical to those in Examples 2.8.3 and 2.8.6.

    Although the computed solution $\hat{x} = [1.67, \, 199.]^T$ has an inaccurate first component, it should not necessarily be viewed as a bad result. The inaccurate entry is much smaller than the accurate one, and in fact $\|\delta x\|_\infty/\|x\|_\infty = .005$, where $\delta x = \hat{x} - x$. This is an excellent outcome for three-digit decimal arithmetic. The small value of $\|\delta x\|_\infty/\|x\|_\infty$ is guaranteed by the well-conditioned coefficient matrix, together with the fact that the computed $\hat{L}$ and $\hat{U}$ do not have large entries (cf. Theorems 2.7.14 and 2.3.6).

It is easy to imagine situations in which the computed result $\hat{x} = [1.67 \ \ 199.]^T$ is acceptable. Suppose for example that x_1 and x_2 represent voltages expressed in the same units. If all that matters is the voltage difference, then the result is okay, since the computed difference $\hat{x}_2 - \hat{x}_1 = 197.33$ differs from the correct difference 199 by only about one percent. $\square$

Exercise 2.8.12

   (a) Calculate $\kappa_\infty(A)$, where A is the coefficient matrix of (2.8.11).

   (b) Perform Gaussian elimination on (2.8.11) with the rows interchanged, using three-digit decimal arithmetic.

   (c) Perform Gaussian elimination on (2.8.11) with the columns interchanged, using three-digit decimal arithmetic.

                                                                                              $\square$

2.9 COMPONENTWISE SENSITIVITY ANALYSIS

In this chapter we have taken the oldest and simplest approach to sensitivity analysis, in which everything is measured by norms. It is called *normwise sensitivity analysis*, and it is accompanied by *normwise backward error analysis*. This style of error analysis has been very successful, but there are some situations in which a different type of analysis, *componentwise sensitivity analysis*, is more appropriate. In the normwise analysis, δA is considered a small perturbation of A if $\|\delta A\|/\|A\|$ is small. This criterion does not force all of the ratios $|\delta a_{ij}|/|a_{ij}|$, which are the perturbations of the components, to be small.

Example 2.9.1 Suppose

$$A = \begin{bmatrix} 1.04 & 2.35 \\ 4.26 \times 10^{-6} & 6.32 \end{bmatrix} \quad \text{and} \quad \delta A = \begin{bmatrix} 1.32 \times 10^{-5} & 5.46 \times 10^{-6} \\ 1.02 \times 10^{-5} & 8.29 \times 10^{-6} \end{bmatrix}.$$

Then

$$A + \delta A = \begin{bmatrix} 1.0400132 & 2.35000546 \\ 1.446 \times 10^{-5} & 6.32000829 \end{bmatrix}.$$

We have $\|\delta A\|_\infty/\|A\|_\infty < 10^{-5}$, but this does not force $|\delta a_{21}|/|a_{21}|$ to be small. Obviously this can happen to any entry for which $|a_{ij}| \ll \|A\|$. $\square$

In componentwise sensitivity analysis, perturbations are considered small only if the perturbation in each component is small relative to that component, that is,

$$\max_{i,j} \frac{|\delta a_{ij}|}{|a_{ij}|}$$

is small. Since we often encounter matrices with zero entries, we prefer the following reformulation, in which a_{ij} does not appear in the denominator: The perturbation δA is *componentwise ϵ-small* with respect to A if there is a positive $\epsilon \ll 1$ such that

$$|\delta a_{ij}| \leq \epsilon |a_{ij}| \quad \text{for } i, j = 1, \dots, n. \tag{2.9.2}$$

Notice that under this condition, if $a_{ij} = 0$, then $\delta a_{ij} = 0$. Thus sparse matrices stay sparse and sparsity patterns are preserved under perturbations of this type.

Recall the following notation, which we introduced in Section 2.7. If B is a matrix (or vector) with (i, j) entry b_{ij}, then $|B|$ is the matrix with the same dimensions whose (i, j) entry is $|b_{ij}|$. We write $|B| \leq |C|$ to mean that $|b_{ij}| \leq |c_{ij}|$ for all i and j. With these notational conventions we can rewrite the condition (2.9.2) as

$$|\delta A| \leq \epsilon |A|. \tag{2.9.3}$$

In the componentwise sensitivity analysis we can ask the same sort of questions as we do in the normwise analysis. For example, if $Ax = b$ and $(A + \delta A)(x + \delta x) = b$, what is the largest δx can be relative to x? Before considering this and related questions, we pause to establish some basic facts about the matrix absolute value notation and matrix inequalities.

rcise 2.9.4

   (a) Show that if $A = BC$, then $|A| \leq |B| |C|$. (This is a matrix inequality.) In particular, $|Ax| \leq |A| |x|$.

   (b) (Review) Show that if $y = |x|$, then $\|y\|_\infty = \|x\|_\infty$.

$\square$

If $A \in \mathbb{R}^{n \times n}$ is nonsingular, we can build various other $n \times n$ matrices from A, for example, $|A|$, A^{-1}, $|A^{-1}|$, and $K = |A^{-1}| |A|$. This last one appears in various error bounds, as we shall see, and we use it to define a new type of condition number, the *Skeel condition number* of A:

$$\text{skeel}(A) = \|K\|_\infty = \| |A^{-1}| |A| \|_\infty.$$

Now we are ready to prove some theorems. The first is the componentwise analogue of Theorem 2.2.4

Theorem 2.9.5 *Let $A \in \mathbb{R}^{n \times n}$ be nonsingular, let $b \in \mathbb{R}^n$ be nonzero, and let x be the unique solution of $Ax = b$. Suppose $\hat{x} = x + \delta x$ is the solution of $A\hat{x} = b + \delta b$, where*

$$|\delta b| \leq \epsilon |b|.$$

Then

$$|\delta x| \leq \epsilon |A^{-1}| |A| |x| \tag{2.9.6}$$

and

$$\frac{\|\delta x\|_\infty}{\|x\|_\infty} \leq \epsilon \, \text{skeel}(A). \tag{2.9.7}$$

Exercise 2.9.8 Prove Theorem 2.9.5 as follows (or do it your own way).

   (a) Prove that $|\delta x| \leq |A^{-1}| \, |\delta b|$ and $|b| \leq |A| \, |x|$. Then deduce (2.9.6).

   (b) Deduce (2.9.7) from (2.9.6).

$\square$

Now consider this componentwise analogue of Theorems 2.3.3 and 2.3.6.

Theorem 2.9.9 *Let* $A \in \mathbb{R}^{n \times n}$ *be nonsingular, let* $b \in \mathbb{R}^n$ *be nonzero, and let* x *be the unique solution of* $Ax = b$. *Suppose* $\hat{x} = x + \delta x$ *satisfies* $(A + \delta A)\hat{x} = b$, *where*

$$|\delta A| \leq \epsilon \, |A|.$$

Then

$$|\delta x| \leq \epsilon \, |A^{-1}| \, |A| \, |\hat{x}| \tag{2.9.10}$$

and

$$\frac{\|\delta x\|_\infty}{\|\hat{x}\|_\infty} \leq \epsilon \, \mathrm{skeel}(A). \tag{2.9.11}$$

If $\epsilon \, \mathrm{skeel}(A) < 1$, *then also*

$$\frac{\|\delta x\|_\infty}{\|x\|_\infty} \leq \frac{\epsilon \, \mathrm{skeel}(A)}{1 - \epsilon \, \mathrm{skeel}(A)}. \tag{2.9.12}$$

Exercise 2.9.13 Prove Theorem 2.9.9 as follows (or do it your own way).

   (a) Prove that $\delta x = -A^{-1} \delta A \hat{x}$ and $|\delta x| \leq |A^{-1}| \, |\delta A| \, |\hat{x}|$. Then deduce (2.9.10).

   (b) Deduce (2.9.11) from (2.9.10).

   (c) Multiply (2.9.11) through by $\|\hat{x}\|_\infty$, apply the triangle inequality to break $\|\hat{x}\|_\infty$ into two parts, and deduce (2.9.12), remembering to point out where you are using the added hypothesis $\epsilon \, \mathrm{skeel}(A) < 1$.

$\square$

Componentwise Backward Error Analysis

Componentwise sensitivity analysis can be combined with componentwise backward stability analysis, whenever the latter is successful. One example of componentwise backward stability that we have already encountered is the forward substitution algorithm for solving triangular systems. Theorem 2.7.2 states that if we solve the triangular system $Gy = b$ by forward substitution using floating-point arithmetic, then the computed solution $\hat{y}$ satisfies $(G + \delta G)\hat{y} = b$, where $|\delta G| \leq 2nu|G| + O(u^2)$. This is a componentwise backward stability result, since it says that each component of δG is tiny relative to the corresponding component of G. As a practical matter we have $|\delta G| \leq Cu|G|$, where C is a modest multiple of 1. We can now apply Theorem 2.9.9 with $\epsilon = Cu$, to conclude that $\hat{y}$ is accurate (in the sense that $\|\delta y\|_\infty / \|y\|_\infty$ is tiny) if the Skeel condition number $\mathrm{skeel}(G)$ is not too large.

Iterative Refinement

Another process that yields componentwise backward stability is iterative refinement. This is an old procedure that was originally used to improve the accuracy of solutions to ill-conditioned systems. Let $\hat{x}$ denote an approximation to the solution of the system $Ax = b$, and let $\hat{r}$ be the associated residual: $\hat{r} = b - A\hat{x}$. The approximation $\hat{x}$ may have been obtained by Gaussian elimination, for example. If we could solve the residual system $Az = \hat{r}$ exactly, then the vector $x = \hat{x} + z$ would be the exact solution of $Ax = b$, as you can easily check. If we did obtain $\hat{x}$ by Gaussian elimination, then an LU decomposition is available, so we can solve $Az = \hat{r}$ inexpensively. Of course the computed solution $\hat{z}$ is not exact. If A is somewhat ill conditioned, it may be far from exact. Nevertheless it is not unreasonable to hope that $\hat{x} + \hat{z}$ will be an improvement over $\hat{x}$. If this is the case, then perhaps we can improve the solution even more by calculating the residual associated with $\hat{x} + \hat{z}$ and repeating the process. In fact we can repeat it as many times as we wish. This gives the following *iterative refinement* algorithm.

Iterative Refinement Algorithm

for $k = 1, \ldots, m$

$$
\begin{array}{l}
\hat{r} \leftarrow b - A\hat{x} \\
\text{Calculate } \hat{z}, \text{ an approximate solution of } Az = \hat{r}. \\
\quad (\text{Use the } LU \text{ decomposition that was computed previously.}) \\
\hat{x} \leftarrow \hat{x} + \hat{z} \\
\text{if } (\|\hat{z}\|/\|\hat{x}\| \text{ is sufficiently small}) \text{ exit} \quad (\text{successful completion})
\end{array}
$$

(2.9.14)

Set flag indicating failure.

This is called an *iterative* algorithm because the number of steps or *iterations* to be performed is not known in advance. The iterations are terminated as soon as the corrections become sufficiently small. Any iterative algorithm should have an upper bound on the number of iterations it is willing to attempt before abandoning the process as a failure. In (2.9.14) that number is denoted by m. Notice that in order to carry out this procedure, we must save copies of A and b to use in the computation of the residuals.

Up until about 1980 it was thought that (2.9.14) cannot hope to succeed unless the residuals are calculated in extended-precision arithmetic. This means that if we are using single-precision arithmetic, the step $\hat{r} \leftarrow b - A\hat{x}$ should be done in double precision. The reason for this is that severe cancellation occurs when $A\hat{x}$ is subtracted from b. The smaller $\hat{r}$ becomes, the worse the cancellation is. The objective of the extended-precision computation is to preserve as many significant digits as possible in the face of this cancellation.

If iterative refinement is carried out in this way and the system is not too badly conditioned (say $\kappa(A) \ll 1/u$), (2.9.14) actually does converge to the true solution of $Ax = b$. Thus iterative refinement can be used to solve the problem to full precision ($\|x - \hat{x}\| \approx u\|x\|$). There is a catch, however. The system that is solved so precisely is the one whose coefficient matrix A and right-hand-side vector b are exactly what is stored in the computer. Because of measurement and representation errors, the stored A and b are mere approximations to the true data for the physical problem that we are trying to solve. If the problem is ill conditioned, the exact solution of $Ax = b$ can be a very unsatisfactory approximation to the solution of the problem we really wish to solve.

Around 1980 it was realized that iterative refinement is useful even if the residuals are not computed with extended precision arithmetic. Full precision ($\| x - \hat{x} \| \approx u \| x \|$) cannot be attained, but some improvement in accuracy is possible. Furthermore, iterative refinement has a good effect on the backward error. Skeel [78] showed that if the system is not too badly conditioned and not too badly out of scale, then *one step* of iterative refinement is usually enough to ensure a *componentwise* backward stable solution. See [52] or [78] for details.

Componentwise backward stability means that the computed solution satisfies $(A + \delta A)\hat{x} = b$, where $|\delta A| \leq Cu|A|$, with C not too big. This can then be combined with Theorem 2.9.9 to get the bound

$$\frac{\| \delta x \|_\infty}{\| \hat{x} \|_\infty} \leq C\, u\, \mathrm{skeel}(A).$$

This is sometimes significantly better than bounds that can be obtained using the normwise error analysis, because $\mathrm{skeel}(A)$ can be much smaller than $\kappa_\infty(A)$ (See Exercise 2.9.15). The greatest advantage comes with matrices that are ill-conditioned simply because the rows are out of scale. The Skeel condition number $\mathrm{skeel}(A)$ is insensitive to row scaling, so it remains small while the normwise condition number $\kappa_\infty(A)$ becomes large in proportion to the badness of the scaling.

There are inexpensive methods for estimating $\mathrm{skeel}(A)$ that work on the same principal as condition estimators for $\kappa_1(A)$. See [52].

Exercise 2.9.15

(a) Show that for every nonsingular matrix A, $\mathrm{skeel}(A) \leq \kappa_\infty(A)$.

(b) Show that if D is a nonsingular diagonal matrix, then $|DA| = |D|\,|A|$ and $|D|^{-1} = |D^{-1}|$.

(c) Show that the Skeel condition number is invariant under row scaling; that is, $\mathrm{skeel}(DA) = \mathrm{skeel}(A)$ for all nonsingular diagonal matrices D.

(d) From part (a) we know that the ratio $\kappa_\infty(A)/\mathrm{skeel}(A)$ is always at least one. Show by example that it can be made arbitrarily large.

$\square$

rcise 2.9.16

(a) Prove that if A is nonsingular, $A + \delta A$ is singular, and $|\delta A| \leq \epsilon |A|$, then $\epsilon \, \text{skeel}(A) \geq 1$ (cf. Theorem 2.3.1).

(b) Prove that if A is nonsingular and $|\delta A| \leq \epsilon |A|$, where $\epsilon \, \text{skeel}(A) < 1$, then $A + \delta A$ is nonsingular.

(c) Discuss the relationship between the result in part (b) and Theorem 2.3.1. Is one stronger than the other?

$\square$

CHAPTER 3

THE LEAST SQUARES PROBLEM

In this chapter we study the least squares problem, which arises repeatedly in scientific and engineering computations. After describing the problem in Section 3.1, we immediately begin to develop the tools that we need to solve the problem: rotators, reflectors, and the QR decomposition, in Section 3.2. In Section 3.3 we show how to use a QR decomposition to solve the least squares problem. In Section 3.4 we introduce the Gram-Schmidt orthonormalization process, demonstrate its relationship to the QR decomposition, and study some of its computational variants. In the interest of getting to the algorithms quickly, we have postponed discussion of some important theoretical issues to Section 3.5. Section 3.6 addresses the important question of updating the QR decomposition when a row or column is added to or deleted from the data matrix.

3.1 THE DISCRETE LEAST SQUARES PROBLEM

A task that occurs frequently in scientific investigations is that of finding a straight line that "fits" some set of data points. Typically we have a fairly large number of points (t_i, y_i), $i = 1, \ldots, n$, collected from some experiment, and often we have some theoretical reason to believe that these points should lie on a straight line. Thus

Fundamentals of Matrix Computations, Third Edition. By David S. Watkins
Copyright © 2010 John Wiley & Sons, Inc.

we seek a linear function $p(t) = a_0 + a_1 t$ such that $p(t_i) = y_i$, $i = 1, \ldots, n$. In practice of course the points will deviate from a straight line, so it is impossible to find a linear $p(t)$ that passes through all of them. Instead we settle for a line that fits the data well, in the sense that the errors

$$|y_i - p(t_i)| \qquad i = 1, \ldots, n$$

are made as small as possible.

It is generally impossible to find a p for which all of the numbers $|y_i - p(t_i)|$ are simultaneously minimized, so instead we seek a p that strikes a good compromise. Specifically, let $r = [r_1, \cdots, r_n]^T$ denote the vector of residuals $r_i = y_i - p(t_i)$. We can solve our problem by choosing a vector norm $\| \cdot \|$ and taking our compromise function to be that p for which $\|r\|$ is made as small as possible. Of course the solution depends on the choice of norm. For example, if we choose the Euclidean norm, we minimize the quantity

$$\|r\|_2 = \left(\sum_{i=1}^{n} |y_i - p(t_i)|^2 \right)^{1/2},$$

whereas if we choose the 1-norm or the ∞-norm, we minimize respectively the quantities

$$\|r\|_1 = \sum_{i=1}^{n} |y_i - p(t_i)| \qquad \text{and} \qquad \|r\|_\infty = \max_{1 \le i \le n} |y_i - p(t_i)|.$$

The problem of finding the minimizing p has been studied for a variety of norms, including those we have just mentioned. By far the nicest theory is that based on the Euclidean norm, and it is that theory that we will study in this chapter. To minimize $\|r\|_2$ is the same as to minimize

$$\|r\|_2^2 = \sum_{i=1}^{n} |y_i - p(t_i)|^2.$$

Thus we are minimizing the sum of the squares of the residuals. For this reason the problem of minimizing $\|r\|_2$ is called the *least squares problem*.[15]

The choice of the 2-norm can be justified on statistical grounds. Suppose the data fail to lie on a straight line because of errors in the measured y_i. If the errors are independent and normally distributed with mean zero and variance σ^2, then the solution of the least squares problem is the maximum likelihood estimator of the true solution.

A straight line is the graph of a polynomial of first degree. Sometimes it is desirable to fit a data set with a polynomial of higher degree. For example, it might be useful to approximate the data of Figure 3.1 by a polynomial of degree 2. If we

---

[15]It is called the *discrete* least squares problem because a finite (discrete) data set (t_i, y_i) is being approximated. The *continuous* least squares problem, in which a continuum of data points is approximated, will be discussed briefly in Section 3.5.

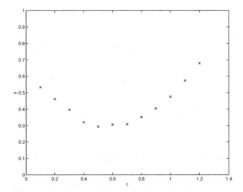

Figure 3.1   Data that can be approximated well by a quadratic polynomial

decide to approximate our data by a polynomial of degree $\leq m - 1$, then the task is to seek $p(t) = a_0 + a_1 t + a_2 t^2 + \cdots + a_{m-1} t^{m-1}$ such that

$$p(t_i) = y_i \qquad i = 1, \ldots, n. \tag{3.1.1}$$

Since the number of data points is typically large and the degree of the polynomial fairly low, it will usually be the case that $n \gg m$. In this case it is too much to ask for a p that satisfies (3.1.1) exactly, but for the moment let us act as if that were our goal. The set of polynomials of degree $\leq m - 1$ is a vector space of dimension m. If $\phi_1, \phi_2, \ldots, \phi_m$ is a basis of this space, then each polynomial p in the space can be expressed in the form

$$p(t) = \sum_{j=1}^{m} x_j \phi_j(t) \tag{3.1.2}$$

for some unique choice of coefficients x_1, x_2, ..., x_m. The obvious basis is $\phi_1(t) = 1$, $\phi_2(t) = t$, $\phi_3(t) = t^2$, ..., $\phi_m(t) = t^{m-1}$, but there are many others, some of which may be better from a computational standpoint. See Example 4.4.15.

Substituting the expression (3.1.2) into the equations (3.1.1), we obtain a set of n linear equations in the m unknowns $x_1, \ldots, x_m$:

$$\sum_{j=1}^{m} x_j \phi_j(t_i) = y_i, \qquad i = 1, \ldots, n,$$

which can be written in matrix form as

$$\begin{bmatrix} \phi_1(t_1) & \phi_2(t_1) & \cdots & \phi_m(t_1) \\ \phi_1(t_2) & \phi_2(t_2) & \cdots & \phi_m(t_2) \\ \phi_1(t_3) & \phi_2(t_3) & \cdots & \phi_m(t_3) \\ \vdots & \vdots & & \vdots \\ \phi_1(t_n) & \phi_2(t_n) & \cdots & \phi_m(t_n) \end{bmatrix} \begin{bmatrix} x_1 \\ x_2 \\ \vdots \\ x_m \end{bmatrix} = \begin{bmatrix} y_1 \\ y_2 \\ y_3 \\ \vdots \\ y_n \end{bmatrix}. \tag{3.1.3}$$

If $n > m$, as is usually the case, this is an *overdetermined system*; that is, it has more equations than unknowns. Thus we cannot expect to find an x that satisfies (3.1.3) exactly. Instead we might seek an x for which the sum of the squares of the residuals is minimized.

It is easy to imagine further generalizations of this problem. For example, the functions $\phi_1, \ldots, \phi_m$ could be taken to be trigonometric or exponential or some other kind of functions. More generally we can consider the overdetermined system

$$Ax = b, \qquad (3.1.4)$$

where $A \in \mathbb{R}^{n \times m}$, $n > m$, and $b \in \mathbb{R}^n$. The *least squares problem* for the system (3.1.4) is to find $x \in \mathbb{R}^m$ for which $\|r\|_2$ is minimized, where $r = b - Ax$ is the vector of residuals. With the help of the orthogonal matrices introduced in the next section, we will develop algorithms for solving the least squares problem (3.1.4), which includes (3.1.3) as a special case.

Exercises

Exercise 3.1.5 Consider the following data.

t_i	1.0	1.5	2.0	2.5	3.0
y_i	1.1	1.2	1.3	1.3	1.4

(a) Set up an overdetermined system of the form (3.1.3) for a straight line passing through the data points. Use the standard basis polynomials $\phi_1(t) = 1$, $\phi_2(t) = t$.

(b) Use MATLAB to calculate the least-squares solution of the system from part (a). This is a simple matter. Given an overdetermined system $Ax = b$, the MATLAB command x = A\b computes the least squares solution. Recall that this is exactly the same command as would be used to tell MATLAB to solve a square system $Ax = b$ by Gaussian elimination.[16] Some useful MATLAB commands:

        t = 1:.5:3; t = t'; s = ones(5,1); A = [s t];

We already know that MATLAB uses Gaussian elimination with partial pivoting in the square case. In the next two sections you will find out what MATLAB does in the overdetermined case.

(c) Use the MATLAB plot command to plot the five data points and your least squares straight line. Type help plot for information about using the plot command, or search in MATLAB's help browser.

(d) Use MATLAB to compute $\|r\|_2$, the norm of the residual.

$\square$

---

[16] In the square case A is expected to be nonsingular. In the overdetermined case it is expected to have *full rank*. Rank will be discussed in Section 3.3 and in subsequent sections.

rcise 3.1.6 Repeat Exercise 3.1.5, but this time compute the best least squares polynomial of degree ≤ 2. Notice that in this case the norm of the residual is smaller. This is to be expected; the space of quadratic polynomials contains the space of linear polynomials, so the quadratic fit should be better than—or in any case no worse than—the linear polynomial fit.   □

rcise 3.1.7 Repeat Exercise 3.1.5, but this time compute the best least squares polynomial of degree ≤ 4. The space of quartic polynomials has dimension 5. Thus we have $n = m = 5$; the system is not overdetermined. The solution interpolates the data exactly (except for roundoff errors). Plot the data points and the solution on the same set of axes. Make the spacing between points small enough that the curve appears smooth. Sample code:

```
tt = .08:.01:3.2;
p4 = x(1)+tt.*(x(2)+tt.*(x(3)+tt.*(x(4)+tt.*x(5))));
plot(...,tt,p4,'k-',...)
```

In the same plot include your least squares linear and quadratic polynomials from the previous two exercises. Notice that the latter are much less oscillatory than the fourth-degree interpolant is. They seem to represent the trend of the data better.   □

rcise 3.1.8 Set up and solve a system of equations for the best least squares fit to the data of Exercise 3.1.5 from the space spanned by the three functions $\phi_1(t) = 1$, $\phi_2(t) = e^t$, and $\phi_3(t) = e^{-t}$.   □

rcise 3.1.9 Consider the following data.

t_i	1	2	3
y_i	1.1	1.3	1.4

(a) Set up a system of three equations in three unknowns that could be solved to find the unique quadratic polynomial that interpolates the data. (As you may know, this is just one of several ways to solve this problem.)

(b) Solve the system using MATLAB to find the interpolant.

(c) Set up a system of three equations in three unknowns that could be solved to find the unique linear combination of the functions $\phi_1(t) = 1$, $\phi_2(t) = e^t$, and $\phi_3(t) = e^{-t}$ that interpolates the data.

(d) Solve the system using MATLAB to find this other interpolant.

(e) Plot the two interpolants on the same set of axes.

□

3.2 ORTHOGONAL MATRICES, ROTATORS, AND REFLECTORS

In this section we will develop powerful tools for solving the least squares problem. As these tools will also be used heavily in the chapters that follow, this is one of the most important sections of the book.

As we did in the first two chapters, we will restrict our attention to real vectors and matrices. However, the entire theory can be carried over to the complex case, and this is done in part in exercises at the end of the section. The results of those exercises will be used in Chapter 5.

We begin by introducing an inner product in $\mathbb{R}^n$. Given two vectors $x = [x_1, \ldots, x_n]^T$ and $y = [y_1, \ldots, y_n]^T$ in $\mathbb{R}^n$, we define the *inner product* of x and y, denoted $\langle x, y \rangle$ by

$$\langle x, y \rangle = \sum_{i=1}^{n} x_i y_i.$$

Although the inner product is a real number, it can also be expressed as a matrix product:

$$\langle x, y \rangle = y^T x = x^T y.$$

The inner product has the following properties, which you can easily verify:

$$
\begin{aligned}
\langle x, y \rangle &= \langle y, x \rangle \\
\langle \alpha_1 x_1 + \alpha_2 x_2, y \rangle &= \alpha_1 \langle x_1, y \rangle + \alpha_2 \langle x_2, y \rangle \\
\langle x, \alpha_1 y_1 + \alpha_2 y_2 \rangle &= \alpha_1 \langle x, y_1 \rangle + \alpha_2 \langle x, y_2 \rangle \\
\langle x, x \rangle &\geq 0 \qquad \text{with equality if and only if } x = 0.
\end{aligned}
$$

for all $x, x_1, x_2, y, y_1, y_2 \in \mathbb{R}^n$ and all $\alpha_1, \alpha_2 \in \mathbb{R}$.

Note also the close relationship between the inner product and the Euclidean norm:

$$\|x\|_2 = \sqrt{\langle x, x \rangle}.$$

The Cauchy-Schwarz inequality (Theorem 2.1.6) can be stated more concisely in terms of the inner product and the Euclidean norm: For every $x, y \in \mathbb{R}^n$,

$$|\langle x, y \rangle| \leq \|x\|_2 \|y\|_2. \tag{3.2.1}$$

When $n = 2$ (or 3) the inner product coincides with the familiar dot product from analytic geometry. Recall that if x and y are two nonzero vectors in a plane and θ is the angle between them, then

$$\cos \theta = \frac{\langle x, y \rangle}{\|x\|_2 \|y\|_2}.$$

It is not unreasonable to employ this formula to *define* the angle between two vectors in $\mathbb{R}^n$. Note first that (3.2.1) guarantees that $\langle x, y \rangle / \|x\|_2 \|y\|_2$ always lies between -1 and 1, so it is the cosine of some angle. We now define the *angle* between (nonzero) x and $y \in \mathbb{R}^n$ to be

$$\theta = \arccos \frac{\langle x, y \rangle}{\|x\|_2 \|y\|_2}. \tag{3.2.2}$$

If $x = 0$ or $y = 0$, we define $\theta = \pi/2$ radians. Two vectors x and y are said to be *orthogonal* if the angle between them is $\pi/2$ radians. Clearly x and y are orthogonal if and only if $\langle x, y \rangle = 0$.

The angle between two vectors in, say, $\mathbb{R}^{100}$ is just as real as the angle between two vectors in a plane. In fact x and y span a two-dimensional subspace of $\mathbb{R}^{100}$. This subspace is nothing but a copy of $\mathbb{R}^2$, that is, a plane. Viewed in this plane, x and y have an angle between them. It is this angle that is produced by the formula (3.2.2).

Orthogonal Matrices

A matrix $Q \in \mathbb{R}^{n \times n}$ is said to be *orthogonal* if $QQ^T = I$. This equation says that Q has an inverse, and $Q^{-1} = Q^T$. Since a matrix always commutes with its inverse, we have $Q^T Q = I$ as well. For square matrices the equations

$$QQ^T = I \qquad Q^T Q = I \qquad Q^T = Q^{-1}$$

are all equivalent, and any one of them could be taken as the definition of an orthogonal matrix.

Exercise 3.2.3 (a) Show that if Q is orthogonal, then Q^{-1} is orthogonal. (b) Show that if Q_1 and Q_2 are orthogonal, then $Q_1 Q_2$ is orthogonal. ☐

Exercise 3.2.4 Show that if Q is orthogonal, then $\det(Q) = \pm 1$. ☐

Exercise 3.2.5 Recall that a *permutation matrix* is a matrix that has exactly one 1 in each row and in each column, all other entries being zero. In Section 1.8 we introduced permutation matrices as a way to represent row interchanges in Gaussian elimination. Show that every permutation matrix is orthogonal. ☐

Theorem 3.2.6 *If $Q \in \mathbb{R}^{n \times n}$ is orthogonal, then for all x, $y \in \mathbb{R}^n$,*

$$(a) \ \langle Qx, Qy \rangle = \langle x, y \rangle, \qquad (b) \ \| Qx \|_2 = \| x \|_2.$$

Proof. (a) $\langle Qx, Qy \rangle = (Qy)^T (Qx) = y^T Q^T Q x = y^T I x = y^T x = \langle x, y \rangle$.
(b) Set $y = x$ in part (a) and take square roots. ☐

Part (b) of the theorem says that Qx and x have the same length. Thus *orthogonal transformations preserve lengths.* Combining parts (a) and (b), we see that

$$\arccos \frac{\langle Qx, Qy \rangle}{\| Qx \|_2 \| Qy \|_2} = \arccos \frac{\langle x, y \rangle}{\| x \|_2 \| y \|_2}.$$

Thus the angle between Qx and Qy is the same as the angle between x and y. We conclude that *orthogonal transformations preserve angles.*

In the least squares problem for an overdetermined system we wish to find x such that $\| b - Ax \|_2$ is minimized. Theorem 3.2.6, part (b), shows that for every orthogonal matrix Q, $\| b - Ax \|_2 = \| Qb - QAx \|_2$. Therefore the solution of the least squares problem is unchanged when A and b are replaced by QA and Qb, respectively. We will eventually solve the least squares problem by finding a Q for which QA has a very simple form, from which the solution of the least squares problem will be easy to determine.

Exercise 3.2.7 Prove the converse of part (a) of Theorem 3.2.6: If Q satisfies $\langle Qx, Qy \rangle = \langle x, y \rangle$ for all x and y in $\mathbb{R}^n$, then Q is orthogonal. (The condition of part (b) also implies that Q is orthogonal, but this is harder to prove.)   □

Exercise 3.2.8 Show that if Q is orthogonal, then $\|Q\|_2 = 1$, $\|Q^{-1}\|_2 = 1$, and $\kappa_2(Q) = 1$. thus Q is perfectly conditioned with respect to the 2-condition number. This suggests that orthogonal matrices will have good computational properties.   □

There are two types of orthogonal transformations that are widely used in matrix computations: rotators and reflectors. Along with Gaussian elimination operations, these are fundamental building blocks of matrix computations. All matrix computations built upon rotators and/or reflectors, as described in this section, are normwise backward stable. This claim will be discussed later in this section and in Exercises 3.2.71–3.2.74.

We will introduce rotators first because they are simpler.

Rotators

Consider vectors in the plane $\mathbb{R}^2$. The operator that rotates each vector through a fixed angle θ is a linear transformation, so it can be represented by a matrix. Let

$$Q = \begin{bmatrix} q_{11} & q_{12} \\ q_{21} & q_{22} \end{bmatrix}$$

be this matrix. Then Q is completely determined by its action on the two vectors $\begin{bmatrix} 1 \\ 0 \end{bmatrix}$ and $\begin{bmatrix} 0 \\ 1 \end{bmatrix}$, because

$$\begin{bmatrix} q_{11} & q_{12} \\ q_{21} & q_{22} \end{bmatrix} \begin{bmatrix} 1 \\ 0 \end{bmatrix} = \begin{bmatrix} q_{11} \\ q_{21} \end{bmatrix} \quad \text{and} \quad \begin{bmatrix} q_{11} & q_{12} \\ q_{21} & q_{22} \end{bmatrix} \begin{bmatrix} 0 \\ 1 \end{bmatrix} = \begin{bmatrix} q_{12} \\ q_{22} \end{bmatrix}.$$

Since the action of Q is to rotate each vector through the angle θ, clearly (see Figure 3.2)

$$Q \begin{bmatrix} 1 \\ 0 \end{bmatrix} = \begin{bmatrix} \cos\theta \\ \sin\theta \end{bmatrix} \quad \text{and} \quad Q \begin{bmatrix} 0 \\ 1 \end{bmatrix} = \begin{bmatrix} -\sin\theta \\ \cos\theta \end{bmatrix}.$$

Thus $\begin{bmatrix} q_{11} \\ q_{21} \end{bmatrix} = \begin{bmatrix} \cos\theta \\ \sin\theta \end{bmatrix}$ and $\begin{bmatrix} q_{12} \\ q_{22} \end{bmatrix} = \begin{bmatrix} -\sin\theta \\ \cos\theta \end{bmatrix}$; that is,

$$Q = \begin{bmatrix} \cos\theta & -\sin\theta \\ \sin\theta & \cos\theta \end{bmatrix}.$$

A matrix of this form is called a *rotator* or *rotation*.

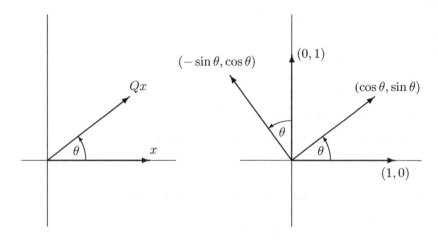

Figure 3.2 Rotation through angle θ

rcise 3.2.9 Verify that every rotator is an orthogonal matrix with determinant 1. What does the inverse of a rotator look like? What transformation does it represent?   □

Rotators can be used to create zeros in a vector or matrix. For example, if $x = \begin{bmatrix} x_1 \\ x_2 \end{bmatrix}$ is a vector with $x_2 \neq 0$, let us see how to find a rotator

$$Q = \begin{bmatrix} \cos\theta & -\sin\theta \\ \sin\theta & \cos\theta \end{bmatrix}$$

such that $Q^T x$ has a zero in its second component: $Q^T x = \begin{bmatrix} y \\ 0 \end{bmatrix}$ for some y. Now

$$Q^T x = \begin{bmatrix} \cos\theta & \sin\theta \\ -\sin\theta & \cos\theta \end{bmatrix} \begin{bmatrix} x_1 \\ x_2 \end{bmatrix} = \begin{bmatrix} (\cos\theta)x_1 + (\sin\theta)x_2 \\ -(\sin\theta)x_1 + (\cos\theta)x_2 \end{bmatrix},$$

which has the form $\begin{bmatrix} y \\ 0 \end{bmatrix}$ if and only if

$$x_1 \sin\theta = x_2 \cos\theta. \tag{3.2.10}$$

Thus θ can be taken to be $\arctan(x_2/x_1)$ or any other angle satisfying $\tan\theta = x_2/x_1$. But notice that we can determine Q without calculating θ itself; we need only $\cos\theta$ and $\sin\theta$. Clearly the choice $\cos\theta = x_1$ and $\sin\theta = x_2$ would satisfy (3.2.10). However, this choice generally violates the basic trigonometric identity $\cos^2\theta + \sin^2\theta = 1$. Therefore we take instead

$$\cos\theta = \frac{x_1}{\sqrt{x_1^2 + x_2^2}} \quad \text{and} \quad \sin\theta = \frac{x_2}{\sqrt{x_1^2 + x_2^2}}, \tag{3.2.11}$$

obtained by dividing our original choice by $\| x \|_2 = \sqrt{x_1^2 + x_2^2}$. Now the basic trigonometric identity is satisfied, and there is a unique $\theta \in [0, 2\pi)$ for which (3.2.11) holds.

Exercise 3.2.12 We have just shown that for every $x \in \mathbb{R}^2$ there exists a rotator Q such that $Q^T x = \begin{bmatrix} y \\ 0 \end{bmatrix}$ for some y.

    (a) Give the geometric interpretation of this fact.

    (b) Show that if $\cos \theta$ and $\sin \theta$ are given by (3.2.11), then $y = \sqrt{x_1^2 + x_2^2} = \| x \|_2$.

$\square$

Now let us see how we can use a rotator to simplify a matrix

$$A = \begin{bmatrix} a_{11} & a_{12} \\ a_{21} & a_{22} \end{bmatrix}.$$

We have seen that there is a rotator Q such that

$$Q^T \begin{bmatrix} a_{11} \\ a_{21} \end{bmatrix} = \begin{bmatrix} r_{11} \\ 0 \end{bmatrix},$$

where $r_{11} = \sqrt{a_{11}^2 + a_{21}^2}$. Define r_{12} and r_{22} by

$$\begin{bmatrix} r_{12} \\ r_{22} \end{bmatrix} = Q^T \begin{bmatrix} a_{12} \\ a_{22} \end{bmatrix},$$

and let

$$R = \begin{bmatrix} r_{11} & r_{12} \\ 0 & r_{22} \end{bmatrix}.$$

Then $Q^T A = R$. This shows that we can transform A to an upper triangular matrix by multiplying it by the orthogonal matrix Q^T. As we shall soon see, it is possible to carry out such a transformation not just for 2-by-2 matrices, but for all $A \in \mathbb{R}^{n \times n}$. That is, for every $A \in \mathbb{R}^{n \times n}$ there is an orthogonal matrix $Q \in \mathbb{R}^{n \times n}$ and an upper triangular matrix $R \in \mathbb{R}^{n \times n}$ such that $Q^T A = R$.

A transformation of this type can be used to solve a system of linear equations $Ax = b$. Multiplying on the left by Q^T, we transform the system to the equivalent system $Q^T Ax = Q^T b$, or $Rx = c$, where $c = Q^T b$. It is a simple matter to calculate c, given b. Then x can be obtained by solving the upper triangular system $Rx = c$ by back substitution.

Another useful viewpoint is reached by rewriting the equations $Q^T A = R$ as

$$A = QR.$$

This shows that A can be expressed as a product QR, where Q is orthogonal and R is upper triangular. The *QR decomposition*, as it is called, can be used in much the same way as an LU decomposition to solve a linear system $Ax = b$. If we have a QR decomposition of A, we can rewrite the system as $QRx = b$. Letting $c = Rx$, we can find c by solving $Qc = b$. This doesn't cost much; Q is orthogonal, so $c = Q^T b$. We can then obtain x by solving $Rx = c$ by back substitution. The computations required here are no different from those required by the method outlined in the previous paragraph. Thus we have actually derived a single method from two different points of view.

Example 3.2.13 We will use a QR decomposition to solve the system

$$\begin{bmatrix} 1 & 2 \\ 1 & 3 \end{bmatrix} \begin{bmatrix} z_1 \\ z_2 \end{bmatrix} = \begin{bmatrix} 1 \\ 2 \end{bmatrix}.$$

First we require Q such that $Q^T \begin{bmatrix} 1 \\ 1 \end{bmatrix} = \begin{bmatrix} * \\ 0 \end{bmatrix}$. We use (3.2.11) with $x_1 = 1$ and $x_2 = 1$ to get $c = \cos\theta = 1/\sqrt{2}$ and $s = \sin\theta = 1/\sqrt{2}$. Then

$$Q = \begin{bmatrix} c & -s \\ s & c \end{bmatrix} = \frac{1}{\sqrt{2}} \begin{bmatrix} 1 & -1 \\ 1 & 1 \end{bmatrix}$$

and

$$R = Q^T A = \frac{1}{\sqrt{2}} \begin{bmatrix} 1 & 1 \\ -1 & 1 \end{bmatrix} \begin{bmatrix} 1 & 2 \\ 1 & 3 \end{bmatrix} = \frac{1}{\sqrt{2}} \begin{bmatrix} 2 & 5 \\ 0 & 1 \end{bmatrix}.$$

Solving $Qc = b$, we have

$$c = Q^T b = \frac{1}{\sqrt{2}} \begin{bmatrix} 1 & 1 \\ -1 & 1 \end{bmatrix} \begin{bmatrix} 1 \\ 2 \end{bmatrix} = \frac{1}{\sqrt{2}} \begin{bmatrix} 3 \\ 1 \end{bmatrix}.$$

Finally we solve $Rz = c$ by back substitution and get $z_2 = 1$ and $z_1 = -1$. $\square$

rcise 3.2.14 Use a QR decomposition to solve the linear system

$$\begin{bmatrix} 2 & 3 \\ 5 & 7 \end{bmatrix} \begin{bmatrix} x_1 \\ x_2 \end{bmatrix} = \begin{bmatrix} 12 \\ 29 \end{bmatrix}.$$

$\square$

We now turn our attention to $n \times n$ matrices. A *plane rotator* is a matrix of the form

$$Q = \begin{bmatrix}
1 & & & & & & & & & \\
& \ddots & & & & & & & & \\
& & 1 & & & & & & & \\
& & & c & & & -s & & & & \leftarrow i \\
& & & & 1 & & & & & \\
& & & & & \ddots & & & & \\
& & & & & & 1 & & & \\
& & & s & & & c & & & & \leftarrow j \\
& & & & & & & 1 & & \\
& & & & & & & & \ddots & \\
& & & & & & & & & 1
\end{bmatrix}$$
$$\quad\quad\quad\quad \underset{i}{\uparrow} \quad\quad\quad \underset{j}{\uparrow}$$

(3.2.15)

$$c = \cos\theta, \qquad s = \sin\theta.$$

All of the entries that have not been filled in are zeros. Thus a plane rotator looks like an identity matrix, except that one pair of rows and columns contains a rotator. Plane rotators are used extensively in matrix computations. They are sometimes called *Givens rotators* or *Jacobi rotators*, depending on the context[17]. We will usually drop the adjective "plane" and refer simply to *rotators*.

Exercise 3.2.16 Prove that every plane rotator is orthogonal and has determinant 1. ☐

Exercise 3.2.17 Let Q be the plane rotator (3.2.15). Show that the transformations $x \to Qx$ and $x \to Q^T x$ alter only the ith and jth entries of x and that the effect on these entries is the same as that of the 2-by-2 rotators $\hat{Q} = \begin{bmatrix} c & -s \\ s & c \end{bmatrix}$ and $\hat{Q}^T = \begin{bmatrix} c & s \\ -s & c \end{bmatrix}$ on the vector $\begin{bmatrix} x_i \\ x_j \end{bmatrix}$. ☐

From Exercise 3.2.17 and (3.2.11) we see that we can transform any vector x to one whose jth entry is zero by applying the plane rotator Q^T, where

$$c = \frac{x_i}{\sqrt{x_i^2 + x_j^2}}, \qquad s = \frac{x_j}{\sqrt{x_i^2 + x_j^2}}. \qquad (3.2.18)$$

---

[17]Rotators applied as described in this section are Givens rotators. Jacobi rotators are discussed in Section 7.2, especially Exercise 7.2.47

If $x_i = x_j = 0$, take $c = 1$ and $s = 0$. In computer programs that use rotators, the norms of the form $\sqrt{x_i^2 + x_j^2}$ should be computed by the method shown in (2.5.2) to avoid any possibility of overflow or harmful underflow. Thus (3.2.18) could be implemented as follows:

$$
\begin{aligned}
&\beta \leftarrow \max\{|x_i|, |x_j|\} \\
&\text{if } \beta = 0 \\
&\quad \left[\; c \leftarrow 1, \;\; s \leftarrow 0 \right. \\
&\text{else} \\
&\quad \left[\begin{array}{l}
\hat{x}_i \leftarrow x_i/\beta, \;\; \hat{x}_j \leftarrow x_j/\beta \\
\nu \leftarrow \sqrt{\hat{x}_i^2 + \hat{x}_j^2} \\
c \leftarrow \hat{x}_i/\nu, \;\; s \leftarrow \hat{x}_j/\nu
\end{array}\right.
\end{aligned}
\qquad (3.2.19)
$$

Another way of organizing these calculations is worked out in Exercise 3.2.62.

We now consider the effect of a rotator on a matrix. Let $A \in \mathbb{R}^{n \times m}$, and consider the transformations $A \to QA$ and $A \to Q^T A$, where Q is as in (3.2.15). It follows easily from Exercise 3.2.17 that these transformations alter only the ith and jth rows of A. Transposing these results, we see that for $B \in \mathbb{R}^{m \times n}$ the transformations $B \to BQ$ and $B \to BQ^T$ alter only the ith and jth columns of B.

rcise 3.2.20

(a) Show that the ith and jth rows of QA are linear combinations of the ith and jth rows of A.

(b) Show that the ith and jth columns of BQ are linear combinations of the ith and jth columns of B.

$\square$

The geometric interpretation of the action of a plane rotator is clear. All vectors lying in the $x_i x_j$ plane are rotated through an angle θ. All vectors orthogonal to the $x_i x_j$ plane are left fixed. A typical vector x is neither in the $x_i x_j$ plane nor orthogonal to it but can be expressed uniquely as a sum $x = p + p^{\perp}$, where p is in the $x_i x_j$ plane, and $p^{\perp}$ is orthogonal to it. The plane rotator rotates p through an angle θ and leaves $p^{\perp}$ fixed.

Theorem 3.2.21 Let $A \in \mathbb{R}^{n \times n}$. Then there exists an orthogonal matrix Q and an upper triangular matrix R such that $A = QR$.

Proof. We will sketch a proof in which Q is taken to be a product of rotators. A more detailed proof using reflectors will be given later. Let Q_{21} be a rotator acting on the $x_1 x_2$ plane, such that Q_{21}^T makes the transformation

$$
\begin{bmatrix}
a_{11} \\
a_{21} \\
a_{31} \\
a_{41} \\
\vdots \\
a_{n1}
\end{bmatrix}
\rightarrow
\begin{bmatrix}
* \\
0 \\
a_{31} \\
a_{41} \\
\vdots \\
a_{n1}
\end{bmatrix} .
$$

Then $Q_{21}^T A$ has a zero in the $(2,1)$ position. Similarly we can find a plane rotator Q_{31}, acting in the $x_1 x_3$ plane, such that $Q_{31}^T(Q_{21}^T A)$ has a zero in the $(3,1)$ position. This rotator does not disturb the zero in the $(2,1)$ position because Q_{31}^T leaves the second row of $Q_{21}^T A$ unchanged. Continuing in this manner, we create rotators Q_{41}, $Q_{51}, \ldots, Q_{n1}$, such that $Q_{n1}^T \cdots Q_{21}^T A$ has zeros in the entire first column, except for the $(1,1)$ position.

Now we go to work on the second column. Let Q_{32} be a plane rotator acting in the $x_2 x_3$ plane, such that the $(3,2)$ entry of $Q_{32}^T(Q_{n1}^T \cdots Q_{21}^T A)$ is zero. This rotator does not disturb the zeros that were created previously in the first column. (Why not?) Continuing on the second column, let $Q_{42}, Q_{52}, \ldots, Q_{n2}$ be rotators such that $Q_{n2}^T \cdots Q_{32}^T Q_{n1}^T \cdots Q_{21}^T A$ has zeros in columns 1 and 2 below the main diagonal.

Next we take care of the third column, fourth column, and so on. In all we create rotators $Q_{21}, Q_{31}, \ldots, Q_{n1}, Q_{32}, \ldots, Q_{n,n-1}$, such that

$$
R = Q_{n,n-1}^T Q_{n,n-2}^T \cdots Q_{21}^T A
$$

is upper triangular. Let $Q = Q_{21} Q_{31} \cdots Q_{n,n-1}$. Then Q, being a product of orthogonal matrices, is itself orthogonal, and $R = Q^T A$; that is, $A = QR$. □

The proof of Theorem 3.2.21 is constructive; it gives us an algorithm for computing Q and R.

Exercise 3.2.22 Show that the algorithm sketched in the proof of Theorem 3.2.21 takes $O(n^3)$ flops to transform A to R.[18] □

Exercise 3.2.23 A mathematically precise proof of Theorem 3.2.21 would use induction on n. Sketch an inductive proof of Theorem 3.2.21. □

Reflectors

We begin with the case $n = 2$, just as we did for rotators. Let $\mathcal{L}$ be any line in $\mathbb{R}^2$ that passes through the origin. The operator that reflects each vector in $\mathbb{R}^2$ through

---

[18] If you count carefully, you will find that $2n^3 + O(n^2)$ flops are needed. The number of multiplications is about twice the number of additions. However, there are clever ways to implement rotators so that the number of multiplications is halved. Rotators implemented in this special way are called *fast Givens rotations* or simply *fast rotators*. See [43] and other references cited there for a discussion of fast rotators.

the line $\mathcal{L}$ is a linear transformation, so it can be represented by a matrix. Let us determine that matrix. Let v be a nonzero vector lying in $\mathcal{L}$. Then every vector that lies in $\mathcal{L}$ is a multiple of v. Let u be a nonzero vector orthogonal to $\mathcal{L}$. Then $\{u, v\}$ is a basis of $\mathbb{R}^2$, so every $x \in \mathbb{R}^2$ can be expressed as a linear combination of u and v: $x = \alpha u + \beta v$. The reflection of x through $\mathcal{L}$ is $-\alpha u + \beta v$ (see Figure 3.3), so the matrix Q of the reflection must satisfy $Q(\alpha u + \beta v) = -\alpha u + \beta v$ for all α and β. For this it is necessary and sufficient that

$$Qu = -u \qquad \text{and} \qquad Qv = v.$$

Without loss of generality we can assume that u was chosen to be a unit vector: $\|u\|_2 = 1$.

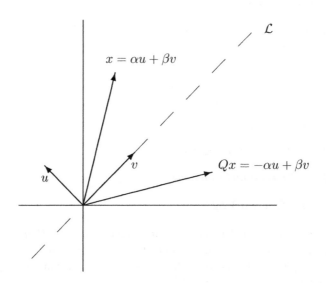

Figure 3.3 Reflection through a line

Consider the matrix $P = uu^T$. P does not have the desired properties, but it brings us closer to our goal. We have $Pu = (uu^T)u = u(u^Tu) = u\|u\|_2^2 = u$ and $Pv = (uu^T)v = u(u^Tv) = 0$, because u and v are orthogonal. With a little thought we can produce a Q with the desired properties by combining P and the identity matrix. Indeed let $Q = I - 2P$. Then $Qu = u - 2Pu = -u$ and $Qv = v - 2Pv = v$. To summarize, the matrix $Q \in \mathbb{R}^{2 \times 2}$ that reflects vectors through the line $\mathcal{L}$ is given by $Q = I - 2uu^T$, where u is a unit vector orthogonal to $\mathcal{L}$.

Now we are ready to consider n-by-n reflectors.

Theorem 3.2.24 *Let $u \in \mathbb{R}^n$ with $\| u \|_2 = 1$, and define $P \in \mathbb{R}^{n \times n}$ by $P = uu^T$. Then*

(a) $Pu = u$.

(b) $Pv = 0$ *if* $\langle u, v \rangle = 0$.

(c) $P^2 = P$.

(d) $P^T = P$.

Remark 3.2.25 A matrix satisfying $P^2 = P$ is called a *projector* or *idempotent*. A projector that is also symmetric ($P^T = P$) is called an *orthoprojector*. The matrix $P = uu^T$ has rank 1, since its range consists of multiples of u. Thus the properties of P can be summarized by saying that P is a rank-1 orthoprojector.   □

Exercise 3.2.26 Prove Theorem 3.2.24.   □

Theorem 3.2.27 *Let $u \in \mathbb{R}^n$ with $\| u \|_2 = 1$, and define $Q \in \mathbb{R}^{n \times n}$ by $Q = I - 2uu^T$. Then*

(a) $Qu = -u$.

(b) $Qv = v$ *if* $\langle u, v \rangle = 0$.

(c) $Q = Q^T$ *(Q is symmetric)*.

(d) $Q^T = Q^{-1}$ *(Q is orthogonal)*.

(e) $Q^{-1} = Q$ *(Q is an involution)*.

Exercise 3.2.28 Prove Theorem 3.2.27.   □

Matrices $Q = I - 2uu^T$ ($\| u \|_2 = 1$) are called *reflectors* or *Householder transformations*, after A. S. Householder, who first employed them in matrix computations. Reflectors, like Gauss transforms (Exercise 1.7.49), are rank-one updates of the identity matrix. See Exercise 3.2.65.

The set $\mathcal{H} = \{ v \in \mathbb{R}^n \mid \langle u, v \rangle = 0 \}$ is an $(n-1)$-dimensional subspace of $\mathbb{R}^n$ known as a *hyperplane*. The matrix Q maps each vector x to its reflection through the hyperplane $\mathcal{H}$. This can be visualized by thinking of the case $n = 3$, in which $\mathcal{H}$ is just an ordinary plane through the origin.

In Theorems 3.2.24 and 3.2.27, u was taken to be a unit vector. This is a convenient choice for the statements and proofs of those theorems, but in computations it is usually more convenient not to normalize the vector. The following theorem makes it possible to skip the normalization.

Proposition 3.2.29 *Let u be a nonzero vector in $\mathbb{R}^n$, and define $\gamma = 2/\| u \|_2^2$ and $Q = I - \gamma uu^T$. Then Q is a reflector satisfying*

(a) $Qu = -u$.

(b) $Qv = v$ if $\langle u, v \rangle = 0$.

Proof. Let $\hat{u} = u/\|u\|_2$. Then $\|\hat{u}\| = 1$, and it is a simple matter to check that $Q = I - 2\hat{u}\hat{u}^T$. Properties (a) and (b) follow easily. $\square$

rcise 3.2.30 Fill in the details of the proof of Proposition 3.2.29. $\square$

Theorem 3.2.31 *Let x, $y \in \mathbb{R}^n$ with $x \neq y$ but $\|x\|_2 = \|y\|_2$. Then there is a unique reflector Q such that $Qx = y$.*

Proof. We will not prove uniqueness because it is not important to our development (but see Exercise 3.2.64). To establish the existence of Q we must find u such that $(I - \gamma uu^T)x = y$, where $\gamma = 2/\|u\|_2^2$. To see how to proceed, consider the case $n = 2$, which is depicted in part (a) of Figure 3.4. Let $\mathcal{L}$ denote the line that bisects

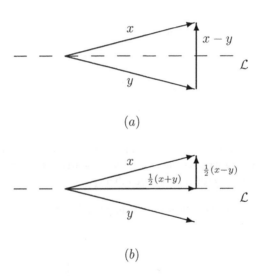

(a)

(b)

Figure 3.4 Reflecting x to y.

the angle between x and y. The reflection of x through this (and only this) line is y. Thus we require a vector u that is orthogonal to $\mathcal{L}$. It appears that $u = x - y$, or any multiple thereof, is the right choice.

Let $u = x - y$, $\gamma = 2/\|u\|_2^2$, and $Q = I - \gamma uu^T$. To prove that $Qx = y$ we first decompose x into a sum:

$$x = \frac{1}{2}(x - y) + \frac{1}{2}(x + y).$$

By part (a) of Proposition 3.2.29, $Q(x - y) = y - x$. Part (b) of Figure 3.4 suggests that $x + y$ is orthogonal to u. To check that this is so, we simply compute the inner product:

$$
\begin{aligned}
\langle x + y, x - y \rangle &= \langle x, x \rangle + \langle y, x \rangle - \langle x, y \rangle - \langle y, y \rangle \\
&= \| x \|_2^2 + 0 - \| y \|_2^2 = 0,
\end{aligned}
$$

because $\| x \|_2 = \| y \|_2$. It follows by part (b) of Proposition 3.2.29 that $Q(x + y) = x + y$. Finally

$$
\begin{aligned}
Qx &= \frac{1}{2} Q(x - y) + \frac{1}{2} Q(x + y) \\
&= \frac{1}{2}(y - x) + \frac{1}{2}(x + y) = y.
\end{aligned}
$$

$\square$

It follows easily from this theorem that reflectors can be used to create zeros in vectors and matrices.

Corollary 3.2.32 *Let $x \in \mathbb{R}^n$ be any nonzero vector. Then there exists a reflector Q such that*

$$
Q \begin{bmatrix} x_1 \\ x_2 \\ \vdots \\ x_n \end{bmatrix} = \begin{bmatrix} * \\ 0 \\ \vdots \\ 0 \end{bmatrix}.
$$

Proof. Let $y = [-\tau, \ 0, \ \ldots, \ 0]^T$, where $\tau = \pm \| x \|_2$. By choosing the sign appropriately we can guarantee that $x \neq y$. Clearly $\| x \|_2 = \| y \|_2$. Thus by Theorem 3.2.31, there is a reflector Q such that $Qx = y$. $\square$

Let us take a look at the construction suggested in this proof. The reflector is $Q = I - \gamma u u^T$, where

$$
u = x - y = \begin{bmatrix} \tau + x_1 \\ x_2 \\ \vdots \\ x_n \end{bmatrix}
$$

and $\gamma = 2/\| u \|_2^2$. Any multiple of this u will generate the same reflector. It turns out to be convenient to normalize u so that its first entry is 1. Thus we will take

$$
u = (x - y)/(\tau + x_1) = \begin{bmatrix} 1 \\ x_2/(\tau + x_1) \\ \vdots \\ x_n/(\tau + x_1) \end{bmatrix} \tag{3.2.33}
$$

instead. We still have $\gamma = 2/\| u \|_2^2$, but now we are referring to the rescaled u (3.2.33). It is a simple matter to calculate u and γ. We require that $\tau = \pm \| x \|_2$, but we did not specify the sign. In theory either choice can be used (so long as $x \neq y$), but in practice the easiest course is to choose τ so that its sign is the same as that of x_1. This ensures that cancellation does not occur in the computation of $\tau + x_1$. It also guarantees that when we normalize u by dividing by $\tau + x_1$, we do not divide by zero, nor do we generate dangerously large entries in u.

rcise 3.2.34 Show that if $\tau = \pm \| x \|_2$ is chosen so that τ and x_1 have the same sign, then all entries of u in (3.2.33) satisfy $|u_i| \leq 1$. Show further that $1 \leq \| u \|_2 \leq 2$. □

The calculation of γ is even easier than it looks. Notice that

$$\| u \|_2^2 = \frac{(\tau + x_1)^2 + x_2^2 + \cdots + x_n^2}{(\tau + x_1)^2}$$
$$= \frac{\tau^2 + 2\tau x_1 + \| x \|_2^2}{(\tau + x_1)^2}.$$

But $\tau^2 = \| x \|_2^2$, regardless of which sign was chosen for τ, so

$$\| u \|_2^2 = 2\tau(\tau + x_1)/(\tau + x_1)^2 = 2\tau/(\tau + x_1),$$

and

$$\gamma = (\tau + x_1)/\tau.$$

The whole procedure for calculating Q is summarized in (3.2.35). In the computation of the Euclidean norm, we have incorporated the procedure (2.5.2) to avoid overflow and harmful underflow. To keep the description brief we have used the MATLAB-like notation $x_{i:j}$ to denote entries i through j of vector x.

Given $x \in \mathbb{R}^n$, this algorithm calculates τ, γ, and u such that $Q = (I - \gamma u u^T)$ is a reflector for which $Qx = [-\tau \ 0 \ \cdots \ 0]^T$. If $x = 0$, γ is set to zero, giving $Q = I$. Otherwise, γ and u are produced, and u is stored over x. Since u_1 is known to be 1, the operation $x_1 \leftarrow 1$ could be skipped, depending on how the output is to be used.

$$
\begin{aligned}
&\beta \leftarrow \max_{1 \leq i \leq n} |x_i| \\
&\text{if } (\beta = 0) \\
&\text{then} \\
&\quad [\ \gamma \leftarrow 0 \\
&\text{else} \\
&\left[
\begin{aligned}
&x_{1:n} \leftarrow x_{1:n}/\beta \\
&\tau \leftarrow \sqrt{x_1^2 + \cdots + x_n^2} \\
&\text{if } (x_1 < 0) \ \tau \leftarrow -\tau \\
&x_1 \leftarrow \tau + x_1 \\
&\gamma \leftarrow x_1/\tau \\
&x_{2:n} \leftarrow x_{2:n}/x_1 \\
&x_1 \leftarrow 1 \\
&\tau \leftarrow \tau\beta
\end{aligned}
\right.
\end{aligned}
$$

(3.2.35)

You can readily see that cancellation cannot occur in this algorithm: The sum $x_1^2 + \cdots + x_n^2$ involves positive numbers only, and we have arranged the computation so that cancellation does not occur in the sum $\tau + x_1$. Therefore the computed τ, γ, and u are accurate.

Exercise 3.2.36 Check that the total work to execute Algorithm (3.2.35) is $O(n)$.   ☐

Algorithm (3.2.35) does not calculate Q explicitly, but then Q is seldom needed; it suffices to save γ and u. For example, when reflectors are used to compute the QR decomposition of a matrix, we need to multiply each reflector Q by a matrix B that is a submatrix of the matrix that is undergoing a transformation to upper-triangular form. Let us see how we can compute QB economically. Suppose $Q \in \mathbb{R}^{n \times n}$, and $B \in \mathbb{R}^{n \times m}$. We have $QB = (I - \gamma u u^T)B = B - \gamma u u^T B$, so the main task is to calculate the product $\gamma u u^T B$. There are good and bad ways to do this. The first good thing we can do is absorb the scalar γ into one of the vectors. Let $v^T = \gamma u^T$, so that $QB = B - u v^T B$.

Exercise 3.2.37 The amount of work required to compute $u v^T B$ depends dramatically upon the order in which the operations are performed. Suppose, as above, that $u \in \mathbb{R}^n$, $v \in \mathbb{R}^n$, and $B \in \mathbb{R}^{n \times m}$.

(a) Show that if we compute $(u v^T)B$, the intermediate result is an $n \times n$ matrix, and the total computation requires about $2n^2 m$ flops.

(b) Show that if we compute $u(v^T B)$, the intermediate result is a $1 \times m$ matrix, and the total computation requires about $3nm$ flops. This arrangement clearly requires much less storage space and arithmetic than the arrangement of part (a).

(c) Show that the computation $QB = B - u v^T B$ requires about $4nm$ flops in all if it is done in the economical way.

(d) How many flops are needed to compute QB if Q is stored in the conventional way as an $n \times n$ matrix?

☐

From this exercise we see that the economical approach to computing QB is to compute $B - u(v^T B)$. The operations are summarized in the following algorithm.

Algorithm to calculate QB and store it over B, where $B \in \mathbb{R}^{n \times m}$ and $Q = (I - \gamma u u^T)$. An auxiliary vector $v \in \mathbb{R}^n$ is used.

$$\begin{aligned} v^T &\leftarrow \gamma u^T \\ v^T &\leftarrow v^T B \\ B &\leftarrow B - u v^T \end{aligned} \qquad (3.2.38)$$

The total flop count is about $4nm$, which is dramatically less expensive than assembling Q explicitly and multiplying Q by B in the conventional manner. This saving is possible because the reflector Q is a rank-one update of the identity (Exercise 3.2.65).

Algorithm (3.2.38) can be implemented as either column-oriented or row-oriented code, depending upon the loop order in the two matrix-matrix multiplications. If column-oriented code is chosen, the auxiliary vector v can be replaced by a single auxiliary scalar, and the flop count can be lowered marginally, by careful organization. However, these modifications will not necessarily result in faster code.

rcise 3.2.39

(a) Find a reflector Q that maps the vector $[3,\ 4,\ 1,\ 3,\ 1]^T$ to a vector of the form $[-\tau,\ 0,\ 0,\ 0,\ 0]^T$. (You need not concern yourself with precautions to avoid over/underflow.) Write Q two ways: (i) in the form $I - \gamma u u^T$ and (ii) as a completely assembled matrix.

(b) Let $a = [0,\ 2,\ 1,\ -1,\ 0]^T$. Calculate Qa two different ways: (i) the efficient way, using $I - \gamma u u^T$, and (ii) using the assembled matrix Q.

$\square$

QR Decomposition by Reflectors

Theorem 3.2.21 states that every $A \in \mathbb{R}^{n \times n}$ can be expressed as a product $A = QR$, where Q is orthogonal and R is upper triangular. We used rotators to prove Theorem 3.2.21, but we promised a second proof based on reflectors.

Proof of Theorem 3.2.21 using reflectors. The proof is by induction on n. When $n = 1$, take $Q = [1]$ and $R = [a_{11}]$ to get $A = QR$. Now we will take an arbitrary $n \geq 2$ and show that the theorem holds for n-by-n matrices if it holds for $(n-1)$-by-$(n-1)$ matrices. Let $Q_1 \in \mathbb{R}^{n \times n}$ be a reflector that creates zeros in the first column of A:

$$Q_1 \begin{bmatrix} a_{11} \\ a_{21} \\ \vdots \\ a_{n1} \end{bmatrix} = \begin{bmatrix} -\tau_1 \\ 0 \\ \vdots \\ 0 \end{bmatrix}. \tag{3.2.40}$$

Recalling that Q_1 is symmetric, we see that $Q_1^T A$ has the form

$$Q_1^T A = Q_1 A = \left[\begin{array}{c|ccc} -\tau_1 & \hat{a}_{12} & \cdots & \hat{a}_{1n} \\ \hline 0 & & & \\ \vdots & & \hat{A}_2 & \\ 0 & & & \end{array} \right]. \tag{3.2.41}$$

By the induction hypothesis $\hat{A}_2$ has a QR decomposition: $\hat{A}_2 = \hat{Q}_2 \hat{R}_2$, where $\hat{Q}_2$ is orthogonal and $\hat{R}_2$ is upper triangular. Define $\tilde{Q}_2 \in \mathbb{R}^{n \times n}$ by

$$\tilde{Q}_2 = \left[\begin{array}{c|ccc} 1 & 0 & \cdots & 0 \\ \hline 0 & & & \\ \vdots & & \hat{Q}_2 & \\ 0 & & & \end{array} \right].$$

Then obviously $\tilde{Q}_2$ is orthogonal, and

$$
\tilde{Q}_2^T Q_1^T A =
\begin{bmatrix}
1 & 0 \cdots 0 \\
\hline
0 & \\
\vdots & \hat{Q}_2 \\
0 &
\end{bmatrix}
\begin{bmatrix}
-\tau_1 & \hat{a}_{12} \cdots \hat{a}_{1n} \\
\hline
0 & \\
\vdots & \hat{A}_2 \\
0 &
\end{bmatrix}
\qquad (3.2.42)
$$

$$
=
\begin{bmatrix}
-\tau_1 & \hat{a}_{12} \cdots \hat{a}_{1n} \\
\hline
0 & \\
\vdots & \hat{R}_2 \\
0 &
\end{bmatrix} .
$$

This matrix is upper triangular; let us call it R. Let $Q = Q_1 \tilde{Q}_2$. Then Q is orthogonal, and $Q^T A = R$. Therefore $A = QR$. □

This proof can be turned into an algorithm for constructing Q and R. Step 1 calculates a reflector $Q_1 = I - \gamma_1 u^{(1)} u^{(1)T}$ to perform the task (3.2.40) and then uses Q_1 to make the transformation (3.2.41). In practice we need not form Q_1 explicitly; we can just save $-\tau_1$, γ_1, and $u^{(1)}$, which were calculated by (3.2.35) with x taken to be the first column of A. There is no need to transform the first column to the form $[-\tau_1 \ 0 \ \cdots \ 0]^T$, as shown in (3.2.41); at least the zeros need not be stored. We can store $-\tau_1 = r_{11}$ in the $(1, 1)$ position, and the rest of the first column can be used to store $u^{(1)}$. Since $u_1^{(1)} = 1$, there is no need to store it, and there is just enough room for the rest of $u^{(1)}$ in the rest of the first column of A. The transformation is achieved simply by dividing each of $a_{21}, \ldots a_{n1}$ by $\tau_1 + a_{11}$. The rest of A can be transformed as indicated in (3.2.41). This is done by applying reflector Q_1 to the submatrix of A consisting of columns 2 through n, using algorithm (3.2.38). This is the expensive part of the computation. Since $m = n - 1$ columns are involved here, the total flop count for step 1 is about $4n^2$. Since the cost of building the reflector is $O(n)$, we ignore it.

The rest of the algorithm consists of reducing $\hat{A}_2$ to upper triangular form. Comparing (3.2.41) with (3.2.42), we see that the first row and column remain unchanged after step 1, so they can be ignored. Step 2 is identical to step 1, except that it operates on $\hat{A}_2$ to convert it to the form

$$
\begin{bmatrix}
-\tau_2 & \hat{a}_{23} \cdots \hat{a}_{2n} \\
\hline
0 & \\
\vdots & \hat{A}_3 \\
0 &
\end{bmatrix} .
$$

In practice we use the zero part of the first column to store $u^{(2)}$. The cost of step 2 is about $4(n - 1)^2$ flops, since it is identical to step 1, except that it acts on an $(n - 1) \times (n - 1)$ matrix. Step 3 is identical to step 2, except that it acts on the smaller matrix $\hat{A}_3$. The cost of the third step is about $4(n - 2)^2$ flops.

After $n - 1$ steps the matrix has been transformed to upper triangular form. The array that originally held A now holds R and the vectors $u^{(1)}, \ldots, u^{(n-1)}$, stripped of their leading 1's. Another array holds $\gamma_1, \ldots, \gamma_{n-1}$. Clearly $R = Q_{n-1}Q_{n-2}\cdots Q_1 A$, where $Q_1 = I - \gamma_1 u^{(1)} u^{(1)T}$,

$$
Q_2 = \left[\begin{array}{c|ccc}
1 & 0 & \cdots & 0 \\
\hline
0 & & & \\
\vdots & & I - \gamma_2 u^{(2)} u^{(2)T} & \\
0 & & &
\end{array}\right],
$$

and in general

$$
Q_i = \left[\begin{array}{c|c}
I_{i-1} & 0 \\
\hline
0 & I - \gamma_i u^{(i)} u^{(i)T}
\end{array}\right],
$$

where I_{i-1} denotes the identity matrix of dimension $i - 1$. Letting

$$
Q = Q_1 Q_2 \cdots Q_{n-1},
$$

we have $Q^T = Q_{n-1}^T \cdots Q_1^T = Q_{n-1} \cdots Q_1$, $R = Q^T A$, and finally $A = QR$. As we shall see, there is no need to assemble Q explicitly.

The algorithm can be summarized as follows.

Algorithm to calculate the QR decomposition of $A \in \mathbb{R}^{n \times n}$ using reflectors

> for $k = 1, \ldots, n-1$
> > Determine a reflector $Q_k = I - \gamma_k u^{(k)} u^{(k)T}$ such that
> > $Q_k[a_{kk} \cdots a_{nk}]^T = [-\tau_k \; 0 \; \cdots \; 0]^T$ (see (3.2.35))
> > Store $u^{(k)}$ over $a_{k:n,k}$, as in (3.2.35).   (3.2.43)
> > $a_{k:n,k+1:n} \leftarrow Q_k \, a_{k:n,k+1:n}$ (see (3.2.38))
> > $a_{kk} \leftarrow -\tau_k$
> $\gamma_n \leftarrow a_{nn}$

We have used the MATLAB-like notation $a_{i:j,k:m}$ to denote the submatrix consisting of the intersection of rows i through j with columns k through m. γ_n functions as a flag. If γ_n or any other γ_i is zero, A is singular. (Why?)

Recalling that the flop count for step one is $4n^2$, for step two is $4(n-1)^2$, and so on, we see that the total flop count for (3.2.43) is about $4n^2 + 4(n-1)^2 + 4(n-2)^2 + \cdots \approx 4n^3/3$, twice that of an LU decomposition.

Most of the work is in the operations $a_{k:n,k+1:n} \leftarrow Q_k \, a_{k:n,k+1:n}$, which should be organized as indicated in (3.2.38). These operations can be done in a column-oriented or row-oriented way.

rcise 3.2.44 If the assembled Q matrix is wanted, for one reason or another, it can be obtained by starting with an array that contains the identity matrix and then applying $Q_1, Q_2, \ldots, Q_{n-1}$ to it in an appropriate order. Show that this costs about $4n^3/3$ additional flops. Thus this doubles the flop count.   □

There are also block variants of the algorithm that are suitable for application to large matrices. These delay the application of the reflectors. Once a certain number of reflectors have been constructed, the whole batch is applied at once in a way that is rich in matrix-matrix multiplications. These operations can be organized into blocks for efficient use of high-speed cache memory and parallel processors. The main idea is worked out in Exercises 3.2.68 and 3.2.69. The LAPACK [1] routine DGEQRF is a Fortran implementation of the QR decomposition by blocks.

Let us see how the QR decomposition, as calculated using reflectors, can be used to solve the system of linear equations $Ax = b$. This is admittedly an expensive way to solve a linear system, but it is a good illustration because the operations are similar to those that we will use to solve least squares problems for overdetermined systems in Section 3.3. Since $A = QR$, our task is to solve $QRx = b$. Letting $c = Rx$, we must first solve $Qc = b$ for c. Then we can solve $Rx = c$ for x. First of all $c = Q^T b = Q_{n-1} \cdots Q_2 Q_1 b$, so we can apply the reflectors $Q_1, Q_2, \ldots, Q_{n-1}$ to b successively to obtain c. Naturally we use (3.2.38) with $m = 1$. Finally we solve $Rx = c$ for x by back substitution.

Applying Q_1 costs $4n$ flops. Applying Q_2 costs only $4(n - 1)$ flops because it is really only an $(n - 1) \times (n - 1)$ reflector; it does not touch the first entry of $Q_1 b$. Similarly Q_3 does not touch the first two entries of $Q_2 Q_1 b$, and the cost of applying it is $4(n - 2)$ flops. The total cost of calculating c is therefore $4n + 4(n - 1) + 4(n - 2) + \cdots \approx 2n^2$. This is the same as it would have cost to calculate $c = Q^T b$ if Q had been available explicitly. The cost of solving for x by back substitution is about n^2 flops. Thus these steps are inexpensive, compared with the cost of computing the QR decomposition.

Uniqueness of the QR Decomposition

For nonsingular matrices the QR decomposition theorem can be strengthened to include the uniqueness of Q and R.

Theorem 3.2.45 *Let $A \in \mathbb{R}^{n \times n}$ be nonsingular. Then there exist unique Q, $R \in \mathbb{R}^{n \times n}$ such that Q is orthogonal, R is upper triangular with positive main-diagonal entries, and $A = QR$.*

Proof. By Theorem 3.2.21, $A = \hat{Q}\hat{R}$, where $\hat{Q}$ is orthogonal and $\hat{R}$ is upper triangular, but $\hat{R}$ does not necessarily have positive main-diagonal entries. Since A is nonsingular, $\hat{R}$ must also be nonsingular, so its main-diagonal entries are nonzero, at least. Let D be the diagonal matrix given by

$$d_{ii} = \begin{cases} 1 & \text{if } \hat{r}_{ii} > 0, \\ -1 & \text{if } \hat{r}_{ii} < 0. \end{cases}$$

Then $D = D^T = D^{-1}$ is orthogonal. Let $Q = \hat{Q}D^{-1}$ and $R = D\hat{R}$. Then Q is orthogonal, R is upper triangular with $r_{ii} = d_{ii}\hat{r}_{ii} > 0$, and $A = QR$. This establishes existence.

Uniqueness can be proved using the same ideas (see Exercise 3.2.46). For variety we will take a different approach, which exploits an interesting connection between the QR and Cholesky decompositions. Suppose $A = Q_1 R_1 = Q_2 R_2$, where Q_1 and

Q_2 are orthogonal, and R_1 and R_2 are upper triangular with positive main-diagonal entries. $A^T A$ is a positive definite matrix, and $A^T A = R_1^T R_1$, since $Q_1^T Q_1 = I$. Thus R_1 is the Cholesky factor of $A^T A$. The same arguments can be applied to R_2, so R_2 is also the Cholesky factor of A. By uniqueness of the Cholesky decomposition, $R_1 = R_2$. Finally $Q_1 = AR_1^{-1} = AR_2^{-1} = Q_2$. $\qquad\square$

rcise 3.2.46 This exercise leads to another uniqueness proof for Theorem 3.2.45.

(a) Suppose $B \in \mathbb{R}^{n \times n}$ is both orthogonal and upper triangular. Prove that B must be a diagonal matrix whose main-diagonal entries are ± 1.

(b) Suppose $Q_1 R_1 = Q_2 R_2$, where Q_1 and Q_2 are both orthogonal, and R_1 and R_2 are both upper triangular and nonsingular. Show that there is a diagonal matrix D with main diagonal entries ± 1 such that $R_2 = DR_1$ and $Q_1 = Q_2 D$.

$\qquad\square$

rcise 3.2.47

(a) Let $A = \begin{bmatrix} 1 & 2 \\ 1 & 3 \end{bmatrix}$. Find a reflector $\hat{Q}$ and an upper triangular $\hat{R}$ such that $A = \hat{Q}\hat{R}$. Assemble $\hat{Q}$ and simplify it.

$$\left(Solution: \hat{Q} = \frac{1}{\sqrt{2}} \begin{bmatrix} -1 & -1 \\ -1 & 1 \end{bmatrix}, \quad \hat{R} = \frac{1}{\sqrt{2}} \begin{bmatrix} -2 & -5 \\ 0 & 1 \end{bmatrix}. \right)$$

(b) Compare your solution from part (a) with the QR decomposition of the same matrix (obtained using a rotator) in Example 3.2.13. Find a diagonal matrix D with main diagonal entries ± 1 such that $Q = \hat{Q}D$ and $\hat{R} = DR$.

$\qquad\square$

rcise 3.2.48 The MATLAB command `qr` performs the QR decomposition of a matrix. In the interest of keeping the output uncomplicated, MATLAB delivers the assembled Q matrix. Try out the following MATLAB commands, for example.

```
n = 7
A = randn(n)
[Q,R] = qr(A)
Q'*Q
norm(eye(n)-Q'*Q)
norm(A-Q*R)
```

Notice that MATLAB does not bother to make the main-diagonal entries of R positive, since this is not essential for most applications. A number of other ways to use the `qr` command will be discussed in the coming section. For further information type `help qr` or search in MATLAB's help browser. $\qquad\square$

Stability of Computations with Rotators and Reflectors

The numerical properties of both rotators and reflectors are excellent. A detailed analysis was carried out by Wilkinson [103, pp. 126–162]. See also [52]. We will content ourselves with a brief summary; some details are worked out in Exercises 3.2.71–3.2.74.

Suppose we wish to convert A to QA, where Q is either a rotator determined by (3.2.18) or a reflector determined by (3.2.35). In the case of the reflector, (3.2.38) should be used to calculate QA. In both cases Wilkinson has shown that

$$\mathrm{fl}(QA) = Q(A + E),$$

where $\|E\|_2/\|A\|_2$ is tiny. In other words, the computed result is the exact result of applying Q to a matrix that is a very slight perturbation of A. That is, the computation of QA is normwise backward stable.

The stability is preserved under repeated applications of rotators and/or reflectors. Consider, for example, the computation of Q_2Q_1A. First of all, $\mathrm{fl}(Q_1A) = Q_1(A + E_1)$. Thus $\mathrm{fl}(Q_2Q_1A) = \mathrm{fl}(Q_2(Q_1(A + E_1))) = Q_2(Q_1(A + E_1) + E_2) = Q_2Q_1(A + E_1 + Q_1^T E_2) = Q_2Q_1(A + E)$, where $E = E_1 + Q_1^T E_2$. Since $\|E\|_2 \leq \|E_1\|_2 + \|E_2\|_2$, we can assert that $\|E\|_2/\|A\|_2$ is tiny.

Many of the algorithms in this book consist of repeated modifications of a matrix by reflectors or rotators. The two QR decomposition algorithms discussed in this section are of this type, and so are many of the eigenvalue algorithms that are discussed in subsequent chapters. The preceding remarks imply that all algorithms of this type are normwise backward stable.

The Complex Case

The following exercises develop the complex analogues of the results obtained in this section. The *inner product* on $\mathbb{C}^n$ is defined by

$$\langle x, y \rangle = \sum_{i=1}^{n} x_i \overline{y}_i, \qquad \left\{ x = [x_1 \ \cdots \ x_n]^T, \ y = [y_1 \ \cdots \ y_n]^T \right\}$$

where the bar denotes complex conjugation. Clearly $\langle x, y \rangle = y^* x = \overline{x^* y}$, where y^* denotes the conjugate transpose of y.

Exercise 3.2.49 Show that the inner product on $\mathbb{C}^n$ satisfies the following properties.

(a) $\langle x, y \rangle = \overline{\langle y, x \rangle}$.

(b) $\langle \alpha_1 x_1 + \alpha_2 x_2, y \rangle = \alpha_1 \langle x_1, y \rangle + \alpha_2 \langle x_2, y \rangle$.

(c) $\langle x, \alpha_1 y_1 + \alpha_2 y_2 \rangle = \overline{\alpha}_1 \langle x, y_1 \rangle + \overline{\alpha}_2 \langle x, y_2 \rangle$.

(d) $\langle x, x \rangle$ is real, $\langle x, x \rangle \geq 0$, and $\langle x, x \rangle = 0$ if and only if $x = 0$.

(e) $\sqrt{\langle x, x \rangle} = \|x\|_2$.

$\square$

The complex analogues of orthogonal matrices are unitary matrices. A matrix $U \in \mathbb{C}^{n \times n}$ is *unitary* if $UU^* = I$, where U^* is the conjugate transpose of U. Equivalent statements are $U^*U = I$ and $U^* = U^{-1}$. Notice that the orthogonal matrices are just the real unitary matrices.

rcise 3.2.50

(a) Show that if U is unitary, then U^{-1} is unitary.

(b) Show that is U_1 and U_2 are unitary, then U_1U_2 is unitary.

□

rcise 3.2.51 Show that if U is unitary, then $|\det(U)| = 1$.   □

rcise 3.2.52 Show that if $U \in \mathbb{C}^{n \times n}$ is unitary and $x, y \in \mathbb{C}^n$, then

(a) $\langle Ux, Uy \rangle = \langle x, y \rangle$.

(b) $\|Ux\|_2 = \|x\|_2$.

(c) $\|U\|_2 = \|U^{-1}\|_2 = \kappa_2(U) = 1$.

Thus unitary matrices preserve the 2-norm and the inner product.   □

rcise 3.2.53 *(Complex Rotators)* Given a nonzero $\begin{bmatrix} a \\ b \end{bmatrix} \in \mathbb{C}^2$ define $U \in \mathbb{C}^{2 \times 2}$ by

$$U = \frac{1}{\gamma} \begin{bmatrix} a & -\bar{b} \\ b & \bar{a} \end{bmatrix}, \qquad \text{where } \gamma = \sqrt{|a|^2 + |b|^2}.$$

Verify that

(a) U is unitary.

(b) $\det(U) = 1$.

(c) $U^* \begin{bmatrix} a \\ b \end{bmatrix} = \begin{bmatrix} \gamma \\ 0 \end{bmatrix}$.

□

The extension to complex plane rotators in $\mathbb{C}^{n \times n}$ is obvious.

rcise 3.2.54 *(Complex Reflectors)* Let $u \in \mathbb{C}^n$ with $\|u\|_2 = 1$, and define $Q \in \mathbb{C}^{n \times n}$ by $Q = I - 2uu^*$. Verify that

(a) $Qu = -u$.

(b) $Qv = v$ if $\langle u, v \rangle = 0$.

(c) $Q = Q^*$ (Q is Hermitian).

(d) $Q^* = Q^{-1}$ (Q is unitary).

(e) $Q^{-1} = Q$ (Q is an involution).

$\square$

Exercise 3.2.55

(a) Prove that if $x, y \in \mathbb{C}^n$, $x \neq y$, $\|x\|_2 = \|y\|_2$, and $\langle x, y \rangle$ is real, then there exists a complex reflector Q such that $Qx = y$.

(b) Let $x \in \mathbb{C}^n$ be a nonzero vector. Express x_1 in polar form as $x_1 = re^{i\theta}$ (with $\theta = 0$ if $x_1 = 0$). Let $\tau = \|x\|_2 e^{i\theta}$ and $y = [-\tau \ 0 \ \cdots \ 0]^T$. Show that $\|x\| = \|y\|$, $x \neq y$, and $\langle x, y \rangle$ is real. Write an algorithm that determines a reflector Q such that $Qx = y$. This is the complex analogue of (3.2.35).

$\square$

Exercise 3.2.56 Show that for every $A \in \mathbb{C}^{n \times n}$ there exist unitary Q and upper triangular R such that $A = QR$. Write a constructive proof using either rotators or reflectors. $\square$

If A is nonsingular, then R can be chosen so that its main-diagonal entries are (real and) positive, in which case Q and R are unique. This is the complex analogue of Theorem 3.2.45.

Theorem 3.2.57 *Let* $A \in \mathbb{C}^{n \times n}$ *be nonsingular. Then there exist unique* $Q, R \in \mathbb{C}^{n \times n}$ *such that* Q *is unitary,* R *is upper triangular with real, positive main-diagonal entries, and* $A = QR$.

Exercise 3.2.58 Prove Theorem 3.2.57. $\square$

Exercise 3.2.59 (Essential uniqueness of the QR decomposition) Let $A \in \mathbb{C}^{n \times n}$ be nonsingular. Suppose $A = QR = \tilde{Q}\tilde{R}$, where Q and $\tilde{Q}$ are unitary, and R and $\tilde{R}$ are upper triangular. (Here we make no assumption about the main-diagonal entries.)

1. Show that $Q = \tilde{Q}D$, where D is a unitary, diagonal matrix.

2. Show that the jth column of $\tilde{Q}$ is a multiple of the jth column of Q for $j = 1, \ldots, n$.

Thus Q and $\tilde{Q}$ are essentially the same, and we can say that the QR decomposition of A is *essentially unique*. $\square$

Exercise 3.2.60 MATLAB will happily compute QR decompositions of complex matrices. Try the following commands, for example.

```
n = 4
A = randn(n) + i*randn(n)
[Q,R] = qr(A)
Q'
Q'*Q
```

```
norm(eye(n)-Q'*Q)
norm(A-Q*R)
```

MATLAB does not bother to make the main-diagonal entries of R positive or even real. Notice that Q' is the *conjugate* transpose of Q. □

rcise 3.2.61 Let $u \in \mathbb{C}^n$ with $\|u\|_2 = 1$. Then $I - \gamma uu^*$ is unitary if $\gamma = 2$ or 0. Find the set of all $\gamma \in \mathbb{C}$ for which $I - \gamma uu^*$ is unitary. □

Additional Exercises

rcise 3.2.62 This exercise works out a slick method of computing a rotator such that

$$\begin{bmatrix} c & s \\ -s & c \end{bmatrix} \begin{bmatrix} x_i \\ x_j \end{bmatrix} = \begin{bmatrix} * \\ 0 \end{bmatrix}. \tag{3.2.63}$$

If $x_j = 0$, take $c = 1$ and $s = 0$. If $x_i = 0$, take $c = 0$ and $s = 1$. Now assume both x_i and x_j are nonzero. The c and s that are to be computed are $c = \cos\theta$ and $s = \sin\theta$ for some unknown angle θ.

(a) Focusing on (3.2.63), show that the desired zero is obtained if and only if $\tan\theta = x_j/x_i$. (This does not determine θ uniquely, but any such θ will do.)

(b) Using elementary trigonometric identities, show that c and s can be obtained by the computations

$$t = \frac{x_j}{x_i}, \quad c = \frac{1}{\sqrt{1 + t^2}}, \quad s = ct.$$

These could be different from the c and s given by (3.2.18); they could be off by a sign. But the sign is unimportant.

(c) The equation $\tan\theta = x_j/x_i$ can equally well be expressed as $\cot\theta = x_i/x_j$. Show that c and s can be obtained by the equations

$$k = \frac{x_i}{x_j}, \quad s = \frac{1}{\sqrt{1 + k^2}}, \quad c = sk.$$

These could be off by a sign from the c and s computed by (3.2.18) or from the formulas of part (b).

(d) Show that all possibility of overflow is avoided if the method of part (b) is used when $|x_j| \le |x_i|$ and the method of part (c) is used when $|x_i| < |x_j|$.

(e) Show that harmful underflows are also eliminated.

□

rcise 3.2.64 Show that if $Qx = y$ ($x \ne y$), where $Q = I - \gamma uu^T \in \mathbb{R}^{n \times n}$, then u must be a multiple of $x - y$. Since all multiples of $x - y$ generate the same reflector, this establishes the uniqueness part of Theorem 3.2.31. □

Exercise 3.2.65 Some of the most important transforming matrices are rank-one updates of the identity matrix.

(a) Recall that the rank of a matrix is equal to the number of linearly independent columns. Prove that $A \in \mathbb{R}^{n \times n}$ has rank one if and only if there exist nonzero vectors $u, v \in \mathbb{R}^n$ such that $A = uv^T$. To what extent is there flexibility in the choice of u and v?

(b) A matrix of the form $G = I - uv^T$ is called a *rank-one update of the identity* for obvious reasons. Prove that if G is singular, then $Gx = 0$ if and only if x is a multiple of u. Prove that G is singular if and only if $v^T u = 1$.

(c) Show that if $G = I - uv^T$ is nonsingular, then G^{-1} has the form $I - \beta uv^T$. Give a formula for β.

(d) Prove that if G is orthogonal, then u and v are proportional, i.e. $v = \rho u$ for some nonzero scalar ρ. Show that if $\| u \|_2 = 1$, then $\rho = 2$.

(e) Show that if $G = I - uv^T$ and $W \in \mathbb{R}^{n \times m}$, then GW can be computed in about $4nm$ flops if the arithmetic is done in an efficient way.

(f) Consider the task of introducing zeros into a vector: Given $x \in \mathbb{R}^n$, we want to find a $G = I - uv^T$ such that

$$G \begin{bmatrix} x_1 \\ x_2 \\ \vdots \\ x_n \end{bmatrix} = \begin{bmatrix} \beta \\ 0 \\ \vdots \\ 0 \end{bmatrix} \qquad (3.2.66)$$

for some nonzero β. Show that G performs the transformation (3.2.66) if and only if

$$u = (1/\gamma)(x - \beta e_1), \quad \text{where} \quad \gamma = v^T x \neq 0, \qquad (3.2.67)$$

and $e_1 = [1, 0, \cdots, 0]^T$.

(g) In (3.2.66) β can be chosen freely. In this section we have already seen how to choose β, u, and v to build a reflector to do this task. Now consider another possibility, assuming $x_1 \neq 0$. Take $\beta = x_1$ and $v = e_1$. What form must u now have, if (3.2.67) is to be satisfied? Show that the resulting G is a Gauss transform (cf. Exercise 1.7.47).

□

Exercise 3.2.68 . This exercise shows how to accumulate reflectors so that they can be applied in a batch using operations that are rich in matrix-matrix multiplications.

(a) Let $G = I + WY^T$, where $W, Y \in \mathbb{R}^{n \times j}$. Show that $G = I + w_1 y_1^T + w_2 y_2^T + \cdots + w_j y_j^T$, where $w_1, \ldots, w_j$ and $y_1, \ldots, y_j$ are column vectors. Thus G is a *rank j update* of the identity.

(b) Let G be as in part (a), with $1 \ll j \ll n$, and let $B \in \mathbb{R}^{n \times m}$. Show that the product GB can be calculated in about $4njm$ flops if the operations are organized carefully. What size matrix-matrix products and sums are performed in the process? (The bigger they are, the more efficiently we can use the computer's cache.) How many flops would it take to compute GB if G were stored as an $n \times n$ matrix in the standard way?

(c) Show that if $Q = I - \gamma u u^T$ and $G = I + W Y^T$, then $QG = I + W Y^T + u v^T$, where $v = -\gamma(u + Y W^T u)$. How many flops does it take to compute v? Show that $QG = I + \tilde{W} \tilde{Y}^T$, where $\tilde{W}, \tilde{Y} \in \mathbb{R}^{n \times (j+1)}$.

(d) Prove that any product of j elementary reflectors can be expressed in the form $I + W Y^T$, where $W, Y \in \mathbb{R}^{n \times j}$. Use induction on j.

(e) Let G be a product of j elementary reflectors. How many flops are needed to compute GB, where $B \in \mathbb{R}^{n \times m}$, if G is applied as a series of j reflectors, one after the other? How does this compare with the cost of applying G in the form $I + W Y^T$? How many additional flops are needed to construct $I + W Y^T$ via the operations indicated in part (c)? (This shows that the extra work is insignificant if $j \ll n$ but can become significant if j is made too large.)

(f) State briefly how (3.2.43) would have to be modified so that the reflectors are applied in batches of j.

(g) Give a simple expression for G^{-1}, if $G = I + W Y^T$ is orthogonal.

$\square$

Exercise 3.2.69 The LAPACK [1] routines use a slightly different representation than the one developed in the previous exercise. Suppose $Q = Q_j Q_{j-1} \cdots Q_1$ is a product of reflectors $Q_i = I - \gamma_i u_i u_i^T$. Proceeding more or less as in the previous exercise, prove by induction on j that $Q = I - U L U^T$, where $U = [u_1 \cdots u_j] \in \mathbb{R}^{n \times j}$, and $L \in \mathbb{R}^{j \times j}$ is lower triangular. How can Q^{-1} be represented? $\square$

Exercise 3.2.70 Show that every 2-by-2 reflector has the form $\begin{bmatrix} c & s \\ s & -c \end{bmatrix}$, where $c^2 + s^2 = 1$.

How does the reflector $Q = \begin{bmatrix} c & s \\ s & -c \end{bmatrix}$ act on the unit vectors $\begin{bmatrix} 1 \\ 0 \end{bmatrix}$ and $\begin{bmatrix} 0 \\ 1 \end{bmatrix}$?

Contrast the action of Q with that of the rotator $\tilde{Q} = \begin{bmatrix} c & -s \\ s & c \end{bmatrix}$. Draw a picture. By examining your picture, determine the line through which the vectors are reflected by Q. Assuming $c > 0$ and $s > 0$, show that this line has slope $\tan(\theta/2)$, where θ is the angle for which $\cos \theta = c$ and $\sin \theta = s$. $\square$

Exercise 3.2.71 This and subsequent exercises demonstrate the backward stability of multiplication by rotators, reflectors, and other orthogonal transformations. We begin with matrix multiplication in general. Let $V \in \mathbb{R}^{n \times n}$ be nonsingular, and let $x \in \mathbb{R}^n$.

(a) Suppose we calculate $\mathrm{fl}(Vx)$ on an ideal computer (Section 2.5) with unit roundoff u. Using (2.5.6) and Proposition 2.7.1, show that $\mathrm{fl}(Vx) = Vx + b$, where b is tiny; specifically, the ith component of b satisfies

$$b_i = \sum_{j=1}^{n} v_{ij}x_j\gamma_{ij},$$

where $|\gamma_{ij}| \leq nu + O(u^2)$.

(b) Continuing from part (a), show that $|b| \leq \gamma|V||x|$, where $\gamma = \max|\gamma_{ij}|$, and $\|b\|_1 \leq \|V\|_1\|x\|_1(nu + O(u^2))$.

(c) Show that $\mathrm{fl}(Vx) = V(x + \delta x)$, where $\delta x \ (= V^{-1}b)$ satisfies $\|\delta x\|_1/\|x\|_1 \leq \kappa_1(V)(nu + O(u^2))$.

(d) Let $A \in \mathbb{R}^{n \times n}$. Using the result of part (c), show that $\mathrm{fl}(VA) = V(A + E)$, where $\|E\|_1/\|A\|_1 \leq \kappa_1(V)(nu + O(u^2))$.

We conclude that multiplication by V is backward stable if V is well conditioned. Orthogonal matrices are the ultimate well-conditioned matrices.   □

Exercise 3.2.72 Use inequalities from Exercise 2.1.33 to prove parts (a) and (b). Then use these inequalities along with the result of Exercise 3.2.71 to work part (c).

(a) For all $E \in \mathbb{R}^{n \times n}$ and nonzero $A \in \mathbb{R}^{n \times n}$, $\|E\|_2/\|A\|_2 \leq n\|E\|_1/\|A\|_1$.

(b) For all nonsingular $V \in \mathbb{R}^{n \times n}$, $\kappa_1(V) \leq n\kappa_2(V)$.

(c) Show that if $Q \in \mathbb{R}^{n \times n}$ is orthogonal, then $\mathrm{fl}(QA) = Q(A + E)$, where

$$\|E\|_2/\|A\|_2 \leq n^3u + O(u^2).$$

Of course, the factor n^3 is very pessimistic. Thus multiplication by orthogonal matrices is backward stable.   □

Exercise 3.2.73 How can the results of Exercises 3.2.71 and 3.2.72 be improved in the case when Q (or V) is a plane rotator?   □

Exercise 3.2.74 The product Qx, where $Q = I - \gamma vv^T$ (with $\gamma = 2/\|v\|_2^2$) is a reflector. In practice there will be slight errors in the computed γ and v, but we will ignore those errors for simplicity. In principle $Qx = y$, where $y = x - \gamma(v^Tx)v$.

(a) Let $\rho = \gamma v^Tx$. Using (2.5.6) and Proposition 2.7.1, show that $\mathrm{fl}(\rho) = \rho + \beta$, where $|\beta| \leq 2(n + 1)u\|x\|_2/\|v\|_2 + O(u^2)$.

(b) $\mathrm{fl}(Qx) = \mathrm{fl}(y) = \mathrm{fl}(x - \mathrm{fl}(\rho)v)$. Show that the ith component satisfies $\mathrm{fl}(y_i) = y_i + d_i$, where $|d_i| \leq (2n + 7)u\|x\|_2 + O(u^2)$.

(c) Show that $\mathrm{fl}(Qx) = Q(x + \delta x)$, where $\|\delta x\|_2/\|x\|_2 \leq \sqrt{n}(2n+7)u + O(u^2)$.

(d) Suppose $A \in \mathbb{R}^{n \times n}$, and we calculate QA by applying the reflector Q to one column of A at a time. Show that $\mathrm{fl}(QA) = Q(A+E)$, where $\|E\|_F/\|A\|_F \leq \sqrt{n}(2n + 7)u + O(u^2)$.

$\square$

3.3 SOLUTION OF THE LEAST SQUARES PROBLEM

Consider an overdetermined system

$$Ax = b, \quad A \in \mathbb{R}^{n \times m}, \ b \in \mathbb{R}^n, \ n > m. \qquad (3.3.1)$$

Our task is to find $x \in \mathbb{R}^m$ such that $\|r\|_2 = \|b - Ax\|_2$ is minimized. So far we do not know whether or not this problem has a solution. If it does have one, we do not know whether the solution is unique. These fundamental questions will be answered in this section. We will settle the existence question affirmatively by constructing a solution. The solution is unique if and only if A has full rank.

Let $Q \in \mathbb{R}^{n \times n}$ be any orthogonal matrix, and consider the transformed system

$$Q^T Ax = Q^T b. \qquad (3.3.2)$$

Let s be the residual for the transformed system. Then $s = Q^T b - Q^T Ax = Q^T(b - Ax) = Q^T r$. Since Q^T is orthogonal, $\|s\|_2 = \|r\|_2$. Thus, for a given x, the residual of the transformed system has the same norm as the residual for the original system. Therefore $x \in \mathbb{R}^m$ minimizes $\|r\|_2$ if and only if it minimizes $\|s\|_2$; that is, the two overdetermined systems have the same least squares solution(s).

It seems reasonable to try to find an orthogonal $Q \in \mathbb{R}^{n \times n}$ for which the system (3.3.2) has a particularly simple form [40]. In the previous section we learned that given a square matrix A there exists an orthogonal Q such that $Q^T A = R$, where R is upper triangular. Thus the linear system $Ax = b$ can be transformed to $Q^T Ax = Q^T b$ ($Rx = c$), which can easily be solved by back substitution. The same approach works for the least squares problem, but first we need a QR decomposition theorem for nonsquare matrices.

Theorem 3.3.3 *Let $A \in \mathbb{R}^{n \times m}$, $n > m$. Then there exist $Q \in \mathbb{R}^{n \times n}$ and $R \in \mathbb{R}^{n \times m}$, such that Q is orthogonal and $R = \begin{bmatrix} \hat{R} \\ 0 \end{bmatrix}$, where $\hat{R} \in \mathbb{R}^{m \times m}$ is upper triangular, and $A = QR$.*

Proof. Let $\breve{A} = [A \ \tilde{A}] \in \mathbb{R}^{n \times n}$, where $\tilde{A} \in \mathbb{R}^{n \times (n-m)}$ is chosen arbitrarily. Then by Theorem 3.2.21 there exist $Q, \breve{R} \in \mathbb{R}^{n \times n}$ such that Q is orthogonal, $\breve{R}$ is upper triangular, and $\breve{A} = Q\breve{R}$. We now partition $\breve{R}$: $\breve{R} = [R \ \tilde{T}]$, where $R \in \mathbb{R}^{n \times m}$. Then $A = QR$. Since $\breve{R}$ is upper triangular, R has the form $\begin{bmatrix} \hat{R} \\ 0 \end{bmatrix}$, where $\hat{R} \in \mathbb{R}^{m \times m}$ is upper triangular. $\square$

Since this QR decomposition is obtained by retaining a portion of an n-by-n QR decomposition, any algorithm that computes a square QR decomposition can easily be modified to compute an n-by-m QR decomposition. In practice there is no need to choose a matrix $\tilde{A}$ to augment A, since that portion of the matrix is thrown away in the end anyway. One simply operates on A alone and quits upon running out of columns, that is, after m steps. For example, Algorithm (3.2.43) can be adapted to n-by-m matrices by changing the upper limit on the loop indices to m and deleting the last instruction, which is irrelevant in this case. Q is produced implicitly as a product of m reflectors: $Q = Q_1 Q_2 \cdots Q_m$.

Exercise 3.3.4 Show that the flop count for a QR decomposition of an n-by-m matrix using reflectors is approximately $2nm^2 - 2m^3/3$. (Thus if $n \gg m$, the flop count is about $2nm^2$.)   □

The QR decomposition of Theorem 3.3.3 may or may not be useful for solving the least squares problem, depending upon whether or not A has full rank. Recall that the *rank* of a matrix is the number of linearly independent columns, which is the same as the number of linearly independent rows. The matrix $A \in \mathbb{R}^{n \times m}$ $(n \geq m)$ has *full rank* if its rank is m; that is, if its columns are linearly independent. In Theorem 3.3.3 the equation $A = QR$ implies that $\mathrm{rank}(A) \leq \mathrm{rank}(R)$.[19] On the other hand the equation $R = Q^T A$ implies that $\mathrm{rank}(R) \leq \mathrm{rank}(A)$. Thus $\mathrm{rank}(R) = \mathrm{rank}(A)$, and A has full rank if and only if R does. Clearly $\mathrm{rank}(R) = \mathrm{rank}(\hat{R})$, and $\hat{R}$ has full rank if and only if it is nonsingular. Thus $\hat{R}$ is nonsingular if and only if A has full rank.

Full-Rank Case

Now consider an overdetermined system $Ax = b$ for which A has full rank. Using the QR decomposition, we can transform the system to the form $Q^T Ax = Q^T b$, or $Rx = c$, where $c = Q^T b$. Writing $c = \begin{bmatrix} \hat{c} \\ d \end{bmatrix}$, where $\hat{c} \in \mathbb{R}^m$, we can express the residual $s = c - Rx$ as

$$s = \begin{bmatrix} \hat{c} \\ d \end{bmatrix} - \begin{bmatrix} \hat{R} \\ 0 \end{bmatrix} x = \begin{bmatrix} \hat{c} - \hat{R}x \\ d \end{bmatrix}.$$

Thus

$$\| s \|_2^2 = \sum_{i=1}^{n} |s_i|^2 = \| \hat{c} - \hat{R}x \|_2^2 + \| d \|_2^2. \tag{3.3.5}$$

Since the term $\| d \|_2^2$ is independent of x, $\| s \|_2$ is minimized exactly when $\| \hat{c} - \hat{R}x \|_2^2$ is minimized. Obviously $\| \hat{c} - \hat{R}x \|_2^2 \geq 0$ with equality if and only if $\hat{R}x = \hat{c}$. Since A has full rank, $\hat{R}$ is nonsingular. Thus the system $\hat{R}x = \hat{c}$ has a unique solution,

---

[19]We recall from elementary linear algebra that $\mathrm{rank}(ST) \leq \mathrm{rank}(T)$ for any matrices S and T for which the product ST is defined. As an exercise, you might like to prove this result yourself.

which is then the unique minimizer of $\|s\|$. We summarize these findings as a theorem.

Theorem 3.3.6 *Let $A \in \mathbb{R}^{n \times m}$ and $b \in \mathbb{R}^n$, $n > m$, and suppose A has full rank. Then the least squares problem for the overdetermined system $Ax = b$ has a unique solution, which can be found by solving the nonsingular system $\hat{R}x = \hat{c}$, where*
$$\begin{bmatrix} \hat{c} \\ d \end{bmatrix} = c = Q^T b, \ \hat{R} \in \mathbb{R}^{m \times m} \ \text{and } Q \in \mathbb{R}^{n \times n} \ \text{are as in Theorem 3.3.3.}$$

rcise 3.3.7 Work this problem by hand. Consider the overdetermined system

$$\begin{bmatrix} 1 \\ 1 \end{bmatrix} [\ x \ ] = \begin{bmatrix} 9 \\ 5 \end{bmatrix},$$

whose coefficient matrix obviously has full rank.

(a) Before you do anything else, guess the least squares solution of the system.

(b) Calculate a QR decomposition of the coefficient matrix, where Q is a 2×2 rotator, and R is 2×1. Use the QR decomposition to calculate the least squares solution. Deduce the norm of the residual without calculating the residual directly.

$\square$

When you solve a least squares problem on a computer, the QR decomposition will normally be calculated using reflectors, in which case Q is stored as a product of reflectors $Q_1 Q_2 \cdots Q_m$. Thus the computation $c = Q^T b = Q_m \cdots Q_1 b$ is accomplished by applying reflectors to b using (3.2.38). Recalling that each reflector does a bit less work than the previous one, we find that the total flop count for this step is about $4n + 4(n - 1) + 4(n - 2) + \cdots + 4(n - m + 1) \approx 4nm - 2m^2$. The equation $\hat{R}x = \hat{c}$ can be solved by back substitution at a cost of m^2 flops. It is also advisable to compute $\|d\|_2$ ($2(n - m)$ flops), as this is the norm of the residual associated with the least squares solution (see (3.3.5)). Therefore $\|d\|_2$ is a measure of goodness of fit. Comparing these flop counts with the result of Exercise 3.3.4, we see that the most expensive step is the computation of the QR decomposition.

rcise 3.3.8 It is a well-known fact from algebra that every nonzero polynomial of degree $\leq m - 1$ is zero at at most $m - 1$ distinct points. Use this fact to prove that the matrix in (3.1.3) has full rank, assuming the points $t_1, \ldots, t_n$ are distinct. $\square$

Exercise 3.3.8 demonstrates that the least squares problem by a polynomial of degree $\leq m - 1$, with $m < n$, always has a unique solution that can be computed by the QR decomposition method.

rcise 3.3.9 Write a pair of Fortran subroutines to solve the least squares problem for the overdetermined system $Ax = b$, where $A \in \mathbb{R}^{n \times m}$, $n > m$, and A has full rank.

The first subroutine uses reflectors to carry out a QR decomposition, $A = QR$, where $Q \in \mathbb{R}^{n \times n}$ is a product of reflectors: $Q = Q_1 Q_2 \cdots Q_m$, and $R = \begin{bmatrix} \hat{R} \\ 0 \end{bmatrix} \in$
$\mathbb{R}^{n \times m}$, where $\hat{R}$ is m-by-m and upper triangular. All of these operations should be carried out in a single n-by-m array plus one or two auxiliary vectors. If $\hat{R}$ has any zero pivots, the subroutine should set an error flag to indicate that A does not have full rank.

The second subroutine uses the QR decomposition to find x. First $c = Q^T b = Q_m Q_{m-1} \cdots Q_1 b$ is calculated by applying the reflectors successively. An additional one-dimensional array is needed for b. This array can also be used for c (and intermediate results). The solution x is found by solving $\hat{R}x = \hat{c}$ by back substitution, where $c = \begin{bmatrix} \hat{c} \\ d \end{bmatrix}$. The solution should calculate $\|d\|_2$, which is the Euclidean norm of the residual.

Use your subroutines to solve the following problems.

(a) Find the least squares quadratic polynomial for the data

t_i	-1	-0.75	-0.5	0	0.25	0.5	0.75
y_i	1.00	0.8125	0.75	1.00	1.3125	1.75	2.3125

(The correct solution is $\phi(t) = 1 + t + t^2$, which fits the data exactly.)

(b) Find the least squares linear polynomials for the two sets of data

t_i	1000	1050	1060	1080	1110	1130
$(\alpha)\ y_i$	6010	6153	6421	6399	6726	6701
$(\beta)\ y_i$	9422	9300	9220	9150	9042	8800

using the basis $\phi_1(t) = 50$, $\phi_2(t) = t - 1065$. Notice that the QR decomposition only needs to be done once. Plot your solutions and the data points.

(c)

$$\begin{bmatrix} 1 & 2 & 3 & 4 \\ 5 & 6 & 7 & 8 \\ 9 & 10 & 11 & 12 \\ 1 & 1 & 1 & 1 \\ 3 & 2 & 1 & 0 \end{bmatrix} \begin{bmatrix} x_1 \\ x_2 \\ x_3 \\ x_4 \end{bmatrix} = \begin{bmatrix} 10 \\ 26 \\ 42 \\ 4 \\ 6 \end{bmatrix}$$

Examine r_{11}, r_{22}, r_{33}, and r_{44}. You will find that some of them are nearly zero, which suggests that the matrix might not have full rank. In fact its rank is 2. This problem illustrates the fact that your program is unlikely to detect rank deficiency. While two of the r_{ii} should equal zero in principle, roundoff errors have made them all nonzero in practice. On the brighter side, you will notice that the "solution" that our program returned does fit the equations remarkably well. In fact it is (up to roundoff error) a solution. As we shall see, the least squares problem for a rank-deficient matrix has infinitely many solutions, of which your program calculated one.

☐

rcise 3.3.10   MATLAB's `qr` command, which we introduced in Exercise 3.2.48, can also be used to compute QR decompositions of non-square matrices.

(a) Try out the following commands, for example.

```
n = 6
m = 3
A = randn(n,m)
[Q,R] = qr(A)
Q'*Q
norm(eye(n)-Q'*Q)
norm(A-Q*R)
```

In the interest of convenience, MATLAB returns Q in assembled form.

(b) Write a short MATLAB script (m-file) that solves least squares problems using the `qr` command. (The submatrix $\hat{R}$ can be accessed by writing R(1:m,1:m) or R(1:m,:), for example.) Use your script to work parts (a) and (b) of Exercise 3.3.9.

This is not the simplest way to solve least squares problems using MATLAB. Recall from Exercise 3.1.5 that the command x = A\b works as well. This gives you a simple way to check your work.

☐

Rank-Deficient Case

While most least squares problems have full rank, it is obviously desirable to be able to handle problems that are not of full rank as well. Most such problems can be solved by a variant of the QR method called the QR decomposition with column pivoting, which we will discuss here. A more reliable (and more expensive) method based on the singular value decomposition will be discussed in Chapter 4.

If A does not have full rank, the basic QR method breaks down in principle because $\hat{R}$ is singular; at least one of its main-diagonal entries $r_{11}, \ldots, r_{mm}$ is zero. The QR decomposition with column pivoting makes column interchanges so that the zero pivots are moved to the lower right hand corner of $\hat{R}$. The resulting decomposition is suitable for solving the rank-deficient least squares problem.

Our initial development will ignore the effects of roundoff errors. At step 1 the 2-norm of each column of A is computed. If the jth column has the largest norm, then columns 1 and j are interchanged. The rest of step 1 is the same as before; a reflector that transforms the first column to the form $[-\tau_1, 0, \cdots, 0]^T$ is determined. This reflector is then applied to columns 2 through m. Since $|\tau_1|$ equals the 2-norm of the first column, the effect of the column interchange is to make $|\tau_1|$ as large as possible. In particular, $\tau_1 \neq 0$ unless $A = 0$.

The second step operates on the submatrix obtained by ignoring the first row and column. Otherwise it is identical to the first step, except that when the columns are interchanged, the full columns should be swapped, not just the portions that lie in the submatrix. This implies that the effect of the column interchange is the same as that of making the corresponding interchange before the QR decomposition is begun.

If the matrix has full rank, the algorithm terminates after m steps. The result is a decomposition $\hat{A} = QR$, where $\hat{A}$ is a matrix obtained from A by permuting the columns. $R = \begin{bmatrix} \hat{R} \\ 0 \end{bmatrix}$, where $\hat{R}$ is upper triangular and nonsingular. If A does not have full rank, there will come a step at which we are forced to take $\tau_i = 0$. That will happen when and only when all of the entries of the remaining submatrix are zero. Suppose this occurs after r steps have been completed. Letting $Q_i \in \mathbb{R}^{n \times n}$ denote the reflector used at step i, we have

$$Q_r Q_{r-1} \cdots Q_1 \hat{A} = \begin{bmatrix} R_{11} & R_{12} \\ 0 & 0 \end{bmatrix} = R,$$

where R_{11} is $r \times r$, upper triangular, and nonsingular. Its main diagonal entries are $-\tau_1$, $-\tau_2$, $\ldots$, $-\tau_r$, which are all nonzero. Clearly $\operatorname{rank}(R) = r$. Letting $Q = Q_1 Q_2 \cdots Q_r$, we see that $Q^T = Q_r Q_{r-1} \cdots Q_1$, $Q^T \hat{A} = R$, and $\hat{A} = QR$. Since $\operatorname{rank}(A) = \operatorname{rank}(\hat{A}) = \operatorname{rank}(R)$, we have $\operatorname{rank}(A) = r$. We summarize these findings as a theorem.

Theorem 3.3.11 *Let $A \in \mathbb{R}^{n \times m}$ with* $\operatorname{rank}(A) = r > 0$. *Then there exist matrices $\hat{A}$, Q, and R, such that $\hat{A}$ is obtained from A by permuting its columns, $Q \in \mathbb{R}^{n \times n}$ is orthogonal, $R = \begin{bmatrix} R_{11} & R_{12} \\ 0 & 0 \end{bmatrix} \in \mathbb{R}^{n \times m}$, $R_{11} \in \mathbb{R}^{r \times r}$ is nonsingular and upper triangular, and*

$$\hat{A} = QR.$$

Let us see how we can use this decomposition to solve the least squares problem. Given $x \in \mathbb{R}^m$, let $\hat{x}$ denote the vector obtained from x by making the same sequence of interchanges to its entries as were made to the columns of A in transforming A to $\hat{A}$. Then $\hat{A}\hat{x} = Ax$, so the problem of minimizing $\| b - Ax \|_2$ is the same as that of minimizing $\| b - \hat{A}\hat{x} \|_2$. An application of Q^T to the overdetermined system $\hat{A}\hat{x} = b$ transforms it to the form $R\hat{x} = Q^T b = c$, or

$$\begin{bmatrix} R_{11} & R_{12} \\ 0 & 0 \end{bmatrix} \begin{bmatrix} \hat{x}_1 \\ \hat{x}_2 \end{bmatrix} = \begin{bmatrix} \hat{c} \\ d \end{bmatrix},$$

where $\hat{x}_1 \in \mathbb{R}^r$ and $\hat{c} \in \mathbb{R}^r$. The residual for this transformed system is

$$s = \begin{bmatrix} \hat{c} - R_{11}\hat{x}_1 - R_{12}\hat{x}_2 \\ d \end{bmatrix},$$

whose norm is

$$\| s \|_2 = \sqrt{\| \hat{c} - R_{11}\hat{x}_1 - R_{12}\hat{x}_2 \|_2^2 + \| d \|_2^2}.$$

Clearly there is nothing we can do with the term $\| d \|_2^2$; We minimize $\| s \|_2$ by minimizing $\| \hat{c} - R_{11}\hat{x}_1 - R_{12}\hat{x}_2 \|_2^2$. This term can never be negative, but there are many choices of $\hat{x}$ for which it is zero. Each of these $\hat{x}$ is a solution of the least squares problem for the overdetermined system $\hat{A}\hat{x} = b$.

To see how to compute these $\hat{x}$, recall that R_{11} is nonsingular. Thus for any choice of $\hat{x}_2 \in \mathbb{R}^{m-r}$ there exists a unique $\hat{x}_1 \in \mathbb{R}^r$ such that

$$R_{11}\hat{x}_1 = \hat{c} - R_{12}\hat{x}_2.$$

Since R_{11} is upper triangular, $\hat{x}_1$ can be calculated by back substitution. Then $\hat{c} - R_{11}\hat{x}_1 - R_{12}\hat{x}_2 = 0$, and $\hat{x} = \begin{bmatrix} \hat{x}_1 \\ \hat{x}_2 \end{bmatrix}$ is a solution to the least squares problem for $\hat{A}\hat{x} = b$. We have thus established the following theorem.

Theorem 3.3.12 *Let* $A \in \mathbb{R}^{n \times m}$ *and* $b \in \mathbb{R}^n$, $n > m$. *Then the least squares problem for the overdetermined system* $Ax = b$ *always has a solution. If* rank$(A) <$ m, *there are infinitely many solutions.*

Not only have we proved this theorem, we have also established an algorithm that can be used to calculate any solution of the least squares problem. Thus we have solved the problem in principle.

In practice we often do not know the rank of A. Then part of the problem is to determine the rank, a task that is complicated by roundoff errors. After r steps of the QR decomposition with column pivoting, A will have been transformed to the form

$$Q_r \cdots Q_1 \hat{A} = \begin{bmatrix} R_{11} & R_{12} \\ 0 & R_{22} \end{bmatrix},$$

where $R_{11} \in \mathbb{R}^{r \times r}$ is nonsingular. If rank$(A) = r$, then in principle $R_{22} = 0$ (telling us that the rank is r), and the algorithm terminates. In practice R_{22} will have been contaminated by rounding errors and will not be exactly zero. Our criterion for determining the rank must take this into account. Thus, for example, we might decide that R_{22} is "numerically zero" if the norm of its largest column is less than $\epsilon \| A \|$, where ϵ is some small parameter depending on the machine precision and the accuracy of the data. This approach generally works well, but unfortunately it is not 100 percent reliable. There exist matrices of the form

$$R = \begin{bmatrix} -\tau_1 & & & * \\ & -\tau_2 & & \\ & & \ddots & \\ 0 & & & -\tau_m \end{bmatrix}$$

that are "nearly" rank deficient, for which none of the $|\tau_i|$ is extremely small. A class of examples due to Kahan is given in [63, p. 31] and also in Exercise 4.2.21. The near rank deficiency of these matrices would not be detected by our simple criterion. A more reliable approach to the detection of rank deficiency is to use the singular value decomposition (Chapter 4).

A few other implementation details need to be mentioned. At each step we need to know the norms of the columns of the remaining submatrix. If these calculations are not done in an economical manner, they can add substantially to the cost of the algorithm.

Exercise 3.3.13 Show that if the norms of the columns are computed in the straightforward manner at each step, the total cost of the norm computations is about $nm^2 - \frac{1}{3}m^3$ flops. How does this compare with the cost of the rest of the algorithm?   □

Fortunately the cost can be decreased substantially for steps $2, 3, \ldots, m$ by using information from previous steps rather than recomputing the norms from scratch. For example, let us see how we can get the norm information for the second step by using the information from the first step. Let $\nu_1, \nu_2, \ldots, \nu_m$ denote the squares of the norms of the columns of A (calculated at a cost of $2nm$ flops). We work with the squares for convenience. For notational simplicity let us assume that ν_1 is the largest (or that the column interchange for step 1 has already been made). After step 1 we have

$$
Q_1 A = \left[
\begin{array}{c|ccc}
-\tau_1 & \tilde{a}_{12} & \cdots & \tilde{a}_{1m} \\
\hline
0 & \tilde{a}_{22} & \cdots & \tilde{a}_{2m} \\
\vdots & \vdots & & \vdots \\
0 & \tilde{a}_{n2} & \cdots & \tilde{a}_{nm}
\end{array}
\right].
$$

Since Q_1 is orthogonal, it preserves the lengths of vectors. Therefore the norms of the columns of $Q_1 A$ are the same as those of the columns of A. For the second step we need the squares of the norms of the columns of the submatrix

$$
\left[
\begin{array}{ccc}
\tilde{a}_{22} & \cdots & \tilde{a}_{2m} \\
\vdots & & \vdots \\
\tilde{a}_{n2} & \cdots & \tilde{a}_{nm}
\end{array}
\right].
$$

These can clearly be obtained by the operations

$$
\nu_j \leftarrow \nu_j - \tilde{a}_{1j}^2 \qquad j = 2, \ldots, m.
$$

This costs a mere $2(m-1)$ flops instead of the $2(n-1)(m-1)$ flops that would have been required to calculate the squared norms from scratch.

Since the calculation of 2-norms is required, there is some danger of overflow or underflow. This danger can be virtually eliminated by rescaling the entire problem in advance. For example, one can calculate $\nu = \max|a_{ij}|$ and multiply all of the entries of A and b by $1/\nu$. The cost of this scaling operation is $O(nm)$. If this scaling operation is done, it is not necessary to perform scaling operations in the calculation of the reflectors (3.2.35).

Finally we make some observations concerning column interchanges. We have already noted that each column swap requires the interchange of the entire columns. Of course the norm information has to be interchanged too. It is also necessary to keep a record of the interchanges, since the computed solution $\hat{x}$ is the least squares solution of the permuted system $\hat{A}\hat{x} = b$. To get the solution x of the original problem, we must apply the inverse permutation to the entries of $\hat{x}$. This is accomplished by performing the interchanges in the reverse order.

rcise 3.3.14 Show that after the QR decomposition with column pivoting, the main-diagonal entries of R_{11} satisfy $|\tau_1| \geq |\tau_2| \geq \cdots \geq |\tau_r|$. □

rcise 3.3.15 To get MATLAB's qr command to calculate a QR decomposition with column pivoting, simply invoke the command with three output arguments:

```
[Q,R,P] = qr(A)
```

This gives the Q and R factors plus a permutation matrix P such that $AP = QR$. The effect of the permutation matrix (applied on the right!) is to scramble the columns. Perform the following sequence of MATLAB commands.

```
A = randn(5,3)
A = A*diag([1 3 9])
[Q,R,P] = qr(A)
```

What was the point of the diag command? What does the permutation matrix P tell you? □

3.4 THE GRAM-SCHMIDT PROCESS

In this section we introduce the idea of an orthonormal set, a new formulation of the QR decomposition, and the Gram-Schmidt process for orthonormalizing a linearly independent set. The main result is that performing a Gram-Schmidt orthonormalization is equivalent to calculating a QR decomposition. It follows that the Gram-Schmidt process can be used to solve least squares problems.

This neat summary overlooks one important fact: the classical Gram-Schmidt process performs badly in the presence of roundoff errors. As we shall see, a simple and effective remedy is to do the orthogonalization twice.

A set of vectors $q_1, q_2, \ldots, q_k \in \mathbb{R}^n$ is said to be *orthonormal* if the vectors are pairwise orthogonal, and each vector has Euclidean norm 1; that is,

$$\langle q_i, q_j \rangle = \begin{cases} 0 & \text{if } i \neq j \\ 1 & \text{if } i = j. \end{cases}$$

Example 3.4.1 Let $e_1, e_2, \ldots, e_n$ be the columns of the identity matrix:

$$e_1 = \begin{bmatrix} 1 \\ 0 \\ 0 \\ \vdots \\ 0 \end{bmatrix} \quad e_2 = \begin{bmatrix} 0 \\ 1 \\ 0 \\ \vdots \\ 0 \end{bmatrix} \quad \cdots \quad e_n = \begin{bmatrix} 0 \\ \vdots \\ 0 \\ 0 \\ 1 \end{bmatrix}.$$

It is evident that $e_1, e_2, \ldots, e_n$ form an orthonormal set. In fact, they form an *orthonormal basis*, since they are also a basis of $\mathbb{R}^n$. From now on we will call e_1, $e_2, \ldots, e_n$ the *standard basis* of $\mathbb{R}^n$, and the notation $e_1, \ldots, e_n$ will be reserved for this basis. □

Theorem 3.4.2 *Let $Q \in \mathbb{R}^{n \times n}$. Then Q is an orthogonal matrix if and only if its columns (rows) form an orthonormal set.*

Proof. Let $q_1, q_2, \ldots, q_n$ denote the columns of Q. Then

$$
Q^T Q = \begin{bmatrix} q_1^T \\ q_2^T \\ \vdots \\ q_n^T \end{bmatrix} \begin{bmatrix} q_1 & q_2 & \cdots & q_n \end{bmatrix} = \begin{bmatrix} q_1^T q_1 & q_1^T q_2 & \cdots & q_1^T q_n \\ q_2^T q_1 & q_2^T q_2 & \cdots & q_2^T q_n \\ \vdots & \vdots & & \vdots \\ q_n^T q_1 & q_n^T q_2 & \cdots & q_n^T q_n \end{bmatrix}.
$$

Thus the entries of $Q^T Q$ are the inner products $\langle q_j, q_i \rangle$. Clearly $Q^T Q = I$ if and only if $q_1, q_2, \ldots, q_n$ form an orthonormal set. The analogous theorem for the rows follows from considering the product QQ^T or from the simple observation that Q^T is orthogonal if and only if Q is. $\qquad \square$

Exercise 3.4.3

(a) Let $A \in \mathbb{R}^{n \times m}$, and let $e_1, \ldots, e_m$ denote the standard basis of $\mathbb{R}^m$. Verify that the ith column of A is Ae_i, $i = 1, \ldots, m$. Thus $A = [Ae_1 \; Ae_2 \; \cdots \; Ae_m]$. This simple observation will be used repeatedly.

(b) Use the observation of part (a) together with the inner product preserving property of orthogonal matrices to obtain a second proof that the columns of an orthogonal matrix are orthonormal.

$\qquad \square$

Using Theorem 3.4.2 as a guide, we introduce a new class of nonsquare matrices that possess some of the properties of orthogonal matrices. The matrix $Q \in \mathbb{R}^{n \times m}$, $n \geq m$ will be called *isometric* (or an *isometry*) if its columns are orthonormal.

Exercise 3.4.4 Prove that $Q \in \mathbb{R}^{n \times m}$ is isometric if and only if $Q^T Q = I$ (in $\mathbb{R}^{m \times m}$). $\quad \square$

The result of this exercise does *not* imply that Q^T is Q^{-1}. Only square matrices can have inverses. It is also *not* true that QQ^T is the identity matrix.

Exercise 3.4.5 Let $Q \in \mathbb{R}^{n \times m}$ ($n > m$) be an isometry with columns $q_1, q_2, \ldots, q_m$.

(a) Show that $QQ^T v = 0$ if v is orthogonal to $q_1, q_2, \ldots, q_m$.

(b) Show that $QQ^T q_i = q_i$, $i = 1, \ldots, m$. Therefore QQ^T behaves like the identity matrix on a proper subspace of $\mathbb{R}^n$.

(c) Show that $(QQ^T)^2 = QQ^T$. Thus QQ^T is a projector. In fact it is an orthoprojector. (See Remark 3.2.25.)

$\qquad \square$

Exercise 3.4.6 Show that if $Q \in \mathbb{R}^{n \times m}$ is an isometry, then

(a) $\langle Qx, Qy \rangle = \langle x, y \rangle$ for all $x, y \in \mathbb{R}^m$,

(b) $\| Qx \|_2 = \| x \|_2$ for all $x \in \mathbb{R}^m$.

(Notice that the norm and inner product on the left-hand side of these equations are from $\mathbb{R}^n$, not $\mathbb{R}^m$.) $\qquad\square$

Thus isometries preserve inner products, norms, and angles. The converses of both parts of Exercise 3.4.6 hold as well.[20]

The Condensed QR Decomposition

The QR Decomposition Theorem for nonsquare matrices can be restated more elegantly in terms of an isometry.

Theorem 3.4.7 *Let* $A \in \mathbb{R}^{n \times m}$, $n \geq m$. *Then there exist matrices* $\hat{Q}$ *and* $\hat{R}$ *such that* $\hat{Q} \in \mathbb{R}^{n \times m}$ *is an isometry,* $\hat{R} \in \mathbb{R}^{m \times m}$ *is upper triangular, and*

$$A = \hat{Q}\hat{R}.$$

Proof. If $n = m$, this is just Theorem 3.2.21. If $n > m$, we know from Theorem 3.3.3 that there exist matrices $Q \in \mathbb{R}^{n \times n}$ and $R \in \mathbb{R}^{n \times m}$ such that Q is orthogonal, $R = \begin{bmatrix} \hat{R} \\ 0 \end{bmatrix}$, $\hat{R} \in \mathbb{R}^{m \times m}$ is upper triangular, and $A = QR$. Let $\hat{Q} \in \mathbb{R}^{n \times m}$ be the matrix consisting of the first m columns of Q. Clearly $\hat{Q}$ is isometric. Let $\tilde{Q} \in \mathbb{R}^{n \times (n-m)}$ be the last $n - m$ columns of Q. Then

$$A = QR = \begin{bmatrix} \hat{Q} & \tilde{Q} \end{bmatrix} \begin{bmatrix} \hat{R} \\ 0 \end{bmatrix} = \hat{Q}\hat{R} + \tilde{Q}0,$$

That is, $A = \hat{Q}\hat{R}$. Since $\hat{Q}$ and $\hat{R}$ have the desired properties, the proof is complete. $\qquad\square$

If A has full rank, Theorem 3.4.7 can be strengthened to include the uniqueness of $\hat{Q}$ and $\hat{R}$.

Theorem 3.4.8 *Let* $A \in \mathbb{R}^{n \times m}$, $n \geq m$, *and suppose* $\mathrm{rank}(A) = m$. *Then there exist unique* $\hat{Q} \in \mathbb{R}^{n \times m}$ *and* $\hat{R} \in \mathbb{R}^{m \times m}$, *such that* $\hat{Q}$ *is isometric,* $\hat{R}$ *is upper triangular with positive entries on the main diagonal, and*

$$A = \hat{Q}\hat{R}.$$

The proof is similar to that of Theorem 3.2.45 and is left as an exercise.

---

[20]The name isometry comes from the norm-preserving property: An isometric operator is one that preserves the metric (i.e. the norm in this case).

Exercise 3.4.9 Prove Theorem 3.4.8. ☐

In the full-rank case the decomposition of Theorem 3.4.7 or Theorem 3.4.8 can be used to solve the least squares problem for the overdetermined system $Ax = b$. Let $\hat{c} = \hat{Q}^T b$. You can easily check that this vector is the same as the vector $\hat{c}$ of Theorem 3.3.6, and the least squares problem can be solved by solving $\hat{R}x = \hat{c}$ by back substitution. When we use this type of QR decomposition, we do not get the norm of the residual in the form $\| d \|_2$.

A Bit of Review

Before we discuss the Gram-Schmidt process, we should review some of the elementary facts about subspaces of $\mathbb{R}^n$. Recall that a *subspace* of $\mathbb{R}^n$ is a nonempty subset S of $\mathbb{R}^n$ that is closed under the operations of addition and scalar multiplication. That is, S is a subspace of $\mathbb{R}^n$ if and only if whenever $v, w \in S$ and $c \in \mathbb{R}$, then $v + w \in S$ and $cv \in S$. Given vectors $v_1, \ldots, v_m \in \mathbb{R}^n$, a *linear combination* of v_1, $\ldots, v_m$ is a vector of the form $c_1 v_1 + c_2 v_2 + \cdots + c_m v_m$, where $c_1, c_2, \ldots, c_m \in \mathbb{R}$. The numbers $c_1, \ldots, c_m$ are called the *coefficients* of the linear combination. In sigma notation a linear combination looks like $\sum_{k=1}^{m} c_k v_k$. The *span* of $v_1, \ldots, v_m$, denoted span$\{v_1, \ldots, v_m\}$, is the set of all linear combinations of $v_1, \ldots, v_m$. It is clear that span$\{v_1, \ldots, v_m\}$ is closed under the operations of addition and scalar multiplication; that is, it is a subspace of $\mathbb{R}^n$.

We have already used terms such as *linear independence* and *basis*, but let us now take a moment to review their precise meaning. The vectors $v_1, \ldots, v_m$ are *linearly independent* if the equation $c_1 v_1 + c_2 v_2 + \cdots + c_m v_m = 0$ has no solution other than $c_1 = c_2 = \cdots = c_m = 0$. In other words, the only linear combination of $v_1, \ldots, v_m$ that equals zero is the one whose coefficients are all zero. If the set $v_1, \ldots, v_m$ is not linearly independent, then it is called *linearly dependent*.

Every orthonormal set is linearly independent: Let $q_1, \ldots, q_m$ be an orthonormal set, and suppose $c_1 q_1 + c_2 q_2 + \cdots + c_m q_m = 0$. Taking inner products of both sides of this equation with q_j, we find that

$$c_1 \langle q_1, q_j \rangle + c_2 \langle q_2, q_j \rangle + \cdots + c_m \langle q_m, q_j \rangle = \langle 0, q_j \rangle = 0.$$

Since $\langle q_i, q_j \rangle = 0$ except when $i = j$, and $\langle q_j, q_j \rangle = 1$, this equation collapses to $c_j = 0$. This is true for all j, so $c_1 = c_2 = \cdots = c_m = 0$. Thus $q_1, \ldots, q_m$ are linearly independent. Of course the converse is false; there are lots of linearly independent sets that are not orthonormal.

Exercise 3.4.10 Give examples of sets of two vectors in $\mathbb{R}^2$ that are

    (a) orthonormal. (Think of something other than e_1, e_2.)

    (b) linearly independent but not orthogonal (hence not orthonormal).

    (c) linearly dependent.

☐

rcise 3.4.11 Prove that the vectors $v_1, \ldots, v_m$ are linearly independent if and only if none of them can be expressed as a linear combination of the others; in symbols, $v_j \notin \text{span}\{v_1, \ldots, v_{j-1}, v_{j+1}, \ldots, v_m\}$ for $j = 1, \ldots, m$. □

Let S be a subspace of $\mathbb{R}^n$, and let $v_1, \ldots, v_m \in S$. Then clearly

$$\text{span}\{v_1, \ldots, v_m\} \subseteq S.$$

We say the $v_1, \ldots, v_m$ *span* S if $\text{span}\{v_1, \ldots, v_m\} = S$. This means that every member of S can be expressed as a linear combination of $v_1, \ldots, v_m$; we say that $v_1, \ldots, v_m$ form a *spanning set* for S. A *basis* of S is a spanning set that is linearly independent. If $v_1, \ldots, v_m$ are linearly independent, then they form a basis of $\text{span}\{v_1, \ldots, v_m\}$, since they are a spanning set by definition. Recall from your elementary linear algebra course that every subspace of $\mathbb{R}^n$ has a basis; in fact it has many bases. Any two bases of S have the same number of elements. This number is called the *dimension* of S. Thus, for example, if $v_1, \ldots, v_m$ are linearly independent, then $\text{span}\{v_1, \ldots, v_m\}$ has dimension m.

The Classical Gram-Schmidt Process

The Gram-Schmidt process is an algorithm that produces orthonormal bases. Let S be a subspace of $\mathbb{R}^n$, and let $v_1, v_2, \ldots, v_m$ be a basis of S. The Gram-Schmidt process takes $v_1, \ldots, v_m$ as input and uses them to produce orthonormal vectors $q_1, \ldots, q_m$ that form a basis of S. Thus $S = \text{span}\{v_1, \ldots, v_m\} = \text{span}\{q_1, \ldots, q_m\}$. In fact, more is true: The vectors also satisfy

$$\begin{aligned}
\text{span}\{q_1\} &= \text{span}\{v_1\} \\
\text{span}\{q_1, q_2\} &= \text{span}\{v_1, v_2\} \\
\text{span}\{q_1, q_2, q_3\} &= \text{span}\{v_1, v_2, v_3\} \\
\vdots \quad & \quad \vdots \\
\text{span}\{q_1, q_2, \ldots, q_m\} &= \text{span}\{v_1, v_2, \ldots, v_m\}.
\end{aligned} \tag{3.4.12}$$

These relationships are important.

We are given linearly independent vectors $v_1, \ldots, v_m \in \mathbb{R}^n$, and we seek orthonormal $q_1, \ldots, q_m$ satisfying (3.4.12). In order to satisfy $\text{span}\{q_1\} = \text{span}\{v_1\}$ we must choose q_1 to be a multiple of v_1. Since we also require $\| q_1 \|_2 = 1$, we define $q_1 = (1/r_{11})v_1$, where $r_{11} = \| v_1 \|_2$. We know that $r_{11} \neq 0$ because $v_1, \ldots, v_m$ are linearly independent, so $v_1 \neq 0$. The equation $q_1 = (1/r_{11})v_1$ implies that $q_1 \in \text{span}\{v_1\}$; hence $\text{span}\{q_1\} \subseteq \text{span}\{v_1\}$. Conversely, the equation $v_1 = r_{11}q_1$ implies $v_1 \in \text{span}\{q_1\}$, and therefore $\text{span}\{v_1\} \subseteq \text{span}\{q_1\}$. Thus $\text{span}\{v_1\} = \text{span}\{q_1\}$.

The second step of the algorithm is to find q_2 such that q_2 is orthogonal to q_1, $\| q_2 \|_2 = 1$, and $\text{span}\{q_1, q_2\} = \text{span}\{v_1, v_2\}$. We will use Figure 3.5 as a guide. In this figure the space $\text{span}\{v_1, v_2\}$ is represented by the plane of the page. We require q_2 such that q_1, q_2 is an orthonormal set that spans this plane. The figure suggests that we can produce a vector $\hat{q}_2$ that lies in the plane and is orthogonal to q_1 by subtracting

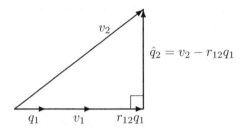

Figure 3.5   Construction of $\hat{q}_2$.

just the right multiple of q_1 from v_2. We can then obtain q_2 by multiplying $\hat{q}_2$ by the appropriate scaling factor. Thus let

$$\hat{q}_2 = v_2 - r_{12}q_1, \tag{3.4.13}$$

where r_{12} is a scalar that should be chosen so that $\langle \hat{q}_2, q_1 \rangle = 0$.

Exercise 3.4.14   Demonstrate that the vector $\hat{q}_2$ given by (3.4.13) satisfies $\langle \hat{q}_2, q_1 \rangle = 0$ if and only if $r_{12} = \langle v_2, q_1 \rangle$.   □

Taking $r_{12} = \langle v_2, q_1 \rangle$, we have $\langle \hat{q}_2, q_1 \rangle = 0$. It is clear that $\hat{q}_2 \neq 0$, for if this were not the case, we would have $v_2 = r_{12}q_1 \in \text{span}\{q_1\}$. But the linear independence of v_1 and v_2 implies that $v_2 \notin \text{span}\{v_1\} = \text{span}\{q_1\}$. Let $r_{22} = \| \hat{q}_2 \|_2 \neq 0$, and define $q_2 = (1/r_{22})\hat{q}_2$. Then clearly $\| q_2 \|_2 = 1$, and $\langle q_1, q_2 \rangle = 0$. It is intuitively clear from Figure 3.5 that $\text{span}\{q_1, q_2\} = \text{span}\{v_1, v_2\}$, and we can easily prove that it is so. First $q_1 \in \text{span}\{v_1\} \subseteq \text{span}\{v_1, v_2\}$. Also $q_2 \in \text{span}\{q_1, v_2\} = \text{span}\{v_1, v_2\}$. Since $q_1, q_2 \in \text{span}\{v_1, v_2\}$, we have $\text{span}\{q_1, q_2\} \subseteq \text{span}\{v_1, v_2\}$. Conversely, $v_1 \in \text{span}\{q_1\} \subseteq \text{span}\{q_1, q_2\}$, and $v_2 = r_{12}q_1 + \hat{q}_2 = r_{12}q_1 + r_{22}q_2 \in \text{span}\{q_1, q_2\}$. Since $v_1, v_2 \in \text{span}\{q_1, q_2\}$, we have $\text{span}\{v_1, v_2\} \subseteq \text{span}\{q_1, q_2\}$. Thus $\text{span}\{q_1, q_2\} = \text{span}\{v_1, v_2\}$.

Now consider the kth step. Suppose we have found orthonormal vectors q_1, q_2, ..., q_{k-1} such that $\text{span}\{q_1, \ldots, q_i\} = \text{span}\{v_1, \ldots, v_i\}$, $i = 1, \ldots, k - 1$. Let us see how to determine q_k. By analogy with (3.4.13) we seek $\hat{q}_k$ of the form

$$\hat{q}_k = v_k - \sum_{j=1}^{k-1} r_{jk}q_j \tag{3.4.15}$$

that is orthogonal to q_1, ..., q_{k-1}. The equations $\langle \hat{q}_k, q_i \rangle = 0$, $i = 1, \ldots, k - 1$, imply that

$$\langle v_k, q_i \rangle - \sum_{j=1}^{k-1} r_{jk}\langle q_j, q_i \rangle = 0 \qquad i = 1, \ldots, k - 1.$$

Since $\langle q_j, q_i \rangle = 0$ when $i \neq j$, and $\langle q_i, q_i \rangle = 1$, these equations reduce to

$$r_{ik} = \langle v_k, q_i \rangle \qquad i = 1, \ldots, k - 1. \tag{3.4.16}$$

Conversely, if r_{ik} are defined by (3.4.16), then $\hat{q}_k$ is orthogonal to $q_1, \ldots, q_{k-1}$.

As in the case $k = 2$, it is easy to show that $\hat{q}_k \neq 0$. Otherwise we would have $v_k = \sum_{j=1}^{k-1} r_{jk} q_j \in \text{span}\{q_1, \ldots, q_{k-1}\}$. But the linear independence of $v_1, \ldots, v_k$ implies that $v_k \notin \text{span}\{v_1, \ldots, v_{k-1}\} = \text{span}\{q_1, \ldots, q_{k-1}\}$. Let

$$r_{kk} = \| \hat{q}_k \|_2 > 0, \qquad (3.4.17)$$

and define

$$q_k = \frac{1}{r_{kk}} \hat{q}_k. \qquad (3.4.18)$$

Then clearly $\| q_k \|_2 = 1$ and $\langle q_k, q_i \rangle = 0$, $i = 1, \ldots, k-1$.

It is also easy to verify that $\text{span}\{q_1, \ldots, q_k\} = \text{span}\{v_1, \ldots, v_k\}$. See Exercises 3.4.30 and 3.4.31.

Equations 3.4.16, 3.4.15, 3.4.17, and 3.4.18 express the kth step of the classical Gram-Schmidt process. Performing this step for $k = 1, \ldots, m$ produces $q_1, \ldots, q_m$.

Before summarizing the algorithm let us note that some of the operations can be expressed more succinctly in terms of matrix-vector multiplication operations. Since $\langle v_k, q_i \rangle = q_i^T v_k$, the $k - 1$ equations of (3.4.16) can be written as the single vector equation

$$\begin{bmatrix} r_{1k} \\ \vdots \\ r_{k-1,k} \end{bmatrix} = \begin{bmatrix} q_1^T \\ \vdots \\ q_{k-1}^T \end{bmatrix} v_k = \begin{bmatrix} q_1 & \cdots & q_{k-1} \end{bmatrix}^T v_k.$$

Introducing MATLAB-like notation, this can be condensed to $r_{1:k-1,k} = (q_{1:k-1})^T v_k$. Applying similar treatment to (3.4.15), we obtain $\hat{q}_k = v_k - (q_{1:k-1})r_{1:k-1,k}$. Now here is the algorithm.

Classical Gram-Schmidt Algorithm. Given linearly independent $v_1, v_2, \ldots,$ v_m, this algorithm produces orthonormal $q_1, q_2, \ldots, q_m$ such that $\text{span}\{q_1, \ldots, q_i\}$ $= \text{span}\{v_1, \ldots, v_i\}$ for $i = 1, \ldots, m$. $q_1, \ldots, q_m$ are stored over $v_1, \ldots, v_m$.

$$\begin{aligned}
&\text{for } k = 1, \ldots, m \\
&\begin{bmatrix} r_{1:k-1,k} \leftarrow (v_{1:k-1})^T v_k \quad (v_{1:k-1} \text{ is really } q_{1:k-1}) \\ v_k \leftarrow v_k - (v_{1:k-1})r_{1:k-1,k} \quad (\text{skipped when } k = 1) \\ (v_k \text{ now contains } \hat{q}_k.) \\ r_{kk} \leftarrow \| v_k \|_2 \\ \text{if } (r_{kk} = 0) \text{ set flag } (v_1, \ldots, v_k \text{ are dependent), exit} \\ v_k \leftarrow (1/r_{kk})v_k \end{bmatrix} \quad (3.4.19)
\end{aligned}$$

Gram-Schmidt equals QR

The promised connection with the QR decomposition follows directly from the Gram-Schmidt equations. Combining (3.4.15) and (3.4.18), and solving for v_k, we obtain

$$v_k = \sum_{j=1}^{k-1} r_{jk} q_j + r_{kk} q_k, \quad k = 1, \ldots, m, \qquad (3.4.20)$$

or

$$
\begin{aligned}
v_1 &= q_1 r_{11} \\
v_2 &= q_1 r_{12} + q_2 r_{22} \\
v_3 &= q_1 r_{13} + q_2 r_{23} + q_3 r_{33} \\
&\ \ \vdots \\
v_m &= q_1 r_{1m} + q_2 r_{2m} + \cdots + q_m r_{mm}.
\end{aligned}
$$

These can be packed into a single matrix equation

$$
\begin{bmatrix} v_1 & v_2 & \cdots & v_m \end{bmatrix} = \begin{bmatrix} q_1 & q_2 & \cdots & q_m \end{bmatrix}
\begin{bmatrix}
r_{11} & r_{12} & \cdots & r_{1m} \\
0 & r_{22} & \cdots & r_{2m} \\
\vdots & \vdots & \ddots & \vdots \\
0 & 0 & \cdots & r_{mm}
\end{bmatrix}.
$$

Defining

$$
V = \begin{bmatrix} v_1 & v_2 & \cdots & v_m \end{bmatrix} \in \mathbb{R}^{n \times m}
$$

$$
Q = \begin{bmatrix} q_1 & q_2 & \cdots & q_m \end{bmatrix} \in \mathbb{R}^{n \times m}
$$

$$
R = \begin{bmatrix}
r_{11} & r_{12} & \cdots & r_{1m} \\
0 & r_{22} & \cdots & r_{2m} \\
\vdots & \vdots & \ddots & \vdots \\
0 & 0 & \cdots & r_{mm}
\end{bmatrix} \in \mathbb{R}^{m \times m},
$$

(3.4.21)

we see that V has full rank, Q is isometric, R is upper triangular with positive entries on the main diagonal (by (3.4.17)), and

$$
V = QR.
$$

Thus Q and R are the unique factors of V guaranteed by Theorem 3.4.8.

This proves that the Gram-Schmidt process provides another means of calculating a QR decomposition: Given a matrix $V \in \mathbb{R}^{n \times m}$ ($n \geq m$) of full rank, let $v_1, \ldots, v_m$ denote the columns of V. Carry out the Gram-Schmidt procedure on $v_1, \ldots, v_m$ to produce an orthonormal set $q_1, \ldots, q_m$ and the coefficients r_{jk}. Then define Q and R by (3.4.21) to get the QR decomposition of V. Conversely, any method for computing the QR decomposition can be used to orthogonalize vectors: Let $v_1, \ldots, v_m$ denote a linearly independent set that is to be orthonormalized. Define $V = [v_1 \cdots v_m] \in \mathbb{R}^{n \times m}$ and use reflectors, rotators, or any other means to produce the unique QR decomposition of V guaranteed by Theorem 3.4.8. Let $q_1, \ldots, q_m$ denote the columns of the resulting isometry Q. Then by the uniqueness of the QR decomposition, $q_1, \ldots, q_m$ are exactly the vectors that would be produced by the Gram-Schmidt process. Thus we have the main result of this section: the Gram-Schmidt orthogonalization is the same as the QR decomposition.

rcise 3.4.22

(a) Let $v_1 = [\ 3,\ -3,\ \ 3,\ -3]^T$, $v_2 = [1,\ 2,\ 3,\ 4]^T$, and $S = \mathrm{span}\{v_1, v_2\} \subseteq \mathbb{R}^4$. Apply the Gram-Schmidt process to S to obtain an orthonormal basis of S. Save the coefficients r_{jk}.

(b) Let

$$V = \begin{bmatrix} 3 & 1 \\ -3 & 2 \\ 3 & 3 \\ -3 & 4 \end{bmatrix} \in \mathbb{R}^{4 \times 2}.$$

Use the result of part (a) to build an isometric $Q \in \mathbb{R}^{4 \times 2}$ and an upper triangular $R \in \mathbb{R}^{2 \times 2}$ such that $V = QR$.

$\square$

The Gram-Schmidt Process and Roundoff Errors

The Gram-Schmidt process produces orthonormal vectors (and a QR decomposition) in principle, but in practice it is vulnerable to roundoff errors. Continuing to let $Q = [\ q_1 \ \cdots \ q_m\ ]$, we know that $q_1, \ldots, q_m$ are truly orthonormal if and only if Q is an isometry, i.e. $I - Q^T Q = 0$. Any method of computing Q will make roundoff errors, so we will have $\| I - Q^T Q \|_2 > 0$. The size of this norm is a good measure of deviation from orthonormality. We would hope to have

$$\| I - Q^T Q \|_2 \approx cu,$$

a modest multiple of the unit roundoff.

Example 3.4.23 Let V be the $n \times m$ matrix whose (i, j) entry is $v_{ij} = p_j^{i-1}$, where $p_j = j/m$. Matrices with entries of the form p_j^{i-1} are called *Vandermonde* matrices. We orthonormalized the columns of V by two different methods: the classical Gram-Schmidt process and the QR decomposition by reflectors. The results for several different choices of n and m are given in the following table.

deviation from orthonormality $\left(\| I - Q^T Q \|_2 \right)$

$n \times m$	classical G-S	QR by reflectors
6×4	1.3×10^{-14}	6.9×10^{-16}
9×6	3.5×10^{-11}	5.8×10^{-16}
12×8	2.5×10^{-06}	7.5×10^{-16}
15×10	8.8×10^{-02}	7.8×10^{-16}
18×12	2.1×10^{-00}	5.4×10^{-16}

With the classical Gram-Schmidt process there is a steady deterioration of orthonormality as the size of the matrix increases. In the 18×12 case (orthonormalizing

12 vectors in $\mathbb{R}^{18}$), we have $\| I - Q^T Q \|_2 = 2.1$, indicating that the vectors are not orthogonal at all. In contrast, the QR decomposition using reflectors yields numerically orthonormal columns in all cases. Recall that for double-precision IEEE floating-point arithmetic, $u \approx 1.11 \times 10^{-16}$. Thus in all cases the QR method gives $\| I - Q^T Q \|_2$ equal to a modest multiple of the unit roundoff. The good results can be attributed to the excellent numerical properties of reflectors.

The problem with the classical Gram-Schmidt process can be traced to the increasing ill conditioning of the Vandermonde matrices as their size increases.   □

We continue to use the notation $V = [v_1 \cdots v_m] \in \mathbb{R}^{n \times m}$. The condition number of a nonsquare matrix with linearly independent columns can be defined by

$$\kappa(V) = \frac{\max\limits_{x \neq 0} \| Vx \| / \| x \|}{\min\limits_{x \neq 0} \| Vx \| / \| x \|} = \frac{\text{maxmag}(V)}{\text{minmag}(V)}.$$

Intuitively $\kappa(V)$ is large whenever the columns $v_1, \ldots, v_m$ are nearly linearly dependent.

It is not hard to see how ill conditioning (large $\kappa(V)$) can be detrimental to the classical Gram-Schmidt process. If the columns are nearly linearly dependent, then at some step of the process v_k will nearly be a linear combination of the previous columns. Then in (3.4.15) v_k will be nearly expressible as a linear combination of $q_1, \ldots, q_{k-1}$, and there will be a severe cancellation in the computation of $\hat{q}_k$. This vector will have a very small norm, and its entries will be inaccurate. These inaccurate numbers are amplified at the step (3.4.18).

One solution to this problem that has been much used over time is the *modified Gram-Schmidt process*. This slight variation of classical Gram-Schmidt is derived in Exercise 3.4.33.

Example 3.4.24 The following table shows that the modified Gram-Schmidt process can indeed improve on classical Gram-Schmidt. Here the matrices are the same as in Example 3.4.23, but this time we have listed the condition numbers instead of the matrix dimensions.

deviation from orthonormality $\left(\| I - Q^T Q \|_2 \right)$

$\kappa_2(V)$	classical G-S	modified G-S
1.1×10^2	1.3×10^{-14}	3.9×10^{-15}
2.8×10^3	3.5×10^{-11}	1.9×10^{-13}
7.3×10^4	2.5×10^{-06}	2.8×10^{-12}
2.0×10^6	8.8×10^{-02}	6.3×10^{-11}
5.3×10^7	2.1×10^{-00}	9.0×10^{-10}

We see that the modified algorithm does a much better job than classical Gram-Schmidt, but still it suffers some deterioration of orthogonality as the condition number increases. Indeed, it is known [8, 52] that the vectors produced by the modified Gram-Schmidt process satisfy

$$\| I_m - Q^T Q \|_2 \approx u\kappa_2(V),$$

and that pattern is seen here. □

Looking back at the numbers in Example 3.4.23, it is evident that the QR decomposition by reflectors does a better job of orthonormalizing vectors than modified Gram-Schmidt does. Fortunately there is a different modification of Gram-Schmidt that is more effective: do the orthogonalization twice.

Reorthogonalization

If one runs the Gram-Schmidt process on a certain ill-conditioned $V = [\, v_1 \cdots v_m \,]$ and finds that the output Q fails to have orthonormal columns due to roundoff errors, one might consider doing a second run of Gram-Schmidt using the columns of Q as input. The chances for success are good. The columns of Q may not be orthogonal, but they are likely to be much less nearly linear dependent than the columns of V were. If Q is sufficiently well conditioned, the output of the second run will be orthonormal.

If one is committed to reorthogonalization from the outset, then a more efficient strategy would get each vector right to begin with. In (3.4.15) one transforms v_k to $\hat{q}_k$, which should be orthogonal to $q_1, \ldots, q_{k-1}$. In practice one can check whether it is or not by calculating the inner products $\langle \hat{q}_k, q_i \rangle$ for $i = 1, \ldots, k-1$. But if one is going to take the time to make these (or similar) computations, then one might as well use them to make further corrections to $\hat{q}_k$. In other words, one might as well do the orthogonalization twice at each step. We could even think of doing a third or fourth orthogonalization if necessary, but it turns out that (usually) twice is enough. Here is the simple algorithm:

> **Classical Gram-Schmidt Algorithm with Reorthogonalization.** Given linearly independent $v_1, v_2, \ldots, v_m$, this algorithm produces orthonormal $q_1, q_2, \ldots, q_m$ such that $\mathrm{span}\{q_1, \ldots, q_i\} = \mathrm{span}\{v_1, \ldots, v_i\}$ for $i = 1, \ldots, m$. $q_1, \ldots, q_m$ are stored over $v_1, \ldots, v_m$.
>
> for $k = 1, \ldots, m$
> $$\begin{aligned}
> &r_{1:k-1,k} \leftarrow (v_{1:k-1})^T v_k \\
> &v_k \leftarrow v_k - (v_{1:k-1})r_{1:k-1,k} \quad \text{(orthogonalization)} \\
> &s_{1:k-1} \leftarrow (v_{1:k-1})^T v_k \\
> &v_k \leftarrow v_k - (v_{1:k-1})s_{1:k-1} \quad \text{(reorthogonalization)} \\
> &r_{1:k-1,k} \leftarrow r_{1:k-1,k} + s_{1:k-1} \quad \text{(only if } R \text{ is needed)} \\
> &r_{kk} \leftarrow \| v_k \|_2 \\
> &\text{if } (r_{kk} = 0) \text{ set flag } (v_1, \ldots, v_k \text{ are dependent}), \text{ exit} \\
> &v_k \leftarrow (1/r_{kk})v_k
> \end{aligned}$$
> (3.4.25)

The folklore that "twice is enough" has been around for many years, but it was proved by Giraud et. al. [39] in 2005. They showed that if V is not numerically singular, i.e. $\kappa_2(V) \ll 1/u$, then (3.4.25) will return a $Q = [\, q_1 \cdots q_m \,]$ that satisfies $\| I - Q^T Q \|_2 \approx cu$, where c is not too large.

Example 3.4.26 We applied classical Gram-Schmidt process with reorthogonalization (3.4.25) to the same five Vandermonde matrices as in Examples 3.4.23 and 3.4.24 with the following results.

deviation from orthonormality $\left(\| I - Q^T Q \|_2 \right)$

$\kappa_2(V)$	reorthogonalization	QR by reflectors
1.1×10^2	1.8×10^{-16}	6.9×10^{-16}
2.8×10^3	2.7×10^{-16}	5.8×10^{-16}
7.3×10^4	4.5×10^{-16}	7.5×10^{-16}
2.0×10^6	2.4×10^{-16}	7.8×10^{-16}
5.3×10^7	4.6×10^{-16}	5.4×10^{-16}

We observe $\| I - Q^T Q \|_2 \approx cu$, as predicted, and there is no deterioration as the condition number increases. The accuracy is comparable to that obtained using the QR decomposition by reflectors. Tests on larger Vandermonde matrices showed that these results held up for condition numbers up to about 10^{16} ($\approx 1/u$). $\square$

The flop counts for the two excellent algorithms featured in Example 3.4.26 are comparable. In Exercise 3.3.4 you showed that the QR decomposition of an $n \times m$ matrix takes about $2nm^2 - 2m^3/3$ flops. However, for this application additional work needs to be done: The orthogonal matrix Q is provided implicitly, as a product of reflectors. If we want the orthonormal vectors, we need to assemble Q, and this doubles the flop count (Exercise 3.4.37). Thus the total is about $4nm^2 - 4m^3/3$ flops.

Exercise 3.4.27   Count the flops for the Gram-Schmidt algorithm with reorthogonalization (3.4.25).

(a) Show that on the kth time through the main loop the flop count is about $8nk$. Notice that the reorthogonalization almost doubles the flop count (but it's worth it).

(b) Deduce that the total flop count is about $4nm^2$.

$\square$

Thus the flop count for the QR method by rotators is somewhat less. We should always bear in mind, however, that the flop count isn't everything. An advantage of the Gram-Schmidt algorithm (3.4.25) is that it makes for fairly efficient code because it is rich in matrix-vector multiplications. A second advantage that should not be overlooked is that it is extremely easy to program.

In some applications the vectors $v_1, v_2, \dots, v_m$ are not all available from the beginning. For example, in the Arnoldi process for computing eigenvectors (Section 6.4), v_k cannot be determined until after $q_1, \dots, q_{k-1}$ have been computed. In this situation the Gram-Schmidt algorithm (3.4.25) is much more convenient than the QR decomposition by reflectors.

rcise 3.4.28 Use several different methods to orthogonalize the columns of the $m \times m$ Hilbert matrix (V = hilb(m)). Try $m = 3, \ldots, 10$, for example. In each case check the effectiveness of the procedure by computing $\| I_m - Q^T Q \|_2$ (norm(eye(m)-Q'*Q)). Check the residual $\| V - QR \|_2$ as well. This exercise is easier than it looks. Just make MATLAB m-files implementing algorithms (3.4.19), (3.4.34), and (3.4.25). They are all quite similar.

(a) Classical Gram-Schmidt (3.4.19).

(b) Modified Gram-Schmidt (Exercise 3.4.33).

(c) Classical Gram-Schmidt with reorthogonalization (3.4.25).

(d) QR decomposition by reflectors. (For this one just use MATLAB's qr command: [Q,R] = qr(V,0).)

□

Additional Exercises

rcise 3.4.29 State and prove a complex version of Theorem 3.4.2.   □

rcise 3.4.30 In this exercise you will show that the subspace equalities

$$\text{span}\{q_1, \ldots, q_k\} = \text{span}\{v_1, \ldots, v_k\}$$

hold in the Gram-Schmidt process.

(a) Use (3.4.15) and (3.4.18) to show that $q_k \in \text{span}\{q_1, \ldots, q_{k-1}, v_k\}$.

(b) Show that if $\text{span}\{q_1, \ldots, q_{k-1}\} = \text{span}\{v_1, \ldots, v_{k-1}\}$, then

$$\text{span}\{q_1, \ldots, q_{k-1}, v_k\} = \text{span}\{v_1, \ldots, v_{k-1}, v_k\}.$$

(c) Verify (3.4.20), and conclude that $v_k \in \text{span}\{q_1, \ldots, q_k\}$.

(d) Using the observations from parts (a), (b), and (c), prove by induction on k that $\text{span}\{q_1, \ldots, q_k\} = \text{span}\{v_1, \ldots, v_k\}$.

□

rcise 3.4.31 Another way to demonstrate the subspace equalities is to make use of the QR decomposition. Let $v_1, \ldots, v_m$ and $q_1, \ldots, q_m$ be two linearly independent sets of vectors, let $V = [v_1 \cdots v_m] \in \mathbb{R}^{n \times m}$, and let $Q = [q_1 \cdots q_m] \in \mathbb{R}^{n \times m}$.

(a) Show that if $V = QR$, where $R \in \mathbb{R}^{m \times m}$ is upper triangular, then

$$\text{span}\{v_1, \ldots, v_k\} \subseteq \text{span}\{q_1, \ldots, q_k\} \quad \text{for} \quad k = 1, \ldots, m.$$

(b) (Review) Show that if R is upper triangular and nonsingular, then R^{-1} is also upper triangular.

(c) Show that if $V = QR$, where R is upper triangular and nonsingular, then

$$\text{span}\{q_1, \ldots, q_k\} \subseteq \text{span}\{v_1, \ldots, v_k\} \quad \text{for} \quad k = 1, \ldots, m.$$

(d) From (a) and (c), deduce that the vectors produced by the Gram-Schmidt process satisfy

$$\text{span}\{q_1, \ldots, q_k\} = \text{span}\{v_1, \ldots, v_k\} \quad \text{for} \quad k = 1, \ldots, m.$$

(e) Conversely, show that if

$$\text{span}\{q_1, \ldots, q_k\} = \text{span}\{v_1, \ldots, v_k\} \quad \text{for} \quad k = 1, \ldots, m,$$

then there is a nonsingular, upper triangular matrix $R \in \mathbb{R}^{m \times m}$ such that $V = QR$.

□

Exercise 3.4.32 When one computes a QR decomposition in practice, one does not always bother to force the main-diagonal elements of R to be positive. In this case the columns of Q may differ from the Gram-Schmidt vectors, but it is not hard to show that the difference is trivial: Let $v_1, \ldots, v_m$ be linearly independent vectors, and let $V = [v_1 \cdots v_m]$. Suppose $V = QR$, where $Q \in \mathbb{R}^{n \times m}$ is an isometry and $R \in \mathbb{R}^{m \times m}$ is upper triangular but does not necessarily have positive main-diagonal entries. Let $q_1, \ldots, q_m$ denote the columns of Q, and let $\hat{q}_1, \ldots, \hat{q}_m$ denote the vectors obtained from $v_1, \ldots, v_m$ by the Gram-Schmidt process. Show that $q_i = \pm\hat{q}_i$ for $i = 1, \ldots, m$.   □

Exercise 3.4.33 (modified Gram-Schmidt process)

(a) Rewrite the classical Gram-Schmidt algorithm (3.4.19) so that the two matrix-vector multiplication operations are written out as loops.

Solution:

$$
\begin{aligned}
&\text{for } k = 1, \ldots, m \\
&\qquad \left[
\begin{aligned}
&\text{for } i = 1, \ldots, k-1 \\
&\qquad \left[\; r_{ik} \leftarrow \langle v_k, v_i \rangle \right. \\
&\text{for } i = 1, \ldots, k-1 \\
&\qquad \left[\; v_k \leftarrow v_k - v_i r_{ik} \right. \\
&r_{kk} \leftarrow \| v_k \|_2 \\
&\text{if } (r_{kk} = 0) \text{ set flag } (v_1, \ldots, v_k \text{ are dependent}), \text{ exit} \\
&v_k \leftarrow (1/r_{kk}) v_k
\end{aligned}
\right.
\end{aligned}
$$

(b) Now merge the two newly-created loops. This yields the

Modified Gram-Schmidt Process

$$
\begin{aligned}
&\text{for } k = 1, \ldots, m \\
&\left[\begin{array}{l}
\text{for } i = 1, \ldots, k - 1 \\
\quad \left[\begin{array}{l}
r_{ik} \leftarrow \langle v_k, v_i \rangle \\
v_k \leftarrow v_k - v_i r_{ik}
\end{array}\right. \\
r_{kk} \leftarrow \| v_k \|_2 \\
\text{if } (r_{kk} = 0) \text{ set flag } (v_1, \ldots, v_k \text{ are dependent}), \text{ exit} \\
v_k \leftarrow (1/r_{kk}) v_k
\end{array}\right.
\end{aligned}
\tag{3.4.34}
$$

How does this change affect the order in which the operations are done?

(c) Prove that in exact arithmetic the classical and modified Gram-Schmidt processes produce exactly the same output.

In the presence of roundoff errors, modified Gram-Schmidt does better than classical Gram-Schmidt because it waits until the last possible moment to compute the coefficient r_{ik}. Since all roundoff errors made up to that point are taken into account, it does a better job of making q_k orthogonal to q_i. □

rcise 3.4.35 In Chapter 1 we saw that the various ways of calculating the LU decomposition can be derived by partitioning the equation $A = LU$ in different ways. It is natural to try to do the same thing with the QR decomposition. Consider a decomposition $V = QR$, where $V = [v_1 \ \cdots \ v_k] \in \mathbb{R}^{n \times k}$, $n \geq k$, and V has full rank; $Q = [q_1 \ \cdots \ q_k] \in \mathbb{R}^{n \times k}$ is an isometry; and $R \in \mathbb{R}^{k \times k}$ is upper triangular with positive entries on the main diagonal. Partition the matrices as

$$
V = [\ \tilde{V} \mid v_k \ ], \qquad Q = [\ \tilde{Q} \mid q_k \ ], \qquad \text{and} \qquad R = \left[\begin{array}{c|c} \tilde{R} & s \\ \hline 0 & r_{kk} \end{array}\right].
$$

Derive an algorithm to calculate q_k, given $q_1, \ldots, q_{k-1}$. You may take the following steps:

(a) Use the equation $V = QR$ in partitioned form to derive a formula for $q_k r_{kk}$ in terms of known quantities and s.

(b) The condition that q_k is orthogonal to $q_1, \ldots, q_{k-1}$ can be written as $\tilde{Q}^T q_k = 0$. Use this equation to derive a formula for s.

(c) Show that q_k and r_{kk} are uniquely determined by $q_k r_{kk}$ and the conditions $\| q_k \|_2 = 1$ and $r_{kk} > 0$.

(d) Parts (a), (b), and (c) can be combined to yield an algorithm to calculate q_k, given $q_1, \ldots, q_{k-1}$. Show that this is exactly the classical Gram-Schmidt algorithm. □

rcise 3.4.36 Let V, Q, and R be as in the previous exercise, and consider the partition

$$
V = [\ v_1 \mid \tilde{V} \ ], \qquad Q = [\ q_1 \mid \tilde{Q} \ ], \qquad \text{and} \qquad R = \left[\begin{array}{c|c} r_{11} & r^T \\ \hline 0 & \tilde{R} \end{array}\right].
$$

(a) Use this partition to derive an algorithm for calculating Q and R.

(b) Show that this algorithm is equivalent to the modified Gram-Schmidt process.

One can also derive a block modified Gram-Schmidt algorithm by this method. ☐

Exercise 3.4.37 If we want to use the QR decomposition by reflectors as a method to orthonormalize vectors, we need to assemble the matrix Q, which is a product of m reflectors: $Q = Q_1 Q_2 \cdots Q_m$.

(a) Clearly $q_1 = Qe_1 = Q_1 Q_2 \cdots Q_m e_1$. Taking into consideration the form of the reflectors $Q_2, \ldots, Q_m$, show that $q_1 = Q_1 e_1$. Thus it costs $4n$ flops to compute q_1.

(b) Generalizing part (a), show that $q_i = Q_1 \cdots Q_i e_i$ for $i = 1, \ldots, m$. How many flops does this require?

(c) Show that the operation of computing $q_1, \ldots, q_m$ costs $2nm^2 - \frac{2}{3}m^2$ flops in all, which is as much as the QR decomposition cost to begin with.

☐

Exercise 3.4.38 Let $A \in \mathbb{R}^{n \times m}$, $n \geq m$, and suppose $\operatorname{rank}(A) = r$. Let $\hat{A}$ be any matrix obtained from A by permuting the columns in such a way that the first r columns are linearly independent. Prove that there exist unique $\hat{Q} \in \mathbb{R}^{n \times r}$ and $R \in \mathbb{R}^{r \times m}$ such that $\hat{Q}$ is an isometry, $R = [\hat{R} \; \tilde{R}]$, $\hat{R} \in \mathbb{R}^{r \times r}$ is upper triangular with positive entries on the main diagonal, and $\hat{A} = \hat{Q}R$. ☐

3.5 GEOMETRIC APPROACH TO THE LEAST SQUARES PROBLEM

In this section we introduce a few basic concepts and prove some fundamental theorems that yield a clear geometric picture of the least squares problem. The tools developed here will also be used later on in the book. It is traditional and possibly more logical to place this material at the beginning of the chapter, but that arrangement would have caused an unnecessary delay in the introduction of the algorithms.

Let S be any subset of $\mathbb{R}^n$. The *orthogonal complement* of S, denoted $S^\perp$ (pronounced S *perp*), is defined to be the set of vectors in $\mathbb{R}^n$ that are orthogonal to S. That is,

$$S^\perp = \{x \in \mathbb{R}^n \mid \langle x, y \rangle = 0 \text{ for all } y \in S\}.$$

The set $S^\perp$ is nonempty since it contains at least the vector 0.

Exercise 3.5.1 Show that $S^\perp$ is a subspace of $\mathbb{R}^n$; that is,

(a) show that the sum of two members of $S^\perp$ is also in $S^\perp$, and

(b) show that every scalar multiple of a member of $S^\perp$ is also a member of $S^\perp$.

☐

rcise 3.5.2 Let $q_1, \ldots, q_n$ be an orthonormal basis of $\mathbb{R}^n$, and let

$$S = \text{span}\{q_1, \ldots, q_k\},$$

where $1 \leq k \leq n - 1$. Show that $S^\perp = \text{span}\{q_{k+1}, \ldots, q_n\}$.   □

Theorem 3.5.3 *Let S be any subspace of $\mathbb{R}^n$. Then for every $x \in \mathbb{R}^n$, there exist unique elements $s \in S$ and $s^\perp \in S^\perp$ for which $x = s + s^\perp$.*

Proof. Let $v_1, \ldots, v_k$ be a basis for S. From elementary linear algebra we know that there exist vectors $v_{k+1}, \ldots, v_n$ such that $v_1, \ldots, v_k, v_{k+1}, \ldots, v_n$ is a basis of $\mathbb{R}^n$. Let $q_1, \ldots, q_n$ be the orthonormal basis of $\mathbb{R}^n$ obtained by applying the Gram-Schmidt procedure to $v_1, \ldots, v_n$. Then the nested subspace property (3.4.12) of the Gram-Schmidt process guarantees that $\text{span}\{q_1, \ldots, q_k\} = \text{span}\{v_1, \ldots, v_k\}$. Thus, by Exercise 3.5.2,

$$S = \text{span}\{q_1, \ldots, q_k\} \quad \text{and} \quad S^\perp = \text{span}\{q_{k+1}, \ldots, q_n\}.$$

Let $x \in \mathbb{R}^n$. We must find $s \in S$ and $s^\perp \in S^\perp$ such that $x = s + s^\perp$. With the help of the basis $q_1, \ldots, q_n$, this is easy. x can be expressed uniquely as a linear combination of these vectors: $x = c_1 q_1 + \cdots + c_n q_n$. Let $s = c_1 q_1 + \cdots + c_k q_k$ and $s^\perp = c_{k+1} q_{k+1} + \cdots + c_n q_n$. Then $s \in S$, $s^\perp \in S^\perp$, and $x = s + s^\perp$.

To see that this decomposition is unique, suppose we have another decomposition $x = \hat{s} + \hat{s}^\perp$, where $\hat{s} \in S$ and $\hat{s}^\perp \in S^\perp$. We will show that $\hat{s} = s$ and $\hat{s}^\perp = s^\perp$. Let $r = s - \hat{s} \in S$. Then the equation $s + s^\perp = \hat{s} + \hat{s}^\perp$ implies that $r = \hat{s}^\perp - s^\perp \in S^\perp$. Thus $r \in S \cap S^\perp$. As you can easily verify, $S \cap S^\perp = \{0\}$. Therefore $r = 0$, which implies $\hat{s} = s$ and $\hat{s}^\perp = s^\perp$.   □

The unique elements s and $s^\perp$ whose existence is guaranteed by Theorem 3.5.3 are called the *orthogonal projections* of x into S and $S^\perp$, respectively.

Given two subspaces S_1 and S_2, the *sum* $S_1 + S_2$ is defined by

$$S_1 + S_2 = \{s_1 + s_2 \mid s_1 \in S_1, \ s_2 \in S_2\}.$$

You can easily verify that $S_1 + S_2$ is a subspace of $\mathbb{R}^n$. If $\mathcal{U} = S_1 + S_2$, then every $u \in \mathcal{U}$ can be expressed as a sum $u = s_1 + s_2$, where $s_1 \in S_1$ and $s_2 \in S_2$. The sum $\mathcal{U} = S_1 + S_2$ is said to be a *direct sum* if for every $u \in \mathcal{U}$ the decomposition $u = s_1 + s_2$ is unique. Direct sums are denoted $S_1 \oplus S_2$. Theorem 3.5.3 states that for any subspace S, $\mathbb{R}^n$ is the direct sum of S and $S^\perp$:

$$\mathbb{R}^n = S \oplus S^\perp.$$

cise 3.5.4 Let S_1 and S_2 be subspaces of $\mathbb{R}^n$.

   (a) Prove that $S_1 + S_2$ is a subspace of $\mathbb{R}^n$.

   (b) Prove that $S_1 + S_2$ is a direct sum if and only if $S_1 \cap S_2 = \{0\}$.

□

Exercise 3.5.5 Let $v_1 = [1, 0, 0]^T$, $v_2 = [0, 1, 1]^T$, and $\mathcal{S} = \mathrm{span}\{v_1, v_2\} \subseteq \mathbb{R}^3$.

    (a) Find (a basis for) $\mathcal{S}^\perp$.

    (b) Find (a basis for) a subspace $\mathcal{U}$ of $\mathbb{R}^3$ such that $\mathcal{U} \neq \mathcal{S}^\perp$ but $\mathcal{S} \oplus \mathcal{U} = \mathbb{R}^3$.

        □

Exercise 3.5.6 Let $\mathcal{S}$ be a subspace of $\mathbb{R}^n$, and let $\mathcal{U} = \mathcal{S}^\perp$. Show that $\mathcal{U}^\perp = \mathcal{S}$. In other words, $\mathcal{S}^{\perp\perp} = \mathcal{S}$.   □

Let $A \in \mathbb{R}^{n \times m}$. Then A can be viewed as a linear transformation mapping $\mathbb{R}^m$ into $\mathbb{R}^n$: The vector $x \in \mathbb{R}^m$ is mapped to $Ax \in \mathbb{R}^n$. Two fundamental subspaces associated with a linear transformation are its null space (also known as the kernel) and its range. The *null space* of A, denoted $\mathcal{N}(A)$, is a subspace of $\mathbb{R}^m$ defined by

$$\mathcal{N}(A) = \{x \in \mathbb{R}^m \mid Ax = 0\}.$$

The *range* of A, denoted $\mathcal{R}(A)$, is a subspace of $\mathbb{R}^n$ defined by

$$\mathcal{R}(A) = \{Ax \mid x \in \mathbb{R}^m\}.$$

Exercise 3.5.7

    (a) Show that $\mathcal{N}(A)$ is a subspace of $\mathbb{R}^m$.

    (b) Show that $\mathcal{R}(A)$ is a subspace of $\mathbb{R}^n$.

        □

Exercise 3.5.8 Show that the rank of A is the dimension of $\mathcal{R}(A)$.   □

Lemma 3.5.9 *Let $A \in \mathbb{R}^{n \times m}$. Then for all $x \in \mathbb{R}^m$ and $y \in \mathbb{R}^n$*

$$\langle Ax, y \rangle = \langle x, A^T y \rangle.$$

Proof. $\langle Ax, y \rangle = y^T(Ax) = (y^T A)x = (A^T y)^T x = \langle x, A^T y \rangle$.   □

Note that the inner product on the left is the inner product on $\mathbb{R}^n$, while that on the right is the inner product in $\mathbb{R}^m$.

The matrix $A^T \in \mathbb{R}^{m \times n}$ can be viewed as a linear transformation mapping $\mathbb{R}^n$ into $\mathbb{R}^m$. Thus it has a null space $\mathcal{N}(A^T) \subseteq \mathbb{R}^n$ and a range $\mathcal{R}(A^T) \subseteq \mathbb{R}^m$. There is an important relationship between these spaces and the corresponding spaces associated with A.

Theorem 3.5.10 $\mathcal{R}(A)^\perp = \mathcal{N}(A^T)$.

Proof. If $y \in \mathcal{R}(A)^\perp$, then $\langle Ax, y \rangle = 0$ for all $x \in \mathbb{R}^m$, so by Lemma 3.5.9, $\langle x, A^T y \rangle = 0$ for all $x \in \mathbb{R}^m$. Thus in particular we can take $x = A^T y$ and get $\langle A^T y, A^T y \rangle = 0$, which implies $\| A^T y \|_2^2 = 0$, $A^T y = 0$, and $y \in \mathcal{N}(A^T)$. Therefore $\mathcal{R}(A)^\perp \subseteq \mathcal{N}(A^T)$. The reverse inclusion $\mathcal{N}(A^T) \subseteq \mathcal{R}(A)^\perp$ is left as an exercise for you.   □

rcise 3.5.11 Prove that $\mathcal{N}(A^T) \subseteq \mathcal{R}(A)^{\perp}$. ☐

Applying the result of Exercise 3.5.6, we see that an alternative statement of Theorem 3.5.10 is $\mathcal{N}(A^T)^{\perp} = \mathcal{R}(A)$. Since $(A^T)^T = A$, the roles of A and A^T can be reversed in Theorem 3.5.10 to yield the following corollary.

Corollary 3.5.12 $\mathcal{N}(A)^{\perp} = \mathcal{R}(A^T)$.

An equivalent statement is $\mathcal{R}(A^T)^{\perp} = \mathcal{N}(A)$. Finally

$$\mathbb{R}^n = \mathcal{R}(A) \oplus \mathcal{N}(A^T),$$

$$\mathbb{R}^m = \mathcal{N}(A) \oplus \mathcal{R}(A^T).$$

In the important special case in which $A \in \mathbb{R}^{n \times n}$ is symmetric, we have $\mathcal{N}(A)^{\perp} = \mathcal{R}(A)$ and $\mathbb{R}^n = \mathcal{N}(A) \oplus \mathcal{R}(A)$.

rcise 3.5.13

(a) The *column space* of A is defined to be the subspace of $\mathbb{R}^n$ spanned by the columns of A. Obviously the dimension of the column space is just the rank of A. Prove that the column space of A is $\mathcal{R}(A)$.

(b) The *row space* of A is defined to be the subspace of $\mathbb{R}^m$ spanned by the rows of A. Prove that the row space of A is just $\mathcal{N}(A)^{\perp}$. Use the equation $Ax = 0$; do not use Theorem 3.5.10 or Corollary 3.5.12.

(c) Obviously the column space of A is the same as the row space of A^T. Use this observation, along with the results of parts (a) and (b), to obtain a second proof of Theorem 3.5.10 and Corollary 3.5.12.

☐

The Discrete Least Squares Problem

Let $A \in \mathbb{R}^{n \times m}$ and $b \in \mathbb{R}^n$, $n \geq m$, and consider the least squares problem for the overdetermined system $Ax = b$.[21] The problem is to find $x \in \mathbb{R}^m$ such that

$$\|b - Ax\|_2 = \min_{w \in \mathbb{R}^m} \|b - Aw\|_2.$$

The set of all Aw such that $w \in \mathbb{R}^m$ is $\mathcal{R}(A)$, so this problem is obviously closely related to that of finding $y \in \mathcal{R}(A)$ such that

$$\|b - y\|_2 = \min_{s \in \mathcal{R}(A)} \|b - s\|_2.$$

[21] In least squares problems we typically have $n \gg m$. However, the results that follow are also valid for $n < m$.

The next theorem shows that this problem has a unique solution, and it gives a simple characterization of the solution. First we have to prove a lemma.

Lemma 3.5.14 *(Pythagorean Theorem) Let u and v be orthogonal vectors in $\mathbb{R}^n$. Then*

$$\|u + v\|_2^2 = \|u\|_2^2 + \|v\|_2^2.$$

Proof. $\|u + v\|_2^2 = \langle u + v, u + v \rangle = \langle u, u \rangle + \langle v, u \rangle + \langle u, v \rangle + \langle v, v \rangle = \langle u, u \rangle + \langle v, v \rangle$, because $\langle u, v \rangle = \langle v, u \rangle = 0$. Thus $\|u + v\|_2^2 = \|u\|_2^2 + \|v\|_2^2$. $\qquad\square$

Theorem 3.5.15 *Let S be a subspace of $\mathbb{R}^n$, and let $b \in \mathbb{R}^n$. Then there exists a unique $y \in S$ such that*

$$\|b - y\|_2 = \min_{s \in S} \|b - s\|_2. \tag{3.5.16}$$

y is the unique element in S such that $b - y \in S^\perp$. In other words, y is the orthogonal projection of b into S.

Exercise 3.5.17 We all learned in school that the shortest distance from a point to a straight line is along the perpendicular. Theorem 3.5.15 generalizes this simple fact. Draw a picture that illustrates Theorem 3.5.15 in the case when the dimension of S is 1. Then the vectors in S all lie in a line. The shortest distance from a point (the tip of b) to a straight line (S) is along the perpendicular ($b - y$). $\qquad\square$

Proof of Theorem 3.5.15. By Theorem 3.5.3 there exist unique elements $y \in S$ and $z \in S^\perp$ such that $b = y + z$. y is the orthogonal projection of b into S. Notice that $b - y = z \in S^\perp$. There can be no other $w \in S$ such that $b - w \in S^\perp$, because then the decomposition $b = w + (b - w)$ would violate the uniqueness part of Theorem 3.5.3. To see that y satisfies (3.5.16), let $s \in S$, and consider $\|b - s\|_2^2$. Since $b - s = (b - y) + (y - s)$, where $b - y \in S^\perp$ and $y - s \in S$, we have by Lemma 3.5.14

$$\|b - s\|_2^2 = \|b - y\|_2^2 + \|y - s\|_2^2.$$

As s runs through S, the term $\|b - y\|_2^2$ remains constant, while the term $\|y - s\|_2^2$ varies but remains strictly positive, except that it equals zero when $y = s$. Thus $\|b - s\|_2^2$, and hence also $\|b - s\|_2$, is minimized when and only when $s = y$. $\qquad\square$

Now let us see what Theorem 3.5.15 tells us about the least squares problem. Taking $S = \mathcal{R}(A)$, we find that there is a unique $y \in \mathcal{R}(A)$ such that $\|b - y\|_2 = \min_{s \in \mathcal{R}(A)} \|b - s\|_2 = \min_{w \in \mathbb{R}^m} \|b - Aw\|_2$. Any $x \in \mathbb{R}^m$ that satisfies $Ax = y$ will be a solution to the least squares problem for the system $Ax = b$. Since $y \in \mathcal{R}(A)$, there must be at least one such x. This proves that the least squares problem always has at least one solution—a fact that we already proved in Section 3.3 by other means.

rcise 3.5.18

(a) Suppose $x \in \mathbb{R}^m$ satisfies $Ax = y$. Show that $A\hat{x} = y$ if and only if $x - \hat{x} \in \mathcal{N}(A)$.

(b) Show that the least squares problem has a unique solution if and only if $\mathcal{N}(A) = \{0\}$.

$\square$

We now have two necessary and sufficient conditions for uniqueness of a solution to the least squares problem. In Section 3.3 we saw that the solution is unique if and only if A has full rank, and now we see that it is unique if and only if $\mathcal{N}(A) = \{0\}$. These two conditions must therefore be equivalent. In the following exercise you are asked to prove directly that they are.

rcise 3.5.19 Let $A \in \mathbb{R}^{n \times m}$. Prove that $\text{rank}(A) = m$ if and only if $\mathcal{N}(A) = \{0\}$. $\square$

The part of Theorem 3.5.15 that characterizes the minimizing vector yields the following corollary.

Corollary 3.5.20 *Let $x \in \mathbb{R}^m$. Then $\| b - Ax \|_2 = \min\limits_{w \in \mathbb{R}^m} \| b - Aw \|_2$ if and only if $b - Ax \in \mathcal{R}(A)^{\perp}$.*

Combining this corollary with Theorem 3.5.10, we see that x solves the least squares problem if and only if $b - Ax \in \mathcal{N}(A^T)$; that is, $A^T(Ax - b) = 0$. Rewriting this last equation, we obtain the following important result.

Theorem 3.5.21 *Let $x \in \mathbb{R}^m$. Then x solves the least squares problem for the system $Ax = b$ if and only if*

$$A^T Ax = A^T b. \tag{3.5.22}$$

The matrix $A^T A$ is in $\mathbb{R}^{m \times m}$, and the vector $A^T b$ is in $\mathbb{R}^m$, so (3.5.22) is a system of m linear equations in m unknowns. Since these are the equations that hold when $b - Ax$ is normal (meaning orthogonal) to $\mathcal{R}(A)$, they are known as the *normal equations*.

The coefficient matrix of the normal equations is *positive semidefinite*; that is, it is symmetric, and $x^T(A^T A)x \geq 0$ for all $x \in \mathbb{R}^m$. If $\text{rank}(A) = m$, then $A^T A$ is positive definite.

rcise 3.5.23 Work this problem by hand. Consider the overdetermined system

$$\begin{bmatrix} 1 \\ 1 \end{bmatrix} [\, x \,] = \begin{bmatrix} 9 \\ 5 \end{bmatrix},$$

which we considered previously in Exercise 3.3.7. Solve the least-squares problem again, but this time do it by setting up and solving the normal equations. $\square$

Exercise 3.5.24

(a) Prove that $A^T A$ is positive semidefinite.

(b) Suppose $A \in \mathbb{R}^{n \times m}$ with $n < m$. Show that $A^T A$ is not positive definite.

(c) Suppose $A \in \mathbb{R}^{n \times m}$ with $n \geq m$. Show that $A^T A$ is positive definite if and only if A has full rank. (We already covered the case $n = m$ in Theorem 1.4.4.)

$\square$

In the full-rank case (with $n \geq m$) the unique solution of the least squares problem can be found by constructing the normal equations (3.5.22) and solving this positive definite system by Cholesky's method. Indeed, up until about 1970 this was the standard technique for solving least squares problems. Its advantages are that it is simple and inexpensive. The system that has to be solved is $m \times m$, and m is usually small. The biggest expense is that of computing the coefficient matrix $A^T A$.

The disadvantage of the normal equations approach is that it is sometimes less accurate than the QR approach. Critical information can be lost when $A^T A$ is formed.

Example 3.5.25 Let

$$A = \begin{bmatrix} 1 & 1 \\ \epsilon & 0 \\ 0 & \epsilon \end{bmatrix},$$

where $\epsilon > 0$ is small. Clearly A has full rank, and

$$A^T A = \begin{bmatrix} 1 + \epsilon^2 & 1 \\ 1 & 1 + \epsilon^2 \end{bmatrix},$$

which is positive definite. However, if ϵ is small enough that ϵ^2 is less than the unit roundoff u, then the computed $A^T A$ will be $\begin{bmatrix} 1 & 1 \\ 1 & 1 \end{bmatrix}$, which is singular. $\square$

In spite of its inferior numerical properties, the normal equation approach is still sometimes used to solve least squares problems. It can be used safely whenever A is well conditioned. This issue will be covered in more detail in Section 4.4, in which we discuss the sensitivity of the least squares problem.

Exercise 3.5.26

(a) Let $B \in \mathbb{R}^{n \times m}$ be any matrix such that $\mathcal{R}(B) = \mathcal{R}(A)$. Show that x is a solution of the least squares problem for the overdetermined system $Ax = b$ if and only if $B^T A x = B^T b$.

(b) Show that if A has full rank, then $\mathcal{R}(A) = \mathcal{R}(B)$ if and only if there exists a nonsingular $C \in \mathbb{R}^{m \times m}$ such that $A = BC$.

$\square$

The most obvious instance of the situation in Exercise 3.5.26 is the case $B = A$, for which the system $B^T A x = B^T b$ is just the system of normal equations (3.5.22). Another instance stems from the QR decomposition $A = \hat{Q}\hat{R}$, where $\hat{Q} \in \mathbb{R}^{n \times m}$ is an isometry and $\hat{R} \in \mathbb{R}^{m \times m}$ is upper triangular (Theorem 3.4.7). If A has full rank, then $\hat{R}$ is nonsingular, and by part (b) of Exercise 3.5.26, $\mathcal{R}(A) = \mathcal{R}(\hat{Q})$. Therefore, by part (a), the unique solution of the least squares problem satisfies

$$\hat{Q}^T A x = \hat{Q}^T b. \tag{3.5.27}$$

This is exactly the system that is solved, in the guise $\hat{R}x = \hat{c}$, when the least squares problem is solved by the QR decomposition method.

rcise 3.5.28 Suppose we solve the normal equations $A^T A x = A^T b$ using a Cholesky decomposition $A^T A = R^T R$. In this approach the last step is to solve an upper triangular system $Rx = y$ by back substitution. Prove that this system is exactly the same as the system $\hat{R}x = \hat{c}$ that arises in the QR decomposition method. (However, the two methods arrive at this system by radically different approaches.) $\quad\square$

rcise 3.5.29 (Derivation of the normal equations using calculus). The function

$$f(x_1, \ldots, x_m) = f(x) = \| b - Ax \|_2^2$$

is a differentiable function of m variables. It has a minimum only when $\nabla f = (\partial f / \partial x_1, \ldots, \partial f / \partial x_m)^T = 0$. Calculate ∇f, and note that the equations $\nabla f = 0$ are just the normal equations. $\quad\square$

The Continuous Least Squares Problem

We introduced the discrete least squares problem by considering the problem of approximating a discrete point set $\{(t_i, y_i) \mid i = 1, \ldots, n\}$ by a simple curve such as a straight line. In the continuous least squares problem, the discrete point set is replaced by continuous data $\{(t, f(t)) \mid t \in [a, b]\}$. Thus, given a function f defined on some interval $[a, b]$, we seek a simple function ϕ (e.g. a linear polynomial) such that ϕ approximates f well on $[a, b]$. The goodness of approximation is measured, not by calculating a *sum* of squares, but by calculating the *integral* of the square of the error:

$$\int_a^b |f(x) - \phi(x)|^2 dx. \tag{3.5.30}$$

The continuous least squares problem is solved by minimizing this integral over whichever set of functions ϕ we are allowing as approximations of f. For example, if the approximating function is to be a first-degree polynomial, we minimize (3.5.30) over the set of functions $\{\phi \mid \phi(t) = a_0 + a_1 t; \; a_0, a_1 \in \mathbb{R}\}$. More generally, if the approximating function is to be a polynomial of degree less than m, we minimize (3.5.30) over the set of functions

$$\mathcal{P}_{m-1} = \left\{ \phi \mid \phi(t) = \sum_{k=0}^{m-1} a_k t^k; \; a_k \in \mathbb{R}, \; k = 0, \ldots, m-1 \right\}.$$

The set $\mathcal{P}_{m-1}$ is an m-dimensional vector space of functions. Still more generally we can let $\mathcal{S}$ be any m-dimensional vector space of functions defined on $[a, b]$ and minimize (3.5.30) over $\mathcal{S}$.

We can solve the continuous least squares problem by introducing an inner product and a norm for functions and utilizing the geometric ideas introduced in this section. The *inner product* of two functions f and g on $[a, b]$ is defined by

$$\langle f, g \rangle = \int_a^b f(x)g(x)\, dx.$$

This inner product enjoys the same algebraic properties as the inner product on $\mathbb{R}^n$. For example, $\langle f, g \rangle = \langle g, f \rangle$, and $\langle f_1 + f_2, g \rangle = \langle f_1, g \rangle + \langle f_2, g \rangle$. Two functions f and g are said to be *orthogonal* if $\langle f, g \rangle = 0$. The *norm* of f is defined by

$$\| f \| = \left(\int_a^b |f(x)|^2 dx \right)^{1/2}.$$

Notice that the norm and inner product are related by $\| f \| = \sqrt{\langle f, f \rangle}$. Furthermore (3.5.30) can be expressed in terms of the norm as $\| f - \phi \|^2$. Thus the continuous least squares problem is to find $\phi \in \mathcal{S}$ such that

$$\| f - \phi \| = \min_{\psi \in \mathcal{S}} \| f - \psi \|. \tag{3.5.31}$$

We proved Theorem 3.5.15 only for subspaces of $\mathbb{R}^n$, but it is valid for function spaces as well. See, for example, [90, Theorem II.7.2]. Thus the continuous least squares problem (3.5.31) has a unique solution, which is characterized by $f - \phi \in \mathcal{S}^\perp$; that is,

$$\langle f - \phi, \psi \rangle = 0 \qquad \text{for all } \psi \in \mathcal{S}. \tag{3.5.32}$$

Let $\phi_1, \ldots, \phi_m$ be a basis for $\mathcal{S}$. Then $\phi = \sum_{j=1}^m x_j \phi_j$ for some unknown coefficients $x_1, \ldots, x_m$. Substituting this expression for ϕ into (3.5.32), setting $\psi = \phi_i$, and applying some of the basic properties of inner products, we obtain

$$\sum_{j=1}^m \langle \phi_j, \phi_i \rangle x_j = \langle f, \phi_i \rangle \qquad i = 1, \ldots, m. \tag{3.5.33}$$

This is a system of m linear equations in m unknowns, which can be written as a matrix equation

$$Cx = d, \tag{3.5.34}$$

where

$$C = \begin{bmatrix} \langle \phi_1, \phi_1 \rangle & \cdots & \langle \phi_m, \phi_1 \rangle \\ \vdots & & \vdots \\ \langle \phi_1, \phi_m \rangle & \cdots & \langle \phi_m, \phi_m \rangle \end{bmatrix} \qquad x = \begin{bmatrix} x_1 \\ \vdots \\ x_m \end{bmatrix} \qquad \text{and} \quad d = \begin{bmatrix} \langle f, \phi_1 \rangle \\ \vdots \\ \langle f, \phi_m \rangle \end{bmatrix}.$$

The matrix C is clearly symmetric. In fact, it is positive definite, so (3.5.34) can be solved by Cholesky's method to yield the solution of the continuous least squares problem.

rcise 3.5.35 With each nonzero $y = [y_1 \cdots y_m]^T \in \mathbb{R}^m$ associate the nonzero function $\psi = \sum_{j=1}^m y_j \phi_j$. Prove that $y^T C y = \langle \psi, \psi \rangle$. Combining this with the fact that $\langle \psi, \psi \rangle > 0$, we conclude that C is positive definite.   □

The next exercise shows that the equations (3.5.34) are analogous to the normal equations of the discrete least squares problem.

rcise 3.5.36 Find $v_1, \ldots, v_m \in \mathbb{R}^m$ for which the normal equations (3.5.22) have the form

$$\sum_{j=1}^m \langle v_j, v_i \rangle x_j = \langle f, v_i \rangle \qquad i = 1, \ldots, m.$$

Thus the normal equations have the same general form as (3.5.33).   □

rcise 3.5.37 Find the polynomial $\phi(t) = x_1 + x_2 t$ of degree 1 that best approximates $f(t) = t^2$ in the least squares sense on $[0, 1]$. Check your answer by verifying that $f - \phi$ is orthogonal to $\mathcal{S} = \{a_0 + a_1 t \mid a_0, a_1 \in \mathbb{R}\}$.   □

rcise 3.5.38 Let $[a, b] = [0, 1]$, $\mathcal{S} = \mathcal{P}_{m-1}$, and let $\phi_1, \ldots, \phi_m$ be the basis of $\mathcal{P}_{m-1}$ defined by $\phi_1(t) = 1$, $\phi_2(t) = t$, $\phi_3(t) = t^2$, ..., $\phi_m(t) = t^{m-1}$. Calculate the matrix C that would appear in (3.5.34) in this case. Note that C is just the $m \times m$ member of the family of Hilbert matrices introduced in Section 2.2. This is the context in which the Hilbert matrices first arose.   □

In Section 2.2 the Hilbert matrices served as an example of a family of ill-conditioned matrices $H_m \in \mathbb{R}^{m \times m}$, $m = 1, 2, 3, \ldots$, whose condition numbers get rapidly worse with increasing m. Now we can observe, at least intuitively, that the ill conditioning originates with the basis $\phi_1(t) = 1$, $\phi_2(t) = t$, $\phi_3(t) = t^2$, ... If you plot the functions t^k on $[0, 1]$, you will see that with increasing k they look more and more alike. That is, they come closer and closer to being linearly dependent. Thus the basis $\phi_1, \ldots, \phi_m$ is (in some sense) ill conditioned and becomes increasingly ill conditioned as m is increased. This ill conditioning is inherited by the Hilbert matrices.

3.6 UPDATING THE QR DECOMPOSITION

There are numerous applications in which the data matrix is updated repeatedly. For example, in a signal processing application, each row of the matrix corresponds to measurements made at a given time. Each time new measurements come in, a new row is added. Now suppose we have solved a least-squares problem (to estimate the trajectory of an object, say) at a particular time, but now we want to solve the problem again, incorporating additional data that has just arrived. To solve the first problem we computed a QR decomposition. Must we now compute a new QR decomposition from scratch, or can we somehow update the old one? It turns out that we can update the old QR decomposition much more cheaply than we can compute a new one from scratch.

This operation and the others discussed in this section are implemented in the MATLAB commands `qrinsert` and `qrdelete`.

Adding a Row

Suppose $A \in \mathbb{R}^{n \times m}$, $n \geq m$, $\text{rank}(A) = m$, and we have a decomposition $A = QR$, where $Q \in \mathbb{R}^{n \times n}$ is orthogonal, and $R \in \mathbb{R}^{n \times m}$ is upper triangular. That is, the upper $m \times m$ block of R is upper triangular, and the rest of R is zero. Now let

$$\tilde{A} = \begin{bmatrix} A_1 \\ z^T \\ A_2 \end{bmatrix} \in \mathbb{R}^{n+1 \times m}, \text{ where } A = \begin{bmatrix} A_1 \\ A_2 \end{bmatrix}, \text{ and } z^T \text{ is a new row, which we}$$

wish to insert somewhere among the rows of A. (It might seem simpler just to adjoin z^T at the bottom of the matrix A, but, as we shall show, it is no harder to insert it anywhere.) Let us figure out how to derive the QR decomposition $\tilde{A} = \tilde{Q}\tilde{R}$ from the QR decomposition of A.

We begin by partitioning Q in the same way as we have partitioned A. Then the QR decomposition takes the form

$$\begin{bmatrix} A_1 \\ A_2 \end{bmatrix} = \begin{bmatrix} Q_1 \\ Q_2 \end{bmatrix} R.$$

It follows immediately that

$$\tilde{A} = \begin{bmatrix} A_1 \\ z^T \\ A_2 \end{bmatrix} = \begin{bmatrix} 0 & Q_1 \\ 1 & 0^T \\ 0 & Q_2 \end{bmatrix} \begin{bmatrix} z^T \\ R \end{bmatrix}. \tag{3.6.1}$$

This is nearly a QR decomposition. The only problem is that $\begin{bmatrix} z^T \\ R \end{bmatrix}$ is Hessenberg, not triangular. We just need to transform it to upper triangular form. We have

$$\begin{bmatrix} z \\ R \end{bmatrix} = \begin{bmatrix} z_1 & z_2 & \cdots \\ r_{11} & r_{12} & \cdots \\ 0 & r_{22} & \cdots \\ 0 & 0 & \ddots \\ \vdots & \vdots & \end{bmatrix}.$$

There is a plane rotator U_1^T acting on rows 1 and 2 such that $U_1^T \begin{bmatrix} z^T \\ R \end{bmatrix}$ has a zero where r_{11} was. This transformation obviously does not destroy any of the zeros that were already present in the matrix. Once we have annihilated r_{11}, we can annihilate r_{22} by applying a plane rotator U_2^T that acts on rows 2 and 3. This does not disturb any zeros that were already present. In particular, the zero in the r_{11} position stays zero because $r_{21} = 0$. The next step is to apply a rotator U_3^T, acting on rows 3 and 4 that annihilates r_{33}. If we continue in this manner for m steps, we obtain an upper triangular matrix

$$\tilde{R} = U_m^T \cdots U_2^T U_1^T \begin{bmatrix} z^T \\ R \end{bmatrix} \in \mathbb{R}^{(n+1) \times m}.$$

Letting

$$\tilde{Q} = \begin{bmatrix} 0 & Q_1 \\ 1 & 0^T \\ 0 & Q_2 \end{bmatrix} U_1 U_2 \cdots U_m,$$

we have $\tilde{A} = \tilde{Q}\tilde{R}$, with $\tilde{Q} \in \mathbb{R}^{(n+1) \times (n+1)}$ orthogonal and $\tilde{R} \in \mathbb{R}^{(n+1) \times m}$ upper triangular.

The cost of producing $\tilde{R}$ is quite low. There are m plane rotators, each acting on rows of effective length m or less. Thus the flop count is clearly $O(m^2)$. You can easily check that it is $3m^2 + O(m)$, to be more precise. Depending on the application, we may or may not wish to update Q. If we do wish to make that update and we have Q stored explicitly, we can obtain $\tilde{Q}$ by applying m plane rotators on the right. This costs $6nm + O(m)$ flops. Recalling that the cost of a QR decomposition from scratch is around $2nm^2$ flops, we see that the update is much cheaper, especially the update of R.

rcise 3.6.2

(a) Show that about $3m^2$ flops are needed to produce $\tilde{R}$.

(b) Show that $\tilde{Q}$ can be obtained from Q in about $6nm$ flops if Q is stored explicitly.

$\square$

rcise 3.6.3 The heart of the procedure that we have just outlined is to reduce the matrix

$$\begin{bmatrix} z^T \\ R \end{bmatrix}$$

to upper triangular form. Reformulate the algorithm so that the matrix that is to be reduced to triangular form is

$$\begin{bmatrix} R \\ z^T \end{bmatrix}.$$

Show how to return this matrix to upper triangular form by a sequence of m plane rotators.

$\square$

rcise 3.6.4 Suppose we have a condensed QR decomposition $A = \hat{Q}\hat{R}$, where $\hat{Q} \in \mathbb{R}^{n \times m}$ is isometric and $\hat{R} \in \mathbb{R}^{m \times m}$ is upper triangular. Develop a procedure for updating this decomposition when a row is adjoined to A. Make sure you get the dimensions of the matrices right. (You'll have to throw something away in the end.) ☐

The procedure we have just outlined also serves as a method for updating the Cholesky decomposition. If we use the normal equations to solve the least-squares problem, we need the Cholesky decomposition of $A^T A$, i.e. $A^T A = \hat{R}^T \hat{R}$. If we now add a row to A, we need the Cholesky decomposition of the updated matrix $\tilde{A}^T \tilde{A}$. It is easy to show that the Cholesky factor $\hat{R}$ is the same as the matrix $\hat{R}$ in the decomposition $A = \hat{Q}\hat{R}$ (cf. proof of Theorem 3.2.45 and Exercise 3.5.28). The updating procedure that we have outlined above (and, in particular, in Exercise 3.6.4), does not require knowledge of the orthogonal matrix Q (or $\hat{Q}$), so it can be used to update the Cholesky factor in $3m^2$ flops.

Exercise 3.6.5 Let $M \in \mathbb{R}^{m \times m}$ be a positive definite matrix, and let R be its Cholesky factor. Let $z \in \mathbb{R}^m$, and let $\tilde{M} = M + zz^T$. Since the matrix zz^T has rank one, $\tilde{M}$ is called a *rank-one update* of M.

(a) Show that $\tilde{M}$ is positive definite.

(b) Let $\tilde{R}$ be the Cholesky factor of $\tilde{M}$. Outline a procedure for obtaining $\tilde{R}$ from R and z in $O(m^2)$ flops.

☐

Deleting a Row

If we are constantly adding new rows of data to a matrix, we might also sometimes wish to delete old, "out-of-date" rows. We thus consider the problem of updating the QR decomposition when we have deleted a row. This procedure is commonly known as *downdating*. Suppose we have the decomposition $\tilde{A} = \tilde{Q}\tilde{R}$, where $\tilde{A} = \begin{bmatrix} A_1 \\ z^T \\ A_2 \end{bmatrix}$, and we seek the QR decomposition of $A = \begin{bmatrix} A_1 \\ A_2 \end{bmatrix}$. We can essentially reverse the process we used for adding a row. However, the reversed process requires knowledge of $\tilde{Q}$ and generally requires more work. A downdating procedure that acts only on $\tilde{R}$ and makes no use of $\tilde{Q}$ is outlined in Exercises 3.6.12 through 3.6.15.

The objective is to transform $\tilde{Q}$ essentially to the form

$$\begin{bmatrix} 0 & Q_1 \\ 1 & 0^T \\ 0 & Q_2 \end{bmatrix},$$

as in (3.6.1). Let

$$\tilde{Q} = \begin{bmatrix} \tilde{Q}_1 \\ w^T \\ \tilde{Q}_2 \end{bmatrix},$$

where the row w^T is in the same position as the row that is to be removed from $\tilde{A}$. We need to transform w^T to the form $\begin{bmatrix} \gamma & 0 & \cdots & 0 \end{bmatrix}$ (where $|\gamma| = 1$). Let U_n be a plane rotator acting on positions n and $n + 1$ such that $w^T U_n$ has a zero in its last $((n + 1)$st) position. Then let U_{n-1} be a rotator acting on positions $n - 1$ and n such that $w^T U_n U_{n-1}$ has zeros in its last two positions. Continuing in this way, we produce plane rotators $U_{n-2}, \ldots U_1$ such that $w^T U_n U_{n-1} \ldots U_1$ has the form $\begin{bmatrix} \gamma & 0 & \cdots & 0 \end{bmatrix}$. If we apply this sequence of rotators to $\hat{Q}$, we obtain a matrix $\check{Q} = \hat{Q} U_n \cdots U_1$ of the form

$$\check{Q} = \begin{bmatrix} 0 & Q_1 \\ \gamma & 0^T \\ 0 & Q_2 \end{bmatrix}. \tag{3.6.6}$$

The first column, apart from the γ, is necessarily zero because the matrix is orthogonal and therefore has orthonormal rows.

Let $\check{R} = U_1^T \cdots U_n^T \tilde{R}$. Then $\tilde{A} = \hat{Q}\tilde{R} = (\hat{Q} U_n \cdots U_1)(U_1^T \cdots U_n^T \tilde{R}) = \check{Q}\check{R}$, so let us examine $\check{R}$. $\check{R}$ is an $(n + 1) \times m$ matrix whose top m rows form an upper triangular matrix and whose last $n + 1 - m$ rows are zero. Now consider how $\tilde{R}$ is altered when we apply The plane rotators $U_n^T, U_{n-1}^T, \ldots, U_1^T$. The first rotators, $U_n^T, \ldots, U_{m+1}^T$, have no effect at all on $\tilde{R}$, because they operate on rows that are identically zero. The first to have any effect is U_m^T, which acts on rows m and $m + 1$. Row m has one nonzero element, namely $\tilde{r}_{mm}$. When U_m^T is applied, this nonzero entry is combined with the zero in position $(m + 1, m)$ to produce a nonzero entry in that position. This is clearly the only new nonzero entry produced by this operation. Similarly, application of U_{m-1}^T acts on rows $m - 1$ and m and creates a new nonzero entry in position $(m, m - 1)$ (and nowhere else), because $\tilde{r}_{m-1,m-1}$ is nonzero. The pattern is now clear. Application of $U_m^T, \ldots, U_1^T$ produces new nonzero entries in positions $(m + 1, m), (m, m - 1), \ldots, (2, 1)$, and nowhere else. This means that $\check{R}$ has upper Hessenberg form. Thus we can partition $\check{R}$ as

$$\check{R} = \begin{bmatrix} v^T \\ R \end{bmatrix},$$

where v^T is the first row of $\check{R}$, and R is upper triangular. Combining this with (3.6.6), we have

$$\begin{bmatrix} A_1 \\ z^T \\ A_2 \end{bmatrix} = \tilde{A} = \check{Q}\check{R} = \begin{bmatrix} 0 & Q_1 \\ \gamma & 0^T \\ 0 & Q_2 \end{bmatrix} \begin{bmatrix} v^T \\ R \end{bmatrix}.$$

Focusing on the top block, we have $A_1 = 0v^T + Q_1 R = Q_1 R$. Now looking at the bottom block, we have $A_2 = 0v^T + Q_2 R$. Putting these blocks together, leaving out the middle row, and letting $Q = \begin{bmatrix} Q_1 \\ Q_2 \end{bmatrix}$, we obtain

$$A = \begin{bmatrix} A_1 \\ A_2 \end{bmatrix} = \begin{bmatrix} Q_1 \\ Q_2 \end{bmatrix} R = QR.$$

Since the rows of Q are orthonormal, Q is an orthogonal matrix. Thus we now have the QR decomposition of A.

Exercise 3.6.7 How much does this downdating procedure cost . . .

(a) . . . if only R is wanted?

(b) . . . if Q is wanted?

$\square$

Exercise 3.6.8 Why does the downdating procedure take more work than the updating procedure? $\square$

Since this procedure requires knowledge of the orthogonal matrix $\tilde{Q}$, we cannot use it for downdating the Cholesky decomposition $\tilde{A}^T \tilde{A} = \tilde{R}^T \tilde{R}$, when a row is removed from $\tilde{A}$. Fortunately a procedure for downdating the Cholesky decomposition does exist. It is quite economical ($O(m^2)$), but it is more delicate, involving the use of hyperbolic transformations. See, for example, [43, §12.5.4] and Exercises 3.6.12 through 3.6.15.

Adding a Column

Consider an experiment in which a large number of individuals, e.g. fruit flies or guinea pigs, are measured in various ways, e.g. wingspan, weight, length of ears. If m characteristics are measured for each of n individuals, the data can be assembled in an $n \times m$ matrix A in which each row corresponds to an individual and each column corresponds to a characteristic. We may be able to learn something about our population by solving some least squares problems using this data. If we want to add a characteristic to our study, we must add a column to the matrix A. Suppose we have a QR decomposition of A. If we adjoin a column to A, can we obtain the QR decomposition of the new matrix efficiently by updating the QR decomposition of A?

Let $A = \begin{bmatrix} A_1 & A_2 \end{bmatrix}$ and $\tilde{A} = \begin{bmatrix} A_1 & z & A_2 \end{bmatrix}$, where z is the column that is to be adjoined. We have a decomposition $A = QR$, with $Q \in \mathbb{R}^{n \times n}$ orthogonal and $R \in \mathbb{R}^{n \times m}$ upper triangular. Partitioning R conformably with A, we have

$$\begin{bmatrix} A_1 & A_2 \end{bmatrix} = Q \begin{bmatrix} R_1 & R_2 \end{bmatrix}.$$

Let us see how to update this decomposition to obtain the QR decomposition of $\tilde{A}$. If we let $w = Q^T z = Q^{-1} z$, then

$$\tilde{A} = \begin{bmatrix} A_1 & z & A_2 \end{bmatrix} = Q \begin{bmatrix} R_1 & w & R_2 \end{bmatrix}.$$

We have disturbed the triangular form of R; our task is to restore it. Suppose w is the kth column. To obtain triangular form we must annihilate the bottom $n - k$ entries of w. Let U_n be a plane rotator acting on rows n and $n - 1$ such that $U_n w$ has a zero in the nth position. Let U_{n-1} be a plane rotator on rows $n - 1$ and $n - 2$ such that $U_{n-1}U_n w$ has a zero in position $n - 1$. The zero that was created on the previous step remains untouched. Continuing in this fashion, we can produce plane rotators $U_{n-2}, \ldots, U_{k+1}$ such that $U_{k+1} \cdots U_n w$ has zeros in positions $k + 1, \ldots, n$. Let

$$\tilde{R} = U_{k+1} \cdots U_n \begin{bmatrix} R_1 & w & R_2 \end{bmatrix}.$$

It is easy to check that $\tilde{R}$ has upper triangular form. The rotators do not alter R_1 at all. They do affect R_2, but only to the extent that they fill out $\tilde{R}$ to a full upper-triangular matrix. Let $\tilde{Q} = Q U_n^T \cdots U_{k+1}^T$. Then $\tilde{A} = \tilde{Q}\tilde{R}$. This is our desired decomposition.

rcise 3.6.9 Verify that $\tilde{R}$ is an upper-triangular matrix. $\qquad\qquad \square$

Unfortunately this updating procedure is not particularly economical. The operation $w = Q^T z$ already costs $2n^2$ flops, and it requires knowledge of Q. If $n \geq m^2$, it is not worthwhile.

Deleting a Column

Now suppose we wish to delete a characteristic from our sample. This corresponds to deleting a column from A. How do perform the update, or perhaps we should say *downdate*, in this case? Fortunately it turns out that removing a column is much cheaper than adding one.

Let $\tilde{A} = \begin{bmatrix} A_1 & z & A_2 \end{bmatrix}$ and $A = \begin{bmatrix} A_1 & A_2 \end{bmatrix}$. Suppose we have the decomposition $\tilde{A} = \tilde{Q}\tilde{R}$, and we would like the QR decomposition of A. Let $\tilde{R} = \begin{bmatrix} \tilde{R}_1 & w & \tilde{R}_2 \end{bmatrix}$. Removing a column from the equation $\tilde{A} = \tilde{Q}\tilde{R}$, we obtain

$$A = \begin{bmatrix} A_1 & A_2 \end{bmatrix} = \tilde{Q} \begin{bmatrix} \tilde{R}_1 & \tilde{R}_2 \end{bmatrix}. \tag{3.6.10}$$

This is not quite a QR decomposition, because the "R" matrix is not upper triangular. For example, if we remove the third column from a 7×5 upper triangular matrix, we obtain a matrix of the form

$$\begin{bmatrix} * & * & * & * \\ 0 & * & * & * \\ 0 & 0 & * & * \\ 0 & 0 & * & * \\ 0 & 0 & 0 & * \\ 0 & 0 & 0 & 0 \\ 0 & 0 & 0 & 0 \end{bmatrix}.$$

It is upper Hessenberg but can be reduced to triangular form by two rotators, U_3^T, acting on rows 3 and 4, followed by U_4^T, acting on rows 4 and 5. More generally, in (3.6.10), if the row w that was removed from $\tilde{R}$ was the kth row, $m - k + 1$ plane rotators, $U_k^T, \ldots, U_m^T$ will be needed to return the matrix to triangular form. Letting $R = U_m^T \cdots U_k^T \begin{bmatrix} \tilde{R}_1 & \tilde{R}_2 \end{bmatrix}$ and $Q = \tilde{Q} U_k \cdots U_m$, we have our QR decomposition of A.

This procedure can be carried out without knowledge of the orthogonal matrix $\tilde{Q}$, so it also serves as a means of downdating the Cholesky decomposition of $\tilde{A}^T \tilde{A}$, when a column is removed from $\tilde{A}$. The cost of obtaining R from $\tilde{R}$ depends on which column is removed. In the worst case (first column is removed) m rotators are needed, and they operate on columns of effective length m or less. Therefore the flop count is $O(m^2)$. In the best case (last column is removed) no flops are needed.

Exercise 3.6.11

(a) Suppose the kth column is deleted from $\tilde{A}$. Show that the cost of producing R is about $3(m - k)^2$ flops.

(b) Show that the cost of computing Q from $\tilde{Q}$ is about $6n(m - k)$ flops if $\tilde{Q}$ is given explicitly.

$\square$

Additional Exercises

Exercise 3.6.12 A 2×2 matrix H is called a *hyperbolic transformation* if is has the form

$$H = \begin{bmatrix} c & s \\ s & c \end{bmatrix},$$

where $c > 0$ and

$$c^2 - s^2 = 1. \tag{3.6.13}$$

The set of all (c, s) that satisfy (3.6.13) is a hyperbola in the c–s plane. For any pair (c, s) satisfying (3.6.13) there is a number α such that $c = \cosh \alpha$ and $s = \sinh \alpha$.

(a) Show that every hyperbolic transformation H is nonsingular. Find the determinant of H. What is H^{-1}? Note that H^{-1} is also hyperbolic.

(b) Let $J = \begin{bmatrix} 1 & 0 \\ 0 & -1 \end{bmatrix}$. Show that if H is hyperbolic, then $H^T J H = J$. Of course, $H = H^T$, but it turns out be be useful to write the identity in terms of the transpose.

(c) Show that if $\begin{bmatrix} a \\ b \end{bmatrix} \in \mathbb{R}^2$ with $|a| > |b|$, then there is a unique hyperbolic transformation H such that

$$\begin{bmatrix} c & s \\ s & c \end{bmatrix} \begin{bmatrix} a \\ b \end{bmatrix} = \begin{bmatrix} \sqrt{a^2 - b^2} \\ 0 \end{bmatrix}.$$

Obtain formulas (resembling (3.2.11)) for c and s in terms of a and b.[22] The condition $c^2 - s^2 = 1$ does not put any bound on c and s; they can be arbitrarily

---

[22] If $|a| = |b|$, there is no hyperbolic transformation that can transform b to zero. If $|a| < |b|$, a different type of hyperbolic transformation can do the job. Since we will not need that type of transformation, we do not introduce it here.

large. (Graphically, the hyperbola is an unbounded figure.) It follows that hyperbolic transformations can be ill conditioned and lack the unconditional stability of rotators. See Exercise 3.6.16.

(d) We can embed hyperbolic transformations in larger matrices, just as we did for rotators. Let

$$
H = \begin{bmatrix} I & & & & \\ & c & & s & \\ & & I & & \\ & s & & c & \\ & & & & I \end{bmatrix},
$$

where $c > 0$ and $c^2 - s^2 = 1$. Suppose the rows and columns in which the hyperbolic transformation is embedded are i and j ($i < j$). Let J be any diagonal matrix with the entries 1 and -1 in positions (i, i) and (j, j), respectively. Show that $H^T J H = J$. (Again the transpose is unnecessary. However, this identity also holds for products of matrices of this type, and then the transpose really is needed.) Show that if $\hat{S} = HS$, then $\hat{S}^T J \hat{S} = S^T J S$. Show that $\hat{S}$ and S differ only in the ith and jth rows.

□

rcise 3.6.14 In this exercise we show how to use hyperbolic transformations to downdate the Cholesky decomposition. Let

$$
\tilde{A} = \begin{bmatrix} A_1 \\ z^T \\ A_2 \end{bmatrix} \in \mathbb{R}^{(n+1)\times m} \quad \text{and} \quad A = \begin{bmatrix} A_1 \\ A_2 \end{bmatrix} \in \mathbb{R}^{n\times m},
$$

and suppose A has rank m. Suppose we have $\tilde{R} \in \mathbb{R}^{m\times m}$, the Cholesky factor of $\tilde{A}^T \tilde{A}$, and we would like to obtain $R \in \mathbb{R}^{m\times m}$, the Cholesky factor of $A^T A$.

(a) Let

$$
J = \begin{bmatrix} I_m & 0 \\ 0^T & -1 \end{bmatrix} \in \mathbb{R}^{(m+1)\times(m+1)}.
$$

Show that

$$
A^T A = \tilde{A}^T \tilde{A} - zz^T = \tilde{R}^T \tilde{R} - zz^T = \begin{bmatrix} \tilde{R}^T & z \end{bmatrix} J \begin{bmatrix} \tilde{R} \\ z^T \end{bmatrix}.
$$

Our plan is to obtain the Cholesky factor of $A^T A$ by annihilating the entries of z^T in the matrix

$$
\tilde{S} = \begin{bmatrix} \tilde{R} \\ z^T \end{bmatrix} = \begin{bmatrix} \tilde{r}_{11} & \tilde{r}_{12} & \cdots \\ 0 & \tilde{r}_{22} & \cdots \\ 0 & 0 & \ddots \\ & \vdots & \\ z_1 & z_2 & \cdots \end{bmatrix}.
$$

For this purpose we will use hyperbolic transformations.

(b) Using the fact that $A^T A$ is positive definite, demonstrate that $\tilde{r}_{11}^2 - z_1^2 > 0$, whence $|\tilde{r}_{11}| > |z_1|$. Construct a hyperbolic transformation $H_1 \in \mathbb{R}^{(m+1) \times (m+1)}$ that acts on rows 1 and $m + 1$, such that

$$
S_1 = H_1 \tilde{S} =
\begin{bmatrix}
r_{11} & r_{12} & \cdots \\
0 & \tilde{r}_{22} & \cdots \\
0 & 0 & \ddots \\
& \vdots & \\
0 & \hat{z}_2 & \cdots
\end{bmatrix}.
$$

The entry z_1 has been annihilated. The entry $\hat{z}_2$ differs from z_2, but it is, in general, nonzero.

(c) Show that $A^T A = \tilde{S}^T J \tilde{S} = S_1^T J S_1$.

(d) In Exercise 3.6.15 we will show that $|\tilde{r}_{22}| > |\hat{z}_2|$, and appropriate corresponding inequalities hold at all subsequent steps. Assuming these inequalities to be true, sketch an algorithm that applies a sequence of m hyperbolic transformations $H_1, \ldots, H_m$ to $\tilde{S}$, thereby transforming it to a matrix $S = S_m = H_m \cdots H_1 \tilde{S}$ of the form $S = \begin{bmatrix} R \\ 0^T \end{bmatrix}$, where R is upper triangular and has positive main-diagonal entries. Show that $A^T A = S^T J S = R^T R$. Thus R is the Cholesky factor of $A^T A$.

(d) Show that this downdating procedure requires about $3m^2$ flops in all.

$\square$

Exercise 3.6.15 After k steps of the algorithm sketched in the previous exercise, we have transformed $\tilde{S}$ to the form

$$
S_k =
\begin{bmatrix}
R_{11} & R_{12} \\
0 & \tilde{R}_{22} \\
0^T & \hat{z}^T
\end{bmatrix},
$$

where R_{11} is $k \times k$. We have $A^T A = S_k^T J S_k$, where J is as defined in the previous exercise. Let $A = \begin{bmatrix} A^{(1)} & A^{(2)} \end{bmatrix}$, where $A^{(1)}$ has k columns.

(a) Show that $A^{(1)T} A^{(1)} = R_{11}^T R_{11}$ and $A^{(1)T} A^{(2)} = R_{11}^T R_{12}$, and deduce that R_{11} and R_{12} are blocks of the Cholesky factor of $A^T A$.

(b) Show that $A^{(2)T} A^{(2)} - R_{12}^T R_{12} = \tilde{R}_{22}^T \tilde{R}_{22} - \hat{z}\hat{z}^T$. Show that the matrix $A^{(2)T} A^{(2)} - R_{12}^T R_{12}$ is the Schur complement of $A^{(1)T} A^{(1)}$ in $A^T A$ and is positive definite. Here we essentially are retreading ground covered in Section 1.4, especially Exercise 1.4.58 and surrounding material.

(c) Use the positive definiteness of $A^{(2)T}A^{(2)} - R_{12}^T R_{12}$ to prove that the $(1,1)$ entry of $\tilde{R}_{22}$, which we call $\tilde{r}_{k+1,k+1}$ and the leading entry of $\hat{z}^T$, which we call $\hat{z}_{k+1}$, satisfy $|\tilde{r}_{k+1,k+1}| > |\hat{z}_{k+1}|$. Thus we can set up the hyperbolic rotator H_{k+1} for the $(k+1)$st step.

$\square$

cise 3.6.16 In this exercise we show that hyperbolic transformations can be arbitrarily ill conditioned.

(a) Let L be a huge positive number (as huge as you please). Find two positive numbers c and s such that $c \geq L$, $s \geq L$, and $c^2 - s^2 = 1$. Then

$$H = \begin{bmatrix} c & s \\ s & c \end{bmatrix}$$

is a hyperbolic transformation with large norm. Show that $\| H \|_\infty = \| H^{-1} \|_\infty$ $= c + s \geq 2L$ and $\kappa_\infty(H) \geq 4L^2$.

(b) Let $v = \begin{bmatrix} 1 \\ 1 \end{bmatrix}$ and $w = \begin{bmatrix} 1 \\ -1 \end{bmatrix}$. Show that $Hv = (c+s)v$ and $Hw = (c-s)w$. The vectors v and w are special vectors called eigenvectors of H. We will study eigenvectors in Chapters 5 and 6. Show that $H^{-1}w = (c+s)w$.

(c) Using results of part (b), show that $\| H \|_2 \geq (c+s)$, $\| H^{-1} \|_2 \geq (c+s)$, and $\kappa_2(H) \geq (c+s)^2$. In Chapter 5 we will find that all three of these inequalities are in fact equations.

(d) Let H be the hyperbolic transformation that maps

$$\begin{bmatrix} a \\ b \end{bmatrix} \quad \text{to} \quad \begin{bmatrix} \sqrt{a^2 - b^2} \\ 0 \end{bmatrix},$$

where $|a| > |b|$. Under what conditions on a and b will H be ill conditioned?

$\square$

In light of the fact that hyperbolic transformations can be ill conditioned, we ought to be suspicious about the stability of any algorithm that uses them. It turns out that the downdating procedure used in Exercise 3.6.14 works well as long as $A^T A$ is "safely" positive definite, meaning well conditioned. However, if we do something reckless, such as removing so many rows that the rank of A becomes less than m, we can expect the algorithm to fail.

CHAPTER 4

THE SINGULAR VALUE DECOMPOSITION (SVD)

The QR decomposition is a fine tool for solving least squares problems when the coefficient matrix is known to have full rank. However, if the matrix does not have full rank, or if the rank is unknown, a more powerful tool is needed. One such tool is the QR decomposition with column pivoting, which we discussed in Section 3.3. In this chapter we introduce an even more powerful tool: the singular value decomposition (SVD). This may be the most important matrix decomposition of all, for both theoretical and computational purposes.

This chapter differs from Chapters 1 and 3 in an important way. Those chapters focused on computation and introduced computations as early as possible. At this point we are not yet ready to explain how to compute the SVD. That will have to wait until we have discussed eigenvalue computations and demonstrated that the SVD problem is an eigenvalue problem (Section 5.8). But the SVD is too important to be left to the end of the book. A great deal can be said about the characteristics and uses of this important decomposition without saying anything about how to compute it.

We begin by introducing the SVD and showing that it can take a variety of forms. Then, in Section 4.2, we establish the connection between singular values and the norm and condition number of a matrix. We also show how to use the SVD to detect the (numerical) rank of a matrix in the presence of roundoff errors and

Fundamentals of Matrix Computations, Third Edition. By David S. Watkins
Copyright © 2010 John Wiley & Sons, Inc.

other uncertainties in the data, we show that rank-deficient matrices are in some sense scarce, and we show how to compute the distance to the nearest rank-deficient matrix. In Section 4.3 we show how to use the SVD to solve least squares problems, even if the coefficient matrix does not have full rank. We also introduce the pseudoinverse, an interesting generalization of the inverse of a matrix. Finally, in Section 4.4, we analyze the sensitivity of the least squares problem in the full-rank case, making use of results proved in Section 4.2.

We will continue to focus on real matrices. However, all of the developments of this chapter can be extended to complex matrices in a straightforward way.

4.1 INTRODUCTION

Let $A \in \mathbb{R}^{n \times m}$, where n and m are positive integers. We make no assumption about which of n and m is larger. Recall that the *range* of A is the subspace of $\mathbb{R}^n$ defined by $\mathcal{R}(A) = \{Ax \mid x \in \mathbb{R}^m\}$. The *rank* of A is the dimension of $\mathcal{R}(A)$.

Theorem 4.1.1 *(SVD Theorem) Let $A \in \mathbb{R}^{n \times m}$ be a nonzero matrix with rank r. Then A can be expressed as a product*

$$A = U\Sigma V^T, \tag{4.1.2}$$

where $U \in \mathbb{R}^{n \times n}$ and $V \in \mathbb{R}^{m \times m}$ are orthogonal, and $\Sigma \in \mathbb{R}^{n \times m}$ is a rectangular "diagonal" matrix

$$\Sigma = \begin{bmatrix} \sigma_1 & & & & \\ & \sigma_2 & & & \\ & & \ddots & & \\ & & & \sigma_r & \\ & & & & 0 \\ & & & & & \ddots \end{bmatrix} \qquad \sigma_1 \geq \sigma_2 \geq \cdots \geq \sigma_r > 0.$$

To prove Theorem 4.1.1, work Exercise 4.1.16. The decomposition (4.1.2) is called the *singular value decomposition* of A. We will usually use the abbreviation SVD. The discussion of many aspects of the SVD will have to be deferred until after we have discussed eigenvalues and eigenvectors. For example, Theorem 4.1.1 says nothing about the uniqueness of the decomposition. It turns out that U and V are not uniquely determined, but they have some partial uniqueness properties that we will be able to discuss later. The entries $\sigma_1, \ldots, \sigma_r$ of Σ *are* uniquely determined, and they are called the *singular values* of A. The columns of U are orthonormal vectors called *left singular vectors* of A, and the columns of V are called *right singular vectors* for reasons that will become apparent. The transpose of A has the SVD $A^T = V\Sigma^T U^T$.

In some contexts it is convenient to fill out the main diagonal of Σ by introducing singular values $\sigma_{r+1} = \sigma_{r+2} = \cdots = \sigma_k = 0$, where $k = \min\{m, n\}$. We will do this whenever it suits us. When this terminology is used, the rank of A is equal to the number of nonzero singular values. The matrix 0 has no nonzero singular values.

Once we have discussed eigenvalues and eigenvectors, we will be able to provide a second proof of the SVD theorem, and we will develop algorithms for computing the SVD. See Section 5.8.

Two matrices A, $B \in \mathbb{R}^{n \times m}$ are said to be *equivalent* if there exist nonsingular matrices $X \in \mathbb{R}^{n \times n}$ and $Y \in \mathbb{R}^{m \times m}$ such that $A = XBY$. If the matrices X and Y are orthogonal, then A and B are said to be *orthogonally equivalent*. Theorem 4.1.1 shows that every $A \in \mathbb{R}^{n \times m}$ is orthogonally equivalent to a diagonal matrix.

Other Forms of the SVD Theorem

The SVD has a simple geometric interpretation, which is a consequence of the following restatement of the SVD theorem.

Theorem 4.1.3 (*Geometric SVD Theorem*) *Let $A \in \mathbb{R}^{n \times m}$ be a nonzero matrix with rank r. Then $\mathbb{R}^m$ has an orthonormal basis $v_1, \ldots, v_m$, $\mathbb{R}^n$ has an orthonormal basis $u_1, \ldots, u_n$, and there exist $\sigma_1 \geq \sigma_2 \geq \ldots \geq \sigma_r > 0$ such that*

$$
Av_i = \begin{cases} \sigma_i u_i & i = 1, \ldots, r \\ 0 & i = r+1, \ldots, m \end{cases} \qquad A^T u_i = \begin{cases} \sigma_i v_i & i = 1, \ldots, r \\ 0 & i = r+1, \ldots, n. \end{cases}
$$
$$(4.1.4)$$

rcise 4.1.5 Show that all of the equations (4.1.4) follow from the SVD $A = U\Sigma V^T$ (and its transpose), where $u_1, \ldots, u_n$ and $v_1, \ldots, v_m$ are the columns of U and V, respectively. □

rcise 4.1.6 Show that the equations in the left half of (4.1.4) imply the matrix equation $AV = U\Sigma$, the equations in the right half of (4.1.4) imply $A^T U = V\Sigma^T$, and either one of these implies the SVD $A = U\Sigma V^T$. □

Think of A as a linear transformation that maps vectors $x \in \mathbb{R}^m$ to vectors $Ax \in \mathbb{R}^n$. Theorem 4.1.3 shows that $\mathbb{R}^m$ and $\mathbb{R}^n$ have orthonormal bases such that A maps the ith basis vector in $\mathbb{R}^m$ to a multiple of the ith basis vector in $\mathbb{R}^n$ (and A^T acts similarly). From elementary linear algebra we know that if we choose bases in $\mathbb{R}^m$ and $\mathbb{R}^n$, then we can represent this or any linear transformation by a matrix with respect to these bases. Theorem 4.1.3 says simply that the diagonal matrix Σ is the matrix of the transformation A with respect to the orthonormal bases $v_1 \ldots, v_m$ and $u_1, \ldots, u_n$ (and Σ^T is the matrix of the transformation A^T). The action of A is depicted in a simple way by the following diagram.

$$
\begin{array}{ccc}
& A & \\
v_1 & \xrightarrow{\ \sigma_1\ } & u_1 \\
v_2 & \xrightarrow{\ \sigma_2\ } & u_2 \\
\vdots & & \vdots \\
v_r & \xrightarrow{\ \sigma_r\ } & u_r \\
v_{r+1} & & \\
\vdots & \Big\} \longrightarrow 0 & \\
v_m & &
\end{array}
$$

A similar diagram holds for A^T. If we set the two diagrams side by side, we obtain

$$
\begin{array}{ccccc}
& A & & A^T & \\
v_1 & \xrightarrow{\ \sigma_1\ } & u_1 & \xrightarrow{\ \sigma_1\ } & v_1 \\
v_2 & \xrightarrow{\ \sigma_2\ } & u_2 & \xrightarrow{\ \sigma_2\ } & v_2 \\
\vdots & \vdots & \vdots & \vdots & \vdots \\
v_r & \xrightarrow{\ \sigma_r\ } & u_r & \xrightarrow{\ \sigma_r\ } & v_r \\
v_{r+1} & & u_{r+1} & & \\
\vdots & \Big\} \longrightarrow 0 & \vdots & \Big\} \longrightarrow 0 & \\
v_m & & u_n & &
\end{array}
\tag{4.1.7}
$$

which serves as a pictorial representation of the SVD Theorem.

The SVD displays orthonormal bases for the four fundamental subspaces $\mathcal{R}(A)$, $\mathcal{N}(A)$, $\mathcal{R}(A^T)$, and $\mathcal{N}(A^T)$. It is clear from (4.1.7) that

$$
\begin{aligned}
\mathcal{R}(A) &= \operatorname{span}\{u_1, \ldots, u_r\} \\
\mathcal{N}(A) &= \operatorname{span}\{v_{r+1}, \ldots, v_m\} \\
\mathcal{R}(A^T) &= \operatorname{span}\{v_1, \ldots, v_r\} \\
\mathcal{N}(A^T) &= \operatorname{span}\{u_{r+1}, \ldots, u_n\}.
\end{aligned}
\tag{4.1.8}
$$

From these representations we see that $\mathcal{R}(A^T) = \mathcal{N}(A)^\perp$ and $\mathcal{R}(A) = \mathcal{N}(A^T)^\perp$; we proved these equations by other means in Theorem 3.5.10.

Another immediate consequence of (4.1.8) is the following fact, which can also be deduced by more elementary means. We record it here for future reference.

Corollary 4.1.9 Let $A \in \mathbb{R}^{n \times m}$. Then $\dim(\mathcal{R}(A)) + \dim(\mathcal{N}(A)) = m$.

The next theorem gives a more condensed version of the SVD.

Theorem 4.1.10 (*Condensed SVD Theorem*) Let $A \in \mathbb{R}^{n \times m}$ be a nonzero matrix of rank r. Then there exist $\hat{U} \in \mathbb{R}^{n \times r}$, $\hat{\Sigma} \in \mathbb{R}^{r \times r}$, and $\hat{V} \in \mathbb{R}^{m \times r}$ such that $\hat{U}$ and $\hat{V}$

are isometries, $\hat{\Sigma}$ is a diagonal matrix with main-diagonal entries $\sigma_1 \geq \ldots \geq \sigma_r > 0$, and

$$A = \hat{U}\hat{\Sigma}\hat{V}^T.$$

rcise 4.1.11 Prove Theorem 4.1.10 by writing the SVD $A = U\Sigma V^T$ in block form in an appropriate way and then chopping off the unneeded blocks. □

Finally we present one more useful form of the SVD.

Theorem 4.1.12 *Let $A \in \mathbb{R}^{n \times m}$ be a nonzero matrix with rank r. Let $\sigma_1, \ldots, \sigma_r$ be the singular values of A, with associated right and left singular vectors $v_1, \ldots, v_r$ and $u_1, \ldots, u_r$, respectively. Then*

$$A = \sum_{j=1}^{r} \sigma_j u_j v_j^T.$$

rcise 4.1.13 Prove Theorem 4.1.12 by partitioning the decomposition $A = \hat{U}\hat{\Sigma}\hat{V}^T$ from Theorem 4.1.10 in an appropriate way. □

rcise 4.1.14 Let

$$A = \begin{bmatrix} 3 & 6 & 9 \\ 4 & 8 & 12 \end{bmatrix}.$$

By inspection, write down the SVD in the condensed form of Theorem 4.1.10 and in the form given by Theorem 4.1.12. How do the two compare? □

rcise 4.1.15 In MATLAB you can use the command `svd` to compute either the singular values or the singular value decomposition of a matrix. Type `help svd` to find out how to use this command. Use MATLAB's `svd` command to check your result from the previous exercise. □

rcise 4.1.16 In this exercise you will prove Theorem 4.1.1 by induction on r, the rank of A.

(a) Suppose $A \in \mathbb{R}^{n \times m}$ has rank 1. Let $u_1 \in \mathbb{R}^n$ be a vector in $\mathcal{R}(A)$ such that $\|u_1\|_2 = 1$. Show that every column of A is a multiple of u_1. Show that A can be written in the form $A = \sigma_1 u_1 v_1^T$, where $v_1 \in \mathbb{R}^m$, $\|v_1\|_2 = 1$, and $\sigma_1 > 0$.

(b) Continuing from part (a), demonstrate that there is an orthogonal matrix $U \in \mathbb{R}^{n \times n}$ whose first column is u_1. (For example, U can be a reflector that maps the unit vector e_1 to u_1.) Similarly there is an orthogonal $V \in \mathbb{R}^{m \times m}$ whose first column is v_1. Show that $A = U\Sigma V^T$, where $\Sigma \in \mathbb{R}^{n \times m}$ has only one nonzero entry, σ_1, in position $(1, 1)$. Thus every matrix of rank 1 has an SVD.

(c) Now suppose $A \in \mathbb{R}^{n \times m}$ has rank $r > 1$. Let v_1 be a unit vector in the direction of maximum magnification by A, i.e. $\|v_1\|_2 = 1$, and $\|Av_1\|_2 = $

$\max_{\|v\|_2=1} \| Av \|_2$. Let $\sigma_1 = \| Av_1 \|_2 = \| A \|_2$, and let $u_1 = \sigma_1^{-1} Av_1$. Let $\tilde{U} \in \mathbb{R}^{n \times n}$ and $\tilde{V} \in \mathbb{R}^{m \times m}$ be orthogonal matrices with first column u_1 and v_1, respectively. Let $\tilde{A} = \tilde{U}^T A \tilde{V}$, so that $A = \tilde{U} \tilde{A} \tilde{V}^T$. Show that $\tilde{A}$ has the form

$$\tilde{A} = \begin{bmatrix} \sigma_1 & z^T \\ 0 & \hat{A} \end{bmatrix}, \qquad (4.1.17)$$

where $z \in \mathbb{R}^{m-1}$ and $\hat{A} \in \mathbb{R}^{(n-1) \times (m-1)}$.

(d) Show that the vector z in (4.1.17) must be zero. You may do this as follows: Let $w = \begin{bmatrix} \sigma_1 \\ z \end{bmatrix} \in \mathbb{R}^m$. Show that $\| \tilde{A}w \|_2 / \| w \|_2 \geq \sqrt{\sigma_1^2 + z^T z}$. Then show that this inequality forces $z = 0$. Thus

$$\tilde{A} = \begin{bmatrix} \sigma_1 & 0 \\ 0 & \hat{A} \end{bmatrix}.$$

(e) Show that $\hat{A}$ has rank $r - 1$. By the induction hypothesis $\hat{A}$ has an SVD $\hat{A} = \hat{U} \hat{\Sigma} \hat{V}^T$. Let $\sigma_2 \geq \sigma_3 \geq \cdots \geq \sigma_r$ denote the positive main-diagonal entries of $\hat{\Sigma}$. Show that $\sigma_1 \geq \sigma_2$. Embed the SVD $\hat{A} = \hat{U} \hat{\Sigma} \hat{V}^T$ in larger matrices to obtain an SVD of $\tilde{A}$. Then use the equation $A = \tilde{U} \tilde{A} \tilde{V}^T$ to obtain an SVD of A.

$\square$

4.2   SOME BASIC APPLICATIONS OF SINGULAR VALUES

Relationship to Norm and Condition Number

In Section 2.1 we defined the *spectral* norm to be the norm induced by the Euclidean vector norm:

$$\| A \|_2 = \max_{x \neq 0} \frac{\| Ax \|_2}{\| x \|_2}.$$

The discussion in Section 2.1 was restricted to square matrices, but this definition makes sense for nonsquare matrices as well. Geometrically $\| A \|_2$ represents the maximum magnification that can be undergone by any vector $x \in \mathbb{R}^m$ when acted on by A. In light of (4.1.7) (and Exercise 4.1.16), it should not be surprising that $\| A \|_2$ equals the maximum singular value of A.

Theorem 4.2.1 *Let $A \in \mathbb{R}^{n \times m}$ have singular values $\sigma_1 \geq \sigma_2 \geq \ldots \geq 0$. Then $\| A \|_2 = \sigma_1$.*

Proof. We must show that $\max\limits_{x \neq 0} \| Ax \|_2 / \| x \|_2 = \sigma_1$. First notice that since $Av_1 = \sigma_1 u_1$,

$$\frac{\| Av_1 \|_2}{\| v_1 \|_2} = \sigma_1 \frac{\| u_1 \|_2}{\| v_1 \|_2} = \sigma_1,$$

so $\max\limits_{x\neq 0} \|Ax\|_2/\|x\|_2 \geq \sigma_1$. Now we must show that no other vector is magnified by more than σ_1.

Let $x \in \mathbb{R}^m$. Then x can be expressed as a linear combination of the right singular vectors of A: $x = c_1 v_1 + c_2 v_2 + \cdots + c_m v_m$. Since $v_1, \ldots, v_m$ are orthonormal, $\|x\|_2^2 = |c_1|^2 + \cdots + |c_m|^2$. Now $Ax = c_1 A v_1 + \cdots + c_r A v_r + \cdots + c_m A v_m = \sigma_1 c_1 u_1 + \cdots + \sigma_r c_r u_r + 0 + \cdots + 0$, where r is the rank of A. Since $u_1, \ldots, u_r$ are also orthonormal, $\|Ax\|_2^2 = |\sigma_1 c_1|^2 + \cdots + |\sigma_r c_r|^2$. Thus $\|Ax\|_2^2 \leq \sigma_1^2(|c_1|^2 + \cdots + |c_r|^2) \leq \sigma_1^2 \|x\|_2^2$; that is, $\|Ax\|_2/\|x\|_2 \leq \sigma_1$.   □

Since A and A^T have the same singular values, we have the following corollary.

Corollary 4.2.2 $\|A\|_2 = \|A^T\|_2$.

rcise 4.2.3 Recall that the Frobenius matrix norm is defined by

$$\|A\|_F = \left(\sum_{i=1}^n \sum_{j=1}^m |a_{ij}|^2 \right)^{1/2}.$$

Show that $\|A\|_F = (\sigma_1^2 + \sigma_2^2 + \cdots + \sigma_r^2)^{1/2}$. (*Hint:* Show that if $B = UC$, where U is orthogonal, then $\|B\|_F = \|C\|_F$.)   □

Now suppose A is square, say $A \in \mathbb{R}^{n \times n}$, and nonsingular. The spectral condition number of A is defined to be

$$\kappa_2(A) = \|A\|_2 \|A^{-1}\|_2.$$

Let us see how $\kappa_2(A)$ can be expressed in terms of the singular values of A. Since A has rank n, it has n strictly positive singular values, and its action is described completely by the following diagram:

$$
\begin{array}{ccc}
 & A & \\
v_1 & \xrightarrow{\sigma_1} & u_1 \\
v_2 & \xrightarrow{\sigma_2} & u_2 \\
\vdots & & \vdots \\
v_n & \xrightarrow{\sigma_n} & u_n
\end{array}
$$

It follows that the corresponding diagram for A^{-1} is

$$
\begin{array}{ccc}
 & A^{-1} & \\
u_1 & \xrightarrow{\sigma_1^{-1}} & v_1 \\
u_2 & \xrightarrow{\sigma_2^{-1}} & v_2 \\
\vdots & & \vdots \\
u_n & \xrightarrow{\sigma_n^{-1}} & v_n
\end{array}
$$

In terms of matrices we have $A = U\Sigma V^T$ and $A^{-1} = V^{-T}\Sigma^{-1}U^{-1} = V\Sigma^{-1}U^T$. Either way we see that the singular values of A^{-1}, in descending order, are $\sigma_n^{-1} \geq \sigma_{n-1}^{-1} \geq \ldots \geq \sigma_1^{-1} > 0$. Applying Theorem 4.2.1 to A^{-1}, we conclude that $\|A^{-1}\|_2 = \sigma_n^{-1}$. These observations imply the following theorem.

Theorem 4.2.4 *Let $A \in \mathbb{R}^{n \times n}$ be a nonsingular matrix with singular values $\sigma_1 \geq \ldots \geq \sigma_n > 0$. Then*

$$\kappa_2(A) = \frac{\sigma_1}{\sigma_n}.$$

Another expression for the condition number that was given in Chapter 2 is

$$\kappa_2(A) = \frac{\text{maxmag}(A)}{\text{minmag}(A)},$$

where

$$\text{maxmag}(A) = \max_{x \neq 0} \frac{\|Ax\|_2}{\|x\|_2},$$

$$\text{minmag}(A) = \min_{x \neq 0} \frac{\|Ax\|_2}{\|x\|_2}.$$

This gives a slightly different view of the condition number. From Theorem 4.2.1 we know that $\text{maxmag}(A) = \sigma_1$. It must therefore be true that $\text{minmag}(A) = \sigma_n$.

Exercise 4.2.5 Prove that $\text{minmag}(A) = \sigma_n$. Show that the minimum magnification is obtained by taking $x = v_n$. ☐

In Section 3.4 we observed that the equation

$$\kappa_2(A) = \frac{\text{maxmag}(A)}{\text{minmag}(A)} \tag{4.2.6}$$

can be used to extend the definition of κ_2 to certain nonsquare matrices. Specifically, if $A \in \mathbb{R}^{n \times m}$, $n \geq m$, and $\text{rank}(A) = m$, then $\text{minmag}(A) > 0$, and we can take (4.2.6) as the definition of the condition number of A. If A is nonzero but does not have full rank, then (still assuming $n \geq m$) $\text{minmag}(A) = 0$, and it is reasonable to define $\kappa_2(A) = \infty$. With this convention the following theorem holds, regardless of whether or not A has full rank.

Theorem 4.2.7 *Let $A \in \mathbb{R}^{n \times m}$, $n \geq m$ be a nonzero matrix with singular values $\sigma_1 \geq \sigma_2 \geq \ldots \geq \sigma_m \geq 0$. (Here we allow some σ_i equal to zero if $\text{rank}(A) < m$.) Then $\text{maxmag}(A) = \sigma_1$, $\text{minmag}(A) = \sigma_m$, and $\kappa_2(A) = \sigma_1/\sigma_m$.*

The proof is left as an easy exercise for you.

Exercise 4.2.8 MATLAB's command `cond` computes the condition number $\kappa_2(A)$. This works for both square and nonsquare matrices. Generate a random 3×3 matrix (A = `randn(3)`) and use MATLAB to compute $\kappa_2(A)$ three different ways: (i) using cond, (ii) taking the ratio of largest to smallest singular value, and (iii) computing $\|A\|_2 \|A^{-1}\|_2$ (norm(A)*norm(inv(A))).   □

The next two theorems establish other important results, which are also easy to prove.

Theorem 4.2.9 *Let* $A \in \mathbb{R}^{n \times m}$ *with* $n \geq m$. *Then* $\|A^T A\|_2 = \|A\|_2^2$ *and* $\kappa_2(A^T A) = \kappa_2(A)^2$.

Recall that $A^T A$ is the coefficient matrix of the normal equations (3.5.22), which can be used to solve the least-squares problem. Theorem 4.2.9 shows that the normal equations can be seriously ill conditioned even if the matrix A is only mildly ill conditioned. For example, if $\kappa_2(A) \approx 10^3$, then $\kappa_2(A^T A) \approx 10^6$.

rcise 4.2.10

(a) Use the SVD of A to deduce the SVD of $A^T A$ and prove Theorem 4.2.9.

(b) Let $M \in \mathbb{R}^{n \times n}$ be positive definite, and let R be the Cholesky factor of M, so that $M = R^T R$. Show that $\|M\|_2 = \|R\|_2^2$ and $\kappa_2(M) = \kappa_2(R)^2$.

□

The results of the next theorem will be used in the analysis of the sensitivity of the least squares problem in Section 4.4.

Theorem 4.2.11 *Let* $A \in \mathbb{R}^{n \times m}$, $n \geq m$, rank$(A) = m$, *with singular values* $\sigma_1 \geq \ldots \geq \sigma_m > 0$. *Then* $\|(A^T A)^{-1}\|_2 = \sigma_m^{-2}$, $\|(A^T A)^{-1} A^T\|_2 = \sigma_m^{-1}$, $\|A(A^T A)^{-1}\|_2 = \sigma_m^{-1}$, *and* $\|A(A^T A)^{-1} A^T\|_2 = 1$.

The matrix $(A^T A)^{-1} A^T$ is called the *pseudoinverse* of A. $A(A^T A)^{-1}$ is the pseudoinverse of A^T. Pseudoinverses will be discussed in greater generality in the next section.

rcise 4.2.12 Let A be as in Theorem 4.2.11 with SVD $A = U \Sigma V^T$.

(a) Determine the singular value decompositions of the matrices

$$(A^T A)^{-1}, \quad (A^T A)^{-1} A^T, \quad A(A^T A)^{-1}, \quad \text{and} \quad A(A^T A)^{-1} A^T$$

in terms of the SVD of A. Use the orthogonality of U and V whenever possible. Pay attention to the dimensions of the various matrices.

(b) Use the results of part (a) to prove Theorem 4.2.11.

□

Numerical Rank Determination

In the absence of roundoff errors and uncertainties in the data, the singular value decomposition reveals the rank of the matrix. Unfortunately the presence of errors makes rank determination problematic. For example, consider the matrix

$$A = \begin{bmatrix} 1/3 & 1/3 & 2/3 \\ 2/3 & 2/3 & 4/3 \\ 1/3 & 2/3 & 3/3 \\ 2/5 & 2/5 & 4/5 \\ 3/5 & 1/5 & 4/5 \end{bmatrix}. \tag{4.2.13}$$

A is obviously of rank 2, as its third column is the sum of the first two. However, if we fail to notice this relationship and decide to use, say, MATLAB to calculate its rank, we have to begin by storing the matrix in the computer. This simple act will result in roundoff errors that destroy the relationship between the columns. Technically speaking, the perturbed matrix has rank 3. If we now use MATLAB's svd command to compute the singular values of A, using IEEE standard double precision floating point arithmetic, we obtain

$$\sigma_1 = 2.5987 \qquad \sigma_2 = 0.3682 \qquad \text{and} \qquad \sigma_3 = 8.6614 \times 10^{-17}.$$

Since there are three nonzero singular values, we must conclude that the matrix has rank 3. However, we cannot fail to notice that one of the singular values is tiny, on the order of the unit roundoff for IEEE double precision. Perhaps we should consider it a zero. For this reason we introduce the notion of numerical rank.

Roughly speaking, a matrix that has k "large" singular values, the others being "tiny," has numerical rank k. For the purpose of determining which singular values are "tiny," we need to introduce a threshold or tolerance ϵ that is roughly on the level of uncertainty in the data in the Matrix. For example, if the only errors are roundoff errors, as in the above example, we might take $\epsilon = 10u\| A \|$, where u is the unit roundoff error. We then say that A has *numerical rank* k if A has k singular values that are substantially larger than ϵ, and all other singular values are smaller than ϵ, that is,

$$\sigma_1 \geq \sigma_2 \geq \cdots \geq \sigma_k \gg \epsilon \geq \sigma_{k+1} \geq \cdots$$

MATLAB has a rank command, which computes the numerical rank of the matrix. When applied to the matrix A above, it yields the answer 2. MATLAB's rank command uses the default threshold $2 \max\{n, m\} u \| A \|_2$, which can be overridden by the user. For more information type help rank in MATLAB.

Sometimes it is not possible to specify the numerical rank exactly. For example, imagine a 2000×1000 matrix with singular values $\sigma_j = (.9)^j$, $j = 1, \ldots, 1000$. Then $\sigma_1 = .9$ and $\sigma_{1000} = 1.75 \times 10^{-46}$, so the numerical rank is definitely less than 1000. However, it is impossible to specify the numerical rank exactly, because there are no gaps in the singular values. For example

$$\sigma_{261} = 1.14 \times 10^{-12} \quad \sigma_{262} = 1.03 \times 10^{-12}$$
$$\sigma_{263} = 9.24 \times 10^{-13} \quad \sigma_{264} = 8.31 \times 10^{-13}$$

If $\epsilon = 10^{-12}$, say, it might be reasonable to say that the numerical rank is approximately 260, but it is certainly not possible to specify it exactly.

The following exercises and theorem provide justification for the use of singular values to define the numerical rank of a matrix. We begin with an exercise that shows that a small perturbation in a matrix that is not of full rank can (and typically will) increase the rank. Here we refer to the exact rank, not the numerical rank.

rcise 4.2.14 Let $A \in \mathbb{R}^{n \times m}$ with rank $r < \min\{n, m\}$. Use the SVD of A to show that for every $\epsilon > 0$ (no matter how small), there exists a full-rank matrix $A_\epsilon \in \mathbb{R}^{n \times m}$ such that $\| A - A_\epsilon \|_2 = \epsilon$. $\qquad\qquad\square$

The nonnegative number $\| A - A_\epsilon \|_2$ is a measure of the distance between the matrices A and A_ϵ. Exercise 4.2.14 shows that every rank-deficient matrix has full-rank matrices arbitrarily close to it. This suggests that matrices of full rank are abundant. This impression is strengthened by the next theorem and its corollary.

Theorem 4.2.15 *Let* $A \in \mathbb{R}^{n \times m}$ *with* $\operatorname{rank}(A) = r > 0$. *Let* $A = U\Sigma V^T$ *be the SVD of* A, *with singular values* $\sigma_1 \geq \sigma_2 \geq \cdots \geq \sigma_r > 0$. *For* $k = 1, \ldots, r - 1$, *define* $A_k = U\Sigma_k V^T$, *where* $\Sigma_k \in \mathbb{R}^{n \times m}$ *is the diagonal matrix* $\operatorname{diag}\{\sigma_1, \ldots, \sigma_k, 0, \ldots, 0\}$. *Then* $\operatorname{rank}(A_k) = k$, *and*

$$\sigma_{k+1} = \| A - A_k \|_2 = \min \left\{ \| A - B \|_2 \mid \operatorname{rank}(B) \leq k \right\}.$$

That is, of all matrices of rank k or less, A_k is closest to A.

Proof. It is obvious that $\operatorname{rank}(A_k) = k$. Since $A - A_k = U(\Sigma - \Sigma_k)V^T$, it is clear that the largest singular value of $A - A_k$ is σ_{k+1}. Therefore $\| A - A_k \|_2 = \sigma_{k+1}$. It remains to be shown only that for any other matrix B of rank k or less, $\| A - B \| \geq \sigma_{k+1}$.

Given any such B, note first that $\mathcal{N}(B)$ has dimension at least $m - k$, for $\dim(\mathcal{N}(B)) = m - \dim(\mathcal{R}(B)) = m - \operatorname{rank}(B) \geq m - k$ by (4.1.9). Also, the space $\operatorname{span}\{v_1, \ldots, v_{k+1}\}$ has dimension $k + 1$. (As usual, $v_1, \ldots, v_m$ denote the columns of V.) Since $\mathcal{N}(B)$ and $\operatorname{span}\{v_1, \ldots, v_{k+1}\}$ are two subspaces of $\mathbb{R}^m$, the sum of whose dimensions exceeds m, they must have a nontrivial intersection. Let $\hat{x}$ be a nonzero vector in $\mathcal{N}(B) \cap \operatorname{span}\{v_1, \ldots, v_{k+1}\}$. We can and will assume that $\| \hat{x} \|_2 = 1$. Since $\hat{x} \in \operatorname{span}\{v_1, \ldots, v_{k+1}\}$, there exist scalars $c_1, \ldots, c_{k+1}$ such that $\hat{x} = c_1 v_1 + \cdots + c_{k+1} v_{k+1}$. Because $v_1, \ldots, v_{k+1}$ are orthonormal, $|c_1|^2 + \cdots + |c_{k+1}|^2 = \| \hat{x} \|_2^2 = 1$. Since $\hat{x} \in \mathcal{N}(B)$, $B\hat{x} = 0$. Thus

$$(A - B)\hat{x} = A\hat{x} = \sum_{i=1}^{k+1} c_i A v_i = \sum_{i=1}^{k+1} \sigma_i c_i u_i.$$

Since $u_1, \ldots, u_{k+1}$ are also orthonormal,

$$\| (A - B)\hat{x} \|_2^2 = \sum_{i=1}^{k+1} |\sigma_i c_i|^2 \geq \sigma_{k+1}^2 \sum_{i=1}^{k+1} |c_i|^2 = \sigma_{k+1}^2.$$

Therefore

$$\| A - B \|_2 \geq \frac{\| (A - B)\hat{x} \|_2}{\| \hat{x} \|_2} \geq \sigma_{k+1}.$$

□

Corollary 4.2.16 *Suppose* $A \in \mathbb{R}^{n \times m}$ *has full rank. Thus* $\operatorname{rank}(A) = r$, *where* $r = \min\{n, m\}$. *Let* $\sigma_1 \geq \cdots \geq \sigma_r$ *be the singular values of* A. *Let* $B \in \mathbb{R}^{n \times m}$ *satisfy* $\| A - B \|_2 < \sigma_r$. *Then* B *also has full rank.*

Exercise 4.2.17 Deduce Corollary 4.2.16 from Theorem 4.2.15.   □

Exercise 4.2.18 In the first part of the proof of Theorem 4.2.15, we used the SVD of A, A_k, and $A - A_k$ in the form given by Theorem 4.1.1. Write down the other forms of the SVD of A_k and $A - A_k$: (a) Diagram (4.1.7), (b) Theorem 4.1.10, (c) Theorem 4.1.12.

□

From Corollary 4.2.16 we see that if A has full rank, then all matrices sufficiently close to A also have full rank. From Exercise 4.2.14 we know that every rank-deficient matrix has full-rank matrices arbitrarily close to it. By Corollary 4.2.16, each of these full-rank matrices is surrounded by other matrices of full rank. In topological language, the set of matrices of full rank is an open, dense subset of $\mathbb{R}^{n \times m}$. Its complement, the set of rank-deficient matrices, is therefore closed and nowhere dense. Thus, in a certain sense, almost all matrices have full rank.

If a matrix does not have full rank, any small perturbation is almost certain to transform it to a matrix that does have full rank. It follows that in the presence of uncertainty in the data, it is impossible to calculate the (exact, theoretical) rank of a matrix or even detect that it is rank deficient. This is a generalization of the assertion, made in Chapters 1 and 2, that it is (usually) impossible to determine whether a square matrix is singular.

Nevertheless, it is reasonable to call a matrix *numerically rank deficient* if it is very close to a rank-deficient matrix, since it could have been rank deficient except for a small perturbation, as was the case for the perturbed version of (4.2.13). Let ϵ be some positive number that represents the magnitude of the data uncertainties in the matrix A. If there exist matrices B of rank k such that $\| A - B \|_2 < \epsilon$ and, on the other hand, for every matrix C of rank $\leq k - 1$ we have $\| A - C \|_2 \gg \epsilon$, then it makes sense to say that the numerical rank of A is k. From Theorem 4.2.15 we know that this condition is satisfied if and only if

$$\sigma_1 \geq \sigma_2 \geq \cdots \geq \sigma_k \gg \epsilon \geq \sigma_{k+1} \geq \cdots$$

This justifies the use of singular values to determine numerical rank.

Exercise 4.2.19 Use MATLAB to generate a random matrix. For example, the command A = randn(40,17); produces a 40×17 matrix of independent normally distributed random numbers with mean zero and variance 1. Use MATLAB's svd command to determine the singular values of A. Notice that (unless you are really (un)lucky), all of the singular values are large. Repeat the experiment several times. Try matrices of various sizes. This shows that a matrix chosen at random (nearly always) has full rank. □

rcise 4.2.20 Use MATLAB to generate a random 8×6 matrix with rank 4. For example, you can proceed as follows.

```
A = randn(8,4);
A(:,5:6) = A(:,1:2) + A(:,3:4);
[Q,R] = qr(randn(6));
A = A*Q;
```

The first command produces a random 8×4 matrix (of rank 4). The second command adjoins two columns to A by adding columns 1 and 2 to columns 3 and 4, respectively. Now A is 8×6 and still has (numerical) rank 4. The third line generates a 6×6 random orthogonal matrix Q by performing a QR decomposition of a random 6×6 matrix. The last line recombines the columns of A by multiplying on the right by Q. The resulting matrix has numerical rank 4.

(a) Print out A on your computer screen. Can you tell by looking at it that it has (numerical) rank 4?

(b) Use MATLAB's `svd` command to obtain the singular values of A. How many are "large?" How many are "tiny?" (Use the command `format short e` to get a more accurate view of the singular values. Type `help format` for more information on the various output formats that are available.)

(c) Use MATLAB's `rank` command to confirm that the numerical rank is 4.

(d) Use the `rank` command with a low enough threshold that it returns the value 6. (Type `help rank` for information about how to do this.)

□

rcise 4.2.21 The *Kahan matrix* $R_n(\theta)$ is an $n \times n$ upper triangular matrix depending on a parameter θ. Let $c = \cos\theta$ and $s = \sin\theta$. Then

$$
R_n(\theta) =
\begin{bmatrix}
1 & & & & \\
& s & & & \\
& & s^2 & & \\
& & & \ddots & \\
& & & & s^{n-1}
\end{bmatrix}
\begin{bmatrix}
1 & -c & -c & \cdots & -c \\
& 1 & -c & \cdots & -c \\
& & 1 & & -c \\
& & & \ddots & \vdots \\
& & & & 1
\end{bmatrix}.
$$

If θ and n are chosen so that s is close to 1 and n is modestly large, then none of the main diagonal entries is extremely small. It thus appears that the matrix is far from rank deficient, but in this case appearances are deceiving. Consider $A = R_n(\theta)$ when $n = 90$ and $\theta = 1.2$ radians.

(a) Show that the largest main-diagonal entry of A is 1 and the smallest is greater than .001.

(b) To generate A in MATLAB and find its singular values, type

```
A = gallery('kahan',90,1.2,0);
sig = svd(A);
```

Examine σ_1, σ_{89}, and σ_{90} (It's not zero; try format short e). What is the numerical rank of A? Type rank(A) to get MATLAB's opinion.

(c) Type A = gallery('kahan',90,1.2,25); to get a slightly perturbed version of the Kahan matrix. (Type help private/kahan for more details.) Repeat part (b) for this perturbed matrix. Then get a QR decomposition with column pivoting: [Q,R,E] = qr(A); . Note that the permutation matrix E is the identity matrix: dif = norm(eye(90)-E), so no pivoting was done. Examine $R(90,90)$, and deduce that the QR decomposition with column pivoting failed to detect the numerical rank deficiency of A.

The Kahan matrix is just one of many matrices in Higham's gallery of test matrices for MATLAB. Type help gallery for a complete list.   □

Orthogonal Decompositions

The QR decomposition with column pivoting gives $AE = QR$ or equivalently $A = QRE^T$, where E is a permutation matrix, a special type of orthogonal matrix. The SVD gives $A = U\Sigma V^T$. Both are examples of orthogonal decompositions $A = YTZ^T$, where Y and Z are orthogonal, and T has a simple form. The QR decomposition is much cheaper to compute than the SVD. However, the SVD always reveals the numerical rank of the matrix, whereas the QR decomposition may sometimes fail to do so, as we have seen in Exercise 4.2.21. Therefore there has been considerable interest in producing an orthogonal decomposition that is cheap to compute yet reveals the rank of the matrix. Particularly noteworthy are the ULV and URV decompositions, which have the attractive feature that they can be updated inexpensively if A is modified by a matrix of rank one, as happens in signal processing applications. See [43, §12.5.5] for details.

Distance to Nearest Singular Matrix

We conclude this section by considering the implications of Theorem 4.2.15 for square, nonsingular matrices. Let $A \in \mathbb{R}^{n \times n}$ be nonsingular, and let A_s denote the singular matrix that is closest to A, in the sense that $\| A - A_s \|_2$ is as small as possible. In Theorem 2.3.1 we showed that

$$\frac{\| A - A_s \|}{\| A \|} \geq \frac{1}{\kappa(A)}$$

for any induced matrix norm, and we mentioned that for the 2-norm, equality holds. We now have the tools to prove this.

Corollary 4.2.22 *Let* $A \in \mathbb{R}^{n \times n}$ *be nonsingular. (Thus A has singular values* $\sigma_1 \geq \sigma_2 \geq \cdots \geq \sigma_n > 0$.) *Let* A_s *be the singular matrix that is closest to A, in the sense that* $\| A - A_s \|_2$ *is as small as possible. Then* $\| A - A_s \| = \sigma_n$, *and*

$$\frac{\| A - A_s \|_2}{\| A \|_2} = \frac{1}{\kappa_2(A)}.$$

These results are immediate consequences of Theorems 4.2.1, 4.2.4, and 4.2.15. In words, the distance from A to the nearest singular matrix is equal to the smallest singular value of A, and the "relative distance" to the nearest singular matrix is equal to the reciprocal of the condition number.

Additional Exercises

rcise 4.2.23 Formulate and prove an expression for the Frobenius condition number $\kappa_F(A)$ for nonsingular $A \in \mathbb{R}^{n \times n}$ in terms of the singular values of A. □

rcise 4.2.24 Let $A \in \mathbb{R}^{n \times m}$ with singular values $\sigma_1 \geq \cdots \geq \sigma_m$ and right singular vectors $v_1, \ldots, v_m$. We have seen that $\| Ax \|_2 / \| x \|_2$ is maximized when $x = v_1$ and minimized when $x = v_m$. More generally show that for $k = 1, \ldots, m$

$$
\begin{aligned}
\sigma_k &= \frac{\| Av_k \|_2}{\| v_k \|_2} = \max \left\{ \frac{\| Ax \|_2}{\| x \|_2} \mid x \neq 0, \ x \in \mathrm{span}\{v_1, \ldots, v_{k-1}\}^{\perp} \right\} \\
&= \min \left\{ \frac{\| Ax \|_2}{\| x \|_2} \mid x \neq 0, \ x \in \mathrm{span}\{v_{k+1}, \ldots, v_m\}^{\perp} \right\}.
\end{aligned}
$$

□

rcise 4.2.25

(a) Let $A \in \mathbb{R}^{2 \times 2}$ with singular values $\sigma_1 \geq \sigma_2 > 0$. Show that the set $\{ Ax \mid \| x \|_2 = 1 \}$ (the image of the unit circle) is an ellipse in $\mathbb{R}^2$ whose major and minor semiaxes have lengths σ_1 and σ_2, respectively.

(b) Let $A \in \mathbb{R}^{n \times m}$, $n \geq m$, $\mathrm{rank}(A) = m$. Show that the set $\{ Ax \mid \| x \|_2 = 1 \}$ is an m-dimensional hyperellipsoid with semiaxes $\sigma_1, \sigma_2, \ldots, \sigma_m$. Notice that the lengths of the longest and shortest semiaxes are $\mathrm{maxmag}(A)$ and $\mathrm{minmag}(A)$, respectively.

□

4.3 THE SVD AND THE LEAST SQUARES PROBLEM

Let $A \in \mathbb{R}^{n \times m}$, $r = \mathrm{rank}(A)$, and $b \in \mathbb{R}^n$, and consider the system of equations

$$Ax = b \tag{4.3.1}$$

with unknown $x \in \mathbb{R}^m$. If $n > m$, then the system is overdetermined, and we cannot expect to find an exact solution. Thus we will seek an x such that $\| b - Ax \|_2$ is minimized. This is exactly the least squares problem, which we studied in Chapter 3. There we found that if $n \geq m$ and $\mathrm{rank}(A) = m$, the least squares problem has a unique solution. If $\mathrm{rank}(A) < m$, the solution is not unique; there are many x for which $\| b - Ax \|_2$ is minimized. Even if $n < m$, it can happen that (4.3.1) does not have an exact solution, so we include that case as well. Thus we will make no assumption about the relative sizes of n and m.

Because the solution of the least squares problem is sometimes not unique, we will consider the following additional problem: of all the $x \in \mathbb{R}^m$ that minimize $\| b - Ax \|_2$, find the one for which $\| x \|_2$ is minimized. As we shall see, this problem always has a unique solution.

Initially we shall assume A and b are known exactly, and all computations are carried out exactly. Once we have settled the theoretical issues, we will discuss the practical questions.

Suppose we have the exact SVD $A = U \Sigma V^T$, where $U \in \mathbb{R}^{n \times n}$ and $V \in \mathbb{R}^{m \times m}$ are orthogonal, and

$$\Sigma = \begin{bmatrix} \hat{\Sigma} & 0 \\ 0 & 0 \end{bmatrix}, \qquad \hat{\Sigma} = \mathrm{diag}\{\sigma_1, \ldots, \sigma_r\},$$

with $\sigma_1 \geq \cdots \geq \sigma_r > 0$. Because U is orthogonal,

$$\| b - Ax \|_2 = \| U^T (b - Ax) \|_2 = \| U^T b - \Sigma (V^T x) \|_2.$$

Letting $c = U^T b$ and $y = V^T x$, we have

$$\| b - Ax \|_2^2 = \| c - \Sigma y \|_2^2 = \sum_{i=1}^{r} |c_i - \sigma_i y_i|^2 + \sum_{i=r+1}^{n} |c_i|^2. \tag{4.3.2}$$

It is clear that this expression is minimized when and only when

$$y_i = \frac{c_i}{\sigma_i}, \qquad i = 1, \ldots, r.$$

Notice that when $r < m$, $y_{r+1}, \ldots, y_m$ do not appear in (4.3.2). Thus they have no effect on the residual and can be chosen arbitrarily. Among all the solutions so obtained, $\| y \|_2$ is clearly minimized when and only when $y_{r+1} = \cdots = y_m = 0$. Since $x = Vy$ and V is orthogonal, $\| x \|_2 = \| y \|_2$. Thus $\| x \|_2$ is minimized when and only when $\| y \|_2$ is. This proves that the least squares problem has exactly one minimum norm solution.

It is useful to repeat the development using partitioned matrices. Let

$$c = \begin{bmatrix} \hat{c} \\ d \end{bmatrix} \qquad \text{and} \qquad y = \begin{bmatrix} \hat{y} \\ z \end{bmatrix},$$

where $\hat{c}, \hat{y} \in \mathbb{R}^r$. Then

$$c - \Sigma y = \begin{bmatrix} \hat{c} \\ d \end{bmatrix} - \begin{bmatrix} \hat{\Sigma} & 0 \\ 0 & 0 \end{bmatrix} \begin{bmatrix} \hat{y} \\ z \end{bmatrix} = \begin{bmatrix} \hat{c} - \hat{\Sigma}\hat{y} \\ d \end{bmatrix},$$

so

$$\|b - Ax\|_2^2 = \|c - \Sigma y\|_2^2 = \|\hat{c} - \hat{\Sigma}\hat{y}\|_2^2 + \|d\|_2^2.$$

This is minimized when and only when $\hat{y} = \hat{\Sigma}^{-1}\hat{c}$, that is, $y_i = c_i/\sigma_i$, $i = 1, \ldots, r$. We can choose z arbitrarily, but we get the minimum norm solution by taking $z = 0$. The norm of the minimal residual is $\|d\|_2$. This solves the problem completely in principle. We summarize the procedure:

1. Calculate $\begin{bmatrix} \hat{c} \\ d \end{bmatrix} = c = U^T b.$

2. Let $\hat{y} = \hat{\Sigma}^{-1}\hat{c}.$

3. If $r < m$, choose $z \in \mathbb{R}^{m-r}$ arbitrarily. (For minimum-norm solution, take $z = 0$.)   (4.3.3)

4. Let $y = \begin{bmatrix} \hat{y} \\ z \end{bmatrix} \in \mathbb{R}^m.$

5. Let $x = Vy.$

Practical Considerations

In practice we do not know the exact rank of A. It is best to use the numerical rank, which we discussed in Section 4.2. All "tiny" singular values should be set to zero.

We have solved the least squares problem under the assumption that we have the matrices U and V at hand. However, you can easily check that the calculation of $\hat{c}$ uses only the first r columns of U, where, in practice, r is the numerical rank. If only the minimum-norm solution is wanted, only the first r columns of V are needed. While the numerical rank is usually not known in advance, it can never exceed $\min\{n, m\}$, so at most $\min\{n, m\}$ columns of U and V are needed.

If $n \gg m$, the computation of U can be expensive, even if we only compute the first m columns. In fact the computation of U can be avoided completely. U is the product of many reflectors and rotators that are generated in the computation of the SVD (discussed in Section 5.8). Since U is needed only so that we can compute $c = U^T b$, we can simply update b instead of assembling U. As each rotator or reflector U_i is generated, we make the update $b \leftarrow U_i^T b$. In the end, b will have been transformed into c. This is much less expensive than computing U explicitly just to get $c = U^T b$. In the process, we get not only $\hat{c}$, but also d, from which we can compute the residual $\|d\|_2$ inexpensively. If several least squares problems with the same A but different right-hand sides $b^{(1)}, b^{(2)}, \ldots$ are to be solved, the updates must be applied to all of the $b^{(j)}$ at once, since the U_i will not be saved.

No matter how the calculations are organized, the SVD is an expensive way to solve the least squares problem. Its principal advantage is that it gives a completely reliable means of determining the numerical rank for rank-deficient least squares problems.

The Pseudoinverse

The pseudoinverse, also known as the Moore-Penrose generalized inverse, is an interesting generalization of the ordinary inverse. Although only square nonsingular matrices have inverses in the ordinary sense, every $A \in \mathbb{R}^{n \times m}$ has a pseudoinverse. Just as the solution of a square nonsingular system $Ax = b$ can be expressed in terms of A^{-1} as $x = A^{-1}b$, the minimum-norm solution to a least squares problem with coefficient matrix $A \in \mathbb{R}^{n \times m}$ can be expressed in terms of the pseudoinverse $A^{\dagger}$ as $x = A^{\dagger}b$.

Given $A \in \mathbb{R}^{n \times m}$ with rank r, the action of A is completely described by the diagram

$$
\begin{array}{ccc}
& A & \\
& \sigma_1 & \\
v_1 & \xrightarrow{} & u_1 \\
& \sigma_2 & \\
v_2 & \xrightarrow{} & u_2 \\
\vdots & & \vdots \\
& \sigma_r & \\
v_r & \xrightarrow{} & u_r \\
v_{r+1} & & \\
\vdots & \left.\right\} \longrightarrow 0 \; , & \\
v_m & &
\end{array}
$$

where $v_1, \ldots, v_m$ and $u_1, \ldots, u_n$ are complete orthonormal sets of right and left singular vectors, respectively, and $\sigma_1 \geq \sigma_2 \geq \cdots \geq \sigma_r > 0$ are the nonzero singular values of A. In matrix form,

$$A = U\Sigma V^T.$$

We wish to define the pseudoinverse $A^{\dagger} \in \mathbb{R}^{m \times n}$ so that it is as much like a true inverse as possible. Therefore we must certainly require $A^{\dagger}u_i = \sigma_i^{-1}v_i$ for $i = 1, \ldots, r$. A reasonable choice for $A^{\dagger}u_{r+1}, \ldots, A^{\dagger}u_n$ is to make them zero. Thus we define the *pseudoinverse* of A to be the matrix $A^{\dagger} \in \mathbb{R}^{m \times n}$ that is uniquely specified by the diagram

$$
\begin{array}{ccc}
& A^{\dagger} & \\
& \sigma_1^{-1} & \\
u_1 & \xrightarrow{} & v_1 \\
& \sigma_2^{-1} & \\
u_2 & \xrightarrow{} & v_2 \\
\vdots & & \vdots \\
& \sigma_r^{-1} & \\
u_r & \xrightarrow{} & v_r \\
u_{r+1} & & \\
\vdots & \left.\right\} \longrightarrow 0 \; . & \\
u_n & &
\end{array}
$$

We see immediately that $\operatorname{rank}\left(A^{\dagger}\right) = \operatorname{rank}(A)$, $u_1, \ldots, u_n$ and $v_1, \ldots, v_m$ are right and left singular vectors of $A^{\dagger}$, respectively, and $\sigma_1^{-1}, \ldots, \sigma_r^{-1}$ are the nonzero singular values. The restricted operators $A : \text{span}\{v_1, \ldots, v_r\} \to \text{span}\{u_1, \ldots, u_r\}$ and $A^\dagger : \text{span}\{u_1, \ldots, u_r\} \to \text{span}\{v_1, \ldots, v_r\}$ are true inverses of one another.

What does $A^\dagger$ look like as a matrix? You can answer this question in the simplest case by working the following exercise.

Exercise 4.3.4 Show that if

$$\Sigma = \begin{bmatrix} \hat{\Sigma} & 0 \\ 0 & 0 \end{bmatrix} \in \mathbb{R}^{n \times m}, \qquad \hat{\Sigma} = \text{diag}\{\sigma_1, \ldots, \sigma_r\},$$

then

$$\Sigma^\dagger = \begin{bmatrix} \hat{\Sigma}^{-1} & 0 \\ 0 & 0 \end{bmatrix} \in \mathbb{R}^{m \times n}.$$

$\square$

To see what $A^\dagger$ looks like in general, simply note that the equations

$$A^\dagger u_i = \begin{cases} v_i \sigma_i^{-1} & i = 1, \ldots, r \\ 0 & i = r+1, \ldots, n \end{cases}$$

can be expressed as a single matrix equation $A^\dagger U = V \Sigma^\dagger$, where $\Sigma^\dagger$ is as given in Exercise 4.3.4. Thus, since U is orthogonal,

$$A^\dagger = V \Sigma^\dagger U^T. \tag{4.3.5}$$

This is the SVD of $A^\dagger$ in matrix form. If we let $\hat{U}$ and $\hat{V}$ denote the first r columns of U and V, respectively, we can rewrite (4.3.5) in the condensed form

$$A^\dagger = \hat{V} \hat{\Sigma}^{-1} \hat{U}^T, \tag{4.3.6}$$

as in Theorem 4.1.10. Equation (4.3.6) gives us a means of calculating $A^\dagger$ by computing the SVD of A. However, there is seldom any reason to compute the pseudoinverse; it is mainly a theoretical tool. In this respect the pseudoinverse plays a role much like that of the ordinary inverse.

It is easy to make the claimed connection between the pseudoinverse and the least squares problem.

Theorem 4.3.7 *Let $A \in \mathbb{R}^{n \times m}$ and $b \in \mathbb{R}^n$, and let $x \in \mathbb{R}^m$ be the minimum-norm solution of*

$$\|b - Ax\|_2 = \min_{w \in \mathbb{R}^m} \|b - Aw\|_2.$$

Then $x = A^\dagger b$.

Proof. By (4.3.3),

$$x = Vy = V \begin{bmatrix} \hat{y} \\ 0 \end{bmatrix} = V \begin{bmatrix} \hat{\Sigma}^{-1} \hat{c} \\ 0 \end{bmatrix} = V \begin{bmatrix} \hat{\Sigma}^{-1} & 0 \\ 0 & 0 \end{bmatrix} \begin{bmatrix} \hat{c} \\ d \end{bmatrix}$$

$$= V \Sigma^\dagger c = V \Sigma^\dagger U^T b = A^\dagger b.$$

$\square$

The pseudoinverse is used in the study of the sensitivity of the rank-deficient least squares problem. See [8].

Exercise 4.3.8 Show that if $A = \sum_{i=1}^{r} \sigma_i u_i v_i^T$, then $A^{\dagger} = \sum_{i=1}^{r} \sigma_i^{-1} v_i u_i^T$.   □

Exercise 4.3.9 Work this exercise using pencil and paper (and exact arithmetic). You might like to use MATLAB to check your work. Let

$$A = \begin{bmatrix} 1 & 2 \\ 2 & 4 \\ 3 & 6 \end{bmatrix} \quad \text{and} \quad b = \begin{bmatrix} 1 \\ 1 \\ 1 \end{bmatrix}.$$

(a) Find the SVD of A. You may use the condensed form given by Theorem 4.1.10 or Theorem 4.1.12.

(b) Calculate $A^{\dagger}$.

(c) Calculate the minimum norm solution of the least-squares problem for the overdetermined system $Ax = b$.

(d) Find a basis for $\mathcal{N}(A)$.

(e) Find all solutions of the least-squares problem.

□

Exercise 4.3.10 MATLAB's `pinv` command returns the pseudoinverse of a matrix. Use the `pinv` command to find the pseudoinverse of

$$A = \begin{bmatrix} 1 & 2 & 3 \\ 2 & 4 & 6 \end{bmatrix}.$$

Use MATLAB's `svd` command to find the singular values of $AA^{\dagger}$ and $A^{\dagger}A$. Explain why these results are what you would expect.   □

Exercise 4.3.11 (Pseudoinverses of full-rank matrices)

(a) Show that if $A \in \mathbb{R}^{n \times m}$, $n \geq m$, and rank$(A) = m$, then $A^{\dagger} = (A^T A)^{-1} A^T$ (cf. Theorem 4.2.11 and Exercise 4.2.12). Compare this result with the normal equations (3.5.22).

(b) Show that if $A \in \mathbb{R}^{n \times m}$, $n \leq m$, and rank$(A) = n$, then $A^{\dagger} = A^T (A^T A)^{-1}$.

□

Additional Exercises

Exercise 4.3.12 Study the effects of treating very small nonzero singular values as nonzeros in the solution of the least squares problem.   □

rcise 4.3.13 Suppose $A \in \mathbb{R}^{n \times m}$, $n < m$, and rank$(A) = n$.

(a) Show that in this case the minimum-norm least squares problem is actually a constrained minimization problem.

(b) This problem can be solved by an SVD. Show that it can also be solved by an LQ decomposition: $A = LQ$, where $L = [\tilde{L}\ 0] \in \mathbb{R}^{n \times m}$, $\tilde{L} \in \mathbb{R}^{n \times n}$ is nonsingular and lower triangular, and $Q \in \mathbb{R}^{m \times m}$ is orthogonal.

(c) How does one calculate an LQ decomposition? Sketch an algorithm for calculating the LQ decomposition and solving the constrained minimization problem.

□

rcise 4.3.14

(a) Let $A \in \mathbb{R}^{n \times m}$ and $B = A^\dagger \in \mathbb{R}^{m \times n}$. Show that the following four relationships hold:

$$BAB = B \qquad\qquad ABA = A$$
$$(BA)^T = BA \qquad (AB)^T = AB. \tag{4.3.15}$$

(b) Conversely, show that if A and B satisfy (4.3.15), then $B = A^\dagger$. Thus equations (4.3.15) characterize the pseudoinverse. In many books these equations are used to define the pseudoinverse.

□

4.4   SENSITIVITY OF THE LEAST SQUARES PROBLEM

In this section we discuss the sensitivity of the solution of the least squares problem under perturbations of A and b. For simplicity we consider only the full-rank case with $n \geq m$. See [8] for more general results.

Given $A \in \mathbb{R}^{n \times m}$ and $b \in \mathbb{R}^n$, with $n \geq m$ and rank$(A) = m$, there is a unique $x \in \mathbb{R}^m$ such that

$$\|b - Ax\|_2 = \min_{w \in \mathbb{R}^m} \|b - Aw\|_2.$$

If we now perturb A and b slightly, the solution x will be altered. We will consider the question of how sensitive x is to perturbations in A and b. We will also consider the sensitivity of the residual $r = b - Ax$. The sensitivity analysis can be combined with backward error analyses to provide an assessment of the accuracy of various methods for solving the least squares problem.

From Section 3.5 it is clear that we can think of the solution of the least squares problem as a two-stage process. First we find a $y \in \mathcal{R}(A)$ whose distance from b is minimal:

$$\|b - y\|_2 = \min_{s \in \mathcal{R}(A)} \|b - s\|_2.$$

Then the least squares solution $x \in \mathbb{R}^m$ is found by solving the equation $Ax = y$ exactly. Because A has full rank, the solution is unique. Even though $A \in \mathbb{R}^{n \times m}$ and $y \in \mathbb{R}^n$, the matrix equation $Ax = y$ is effectively a system of m equations in m unknowns, for y lies in the m-dimensional subspace $\mathcal{R}(A)$, and $A : \mathbb{R}^m \rightarrow \mathcal{R}(A)$ can be viewed as a mapping of one m-dimensional space into another.

Exercise 4.4.1 The m-dimensional nature of the system can be revealed by a judicious change of coordinate system.

> (a) Show that the QR decomposition does just this. How is the vector y determined? How is the system $Ax = y$ solved?
>
> (b) Repeat part (a) using the SVD in place of the QR decomposition.

$\square$

Let $r = b - y = b - Ax$. This is the residual of the least squares problem. By Theorem 3.5.15, r is orthogonal to $\mathcal{R}(A)$. In particular, b, y, and r form a right triangle, as illustrated in Figure 4.1. The angle θ between b and y is the smallest

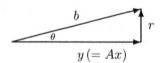

Figure 4.1   Right triangle formed by b, y, and r

angle between b and an element of $\mathcal{R}(A)$. It is clear that

$$\| r \|_2 = \| b \|_2 \sin \theta, \tag{4.4.2}$$

$$\| y \|_2 = \| b \|_2 \cos \theta. \tag{4.4.3}$$

These equations and Figure 4.1 will prove useful in our sensitivity analysis.

The Effect of Perturbations of b

At first we will examine the effects of perturbations of b only. Given a perturbation δb, let $y + \delta y$ denote the element of $\mathcal{R}(A)$ that is closest to $b + \delta b$, and let $\hat{x} = x + \delta x$ be the exact solution of $A\hat{x} = y + \delta y$. Then $\hat{x}$ is the minimizer of $\| (b + \delta b) - Aw \|_2$. We would like to be able to say that if $\| \delta b \|_2 / \| b \|_2$ is small, then $\| \delta x \|_2 / \| x \|_2$ is also small. As we shall see, there are two reasons why this might fail to be the case. The first is that $\| \delta y \|_2 / \| y \|_2$ might fail to be small. To see this, refer to Figure 4.1 and consider what happens when b is orthogonal or nearly orthogonal to $\mathcal{R}(A)$. Then y is either zero or very small. It is clear that a small perturbation of b in a direction parallel to $\mathcal{R}(A)$ will cause a perturbation δy that is large relative to $\| y \|_2$. The second problem is that even if $\| \delta y \|_2 / \| y \|_2$ is small, $\| \delta x \|_2 / \| x \|_2$ can be large if the linear system $Ax = y$ is ill conditioned. Any error bound for $\| \delta x \|_2 / \| x \|_2$ must reflect both of these factors. In the next exercise you will derive such a bound.

rcise 4.4.4

   (a) Defining $\delta r = \delta b - \delta y$, we have $\delta b = \delta y + \delta r$. Show that δr is orthogonal to δy. Deduce that $\| \delta y \|_2 \leq \| \delta b \|_2$.

   (b) Show that, provided $\cos \theta \neq 0$,

$$\frac{\| \delta y \|_2}{\| y \|_2} \leq \frac{1}{\cos \theta} \frac{\| \delta b \|_2}{\| b \|_2}. \tag{4.4.5}$$

   (c) Show that $\| y \|_2 \leq \text{maxmag}(A) \| x \|_2$, $\| \delta y \|_2 \geq \text{minmag}(A) \| \delta x \|_2$, and

$$\frac{\| \delta x \|_2}{\| x \|_2} \leq \kappa_2(A) \frac{\| \delta y \|_2}{\| y \|_2}. \tag{4.4.6}$$

   (d) Conclude that

$$\frac{\| \delta x \|_2}{\| x \|_2} \leq \frac{\kappa_2(A)}{\cos \theta} \frac{\| \delta b \|_2}{\| b \|_2}. \tag{4.4.7}$$

                                                                        $\square$

Inequality (4.4.5) shows that if θ is not close to $\pi/2$, that is, if b is not nearly orthogonal to the range of A, a small perturbation in b results in a small perturbation in y. Of course, if θ is close to $\pi/2$, we can have a disaster. But this almost never happens in real problems; the angle θ, like $\| r \|_2$, is a measure of how well the least squares solution fits the data. It will not be close to $\pi/2$ unless all of the vectors in $\mathcal{R}(A)$ approximate b very poorly. Thus, as long as the problem is formulated so that the least squares solution fits the data reasonably well, (4.4.5) will give a useful bound on $\| \delta y \|_2 / \| y \|_2$.

Inequality (4.4.6) shows the effect of $\kappa_2(A)$. This is really no different from the inequality (2.2.5), which we derived during the discussion of the sensitivity of square linear systems.

Inequality (4.4.7) combines the two effects. We see that if $\kappa_2(A)$ is not large, and $\cos \theta$ is not close to zero, then a small perturbation in b results in a small perturbation in x.

Inequality (4.4.7) can also be derived using the normal equations (3.5.22). We will work through this second derivation as preparation for the discussion of the effects of perturbations of A. Here we will make use of some of our elementary results on singular values.

From Theorem 3.5.21 we know that the solution of the least squares problem satisfies the normal equations

$$A^T A x = A^T b. \tag{4.4.8}$$

The solution of the perturbed problem satisfies the perturbed normal equations

$$A^T A(x + \delta x) = A^T (b + \delta b). \tag{4.4.9}$$

Subtracting (4.4.8) from (4.4.9), we find that $(A^T A)\delta x = A^T \delta b$, or

$$\delta x = (A^T A)^{-1} A^T \delta b.$$

Thus $\| \delta x \|_2 \leq \| (A^T A)^{-1} A^T \|_2 \| \delta b \|_2$. By Theorem 4.2.11, $\| (A^T A)^{-1} A^T \|_2 = \sigma_m^{-1}$, where σ_m is the smallest singular value of A, so

$$\| \delta x \|_2 \leq \sigma_m^{-1} \| \delta b \|_2. \tag{4.4.10}$$

On the other hand, we see from (4.4.3) that $\| b \|_2 \cos \theta = \| y \|_2 = \| Ax \|_2 \leq \| A \|_2 \| x \|_2 = \sigma_1 \| x \|_2$. Thus

$$\frac{1}{\| x \|_2} \leq \frac{\sigma_1}{\cos \theta} \frac{1}{\| b \|_2}. \tag{4.4.11}$$

Multiplying (4.4.10) by (4.4.11), and recalling that $\kappa_2(A) = \sigma_1/\sigma_m$ (Theorem 4.2.4), we get (4.4.7).

The Effect of Perturbations of A

Unfortunately perturbations of A have a more severe effect than perturbations of b. As the following Theorem shows, the sensitivity of x to perturbations of A depends, in part, on the square of the condition number of A. Our approach will be to state the theorem, discuss its implications, then prove it.

Theorem 4.4.12 *Let $A \in \mathbb{R}^{n \times m}$, $b \in \mathbb{R}^n$, $n \geq m$, and $\mathrm{rank}(A) = m$. Let $x \in \mathbb{R}^m$ be the unique solution of*

$$\| b - Ax \| = \min_{w \in \mathbb{R}^m} \| b - Aw \|_2.$$

Let θ denote the angle between b and Ax, and assume $\theta \neq \pi/2$. Let $\delta A \in \mathbb{R}^{n \times m}$, $\delta b \in \mathbb{R}^n$, $\epsilon_A = \| \delta A \|_2 / \| A \|_2$, $\epsilon_b = \| \delta b \|_2 / \| b \|_2$, and $\epsilon = \max\{\epsilon_A, \epsilon_b\}$. Assume $\epsilon \ll 1$, and in particular, $\epsilon_A < 1/\kappa_2(A)$. Let $\hat{x} = x + \delta x \in \mathbb{R}^m$ be the unique solution of

$$\| (b + \delta b) - (A + \delta A)\hat{x} \| = \min_{w \in \mathbb{R}^m} \| (b + \delta b) - (A + \delta A)w \|_2.$$

Then

$$\frac{\| \delta x \|_2}{\| x \|_2} \leq \frac{2\kappa_2(A)}{\cos \theta} \epsilon_b + 2(\kappa_2(A)^2 \tan \theta + \kappa_2(A))\epsilon_A + O(\epsilon^2). \tag{4.4.13}$$

Let $r = b - Ax$ and $\hat{r} = r + \delta r = (b + \delta b) - (A + \delta A)\hat{x}$. Then

$$\frac{\| \delta r \|_2}{\| b \|_2} \leq 2\epsilon_b + 3\kappa_2(A)\epsilon_A + O(\epsilon^2). \tag{4.4.14}$$

In both of the bounds, the term $O(\epsilon^2)$ stands for terms that contain factors ϵ_A^2 or $\epsilon_A \epsilon_b$ and are negligible. If we take $\epsilon_A = 0$ in (4.4.13), we obtain (4.4.7), except for an inessential factor 2.

The most striking feature of (4.4.13) is that it depends on the square of $\kappa_2(A)$. This is not surprising if one considers that a perturbation in A causes a perturbation in $\mathcal{R}(A)$ that turns out to be (at worst) proportional to $\kappa_2(A)$. Thus a perturbation ϵ_A can cause a change of magnitude $\kappa_2(A)\epsilon_A$ in y $(= Ax \in \mathcal{R}(A))$, which can in turn cause a change of magnitude $\kappa_2(A)^2\epsilon$ in x.

The presence of $\kappa_2(A)^2$ in (4.4.13) means that even if A is only mildly ill conditioned, a small perturbation in A can cause a large change in x. An exception is the class of problems for which the least squares solution fits the data exactly; that is, $r = 0$. Then $\tan\theta = 0$ as well, and the offending term disappears. While a perfect fit is unusual, a good fit is not. If the fit is good, $\tan\theta$ will be small and will partially cancel out the factor $\kappa_2(A)^2$.

Since we measure the size of residuals relative to b, it is reasonable to measure the change in the residual relative to $\|b\|_2$, as we have done in (4.4.14). Notice that (4.4.14) depends on $\kappa_2(A)$, not $\kappa_2(A)^2$. This means that the residual, and the goodness of fit, is generally much less sensitive to perturbations than the least squares solution is.

Keeping the condition number under control.

From Theorem 4.4.12 it is clear that it is important to avoid having to solve least squares problems for which the coefficient matrix is ill conditioned. This is something over which you have some control. For the purpose of illustration, consider the problem of fitting a set of data points by a polynomial of low degree. Then the system to be solved has the form (3.1.3); the coefficient matrix is determined by the point abscissae $t_1, t_2, \ldots, t_n$ and the basis functions $\phi_1, \phi_2, \ldots, \phi_m$. In principle any basis for the space of approximating polynomials can be used. However, the choice of basis affects the condition number of the system.

Example 4.4.15 Suppose we need to find the least squares first-degree polynomial for a set of seven points whose abscissae are

t_1	t_2	t_3	t_4	t_5	t_6	t_7
1.01	1.02	1.03	1.04	1.05	1.06	1.07

The obvious choice of basis for the first-degree polynomials is $\phi_1(t) = 1$ and $\phi_2(t) = t$. Let us contrast the behavior of this basis with that of the more carefully constructed basis $\tilde{\phi}_1 = 1$ and $\tilde{\phi}_2 = 30(t - 1.04)$. Using (3.1.3) we find that the two bases give rise to the coefficient matrices

$$A = \begin{bmatrix} 1 & 1.01 \\ 1 & 1.02 \\ 1 & 1.03 \\ 1 & 1.04 \\ 1 & 1.05 \\ 1 & 1.06 \\ 1 & 1.07 \end{bmatrix} \qquad \tilde{A} = \begin{bmatrix} 1 & -.9 \\ 1 & -.6 \\ 1 & -.3 \\ 1 & 0 \\ 1 & .3 \\ 1 & .6 \\ 1 & .9 \end{bmatrix},$$

respectively. According to MATLAB, the condition numbers are $\kappa_2(A) \approx 104$ and $\kappa_2(\tilde{A}) \approx 1.67$. It is not hard to see why $\tilde{A}$ is so much better than A. The abscissae are concentrated in the interval $[1.01, 1.07]$, on which the function ϕ_2 varies very little; it looks a lot like ϕ_1. Consequently the two columns of A are nearly equal, and A is (mildly) ill conditioned. By contrast the basis $\tilde{\phi}_1$, $\tilde{\phi}_2$ was chosen with the interval $[1.01, 1.07]$ in mind. $\tilde{\phi}_2$ is centered on the interval and varies from -0.9 to $+0.9$, while $\tilde{\phi}_1$ remains constant. As a consequence the columns of $\tilde{A}$ are not close to being linearly dependent. (In fact they are orthogonal.) The factor 30 in the definition of $\tilde{\phi}_2$ guarantees that the second column of $\tilde{A}$ has roughly the same magnitude as the first column, so the columns are not out of scale. Thus $\tilde{A}$ is well conditioned.   □

In general one wants to choose functions that are not close to being linearly dependent on the interval of interest. Then the columns of the resulting A will not be close to being linearly dependent. Consequently A will be well conditioned.

Accuracy of techniques for solving the least squares problem.   A standard backward error analysis [52] (or Exercises 3.2.71– 3.2.74) shows that the QR decomposition method using reflectors or rotators is normwise backward stable. Thus the computed solution is the exact minimizer of a perturbed function

$$\| (b + \delta b) - (A + \delta A)w \|_2,$$

where the perturbations are of the same order of magnitude as the unit roundoff u. This means that this method works as well as we could hope. The computed solution and residual satisfy (4.4.13) and (4.4.14) with $\epsilon_A \approx \epsilon_b \approx u$.

As Björck has shown (see, e.g., [8]), the modified Gram-Schmidt method is about as accurate as the QR decomposition by reflectors.

The method of normal equations has somewhat different characteristics. Suppose $A^T A$ and $A^T b$ have been calculated exactly, and the system $A^T A x = A^T b$ has been solved by Cholesky's method. From Section 2.7 (especially Exercise 2.7.29) we know that the computed solution is the exact solution of a perturbed equation $(A^T A + E)\hat{x} = A^T b$, where $\| E \|_2 / \| A^T A \|_2$ is a modest multiple of u. It follows from Theorem 2.3.6 that $\| \delta x \|_2 / \| x \|_2$ is roughly $\kappa_2(A^T A)u$. By Theorem 4.2.9, $\kappa_2(A^T A) = \kappa_2(A)^2$. Given that the factor $\kappa_2(A)^2$ also appears in (4.4.13), it looks as if the normal equations method is about as accurate as the other methods. However, in problems with a small residual, the $\kappa_2(A)^2 \tan \theta$ term in (4.4.13) is diminished. In these cases the other methods will be more accurate.

This analysis has assumed that $A^T A$ and $A^T b$ have been computed accurately. In practice significant errors can occur in the computation of $A^T A$, as was shown in Example 3.5.25. Furthermore, if $\cos \theta$ is near zero, cancellations will occur in the calculation of $A^T b$ and result in large relative errors.

rcise 4.4.16 This exercise demonstrates that the QR method is superior to the normal equations method when the condition number is bad. It also shows how a change of basis can improve the condition number. Consider the following data.

t_i	y_i
0.98765431	2.1
0.98765432	2.1
0.98765433	2.1
0.98765434	2.0
0.98765435	1.9
0.98765436	1.9
0.98765437	1.8
0.98765438	1.7
0.98765439	1.7

(a) Set up an over-determined system $Ax = b$ for a linear polynomial. Use the basis $\phi_1(t) = 1$, $\phi_2(t) = t$. Using MATLAB, find the condition number of A. Find the least-squares solution by the QR decomposition method (e.g. x=A\b in MATLAB). Calculate the norm of the residual.

(b) Obtain the normal equations $A^T Ax = A^T b$. Find the condition number of the coefficient matrix. Solve the normal equations. Calculate the norm of the residual ($\| b - Ax \|_2$).

(c) Compare your solutions from parts (a) and (b). Plot the two polynomials and the data points on a single graph.

(d) Now set up a different over-determined system $\hat{A}\hat{x} = \hat{b}$ for the same data using the basis $\hat{\phi}_1(t) = 1$, $\hat{\phi}_2(t) = 3 \times 10^7 \times (t - .98765435)$. Find the condition number of $\hat{A}$. Find the least-squares solution by the QR decomposition method. Calculate the norm of the residual.

(e) Obtain the normal equations $\hat{A}^T \hat{A}\hat{x} = \hat{A}^T \hat{b}$. Find the condition number of the coefficient matrix. Solve the normal equations. Calculate the norm of the residual ($\| \hat{b} - \hat{A}\hat{x} \|_2$).

(f) Compare your solutions from parts (d) and (e). This time they are identical (to 15 decimal places if you care to check). What polynomial do these solutions represent? You might want to plot it just to make sure you've worked the problem correctly. If you do make a plot, you will see that your line looks identical to the one computed using the QR decomposition in part (a). That one was correct to only about eight decimal places, whereas the solutions computed using the well-conditioned matrices are correct to machine precision, but of course the difference does not show up in a plot (and wouldn't matter in most applications either).

$\square$

Proof of Theorem 4.4.12   First of all, the perturbed problem has a unique solution, since $A + \delta A$ has full rank. This is a consequence of Corollary 4.2.16.

Rather than proving the theorem in full generality, we will examine the case in which only A is perturbed; that is, $\delta_b = 0$. This will simplify the manipulations somewhat without altering the spirit of the proof. The complete proof is left as an exercise.

Furthermore, we shall make the assumption $\epsilon_A + \frac{1}{2}\epsilon_A^2 \leq 1/(4\kappa_2(A)^2)$, which is considerably more restrictive than the hypothesis that was given in the statement of the theorem. With a more careful argument we could remove this restriction.

The solution $\hat{x} = x + \delta x$ of the perturbed least squares problem must satisfy the perturbed normal equations

$$(A + \delta A)^T (A + \delta A)(x + \delta x) = (A + \delta A)^T b.$$

Subtracting the unperturbed normal equations $A^T A x = A^T b$ and making some routine manipulations, we find that

$$A^T A \delta x = \delta A^T (b - Ax) - A^T \delta Ax - (A^T \delta A + \delta A^T A + \delta A^T \delta A)\delta x - \delta A^T \delta Ax$$

or

$$
\begin{aligned}
\delta x \; = \; & (A^T A)^{-1}\delta A^T r - (A^T A)^{-1}A^T \delta Ax \\
& -(A^T A)^{-1}(A^T \delta A + \delta A^T A + \delta A^T \delta A)\delta x \qquad (4.4.17) \\
& -(A^T A)^{-1}\delta A^T \delta Ax.
\end{aligned}
$$

Taking norms of both sides and using the fact that $\|B^T\|_2 = \|B\|_2$ for any B (Corollary 4.2.2), we obtain the inequality

$$
\begin{aligned}
\|\delta x\|_2 \; \leq \; & \|(A^T A)^{-1}\|_2 \|\delta A\|_2 \|r\|_2 + \|(A^T A)^{-1}A^T\|_2 \|\delta A\|_2 \|x\|_2 \\
& + \|(A^T A)^{-1}\|_2 (2\|A\|_2 \|\delta A\|_2 + \|\delta A\|_2^2)\|\delta x\|_2 \\
& + \|(A^T A)^{-1}\|_2 \|\delta A\|_2^2 \|x\|_2.
\end{aligned}
$$

As usual, let σ_1 and σ_m denote the largest and smallest singular values of A, respectively. Then $\|A\|_2 = \sigma_1$, $\|\delta A\|_2 = \sigma_1 \epsilon_A$, and from Theorem 4.2.11 we know that $\|(A^T A)^{-1}\|_2 = \sigma_m^{-2}$ and $\|(A^T A)^{-1}A^T\|_2 = \sigma_m^{-1}$. Furthermore, by (4.4.2) and (4.4.3), $\|r\|_2 = \|b\|_2 \sin\theta = \|Ax\|_2 \tan\theta \leq \sigma_1 \|x\|_2 \tan\theta$. Applying these results, making the substitution $\kappa_2(A) = \sigma_1/\sigma_m$, and moving all terms involving $\|\delta x\|_2$ to the left-hand side of the inequality, we obtain

$$\left[1 - 2\kappa_2(A)^2(\epsilon_A + \tfrac{1}{2}\epsilon_A^2)\right] \|\delta x\|_2 \le$$
$$\left[\kappa_2(A)^2 \tan\theta + \kappa_2(A)\right]\epsilon_A \|x\|_2 + \kappa_2(A)^2 \epsilon_A^2 \|x\|_2.$$

If $\epsilon_A + \tfrac{1}{2}\epsilon_A^2 \le 1/(4\kappa_2(A)^2)$, then the coefficient of $\|\delta x\|_2$ is at least 1/2. Therefore

$$\frac{\|\delta x\|_2}{\|x\|_2} \le 2\left[\kappa_2(A)^2 \tan\theta + \kappa_2(A)\right]\epsilon_A + O(\epsilon_A^2).$$

This is (4.4.13) in the case $\epsilon_b = 0$.

To obtain the bound (4.4.14) on the residual, we note first that since $r + \delta r = b - (A + \delta A)(x + \delta x)$ and $r = b - Ax$,

$$\delta r = -\delta Ax - A\delta x - \delta A\delta x.$$

Using (4.4.17), and noting that by (4.4.13) $\|\delta A\|_2 \|\delta x\|_2 = O(\epsilon^2)$, we obtain

$$\begin{aligned}\|\delta r\|_2 \le\ & \|\delta A\|_2 \|x\|_2 + \|A(A^T A)^{-1}\|_2 \|\delta A\|_2 \|r\|_2 \\ & + \|A(A^T A)^{-1} A^T\|_2 \|\delta A\|_2 \|x\|_2 + O(\epsilon^2).\end{aligned}$$

Since $\|Ax\|_2 \ge \sigma_m \|x\|_2$, we have $\|x\|_2 \le \sigma_m^{-1}\|Ax\|_2 = \sigma_m^{-1}\cos\theta\|b\|_2$. Also, by Theorem 4.2.11, $\|A(A^T A)^{-1}\|_2 = \sigma_m^{-1}$ and $\|A(A^T A)^{-1} A^T\|_2 = 1$. Therefore

$$\begin{aligned}\frac{\|\delta r\|_2}{\|b\|_2} \le\ & (2\cos\theta + \sin\theta)\kappa_2(A)\epsilon_A + O(\epsilon^2) \\ <\ & 3\kappa_2(A)\epsilon_A + O(\epsilon^2).\end{aligned}$$

This is just (4.4.14) in the case $\epsilon_b = 0$. $\square$

·cise 4.4.18 Work out a complete proof of Theorem 4.4.12 with $\epsilon_A > 0$ and $\epsilon_b > 0$. $\square$

CHAPTER 5

EIGENVALUES AND EIGENVECTORS I

Eigenvalues and eigenvectors turn up in stability theory, theory of vibrations, quantum mechanics, statistical analysis, and many other fields. It is therefore important to have efficient, reliable methods for computing these objects. The main business of this chapter is to develop such algorithms, culminating in the powerful and elegant Francis algorithm, also known as the implicitly-shifted QR algorithm.

Before we embark on the development of algorithms, we take the time to illustrate (in Section 5.1) how eigenvalues and eigenvectors arise in the analysis of systems of differential equations. The material is placed here entirely for motivational purposes. It is intended to convince you, the student, that eigenvalues are important. Section 5.1 is not, strictly speaking, a prerequisite for the rest of the chapter.

Section 5.1 also provides an opportunity to introduce MATLAB's eig command. When you use eig to compute the eigenvalues and eigenvectors of a matrix, you are using Francis's algorithm.

5.1 SYSTEMS OF DIFFERENTIAL EQUATIONS

Many applications of eigenvalues and eigenvectors arise from the study of systems of differential equations.

Fundamentals of Matrix Computations, Third Edition. By David S. Watkins
Copyright © 2010 John Wiley & Sons, Inc.

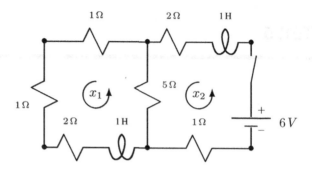

Figure 5.1   Solve for the time-varying loop currents.

Example 5.1.1 The electrical circuit in Figure 5.1 is the same as the one that was featured in Example 1.2.8, except that two inductors and a switch have been added. Whereas resistors resist current, inductors resist changes in current. If we are studying constant, unvarying currents, as in Example 1.2.8, we can ignore the inductors, since their effect is felt only when the currents are changing. However, if the currents are varying in time, we must take the inductances into account.

Once the switch in the circuit is closed, current will begin to flow. The loop currents are functions of time: $x_i = x_i(t)$. Just as before, an equation for each loop can be obtained by applying Kirchhoff's voltage law: the sum of the voltage drops around the loop must be zero. The voltage drop across an inductor is proportional to the rate of change of the current. The constant of proportionality is the *inductance*. Thus if the current flowing through an inductor is $x(t)$ amps at time t seconds, and the inductance is m henries, the voltage drop across the inductor at time t is $m\dot{x}(t)$ volts, where $\dot{x}$ denotes the time derivative $\frac{dx}{dt}$ (amperes/second). Because of the presence of derivative terms, the resulting loop equations are now differential equations. Thus we have a system of two differential equations (one for each loop) in two unknowns (the loop currents).

Let us write down the two equations. First consider the first loop. As you will recall, the voltage drop across the 5 Ω resistor in the direction indicated by the arrow for the first loop is $5(x_1 - x_2)$ volts. The voltage drop across the 1 henry inductor is $1\dot{x}_1$ volts. Summing these voltage drops, together with the voltage drops across the other resistors in loop 1, we obtain the equation

$$5(x_1 - x_2) + 1x_1 + 1x_1 + 2x_1 + 1\dot{x}_1 = 0.$$

Similarly, in loop 2,

$$5(x_2 - x_1) + 1x_2 - 6 + 1\dot{x}_2 + 2x_2 = 0.$$

These are exactly the same as the equations we obtained in Example 1.2.8, except for the derivative terms. Rearranging these equations and employing matrix notation, we can rewrite the system as

$$\left[\begin{array}{c} \dot{x}_1(t) \\ \dot{x}_2(t) \end{array} \right] = \left[\begin{array}{cc} -9 & 5 \\ 5 & -8 \end{array} \right] \left[\begin{array}{c} x_1 \\ x_2 \end{array} \right] - \left[\begin{array}{c} 0 \\ -6 \end{array} \right]. \tag{5.1.2}$$

Suppose the switch is closed at time zero. Then we can find the resulting currents by solving the system (5.1.2) subject to the initial conditions

$$\left[\begin{array}{c} x_1(0) \\ x_2(0) \end{array} \right] = \left[\begin{array}{c} 0 \\ 0 \end{array} \right]. \tag{5.1.3}$$

As we shall see, we can solve this problem with the help of eigenvalues and eigenvectors.　　　　□

The system of differential equations in Example 5.1.1 has the general form $\dot{x} = Ax - b$, where

$$x = \left[\begin{array}{c} x_1(t) \\ x_2(t) \end{array} \right], \quad A = \left[\begin{array}{cc} -9 & 5 \\ 5 & -8 \end{array} \right], \quad \text{and} \quad b = \left[\begin{array}{c} 0 \\ -6 \end{array} \right].$$

A larger electrical circuit will lead to a larger system of equations of the same or similar form. For example, in Exercise 5.1.21 we consider a system of eight differential equations. The form is again $\dot{x} = Ax - b$, but now A is an 8×8 matrix. In general we will consider systems of n differential equations, in which the matrix A is $n \times n$.

Homogeneous Systems

An important step in solving a system of the form $\dot{x} = Ax - b$ is to solve the related *homogeneous* system $\dot{x} = Ax$, obtained by dropping the forcing term b (corresponding to the battery in Example 5.1.1). Thus we consider the problem of solving

$$\frac{dx}{dt} = Ax, \tag{5.1.4}$$

where A is an $n \times n$ matrix. Equation (5.1.4) is shorthand for

$$\begin{array}{rcl} \dfrac{dx_1}{dt} &=& a_{11}x_1 + a_{12}x_2 + \cdots + a_{1n}x_n \\[2mm] \dfrac{dx_2}{dt} &=& a_{21}x_1 + a_{22}x_2 + \cdots + a_{2n}x_n \\[2mm] \vdots & & \vdots \\[2mm] \dfrac{dx_n}{dt} &=& a_{n1}x_1 + a_{n2}x_2 + \cdots + a_{nn}x_n. \end{array}$$

A common approach to solving linear differential equations is to begin by looking for solutions of a particularly simple form. Therefore let us seek solutions of the form

$$x(t) = g(t)v, \tag{5.1.5}$$

where $g(t)$ is a nonzero scalar (real or complex valued) function of t, and v is a nonzero constant vector. The time-varying nature of $x(t)$ is expressed by $g(t)$, while the vector nature of $x(t)$ is expressed by v. Substituting form (5.1.5) into (5.1.4), we obtain the equation

$$\dot{g}(t)v = g(t)Av,$$

or equivalently

$$\frac{\dot{g}(t)}{g(t)}v = Av. \tag{5.1.6}$$

Since v and Av are constant vectors, (5.1.6) implies that $\dot{g}(t)/g(t)$ must be constant. That is, there exists a (real or complex) constant λ such that

$$\frac{\dot{g}(t)}{g(t)} = \lambda. \tag{5.1.7}$$

In addition (5.1.6) implies

$$Av = \lambda v.$$

A nonzero vector v for which there exists a λ such that $Av = \lambda v$ is called an *eigenvector* of A. The number λ is called the *eigenvalue* of A associated with v. So far we have shown that if $x(t)$ is a solution of (5.1.4) of the form (5.1.5), then v must be an eigenvector of A, and $g(t)$ must satisfy the differential equation (5.1.7), where λ is the eigenvalue of A associated with v. The general solution of the scalar differential equation (5.1.7) is $g(t) = ce^{\lambda t}$, where c is an arbitrary constant. Conversely, if v is an eigenvector of A with associated eigenvalue λ, then

$$e^{\lambda t}v$$

is a solution of (5.1.4), as you can easily verify. Thus each eigenvector of A gives rise to a solution of (5.1.4). If A has enough eigenvectors, then every solution of (5.1.4) can be realized as a linear combination of these simple solutions. Specifically, suppose A (which is $n \times n$) has a set of n linearly independent eigenvectors $v_1, \ldots, v_n$ with associated eigenvalues $\lambda_1, \ldots, \lambda_n$. Then for any constants $c_1, \ldots, c_n$,

$$x(t) = c_1 e^{\lambda_1 t}v_1 + c_2 e^{\lambda_2 t}v_2 + \cdots + c_n e^{\lambda_n t}v_n \tag{5.1.8}$$

is a solution of (5.1.4). Since $v_1, \ldots, v_n$ are linearly independent, (5.1.8) turns out to be the general solution of (5.1.4); that is, every solution of (5.1.4) has the form (5.1.8). See Exercise 5.1.26.

An $n \times n$ matrix that possesses a set of n linearly independent eigenvectors is said to be a *semisimple* or *nondefective* matrix. In the next section we will see that in some sense "most" matrices are semisimple. Thus for "most" systems of the form (5.1.4), the general solution is (5.1.8).

It is easy to show (Section 5.2) that λ is an eigenvalue of A if and only if λ is a solution of the *characteristic equation* $\det(\lambda I - A) = 0$. For each eigenvalue λ a corresponding eigenvector can be found by solving the equation $(\lambda I - A)v = 0$. For small enough (e.g. 2-by-2) systems of differential equations it is possible to solve the characteristic equation exactly and thereby solve the system.

Example 5.1.9 Find the general solution of

$$\dot{x}_1 = 2x_1 + 3x_2$$

$$\dot{x}_2 = x_1 + 4x_2.$$

The coefficient matrix is $A = \begin{bmatrix} 2 & 3 \\ 1 & 4 \end{bmatrix}$, whose characteristic equation is

$$\begin{aligned} 0 &= \det(\lambda I - A) = \det \begin{bmatrix} \lambda - 2 & -3 \\ -1 & \lambda - 4 \end{bmatrix} \\ &= (\lambda - 2)(\lambda - 4) - 3 = \lambda^2 - 6\lambda + 5 \\ &= (\lambda - 1)(\lambda - 5). \end{aligned}$$

Therefore the eigenvalues are $\lambda_1 = 1$ and $\lambda_2 = 5$. Solving the equation $(\lambda I - A)v = 0$ twice, once with $\lambda = \lambda_1$ and once with $\lambda = \lambda_2$, we find (nonunique) solutions

$$v_1 = \begin{bmatrix} 3 \\ -1 \end{bmatrix} \quad \text{and} \quad v_2 = \begin{bmatrix} 1 \\ 1 \end{bmatrix}.$$

Since these two ($= n$) vectors are obviously linearly independent, we conclude that the general solution of the system is

$$x(t) = c_1 e^t \begin{bmatrix} 3 \\ -1 \end{bmatrix} + c_2 e^{5t} \begin{bmatrix} 1 \\ 1 \end{bmatrix}. \tag{5.1.10}$$

$\square$

·cise 5.1.11 Check the details of Example 5.1.9. $\square$

The computations in Example 5.1.9 were quite easy because the matrix was only 2×2, and the eigenvalues turned out to be integers. In a more realistic problem the system would bigger, and the eigenvalues would be irrational numbers that we would have to approximate. Imagine trying to solve a system of the form $\dot{x} = Ax$ with, say, $n = 200$. We would need to find the 200 eigenvalues of a 200×200 matrix. Just trying to determine the characteristic equation (a polynomial equation of degree 200) is a daunting task, then we have to solve it. What is more, the problem of finding the zeros of a polynomial, given the coefficients, turns out to be ill conditioned whenever there are clustered roots, even when the underlying eigenvalue problem is well conditioned (Exercises 5.1.24 and 5.1.25). Clearly we need a better method. A major task of this chapter is to develop one.

Nonhomogeneous Systems

We now consider the problem of solving the nonhomogeneous system

$$\frac{dx}{dt} = Ax - b, \tag{5.1.12}$$

where b is a nonzero constant vector. Since b is invariant in time, it is reasonable to try to find a solution that is invariant in time: $x(t) = z$, where z is a constant vector. Substituting this form into (5.1.12), we have $dx/dt = 0$, so $0 = Az - b$, i.e. $Az = b$. Assuming A is nonsingular, we can solve the equation $Az = b$ to obtain a unique time-invariant solution z.

Once we have z in hand, we can make the following simple observation. If $y(t)$ is any solution of the homogeneous problem $\dot{x} = Ax$, then the sum $x(t) = y(t) + z$ is a solution of the nonhomogeneous system. Indeed $\dot{x} = \dot{y} + 0 = Ay = A(x - z) = Ax - Az = Ax - b$, so $\dot{x} = Ax - b$. Moreover, if $y(t)$ is the general solution of the homogeneous problem, then $x(t) = y(t) + z$ is the general solution of the nonhomogeneous problem.

Example 5.1.13 Let us solve the nonhomogeneous differential equation

$$\begin{bmatrix} \dot{x}_1(t) \\ \dot{x}_2(t) \end{bmatrix} = \begin{bmatrix} -9 & 5 \\ 5 & -8 \end{bmatrix} \begin{bmatrix} x_1 \\ x_2 \end{bmatrix} - \begin{bmatrix} 0 \\ -6 \end{bmatrix}.$$

This is just (5.1.2) from Example 5.1.1. The two components of the solution, subject to the initial conditions (5.1.3), represent the two loop currents in the electrical circuit in Figure 5.1.

We must find a time-invariant solution and the general solution of the homogeneous problem, then add them together. First of all, the time-invariant solution must satisfy

$$\begin{bmatrix} -9 & 5 \\ 5 & -8 \end{bmatrix} \begin{bmatrix} z_1 \\ z_2 \end{bmatrix} = \begin{bmatrix} 0 \\ -6 \end{bmatrix}.$$

This is exactly the system we had to solve in Example 1.2.8. The solution is $z_1 = 30/47 = 0.6383$ and $z_2 = 54/47 = 1.1489$. Recall that these numbers represent the loop currents in the circuit if inductances are ignored.

Now let us solve the homogeneous system

$$\begin{bmatrix} \dot{x}_1(t) \\ \dot{x}_2(t) \end{bmatrix} = \begin{bmatrix} -9 & 5 \\ 5 & -8 \end{bmatrix} \begin{bmatrix} x_1 \\ x_2 \end{bmatrix}.$$

We can proceed as in Example 5.1.9 and form the characteristic equation $0 = \det(\lambda I - A) = \lambda^2 + 17\lambda + 47$, but this time the solutions are not integers. Applying the quadratic formula, we find that the solutions are $\lambda = (-17 \pm \sqrt{101})/2$. The rounded eigenvalues are $\lambda_1 = -13.5249$ and $\lambda_2 = -3.4751$. We can now substitute these values back into the equation $(\lambda I - A)v = 0$ to obtain eigenvectors. This is tedious by hand because the numbers are long decimals.

An easier approach is to let MATLAB do the work. Using MATLAB's `eig` command, we obtain the (rounded) eigenvectors

$$v_1 = \begin{bmatrix} 0.7415 \\ -0.6710 \end{bmatrix} \quad \text{and} \quad v_2 = \begin{bmatrix} 0.6710 \\ 0.7415 \end{bmatrix}.$$

Thus the general solution to the homogeneous problems is

$$c_1 e^{-13.5249t} \begin{bmatrix} 0.7415 \\ -0.6710 \end{bmatrix} + c_2 e^{-3.4751t} \begin{bmatrix} 0.6710 \\ 0.7415 \end{bmatrix}.$$

Adding this to the time-invariant solution of the nonhomogeneous problem, we obtain the general solution of the nonhomogeneous problem:

$$x(t) = \begin{bmatrix} 0.6383 \\ 1.1489 \end{bmatrix} + c_1 e^{-13.5249t} \begin{bmatrix} 0.7415 \\ -0.6710 \end{bmatrix} + c_2 e^{-3.4751t} \begin{bmatrix} 0.6710 \\ 0.7415 \end{bmatrix}.$$

$$(5.1.14)$$

To obtain the particular solution that gives the loop currents in the circuit, assuming the switch is closed at time zero, we must apply the initial condition (5.1.3), $x(0) = 0$. Making this substitution in (5.1.14), we obtain

$$0 = \begin{bmatrix} 0.6383 \\ 1.1489 \end{bmatrix} + c_1 \begin{bmatrix} 0.7415 \\ -0.6710 \end{bmatrix} + c_2 \begin{bmatrix} 0.6710 \\ 0.7415 \end{bmatrix},$$

which can be seen to be a system of two linear equations that we can solve for c_1 and c_2. Doing so (using MATLAB) we obtain

$$c_1 = 0.2977 \quad \text{and} \quad c_2 = -1.2802.$$

Using this solution we can determine what the loop currents will be at any time.

We bring this example to a close by considering the long-term behavior of the circuit. Examining (5.1.14), we see that the exponential functions $e^{-13.5249t}$ and $e^{-3.4751t}$ both tend to zero as $t \to \infty$, since the exponents are negative. This is a consequence of both eigenvalues $\lambda_1 = -13.5249$ and $\lambda_2 = -3.4751$ being negative. In fact the convergence to zero is quite rapid in this case. After a short time, the exponential functions become negligible and the solution is essentially constant: $x(t) \approx \begin{bmatrix} 0.6383 \\ 1.1489 \end{bmatrix}$. In other words, after a brief transient phase, the circuit settles down to its steady state. Figure 5.2 shows a plot of the loop currents as a function of time during the first second after the switch has been closed. Notice that after just one second the currents are already quite close to their steady-state values.   □

A larger circuit is studied in Exercise 5.1.21. More interesting circuits can be built by inserting some capacitors and thereby obtaining solutions that oscillate in time. Of course, mass-spring systems also exhibit oscillatory behavior.

Dynamics of Simple Mass-Spring Systems

Figure 5.3 depicts a cart attached to a wall by a spring. At time zero the cart is at rest at its equilibrium position $x = 0$. At that moment a steady force of 2 newtons

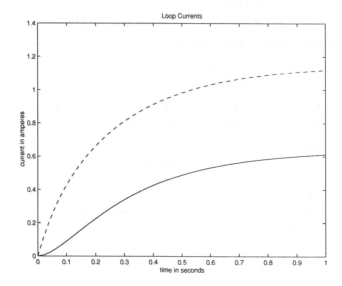

Figure 5.2 Loop currents as a function of time

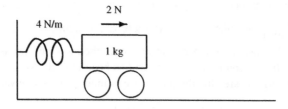

Figure 5.3 Single cart and spring

is applied, pushing the cart to the right and stretching the spring. In Section 1.2 we studied this same system (Figure 1.3) and found that the new equilibrium position is $x = 0.5$ meter. That is, the force moves the cart one half meter to the right. Now we will ask the more specific question of how the cart gets from its old equilibrium position to the new one. The solution of the static problem does not depend on the mass of the cart, but the dynamic problem does. The rolling friction also matters. Let us say that the mass of the cart is 1 kilogram. We'll make the linear assumption that the rolling friction is proportional to the velocity: $-k\dot{x}(t)$ newtons, where k is a positive proportionality constant with units kg/s.

The physical principle that we will apply is Newton's second law, $f = ma$, in the form $a = f/m$: the acceleration of the cart is equal to the sum of the forces on the cart divided by the cart's mass. The choice $m = 1$ simplifies things. The three forces on the cart are the external force $+2$ N, the restoring force of the spring $-4x(t)$ N, and the rolling friction $-k\dot{x}(t)$ N. We recall also that the acceleration is $\ddot{x}$, the second derivative of displacement with respect to time. Thus Newton's second law gives the differential equation

$$\ddot{x} = 2 - 4x - k\dot{x}, \tag{5.1.15}$$

which we can solve subject to the initial conditions $x(0) = \dot{x}(0) = 0$ to find x. It turns out that the nature of the motion depends heavily on the value of k.

The differential equation (5.1.15) is simple enough that we can solve it as it stands using standard techniques from a first course in differential equations, but we will take a different approach. The differential equation is of second order; that is, it involves a second derivative. We will convert it to a system of two first-order differential equations by a standard technique. Let $x_1(t) = x(t)$ and $x_2(t) = \dot{x}(t)$. Then we can rewrite the differential equation as $\dot{x}_2 = 2 - 4x_1 - kx_2$. Combining this with the trivial differential equation $\dot{x}_1 = x_2$, we obtain the following system and initial condition:

$$\begin{bmatrix} \dot{x}_1 \\ \dot{x}_2 \end{bmatrix} = \begin{bmatrix} 0 & 1 \\ -4 & -k \end{bmatrix} \begin{bmatrix} x_1 \\ x_2 \end{bmatrix} - \begin{bmatrix} 0 \\ -2 \end{bmatrix}, \qquad \begin{bmatrix} x_1(0) \\ x_2(0) \end{bmatrix} = \begin{bmatrix} 0 \\ 0 \end{bmatrix}.$$

This system has the general form $\dot{x} = Ax - b$, just as our system in the electrical circuit example did, and it can be solved by the same means. A system of this type is solved in Exercise 5.1.22.

A system of three masses attached by springs is considered in Exercise 5.1.23. In that problem Newton's second law is applied to each of the carts to obtain a system of three second-order differential equations, which is then rewritten as a system of six first-order differential equations.

Stability of Linear and Nonlinear Systems

In stability theory the behavior of systems as $t \to \infty$ is studied. In Example 5.1.13 we noted that as $t \to \infty$ the loop currents tend to their equilibrium or steady-state values. This is so because the exponents in the general solution (5.1.14) are negative, which is in turn a consequence of the eigenvalues being negative. It holds regardless of the values of the constants c_1 and c_2, which means that it holds for all solutions. Thus if the loop currents are somehow perturbed away from their equilibrium values, they will return to those values. This is *stable* behavior.

The system in Example 5.1.9, on the other hand, behaves quite differently. There the eigenvalues are positive, so the exponents in the general solution (5.1.10) are positive and become large as $t \to \infty$. Consequently all solutions become large as time increases. The only exception is the steady-state solution 0, obtained by setting $c_1 = c_2 = 0$. If the system is initially in this equilibrium state but then is somehow perturbed away from it, it will not return to that state. It will blow up, because all nonzero solutions blow up. This is *unstable* behavior.

In general we can assess the stability of homogeneous linear systems (5.1.4) by examining the general solution (5.1.8). (In the nonhomogeneous case we just add a constant.) If all of the eigenvalues λ_k are real and negative, then all solutions will satisfy $x(t) \to 0$ as $t \to \infty$. On the other hand, if at least one of the eigenvalues is positive, then there will be solutions that satisfy $\| x(t) \| \to \infty$ as $t \to \infty$. In the former case the system is said to be *asymptotically stable*; in the latter case it is *unstable*.

As we shall see, eigenvalues need not always be real numbers. Even if A is real, some or all of its eigenvalues can be complex. What happens then? Say $\lambda_k = \alpha_k + i\beta_k$. Then $e^{\lambda_k t} = e^{\alpha_k t}e^{i\beta_k t} = e^{\alpha_k t}(\cos \beta_k t + i \sin \beta_k t)$, which grows if $\alpha_k > 0$ and decays if $\alpha_k < 0$. Since $\alpha_k = \text{Re}(\lambda_k)$ (the *real part* of λ_k), we conclude that (5.1.4) is asymptotically stable if all eigenvalues of A satisfy $\text{Re}(\lambda_k) < 0$, and it is unstable if one or more of the eigenvalues satisfies $\text{Re}(\lambda_k) > 0$. We have come to this conclusion under the assumption that A is a semisimple matrix. However, the conclusion turns out to be valid for all A.

Notice that complex eigenvalues signal oscillations in the solution, since they give rise to sine and cosine terms. The frequency of oscillation of the kth term is governed by $\beta_k = \text{Im}(\lambda_k)$. See Exercises 5.1.22 and 5.1.23 for examples of systems that have complex eigenvalues and therefore exhibit oscillatory behavior.

Now consider a nonlinear system of n first-order differential equations

$$
\begin{aligned}
\dot{x}_1 &= f_1(x_1, x_2, \ldots, x_n) \\
\dot{x}_2 &= f_2(x_1, x_2, \ldots, x_n) \\
&\vdots \qquad\qquad \vdots \\
\dot{x}_n &= f_n(x_1, x_2, \ldots, x_n)
\end{aligned}
$$

or, in vector notation,

$$\dot{x} = f(x). \tag{5.1.16}$$

A vector $\hat{x}$ is called an *equilibrium point* of (5.1.16) if $f(\hat{x}) = 0$, since then the constant function $x(t) = \hat{x}$ is a solution. An equilibrium point $\hat{x}$ is called *asymptotically stable* if there is a neighborhood N of $\hat{x}$, such that for every solution of (5.1.16) for which $x(t) \in N$ for some t, it must happen that $x(t) \to \hat{x}$ as $t \to \infty$. Physically this means that a system whose state is perturbed slightly from an equilibrium point will return to that equilibrium point. We can analyze the stability of each equilibrium point by linearizing (5.1.16) about that point.

Specifically, consider a multivariate Taylor expansion of f centered on $\hat{x}$:

$$f_i(x_1, \ldots, x_n) = f_i(\hat{x}) + \sum_{j=1}^{n} \frac{\partial f_i}{\partial x_j}(\hat{x})(x_j - \hat{x}_j)$$

$$+ \frac{1}{2} \sum_{j=1}^{n} \sum_{k=1}^{n} \frac{\partial^2 f_i}{\partial x_j \, \partial x_k}(\hat{x})(x_j - \hat{x}_j)(x_k - \hat{x}_k) + \cdots$$

for $i = 1, \ldots, n$. In vector notation we have the more compact expression

$$f(x) = f(\hat{x}) + \frac{\partial f}{\partial x}(\hat{x})(x - \hat{x}) + \text{ higher order terms,}$$

where $\partial f / \partial x$ is the *Jacobian* matrix, whose (i, j) entry is $\partial f_i / \partial x_j$. For x near $\hat{x}$, the quadratic and higher terms are very small. Also $f(\hat{x}) = 0$, so $f(x) \approx \frac{\partial f}{\partial x}(\hat{x})(x - \hat{x})$. It is thus reasonable to expect that for x near $\hat{x}$ the solutions of (5.1.16) are approximated well by solutions of

$$\frac{d}{dt}(x - \hat{x}) = \dot{x} = \frac{\partial f}{\partial x}(\hat{x})(x - \hat{x}). \tag{5.1.17}$$

This is a linear system with coefficient matrix $\frac{\partial f}{\partial x}(\hat{x})$. From our discussion of stability of linear systems we know that all solutions of (5.1.17) satisfy $x - \hat{x} \to 0$ as $t \to \infty$ if all eigenvalues of $\frac{\partial f}{\partial x}(\hat{x})$ satisfy $\text{Re}(\lambda_k) < 0$. We conclude that $\hat{x}$ is an asymptotically stable equilibrium point of (5.1.16) if all eigenvalues of the Jacobian $\frac{\partial f}{\partial x}(\hat{x})$ have negative real parts.

cise 5.1.18 The differential equation of a damped pendulum is

$$\ddot{\theta} + k_1 \dot{\theta} + k_2 \sin \theta = 0,$$

where k_1 and k_2 are positive constants, and θ is the angle of the pendulum from its vertical resting position. This differential equation is nonlinear because $\sin \theta$ is a nonlinear function of θ. Introducing new variables $x_1 = \theta$ and $x_2 = \dot{\theta}$, rewrite the second-order differential equation as a system of two first-order differential equations in x_1 and x_2. Find all equilibrium points of the system, and note that they correspond to $\dot{\theta} = 0$, $\theta = n\pi$, $n = 0, \pm 1, \pm 2, \ldots$. Calculate the Jacobian of the system and show that the equilibrium points $\theta = n\pi$ are asymptotically stable for even n and unstable for odd n. Interpret your result physically. $\square$

Additional Exercises

cise 5.1.19 Suppose A is nonsingular and has linearly independent eigenvectors $v_1, \ldots,$ v_n. Let V be the nonsingular $n \times n$ matrix whose columns are $v_1, \ldots, v_n$.

(a) Show that (5.1.8) can be rewritten as $x(t) = V e^{\Lambda t} c$, where c is a column vector, Λt is the diagonal matrix $\text{diag}\{\lambda_1 t, \ldots, \lambda_n t\}$, and $e^{\Lambda t}$ is its *matrix exponential*: $e^{\Lambda t} = \text{diag}\{e^{\lambda_1 t}, \ldots, e^{\lambda_n t}\}$.

(b) Show that the general solution of $\dot{x} = Ax - b$ has the form $x(t) = z + V e^{\Lambda t} c$, where z satisfies $Az = b$.

(c) To solve the initial value problem $\dot{x} = Ax - b$ with initial condition $x(0) = \hat{x}$, we need to solve for the constants in the vector c. Show that c can be obtained by solving the system $Vc = \hat{x} - z$. Since V is nonsingular, this system has a unique solution.

$\square$

Exercise 5.1.20 Using MATLAB, work out the details of Example 5.1.13. The MATLAB command [V,D]=eig(A) returns (if possible) a matrix V whose columns are linearly independent eigenvectors of A and a diagonal matrix D whose main diagonal entries are the eigenvalues of A. Thus V and D are the same as the matrices V and Λ of the previous exercise. You may find the results of the previous exercise helpful as you work through this exercise. Here are some sample plot commands:

```
t = 0:.02:1;
x = z*ones(size(t));
for j=1:2; x = x + V(:,j)*c(j)*exp(t*D(j,j)); end
plot(t,x(1,:),t,x(2,:),'--')
title('Loop Currents')
xlabel('time in seconds')
ylabel('current in amperes')
print
```

Remember that for more information on the usage of these commands, you can type help plot, help print, etc., or look in MATLAB's help browser. $\square$

Exercise 5.1.21 In working through this exercise, do not overlook the advice given in the previous two exercises. The circuit in Figure 5.4 is the same as the one that appeared in Exercise 1.2.19, except that some inductors have been added around the edges. All of the resistances are $1\,\Omega$, and all of the inductances are 1 H. Notice the unusual numbering of the loops. Initially the loop currents are all zero, because the switch is open. Suppose the switch is closed at time 0.

(a) Write down a system of nine equations for the nine unknown loop currents. All except the ninth equation involve derivatives.

(b) Solve the ninth equation for x_9 in terms of other loop currents, and use this expression to eliminate x_9 from the other equations. Obtain thereby a system of eight differential equations $\dot{x} = Ax - b$ with initial condition $x(0) = 0$.

(c) Solve the system $Az = b$ to obtain a steady-state solution of the differential equation. Compute the steady-state value of the loop current z_9 also. (Checkpoint: $z_1 = 0.3$ amperes.)

(d) Get the eigenvalues and eigenvectors of A, and use them to construct the general solution of the homogeneous equation $\dot{z} = Az$.

(e) Deduce the solution of the initial value problem $\dot{x} = Ax - b$, $x(0) = 0$.

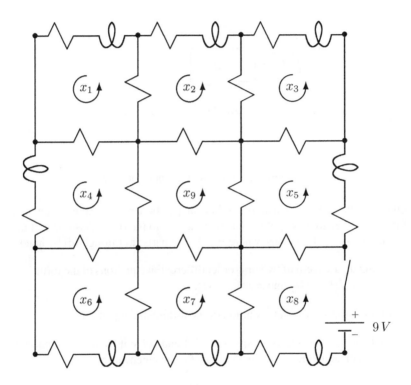

Figure 5.4   Solve for the time-varying loop currents.

(f) On separate graphs, plot the loop currents $x_1(t)$ and $x_8(t)$ for $0 \leq t \leq 6$ seconds. (Checkpoint: Do these look realistic?)

(g) Once the switch is closed, about how long does it take for all of the loop currents to be within (i) ten percent of their final values? (ii) one percent?

□

Exercise 5.1.22 Consider a cart attached to a wall by a spring, as shown in Figure 5.5. At

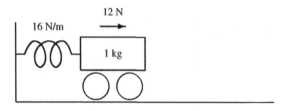

Figure 5.5   Solve for the motion of the cart.

time zero the cart is at rest at its equilibrium position $x = 0$. At that moment a steady force of 12 newtons is applied, pushing the cart to the right. Assume that the rolling friction is $-k\,\dot{x}(t)$ newtons, where $k \geq 0$. Do parts (a) through (d) by hand.

(a) Set up a system of two first-order differential equations of the form $\dot{x} = Ax - b$ for $x_1(t) = x(t)$ and $x_2(t) = \dot{x}(t)$.

(b) Find the steady-state solution of the differential equation.

(c) Find the characteristic equation of A and solve it by the quadratic formula to obtain an expression (involving k) for the eigenvalues of A.

(d) There is a *critical value* of k at which the eigenvalues of A change from real to complex. Find this critical value.

(e) Using MATLAB, solve the initial value problem for the cases (i) $k = 2$, (ii) $k = 6$, (iii) $k = 10$, and (iv) $k = 14$. Rather than reporting your solutions, simply plot $x_1(t)$ for $0 \leq t \leq 3$ for each of your four solutions on a single set of axes. (Do not overlook the help given in Exercises 5.1.19 and 5.1.20.) Comment on your plots (e.g. rate of decay to steady state, presence or absence of oscillations).

(f) What happens when $k = 0$?

□

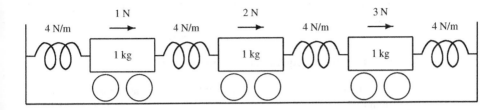

Figure 5.6   Solve for the motion of the carts.

rcise 5.1.23 Consider a system of three carts attached by springs, as shown in Figure 5.6. The carts are initially at rest. At time zero the indicated forces are applied, causing the carts to move toward a new equilibrium. Let x_1, x_2, and x_3 denote the displacements of the carts, and let x_4, x_5, and x_6 denote their respective velocities. Suppose the coefficients of rolling friction of the three carts are k_1, k_2, and k_3, respectively.

(a) Apply Newton's second law to each cart to obtain a system of three second-order differential equations for the displacements x_1, x_2, and x_3. You may find it useful to review Example 1.2.10.

(b) Introducing the velocity variables x_4, x_5, and x_6, rewrite your system as a system of six first-order differential equations. Write your system in the form $\dot{x} = Ax - b$.

(c) Find the steady-state solution of the system.

(d) Solve the initial value problem under each of the conditions listed below. In each case plot x_1, x_2, and x_3 on a single set of axes for $0 \le t \le 20$, and comment on the plot.

   (1) $k_1 = 1$, $k_2 = 0$, and $k_3 = 0$.
   (2) $k_1 = 1$, $k_2 = 8$, and $k_3 = 8$.
   (3) $k_1 = 8$, $k_2 = 8$, and $k_3 = 8$.

$\square$

rcise 5.1.24 Let $A = \begin{bmatrix} 1 & 0 \\ 0 & 1 \end{bmatrix}$, the 2×2 identity matrix.

(a) Show that the characteristic equation of A is

$$\lambda^2 - 2\lambda + 1 = 0.$$

The eigenvalues are obviously $\lambda_1 = 1$ and $\lambda_2 = 1$.

(b) We now perturb one coefficient of the characteristic polynomial slightly and consider the equation

$$\lambda^2 - 2\lambda + (1 - \epsilon) = 0,$$

where $\epsilon \ll 1$. Solve the equation for the roots $\hat{\lambda}_1$ and $\hat{\lambda}_2$.

(c) Show that when $\epsilon = 10^{-12}$, $|\hat{\lambda}_1 - \lambda_1|$ and $|\hat{\lambda}_2 - \lambda_2|$ are one million times bigger than ϵ.

(d) Sketch the graphs of the original and perturbed polynomials (using some ϵ bigger than 10^{-12}, obviously), and give some indication of why the roots are so sensitive to the ϵ perturbation.

This shows that the problem of finding the roots of a polynomial equation can be ill conditioned. Difficulties arise any time there are multiple roots or tightly clustered roots. The more roots there are in the cluster, the worse the situation is. In contrast, the *eigenvalue problem* for the identity matrix is well conditioned. Small perturbations in the entries of A cause small changes in the eigenvalues. This is true not only for the identity matrix but for any symmetric matrix, regardless of whether or not the eigenvalues are clustered, as we shall show in Section 7.1 (Corollary 7.1.8).   ☐

Exercise 5.1.25 Figure out what the following MATLAB commands do. For example, type `help poly` in **MATLAB** to find out what `poly` does.

```
A = eye(2)
lam = eig(A)
c = poly(A)
roots(c)
format long
A2 = A + 200*eps*randn(2)
lam2 = eig(A2)
eigdif = norm(lam2-lam)
c2 = c + [0 200*eps*randn(1,2)]
roots2 = roots(c2)
rootdif = norm(roots2-lam)
```

Run this code several times. The results will vary because random numbers are being used. Explain what the output shows.   ☐

Exercise 5.1.26 In this exercise we show that (5.1.8) is the general solution of (5.1.4).

(a) Show that for any $\hat{x} \in \mathbb{C}^n$ there exist (unique) $c_1, \ldots, c_n$ such that the function $x(t)$ of (5.1.8) satisfies $x(0) = \hat{x}$.

(b) It is a basic fact of the theory of differential equations that for a given $\hat{x} \in \mathbb{C}^n$, the *initial value problem* $dx/dt = Ax$, $x(0) = \hat{x}$, has exactly one solution (see any text on differential equations). Use this fact together with the result of part (a) to show that every solution of (5.1.4) has the form (5.1.8).

☐

5.2 BASIC FACTS ABOUT EIGENVALUES AND EIGENVECTORS

In most scientific and engineering applications that require eigenvalue computations, the matrix has real entries. Nevertheless, as we shall soon see, the natural setting for the study of eigenvalues is the field of complex numbers. Thus let $A \in \mathbb{C}^{n \times n}$. A vector $v \in \mathbb{C}^n$ is called an *eigenvector* of A if $v \neq 0$ and Av is a multiple of v; that is, there exists a $\lambda \in \mathbb{C}$ such that

$$Av = \lambda v. \tag{5.2.1}$$

The scalar λ is called the *eigenvalue* of A *associated with* the eigenvector v. Similarly v is called *an* eigenvector of A *associated with* the eigenvalue λ. The pair (λ, v) is called an *eigenpair* of A. Whereas each eigenvector has a unique eigenvalue associated with it, each eigenvalue is associated with many eigenvectors. For example, if v is an eigenvector of A associated with the eigenvalue λ, then every nonzero multiple of v is also an eigenvector of A associated with λ.

The set of all eigenvalues of A is called the *spectrum* of A.

Frequently we will wish to work with eigenvectors with unit norm. This is always possible, for given any eigenvector v of A, the vector $\hat{v} = (1/\|v\|)v$ is also an eigenvector of A associated with the same eigenvalue, and $\|\hat{v}\| = 1$. This normalization process can be carried out using any vector norm.

rcise 5.2.2 Show that A is nonsingular if and only if 0 is not an eigenvalue of A. (Thus we could add another line to the list in Theorem 1.2.3.)   □

In the previous section we mentioned that λ is an eigenvalue of A if and only if it is a root of the characteristic equation. Let us now see why this is true. The equation (5.2.1) can be rewritten as $Av = \lambda Iv$ or $(\lambda I - A)v = 0$, where I is the $n \times n$ identity matrix. Thus if v is an eigenvector with eigenvalue λ, then v is a nonzero solution of the homogeneous matrix equation $(\lambda I - A)v = 0$. Therefore, by part (b) of Theorem 1.2.3, the matrix $(\lambda I - A)$ is singular. Hence, by part (e) of the same theorem, $\det(\lambda I - A) = 0$. Since each step of this argument can be reversed, we have established the following theorem.

Theorem 5.2.3 λ *is an eigenvalue of A if and only if*

$$\det(\lambda I - A) = 0. \tag{5.2.4}$$

As we mentioned previously, (5.2.4) is called the *characteristic equation* of A. Because of the form of the equation, determinants will play a prominent role in this section. Although it is a useful theoretical device, the determinant is of little value for actual computations. Consequently determinants will appear only rarely in subsequent sections of the book.

It is not hard to see that $\det(\lambda I - A)$ is a polynomial in λ of degree n. It is called the *characteristic polynomial* of A.

Exercise 5.2.5

(a) Show that the characteristic polynomial of

$$B = \begin{bmatrix} 1 & 2 \\ 3 & 4 \end{bmatrix}$$

is $\lambda^2 - 5\lambda - 2$, a polynomial of degree 2.

(b) Show that the characteristic polynomial of

$$C = \begin{bmatrix} 1 & 1 & 1 \\ 2 & 3 & 4 \\ 0 & 1 & 1 \end{bmatrix}$$

is $\lambda^3 - 5\lambda^2 + \lambda + 1$, a polynomial of degree 3.

$\square$

Exercise 5.2.6 Let

$$B = \begin{bmatrix} 1 & 3 \\ 0 & 2 \end{bmatrix} \quad \text{and} \quad C = \begin{bmatrix} 1 & 3 \\ 0 & 1 \end{bmatrix}.$$

(a) Show that the characteristic polynomial of B is $(\lambda - 1)(\lambda - 2)$ and the eigenvalues are 1 and 2. Thus B has two distinct eigenvalues.

(b) Show that the characteristic polynomial of C is $(\lambda - 1)(\lambda - 1)$ and the eigenvalues are 1 and 1. Thus C does not have two distinct eigenvalues. It does have two eigenvalues if you count the eigenvalue 1 twice. The eigenvalue 1 is said to have *algebraic multiplicity* two, since it appears twice in the factorization of the characteristic polynomial of C.

$\square$

Every nth degree polynomial equation has n complex roots, so A has n eigenvalues, some of which may be repeated. If A is a real matrix, the characteristic polynomial has real coefficients. But the zeros of a real polynomial are not necessarily real, so a real matrix can have complex, non-real, eigenvalues. This is why it makes sense to work with complex numbers from the beginning. Notice, however, that if a real matrix has complex eigenvalues, they must occur in complex conjugate pairs; that is, if $\lambda = \alpha + \beta i$ ($\alpha, \beta \in \mathbb{R}$) is an eigenvalue of $A \in \mathbb{R}^{n \times n}$, then so is the complex conjugate $\overline{\lambda} = \alpha - \beta i$.

Exercise 5.2.7 Prove the above assertion in two different ways:

(a) First verify the following two basic facts about complex conjugates. If $z_1 = \alpha_1 + \beta_1 i \in \mathbb{C}$ and $z_2 = \alpha_2 + \beta_2 i \in \mathbb{C}$, then $\overline{z_1 + z_2} = \overline{z}_1 + \overline{z}_2$ and $\overline{z_1 z_2} = \overline{z}_1 \overline{z}_2$.

(b) Let $A \in \mathbb{R}^{n \times n}$. Let λ be a complex eigenvalue of A and let v be an associated eigenvector. Let $\overline{v}$ denote the vector obtained by taking complex conjugates of the entries of v. Show that $\bar{v}$ is an eigenvector of A with associated eigenvalue $\bar{\lambda}$.

(c) Let $p(\lambda) = a_0 + a_1\lambda + a_2\lambda^2 + \cdots + a_n\lambda^n$ be a polynomial with real coefficients $a_0, \ldots, a_n$. Show that for any $z = \alpha + \beta i \in \mathbb{C}$, $p(\bar{z}) = \overline{p(z)}$. Conclude that $p(z) = 0$ if and only if $p(\bar{z}) = 0$.

□

Now is a good time to mention one other important point about real matrices. Let $A \in \mathbb{R}^{n \times n}$ and suppose it has a real eigenvalue λ. Then in the homogeneous equation $(\lambda I - A)v = 0$, the coefficient matrix $\lambda I - A$ is real. The fact that $\lambda I - A$ is singular implies that the equation $(\lambda I - A)v = 0$ has nontrivial *real* solutions. We conclude that every real eigenvalue of a real matrix has a real eigenvector associated with it.

rcise 5.2.8 Let $A \in \mathbb{C}^{n \times n}$.

(a) Use the characteristic equation to show that A and A^T have the same eigenvalues.

(b) Show that λ is an eigenvalue of A if and only if $\bar{\lambda}$ is an eigenvalue of A^*. (A^* denotes the complex conjugate transpose of A.)

□

There are some types of matrices for which the eigenvalues are obvious. The following theorem will prove very useful later on.

Theorem 5.2.9 *Let $T \in \mathbb{C}^{n \times n}$ be a (lower- or upper-) triangular matrix. Then the eigenvalues of T are the main-diagonal entries $t_{11}, \ldots, t_{nn}$.*

Proof. The determinant of a triangular matrix R is $r_{11}r_{22}\cdots r_{nn}$, the product of the main-diagonal entries. Since T is triangular, so is $\lambda I - T$. Therefore the characteristic equation of T is $0 = \det(\lambda I - T) = (\lambda - t_{11})(\lambda - t_{22})\cdots(\lambda - t_{nn})$, whose roots are obviously $t_{11}, \ldots, t_{nn}$.

For a second proof see Exercise 5.2.19.

□

The following generalization of Theorem 5.2.9 is also quite useful.

Theorem 5.2.10 *Let A be a block triangular matrix, say*

$$A = \begin{bmatrix} A_{11} & A_{12} & \cdots & A_{1m} \\ & A_{22} & \cdots & A_{2m} \\ & & \ddots & \vdots \\ & & & A_{mm} \end{bmatrix}.$$

Then the set of eigenvalues of A, counting algebraic multiplicity, equals the union of the sets of eigenvalues of $A_{11}, A_{22}, \ldots, A_{mm}$.

Proof. Because $\lambda I - A$ is block triangular, $\det(\lambda I - A) = \det(\lambda I - A_{11}) \det(\lambda I - A_{22}) \cdots \det(\lambda I - A_{mm})$. Thus the set of roots of the characteristic polynomial of A equals the union of the roots of the characteristic polynomials for $A_{11}, A_{22}, \ldots, A_{mm}$.   □

Although A has been depicted as block upper triangular, the result is obviously valid for block lower-triangular matrices as well.

A second proof, which does not, however, yield the multiplicity information, is outlined in Exercise 5.2.20.

Theorem 5.2.11 *Let $v_1, \ldots, v_k$ be eigenvectors of A associated with distinct eigenvalues $\lambda_1, \ldots, \lambda_k$. Then $v_1, \ldots, v_k$ are linearly independent.*

Proof. See any text on linear algebra, or work through Exercise 5.2.21.   □

Recall from the previous section that a matrix $A \in \mathbb{C}^{n \times n}$ is *semisimple* or *nondefective* if there exists a set of n linearly independent eigenvectors of A. A matrix that is not semisimple is called *defective*.

Corollary 5.2.12 *If $A \in \mathbb{C}^{n \times n}$ has n distinct eigenvalues, then A is semisimple.*

If we pick a matrix "at random," its characteristic equation is almost certain to have distinct roots. After all, the occurrence of repeated roots is an exceptional event. Thus a matrix chosen "at random" is almost certain to be semisimple. This is the basis of the claim made in the previous section that "most" matrices are semisimple. In addition it is not hard to show that every defective matrix has semisimple matrices arbitrarily close to it. Given a defective matrix (which must necessarily have some repeated eigenvalues), a slight perturbation suffices to split the repeated eigenvalues and yield a semisimple matrix (Exercise 5.4.33). In the language of topology, the set of semisimple matrices is *dense* in the set of all matrices.

A defective matrix must have at least one repeated eigenvalue. However, the presence of repeated eigenvalues does not guarantee that the matrix is defective.

Exercise 5.2.13   Find all eigenvalues and eigenvectors of the identity matrix $I \in \mathbb{C}^{n \times n}$. Conclude that (if $n > 1$) the identity matrix has repeated eigenvalues but is not defective.   □

So far we do not even know that defective matrices exist. The next exercise takes care of that.

Exercise 5.2.14

(a) The matrix $\begin{bmatrix} 0 & 1 \\ 0 & 0 \end{bmatrix}$ has only the eigenvalue zero. (Why?) Show that it does not have two linearly independent eigenvectors.

(b) Find a 2×2 matrix that has only the eigenvalue zero but does have two linearly independent eigenvectors. (There is only one such matrix.)

(c) Show that the matrix

$$\begin{bmatrix} c & 1 & & & \\ & c & 1 & & \\ & & \ddots & \ddots & \\ & & & c & 1 \\ & & & & c \end{bmatrix}$$

is defective. A matrix of this form is called a *Jordan block*.

(d) Show that the matrix

$$\begin{bmatrix} 5 & 0 & 0 \\ 0 & 7 & x \\ 0 & 0 & 7 \end{bmatrix}$$

is defective if $x \neq 0$ and semisimple if $x = 0$.

□

Equivalence of Eigenvalue Problems and Polynomial Equations

We have seen that finding the eigenvalues of A is equivalent to finding the roots of the polynomial equation $\det(\lambda I - A) = 0$. Therefore, if we have a way of finding the roots of an arbitrary polynomial equation, then in principle we can find the eigenvalues of an arbitrary matrix. The next theorem shows that the converse is true as well: if we have a way of finding the eigenvalues of an arbitrary matrix, then in principle we can find the zeros of an arbitrary polynomial.

An arbitrary polynomial of degree n has the form $q(\lambda) = b_0 + b_1\lambda + b_2\lambda^2 + \cdots + b_n\lambda^n$, with $b_n \neq 0$. If we divide all coefficients of $q(\lambda)$ by b_n, we obtain a new polynomial $p(\lambda) = a_0 + a_1\lambda + a_2\lambda^2 + \cdots + 1\lambda^n$ ($a_k = b_k/b_n$) that has same zeros and is *monic*; that is, its leading coefficient is $a_n = 1$. Since we can always replace the polynomial equation $q(\lambda) = 0$ by the equivalent equation $p(\lambda) = 0$, it suffices to consider monic polynomials. Given a monic polynomial, we define the *companion matrix* of p to be the $n \times n$ matrix

$$A = \begin{bmatrix} -a_{n-1} & -a_{n-2} & -a_{n-3} & \cdots & -a_0 \\ 1 & & & & \\ & 1 & & & \\ & & \ddots & & \\ & & & 1 & \end{bmatrix}. \tag{5.2.15}$$

This matrix has ones on the subdiagonal, the opposites of the coefficients of p in the first row, and zeros elsewhere.

Theorem 5.2.16 *Let* $p(\lambda) = a_0 + a_1\lambda + a_2\lambda^2 + \cdots + 1\lambda^n$ *be a monic polynomial of degree* n, *and let* A *be its companion matrix (5.2.15). Then the characteristic* *polynomial of A is p. Thus the roots of the equation $p(\lambda) = 0$ are the eigenvalues of A.*[23]

Proof. See Exercise 5.2.23.   □

Necessity of Iterative Methods

Thanks to Theorem 5.2.16 we see that the eigenvalue problem and the problem of finding roots of a polynomial equation are equivalent. Although the algorithms that we will develop make no direct use of this equivalence, it does have an important theoretical implication. The polynomial root-finding problem is an old one that has attracted the interest of many great minds. In particular, early in the nineteenth century, Neils Henrik Abel was able to prove that there is no general formula[24] for the roots of a polynomial equation of degree n if $n > 4$. (See, e.g., [48].) It follows that there is no general formula for the eigenvalues of an $n \times n$ matrix if $n > 4$.

Numerical methods can be divided into two broad categories—direct and iterative. A *direct method* is one that produces the result in a prescribed, finite number of steps. All versions of Gaussian elimination for solving $Ax = b$ are direct methods, and so are all of the methods that we developed in Chapter 3 for solving the least squares problem. By contrast an *iterative method* is one that produces a sequence of approximants that (hopefully) converges to the true solution of the problem. Each step or *iteration* of the algorithm produces a new approximant. In principle the sequence of approximants is infinite, but in practice we cannot run the algorithm forever. We stop once we have produced an approximant that is sufficiently good that we are willing to accept it as the solution. The number of iterations that will be needed to reach this point is usually not known in advance, although it can often be estimated.

Abel's theorem shows there are no direct methods for solving the general eigenvalue problem, for the existence of a finite, prespecified procedure would imply the existence of a (perhaps very complicated) formula for the solutions of an arbitrary polynomial equation. Therefore all eigenvalue algorithms are iterative.

Convergence Issues

In the next section we will study some of the simplest iterative methods. Each of these produces a sequence of vectors q_1, q_2, q_3, ... that (usually) converges to an eigenvector v as $j \to \infty$; that is, $q_j \to v$ as $j \to \infty$. For each iterative method that we introduce, we must concern ourselves with whether or not (or under what

---

[23]MATLAB has a command called `roots` that makes use of this correspondence. If x is a vector containing the coefficients of the polynomial p, the command `roots(x)` returns the roots of the equation $p(\lambda) = 0$. MATLAB does it as follows. The coefficients are normalized to make the polynomial monic. Then the companion matrix is formed. Finally the eigenvalues of the companion matrix are computed by Francis's algorithm. These are the desired roots.

[24]We allow formulas that involve addition, subtraction, multiplication, division, and the extraction of roots.

conditions) it really does converge to an eigenvector. First of all, what do we mean when we write $q_j \to v$? We mean that each component of q_j converges to the corresponding component of v as $j \to \infty$. You can easily check that this componentwise convergence holds if and only if $\|q_j - v\|_\infty \to 0$ or $\|q_j - v\|_1 \to 0$ or $\|q_j - v\|_2 \to 0$, as $j \to \infty$.[25] We will make extensive use of norms in our discussion of the convergence of iterative methods. In principle it does not matter which norm we use, as all norms on $\mathbb{C}^n$ are equivalent. This is so because $\mathbb{C}^n$ has finite dimension; see [90, Theorem II.3.1] or any other text on functional analysis.

Although the yes or no question of convergence is important, the *rate* of convergence is even more important. A convergent iterative method is of no practical interest if it takes millions of iterations to produce a useful result. Convergence rates will be discussed in the coming section in connection with the iterative methods introduced there.

A Variety of Iterative Methods

As we have already indicated, the methods to be introduced in the next section generate sequences of vectors. In subsequent sections of this chapter we will develop more sophisticated methods that generate sequences of matrices A_1, A_2, A_3, ..., such that each matrix in the sequence has the same eigenvalues as the original matrix A, and the matrices converge to a simple form such as upper-triangular form, from which the eigenvalues can be read. In Chapter 6 we will introduce yet another class of iterative methods that generate sequences of subspaces of $\mathbb{C}^n$. By the middle of Chapter 6 it will be clear that all of these methods, whether they generate vectors, matrices, or subspaces, are intimately related.

Additional Exercises

cise 5.2.17 Let $A \in \mathbb{R}^{n \times m}$ have SVD $A = U\Sigma V^T$. The right singular vectors are v_1, ..., v_m, the columns of V; the left singular vectors are $u_1, \ldots, u_n$, the columns of U. Show that $v_1, \ldots, v_m$ are linearly independent eigenvectors of $A^T A$, and $u_1, \ldots, u_n$ are linearly independent eigenvectors of AA^T, associated with eigenvalues $\sigma_1^2, \ldots,$ $\sigma_r^2, 0, 0, \ldots$, where r is the rank of A. □

cise 5.2.18 Let λ be an eigenvalue of A, and consider the set

$$S_\lambda = \{v \in \mathbb{C}^n \mid Av = \lambda v\},$$

which consists of all eigenvectors associated with λ, together with the zero vector.

(a) Show that if $v_1 \in S_\lambda$ and $v_2 \in S_\lambda$, then $v_1 + v_2 \in S_\lambda$.

(b) Show that if $v \in S_\lambda$ and $c \in \mathbb{C}$, then $cv \in S_\lambda$.

---

[25]Our discussion of vector norms in Section 2.1 was restricted to norms on $\mathbb{R}^n$. However, that entire discussion can be carried over to the complex setting.

Thus S_λ is a subspace of $\mathbb{C}^n$. It is called the *eigenspace* of A associated with λ. Even if λ is not an eigenvalue, we can still define the space S_λ, but in this case $S_\lambda = \{0\}$. Clearly λ is an eigenvalue of A if and only if $S_\lambda \neq \{0\}$.   □

Exercise 5.2.19 This exercise gives a second proof that the main-diagonal entries of a triangular matrix are its eigenvalues. Let T be a triangular matrix. For the sake of argument let us assume that it is upper triangular. Let μ be a complex number.

(a) Suppose $\mu = t_{ii}$ for some i. We will show that μ is an eigenvalue by producing an eigenvector associated with μ. That is, we will produce a nonzero $v \in \mathbb{C}^n$ such that $(\mu I - T)v = 0$. It may be the case that $\mu = t_{ii}$ for several different i. Let j be the smallest index for which $\mu = t_{jj}$. Thus $\mu \neq t_{ii}$ for all $i < j$. Show that if $j = 1$, then $(\mu I - T)e_1 = 0$, where $e_1 = \begin{bmatrix} 1, 0, \cdots, 0 \end{bmatrix}^T$. If $j \neq 1$, partition T as follows:

$$T = \begin{bmatrix} T_{11} & x & T_{13} \\ 0 & t_{jj} & y^T \\ 0 & 0 & T_{13} \end{bmatrix},$$

where T_{11} is $(j-1) \times (j-1)$, $x \in \mathbb{C}^{j-1}$, and so on. Prove that there is a unique vector v of the form

$$v = \begin{bmatrix} z \\ 1 \\ 0 \end{bmatrix},$$

where $z \in \mathbb{C}^{j-1}$, such that $(\mu I - T)v = 0$. Thus μ is an eigenvalue of T.

(b) Show that if $\mu \neq t_{ii}$ for all i, and $v \in \mathbb{C}^n$ satisfies $(\mu I - T)v = 0$, then $v = 0$. (Do not invoke determinants; just prove the result directly from the equation.) Thus μ is not an eigenvalue of T.

□

Exercise 5.2.20 Let $A \in \mathbb{C}^{n \times n}$ be a block triangular matrix:

$$A = \begin{bmatrix} A_{11} & A_{12} \\ 0 & A_{22} \end{bmatrix},$$

where $A_{11} \in \mathbb{C}^{j \times j}$ and $A_{22} \in \mathbb{C}^{k \times k}$, $j + k = n$.

(a) Let λ be an eigenvalue of A_{11} with eigenvector v. Show that there exists a $w \in \mathbb{C}^k$ such that $\begin{bmatrix} v \\ w \end{bmatrix}$ is an eigenvector of A with associated eigenvalue λ.

(b) Let λ be an eigenvalue of A_{22} with associated eigenvector w, and suppose λ is not also an eigenvector of A_{11}. Show that there is a unique $v \in \mathbb{C}^j$ such that $\begin{bmatrix} v \\ w \end{bmatrix}$ is an eigenvector of A with associated eigenvalue λ.

(c) Let λ be an eigenvalue of A with associated eigenvector $\begin{bmatrix} v \\ w \end{bmatrix}$. Show that either w is an eigenvector of A_{22} with associated eigenvalue λ of v is an eigenvector of A_{11} with associated eigenvalue λ. (*Hint:* Either $w \neq 0$ or $w = 0$. Consider these two cases.)

(d) Combining parts (a), (b), and (c), show that λ is an eigenvalue of A if and only if it is an eigenvalue of A_{11} or A_{22}.

$\square$

rcise 5.2.21 Prove Theorem 5.2.11 by induction on k. You might like to make use of the following outline. First show that the theorem is true for $k = 1$. Then suppose k is arbitrary and show that if the theorem holds for sets of $k - 1$ vectors (induction hypothesis), then it must also hold for sets of k vectors. Suppose

$$c_1 v_1 + c_2 v_2 + \cdots + c_k v_k = 0. \tag{5.2.22}$$

To establish independence of $v_1, \ldots, v_k$, you need to show that $c_1 = c_2 = \cdots = c_k = 0$. Apply $(A - \lambda_k I)$ to both sides of (5.2.22). Simplify the result by using the equations $A v_j = \lambda_j v_j$ for $j = 1, \ldots, k$, and note that one term is eliminated from the left-hand side. Use the induction hypothesis to conclude that $c_1 = c_2 = \cdots = c_{k-1} = 0$. Point out clearly how you are using the distinctness of the eigenvalues here. Finally, go back to (5.2.22) and show that $c_k = 0$. $\square$

rcise 5.2.23 Parts (a) and (b) of this exercise provide two independent proofs of Theorem 5.2.16. Let $p(\lambda) = a_0 + a_1\lambda + a_2\lambda^2 + \cdots + 1\lambda^n$, and let A be its companion matrix (5.2.15).

(a) Let $\lambda \in \mathbb{C}$ be an eigenvalue of A with associated eigenvector v. Write down and examine the equation $Av = \lambda v$. Demonstrate that v must be a multiple of the vector

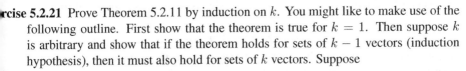

$$\begin{bmatrix} \lambda^{n-1} \\ \vdots \\ \lambda^2 \\ \lambda \\ 1 \end{bmatrix}, \tag{5.2.24}$$

and λ must satisfy $p(\lambda) = 0$. Conversely, show that if $p(\lambda) = 0$, then λ is an eigenvalue of A with eigenvector v given by (5.2.24). Thus the eigenvalues of A are exactly the zeros of p.

(b) For $j = 1, \ldots, n$, let $\chi_j(\lambda) = \lambda^j + a_{n-1}\lambda^{j-1} + \cdots + a_{n-j+1}\lambda + a_{n-j}$. Thus χ_j is a monic polynomial of degree j, and $\chi_n = p$. Let A_j be the companion matrix of χ_j. Prove by induction on j that $\det(\lambda I - A_j) = \chi_j(\lambda)$, $j = 1, \ldots, n$. *Hint:* Expand the last column of the determinant $\det(\lambda I - A_j)$ to obtain the recurrence

$$\det(\lambda I - A_j) = \lambda \det(\lambda I - A_{j-1}) + a_{n-j}.$$

$\square$

5.3   THE POWER METHOD AND SOME SIMPLE EXTENSIONS

Let $A \in \mathbb{C}^{n \times n}$. To avoid complicating the analysis we assume that A is semisimple. This means that A has a set of n linearly independent eigenvectors $v_1, \ldots, v_n$, which must then form a basis for $\mathbb{C}^n$. Let $\lambda_1, \ldots, \lambda_n$ denote the eigenvalues associated with $v_1, \ldots, v_n$, respectively. Let us assume that the vectors are ordered so that $|\lambda_1| \geq |\lambda_2| \geq \cdots \geq |\lambda_n|$. If $|\lambda_1| > |\lambda_2|$, λ_1 is called the *dominant eigenvalue* and v_1 is called a *dominant eigenvector* of A.

If A has a dominant eigenvalue, then we can find it and an associated dominant eigenvector by the *power method*. The basic idea is to pick a vector q and form the sequence

$$q, \ Aq, \ A^2 q, \ A^3 q, \ldots$$

To calculate this sequence, it is not necessary to form the powers of A explicitly. Each vector in the sequence can be obtained by multiplying the previous vector by A, e.g. $A^{j+1} q = A(A^j q)$. It is easy to show that the sequence converges, in a sense, to a dominant eigenvector, for almost all choices of q. Since $v_1, \ldots, v_n$ form a basis for $\mathbb{C}^n$, there exist constants $c_1, \ldots, c_n$ such that

$$q = c_1 v_1 + c_2 v_2 + \cdots + c_n v_n.$$

So far we do not know $v_1, \ldots, v_n$, so we do not know what $c_1, \ldots, c_n$ are either. However, it is clear that for practically any choice of q, c_1 will be nonzero (Exercise 5.3.43). The argument that follows is valid for every q for which $c_1 \neq 0$. Multiplying q by A, we have

$$
\begin{aligned}
Aq &= c_1 A v_1 + c_2 A v_2 + \cdots + c_n A v_n \\
&= c_1 \lambda_1 v_1 + c_2 \lambda_2 v_2 + \cdots + c_n \lambda_n v_n.
\end{aligned}
$$

Similarly

$$A^2 q = c_1 \lambda_1^2 v_1 + c_2 \lambda_2^2 v_2 + \cdots + c_n \lambda_n^2 v_n,$$

and in general

$$A^j q = c_1 \lambda_1^j v_1 + c_2 \lambda_2^j v_2 + \cdots + c_n \lambda_n^j v_n. \tag{5.3.1}$$

Since λ_1 dominates the other eigenvalues, the component of $A^j q$ in the direction of v_1 becomes steadily greater relative to the other components as j increases. This can be made clearer by rewriting (5.3.1) as

$$A^j q = \lambda_1^j \left(c_1 v_1 + c_2 (\lambda_2 / \lambda_1)^j v_2 + \cdots + c_n (\lambda_n / \lambda_1)^j v_n \right). \tag{5.3.2}$$

The fact that every multiple of an eigenvector is also an eigenvector means that the magnitude of an eigenvector is unimportant; only the direction matters. Therefore the factor λ_1^j in (5.3.2) is, in principle, unimportant. Thus, instead of the sequence $(A^j q)$, let us consider the rescaled sequence (q_j), where $q_j = A^j q / \lambda_1^j$. From (5.3.2) it is clear that $q_j \to c_1 v_1$ as $j \to \infty$. Indeed, for any vector norm,

$$
\begin{aligned}
\| q_j - c_1 v_1 \| &= \| c_2 (\lambda_2 / \lambda_1)^j v_2 + \cdots + c_n (\lambda_n / \lambda_1)^j v_n \| \\
&\leq |c_2| \, |\lambda_2 / \lambda_1|^j \, \| v_2 \| + \cdots + |c_n| \, |\lambda_n / \lambda_1|^j \, \| v_n \| \\
&\leq \left(|c_2| \, \| v_2 \| + \cdots + |c_n| \, \| v_n \| \right) |\lambda_2 / \lambda_1|^j.
\end{aligned}
$$

Here we have used the fact that $|\lambda_i| \leq |\lambda_2|$ for $i = 3, \ldots, n$. Letting $C = |c_2| \, \|v_2\| + \cdots + |c_n| \, \|v_n\|$, we have

$$\|q_j - c_1 v_1\| \leq C|\lambda_2/\lambda_1|^j \qquad j = 1, 2, 3, \ldots \qquad (5.3.3)$$

Since $|\lambda_1| > |\lambda_2|$, it follows that $|\lambda_2/\lambda_1|^j \to 0$ as $j \to \infty$. Thus $\|q_j - c_1 v_1\| \to 0$. This means that for sufficiently large j, q_j is a good approximation to the dominant eigenvector $c_1 v_1$. The number $\|q_j - c_1 v_1\|$ gives a measure of the error in this approximation. From (5.3.3) we see that, roughly speaking, the magnitude of the error is decreased by a factor $|\lambda_2/\lambda_1|$ with each iteration. Therefore the ratio $|\lambda_2/\lambda_1|$ is an important indicator of the rate of convergence.

The convergence behavior exhibited by the power method is called linear convergence. In general a sequence (x_j) that converges to x is said to converge *linearly* if there is a number r satisfying $0 < r < 1$, such that

$$\lim_{j \to \infty} \frac{\|x_{j+1} - x\|}{\|x_j - x\|} = r. \qquad (5.3.4)$$

This means that $\|x_{j+1} - x\| \approx r \|x_j - x\|$ for sufficiently large j.[26] The number r is called the *convergence ratio* or *contraction number* of the sequence. The power method generally exhibits linear convergence with convergence ratio $r = |\lambda_2/\lambda_1|$ (Exercise 5.3.44).

In practice the sequence $q_j = A^j q/\lambda_1^j$ is inaccessible because we do not know λ_1 in advance. On the other hand, it is impractical to work with $A^j q$ itself because $\|A^j q\| \to \infty$ if $|\lambda_1| > 1$ and $\|A^j q\| \to 0$ if $|\lambda_1| < 1$. In order to avoid overflow or underflow and to recognize when the iterates are converging, we must employ some kind of scaling. Thus we let $q_0 = q$ and define

$$q_{j+1} = Aq_j/s_{j+1},$$

where s_{j+1} is some convenient scaling factor. The exact choice of scale factor is unimportant, since we are interested in the directions of the vectors, not the lengths. A simple and convenient strategy is to take s_{j+1} to be that entry of Aq_j that is largest in absolute value. The effect of this choice is that the largest component of each q_j is 1, and the sequence converges to a dominant eigenvector whose largest component is 1.

Iterations of the power method are relatively inexpensive. The cost of multiplying the $n \times n$ matrix A by q_j is $2n^2$ flops. The normalizing operations require only $O(n)$ work, as you can easily verify, so the total cost of a power iteration is about $2n^2$ flops. Thus m iterations will cost $2n^2 m$ flops. This count assumes that A is not a sparse matrix. If A is sparse, the cost of calculating Aq_j will be considerably less than $2n^2$ flops. For example, if A has five nonzero entries in each row, the cost of computing Aq_j is only about $10n$ flops.

---

[26] Some authors also allow $r = 1$, but it seems preferable to think of that case as representing *slower than linear* convergence. A sequence that satisfies (5.3.4) with $r = 1$ converges very slowly indeed.

Our development of the power method has assumed that the matrix is semisimple. This turns out not to be crucial; essentially the same conclusions can be drawn for defective matrices.

Example 5.3.5 We will use the power method to calculate a dominant eigenvector of

$$A = \begin{bmatrix} 9 & 1 \\ 1 & 2 \end{bmatrix}.$$

If we start with the vector $q_0 = \begin{bmatrix} 1 & 1 \end{bmatrix}^T$, then on the first step we have $Aq_0 = \begin{bmatrix} 10 & 3 \end{bmatrix}^T$. Dividing by the scale factor $s_1 = 10$, we get $q_1 = \begin{bmatrix} 1 & 0.3 \end{bmatrix}^T$. Then $Aq_1 = \begin{bmatrix} 9.3 & 1.6 \end{bmatrix}^T$, $s_2 = 9.3$, and $q_2 = \begin{bmatrix} 1 & 0.172034 \end{bmatrix}^T$. Subsequent iterates are listed in Table 5.1. Only the second component of each q_j is listed because the

j	s_j	q_j (second component)
3	9.172043	0.146541
4	9.146541	0.141374
5	9.141374	0.140323
6	9.140323	0.140110
7	9.140110	0.140066
8	9.140066	0.140057
9	9.140057	0.140055
10	9.140055	0.140055

Table 5.1   Iterates of power method

first component is always 1. We see that after 10 iterations the sequence of vectors (q_j) has converged to six decimal places. Thus (to six decimal places)

$$v_1 = \begin{bmatrix} 1.0 \\ 0.140055 \end{bmatrix}.$$

The sequence s_j converges to the dominant eigenvalue $\lambda_1 = 9.140055$.   □

In Example 5.3.5 and also in Exercises 5.3.7, 5.3.10, and 5.3.11, the iterates converge reasonably rapidly. This is so because in each case the ratio $|\lambda_2/\lambda_1|$ is fairly small. For most matrices this ratio is not so favorable. Not uncommonly $|\lambda_2/\lambda_1|$ is very close to 1, in which case convergence is slow. See Exercise 5.3.9.

Exercise 5.3.6   Calculate the characteristic polynomial of the matrix of Example 5.3.5, and use the quadratic formula to find the eigenvalues. Calculate the dominant eigenvector as well, and verify that the results of Example 5.3.5 are correct. Calculate the ratios of errors $\| q_{j+1} - v_1 \| / \| q_j - v_1 \|$, $j = 0, 1, 2, \ldots$, and note that they agree very well with the theoretical convergence ratio $|\lambda_2/\lambda_1|$. Why is the agreement so good? Would you expect the agreement to be as good if A were 3×3 or larger?   □

rcise 5.3.7 Let $A = \begin{bmatrix} 8 & 1 \\ -2 & 1 \end{bmatrix}$.

(a) Use the power method with $q_0 = \begin{bmatrix} 1 & 1 \end{bmatrix}^T$ to calculate the dominant eigenvalue and eigenvector of A. Tabulate q_j and s_j at each step. Iterate until the q_j and s_j have converged to at least six decimal places.

(b) Now that you have calculated the eigenvector v, calculate the ratios

$$\| q_{j+1} - v \| / \| q_j - v \|, \qquad j = 0, 1, 2, \ldots$$

to find the observed rate of convergence. Using the characteristic equation, solve for the eigenvalues and calculate the ratio $|\lambda_2/\lambda_1|$, which gives the theoretical rate of convergence. How well does theory agree with practice in this case?

□

rcise 5.3.8 Work this problem by hand. Let $A = \begin{bmatrix} 0 & 1 \\ 1 & 0 \end{bmatrix}$. Carry out the power method with starting vector $q_0 = \begin{bmatrix} a & b \end{bmatrix}^T$, where $a \geq 0, b \geq 0$, and $a \neq b$. Explain why the sequence fails to converge. Why does the convergence argument given above fail in this case?

□

rcise 5.3.9 Work this problem by hand. Only at the very last step will you need a calculator. Let $A = \begin{bmatrix} 0.99 & 0 \\ 0 & 1 \end{bmatrix}$.

(a) Find the eigenvalues of A and associated eigenvectors.

(b) Carry out power iterations starting with $q_0 = \begin{bmatrix} 1 & 1 \end{bmatrix}^T$. Derive a general expression for q_j.

(c) How may iterations are required in order to obtain $\| q_j - v_1 \|_\infty / \| v_1 \|_\infty < 10^{-6}$?

□

rcise 5.3.10 Let $A = \begin{bmatrix} 1 & 1 & 1 \\ -1 & 9 & 2 \\ 0 & -1 & 2 \end{bmatrix}$.

(a) Using MATLAB, apply the power method to A, starting with

$$q_0 = \begin{bmatrix} 1 & 1 & 1 \end{bmatrix}^T.$$

Do at least ten iterations. Here are some sample MATLAB commands.

```
A = [ 1 1 1 ; -1 9 2 ; 0 -1 2]
q = ones(3,1)
```

```
iterate(:,1) = q;
for j = 1:10
  q = A*q;
  [bgst,index] = max(abs(q));
  scale_factor(j+1) = q(index(1));
  q = q/scale_factor(j+1);
  iterate(:,j+1) = q;
end
iterate
scale_factor
```

If you store some of these (or similar) commands in a file named, say, powr.m, then each time you type the command `powr`, MATLAB will execute that sequence of instructions. The numbers are stored to much greater accuracy than what is normally displayed on the screen. If you wish to display more digits, type `format long`.

(b) Use the command `[V,D] = eig(A)` to get the eigenvalues and eigenvectors of A. Compare the dominant eigenvalue from `D` with the scale factors that you computed in part (a).

(c) Locate a dominant eigenvector from `V` and rescale it appropriately, so that you can compare it with the iterates from part (a). Calculate the ratios

$$\|q_{j+1} - v\|_2 / \|q_j - v\|_2 \quad \text{for } j = 1, 2, 3, \ldots.$$

(d) Using `D` from part (b), calculate the ratio $|\lambda_2/\lambda_1|$ and compare it to the ratios computed in part (c).

□

Exercise 5.3.11 Repeat Exercise 5.3.10 using the matrix

$$A = \begin{bmatrix} 1 & 1 & 1 \\ -1 & 9 & 2 \\ -4 & -1 & 2 \end{bmatrix}.$$

What is different here? The ratios of errors are more erratic. Compute

$$\sqrt[j]{\|q_{j+1} - v\| / \|q_1 - v\|}, \qquad j = 1, 2, 3, \ldots,$$

and compare these values with $|\lambda_2/\lambda_1|$. Why would you expect these to be close?

□

Exercise 5.3.12 Repeat Exercise 5.3.10 using the matrix

$$A = \begin{bmatrix} 1 & 1 & 1 \\ -1 & 3 & 2 \\ -4 & -1 & 2 \end{bmatrix}.$$

What happened this time?

□

The Google Matrix

We have seen that the convergence of the power method can be quite slow, and indeed the basic power method, as we have just described it, is not used much in practice. However, there is one spectacular current example that is worth mentioning. The *Google matrix* is a gigantic sparse matrix that models the internet [62]. The entries of the dominant eigenvector are used to rank the importance of web pages, and this information is used in Google searches. Since the Google matrix has $|\lambda_2/\lambda_1| = .85$ by design, the dominant eigenvector can be estimated in a reasonable number of iterations of the power method.

Inverse Iteration and the Shift-and-Invert Strategy

We continue to assume that $A \in \mathbb{C}^{n \times n}$ is semisimple with linearly independent eigenvectors $v_1, \ldots, v_n$ and associated eigenvalues $\lambda_1, \ldots, \lambda_n$, arranged in order of descending magnitude. If A is nonsingular, we can apply the power method to A^{-1}. This is the *inverse power method* or *inverse iteration*. (By contrast, the power method applied to A is sometimes called *direct iteration*.)

Exercise 5.3.13 Suppose $A \in \mathbb{C}^{n \times n}$ is nonsingular. Then all eigenvalues of A are nonzero. Show that if v is an eigenvector of A with associated eigenvalue λ, then v is also an eigenvector of A^{-1} with associated eigenvalue λ^{-1}.   □

From Exercise 5.3.13 we see that A^{-1} has linearly independent eigenvectors v_n, $v_{n-1}, \ldots v_1$ with eigenvalues $\lambda_n^{-1}, \lambda_{n-1}^{-1}, \ldots \lambda_1^{-1}$. If $|\lambda_n^{-1}| > |\lambda_{n-1}^{-1}|$ (i.e. $|\lambda_{n-1}| > |\lambda_n|$) and we start with a vector $q = c_n v_n + \cdots + c_1 v_1$ for which $c_n \neq 0$, then the inverse power iterates will converge to a multiple of v_n, an eigenvector associated with the smallest eigenvalue of A. The convergence ratio is $|\lambda_{n-1}^{-1}/\lambda_n^{-1}| = |\lambda_n/\lambda_{n-1}|$, so convergence will be fast when $|\lambda_{n-1}| \gg |\lambda_n|$. This suggests that we *shift* the eigenvalues so that the smallest one is very close to zero. To understand how this works, we need only the following simple result.

Exercise 5.3.14 Let $A \in \mathbb{C}^{n \times n}$, and let $\rho \in \mathbb{C}$. Show that if v is an eigenvector of A with eigenvalue λ, then v is also an eigenvector of $A - \rho I$ with eigenvalue $\lambda - \rho$.   □

From Exercise 5.3.14 we see that if A has eigenvalues $\lambda_1, \ldots, \lambda_n$, then $A - \rho I$ has eigenvalues $\lambda_1 - \rho, \lambda_2 - \rho, \ldots, \lambda_n - \rho$. The scalar ρ is called a *shift*. If the shift is chosen so that it is a good approximation to λ_n, then $|\lambda_{n-1} - \rho| \gg |\lambda_n - \rho|$, and inverse iteration applied to $A - \rho I$ will converge rapidly to a multiple of v_n. Actually there is nothing special about λ_n; the shift ρ can be chosen to approximate any one of the eigenvalues of A. If ρ is a good enough approximation to λ_i that $\lambda_i - \rho$ is much smaller than any other eigenvalue of $A - \rho I$, then (for almost any starting vector q) inverse iteration applied to $A - \rho I$ will converge to a multiple of the eigenvector v_i. The convergence ratio is $|(\lambda_i - \rho)/(\lambda_k - \rho)|$, where $\lambda_k - \rho$ is the second smallest eigenvalue of $A - \rho I$. The closer ρ is to λ_i, the swifter the convergence will be.

Shifts can be used in conjunction with direct iteration as well, but they are not nearly so effective in that context. The combination of shifting and inverse iteration works well because there exist shifts for which one eigenvalue of the shifted matrix is much smaller than all other eigenvalues. In contrast, there (usually) do not exist shifts for which one eigenvalue is much larger than all other eigenvalues.

To get rapid convergence, we shift first and then apply inverse iteration. This is called the *shift-and-invert* strategy. It is one of the most important ideas for eigenvalue computations.

Exercise 5.3.15 Work this problem by hand. Diagonal matrices such as

$$
A = \begin{bmatrix} 2.99 & 0 & 0 \\ 0 & 1.99 & 0 \\ 0 & 0 & 1.00 \end{bmatrix}
$$

have particularly simple eigensystems.

(a) Find the eigenvalues and eigenvectors of A.

(b) Find the eigenvalues of $A - \rho I$ and $(A - \rho I)^{-1}$, where $\rho = 0.99$. Perform both direct and inverse iteration on $A - \rho I$, starting with $q_0 = [1 \ 1 \ 1]^T$. To which eigenvector does each sequence converge? Which converges faster?

(c) Perform inverse iteration with values (i) $\rho = 2.00$ and (ii) $\rho = 3.00$, using the same starting vector. To which eigenvectors do these sequences converge?

□

Let us consider some of the practical aspects of inverse iteration. The iterates will satisfy

$$
q_{j+1} = (A - \rho I)^{-1} q_j / s_{j+1},
$$

but there is no need to calculate $(A - \rho I)^{-1}$ explicitly. Instead one can solve the linear system $(A - \rho I)\hat{q}_{j+1} = q_j$ and then set $q_{j+1} = \hat{q}_{j+1}/s_{j+1}$, where s_{j+1} equals the component of $\hat{q}_{j+1}$ that is largest in magnitude. If the system is to be solved by Gaussian elimination, then the LU decomposition of $A - \rho I$ has to be done only once. Then each iteration consists of forward substitution, back substitution, and normalization. For a full $n \times n$ matrix the cost is $\frac{2}{3}n^3$ flops for the LU decomposition plus $2n^2$ flops per iteration.

Example 5.3.16 Let us apply inverse iteration with a shift to the matrix

$$
A = \begin{bmatrix} 9 & 1 \\ 1 & 2 \end{bmatrix}.
$$

We know from Example 5.3.5 that 9 is a good approximation to an eigenvalue of A. Even if we did not know this, we might expect as much from the dominant 9 in the $(1, 1)$ position of the matrix. Thus it is reasonable to use the shift $\rho = 9$.

Starting with $q_0 = \begin{bmatrix} 1 & 1 \end{bmatrix}^T$, we solve the system $(A - 9I)\hat{q}_1 = q_0$ to get $\hat{q}_1 = \begin{bmatrix} 8.0 & 1.0 \end{bmatrix}^T$. We then rescale by taking $s_1 = 8$ to get $q_1 = \begin{bmatrix} 1.0 & 0.125 \end{bmatrix}^T$. Solving $(A - 9I)\hat{q}_2 = q_1$, we get $\hat{q}_2 = \begin{bmatrix} 7.125 & 1.0 \end{bmatrix}^T$. Thus $s_2 = 7.125$, and $q_2 = \begin{bmatrix} 1.0 & 0.140351 \end{bmatrix}^T$. Subsequent iterates are listed in Table 5.2. As in

j	s_j	q_j (second component)
3	7.140351	0.140449
4	7.140449	0.140055
5	7.140055	0.140055

Table 5.2 Iterates of inverse power method

Table 5.1, we have listed only the second component of q_j because the first component is always 1. After five iterations we have the eigenvector $v_1 = \begin{bmatrix} 1.0 & 0.140055 \end{bmatrix}^T$ correct to six decimal places. The good choice of shift gave much faster convergence than we had in Example 5.3.5. The scale factors converge to an eigenvalue of $(A - 9I)^{-1}$: $(\lambda_1 - 9)^{-1} = 7.140055$. Solving for λ_1, we obtain $\lambda_1 = 9.140055$. Another way to obtain the eigenvalue is to compute the Rayleigh quotient, which will be discussed below. □

rcise 5.3.17 Let $A = \begin{bmatrix} 8 & 1 \\ -2 & 1 \end{bmatrix}$.

(a) Use inverse iteration with $\rho = 8$ and $q_0 = \begin{bmatrix} 1 & 1 \end{bmatrix}^T$ to calculate an eigenvalue and eigenvector of A. (On this small problem it may be easiest simply to calculate $B = (A - 8I)^{-1}$ and perform direct iterations with B.)

(b) Now that you have calculated an eigenvector v, calculate the ratios

$$\| q_{j+1} - v \| / \| q_j - v \|$$

for $j = 0, 1, 2, \ldots$, to find the observed rate of convergence. Solve for the eigenvalues using the characteristic equation or MATLAB and calculate the theoretical convergence rate $|(\lambda_1 - 8)/(\lambda_2 - 8)|$. How well does theory agree with practice?

□

rcise 5.3.18 Let $A = \begin{bmatrix} 1 & 1 & 1 \\ -1 & 9 & 2 \\ 0 & -1 & 2 \end{bmatrix}$.

(a) Using MATLAB, apply inverse iteration to A with shift $\rho = 9$, starting with $q_0 = \begin{bmatrix} 1 & 1 & 1 \end{bmatrix}^T$. For simplicity, just form $B = (A - 9I)^{-1}$ (B =

inv(A-9*eye(3)) in MATLAB) and iterate with B. Some of the commands shown in Exercise 5.3.10 may be useful here. Use format long to view more digits of your iterates. Do at least ten iterations.

(b) Use [V, D] = eig(A) to get the true dominant eigenvector. Calculate the errors $\|q_j - v\|_2$ and ratios $\|q_{j+1} - v\|_2 / \|q_j - v\|_2$ for $j = 1, 2, 3, \ldots$. Compute the theoretical convergence rate (from the known eigenvalues) and compare it with these ratios. Notice that we have faster convergence than we had when we applied the power method to this same matrix in Exercise 5.3.10.

□

Exercise 5.3.19 The matrix west0479 is a sparse 479×479 matrix that is supplied with MATLAB. We used this matrix previously in Exercise 1.9.2. Some useful commands are

```
load west0479
A = west0479;  %  Just to get a short abbreviation.
issparse(A)    %  = 1 for sparse, = 0 for dense matrices.
size(A)
nnz(A)         %  number of nonzeros of A
disp(A)
spy(A)
```

(a) It is known that A has an eigenvalue near $\rho = 15 + 35i$. Use inverse iteration with shift ρ to calculate this eigenvalue. (*Solution:* $\lambda = 17.548546093831 + 34.237822957500i$.) More useful commands:

```
n = 479;
itnum = 20; % number of iterations
iterate = zeros(n,itnum+1);
scale_factor = zeros(itnum+1,1);
shift = 15 + 35i;
[L,U,P] = lu(A - shift*speye(n));
q = randn(n,1); % random starting vector
[bgst,index] = max(abs(q));
scale_factor(1) = q(index(1));
q = q/scale_factor(1);
iterate(:,1) = q;
for j = 1:itnum
   q = P*q;
   q = L\q;
   q = U\q;
   [bgst,index] = max(abs(q));
   scale_factor(j+1) = q(index(1));
   q = q/scale_factor(j+1);
   iterate(:,j+1) = q;
end
format long
eigval = shift + 1./scale_factor
```

(b) Explain the function of the four commands involving `L`, `U`, and `P` in part (a). How do they help make the algorithm economical?

(c) When the matrix is large and sparse, the speed of the algorithm depends heavily upon how it is implemented. Run three different implementations of inverse iteration, keeping track of the execution time:

   (i) Use the implementation shown in part (a).

  (ii) Let `B = A - shift*speye(n)` and use `q = B\q` to do the inverse iteration. This lazy implementation forces MATLAB to redo the sparse *LU* decomposition on each iteration.

 (iii) Let `C = full(inv(B))`, where `B` is as before, and use `q = C*q` to do the iteration. (Type `spy(C)` to verify that `C` is not at all sparse.)

Time, say, 20 iterations by each method. Also determine separately the time it takes to compute the sparse *LU* decomposition and the time it takes to compute the inverse. Discuss your results. More sample code:

```
t1 = cputime;
[L,U,P] = lu(A - shift*speye(n));
t2 = cputime;
decomp_time = t2 - t1
```

                                                     □

rcise 5.3.20 This exercise is similar to the previous one, except that it features a scalable family of matrices. The MATLAB command `delsq` generates sparse matrices associated with discretized Laplacian ("del squared") operators on various regions. For example, try

```
m = 62
A = delsq(numgrid('S',m));
issparse(A)
size(A)
```

This produces a 3600×3600 matrix. You can make a larger or smaller matrix by adjusting the size of m. In general the matrix A has dimension $(m - 2)^2$. We said more about this family in Exercise 1.6.7.

(a) This matrix is symmetric and positive definite. Its largest eigenvalue is near 8. Use inverse iteration with shift $\rho = 8$ to estimate this eigenvalue to ten decimal places accuracy. (The solution depends on m.) Borrow code from Exercise 5.3.19, part (a). Use a random starting vector.

(b) Repeat part (c) of Exercise 5.3.19 using a `delsq` matrix. Make m large enough that you get impressive results, but not so large that you die of old age before the results come back.

                                                     □

Inverse Iteration with Exact Shifts

The introduction of shifts allows us to find any eigenvector, not just those associated with the largest and smallest eigenvalues. However, in order to find a given eigenvector, we must have a good approximation to the associated eigenvalue to begin with. It follows that a good application for inverse iteration is to find eigenvectors associated with eigenvalues that have been computed by some other means. Using the computed eigenvalue as a shift, we typically obtain an excellent eigenvector in just one or two iterations. We call this *inverse iteration with an exact shift*. It is an important application of inverse iteration.

Exercise 5.3.21 An oracle tells you that $\lambda = 17.548546093831 + 34.237822957500i$ is an eigenvalue of the matrix `west0479` from MATLAB. Perform one step of inverse iteration, using this computed eigenvalue as a shift, to get an estimate v of an associated eigenvector. Borrow MATLAB code from Exercise 5.3.19. To get an idea of the quality of this estimated eigenvector, calculate the residual $\| Av - \lambda v \|_2 / \| v \|_2$. A residual of zero would indicate a perfect eigenpair. If you work this problem correctly, you will get an impressively small residual. (*Warning:* A small residual does not absolutely guarantee an accurate eigenvector. See the discussion of sensitivity of eigenvalues and eigenvectors in Section 7.1.)   □

Inverse iteration with exact shifts works amazingly well when one considers the circumstances. If ρ is an eigenvalue of A, then $A - \rho I$ is singular, so $(A - \rho I)^{-1}$ does not exist. How can we possibly do inverse iteration with this shift? Well, in practice ρ is (almost) never exactly an eigenvalue; there is (almost) always some error. Therefore $A - \rho I$ is (almost) always nonsingular. Furthermore, recall from Chapter 2 that in numerical practice it is impossible to distinguish between singular and nonsingular matrices. Therefore, even if $A - \rho I$ is singular, we will (almost) never be able to detect the fact. Therefore the method (almost) never breaks down in practice.

However, if ρ is nearly an eigenvalue, this means that $A - \rho I$ is nearly singular; that is, it is ill conditioned (Exercise 5.3.47). One might reasonably fear that the ill condition of $A - \rho I$ would spoil the computation, since at each step a system of the form $(A - \rho I)\hat{q}_{j+1} = q_j$ must be solved. Fortunately the ill conditioning turns out not to be a problem in this context. Consider a single step $(A - \rho I)\hat{q}_1 = q_0$. If ρ is nearly an eigenvalue of A, then $\hat{q}_1$ is already very close to an eigenvector of A, if the system has been solved exactly. But suppose instead that the system was solved by Gaussian elimination with partial pivoting. Then we know from Chapter 2 (Theorem 2.7.14 and ensuing discussion) that the computed $\hat{q}_1$ actually satisfies a perturbed system $(A + \delta A - \rho I)\hat{q}_1 = q_0$, where δA is normally tiny relative to A. Since δA is tiny, ρ should also be a good approximation to an eigenvalue of $A + \delta A$, so $\hat{q}_1$ should be nearly an eigenvector of $A + \delta A$, which should then also be close to an eigenvector of A. Thus inverse iteration should be effective.

Whether or not all this is true depends upon whether the small perturbation in A really does cause only small perturbations of the eigenvalue and eigenvector. This depends in turn on condition numbers for the eigenvalue and eigenvector, which are independent of the condition number of $A - \rho I$. Thus the effectiveness of inverse iteration does not depend on the condition number of $A - \rho I$. We will discuss condition numbers for eigenvalues and eigenvectors in Section 7.1.

The Rayleigh Quotient

In using inverse iteration to calculate an eigenvector of A, there is no reason why one could not use a different shift at each step. This might be useful in situations in which the eigenvalue is not known in advance. If q_j is close enough to an eigenvector, it should somehow be possible to use q_j to obtain an estimate of the associated eigenvalue. This estimate could then be used as the shift for the next iteration. In this way, we would get an improved shift, hence an improved convergence ratio, for each iteration.

Suppose $q \in \mathbb{C}^n$ approximates an eigenvector of A. How can we use q to estimate the associated eigenvalue? Our approach will be to minimize a residual. If q is exactly an eigenvector, then there exists exactly one number ρ for which

$$Aq = \rho q. \tag{5.3.22}$$

This number is the eigenvalue. If q is not an eigenvector, then there is no value of ρ for which (5.3.22) is satisfied. Equation (5.3.22) is in fact an overdetermined system of n equations in the single unknown ρ. Letting r denote the residual, $r = Aq - \rho q$, we can find the value of ρ for which the 2-norm $\| r \|_2$ assumes a minimum. In the case when q is an eigenvector, the minimizing ρ is exactly the associated eigenvalue. It therefore seems reasonable that if q merely approximates an eigenvector, then the minimizing ρ should at least yield a good estimate of the associated eigenvalue.

The choice of the 2-norm means that the minimization problem is a least squares problem, so we can solve it by any of the techniques developed in Chapter 3. Thus a QR decomposition or the normal equations, for example. It is simplest to use the normal equations. In Chapter 3 we restricted our attention to real matrices, but all of the developments of that chapter can be carried over to the complex setting. The only modification that must be made is that whenever a transpose, say B^T, occurs, it must be replaced by the conjugate transpose B^*. This is the matrix obtained by transposing B and then taking complex conjugates of all entries. Making this modification to Theorem 3.5.21, we find that the normal equations for the complex overdetermined system $Cz = b$ are $C^*Cz = C^*b$. Rewriting (5.3.22) as

$$q\rho = Aq, \tag{5.3.23}$$

we see that the role of C is played by q, that of z is played by ρ, and that of b is played by Aq. Thus the normal equations for (5.3.23) are

$$(q^*q)\rho = q^*Aq.$$

Actually there is only one normal equation in this case, because there is only one unknown. Its solution is

$$\rho = \frac{q^*Aq}{q^*q}.$$

This number is called the *Rayleigh quotient* of q with respect to A. We have just proved the following theorem.

Theorem 5.3.24 *Let* $A \in \mathbb{C}^{n \times n}$ *and* $q \in \mathbb{C}^n$. *The unique complex number that minimizes* $\| Aq - \rho q \|_2$ *is the Rayleigh quotient* $\rho = q^* A q / q^* q$.

In particular, if q is an eigenvector of A, then the Rayleigh quotient equals the associated eigenvalue. This is a consequence of Theorem 5.3.24, but it is also perfectly obvious from the form of the Rayleigh quotient.

Two other proofs of Theorem 5.3.24 are pursued in Exercise 5.3.48.

The next theorem shows that the Rayleigh quotient of q does indeed approximate an eigenvalue if q approximates an eigenvector. In the theorem the vectors are normalized so that they have Euclidean norm 1. This simplifies both the statement and the proof of the theorem, but it is a matter of convenience, not necessity. Notice that if $\| q \|_2 = 1$, the Rayleigh quotient has the simpler form $\rho = q^* A q$.

Theorem 5.3.25 *Let* $A \in \mathbb{C}^{n \times n}$ *and let* v *be an eigenvector of* A *with associated eigenvalue* λ. *Assume* $\| v \|_2 = 1$. *Let* $q \in \mathbb{C}^n$ *with* $\| q \|_2 = 1$, *and let* $\rho = q^* A q$ *be the Rayleigh quotient of* q. *Then*

$$| \lambda - \rho | \leq 2 \| A \|_2 \| v - q \|_2.$$

Proof. Since $Av = \lambda v$ and $\| v \|_2 = 1$, $\lambda = v^* A v$. Thus $\lambda - \rho = v^* A v - q^* A q = v^* A v - v^* A q + v^* A q - q^* A q = v^* A(v - q) + (v - q)^* A q$. Therefore $| \lambda - \rho | \leq | v^* A(v - q) | + | (v - q)^* A q |$. By the Cauchy-Schwarz inequality (3.2.1), $| v^* A(v - q) | \leq \| v \|_2 \| A(v - q) \|_2 = \| A(v - q) \|_2$, and by Theorem 2.1.24,

$$\| A(v - q) \|_2 \leq \| A \|_2 \| v - q \|_2.$$

Thus $| v^* A(v - q) | \leq \| A \|_2 \| v - q \|_2$. Similarly $| (v - q)^* A q | \leq \| A \|_2 \| v - q \|_2$. The assertion of the theorem follows. $\qquad \square$

Thus if $\| v - q \|_2 < \epsilon$, then $| \lambda - \rho | \leq 2 \| A \|_2 \epsilon$. This can be expressed briefly by saying that if $\| v - q \|_2 = O(\epsilon)$, then $| \lambda - \rho | = O(\epsilon)$.

Rayleigh Quotient Iteration

Rayleigh quotient iteration is that variant of inverse iteration in which the Rayleigh quotient of each q_j is computed and used as the shift for the next iteration. Thus a step of Rayleigh quotient iteration is as follows:

$$\rho_j = \frac{q_j^* A q_j}{q_j^* q_j}, \qquad (A - \rho_j I) \hat{q}_{j+1} = q_j, \qquad q_{j+1} = s_{j+1}^{-1} \hat{q}_{j+1},$$

where s_{j+1} is any convenient scaling factor.

Because a different shift is used at each step, it is difficult to analyze the global convergence properties of Rayleigh quotient iteration. The algorithm is not guaranteed to converge to an eigenvector, but experience suggests that it seldom fails.

When it does converge, it generally converges rapidly. The next example illustrates the swift convergence of Rayleigh quotient iteration.

Example 5.3.26 Consider once again the matrix

$$A = \begin{bmatrix} 9 & 1 \\ 1 & 2 \end{bmatrix}.$$

We used Rayleigh quotient iteration starting with $q_0 = \begin{bmatrix} 1 & 1 \end{bmatrix}^T$ to calculate an eigenvector of A. The results are listed in Table 5.3. Again we have normalized the iterates so that the first component is 1. You can easily check that q_5 is an eigenvector

j	q_j (second component)	ρ_j
0	1.00000000000000	6.50000000000000
1	−0.27272727272727	8.00769230769231
2	0.22155834654310	9.09484426192450
3	0.13955130581106	9.14005316814698
4	0.14005494476317	9.14005494464026
5	0.14005494464026	9.14005494464026

Table 5.3 Rayleigh quotient iteration

with associated eigenvalue ρ_5, correct to fourteen decimal places. Notice the rapid convergence: q_2 has no correct digits, q_3 has (essentially) three correct digits, and q_4 has nine correct digits. □

rcise 5.3.27 Calculate the ratios $\| q_{j+1} - v \| / \| q_j - v \|$, $j = 0$, 1, 2, 3, from Table 5.3, where $v = q_5 = $ the eigenvector. Notice that the ratios decrease dramatically with increasing j. □

Exercise 5.3.27 shows that the rate of convergence observed in Example 5.3.26 is better than linear. This is to be expected since the convergence ratio depends on a ratio of shifted eigenvalues that improves from one step to the next.

Let us take a closer look at the convergence rate of Rayleigh quotient iteration. (However, what follows should not be mistaken for a rigorous analysis.) Let (q_j) be a sequence of vectors generated by Rayleigh quotient iteration. We shall simplify the analysis by assuming that $\| q_j \|_2 = 1$ for all j. Clearly we may choose to normalize so that this is the case. Suppose $q_j \to v_i$ as $j \to \infty$. Then also $\| v_i \|_2 = 1$. Assume further that the associated eigenvalue λ_i is not a multiple eigenvalue, and let λ_k denote the closest eigenvalue to λ_i with $k \neq i$. Since the jth step of Rayleigh quotient iteration is just a power iteration with matrix $(A - \rho_j I)^{-1}$, we know that

$$\| v_i - q_{j+1} \|_2 \approx r_j \| v_i - q_j \|_2, \tag{5.3.28}$$

where r_j is the ratio of the two eigenvalues of $(A - \rho_j I)^{-1}$ of largest absolute value. By Theorem 5.3.25 the Rayleigh quotients ρ_j converge to λ_i. Once ρ_j is close enough to λ_i, the two largest eigenvalues of $(A - \rho_j I)^{-1}$ will be $(\lambda_i - \rho_j)^{-1}$ and $(\lambda_k - \rho_j)^{-1}$. Thus

$$r_j = |(\lambda_k - \rho_j)^{-1}/(\lambda_i - \rho_j)^{-1}| = |(\lambda_i - \rho_j)/(\lambda_k - \rho_j)|.$$

By Theorem 5.3.25, $|\lambda_i - \rho_j| \le 2\|A\|_2 \|v_i - q_j\|_2$. Also, since $\rho_j \approx \lambda_i$, we can make the approximation $|\lambda_k - \rho_j| \approx |\lambda_k - \lambda_i|$. Thus

$$r_j \approx \frac{2\|A\|_2}{|\lambda_k - \lambda_i|}\|v_i - q_j\|_2 = C\|v_i - q_j\|_2,$$

where $C = 2\|A\|_2/|\lambda_k - \lambda_i|$. Substituting this estimate of r_j into (5.3.28), we obtain the estimate

$$\|v_i - q_{j+1}\|_2 \approx C\|v_i - q_j\|_2^2. \tag{5.3.29}$$

Thus the error after $j + 1$ iterations is proportional to the square of the error after j iterations. Another way to express this is to say that if $\|v_i - q_j\|_2 = O(\epsilon)$, then $\|v_i - q_{j+1}\|_2 = O(\epsilon^2)$. If ϵ is small, then ϵ^2 is tiny. A sequence whose convergence rate satisfies (5.3.29) is said to converge *quadratically*. More specifically, we say that $q_j \to v$ *quadratically* as $j \to \infty$ if there exists a nonzero constant C such that

$$\lim_{j \to \infty} \frac{\|v - q_{j+1}\|_2}{\|v - q_j\|_2^2} = C. \tag{5.3.30}$$

The estimates that we have made above indicate that Rayleigh quotient iteration typically converges quadratically when it does converge.

A rule of thumb that one sometimes hears is this: quadratic convergence means that the number of correct digits doubles with each iteration. This is true if $C \approx 1$ in (5.3.29) and (5.3.30), for if q_j agrees with v_i to s_j decimal places, then $\|v_i - q_j\|_2 \approx 10^{-s_j}$. Thus, by (5.3.29) with $C \approx 1$, $\|v_i - q_{j+1}\|_2 \approx 10^{-2s_j}$; that is, q_{j+1} agrees with v_i to about $2s_j$ decimal places. Even if $C \not\approx 1$, the rule of thumb is valid in the limit. If $C \approx 10^t$, then $\|v_i - q_{j+1}\|_2 \approx 10^{t-2s_j}$. Thus q_{j+1} agrees with v_i to about $2s_j - t$ decimal places. As s_j grows, t becomes increasingly insignificant. Once s_j is large enough, t can be ignored.

In Example 5.3.26 the rate of convergence appears to be better than quadratic, since the number of correct digits roughly triples with each iteration. This is not an accident, rather it is a consequence of the special form of the matrix. Note that in this example A is a real symmetric matrix. For matrices of this type, the Rayleigh quotient approximates the eigenvalue better than Theorem 5.3.25 would indicate. In Section 5.4 we will see (Exercise 5.4.34) that for symmetric matrices, if $\|v - q\|_2 = O(\epsilon)$, then $|\lambda - \rho| = O(\epsilon^2)$. (The notation is that of Theorem 5.3.25.) Using this estimate instead of Theorem 5.3.25 to estimate r_j, we find that

$$\|v_i - q_{j+1}\|_2 \approx C\|v_i - q_j\|_2^3.$$

This is called *cubic* convergence. More precisely, $q_j \to v$ *cubically* as $j \to \infty$ if there is a constant $C \ne 0$ such that

$$\lim_{j \to \infty} \frac{\|v - q_{j+1}\|_2}{\|v - q_j\|_2^3} = C. \tag{5.3.31}$$

In this case the number of correct digits roughly triples with each iteration.

rcise 5.3.32   In Example 5.3.26, the computations were performed using IEEE standard double-precision floating point arithmetic, which gives about sixteen decimal digits accuracy. If the computations had been done in quadruple precision arithmetic, about how may correct digits would you expect the iterate q_5 to have?   □

Quite a lot can be said about Rayleigh quotient iteration in the symmetric case that cannot be said in general. Not only is the convergence cubic when it occurs, it is also known that Rayleigh quotient iteration converges for almost all choices of starting vector. Of course the eigenvector that is reached depends on the choice of starting vector. Unfortunately there is no simple characterization of this dependence. See [69, Chapter 4] for details.

Rayleigh quotient iteration can be expensive. Since a different shift is used at each iteration, a new LU decomposition is needed at each step. This costs $O(n^3)$ flops per iteration for a full matrix and thus makes the method too expensive. However, there are some classes of matrices for which Rayleigh quotient iteration is economical. For example, a matrix A is said to have *upper Hessenberg* form if $a_{ij} = 0$ whenever $i > j + 1$. This means that the matrix has the nearly triangular form

$$
\begin{bmatrix}
* & * & * & * & * \\
* & * & * & * & * \\
  & * & * & * & * \\
  &   & * & * & * \\
  &   &   & * & *
\end{bmatrix}.
$$

rcise 5.3.33 Let A be an upper Hessenberg matrix.

(a) Show that an LU decomposition of A (with partial pivoting) can be accomplished in about n^2 flops. What special form does L have in this case?

(b) Show that a QR decomposition of A can be accomplished by applying a sequence of $n - 1$ plane rotators to A. Show that the total flop count is $O(n^2)$, assuming the factor Q is not assembled but left as a product of rotators.

(c) If Q were assembled, what special form would it have?

□

Exercise 5.3.33 shows that we can apply Rayleigh quotient iteration to an upper Hessenberg matrix at a cost of only $O(n^2)$ flops per iteration, which is relatively economical. Hessenberg matrices will play an important role in this chapter. In Section 5.5 we will see that the eigenvalue problem for an arbitrary matrix can be reduced to that of finding the eigenvalues (and eigenvectors, if desired) of a related upper Hessenberg matrix. We can use Rayleigh quotient iteration to attack the Hessenberg matrix at reasonable cost. However, we will not focus on Rayleigh quotient iteration. Instead we will develop a more powerful algorithm, Francis's implicitly shifted QR algorithm. Rayleigh quotient iterations take place implicitly within (certain versions of) Francis's algorithm.

Exercise 5.3.34 Let $A = \begin{bmatrix} 8 & 1 \\ -2 & 1 \end{bmatrix}$. Write a simple MATLAB script to perform Rayleigh quotient iteration on A.

(a) Use $q_0 = \begin{bmatrix} 1 & 1 \end{bmatrix}^T$ as a starting vector. Iterate until the limit of machine precision is reached. (Use `format long` to display 15 digits.) Notice that the iterates converge to a different eigenvector than in Exercises 5.3.7 and 5.3.17. Notice also that once the iterates get close to the limit, the number of correct digits roughly doubles with each iteration. This is true of both the iteration vector and the Rayleigh quotient. Thus both converge quadratically.

(b) Repeat part (a) with the starting vector $q_0 = \begin{bmatrix} 1 & 0 \end{bmatrix}^T$. This time the iterates converge to the same eigenvector as in Exercises 5.3.7 and 5.3.17.

□

Exercise 5.3.35 Work this problem by hand. Consider the real, symmetric matrix

$$A = \begin{bmatrix} 0 & 1 \\ 1 & 0 \end{bmatrix}.$$

(a) Apply Rayleigh quotient iteration to A with starting vector $q_0 = \begin{bmatrix} 1 & 0 \end{bmatrix}^T$. This is an exceptional case in which Rayleigh quotient iteration fails to converge.

(b) Calculate the eigenvalues and eigenvectors by some other means and show that (i) the iterates q_0, q_1, q_2, ... exactly bisect the angle between the two linearly independent eigenvectors and (ii) the Rayleigh quotients ρ_0, ρ_1, ρ_2, ... lie exactly half way between the two eigenvalues. Thus the sequence "cannot decide" which eigenvector to approach.

□

Exercise 5.3.36 Let

$$A = \begin{bmatrix} 1 & 1 & 1 \\ -1 & 9 & 2 \\ 0 & -1 & 2 \end{bmatrix}.$$

Use Rayleigh quotient iteration with various starting vectors to calculate three linearly independent eigenvectors and the associated eigenvalues. You can use the MATLAB script you wrote for Exercise 5.3.34.

□

Additional Exercises

Exercise 5.3.37 Each of the following sequences of positive numbers converges to zero. For each sequence (a_j), determine whether (a_j) converges (i) slower than linearly (i.e. $\lim_{j \to \infty} a_{j+1}/a_j = 1$), (ii) linearly (i.e. $0 < \lim_{j \to \infty} a_{j+1}/a_j = r <$

1), (iii) quadratically (i.e. $\lim_{j\to\infty} a_{j+1}/a_j^2 = C \neq 0$), or (iv) cubically (i.e. $\lim_{j\to\infty} a_{j+1}/a_j^3 = C \neq 0$).

(a) $10^{-1}, 10^{-2}, 10^{-3}, 10^{-4}, 10^{-5}, \ldots$

(b) $10^{-1}, 10^{-2}, 10^{-4}, 10^{-8}, 10^{-16}, \ldots$

(c) $1, 1/2, 1/3, 1/4, 1/5, \ldots$

(d) $10^{-3}, 10^{-6}, 10^{-9}, 10^{-12}, 10^{-15}, \ldots$

(e) $0.9, 0.81, 0.729, (0.9)^4, (0.9)^5, \ldots$

(f) $10^{-1}, 10^{-3}, 10^{-9}, 10^{-27}, 10^{-81}, \ldots$

Notice that quadratic and cubic convergence are qualitatively better than linear convergence. However, linear convergence can be quite satisfactory if the convergence ratio r is small enough. Convergence that is slower than linear is very slow. □

rcise 5.3.38 The intent of this exercise is to show that quadratic convergence is nearly as good as cubic convergence. Suppose algorithm X produces a sequence (q_j) such that $\| q_j - v_i \| = a_j \to 0$ quadratically. Define a new algorithm Y for which one step consists of two steps of algorithm X. Then algorithm Y produces a sequence $\hat{q}_j$ for which $\| \hat{q}_j - v_i \| = b_j = a_{2j}$. Prove that $b_j \to 0$ *quartically*, that is, $\lim_{j\to\infty} b_{j+1}/b_j^4 = M \neq 0$. Thus the convergence of method Y is faster than cubic. □

rcise 5.3.39 Suppose $q_j \to v$ linearly with convergence ratio r. Show that on each iteration the number of correct decimal digits increases by about $m = -\log_{10} r$. (A rigorous argument is not required.) Thus the rate of increase in the number of correct digits is constant. □

rcise 5.3.40 The residents of Pullburg shop for groceries exactly once a week at one of three markets, named 1, 2, and 3. They are not faithful customers; they switch stores frequently. For $i, j = 1, 2, 3$, let p_{ij} be a number between 0 and 1 that represents the fraction of people who shopped at store j in a given week who will switch to store i in the following week. This can also be interpreted as the probability that a given customer will switch from store j to store i from one week to the next. These numbers can be arranged into a matrix P. Suppose the values of the p_{ij} for Pullburg are

$$P = \begin{bmatrix} 0.63 & 0.18 & 0.14 \\ 0.26 & 0.65 & 0.31 \\ 0.11 & 0.17 & 0.55 \end{bmatrix}.$$

Since $p_{23} = .31$, 31 percent of the people who shop at store 3 in a given week will switch to store 2 in the following week. Fifty-five percent of the people who shopped in store 3 will return to store 3 the following week, and so on. Notice that the sum of the entries in each column of P is one.

(a) A matrix $P \in \mathbb{R}^{n \times n}$ whose entries are nonnegative and whose column sums are all 1 is called a *stochastic* matrix. If Pullburg had n stores, the probabilities p_{ij} would form an $n \times n$ matrix. Show that regardless of what the probabilities are, the resulting $P \in \mathbb{R}^{n \times n}$ must be a stochastic matrix. Show that $w^T P = w^T$, where $w^T = [1, 1, \cdots, 1]$, and 1 is an eigenvalue of P. (We call w^T a *left eigenvector* of P.) The Perron-Frobenius theory of non-negative matrices [53], [60] guarantees that, if all (or "enough") of the p_{ij} are strictly positive, then 1 is the dominant eigenvalue of P.

(b) Let $q_k \in \mathbb{R}^3$ be a vector whose three components represent the fraction of people in Pullburg who shopped at stores 1, 2, and 3, respectively, in week k. For example, if $q_{20} = \begin{bmatrix} 0.24 & 0.34 & 0.42 \end{bmatrix}^T$, then 24% of the population shopped in store 1, 34% in store 2, and 42% in store 3 during week 20. Show that

$$q_{k+1} = Pq_k \qquad (5.3.41)$$

for all k. The sequence (q_k) is called a *Markov chain*. Equation (5.3.41) shows that we can compute a Markov chain by applying the power method with the transition matrix P and scaling factor 1.

(c) Assuming that at week 1 one third of the population shopped at each store, calculate the fraction that visited stores 1, 2, and 3 in subsequent weeks by executing the *Markov process* (5.3.41) repeatedly. Use MATLAB to do the calculations. What is the long-term trend?

(d) How could you have determined the long-term trend without executing the Markov process? Use MATLAB's `eig` command to determine the long-term trend without executing the Markov process.

(e) Calculate the ratio $|\lambda_2 / \lambda_1|$ to determine the rate at which the weekly state vectors q_k approach the limiting value.

$\square$

Exercise 5.3.42 Stochastic matrices, introduced in the previous exercise, provide an example of a class of matrices for which the dominant eigenvalue tends to be significantly larger than the others.

(a) Generate a 20×20 random stochastic matrix. For example,

```
P = rand(20);
P = P*inv(diag(sum(P,1)));
```

The first command (`rand`, not `randn`) generates a matrix with random (non-negative) entries uniformly distributed in $[0, 1]$. The second command rescales each column by dividing it by its column sum. Check that the matrix really is stochastic. Compute the eigenvalues and the ratio $|\lambda_2 / \lambda_1|$ $(= \max\limits_{j \neq 1} |\lambda_j|)$.

(b) Starting from a random starting vector

```
q = rand(20,1); q = q/norm(q,1);
```

compute enough iterates of the Markov chain (q_j) to observe that it converges reasonably quickly.

□

rcise 5.3.43 This exercise is intended to convince you that if a vector q is chosen at random, then in the expansion

$$q = c_1 v_1 + c_2 v_2 + \cdots + c_n v_n$$

the coefficient c_1 is almost certain to be nonzero.

(a) Picture the situation in $\mathbb{R}^2$. Let v_1 and v_2 be any two linearly independent vectors in $\mathbb{R}^2$. The set of all vectors $q = c_1 v_1 + c_2 v_2$ such that $c_1 = 0$ is just the subspace span$\{v_2\}$. Sketch this subspace and note that it is a very small subset of $\mathbb{R}^2$. A vector chosen at random is almost certain not to lie in it.

(b) Repeat part (a) in $\mathbb{R}^3$.

In general any proper subspace is a very small subset of the larger space in which it lies.

□

rcise 5.3.44 Show that the sequence $(q_j) = (A^j q / \lambda_1^j)$ generated by the power method converges linearly to $c_1 v_1$ with convergence ratio $r = |\lambda_2/\lambda_1|$, provided that $|\lambda_1| > |\lambda_2| > |\lambda_3|$, $c_1 \neq 0$, and $c_2 \neq 0$.[27]

□

rcise 5.3.45 What happens if the power method is applied with a starting vector $q = c_1 v_1 + c_2 v_2 + \cdots + c_n v_n$ for which $c_1 = 0 \ldots$

(a) ... assuming the arithmetic is done exactly?

(b) ... assuming roundoff errors are made?

□

rcise 5.3.46   Suppose the power method is applied to $A \in \mathbb{C}^{n \times n}$ using some scaling strategy that produces a sequence (q_j) that converges to a dominant eigenvector. Prove that the sequence of scale factors (s_j) converges to the dominant eigenvalue.

□

rcise 5.3.47 Let $A \in \mathbb{C}^{n \times n}$ be a matrix with eigenvalues λ and μ such that $\lambda \neq \mu$. Suppose that ρ is not an eigenvalue of A.

(a) Show that $\| A - \rho I \| \geq |\lambda - \rho|$, where $\| \cdot \|$ denotes any induced matrix norm.

---

[27]It is possible to construct examples (satisfying $|\lambda_2| = |\lambda_3|$) that violate (5.3.4). These examples do, however, satisfy (5.3.3). See Exercise 5.3.11.

(b) Show that $\kappa(A - \rho I) \geq |\lambda - \rho|/|\mu - \rho|$, where κ denotes the condition number with respect to any induced matrix norm.

(c) Let (ρ_j) be a sequence of shifts such that $\rho_j \rightarrow \mu$. Prove that $\kappa(A - \rho_j I) \rightarrow \infty$ as $j \rightarrow \infty$.

$\square$

Exercise 5.3.48 Prove Theorem 5.3.24 two different ways.

(a) Use a complex QR decomposition of $q \in \mathbb{C}^{n \times 1}$ to prove that the Rayleigh quotient minimizes $\| Aq - \rho q \|_2$. (The factor Q in the QR decomposition satisfies $Q^*Q = I$.)

(b) Letting $\rho = \alpha + \beta i$, the function

$$f(\alpha, \beta) = \| Aq - \rho q \|_2^2 = \langle Aq - \rho q, Aq - \rho q \rangle$$

is a smooth function in the two real variables α and β. Use differential calculus to show that f is minimized if and only if $\rho = \alpha + \beta i$ is the Rayleigh quotient.

$\square$

Exercise 5.3.49 Let $A = \begin{bmatrix} a & b \\ -b & a \end{bmatrix}$, where $a, b \in \mathbb{R}$ and $b \neq 0$. Perform Rayleigh quotient iteration with starting vector $q_0 = \begin{bmatrix} c & d \end{bmatrix}^T$, where $c, d \in \mathbb{R}$, and $c^2 + d^2 = 1$. Analyze the problem. $\square$

5.4 SIMILARITY TRANSFORMS

Two matrices $A, B \in \mathbb{C}^{n \times n}$ are said to be *similar* if there exists a nonsingular $S \in \mathbb{C}^{n \times n}$ such that

$$B = S^{-1}AS. \tag{5.4.1}$$

Equation (5.4.1) is called a *similarity transformation*, and S is called the *transforming matrix*. Obviously equation (5.4.1) is equivalent to

$$AS = SB. \tag{5.4.2}$$

This form is often easier to use.

As we shall see, similar matrices have the same eigenvalues, and their eigenvectors are related in a simple way. Some of the most important eigenvalue algorithms employ a sequence of similarity transformations to reduce a matrix to a simpler form. That is, they replace the original matrix by a similar one whose eigenvalues are either obvious or easily determined. This section prepares us for those algorithms by setting out some of the basic facts about similarity transformations. In the process we will cover some important material on special classes of matrices such as symmetric and orthogonal matrices and their complex counterparts.

Theorem 5.4.3 *Similar matrices have the same eigenvalues.*

Proof. Suppose A and B are similar. Then there is a nonsingular S such that $B = S^{-1}AS$. To show that A and B have the same eigenvalues, it suffices to show that they have the same characteristic polynomial. Now $\lambda I - B = S^{-1}\lambda IS - S^{-1}AS = S^{-1}(\lambda I - A)S$, so $\det(\lambda I - B) = \det(S^{-1})\det(\lambda I - A)\det(S) = \det(\lambda I - A)$. To get this last equality we are using the facts that multiplication of complex numbers is commutative and $\det(S^{-1})\det(S) = 1$. Thus A and B have the same characteristic polynomial. $\qquad\square$

This proof shows that the similarity transformation preserves the *algebraic multiplicity* of the eigenvalue. That is, if μ is a root of order k of the equation $\det(\lambda I - A) = 0$, then it is also a root of $\det(\lambda I - B) = 0$ of order k. The next theorem shows how the eigenvector is transformed.

Theorem 5.4.4 *Suppose $B = S^{-1}AS$. Then v is an eigenvector of A with eigenvalue λ if and only if $S^{-1}v$ is an eigenvector of B with associated eigenvalue λ.*

Proof. Suppose $Av = \lambda v$. Then $B(S^{-1}v) = S^{-1}ASS^{-1}v = S^{-1}Av = S^{-1}\lambda v = \lambda(S^{-1}v)$. Thus $B(S^{-1}v) = \lambda(S^{-1}v)$. Conversely, if $B(S^{-1}v) = \lambda(S^{-1}v)$, one easily deduces that $Av = \lambda v$ by a similar argument. $\qquad\square$

rcise 5.4.5 Let D be any diagonal matrix. Find a set of n linearly independent eigenvectors of D, thereby demonstrating that D is semisimple. $\qquad\square$

The next theorem demonstrates that a matrix is semisimple if and only if it is similar to a diagonal matrix.

Theorem 5.4.6 *Let $A \in \mathbb{C}^{n \times n}$ be a semisimple matrix with linearly independent eigenvectors $v_1, v_2, \dots, v_n$ and eigenvalues $\lambda_1, \lambda_2, \dots, \lambda_n$. Define a diagonal matrix D and a nonsingular matrix V by $D = \operatorname{diag}\{\lambda_1, \dots, \lambda_n\}$ and $V = \begin{bmatrix} v_1 & \cdots & v_n \end{bmatrix}$. Then $V^{-1}AV = D$. Conversely, suppose A satisfies $V^{-1}AV = D$, where D is diagonal and V is nonsingular. Then the columns of V are n linearly independent eigenvectors of A, and the main-diagonal entries of D are the associated eigenvalues. In particular, A is semisimple.*

rcise 5.4.7 Prove Theorem 5.4.6. Work with the simpler equation $AV = VD$ instead of $D = V^{-1}AV$.

(a) Suppose A is semisimple, and D and V are as described. Show that the equations $Av_i = v_i\lambda_i$, $i = 1, \dots, n$, imply $AV = VD$.

(b) Conversely, show that the equation $AV = VD$ implies $Av_i = v_i\lambda_i$, $i = 1, \dots, n$.

$\qquad\square$

Theorem 5.4.6 is about semisimple matrices. With considerable additional effort, we could obtain an extension of Theorem 5.4.6 that is valid for all matrices. See the discussion of the *Jordan canonical form* in [36], [53], [60], or [101].

Theorem 5.4.6 tells us that we can solve the eigenvalue problem completely if we can find a similarity that transforms A to diagonal form. Unfortunately the proof is not constructive; it does not show us a way to construct D and V without knowing the eigenvalues and eigenvectors in advance. The theorem does, however, suggest a simple form toward which we might work. For example, we might try to construct a sequence of similar matrices $A = A_0, A_1, A_2, \ldots$, that converges to diagonal form.

It turns out that we will be able to carry out such a program for certain classes of matrices, but in general it is too ambitious. For one thing, it cannot succeed if the matrix is defective. Even if the matrix is semisimple, the eigenvector matrix V can be ill conditioned, which is almost as bad. Similarity transformations by ill-conditioned matrices can cause disastrous growth of errors.

Geometric View of Semisimple Matrices

Recall from elementary linear algebra that every matrix $A \in \mathbb{C}^{n \times n}$ can be viewed as a linear transformation whose action is to map $x \in \mathbb{C}^n$ to $Ax \in \mathbb{C}^n$. If A is semisimple, then its action on $\mathbb{C}^n$ is easily pictured. A has a set of n linearly independent eigenvectors $v_1, \ldots, v_n$, which form a basis for $\mathbb{C}^n$. Every $x \in \mathbb{C}^n$ can be expressed as a linear combination $x = c_1 v_1 + \cdots + c_n v_n$. The action of A on each $c_i v_i$ is simply to multiply it by the scalar λ_i, and the action of A on x is just a sum of such simple actions: $Ax = \lambda_1 c_1 v_1 + \lambda_2 c_2 v_2 + \cdots + \lambda_n c_n v_n$.

Theorem 5.4.6 is actually just a restatement of this fact in matrix form. We will attempt to clarify this without going into detail. Recall again from elementary linear algebra that a linear transformation $A : \mathbb{C}^n \rightarrow \mathbb{C}^n$ can be represented in numerous ways. Given any basis of $\mathbb{C}^n$, there is a unique matrix that represents A with respect to that basis. Two matrices represent the same linear transformation (with respect to different bases) if and only if they are similar. Thus a similarity transformation amounts just to a change of basis, that is, a change of coordinate system. Theorem 5.4.6 says that if A is semisimple, then there exists a coordinate system in which it is represented by a diagonal matrix. This coordinate system has eigenvectors of A as its coordinate axes.

Unitary Similarity Transformations

In Chapter 3 we introduced orthogonal matrices and noted that they have numerous desirable properties. The complex analogue of the orthogonal matrix is the unitary matrix, which was also introduced in Chapter 3. Recall that a matrix $U \in \mathbb{C}^{n \times n}$ is *unitary* if $U^* U = I$, that is, $U^* = U^{-1}$. The class of unitary matrices contains the (real) orthogonal matrices. In Exercises 3.2.50 through 3.2.58 you showed that: (1) The product of unitary matrices is unitary. (2) The inverse of a unitary matrix is unitary. (3) The complex inner product $\langle x, y \rangle = y^* x$ and Euclidean norm are preserved by unitary matrices; that is, $\langle Ux, Uy \rangle = \langle x, y \rangle$ and $\| Ux \|_2 = \| x \|_2$ for all $x, y \in \mathbb{C}^n$. (4) Rotators and reflectors have complex analogues. (5) Every $A \in \mathbb{C}^{n \times n}$ can be expressed as a product $A = QR$, where Q is unitary and R is upper triangular. You also showed (Exercise 3.4.29) that a matrix $U \in \mathbb{C}^{n \times n}$ is unitary if and only if its columns are orthonormal. Of course the orthonormality is with respect to the complex inner product.

Two matrices A, $B \in \mathbb{C}^{n \times n}$ are *unitarily similar* if there is a unitary matrix $U \in \mathbb{C}^{n \times n}$ such that $B = U^{-1}AU$. Since $U^{-1} = U^*$, the similarity can also be expressed as $B = U^*AU$. If A, B, and U are all real, then U is orthogonal, and A and B are said to be *orthogonally similar*. Unitary similarity transformations have some nice properties not possessed by similarity transformations in general.

rcise 5.4.8

(a) Show that if U is unitary, then $\|U\|_2 = 1$ and $\kappa_2(U) = 1$.

(b) Show that if A and B are unitarily similar, then $\|B\|_2 = \|A\|_2$ and $\kappa_2(A) = \kappa_2(B)$.

(c) Suppose $B = U^*AU$, where U is unitary. Show that if A is perturbed slightly, then the resulting perturbation in B is of the same magnitude. Specifically, show that if $B + \delta B = U^*(A + \delta A)U$, then $\|\delta B\|_2 = \|\delta A\|_2$.

$\square$

This exercise shows that any errors that a matrix may contain will not be amplified by subsequent unitary similarity transformations. The same cannot be said of arbitrary similarity transformations (Exercises 5.4.28 and 5.4.29).

The results in Exercise 5.4.8 are related to results from Section 3.2, in which we showed that algorithms built using well-conditioned transformation matrices, in particular rotators and reflectors, are normwise backward stable. Those results are valid for unitary matrices as well. Thus all algorithms that consist of a sequence of transformations by unitary matrices (and in particular the complex analogues of reflectors and plane rotators) are normwise backward stable.

In addition to their favorable error propagation properties, unitary similarity transformations also preserve certain desirable matrix structures. For example, consider the following theorem. (For other examples see Exercises 5.4.36 to 5.4.42.)

Theorem 5.4.9 *If $A = A^*$ and A is unitarily similar to B, then $B = B^*$.*

Proof. Since $B = U^*AU$ for some unitary U, we have $B^* = (U^*AU)^* = U^*A^*U^{**} = U^*AU = B$. $\square$

rcise 5.4.10 Show by example that the conclusion of Theorem 5.4.9 does not hold for general similarity transformations. $\square$

Recall that a matrix that satisfies $A = A^*$ is called *Hermitian*. In words, Theorem 5.4.9 states that the Hermitian property is preserved under unitary similarity transformations. If we take all of the matrices in Theorem 5.4.9 to be real, we see that if A is real and symmetric and B is orthogonally similar to A, then B is also symmetric. In other words, the symmetry property is preserved under orthogonal similarity transformations. Eigenvalue problems for which the coefficient matrix is symmetric occur frequently in applications. Since symmetric and Hermitian matrices have special properties that make them easier to handle than general matrices, it is useful to have at hand a class of similarity transformations that preserve these properties.

The next result, Schur's theorem, is the most important result of this section. It states that every (square) matrix is unitarily similar to a triangular matrix.

Theorem 5.4.11   *(Schur's Theorem) Let $A \in \mathbb{C}^{n \times n}$. Then there exists a unitary matrix $U \in \mathbb{C}^{n \times n}$ and an upper-triangular matrix $T \in \mathbb{C}^{n \times n}$ such that $T = U^* A U$.*

We can equally well write $A = UTU^*$. This is a *Schur decomposition* of A.
Proof. The proof is by induction on n. The result is trivial for $n = 1$. Now let us show that it holds for $n = k$, given that it holds for $n = k - 1$. Let $A \in \mathbb{C}^{k \times k}$. Let λ be an eigenvalue of A and v an associated eigenvector, chosen so that $\| v \|_2 = 1$. Let U_1 be any unitary matrix that has v as its first column. There are many such matrices: just take any orthonormal basis of $\mathbb{C}^k$ whose first member is v, and let U_1 be the matrix whose columns are the members of the basis. Let $W \in \mathbb{C}^{k \times (k-1)}$ denote the submatrix of U_1 consisting of columns 2 through k, so that $U_1 = \begin{bmatrix} v & W \end{bmatrix}$. Since the columns of W are orthogonal to v, $W^* v = 0$. Let $A_1 = U_1^* A U_1$. Then

$$A_1 = \begin{bmatrix} v^* \\ W^* \end{bmatrix} A \begin{bmatrix} v & W \end{bmatrix} = \begin{bmatrix} v^* A v & v^* A W \\ W^* A v & W^* A W \end{bmatrix}.$$

Since $Av = \lambda v$, it follows that $v^* A v = \lambda$ and $W^* A v = \lambda W^* v = 0$. Let $\hat{A} = W^* A W$. Then A_1 has the form

$$A_1 = \begin{bmatrix} \lambda & * & \cdots & * \\ 0 & & & \\ \vdots & & \hat{A} & \\ 0 & & & \end{bmatrix}.$$

$\hat{A} \in \mathbb{C}^{(k-1) \times (k-1)}$, so by the induction hypothesis there exists a unitary matrix $\hat{U}_2$ and an upper-triangular matrix $\hat{T}$ such that $\hat{T} = \hat{U}_2^* \hat{A} \hat{U}_2$. Define $U_2 \in \mathbb{C}^{k \times k}$ by

$$U_2 = \begin{bmatrix} 1 & 0 & \cdots & 0 \\ 0 & & & \\ \vdots & & \hat{U}_2 & \\ 0 & & & \end{bmatrix}.$$

Then U_2 is unitary, and

$$U_2^* A_1 U_2 = \begin{bmatrix} \lambda & * & \cdots & * \\ 0 & & & \\ \vdots & & \hat{U}_2^* \hat{A} \hat{U}_2 & \\ 0 & & & \end{bmatrix} = \begin{bmatrix} \lambda & * & \cdots & * \\ 0 & & & \\ \vdots & & \hat{T} & \\ 0 & & & \end{bmatrix},$$

which is upper triangular. Let us call this matrix T, and let $U = U_1 U_2$. Then $T = U_2^* A_1 U_2 = U_2^* U_1^* A U_1 U_2 = U^* A U$. $\qquad\qquad\qquad\qquad\qquad\qquad\qquad\qquad\square$

The main-diagonal entries of T are the eigenvalues of A. Thus, if we can find the unitary similarity transformation U that transforms A to upper triangular form, then we have the eigenvalues of A. Unfortunately the proof of Schur's theorem is non-constructive; it does not give us a recipe for computing U without knowing eigenvectors in advance. Nevertheless it gives us reason to hope that we will be able to create an algorithm that produces a sequence of unitarily similar matrices $A = A_0$, A_1, A_2, ... that converges to upper-triangular form. Indeed there is an algorithm that does essentially this, namely, Francis's algorithm, which we will introduce in Section 5.6.

It follows easily from the equation $T = U^* A U$ that the first column of U is necessarily an eigenvector of A (regardless of how we may choose to compute U and T). Indeed, rewrite the equation as $AU = UT$, then compare first columns. The first column of AU is Au_1, where u_1 denotes the first column of U. This must equal the first column of UT, which is $u_1 t_{11}$. Thus $Au_1 = u_1 t_{11}$, which means that u_1 is an eigenvector of A with associated eigenvalue t_{11}. In general the other columns of U are not eigenvectors of A. The extraction of the other eigenvectors requires a bit more work. See Exercise 5.4.32 and Section 5.7.

Schur's theorem is comparable in spirit to Theorem 5.4.6, which states, in part, that every semisimple matrix is similar to a diagonal matrix. Schur's theorem is more modest in the sense that the triangular form is not as simple and elegant as the diagonal form of Theorem 5.4.6. On the other hand, Schur's theorem is valid for all matrices, not just the semisimple ones. Moreover, the unitary similarity transformations of Schur's theorem are well behaved from a numerical standpoint.

A class of matrices for which Schur's theorem and Theorem 5.4.6 overlap is that of Hermitian matrices. If A is Hermitian, the matrix $T = U^* A U$ is not only upper triangular but Hermitian as well (Theorem 5.4.9). This obviously implies that T is diagonal. This result is known as the *spectral theorem* for Hermitian matrices.

Theorem 5.4.12 (Spectral Theorem for Hermitian Matrices) *Suppose $A \in \mathbb{C}^{n \times n}$ is Hermitian. Then there exists a unitary matrix $U \in \mathbb{C}^{n \times n}$ and a diagonal matrix $D \in \mathbb{R}^{n \times n}$ such that $D = U^* A U$. The columns of U are eigenvectors and the main-diagonal entries of D are eigenvalues of A.*

We can equally well write $A = UDU^*$. This is a *spectral decomposition* of A.

That the columns of U are eigenvectors follows from the last part of Theorem 5.4.6. The diagonal matrix is real because $D = D^*$. This proves that the eigenvalues of a Hermitian matrix are real.

Corollary 5.4.13 *The eigenvalues of a Hermitian matrix are real. In particular, the eigenvalues of a real symmetric matrix are real.*

Corollary 5.4.14 *Every Hermitian matrix in $\mathbb{C}^{n \times n}$ has a set of n orthonormal eigenvectors. In particular, every Hermitian matrix is semisimple.*

Theorem 5.4.15 *Let A be a Hermitian matrix, and let v and w be eigenvectors of A associated with distinct eigenvalues. then v and w are necessarily orthogonal.*

Proof. Suppose $Av = \lambda v$ and $Aw = \mu w$, where $\lambda \neq \mu$. Then

$$\lambda w^* v = w^* A v = w^* A^* v = (Aw)^* v = \mu w^* v.$$

Since $\lambda \neq \mu$, we must have $w^* v = 0$. Thus v and w are orthogonal. $\square$

There are other classes of matrices for which spectral theorems analogous to Theorem 5.4.12 hold. A matrix is called *skew Hermitian* if $A^* = -A$. We are already familiar with the *unitary* matrices, which satisfy $A^* = A^{-1}$ or, equivalently, $A^*A = AA^* = I$. A matrix is *normal* if $A^*A = AA^*$. Spectral theorems for these classes of matrices are discussed in Exercises 5.4.40 through 5.4.42.

The class of normal matrices is of interest for at least two reasons. For one thing, it contains all Hermitian, skew-Hermitian, and unitary matrices, so every property that holds for normal matrices holds for these classes of matrices as well. Furthermore, this is exactly the class of matrices for which a spectral theorem holds; the following theorem states that a matrix is unitarily similar to a diagonal matrix if *and only if* it is normal.

Theorem 5.4.16   *(Spectral Theorem for Normal Matrices)*   *Let $A \in \mathbb{C}^{n \times n}$. Then A is normal if and only if there exists a unitary matrix $U \in \mathbb{C}^{n \times n}$ and a diagonal matrix $D \in \mathbb{C}^{n \times n}$ such that $D = U^* A U$.*

This theorem is proved in Exercise 5.4.42.

Corollary 5.4.17   *(a) Let $A \in \mathbb{C}^{n \times n}$ be normal. Then A has a set of n orthonormal eigenvectors.*

(b) Conversely, if $A \in \mathbb{C}^{n \times n}$ has a set of n orthonormal eigenvectors, then A is normal.

Exercise 5.4.18   Verify Corollary 5.4.17. $\square$

Corollary 5.4.19   *(a) Every normal matrix is semisimple.*

(b) Every skew-Hermitian matrix has a set of n orthonormal eigenvectors. In particular, every skew-Hermitian matrix is semisimple.

(c) Every unitary matrix has a set of n orthonormal eigenvectors. In particular, every unitary matrix is semisimple.

Real Matrices

If we want to solve the general eigenvalue problem, we have to be prepared to deal with complex numbers, even if the matrix is real. There is, however, one important class of real matrices for which we can solve the entire eigenvalue problem without going outside of the real number system, namely the symmetric matrices.

Theorem 5.4.20 (Spectral Theorem for Real Symmetric Matrices) *Suppose $A \in \mathbb{R}^{n \times n}$ is symmetric. Then there exists an orthogonal matrix $U \in \mathbb{R}^{n \times n}$ and a diagonal matrix $D \in \mathbb{R}^{n \times n}$ such that $D = U^T A U$.*

Proof. The proof is by induction on n. It is the same as the proof of Schur's theorem, except that the matrices are real and we can exploit symmetry. Therefore we just sketch the proof.

Let λ be any eigenvalue of A. Since λ is real, it has a real eigenvector v, which may be chosen so that $\| v \|_2 = 1$, associated with it. Let U_1 be a real orthogonal matrix whose first column is v, and let $A_1 = U_1^T A U_1$. Then A_1 is real and symmetric, and (as in the proof of Schur's theorem)

$$A_1 = \left[\begin{array}{c|ccc} \lambda & 0 & \cdots & 0 \\ \hline 0 & & & \\ \vdots & & \hat{A} & \\ 0 & & & \end{array} \right].$$

Since $\hat{A} \in \mathbb{R}^{(n-1) \times (n-1)}$ is symmetric, we can assume inductively that there is an orthogonal matrix $\hat{U}_2$ and a diagonal matrix $\hat{D}$ such that $\hat{D} = \hat{U}_2^T \hat{A} \hat{U}_2$. Let

$$U_2 = \left[\begin{array}{c|ccc} 1 & 0 & \cdots & 0 \\ \hline 0 & & & \\ \vdots & & \hat{U}_2 & \\ 0 & & & \end{array} \right], \qquad D = \left[\begin{array}{c|ccc} \lambda & 0 & \cdots & 0 \\ \hline 0 & & & \\ \vdots & & \hat{D} & \\ 0 & & & \end{array} \right],$$

and $U = U_1 U_2$. Then U is orthogonal and $D = U^T A U$. □

cise 5.4.21 Write down a detailed proof of Theorem 5.4.20. □

Corollary 5.4.22 *Let $A \in \mathbb{R}^{n \times n}$ be symmetric. Then A has a set of n real, orthonormal eigenvectors.*

When working with nonsymmetric matrices, we must be prepared to deal with complex numbers. Nevertheless it usually pays to delay their introduction for as long as possible, since complex arithmetic is much slower than real arithmetic. It turns out that we can bring a matrix nearly to triangular form without using complex numbers.

A matrix $T \in \mathbb{R}^{n \times n}$ is called *quasi-triangular* if it has the block upper triangular form

$$T = \left[\begin{array}{cccc} T_{11} & T_{12} & \cdots & T_{1m} \\ 0 & T_{22} & & T_{2m} \\ \vdots & \ddots & \ddots & \vdots \\ 0 & \cdots & 0 & T_{mm} \end{array} \right],$$

where each main diagonal block is either 1×1 or 2×2, and each 2×2 block has complex eigenvalues. Thus the 1×1 blocks carry the real eigenvalues of T, and the 2×2 blocks carry the complex conjugate pairs. It is a simple matter to compute the eigenvalues of a 2×2 matrix. One simply applies the quadratic formula (carefully) to its characteristic polynomial. The Wintner-Murnaghan Theorem, also known as the Real Schur Theorem, states that every matrix in $\mathbb{R}^{n \times n}$ is orthogonally similar to a quasi-triangular matrix.

Theorem 5.4.23   *(Wintner-Murnaghan) Let $A \in \mathbb{R}^{n \times n}$. Then there exist an orthogonal matrix $U \in \mathbb{R}^{n \times n}$ and a quasi-triangular matrix $T \in \mathbb{R}^{n \times n}$ such that $T = U^T A U$.*

The proof is like that of Schur's theorem, except that the dimension has to be reduced by two whenever we encounter a pair of complex eigenvalues. See Exercise 5.4.49 or Exercise 6.1.15.

If A is symmetric, the Wintner-Murnaghan Theorem reduces to Theorem 5.4.20. Other classes of real matrices for which the Wintner-Murnaghan Theorem simplifies nicely are the skew-symmetric ($A^T = -A$), orthogonal ($A^T = A^{-1}$), and normal ($A^T A = A A^T$) matrices. See Exercises 5.4.50 through 5.4.53.

Balancing

We end this section with a brief mention of a computational issue related to the minimization of roundoff errors that is relevant to nonsymmetric matrices. Probably the simplest type of similarity transformation is the diagonal scaling $D^{-1}AD$, where $D = \operatorname{diag}\{d_1, \ldots, d_n\}$ is a diagonal matrix. Although A and $D^{-1}AD$ have the same eigenvalues, it does not follow that an algorithm will be able to compute the eigenvalues of each with equal accuracy. On the assumption that the errors in computing the eigenvalues of A are proportional to $u\|A\|_F$, it has been suggested that a diagonal D should be chosen so that $\|D^{-1}AD\|_F$ is minimized or approximately minimized, and the eigenvalue computation should be applied to $D^{-1}AD$ instead of A. This procedure, which is known as *balancing*, has been shown to reduce errors in many instances. A specific procedure for balancing matrices is laid out in detail in [72], and this procedure is used in LAPACK and MATLAB. When MATLAB's `eig` command is used to compute the eigenvalues of a nonsymmetric matrix, the default action is to balance the matrix before doing anything else.

Additional Exercises

Exercise 5.4.24 MATLAB's command `[V,D] = eig(A)` produces a diagonal D whose main-diagonal entries are the eigenvalues of A. The associated eigenvectors are the columns of V. The matrices satisfy $AV = VD$. Normally V is nonsingular, and we have $D = V^{-1}AV$, as in Theorem 5.4.6. However, this is impossible if A is not semisimple. Try out MATLAB's `eig` command on the defective matrix

$$A = \begin{bmatrix} 2 & 0 & 0 \\ 0 & 2 & 1 \\ 0 & 0 & 2 \end{bmatrix}.$$

Comment on the output.   □

rcise 5.4.25 The *geometric multiplicity* of an eigenvalue is defined to be the dimension of the associated eigenspace $\{v \in \mathbb{C}^n \mid Av = \lambda v\}$. Show that a similarity transformation preserves the geometric multiplicity of each eigenvalue.   □

rcise 5.4.26 Let A be a semisimple matrix. How are the geometric and algebraic multiplicities related? Recall that the algebraic multiplicity of an eigenvalue λ is equal to the multiplicity of λ as a root of the characteristic equation.   □

rcise 5.4.27 Suppose A is similar to B. Show that if A is nonsingular, then B is also nonsingular, and A^{-1} is similar to B^{-1}.   □

rcise 5.4.28 Let $S = \begin{bmatrix} 1 & \alpha \\ 0 & 1 \end{bmatrix}$, where α is a real parameter.

(a) Let $A = \begin{bmatrix} 2 & 0 \\ 0 & 1 \end{bmatrix}$ and $B = S^{-1}AS$. Calculate B and conclude that $\|B\|_\infty$ can be made arbitrarily large by taking α large. (The same is true of $\|B\|_2$. We work with the ∞-norm for convenience.)

(b) Let $A = \begin{bmatrix} 1 & 0 \\ 0 & 1 \end{bmatrix}$, and $\delta A = \dfrac{\epsilon}{2}\begin{bmatrix} 1 & 1 \\ 1 & 1 \end{bmatrix}$. Notice that $\|\delta A\|_\infty / \|A\|_\infty = \epsilon$. Let $B = S^{-1}AS$ and $B + \delta B = S^{-1}(A + \delta A)S$. Calculate B and δB, and show that $\|\delta B\|_\infty / \|B\|_\infty$ can be made arbitrarily large by taking α large. Notice that this example is even more extreme than that of part (a), since $\|\delta B\|_\infty$ is (asymptotically) proportional to α^2, not α.

    □

rcise 5.4.29 Suppose $B = S^{-1}AS$ and $B + \delta B = S^{-1}(A + \delta A)S$. Show that

(a) $\dfrac{1}{\kappa(S)}\|A\| \leq \|B\| \leq \kappa(S)\|A\|$,

(b) $\dfrac{1}{\kappa(S)^2}\dfrac{\|\delta A\|}{\|A\|} \leq \dfrac{\|\delta B\|}{\|B\|} \leq \kappa(S)^2\dfrac{\|\delta A\|}{\|A\|}$,

where $\|\cdot\|$ denotes any matrix norm, and κ denotes the condition number associated with that norm. Notice that the results of parts (b) and (c) of Exercise 5.4.8 are special cases of these results.   □

rcise 5.4.30 Let $v \in \mathbb{C}^n$ be any nonzero vector. Show how to build a (complex) reflector whose first column is a multiple of v. Such a reflector could play the role of U_1 in the proof of Schur's theorem.   □

Exercise 5.4.31 Show that in Schur's theorem . . .

    (a) T can be chosen so that the eigenvalues of A appear in any desired order on the main diagonal of T.

    (b) U can be chosen so that its first column equals any desired eigenvector of A for which $\|v\|_2 = 1$. (*Hint:* Show that in the proof of Schur's theorem, the first column of U_1 equals the first column of U.)

    □

Exercise 5.4.32 Let $T \in \mathbb{C}^{n \times n}$ be an upper-triangular matrix with distinct eigenvalues. Sketch an algorithm that calculates a set of n linearly independent eigenvectors of T by finding solutions of equations of the form $(\lambda I - T)v = 0$. About how many flops does the whole operation require? What difficulties can arise when the eigenvalues are not distinct? This exercise will be worked out in Section 5.7.   □

Exercise 5.4.33 Let $A \in \mathbb{C}^{n \times n}$ be a defective matrix. Use Schur's theorem to show that for every $\epsilon > 0$, there is a semisimple matrix $A_\epsilon \in \mathbb{C}^{n \times n}$ such that $\|A - A_\epsilon\|_2 < \epsilon$. Thus the set of semisimple matrices is dense in $\mathbb{C}^{n \times n}$.   □

Exercise 5.4.34 In Section 5.3 it was claimed that for real symmetric matrices, the Rayleigh quotient gives a particularly good approximation to the associated eigenvalue. In this exercise we verify that claim for Hermitian matrices. Let $A \in \mathbb{C}^{n \times n}$ be Hermitian, and let $q \in \mathbb{C}^n$ be a vector satisfying $\|q\|_2 = 1$ that approximates an eigenvector v of A. Assume that $\|v\|_2 = 1$ as well. By Corollary 5.4.14, A has n orthonormal eigenvectors $v_1, \ldots, v_n$, with associated eigenvalues $\lambda_1, \ldots, \lambda_n$. We can choose v_1, $\ldots, v_n$ in such a way that $v_1 = v$. The approximate eigenvector q can be expressed as a linear combination of $v_1, \ldots, v_n$: $q = c_1 v_1 + \cdots + c_n v_n$.

    (a) Show that $\sum_{k=1}^{n} |c_k|^2 = \|q\|_2^2 = 1$.

    (b) Show that $\sum_{k=2}^{n} |c_k|^2 \leq \|v_1 - q\|_2^2$.

    (c) Derive an expression for the Rayleigh quotient $\rho = q^* A q$ in terms of the coefficients c_i and the eigenvalues λ_i.

    (d) Show that $|\lambda_1 - \rho| \leq C \|v_1 - q\|_2^2$, where $C = \max_{2 \leq k \leq n} |\lambda_1 - \lambda_k|$.

    (*Hint:* $\lambda_1 = \sum_{k=1}^{n} \lambda_1 |c_k|^2$.)

Thus if $\|v_1 - q\|_2 = O(\epsilon)$, then $|\lambda_1 - \rho| = O(\epsilon^2)$.   □

Exercise 5.4.35 Use Rayleigh quotients to obtain a second proof that the eigenvalues of a Hermitian matrix are real (cf. Exercise 1.4.63).   □

Exercise 5.4.36 Recall that a Hermitian matrix is called *positive definite* if for all nonzero $x \in \mathbb{C}^n$, $x^* A x > 0$. Prove that if A is positive definite and B is unitarily similar to A, then B is also positive definite.   □

rcise 5.4.37 Use Rayleigh quotients to prove that the eigenvalues of a positive definite matrix are positive. The next exercise indicates a second way to prove this.   □

rcise 5.4.38 Let $A \in \mathbb{C}^{n \times n}$ be a Hermitian matrix. Use Theorem 5.4.12 and the result of Exercise 5.4.36 to prove that A is positive definite if and only if all of its eigenvalues are positive.   □

rcise 5.4.39 A Hermitian matrix $A \in \mathbb{C}^{n \times n}$ is *positive semidefinite* if $x^* A x \geq 0$ for all $x \in \mathbb{C}^n$. Formulate and prove results analogous to those of Exercises 5.4.36 to 5.4.38 for positive semidefinite matrices.   □

rcise 5.4.40 A matrix $A \in \mathbb{C}^{n \times n}$ is *skew Hermitian* if $A^* = -A$.

   (a) Prove that if A is skew Hermitian and B is unitarily similar to A, then B is also skew Hermitian.

   (b) What special form does Schur's theorem (5.4.11) take when A is skew Hermitian?

   (c) Prove that the eigenvalues of a skew Hermitian matrix are purely imaginary; that is, they satisfy $\overline{\lambda} = -\lambda$. Give two proofs, one based on Schur's theorem and one based on the Rayleigh quotient.

                                         □

rcise 5.4.41

   (a) Prove that if A is unitary and B is unitarily similar to A, then B is also unitary.

   (b) Prove that a matrix $T \in \mathbb{C}^{n \times n}$ that is both upper triangular and unitary must be a diagonal matrix. (You will prove a more general result in Exercise 5.4.42.)

   (c) What special form does Schur's theorem (5.4.11) take when A is unitary?

   (d) Prove that the eigenvalues of a unitary matrix satisfy $\overline{\lambda} = \lambda^{-1}$. Equivalently; $\lambda \overline{\lambda} = 1$ or $|\lambda|^2 = 1$; that is, the eigenvalues lie on the unit circle in the complex plane. Give two proofs.

                                         □

rcise 5.4.42 A matrix $A \in \mathbb{C}^{n \times n}$ is *normal* if $AA^* = A^*A$.

   (a) Prove that all Hermitian, skew-Hermitian, and unitary matrices are normal.

   (b) Prove that if A is normal and B is unitarily similar to A, then B is also normal.

   (c) Prove that a matrix $T \in \mathbb{C}^{n \times n}$ that is both upper triangular and normal must be a diagonal matrix. (*Hint:* Use induction on n. Write T in the partitioned form

$$T = \begin{bmatrix} t_{11} & s^* \\ 0 & \hat{T} \end{bmatrix},$$

then write down the equation $TT^* = T^*T$ in partitioned form and deduce that $s = 0$ and $\hat{T}$ is also (triangular and) normal.)

(d) Prove that every diagonal matrix is normal.

(e) Prove Theorem 5.4.16: A is normal if and only if A is unitarily similar to a diagonal matrix.

$\square$

Exercise 5.4.43   Let $D \in \mathbb{C}^{n \times n}$ be a diagonal matrix. Show that ...

(a) D is Hermitian if and only if its eigenvalues are real.

(b) D is positive semidefinite if and only if its eigenvalues are nonnegative.

(c) D is positive definite if and only if its eigenvalues are positive.

(d) D is skew Hermitian if and only if its eigenvalues lie on the imaginary axis of the complex plane.

(e) D is unitary if and only if its eigenvalues lie on the unit circle of the complex plane.

$\square$

Exercise 5.4.44   Let $A \in \mathbb{C}^{n \times n}$ be a normal matrix. Show that ...

(a) A is Hermitian if and only if its eigenvalues lie on the real axis.

(b) A is positive semidefinite if and only if its eigenvalues are nonnegative.

(c) A is positive definite if and only if its eigenvalues are positive.

(d) A is skew Hermitian if and only if its eigenvalues lie on the imaginary axis.

(e) A is unitary if and only if its eigenvalues lie on the unit circle.

$\square$

Exercise 5.4.45   Verify that the results of Exercise 5.4.34 can be carried over verbatim to normal matrices. Thus the good approximation properties of the Rayleigh quotient hold for normal matrices in general, not just Hermitian matrices. $\square$

Exercise 5.4.46   Let $A \in \mathbb{R}^{n \times m}$. Then $A^T A$ and $A A^T$ are real symmetric matrices, so each has a spectral decomposition as guaranteed by Theorem 5.4.20. For example, there is an orthogonal matrix $W \in \mathbb{R}^{m \times m}$ and a diagonal matrix $\Lambda \in \mathbb{R}^{m \times m}$ such that $A^T A = W \Lambda W^T$. How are the spectral decompositions of $A^T A$ and $A A^T$ related to the singular value decomposition $A = U \Sigma V^T$? $\square$

Exercise 5.4.47   Show that if $A \in \mathbb{C}^{n \times n}$ is normal, then its eigenvectors are singular vectors, and the eigenvalues and singular values are related by $\sigma_j = |\lambda_j|$, $j = 1, \ldots, n$, if ordered by decreasing magnitude. In particular, if A is positive definite, its eigenvalues are the same as its singular values. □

cise 5.4.48 Let $A \in \mathbb{R}^{n \times m}$ be a nonzero matrix, and consider the related square real symmetric matrix

$$C = \begin{bmatrix} 0 & A^T \\ A & 0 \end{bmatrix}.$$

Suppose $\begin{bmatrix} v \\ u \end{bmatrix}$ is an eigenvector of C associated with the nonzero eigenvalue σ. Here $v \in$ and $u \in \mathbb{R}^n$.

(a) Show that $Av = \sigma u$ and $A^T u = \sigma v$.

(b) Show that $\begin{bmatrix} v & -u \end{bmatrix}$ is an eigenvector of C associated with the eigenvalue $-\sigma$. Thus the nonzero eigenvalues of C come in $\pm$ pairs.

(c) By Theorem 5.4.15 these two eigenvectors must be orthogonal to each other. Deduce that $\|v\|_2 = \|u\|_2$.

(d) Without loss of generality assume $\sigma > 0$, $\|v\|_2 = 1$, and $\|u\|_2 = 1$. Deduce that v and u are right and left singular vectors of A, respectively, associated with the singular value σ.

(e) Conversely, show that if v and u are right and left singular values of A, respectively, associated with the singular value σ, then $\begin{bmatrix} v \\ u \end{bmatrix}$ and $\begin{bmatrix} v \\ -u \end{bmatrix}$ are eigenvectors of C associated with eigenvalues $\pm\sigma$. Thus the singular values of A are exactly the positive eigenvalues of C.

□

cise 5.4.49 Prove Theorem 5.4.23 (Wintner-Murnaghan) by induction on n, using the same general strategy as in the proof of Schur's theorem. Let λ be an eigenvalue of the matrix $A \in \mathbb{R}^{n \times n}$. If λ is real, reduce the problem exactly as in Schur's Theorem. Now let us focus on the complex case. Suppose $Av = \lambda v$, where $v = x + iy$, x, $y \in \mathbb{R}^n$, $\lambda = \alpha + i\beta$, $\alpha, \beta \in \mathbb{R}$, and $\beta \neq 0$.

(a) Show that $AZ = ZB$, where $Z = \begin{bmatrix} x & y \end{bmatrix}$ and $B = \begin{bmatrix} \alpha & \beta \\ -\beta & \alpha \end{bmatrix}$. Show that the eigenvalues of B are λ and $\overline{\lambda}$.

(b) Show that the matrix Z has rank two (full rank). In other words, x and y are linearly independent. (Show that if they are not independent, then x and y must satisfy $Ax = \lambda x$ and $Ay = \lambda y$, and λ must be real.)

(c) Consider a condensed QR decomposition (Theorem 3.4.7) $Z = VR$, where $V \in \mathbb{R}^{n \times 2}$ has orthonormal columns, and $R \in \mathbb{R}^{2 \times 2}$ is upper triangular. Show that $AV = VC$, where C is similar to B.

(d) Show that there exists an orthogonal matrix $U_1 \in \mathbb{R}^{n \times n}$ whose first two columns are the columns of V. Let $A_1 = U_1^T A U_1$. Show that A_1 has the form

$$
A_1 = \left[
\begin{array}{cc|ccc}
& & * & \cdots & * \\
\multicolumn{2}{c|}{\raisebox{1.5ex}{C}} & * & \cdots & * \\
\hline
0 & 0 & & & \\
\vdots & \vdots & & \hat{A} & \\
0 & 0 & & &
\end{array}
\right],
$$

where $\hat{A} \in \mathbb{R}^{(n-2) \times (n-2)}$.

(e) Apply the induction hypothesis to $\hat{A}$ to complete the proof.

(f) Write out a complete and careful proof of the Wintner-Murnaghan Theorem.

Remarks: The equation $AZ = ZB$ implies that the columns of Z span an *invariant subspace* of A. We will discuss invariant subspaces systematically in Section 6.1. Part (b) shows that this subspace has dimension 2. Part (c) introduces an orthonormal basis for the space. We note finally that the eigenvalues and the 2×2 blocks can be made to appear in any order on the main diagonal of T. □

Exercise 5.4.50 A matrix $A \in \mathbb{R}^{n \times n}$ is *skew symmetric* if $A^T = -A$. Thus a skew-symmetric matrix is just a skew-Hermitian matrix that is real. In particular, the eigenvalues of a skew-symmetric matrix are purely imaginary (Exercise 5.4.40).

(a) Show that if $A \in \mathbb{R}^{n \times n}$ is skew symmetric and n is odd, then A is singular.

(b) What does a 1×1 skew-symmetric matrix look like? What does a 2×2 skew-symmetric matrix look like?

(c) What special form does Theorem 5.4.23 take when A is skew symmetric? Be as specific as you can.

□

Exercise 5.4.51 The *trace* of a matrix $A \in \mathbb{C}^{n \times n}$, denoted $\mathrm{tr}(A)$, is defined to be the sum of the main diagonal entries: $\mathrm{tr}(A) = \sum_{k=1}^{n} a_{kk}$.

(a) Show that for $C, D \in \mathbb{C}^{n \times n}$, $\mathrm{tr}(C + D) = \mathrm{tr}(C) + \mathrm{tr}(D)$.

(b) Show that for $C \in \mathbb{C}^{n \times m}$ and $D \in \mathbb{C}^{m \times n}$, $\mathrm{tr}(CD) = \mathrm{tr}(DC)$.

(c) Show that $\| B \|_F^2 = \mathrm{tr}(B^* B) = \mathrm{tr}(BB^*)$, where $\| \cdot \|_F$ denotes the Frobenius norm (Example 2.1.22).

(d) Suppose $A \in \mathbb{C}^{n \times n}$ is normal, and consider the partition

$$
A = \left[
\begin{array}{cc}
A_{11} & A_{12} \\
A_{21} & A_{22}
\end{array}
\right],
$$

where A_{11} and A_{22} are square matrices. Show that $\|A_{21}\|_F = \|A_{12}\|_F$. (*Hint:* Write the equation $A^*A = AA^*$ in partitioned form, and take the trace of the (1,1) block.)

□

rcise 5.4.52

(a) Suppose $T \in \mathbb{C}^{n \times n}$ is normal and has the block-triangular form

$$T = \begin{bmatrix} T_{11} & T_{12} \\ 0 & T_{22} \end{bmatrix},$$

where T_{11} and T_{22} are square matrices. Show that $T_{12} = 0$, and T_{11} and T_{22} are normal.

(b) Suppose $T \in \mathbb{C}^{n \times n}$ is normal and has the block-triangular form

$$T = \begin{bmatrix} T_{11} & T_{12} & \cdots & T_{1k} \\ 0 & T_{22} & \cdots & T_{2k} \\ \vdots & \ddots & \ddots & \vdots \\ 0 & \cdots & 0 & T_{kk} \end{bmatrix}$$

with square blocks on the main diagonal. Use induction on k and the result of part (a) to prove that T is block diagonal and the main-diagonal blocks are normal.

□

rcise 5.4.53 We now return our attention to real matrices.

(a) Let $A = \begin{bmatrix} a & b \\ c & d \end{bmatrix} \in \mathbb{R}^{2 \times 2}$. Show that A is normal if and only if either A is symmetric or A has the form $A = \begin{bmatrix} a & b \\ -b & a \end{bmatrix}$.

(b) In the symmetric case A has real eigenvalues. Find the eigenvalues of

$$A = \begin{bmatrix} a & b \\ -b & a \end{bmatrix}.$$

(c) What special form does Theorem 5.4.23 take when $A \in \mathbb{R}^{n \times n}$ is normal? Make use of the result from Exercise 5.4.52.

□

rcise 5.4.54 Every orthogonal matrix is normal, so the results of Exercise 5.4.53 are valid for all orthogonal matrices.

(a) What does a 1×1 orthogonal matrix look like?

(b) Give necessary and sufficient conditions on a and b such that the matrix $\begin{bmatrix} a & b \\ -b & a \end{bmatrix} \in \mathbb{R}^{2 \times 2}$ is orthogonal.

(c) What special form does Theorem 5.4.23 take when A is orthogonal? Be as specific as you can.

$\square$

Exercise 5.4.55

(a) Show that if A and B are similar matrices, then $\operatorname{tr}(A) = \operatorname{tr}(B)$. (*Hint:* Apply part (b) of Exercise 5.4.51 with $C = S^{-1}$ and $D = AS$.)

(b) Show that the trace of a matrix equals the sum of its eigenvalues.

$\square$

Exercise 5.4.56 Prove that the determinant of a matrix equals the product of its eigenvalues.

$\square$

Exercise 5.4.57 Use MATLAB's `help` facility to find out about the commands `schur` and `rsf2csf`. Use them to find the forms of Schur (Theorem 5.4.11) and Winter-Murnaghan (Theorem 5.4.23) for the matrix

$$A = \begin{bmatrix} 1 & 2 & 3 \\ -4 & 5 & 6 \\ -7 & -8 & 9 \end{bmatrix}.$$

$\square$

5.5   REDUCTION TO HESSENBERG AND TRIDIAGONAL FORMS

The results of the previous section encourage us to seek algorithms that reduce a matrix to triangular form by similarity transformations, as a means of finding the eigenvalues of the matrix. On theoretical grounds we know that there is no algorithm that accomplishes this task in a finite number of steps; such an algorithm would violate Abel's classical theorem, cited in Section 5.2. It turns out, however, that there are finite algorithms, that is, direct methods, that bring a matrix very close to upper-triangular form.

Recall that an $n \times n$ matrix A is called *upper Hessenberg* if $a_{ij} = 0$ whenever $i > j + 1$. Thus an upper Hessenberg matrix has the form

$$\begin{bmatrix} * & * & * & * & * \\ * & * & * & * & * \\ & * & * & * & * \\ & & * & * & * \\ & & & * & * \end{bmatrix}.$$

In this section we will develop an algorithm that uses unitary similarity transformations to transform a matrix to upper Hessenberg form in $\frac{10}{3}n^3$ flops. This algorithm does not of itself solve the eigenvalue problem, but it is extremely important nevertheless because it reduces the problem to a form that can be manipulated inexpensively. For example, you showed in Exercise 5.3.33 that both the LU and QR decompositions of an upper Hessenberg matrix can be computed in $O(n^2)$ flops. Thus Rayleigh quotient iteration can be performed at a cost of only $O(n^2)$ flops per iteration.

The reduction to Hessenberg form is especially helpful when the matrix is Hermitian. Since unitary similarity transformations preserve the Hermitian property, the reduced matrix is not merely Hessenberg, it is tridiagonal. That is, it has the form

$$
\begin{bmatrix}
* & * & & & \\
* & * & * & & \\
 & * & * & * & \\
 & & * & * & * \\
 & & & * & *
\end{bmatrix}.
$$

Tridiagonal matrices can be manipulated very inexpensively. Moreover, the symmetry of the matrix can be exploited to reduce the cost of the reduction to tridiagonal form to a modest $\frac{4}{3}n^3$ flops.

rcise 5.5.1 Let $A \in \mathbb{C}^{n \times n}$ be tridiagonal.

(a) Show that the cost of an LU decomposition of A (with or without pivoting) is $O(n)$ flops.

(b) Show that the cost of a QR decomposition of A is $O(n)$ flops, provided the matrix Q is not assembled explicitly.

□

Reduction of General Matrices

The algorithm that reduces a matrix to upper Hessenberg form is quite similar to the algorithm that performs a QR decomposition using reflectors. It accomplishes the task in $n - 2$ steps. The first step introduces the desired zeros in the first column, the second step takes care of the second column, and so on.

Let us proceed with the first step. Partition A as

$$
A = \begin{bmatrix} a_{11} & c^T \\ b & \hat{A} \end{bmatrix}.
$$

Let $\hat{Q}_1$ be a (real or complex) reflector such that $\hat{Q}_1 b = [-\tau_1, 0, \cdots, 0]^T$ ($|\tau_1| = \|b\|_2$), and let

$$
Q_1 = \begin{bmatrix} 1 & 0^T \\ 0 & \hat{Q}_1 \end{bmatrix}.
$$

Let

$$
A_{1/2} = Q_1 A = \left[\begin{array}{c|c} a_{11} & c^T \\ \hline -\tau_1 & \\ 0 & \\ \vdots & \hat{Q}_1 \hat{A} \\ 0 & \end{array} \right],
$$

which has the desired zeros in the first column. This operation is just like the first step of the QR decomposition algorithm, except that it is less ambitious. Instead of transforming all but one entry to zero in the first column, it leaves two entries nonzero. The reason for the diminished ambitiousness is that now we must complete a similarity transformation by multiplying on the right by Q_1^{-1}. Letting $A_1 = Q_1 A Q_1^{-1}$ and recalling that $Q_1^{-1} = Q_1^* = Q_1$, we have

$$
A_1 = A_{1/2} Q_1 = \left[\begin{array}{c|c} a_{11} & c^T \hat{Q}_1 \\ \hline -\tau_1 & \\ 0 & \\ \vdots & \hat{Q}_1 \hat{A} \hat{Q}_1 \\ 0 & \end{array} \right] = \left[\begin{array}{c|ccc} a_{11} & * & \cdots & * \\ \hline -\tau_1 & & & \\ 0 & & & \\ \vdots & & \hat{A}_1 & \\ 0 & & & \end{array} \right].
$$

Because of the form of Q_1, this operation does not destroy the zeros in the first column. If we had been more ambitious, we would have failed at this point.

The second step creates zeros in the second column of A_1, that is, in the first column of $\hat{A}_1$. Thus we pick a reflector $\hat{Q}_2 \in \mathbb{C}^{(n-2) \times (n-2)}$ in just the same way as the first step, except that A is replaced by A_1. Let

$$
Q_2 = \left[\begin{array}{cc|ccc} 1 & 0 & 0 & \cdots & 0 \\ 0 & 1 & 0 & \cdots & 0 \\ \hline 0 & 0 & & & \\ \vdots & \vdots & & \hat{Q}_2 & \\ 0 & 0 & & & \end{array} \right].
$$

Then

$$
A_{3/2} = Q_2 A_1 = \left[\begin{array}{c|c|ccc} a_{11} & * & * & \cdots & * \\ \hline -\tau_1 & * & * & \cdots & * \\ 0 & -\tau_2 & & & \\ \vdots & \vdots & & \hat{Q}_2 \hat{A}_2 & \\ 0 & 0 & & & \end{array} \right].
$$

We complete the similarity transformation by multiplying on the right by $Q_2^{-1} = Q_2$. Because the first two columns of Q_2 are equal to the first two columns of the identity matrix, this operation does not alter the first two columns of $A_{3/2}$. Thus

$$
A_2 = A_{3/2}Q_2 = \left[\begin{array}{cc|ccc}
a_{11} & * & * & \cdots & * \\
\hline
-\tau_1 & * & * & \cdots & * \\
\hline
0 & -\tau_2 & & & \\
\vdots & \vdots & & \hat{Q}_2\hat{A}_2\hat{Q}_2 & \\
0 & 0 & & &
\end{array}\right].
$$

The third step creates zeros in the third column, and so on. After $n-2$ steps the reduction is complete. The result is an upper Hessenberg matrix that is unitarily similar to A: $B = Q^*AQ$, where

$$
Q = Q_1Q_2\cdots Q_{n-2} \quad \text{and} \quad Q^* = Q_{n-2}Q_{n-3}\cdots Q_1.
$$

If A is real, then all operations are real, Q is real and orthogonal, and B is orthogonally similar to A.

A computer program to perform the reduction can be organized in exactly the same way as a QR decomposition by reflectors is. For simplicity we will restrict our attention to the real case. Most of the details do not require discussion because they were already covered in Section 3.2. The one way in which the present algorithm is significantly different from the QR decomposition is that it involves multiplication by reflectors on the right as well as the left. As you might expect, the right multiplications can be organized in essentially the same way as the left multiplications are. If we need to calculate CQ, where $Q = I - \gamma uu^T$ is a reflector, we can write $CQ = C - \gamma Cuu^T$. Thus we can compute $v = C(\gamma u)$ and then $CQ = C - vu^T$. This is just a variant of algorithm (3.2.38).

Real Householder Reduction to Upper Hessenberg Form

for $k = 1, \ldots, n - 2$

$$\begin{array}{l} \beta \leftarrow \max\{|a_{ik}| \mid i = k + 1, \ldots, n\} \\ \gamma_k \leftarrow 0 \\ \text{if } (\beta \neq 0) \text{ then} \end{array}$$

$\quad$ % Set up the reflector $\hat{Q}_k$

$\quad a_{k+1:n,k} \leftarrow \beta^{-1} a_{k+1:n,k}$

$\quad \tau_k \leftarrow \sqrt{a_{k+1,k}^2 + \cdots + a_{n,k}^2}$

$\quad$ if $(a_{k+1,k} < 0)\, \tau_k \leftarrow -\tau_k$

$\quad \eta \leftarrow a_{k+1,k} + \tau_k$

$\quad a_{k+1,k} \leftarrow 1$

$\quad a_{k+2:n,k} \leftarrow a_{k+2:n,k}/\eta$

$\quad \gamma_k \leftarrow \eta/\tau_k$

$\quad \tau_k \leftarrow \tau_k \beta$

$\quad$ % Multiply on the left by $\hat{Q}_k$

$\quad b_{k+1:n,1}^T \leftarrow a_{k+1:n,k}^T a_{k+1:n,k+1:n}$

$\quad b_{k+1:n,1}^T \leftarrow -\gamma_k b_{k+1:n,1}^T$

$\quad a_{k+1:n,k+1:n} \leftarrow a_{k+1:n,k+1:n} + a_{k+1:n,k} b_{k+1:n,1}^T$

$\quad$ % Multiply on the right by $\hat{Q}_k$

$\quad b_{1:n,1} \leftarrow a_{1:n,k+1:n} a_{k+1:n,k}$

$\quad b_{1:n,1} \leftarrow -\gamma_k b_{1:n,1}$

$\quad a_{1:n,k+1:n} \leftarrow a_{1:n,k+1:n} + b_{1:n,1} a_{k+1:n,k}^T$

$\quad a_{k+1,k} \leftarrow -\tau_k$

$\tau_{n-1} \leftarrow -a_{n,n-1}$

exit

$$(5.5.2)$$

This algorithm takes as input an array that contains $A \in \mathbb{R}^{n \times n}$. It returns an upper Hessenberg matrix $B = Q^T A Q$, whose nonzero elements are in their natural positions in the array. The portion of the array below the subdiagonal is not set to zero. It is used to store the vectors u_k, used to generate the reflectors $\hat{Q}_k = I - \gamma_k u_k u_k^T$, where $\hat{Q}_k$ is the $(n - k) \times (n - k)$ reflector used at step k. The leading 1 in u_k is not stored. The scalar γ_k $(k = 1, \ldots, n - 2)$ is stored in a separate array γ. Thus the information needed to construct the orthogonal transforming matrix Q is available.

Exercise 5.5.3 Count the flops in (5.5.2). Show that the left multiplication part costs about $\frac{4}{3}n^3$ flops, the right multiplication part costs about $2n^3$ flops, and the rest of the cost of the reflector setups is $O(n^2)$. Thus the total cost is about $\frac{10}{3}n^3$ flops. Why do the right multiplications require more flops than the left multiplications do? $\quad\square$

Suppose we have transformed $A \in \mathbb{R}^{n \times n}$ to upper Hessenberg form $B = Q^T A Q$ using (5.5.2). Suppose further that we have found some eigenvectors of B and we would now like to find the corresponding eigenvectors of A. For each eigenvector v of B, Qv is an eigenvector of A. Since

$$Qv = Q_1 Q_2 \cdots Q_{n-2} v,$$

we can easily calculate Qv by applying $n - 2$ reflectors in succession using (3.2.38) repeatedly. In fact we can process all of the eigenvectors at once. If we have m of them, we can build an $n \times m$ matrix V whose columns are the eigenvectors, then we can compute $QV = Q_1 Q_2 \cdots Q_{n-2} V$, by applying (3.2.38) $n - 2$ times. The cost is about $2n^2 m$ flops for m vectors.

If we wish to generate the transforming matrix Q itself, we can do so by computing $Q = QI = Q_1 Q_2 \cdots Q_{n-2} I$, by applying (3.2.38) $n - 2$ times.

The Symmetric Case

If A is Hermitian, then the matrix B produced by (5.5.2) is not merely Hessenberg, it is tridiagonal. Furthermore it is possible to exploit the symmetry of A to reduce the cost of the reduction to $\frac{4}{3}n^3$ flops, less than half the cost of the non-symmetric case. In the interest of simplicity we will restrict our attention to the real symmetric case. We begin with the matrix

$$A = \begin{bmatrix} a_{11} & b^T \\ b & \hat{A} \end{bmatrix}.$$

In the first step of the reduction we transform A to $A_1 = Q_1 A Q_1$, where

$$A_1 = \left[\begin{array}{c|c} a_{11} & b^T \hat{Q}_1 \\ \hline \hat{Q}_1 b & \hat{Q}_1 \hat{A} \hat{Q}_1 \end{array} \right] = \left[\begin{array}{c|cccc} a_{11} & -\tau_1 & 0 & \cdots & 0 \\ \hline -\tau_1 & & & & \\ 0 & & & \hat{A}_1 & \\ \vdots & & & & \\ 0 & & & & \end{array} \right].$$

We save a little bit here by not performing the computation $b^T \hat{Q}_1$, which duplicates the computation $\hat{Q}_1 b$.

The bulk of the effort in this step is expended in the computation of the symmetric submatrix $\hat{A}_1 = \hat{Q}_1 \hat{A} \hat{Q}_1$. We must do this efficiently if we are to realize significant savings. If symmetry is not exploited, it costs about $4n^2$ flops to calculate $\hat{Q}_1 \hat{A}$ and another $4n^2$ flops to calculate $(\hat{Q}_1 \hat{A}) \hat{Q}_1$, as you can easily verify. Thus the entire computation of $\hat{A}_1$ costs about $8n^2$ flops. It turns out that we can cut this figure in half by carefully exploiting symmetry.

$\hat{Q}_1$ is a reflector given in the form $\hat{Q}_1 = I - \gamma u u^T$. Thus

$$\begin{aligned} \hat{A}_1 &= (I - \gamma u u^T)\hat{A}(I - \gamma u u^T) \\ &= \hat{A} - \gamma \hat{A} u u^T - \gamma u u^T \hat{A} + \gamma^2 u u^T \hat{A} u u^T. \end{aligned}$$

The terms in this expression admit considerable simplification if we introduce the auxiliary vector $v = -\gamma \hat{A} u$. We have $-\gamma \hat{A} u u^T = v u^T$, $-\gamma u u^T \hat{A} = u v^T$, and $\gamma^2 u u^T \hat{A} u u^T = -\gamma u u^T v u^T$. Introducing the scalar $\alpha = -\frac{1}{2}\gamma u^T v$, we can rewrite this last term as $2\alpha u u^T$. Thus

$$\hat{A}_1 = \hat{A} + v u^T + u v^T + 2\alpha u u^T.$$

The final manipulation is to split the last term into two pieces in order to combine one piece with the term $v u^T$ and the other with $u v^T$. In other words, let $w = v + \alpha u$. Then

$$\hat{A}_1 = \hat{A} + w u^T + u w^T.$$

This equation translates into the code segment

```
for j = 2, ..., n
  ⌈ for i = j, ..., n
  ⌊   ⌈ a_ij ← a_ij + w_i u_j + u_i w_j
```

which costs four flops per updated array entry. By symmetry we need only update the main diagonal and lower triangle. Thus the total number of flops in this segment is $4(n-1)n/2 \approx 2n^2$. This does not include the cost of calculating w. First of all, the computation $v = -\gamma \hat{A} u$ costs about $2n^2$ flops. The computation $\alpha = -\frac{1}{2}\gamma u^T v$ costs about $2n$ flops, as does the computation $w = v + \alpha u$. Thus the total flop count for the first step is about $4n^2$, as claimed.

The second step of the reduction is identical to the first step, except that it acts on the submatrix $\hat{A}_1$. In particular, (unlike the nonsymmetric reduction) it has no effect on the first row of A_1. Thus the flop count for the second step is about $4(n-1)^2$ flops. After $n-2$ steps the reduction is complete. The total flop count is approximately $4(n^2 + (n-1)^2 + (n-2)^2 + \cdots) \approx \frac{4}{3}n^3$.

Reduction of a Real Symmetric Matrix to Tridiagonal Form

for $k = 1, \ldots, n - 2$
$$\left[\begin{array}{l} \beta \leftarrow \max\{|a_{ik}| \mid i = k + 1, \ldots, n\} \\ \gamma_k \leftarrow 0 \\ \text{if } (\beta \neq 0) \text{ then} \\ \text{set up the reflector } \hat{Q}_k \text{ exactly as in (5.5.2)} \\ w_{k+1:n} \leftarrow 0 \\ \text{for } j = k + 1, \ldots, n \\ \quad \left[\; w_{j:n} \leftarrow w_{j:n} + a_{j:n,j} a_{jk} \right. \\ \text{for } i = k + 1, \ldots, n \\ \quad \left[\; w_i \leftarrow w_i + a_{i+1:n,i}^T a_{i+1:n,k} \right. \\ w_{k+1:n} \leftarrow -\gamma_k w_{k+1:n} \\ \alpha \leftarrow w_{k+1:n} a_{k+1:n,k} \\ \alpha \leftarrow -\gamma_k \alpha/2 \\ w_{k+1:n} \leftarrow w_{k+1:n} + \alpha a_{k+1:n,k} \\ \text{for } j = k + 1, \ldots, n \\ \quad \left[\; a_{j:n,j} \leftarrow a_{j:n,j} + w_{j:n} a_{jk} + a_{j:n,k} w_j \right. \end{array} \right.$$
$$\tau_{n-1} \leftarrow -a_{n,n-1}$$
for $i = 1, \ldots, n$
$$\left[\; d_i \leftarrow a_{ii} \right.$$
$$s_{1:n-1} \leftarrow -\tau_{1:n-1}$$

(5.5.4)

This algorithm accesses only the main diagonal and lower triangle of A. It stores the main-diagonal entries of the tridiagonal matrix B in a one-dimensional array d ($d_i = b_{ii}$, $i = 1, \ldots, n$) and the off-diagonal entries in a one-dimensional array s ($s_i = b_{i+1,i} = b_{i,i+1}$, $i = 1, \ldots, n - 1$). The information about the reflectors used in the similarity transformation is stored exactly as in (5.5.2).

Additional Exercises

rcise 5.5.5 Write a Fortran subroutine that implements (5.5.2).   ☐

rcise 5.5.6 Write a Fortran subroutine that implements (5.5.4).   ☐

rcise 5.5.7 MATLAB's `hess` command transforms a matrix to upper Hessenberg form. To get just the upper Hessenberg matrix, type `B = hess(A)`. To get both the Hessenberg matrix and the transforming matrix (fully assembled), type `[Q,B] = hess(A)`.

(a) Using MATLAB, generate a random 100×100 (or larger) matrix and reduce it to upper Hessenberg form:

```
n = 100
A = randn(n);   % or A = randn(n) + i*randn(n);
[Q,B] = hess(A);
```

(b) Use Rayleigh quotient iteration to compute an eigenpair of B, starting from a random complex vector. For example,

```
q = randn(n,1) + i*randn(n,1);
q = q/norm(q);
rayquo = q'*B*q;
qq = (B - rayquo*eye(n))\q;
```

and so on. Once you have an eigenvector q of B, transform it to an eigenvector $v = Qq$ of A. Your last Rayleigh quotient is your eigenvalue.

(c) Give evidence that your Rayleigh quotient iteration converged quadratically.

(d) Check that you really do have an eigenpair of A by computing the residual norm $\| Av - \lambda v \|_2$.

□

Exercise 5.5.8 The first step of the reduction to upper Hessenberg form introduces the needed zeros into the vector b, where

$$A = \begin{bmatrix} a_{11} & c^T \\ b & \hat{A} \end{bmatrix}.$$

We used a reflector to do the task, but there are other types of transformations that we could equally well have used. Sketch an algorithm that uses Gaussian elimination with partial pivoting to introduce the zeros at each step. Don't forget that each transformation has to be a similarity transformation. Thus we get a similarity transformation to Hessenberg form that is, however, nonunitary. This is a good algorithm that has gone out of favor. The old library EISPACK includes it, but it has been left out of LAPACK.   □

5.6 FRANCIS'S ALGORITHM

For many years now, the most widely used algorithm for computing the complete set of eigenvalues of a matrix has been *Francis's algorithm*, also known as the implicitly-shifted QR algorithm [32]. The present section is devoted to a description of the algorithm and some numerical experiments that show how well it works. An explanation of why it works is largely postponed to the early sections of Chapter 6. You can read those sections right now if you prefer.

Let A be an $n \times n$ matrix whose eigenvalues we would like to know. In the previous section we observed that A can be transformed to upper Hessenberg form by a unitary similarity transformation. Let us suppose this has been done, and A_0 is an upper Hessenberg matrix unitarily similar to A. The algorithm generates a sequence (A_k) of upper-Hessenberg matrices that converges to upper-triangular form (or perhaps a block-triangular form), allowing the eigenvalues to be read from the main diagonal.

To describe the algorithm it suffices to describe a single iteration, starting with A_k and ending with A_{k+1}. Let us simplify (and recycle) notation by writing $A = A_k$ and $\hat{A} = A_{k+1}$. Thus our Francis iteration will take us from the Hessenberg matrix A to the Hessenberg matrix $\hat{A}$. An upper Hessenberg matrix whose subdiagonal entries

$a_{j+1,j}$ are all nonzero is called an *unreduced* or *proper* upper Hessenberg matrix. I prefer the latter term, as the former is illogical. We can assume without loss of generality that our matrix A is properly upper Hessenberg. If not, say $a_{j+1,j} = 0$ for some j, then A has the block-triangular form

$$\begin{bmatrix} A_{11} & A_{12} \\ 0 & A_{22} \end{bmatrix},$$

where A_{11} is $j \times j$. Thus the eigenvalue problem for A can be solved by solving the eigenvalue problems for the upper Hessenberg matrices A_{11} and A_{22} separately. If either A_{11} or A_{22} is not properly upper Hessenberg, it can be broken down further until we have broken A into several properly upper Hessenberg matrices.

Francis iteration of degree one

Assume now that A is properly upper Hessenberg.[28] We begin with the Francis iteration of degree one (aka. single-shift implicit QR algorithm). Pick a shift[29] ρ and notice that the first column of $A - \rho I$ has the form

$$p = \begin{bmatrix} a_{11} - \rho \\ a_{21} \\ 0 \\ \vdots \\ 0 \end{bmatrix}.$$

Let Q_0 be a rotator (or reflector or other unitary matrix) acting in the 1–2 plane that eliminates the entry a_{21}, that is, such that

$$Q_0^* p = \begin{bmatrix} * \\ 0 \\ 0 \\ \vdots \\ 0 \end{bmatrix}. \tag{5.6.1}$$

Now use Q_0 to transform A to $Q_0^* A Q_0$.[30] The transformation $A \to Q_0^* A$ recombines rows 1 and 2 and leaves the Hessenberg form intact. The transformation $Q_0^* A \to Q_0^* A Q_0$ recombines columns 1 and 2 and disturbs the Hessenberg form, introducing

---

[28] The reason for making this restriction will become evident when we explain why the algorithm works in Chapter 6. See also Exercise 5.6.2.

[29] In Section 5.3 we observed that shifting the matrix can often be beneficial, so that's what we are doing here. One good choice, known as the *Rayleigh-quotient shift*, is $\rho = a_{nn}$. Better shifts will be discussed later, and the effect of shifting will be clarified in Chapter 6.

[30] We have stated that Q_0 should be a unitary transformation that effects a certain action, but we have not specified Q_0 precisely. The same will be true of the subsequent transformations Q_1, Q_2, and so on. Therefore there is some ambiguity in our description of the algorithm. As we shall see, this ambiguity is inconsequential.

a nonzero in the $(3,1)$ position. Using the 4×4 case as an illustration, the matrix now has the form

$$\begin{bmatrix} * & * & * & * \\ * & * & * & * \\ + & * & * & * \\ & & * & * \end{bmatrix}.$$

There is a *bulge* in the Hessenberg form indicated by the plus symbol in the $(3,1)$ position.

The rest of the iteration consists of returning the matrix to upper Hessenberg form by the method described in the previous section. But the procedure is greatly simplified in this case, as there is only one nonzero entry disturbing the Hessenberg form. Let Q_1 be a rotator (or reflector) acting in the 2–3 plane that eliminates the $(3,1)$ entry, so that $Q_1^* Q_0^* A Q_0$ is in upper Hessenberg form. But now we must multiply by Q_1 on the right to complete a similarity transformation. This operation recombines columns 2 and 3 of the matrix and creates a bulge in the $(4,2)$ position. Again using the 4×4 case as an illustration, the matrix $Q_1^* Q_0^* A Q_0 Q_1$ has the form

$$\begin{bmatrix} * & * & * & * \\ * & * & * & * \\ & * & * & * \\ & + & * & * \end{bmatrix}.$$

Next we choose a rotator Q_2 acting in the 3–4 plane such that Q_2^* eliminates the bulge in the $(4,2)$ position. When we apply Q_2 on the right to complete the similarity transformation, columns 3 and 4 are recombined. In the 4×4 case, the matrix is now in upper Hessenberg form, and the iteration is complete. If $n > 4$, there is a new bulge in the $(5,3)$ position. This bulge can be chased to the $(6,4)$ position by applying a transformation Q_3 that acts on rows/columns 4 and 5. Continuing in this way, we chase the bulge downward until it is pushed off the bottom of the matrix. At this point the Francis iteration is complete. In the general $n \times n$ case, the last transformation is Q_{n-2}. We have

$$\hat{A} = Q_{n-2}^* \cdots Q_1^* Q_0^* A Q_0 Q_1 \cdots Q_{n-2},$$

where $\hat{A}$ denotes the next iterate. Since Francis's algorithm creates a bulge in the Hessenberg form and then chases it from the matrix, we call it a *bulge-chasing* algorithm.

Francis's algorithm can be applied to general complex matrices, in which case all the arithmetic is of course complex. However, if the matrix and the shift are real, then the entire iteration can be carried out in real arithmetic.

Exercise 5.6.2 Suppose A is upper Hessenberg but not properly upper Hessenberg. Demonstrate that if $a_{j+1,j} = 0$, the bulge that is supposed to appear in position $(j+1, j-1)$ at step $j - 2$ of the bulge chase does not in fact appear. Thus the bulge chase ends prematurely. □

Symmetric Case

The Francis iteration of degree one is particularly important in the real symmetric case. Symmetry is preserved in the reduction to Hessenberg form and in the bulge chase. Thus the Hessenberg matrix is symmetric and tridiagonal, we have

$$
Q_0^* A Q_0 \;=\; \begin{bmatrix} * & * & + & \\ * & * & * & \\ + & * & * & * \\ & & * & * \end{bmatrix},
$$

$$
Q_1^* Q_0^* A Q_0 Q_1 \;=\; \begin{bmatrix} * & * & & \\ * & * & * & + \\ & * & * & * \\ & + & * & * \end{bmatrix},
$$

and so on.

Each Francis iteration begins by choosing a shift. A good choice is the *Rayleigh quotient shift* $\rho = a_{nn}$ (Exercise 5.6.18), but a better one is the *Wilkinson shift*, described as follows: compute the two real eigenvalues of the submatrix

$$
\begin{bmatrix} a_{n-1,n-1} & a_{n-1,n} \\ a_{n-1,n} & a_{n,n} \end{bmatrix},
$$

and take that one that is closer to a_{nn} as the shift. In case of a tie, take the one that is smaller in magnitude as the shift.

rcise 5.6.3 In this exercise you will put together some rudimentary code to do a Francis iteration on an $n \times n$ symmetric, tridiagonal matrix, and you will use the code to explore the properties of the algorithm.

(a) Review the quadratic formula and convince yourself that the following MAT-LAB code segment computes the Wilkinson shift.

```
htr = .5*(a(n-1,n-1) + a(n,n)); % half a 2x2 trace
dscr = sqrt((.5*(a(n-1,n-1)-a(n,n)))^2 + a(n,n-1)^2);
% discriminant
if htr < 0, dscr = -dscr; end % to avoid cancellation
root1 = htr + dscr;  % quadratic formula
if root1 == 0 % almost never happens
   root2 = 0;
else % almost always happens
   det = a(n-1,n-1)*a(n,n) - a(n,n-1)^2;
   % 2x2 determinant = product of roots
   root2 = det/root1;
end
if abs(a(n,n)-root1) < abs(a(n,n)-root2)
```

```
    shift = root1;
else
    shift = root2;
end
```

(b) Once the shift has been chosen, a rotator that effects the transformation (5.6.1) must be built. Convince yourself that the following code segment does this and creates the bulge.

```
cs = a(1,1) - shift; sn = a(2,1);
r = norm([cs sn]);
cs = cs/r; sn = sn/r;
% normalizing to make cs^2 + sn^2 = 1
q0 = [ cs -sn; sn cs]; % Givens rotator
a(1:2,:) = q0'*a(1:2,:); % left multiplication
a(:,1:2) = a(:,1:2)*q0;  % right multiplication
```

(c) Convince yourself that the following loop chases the bulge .

```
for ii = 1:n-2
    % Chase the bulge from position (ii+2,ii).
    cs = a(ii+1,ii); sn = a(ii+2,ii);
    r = norm([cs sn]);
    cs = cs/r; sn = sn/r;
    a(ii+1,ii) = r; a(ii+2,ii) = 0;
    qi = [cs -sn; sn cs];
    % Givens rotator to chase the bulge.
    a(ii+1:ii+2,ii+1:n) = qi'*a(ii+1:ii+2,ii+1:n);
    a(:,ii+1:ii+2) = a(:,ii+1:ii+2)*qi;
end
```

(d) Put the code fragments together to make a MATLAB m-file that performs a symmetric Francis iteration of degree one. Put some code at the end that prints out some of the matrix entries. For example, the code

```
format short e
subdiag = diag(a,-1)
format long
bottom_entry = a(n,n)
```

causes the subdiagonal entries of A to be printed out in exponential format and the (n, n) entry to be printed out in long format. Our objective is to make the subdiagonal entries go to zero and the diagonal entries converge to eigenvalues. You might also want to print out the shift on each iteration.

(e) Try out your Francis iteration on the famous matrix

$$\begin{bmatrix} 2 & 1 & & & & \\ 1 & 2 & 1 & & & \\ & 1 & 2 & 1 & & \\ & & 1 & 2 & 1 & \\ & & & 1 & 2 & 1 \\ & & & & 1 & 2 \end{bmatrix},$$

which can be generated by the commands

```
n = 6;
row = [ 2 1 zeros(1,n-2)];
t = toeplitz(row);
a = t;
```

for example. Perform several Francis iterations and see what you get. What evidence of convergence do you see? What does the rate of convergence appear to be? (linear?, quadratic?, cubic?) Iterate until $|a_{n,n-1}|$ is below 10^{-16}. At this point it is at the level of roundoff errors and can be set to zero, and the (n, n) entry of A is (extremely close to) an eigenvalue. To check that this really is an eigenvalue, you can use MATLAB's eig command (e.g. `evals = eig(t)`) to find the eigenvalues of the matrix.

(f) Modify your code so that it does the Rayleigh quotient shift $\rho = a_{nn}$ instead of the Wilkinson shift, and try it on the above matrix. This is an example of one of the rare matrices on which the Rayleigh quotient shift performs poorly.

(f) Modify your code so that it uses $\rho = 0$ as a shift on every iteration. Notice that in this case we get rather slow convergence.

(g) Modify your code so that is uses the constant shift $\rho = .75$, which is close to an eigenvalue of A. Notice that we get fairly rapid convergence but not as rapid as we got with the Wilkinson shift. What kind of convergence (linear?, quadratic?, cubic?) are we observing here?

(h) Use the constant shift $\rho = .753020396282533$, which is extremely close to an eigenvalue of A. What happens?

□

rcise 5.6.4 Try out your code from Exercise 5.6.3 on a random symmetric tridiagonal matrix. The following commands show ways of generating two different classes of random matrices.

```
n = 6;
% method 1:
a = randn(n); % random matrix
```

```
a = a + a';  % random symmetric matrix
a = hess(a);
% method 2:
md = randn(1,n);    % random main diagonal
sd = randn(1,n-1);  % random subdiagonal
a = diag(md) + diag(sd,-1) + diag(sd,1);
```

Try at least the following:

  (a) Wilkinson shift

  (b) Rayleigh quotient shift

  (c) zero shift

  (d) constant shift close to an eigenvalue.

What is the same and what is different from the outcome of Exercise 5.6.3?   □

Exercises 5.6.3 and 5.6.4 demonstrate that Francis's method with the Wilkinson shift causes the entry $a_{n,n-1}$ to converge rapidly to zero. Once that entry gets down to the level of roundoff error, it can be set to zero, and a_{nn} can be declared an eigenvalue. Then the algorithm can proceed on the remaining $(n-1) \times (n-1)$ submatrix and go after the next eigenvalue. After a few iterations, the second eigenvalue will have been found, and the search for a third eigenvalue can proceed on a submatrix of order $n-2$. The process of extracting an eigenvalue and then continuing on a smaller matrix is called *deflation*. After $n-1$ deflations, the process is complete; all eigenvalues have been found.

It can also sometimes happen that a matrix breaks apart in the middle. Before each iteration, all of the subdiagonal entries $a_{j+1,j}$ should be inspected to see if any of them can be set to zero. A commonly used criterion is that if

$$|a_{j+1,j}| \le u(|a_{jj}| + |a_{j+1,j+1}|),$$

where u is the unit roundoff (around 10^{-16} in double precision), we set $a_{j+1,j}$ to zero. If we are able to do this for some j that is around $n/2$, we get to break our problem into two half-size problems.

The code in Exercise 5.6.3 is quite primitive and lacks several features that a good implementation of the symmetric Francis iteration ought to have. A symmetric tridiagonal matrix has the form

$$A = \begin{bmatrix} a_1 & b_1 & & & \\ b_1 & a_2 & \ddots & & \\ & \ddots & \ddots & b_{n-1} & \\ & & b_{n-1} & a_n \end{bmatrix}.$$

All of the information about the matrix can be stored in two vectors,

$$a = \begin{bmatrix} a_1 & \cdots & a_n \end{bmatrix} \quad \text{and} \quad b = \begin{bmatrix} b_1 & \cdots & b_{n-1} \end{bmatrix},$$

so there is no point in storing A in the conventional way. Instead one should simply store the vectors a and b and operate on them. In addition a few extra scratch spaces are needed: a few for the shift computation, one for the bulge, and a few for building the rotators or reflectors that build and chase the bulge. But it is clear that the total storage space needed is $O(n)$ rather than the $O(n^2)$ that is needed for a matrix that is stored conventionally.

A good code will fully exploit symmetry and not perform any unnecessary operations on zeros. The use of the compact data structure consisting mainly of the two vectors a and b helps this to happen. See Exercise 5.6.22.

·cise 5.6.5

(a) Show that if the symmetric Francis iteration is implemented efficiently, there are $O(1)$ flops associated with each rotator. (This just means that the number of flops per rotator is independent of n.)

(b) Deduce that the number of flops per symmetric Francis iteration is $O(n)$.

(c) Our experience from Exercises 5.6.3 and 5.6.4 suggests that only a few iterations are needed to find an eigenvalue, assuming the Wilkinson shift is used. Much more extensive experience confirms this; typically only about two iterations per eigenvalue are needed on average. Assuming that this is the case, deduce that the total number of flops needed to find the complete set of eigenvalues of an $n \times n$ symmetric tridiagonal matrix is $O(n^2)$.

(d) Suppose that right at the start it is found that $a_{j+1,j} = 0$ for some $j \approx n/2$. Then the problem becomes one of computing the eigenvalues of two matrices of dimension approximately $n/2$. How much work does this save relative to computing the eigenvalues of a single $n \times n$ matrix?

(e) Now consider the problem of computing the complete set of eigenvalues of a dense symmetric matrix. This can be done in two stages: (1) reduce the matrix to tridiagonal form, (2) apply the symmetric Francis iteration. Which of these steps is more expensive in terms of flop count?

□

The symmetric Francis iteration with the Wilkinson shift is guaranteed to converge [69], and the rate of convergence is normally cubic, as seen in Exercises 5.6.3 and 5.6.4. With the Rayleigh quotient shift we have convergence in all but certain exceptional cases. When convergence occurs, it is cubic in all but exceptional cases [101, 102]. One of those exceptional cases is the matrix in Exercise 5.6.3. See also Exercise 5.6.20. In Exercise 5.6.4 we observed the normal convergence rate of the Rayleigh quotient shift. With a fixed shift the convergence rate is normally linear. We will have more to say about convergence later on in this section and in Sections 6.2 and 6.3.

Francis iteration of degree two

Let A be an $n \times n$ properly upper Hessenberg matrix. Symmetry is not assumed. Recall that the Francis iteration of degree one starts by picking a shift ρ and considering the first column of the matrix $A - \rho I$. The Francis iteration of degree two (aka. double-shift QR algorithm) starts by picking two shifts ρ_1 and ρ_2 and considering the first column of the product $(A - \rho_1 I)(A - \rho_2 I)$. Notice that we have no intention of computing the entire matrix $(A - \rho_1 I)(A - \rho_2 I)$; that would be expensive ($O(n^3)$ flops). The first column, in contrast, is easy to compute. It is

$$
p = (A - \rho_1 I)(A - \rho_2 I)e_1 = \begin{bmatrix} (a_{11} - \rho_1)(a_{11} - \rho_2) + a_{12}a_{21} \\ a_{21}((a_{11} + a_{22}) - (\rho_1 + \rho_2)) \\ a_{32}a_{21} \\ 0 \\ \vdots \\ 0 \end{bmatrix}. \tag{5.6.6}
$$

Exercise 5.6.7 Check that (5.6.6) is correct. □

Let Q_0 be a reflector (or a pair of rotators or some other unitary transformation) acting on rows 1–2–3, such that

$$
Q_0^* p = \begin{bmatrix} * \\ 0 \\ 0 \\ \vdots \\ 0 \end{bmatrix}.
$$

By the way, this equation is equivalent to $Q_0 e_1 = \alpha p$ for some scalar α, which means that the first column of Q_0 is proportional to p.[31] Use Q_0 to transform A to $Q_0^* A Q_0$. The left multiplication by Q_0^* recombines rows 1 through 3 and creates a nonzero entry in position $(3, 1)$. The right multiplication by Q_0 recombines columns 1 through 3 and creates additional nonzero entries in positions $(4, 1)$ and $(4, 2)$. Using the 6×6 case as an illustration, the matrix now has the form

$$
\begin{bmatrix} * & * & * & * & * & * \\ * & * & * & * & * & * \\ + & * & * & * & * & * \\ + & + & * & * & * & * \\ & & & * & * & * \\ & & & & * & * \end{bmatrix}. \tag{5.6.8}
$$

This matrix has a bigger bulge than we saw before. We call this a *bulge of degree two*. The rest of the double Francis iteration consists of returning the matrix to

---

[31] Again there is some ambiguity in the description of the algorithm, but this ambiguity turns out to be inconsequential.

upper Hessenberg form by the method described in Section 5.5. Since there are only three nonzeros disturbing the Hessenberg form, this is much simpler than the general Hessenberg reduction, and it amounts to a bulge chase. Let Q_1 be a reflector acting on rows 2–3–4 that transforms the two bulge entries in the first column to zero. By this we mean that a left multiplication by Q_1^* ($= Q_1$) returns the first column to Hessenberg form. To complete a similarity transformation we must multiply on the right by Q_1. This recombines columns 2 through 4 and creates new nonzeros in positions $(5, 2)$ and $(5, 3)$. The matrix now has the form

$$\begin{bmatrix} * & * & * & * & * & * \\ * & * & * & * & * & * \\ & * & * & * & * & * \\ & + & * & * & * & * \\ & + & + & * & * & * \\ & & & & * & * \end{bmatrix}.$$

The bulge has been pushed down and over one position. In the next step we build a reflector acting on rows 3–4–5 that restores Hessenberg form in the second column, pushing the bulge down and over one more position. Continuing in this manner we chase the bulge downward until it gets pushed off the bottom of the matrix. At this point we have our new iterate

$$\hat{A} = Q_{n-2}^* \cdots Q_1^* Q_0^* A Q_0 Q_1 \cdots Q_{n-2}.$$

Each of the reflectors Q_j acts on three rows (or columns), except for Q_{n-2}, which acts only on rows $n - 1$ and n.

In general the matrix A can be complex, and so can the shifts, in which case all the arithmetic is complex. But Francis's motivation for developing this algorithm came from the real case. A real nonsymmetric matrix can have complex eigenvalues, which come in conjugate pairs. As we observed in Exercise 5.6.3, we can get fast convergence to an eigenvalue if (and only if) the shifts approximate the eigenvalue well. It follows that in the general real case we will often want to use complex shifts. Francis developed his double-shift algorithm as a way of making complex shifts while staying in real arithmetic. Notice that if A is real and the shifts ρ_1 and ρ_2 are either both real or form a complex conjugate pair ($\rho_2 = \bar{\rho}_1$), then the column p in (5.6.6) is real. It follows that the entire Francis iteration can be carried out in real arithmetic.

rcise 5.6.9 Verify that if A is real, and ρ_1 and ρ_2 are complex conjugates, then the column vector p in (5.6.6) is real. (Indeed the entire matrix $(A - \rho_1 I)(A - \rho_2 I)$ is real.) □

How should we choose the shifts? One reasonable choice is to take ρ_1 and ρ_2 to be the two eigenvalues of

$$\begin{bmatrix} a_{n-1,n-1} & a_{n-1,n} \\ a_{n,n-1} & a_{n,n} \end{bmatrix}.$$

This is the *generalized Rayleigh quotient* shift strategy. If A is real, then the shifts are either both real or a complex conjugate pair.

Exercise 5.6.10

(a) Convince yourself that the following quick and dirty MATLAB code performs a Francis iteration of degree two on a real matrix.

```
rho = eig(a(n-1:n,n-1:n));
% We "cheat" by using eig to compute shifts.
p=[ (a(1,1)-rho(1))*(a(1,1)-rho(2))+a(1,2)*a(2,1);  ...
    a(2,1)*(a(1,1)+a(2,2)-rho(1)-rho(2));  ...
    a(3,2)*a(2,1) ];
p = real(p); % should be real anyway
[Q R] = qr(p);  % easy way to build orthogonal matrix
                % with first column proportional to p.
a(1:3,:) = Q'*a(1:3,:);
a(1:4,1:3) = a(1:4,1:3)*Q; % bulge created
a = hess(a); % crude bulge chase by hess
format short e
subdiag = diag(a,-1) % Print out some stuff.
```

(b) Try out the code on the matrix generated by the commands

```
n = 6;
r = ones(1,n);
c = [ 1 -1 zeros(1,n-2) ];
a = toeplitz(c,r);
```

which has three pairs of complex conjugate eigenvalues. Be patient; it takes a few iterations for convergence to set in. What evidence for convergence do you eventually see? What convergence rate do you eventually see?

(c) What pair of eigenvalues have you just found? (You can "cheat" and use `eig(a(n-1:n,n-1:n))`.)

(d) Modify the code so that it uses constant shifts $1.5 \pm 0.53i$ (close to a pair of eigenvalues). What rate of convergence do you observe?

(e) Now try constant shifts $1.595526541172915 \pm 0.530977194534957i$ (extremely close to a pair of eigenvalues). What do you observe now?

$\square$

Exercise 5.6.11 Try out the code from the previous exercise on some random matrices, which you can generate as follows.

```
n = 7; % or as large as you please
a = randn(n);
a = hess(a);
```

Try several matrices with even n and several with odd n.   □

In Exercise 5.6.10 when we used the generalized Rayleigh quotient shift, we observed that $a_{n-1,n-2} \to 0$ quadratically, isolating a pair of complex conjugate eigenvalues in the lower-right-hand 2×2 submatrix. Once $a_{n-1,n-2}$ reaches the level of roundoff error, we can set it zero, deflate out the 2×2 with the pair of complex eigenvalues, and proceed to look for the next pair of eigenvalues in the remaining submatrix of order $n - 2$.

In Exercise 5.6.11 we observed the same thing, except that sometimes $a_{n,n-1} \to 0$ quadratically, yielding a single real eigenvalue. Also, sometimes when $a_{n-1,n-2} \to 0$, the 2×2 lower-right-hand submatrix can have two real eigenvalues instead of a complex conjugate pair.

Unfortunately convergence is not guaranteed, as Exercise 5.6.21 shows. Fortunately convergence failures are rare and can usually be avoided by an exceptional shift strategy: whenever the iterations seem not to be converging, take one step with "random" shifts to shake things up.

The code in Exercise 5.6.10 is very crude. A good implementation of a double Francis iteration will take account of the shape of the bulge and not waste time doing operations on numbers that are known to be zero. If this is done, then the work to apply each reflector will be $O(n)$ flops. Since each iteration uses $O(n)$ reflectors, the total flop count for a Francis iteration of degree 2 is $O(n^2)$. The more precise figure $10n^2$ flops/iteration is derived in Exercise 5.6.28.

Usually it just takes a few iterations $(O(1))$ to find an eigenvalue or a pair of eigenvalues. Assuming this, the total cost in flops for finding all n eigenvalues is n times $O(n^2)$ or $O(n^3)$. Thus Francis's algorithm of degree two is considered to have a complexity of $O(n^3)$.

Francis iteration of arbitrary degree

We continue to assume that A is a properly upper Hessenberg matrix. We have described Francis iterations of degrees one and two, and it now ought to be clear that we can do Francis iterations of any degree m in principal. We first pick m shifts $\rho_1, \ldots, \rho_m$. Then we compute a vector proportional to the first column of $p(A) = (A - \rho_1 I) \cdots (A - \rho_m I)$. This can be computed by the simple iteration

$$
\begin{aligned}
&p \leftarrow e_1 \\
&\text{for } j = 1 \ldots m \\
&\quad \left[\begin{array}{l} p \leftarrow (A - \rho_j I)p \\ p \leftarrow p/\|p\| \quad \text{(optional)} \end{array} \right.
\end{aligned}
\tag{5.6.12}
$$

It is easy to check that the resulting p has only $m + 1$ nonzero entries and can be computed at a cost of $O(m^3)$ flops. This is cheap if m is not too big.

Exercise 5.6.13 Let $p = p(A)e_1 = (A - \rho_1 I) \cdots (A - \rho_m I)e_1$.

    (a) Prove by induction on m that only the first $m + 1$ entries of p can be nonzero.

    (b) Show that the cost of the jth iteration of (5.6.12) is about $2j^2$ flops.

    (c) Deduce that the total flop count for (5.6.12) is about $\frac{2}{3}m^3$. This is negligible if $m \ll n$.

$\square$

Once we have p, we let Q_0 be a unitary transformation (e.g. reflector) such that $Q_0^* p = \beta e_1$ for some β. We then make the similarity transformation $A \to Q_0^* A Q_0$ to create a bulge in the Hessenberg form whose size is directly proportional to m For example, in the case $m = 3$, the bulge looks like this:

$$\begin{bmatrix} * & * & * & * & * & * & * \\ * & * & * & * & * & * & * \\ + & * & * & * & * & * & * \\ + & + & * & * & * & * & * \\ + & + & + & * & * & * & * \\ & & & * & * & * & * \\ & & & & * & * & * \end{bmatrix}.$$

Once the bulge has been formed, the Francis iteration is completed by returning the matrix to Hessenberg form by chasing the bulge.

The following result, which is valid for Francis iterations of any degree, is one of the keys to understanding why the algorithm works. We will use it in Chapter 6.

Theorem 5.6.14 *Suppose $\hat{A} = Q^* A Q$ is produced by one iteration of Francis's algorithm of degree m with shifts $\rho_1, \ldots, \rho_m$. Let $p(A) = (A - \rho_1 I) \cdots (A - \rho_m I)$. Then there is a nonzero constant α such that $Q e_1 = \alpha p(A) e_1$. In words, the first column of Q is proportional to the first column of $p(A)$.*

Proof. $Q = Q_0 Q_1 \cdots Q_{n-2}$, where Q_0 is the transformation that creates the bulge, and $Q_1, Q_2, \ldots, Q_{n-2}$ return the matrix to upper Hessenberg form by a variant of the algorithm described in Section 5.5. Referring back to that section, we recall that Q_1 has the form

$$Q_1 = \begin{bmatrix} 1 & 0^T \\ 0 & \hat{Q}_1 \end{bmatrix},$$

reflecting the fact that left multiplication by Q_1^* does not touch the first row of A, and right multiplication by Q_1 does not touch the first column. A consequence of this form is that $Q_1 e_1 = e_1$. All of the subsequent transformations clearly also have this property: $Q_2 e_1 = e_1, \ldots Q_{n-2} e_1 = e_1$. On the other hand, Q_0 is chosen so that Q_0^* transforms the first column of $p(A)$ to a multiple of e_1. In other words, $Q_0^* p(A) e_1 = \beta e_1$ for some nonzero β. Multiplying through by Q_0, we can rewrite this as $Q_0 e_1 = \alpha p(A) e_1$, where $\alpha = 1/\beta$. Finally, $Q e_1 = Q_0 Q_1 \cdots Q_{n-2} e_1 = Q_0 e_1 = \alpha p(A) e_1$. $\square$

In principle the degree of a Francis iteration can be any positive integer. However it has been found that for large values of m, roundoff errors cause the information about the shifts not to be transmitted accurately. This phenomenon, which is known as *shift blurring* [101, Ch. 7], causes the iterations to be ineffective. For this reason the degree m is seldom taken be to greater than six or so.

We must also ask where one gets m shifts when, say, $m = 6$. The *generalized Rayleigh quotient* shift strategy is to take those shifts to be the eigenvalues of the lower-right-hand $m \times m$ submatrix of A. These could be computed by a supplementary routine that computes the eigenvalues by Francis iterations of degree one or two.

Convergence of Francis's algorithm

The mechanism underlying Francis's algorithm will be elucidated in Chapter 6. Here we state briefly and without justification some basic facts about the convergence of Francis's algorithm and why shifts are chosen as they are.

For simplicity we consider first the *stationary* variant of Francis's algorithm, in which shifts are chosen in advance, and the same shifts are used on every iteration. Let $\rho_1, \ldots, \rho_m$ be the shifts, and suppose we use these shifts to do Francis iterations of degree m to produce a sequence of iterates (A_j). Let $p(z)$ denote the polynomial of degree m given by $p(z) = (z - \rho_1) \cdots (z - \rho_m)$, so that $p(A) = (A - \rho_1 I) \cdots (A - \rho_m I)$. If $\lambda_1, \ldots, \lambda_m$ are the eigenvalues of A, then the eigenvalues of $p(A)$ are $p(\lambda_1), \ldots, p(\lambda_n)$. Suppose we number the eigenvalues so that $|p(\lambda_1)| \geq |p(\lambda_2)| \geq \cdots \geq |p(\lambda_n)|$.

·cise 5.6.15 Show that if (λ, v) is an eigenpair of A, then $(p(\lambda), v)$ is an eigenpair of $p(A)$. □

In Chapter 6 we will show that Francis's algorithm is a variant of nested subspace iteration, which is an easily understood generalization of the power method, driven by $p(A)$. As a consequence, the convergence is governed by the ratios

$$|p(\lambda_{k+1})/p(\lambda_k)| \leq 1, \quad k = 1, \ldots, n - 1.$$

Precisely, if for some given k, the ratio $|p(\lambda_{k+1})/p(\lambda_k)|$ is strictly less than 1, then the kth subdiagonal entry $a_{k+1,k}^{(j)}$ converges to zero linearly with convergence ratio $|p(\lambda_{k+1})/p(\lambda_k)|$ as $j \to \infty$. Therefore, if we can choose p so that any one of the ratios $|p(\lambda_{k+1})/p(\lambda_k)|$ is very small compared to 1, the corresponding subdiagonal entry of (A_j) will converge to zero rapidly, allowing the eigenvalue problem for A to be split into two smaller problems. It will be seen, moreover, that once the split occurs, say

$$A_j = \begin{bmatrix} A_{11} & A_{12} \\ 0 & A_{22} \end{bmatrix},$$

the eigenvalues of A_{11} are $\lambda_1, \ldots, \lambda_k$, and the eigenvalues of A_{22} are $\lambda_{k+1}, \ldots, \lambda_n$.

Consider the case $m = 1$, in which we have $p(z) = z - \rho$. The ratios of interest are $|\lambda_{k+1} - \rho|/|\lambda_k - \rho|$, where $|\lambda_1 - \rho| \geq \cdots \geq |\lambda_n - \rho|$. If it happens that ρ

approximates any one of the eigenvalues particularly well and in particular much better than it approximates any of the other eigenvalues, then the eigenvalue that is approximated well will be numbered λ_n, and the ratio $|\lambda_n - \rho|/|\lambda_{n-1} - \rho|$ will be tiny. Thus $a_{n,n-1}^{(j)}$ will converge rapidly to zero, and the entries $a_{nn}^{(j)}$ will converge rapidly to the eigenvalue λ_n.

In the case $m = 2$, two shifts are chosen, and we have $p(z) = (z - \rho_1)(z - \rho_2)$. If it happens that ρ_1 and ρ_2 are excellent approximations to two different eigenvalues of A, then those will be the two eigenvalues for which $|p(\lambda)|$ is smallest. They will be numbered λ_{n-1} and λ_n, and we will have $|p(\lambda_{n-2})| \gg |p(\lambda_{n-1})|$. Thus $|p(\lambda_{n-1})/p(\lambda_{n-2})| \ll 1$, $a_{n-1,n-2}^{(j)} \to 0$ rapidly, and the eigenvalues of the bottom 2×2 submatrix converge rapidly to the eigenvalues λ_{n-1} and λ_n.

Exercise 5.6.16 Suppose A has eigenvalues $\lambda_1, \ldots, \lambda_n$, and $p(z) = (z - \rho_1)(z - \rho_2)$. Suppose the shifts satisfy $\rho_1 \approx \lambda_{n-1}$ and $\rho_2 \approx \lambda_n$. Let $m_1 = \min\{|\lambda_i - \rho_1| \mid i \neq n-1\}$ and $m_2 = \min\{|\lambda_i - \rho_2| \mid i \neq n\}$. Show that if $|\lambda_{n-1} - \rho_1|/m_1$ and $|\lambda_n - \rho_2|/m_2$, are sufficiently small, then $|p(\lambda_{n-1})/p(\lambda_{n-2})| \ll 1$. A rigorous proof is not required. $\square$

In general we have m shifts, and $p(z) = (z - \rho_1) \cdots (z - \rho_m)$. If each of the shifts is an excellent approximation to an eigenvalue, we will have $|p(\lambda_{n-m+1})| \gg |p(\lambda_{n-m})|$, so $a_{n-m+1,n-m}^{(j)} \to 0$ rapidly, and the eigenvalues of the bottom $m \times m$ submatrix will converge rapidly to $\lambda_{n-m+1}, \ldots, \lambda_n$.

Even if only some of the shifts approximate eigenvalues well (and others don't), we still get a good result. If $\widehat{m}$ of the m shifts approximate $\widehat{m}$ distinct eigenvalues well, then the argument of the previous paragraph remains valid with m replaced by $\widehat{m}$, $a_{n-\widehat{m}+1,n-\widehat{m}}^{(j)} \to 0$ rapidly, and the eigenvalues of the bottom $\widehat{m} \times \widehat{m}$ submatrix converge rapidly to $\lambda_{n-\widehat{m}+1}, \ldots \lambda_n$.

The arguments in the past few paragraphs discuss what happens if some or all of the shifts are excellent approximations to eigenvalues. Of course, if we choose the shifts in advance, we cannot reasonably expect any of them to approximate eigenvalues well. Nevertheless, whatever the shifts may be, we can reasonably expect that in most cases the ratios $|p(\lambda_{k+1})/p(\lambda_k)|$ will be less than 1 for many choices of k, with the consequence that $a_{k+1,k}^{(j)} \to 0$ as $j \to \infty$ for many (perhaps most) values of k. The convergence might be quite leisurely, but after many iterations we can reasonably expect that some of the subdiagonal entries $a_{k+1,k}^{(j)}$ will be quite small. If it happens that $|a_{n-m+1,n-m}^{(j)}|$ is very small, then the eigenvalues of the bottom $m \times m$ submatrix will be excellent approximations to the eigenvalues $\lambda_{n-m+1}, \ldots, \lambda_n$ and would make excellent shifts. So why not use them as shifts for subsequent iterations?

If we decide to choose new shifts in this way, there are a couple of questions that need to be answered. First of all, how many iterations must pass before it makes sense to take the eigenvalues of the lower-right-hand $m \times m$ matrix as shifts? Once we have made the change, convergence will be much faster than before. After a few more iterations we will have much better approximations to the eigenvalues of

$\lambda_{n-m+1}, \ldots, \lambda_n$, so perhaps we should change our shifts again. This brings us to the second question: How often should we change the shifts?

The following answers were arrived at after extensive experimentation and testing: (1) The shifts can be taken as the eigenvalues of the lower-right-hand $m \times m$ submatrix (or some similar strategy) right from the very start. (2) New shifts can be chosen for each iteration. This is the *dynamic* variant of Francis's algorithm, and this is the way it is used in practice. At first the shifts will be poor approximations to eigenvalues, but after a number of iterations, they will begin to home in on the eigenvalues and become increasingly effective as shifts. Improving shifts lead to faster convergence, which leads to still better shifts. This positive feedback mechanism leads to quadratic convergence [101, Ch. 5]. For normal matrices, including real symmetric matrices, the local convergence rate is cubic.

The dynamic shifting strategy, choosing new shifts on each iteration, greatly improves the performance of the algorithm. At the same time it makes the analysis more difficult. So far nobody has been able to devise a good shifting strategy *and* prove that it is globally convergent. In particular, the generalized Rayleigh quotient shift strategy is not globally convergent (Exercise 5.6.21). For this reason, publicly available implementations of Francis's algorithm include *exceptional shifts*, used to shake things up in situations where the algorithm seems not to be converging. Nobody has proved that this strategy always works.

Francis's algorithm for larger matrices

Francis's algorithm is primarily a "small matrix" algorithm. In order to apply it, you must be able to perform similarity transformations on the matrix. This means that you must be able to store the matrix in the conventional manner and operate on its entries. In the next chapter we will introduce some "large matrix" algorithms that assume that the matrix is sparse, and we are not willing to destroy the sparseness by performing similarity transformations.

The definition of *small matrix* has crept upward over the years. Nowadays it is routine to compute the complete set of eigenvalues of a 1000×1000 matrix by Francis's algorithm on a laptop computer, so 1000×1000 is small. Not long before this writing, the complete set of eigenvalues of a matrix of order $n = 10^5$ was computed using a version of Francis's algorithm on a parallel supercomputer [44]. For handling these larger "small" matrices, some useful innovations have been introduced.

Since the Francis iterations of degree 1 and 2 operate on only a few rows and columns of a matrix at a time, they are not able to realize the high performance obtainable through efficient use of cache memory, as discussed in Section 1.1. Our first innovation addresses this issue. Think of a properly upper Hessenberg matrix A that is of order a few thousand, for example. The task of computing the eigenvalues of the lower-right-hand 200×200 submatrix, say, (by Francis iterations of degree 2) is trivial in comparison with the task of computing all of the eigenvalues of A. If we compute these 200 quantities, we can use them as shifts for 100 Francis iterations of degree 2.[32] These iterations need not be done one at a time. Once the bulge for the first iteration has been started and chased a small ways downward, the second bulge can be introduced, then the third, and so on. In this way a long chain of closely-spaced bulges can be chased down through the matrix. Each bulge chase entails a large number of 3×3 reflectors. If a number of these reflectors (from all bulges in the chain) are accumulated into a single larger orthogonal transformation, then most of the work of the algorithm can be executed as matrix-matrix multiplications, yielding high performance through efficient use of cache. For details see [101, § 4.9] and the primary references cited there. It is interesting that this reorganization roughly doubles the total flop count for the algorithm but nevertheless yields a big performance boost for large matrices.

On large parallel supercomputers, on which the matrix is dealt out over a large number of processors, parallelism can be achieved by chasing several bulge chains at once [44].

A final refinement, which has proved to increase performance substantially by decreasing the total number of iterations, is called *aggressive early deflation*. You can read about this interesting procedure in [101, § 4.9] and the primary references cited there.

Additional Exercises

Exercise 5.6.17 To get an idea how well Francis's algorithm works without writing your own program, just try out MATLAB's `eig` command, which balances the matrix (Section 5.4), reduces it to upper Hessenberg form and then applies Francis iterations of degree two (if the matrix is nonsymmetric). For example, try

```
n = 1000;
A = randn(n);
t = cputime;
lambda = eig(A)
elapsed_time = cputime - t
```

Adjust the value of n to suit the speed of your computer. ☐

Exercise 5.6.18 Consider the symmetric tridiagonal matrix

$$A = \begin{bmatrix} a_1 & b_1 & & \\ b_1 & a_2 & \ddots & \\ & \ddots & \ddots & b_{n-1} \\ & & b_{n-1} & a_n \end{bmatrix}. \tag{5.6.19}$$

(a) Show that if $b_{n-1} = 0$, then a_n is an eigenvalue of A with the associated eigenvector e_n. It is thus natural to expect that if $|b_{n-1}|$ is small, then e_n will be close to an eigenvector of A, and a_n will be close to an eigenvalue.

---

[32]We could equally well do one Francis iteration of degree 200, but that would fail due to shift blurring [101, Ch. 7].

(b) Show that a_n is the Rayleigh quotient of A with respect to the vector e_n. This explains why $\rho = a_{nn}$ is called the Rayleigh quotient shift.

□

rcise 5.6.20 This exercise shows that the Rayleigh quotient shift strategy can sometimes fail.

(a) Consider a Francis iteration of degree one using the Rayleigh quotient shift on the matrix

$$A = \begin{bmatrix} 0 & 1 \\ 1 & 0 \end{bmatrix}.$$

Show that the initial vector that determines the first rotator is $p = \begin{bmatrix} 0 \\ 1 \end{bmatrix}$.
What is the first rotator (or reflector if you prefer)? What effect does this transformation have? No bulge is formed because the matrix is too small. Thus the iteration ends after one transformation. Because of the ambiguity in our description of the algorithm, there are two possible outcomes. However, the difference between them is insignificant.

(b) What happens on the second iteration?

(c) Compute (by elementary means) the eigenvalues of A. Is the convergence failure observed here in conflict with the convergence theory outlined in this section?

(d) Think of a change of strategy that would allow Francis's algorithm to find the eigenvalues of this matrix.

□

rcise 5.6.21 This exercise shows that the Francis iteration with the generalized Rayleigh quotient shift can fail to converge. Consider the $n \times n$ matrix with 1's on the subdiagonal, a 1 in the upper-right corner, and zeros elsewhere. The 4×4 version of this matrix has the form

$$A = \begin{bmatrix} & & & 1 \\ 1 & & & \\ & 1 & & \\ & & 1 & \end{bmatrix}.$$

(a) Perform a Francis iteration of degree one with the Rayleigh quotient shift. Because of the ambiguity in the description of the algorithm, the outcome is not quite uniquely determined. For the simplest outcome, use transformations of the form $\begin{bmatrix} 0 & 1 \\ 1 & 0 \end{bmatrix}$ wherever possible. (This is a reflector, not a rotator.)

(b) Taking $m < N$, what shifts do we get if we use the generalized Rayleigh quotient shift (eigenvalues of the lower-right-hand $m \times m$ matrix)? Perform a Francis iteration of degree m with these m shifts. This is easier than it may seem. In case of doubt or confusion, try the case $m = 2$ first.

(c) Show that A is an orthogonal matrix. Deduce that its eigenvalues all satisfy $|\lambda| = 1$.

(d) Is the lack of convergence observed in parts (a) and (b) in conflict with the convergence theory outlined in this section?

(e) Part (c) gave some basic information about the spectrum of A. Actually it is possible to deduce exactly all of the eigenvalues and eigenvectors of this special matrix. Show that if ω is a complex number satisfying $\omega^n = 1$, then $v = \begin{bmatrix} \omega^{n-1} & \omega^{n-2} & \cdots & \omega & 1 \end{bmatrix}^T$ is an eigenvector of A associated with the eigenvalue ω. Deduce that the eigenvalues of A are exactly the nth roots of unity.

$\square$

Exercise 5.6.22

(a) Demonstrate that

$$\begin{bmatrix} c & s \\ -s & c \end{bmatrix} \begin{bmatrix} a_1 & b_1 \\ b_1 & a_2 \end{bmatrix} \begin{bmatrix} c & -s \\ s & c \end{bmatrix} = \begin{bmatrix} \hat{a}_1 & \hat{b}_1 \\ \hat{b}_1 & \hat{a}_2 \end{bmatrix},$$

where

$$\begin{aligned} \hat{a}_1 &= c^2 a_1 + 2cs b_1 + s^2 a_2 \\ \hat{a}_2 &= s^2 a_1 - 2cs b_1 + c^2 a_2 \\ \hat{b}_1 &= (c^2 - s^2) b_1 + cs(a_2 - a_1). \end{aligned}$$

(b) Consider the symmetric matrix

$$\hat{A} = \begin{bmatrix} a_1 & b_1 & & & & & & & \\ b_1 & a_2 & b_2 & & & & & & \\ & b_2 & a_3 & & & & & & \\ & & & \ddots & & & & & \\ & & & & a_{i-1} & b_{i-1} & g & & \\ & & & & b_{i-1} & a_i & b_i & & \\ & & & & g & b_i & a_{i+1} & & \\ & & & & & & & \ddots & b_{n-1} \\ & & & & & & & b_{n-1} & a_n \end{bmatrix},$$

which would be tridiagonal if it did not have the bulge g in positions $(i+1, i-1)$ and $(i-1, i+1)$. Determine a rotator $\tilde{Q} = \begin{bmatrix} c & -s \\ s & c \end{bmatrix}$ such that

$$\tilde{Q}^T \begin{bmatrix} b_{i-1} \\ g \end{bmatrix} = \begin{bmatrix} * \\ 0 \end{bmatrix}. \text{ Define } Q_{i-1} \in \mathbb{R}^{n \times n} \text{ by}$$

$$Q_{i-1} = \begin{bmatrix} I & 0 & 0 \\ 0 & \tilde{Q} & 0 \\ 0 & 0 & I \end{bmatrix},$$

where $\tilde{Q}$ is embedded in positions i and $i+1$. Calculate $Q_{i-1}^T \hat{A} Q_{i-1}$ using the result from part (a), and note that the resulting matrix has a bulge in positions $(i+2, i)$ and $(i, i+2)$. Show that, taking symmetry into account, the transformation from $\hat{A}$ to $Q_{i-1}^T \hat{A} Q_{i-1}$ requires the calculation of only six new entries.

(c) Building on part (b) and Exercise 5.6.3, write MATLAB code for a symmetric Francis iteration in which the tridiagonal matrix (5.6.19) is stored in the form of a pair of vectors

$$a = \begin{bmatrix} a_1 & \cdots & a_n \end{bmatrix} \quad \text{and} \quad b = \begin{bmatrix} b_1 & \cdots & b_{n-1} \end{bmatrix}.$$

(c) Try out your code on the matrices used in Exercises 5.6.3 and 5.6.4 and larger versions of those matrices.

□

rcise 5.6.23 We continue to use the notation of Exercise 5.6.22.

(a) Suppose $c \neq 0$. Show that $\tilde{Q}$ can be chosen so that $c > 0$. Let $t = s/c$. Show that (if $c > 0$)

$$c^2 = \frac{1}{1+t^2} \qquad s^2 = \frac{t^2}{1+t^2} \qquad cs = \frac{t}{1+t^2}$$

$$c = \frac{1}{\sqrt{1+t^2}} \quad \text{and} \quad s = \frac{t}{\sqrt{1+t^2}}.$$

(Keep in mind that the letters s, c, and t stand for sine, cosine, and tangent, respectively. Thus you can use various trigonometric identities such as $c^2 + s^2 = 1$.)

(b) Rewrite the formulas that transform $\hat{A}$ to $Q_{i-1}^T \hat{A} Q_{i-1}$ entirely in terms of t, omitting all reference to c and s. In particular, notice how simple the formula for t itself is.

(c) Now suppose $s \neq 0$. Show that $\tilde{Q}$ can be chosen so that $s > 0$. Let $k = c/s$. (k stands for *kotangent*.) Show that (if $s > 0$)

$$s^2 = \frac{1}{1+k^2} \qquad c^2 = \frac{k^2}{1+k^2} \qquad cs = \frac{k}{1+k^2}$$

$$s = \frac{1}{\sqrt{1 + k^2}} \quad \text{and} \quad c = \frac{k}{\sqrt{1 + k^2}}.$$

(d) Rewrite the formulas that transform $\hat{A}$ to $Q_{i-1}^T \hat{A} Q_{i-1}$ entirely in terms of k, omitting all reference to c and s. In particular, notice how simple the formula for k is.

(e) Depending on the relative sizes of g and b_{i-1}, either t or k can be very large. There is a slight danger of overflow. However, k and t can't both be large at the same time. All danger of overflow can be avoided by choosing the appropriate set of formulas for a given step. Show that if $|g| \leq |b_{i-1}|$, then $|t| \leq 1$, $t^2 \leq 1$, and $1 \leq 1 + t^2 \leq 2$. Show that if $|g| \geq |b_{i-1}|$, then $|k| \leq 1$, $k^2 \leq 1$, and $1 \leq 1 + k^2 \leq 2$.

□

Exercise 5.6.24 Rewrite your MATLAB code from Exercise 5.6.22 so that it uses tangents and cotangents, as shown in Exercise 5.6.23. Try to minimize the number of arithmetic operations. The initial rotator, which creates the first bulge, can be handled in about the same way as the other rotators. □

Numerous ways to organize a symmetric Francis step have been proposed. A number of them are discussed in [69] in the context of the equivalent implicitly-shifted QL algorithm, which chases bulges from bottom to top.

Exercise 5.6.25 Write MATLAB code that calculates the eigenvalues of a real, symmetric, tridiagonal matrix by Francis iterations of degree one.

Since the algorithm requires properly tridiagonal matrices, before each iteration you must check the subdiagonal entries to see whether the problem can be deflated or split. In practice any entry b_k that is very close to zero should be regarded as a zero. Use the following criterion: set b_k to zero whenever

$$|b_k| < u(|a_k| + |a_{k+1}|),$$

where u is the unit roundoff error. In IEEE double precision arithmetic, $u \approx 10^{-16}$. This criterion guarantees that the error that is made when b_k is set to zero is comparable in magnitude to the numerous roundoff errors that are made during the computation. In particular, backward stability is not compromised.

Once the matrix has been broken into two or more pieces, the easiest way to keep track of which portions of the matrix still need to be processed is to work from the bottom up. Thus you should check the subdiagonal entries starting from the bottom, and as soon as you find a zero, perform a QR iteration on the bottom submatrix. If you always work on the bottom matrix first, it will be obvious when you are done.

Take full advantage of symmetry. The only storage area you should need is a pair of one dimensional arrays for the main diagonal and subdiagonal, and a few additional single storage locations for temporary variables such as the bulge, tangent, and cotangent.

The usual choice of shift is the Wilkinson shift, since it guarantees convergence. However, since we wish to use this program as a learning tool, build in three shift options: (i) zero shift, (ii) Rayleigh quotient shift, and (iii) Wilkinson shift. Since we wish to observe the convergence of the algorithm, the subroutine should optionally print out the matrix at each iteration and also print out how many iterations are required for each deflation or reduction. Since we wish to observe the convergence of the eigenvalues, print out the *main-diagonal* using `format long` or `format long e`. We are interested in only the magnitude of the *subdiagonal* entries, so they can be printed out in `format short e`.

This subroutine, like all iterative algorithms, must have some limit on the number of iterations allowed. I suggest that you not allow more than $100n$ iterations. (The limit could be set much lower, say $10n$, if only the Wilkinson shift were being used.) If the iteration limit is reached, the subroutine should set an error flag and return.

Be sure to write clear, structured code, document it with a reasonable number of comments, and document clearly all the variables that are passed to and from the subroutine.

Try out your subroutine on the following two matrices using all three shift options.

$$\begin{bmatrix} 16 & 1 & & & \\ 1 & 8 & 1 & & \\ & 1 & 4 & 1 & \\ & & 1 & 2 & 1 \\ & & & 1 & 1 \end{bmatrix}.$$

Try the above matrix first, since it converges fairly rapidly, even with zero shifts. The eigenvalues are approximately 16.124, 8.126, 4.244, 2.208, and 0.297. Print out the matrix after each iteration and observe the spectacular cubic convergence of the shifted cases. Next try the matrix

$$\begin{bmatrix} 2 & 1 & & & & \\ 1 & 2 & 1 & & & \\ & 1 & 2 & 1 & & \\ & & 1 & 2 & 1 & \\ & & & 1 & 2 & 1 \\ & & & & 1 & 2 \end{bmatrix}$$

from Exercise 5.6.3. This matrix causes problems for the Rayleigh quotient shift, as we already observed. Since this matrix requires many iterations for the unshifted case, don't print out the matrix after each iteration. Just keep track of the number of iterations required per eigenvalue.

Experiment with larger versions of this matrix. The $n \times n$ version has eigenvalues $\lambda_k = 4\sin^2(k\pi/(2n+2))$, $k = 1, \ldots, n$.

You can also test your subroutine's reduction mechanism by concocting some larger matrices with some zeros on the subdiagonal. For example, you can string together three or four copies of the matrices given above. □

If you run the code from Exercise 5.6.24 on some large examples, you will find that it runs fairly slowly. Since MATLAB is an interpreted language, it has to interpret each instruction before it executes it. This process will be slow on codes that contain a large number of instructions that tell the computer to do relatively small tasks. That is the case in the symmetric Francis code of Exercise 5.6.24.

If you want really fast code, you should write in a compiled language such as C or Fortran. The compile step creates machine language code that can be executed rapidly.

Exercise 5.6.26 Write Fortran or C code to solve symmetric tridiagonal eigenvalue problems using Francis's algorithm of degree one with the Wilkinson shift. Race your compiled code against the MATLAB code on some large examples.   □

Exercise 5.6.27 Often the shifts for a Francis iteration of degree two are taken to be the generalized Rayleigh quotient shifts, the two eigenvalues of the lower right hand 2×2 submatrix

$$\begin{bmatrix} a_{n-1,n-1} & a_{n-1,n} \\ a_{n,n-1} & a_{n,n} \end{bmatrix}.$$

Then either ρ_1 and ρ_2 are real or $\rho_2 = \overline{\rho}_1$.

(a) Show that if we choose the shifts this way, we have

$$\begin{aligned} \rho_1 + \rho_2 &= a_{n-1,n-1} + a_{n,n} & (= \text{trace}) \\ \rho_1 \rho_2 &= a_{n-1,n-1} a_{n,n} - a_{n,n-1} a_{n-1,n} & (= \text{determinant}) \end{aligned}$$

(b) Show that the first column of $B = (A - \rho_1 I)(A - \rho_2 I)$ is proportional to

$$\begin{aligned} v_1 &= \frac{(a_{11} - a_{n-1,n-1})(a_{11} - a_{n,n}) - a_{n,n-1} a_{n-1,n}}{a_{21}} + a_{12} \\ v_2 &= a_{11} + a_{22} - a_{n-1,n-1} - a_{n,n} \\ v_3 &= a_{32}. \end{aligned}$$

This is a good formula to use for computing the initial reflector Q_0. Exactly this formula was used in the first official version of Francis's algorithm in Contribution II/14 of the Handbook [105]. It will work well for any matrix you would normally encounter. However, the newer code DLAHQR in LAPACK uses a more careful formula with rescaling to increase the dynamic range of matrices that can be handled.

□

Exercise 5.6.28 Given a nonzero $y \in \mathbb{R}^3$, let $Q \in \mathbb{R}^{3 \times 3}$ be the reflector such that

$$Q \begin{bmatrix} y_1 \\ y_2 \\ y_3 \end{bmatrix} = \begin{bmatrix} -\tau \\ 0 \\ 0 \end{bmatrix} \quad \text{and} \quad \tau = \begin{cases} \|y\|_2 & \text{if } y_1 \geq 0 \\ -\|y\|_2 & \text{if } y_1 < 0. \end{cases}$$

(a) (Review) Show that $Q = I - \gamma u u^T$, where

$$u = \begin{bmatrix} 1 \\ y_2/(y_1 + \tau) \\ y_3/(y_1 + \tau) \end{bmatrix}^T$$

and $\gamma = (y_1 + \tau)/\tau$. Thus $Q = I - v u^T$, where $v = \gamma u$.

(b) Let $B \in \mathbb{R}^{3 \times k}$ be a submatrix to be transformed by Q. Noting that $u_1 = 1$, show that the operation $B \to QB = B - v(u^T B)$ requires only $5k$ multiplications and $5k$ additions. (Since $Q = Q^T = I - u v^T$, operations of the type $C \to CQ$ can be performed in exactly the same way.)

(c) Show that if the operations are carried out as indicated in part (b), an iteration of Francis's algorithm of degree two requires only about $5n^2$ multiplications and $5n^2$ additions, that is, $10n^2$ flops.

$\square$

rcise 5.6.29 Drawing from Exercises 5.6.27 and 5.6.28, write a MATLAB m-file that executes an efficient Francis iteration of degree two. Try it out using examples from Exercises 5.6.10 and 5.6.11. What evidence of quadratic convergence do you see? $\square$

rcise 5.6.30 Write a MATLAB m-file that calculates the eigenvalues of an upper Hessenberg matrix by the Francis algorithm of degree 2. Since the algorithm requires properly upper Hessenberg matrices, before each iteration you must check the subdiagonal entries to see whether the problem can be deflated or split. In practice any entry $a_{k+1,k}$ that is very close to zero should be regarded as a zero. Use the following criterion: set $a_{k+1,k}$ to zero whenever

$$|a_{k+1,k}| < u(|a_{k,k}| + |a_{k+1,k+1}|),$$

where u is the unit roundoff error. In IEEE double precision arithmetic, $u \approx 10^{-16}$. This criterion guarantees that the error that is made when $a_{k+1,k}$ is set to zero is comparable in magnitude to the numerous roundoff errors that are made during the computation. In particular, backward stability is not compromised.

Once the matrix has been broken into two or more pieces, the easiest way to keep track of which portions of the matrix still need to be processed is to work from the bottom up. Thus you should check the subdiagonal entries starting from the bottom, and as soon as you find a zero, perform a QR iteration on the bottom submatrix. If you always work on the bottom matrix first, it will be obvious when you are done. An isolated 1×1 block is a real eigenvalue. An isolated 2×2 block contains a pair of complex or real eigenvalues that can be found (carefully) by the quadratic formula. (See Exercise 5.6.31). Once you get to the top of the matrix, you are done.

Since we wish to observe the quadratic convergence of the algorithm, the subroutine should optionally print out the subdiagonal of the matrix after each iteration. It should also print out how many iterations are required for each deflation or reduction.

Try your code out on examples from Exercises 5.6.10 and 5.6.11. You can check the accuracy of your results by comparing them with the eigenvalues as computed by the `eig` command. Try some large random examples. ☐

Exercise 5.6.31 This exercise discusses the careful calculation of the eigenvalues of a real 2×2 matrix. Suppose we want to find the eigenvalues of

$$A = \begin{bmatrix} a_{11} & a_{12} \\ a_{21} & a_{22} \end{bmatrix} \in \mathbb{R}^{2 \times 2}.$$

A straightforward application of the quadratic formula to the characteristic polynomial can sometimes give inaccurate results. We can avoid difficulties by performing an orthogonal similarity transformation to get A into a special form. Let

$$Q = \begin{bmatrix} c & s \\ -s & c \end{bmatrix}$$

be a rotator. Thus $c = \cos\theta$ and $s = \sin\theta$ for some θ. Let $\hat{A} = Q^T A Q$.

(a) It turns out to be desirable to choose Q so that $\hat{A}$ satisfies $\hat{a}_{11} = \hat{a}_{22}$. Show that if $\hat{a}_{11} = \hat{a}_{22}$, then the eigenvalues of A and $\hat{A}$ are

$$\hat{a}_{11} \pm \sqrt{\hat{a}_{21}\hat{a}_{12}}.$$

Show that the eigenvalues are a complex conjugate pair if $\hat{a}_{21}$ and $\hat{a}_{12}$ have opposite sign. This formula computes the complex eigenvalues with no cancellations, so it computes eigenvalues that are as accurate as the data warrants. A transformation of this type is used by the routines in LAPACK. In part (b) you will show that such a transformation is always possible. The case of real eigenvalues is dealt with further in part (c).

(b) Show that

$$\hat{A} = \begin{bmatrix} c^2 a_{11} + s^2 a_{22} - cs(a_{12} + a_{21}) & c^2 a_{12} - s^2 a_{21} + cs(a_{11} - a_{22}) \\ c^2 a_{21} - s^2 a_{12} + cs(a_{11} - a_{22}) & c^2 a_{22} + s^2 a_{11} + cs(a_{12} + a_{21}) \end{bmatrix}.$$

Using the trigonometric identities $\cos 2\theta = c^2 - s^2$ and $\sin 2\theta = 2sc$, show that $\hat{a}_{11} = \hat{a}_{22}$ if and only if

$$(a_{11} - a_{22})\cos 2\theta = (a_{12} + a_{21})\sin 2\theta.$$

Show that if $a_{12} + a_{21} = 0$, we achieve $\hat{a}_{11} = \hat{a}_{22}$ by taking $c = s = 1/\sqrt{2}$. Otherwise, choose θ so that

$$\hat{t} = \tan 2\theta = \frac{a_{11} - a_{22}}{a_{12} + a_{21}}$$

to get the desired result. Let $t = \tan\theta = s/c$. Use half-angle formulas and other trigonometric identities to show that

$$t = \frac{\sin 2\theta}{1 + \cos 2\theta} = \frac{\hat{t}}{1 + \sqrt{1 + \hat{t}^2}}.$$

Show further that $c = 1/\sqrt{1+t^2}$ and $s = ct$. An alternative formula for t that uses $\cot 2\theta$ instead of $\tan 2\theta$ (useful when $|a_{11} - a_{22}| \gg |a_{12} + a_{21}|$) can be inferred from Exercise 7.2.47.

(c) If the eigenvalues turn out to be real, that is, if $\hat{a}_{21}\hat{a}_{12} \geq 0$ a second rotator can be applied to transform $\hat{A}$ to upper triangular form. To keep the notation simple, let us drop the hat from $\hat{A}$. That is, we assume A already satisfies $a_{11} = a_{22}$ and $a_{21}a_{12} \geq 0$, and we seek a transformation $\hat{A} = Q^T A Q$ for which $\hat{a}_{21} = 0$. Show that the condition $a_{11} = a_{22}$ implies that $\hat{A}$ has the simpler form

$$\hat{A} = \begin{bmatrix} a_{11} - cs(a_{12} + a_{21}) & c^2 a_{12} - s^2 a_{21} \\ c^2 a_{21} - s^2 a_{12} & a_{11} + cs(a_{12} + a_{21}) \end{bmatrix}.$$

Show that if $a_{12} = 0$, we obtain the desired result by taking $c = 0$ and $s = 1$. Otherwise, show that $\sqrt{a_{21}/a_{12}}$ is a real number, and $t = s/c = \sqrt{a_{21}/a_{12}}$ gives the desired transformation. Then $\hat{a}_{11}$ and $\hat{a}_{22}$ are the eigenvalues of A.

□

·cise 5.6.32 Write Fortran or C code to compute the eigenvalues of an upper Hessenberg matrix by Francis's algorithm of degree two with the generalized Rayleigh quotient shift. Race your compiled code against the MATLAB code on some large examples.

□

·cise 5.6.33 In this exercise you will show that if the shift for a Francis iteration is exactly an eigenvalue, then that eigenvalue is deflated in one iteration. This is a theoretical result that is true in exact arithmetic. To see what happens in practice, refer back to Exercises 5.6.3 (h) and 5.6.10 (e), or perform some additional experiments of your own.

This exercise has many steps, but they are all easy. Let A be an $n \times n$ properly upper Hessenberg matrix, and let ρ be an eigenvalue of A. We consider the effect of one Francis iteration of degree one on A with shift ρ. Recall that the iteration transforms A to

$$\hat{A} = Q_{n-2}^* \cdots Q_1^* Q_0^* A Q_0 Q_1 \cdots Q_{n-2}.$$

The transformation by Q_0 creates the bulge, and the subsequent transformations chase the bulge. Let

$$A_k = Q_k^* \cdots Q_1^* Q_0^* A Q_0 Q_1 \cdots Q_k, \quad k = 0, 1, \ldots, n-2.$$

Then

$$A_k = Q_k^* A_{k-1} Q_k,$$

and for $k < n-2$, Q_k has a bulge at position $(k+3, k+1)$. Let $a_{ij}^{(k)}$ denote the (i, j) entry of A_k.

(a) Show that the first $n-1$ columns of $A - \rho I$ are linearly independent. Let v be an eigenvector of A associated with the eigenvalue ρ, so that $(A - \rho I)v = 0$. Show that $v_n \neq 0$, where v_n denotes the last component of v.

(b) Consider the matrix A_{n-3}, which has a bulge in position $(n, n-2)$ (the very bottom!). Let $\tilde{Q} = Q_0 \cdots Q_{n-3}$ and $\tilde{v} = \tilde{Q}^* v$. Show that $(A_{n-3} - \rho I)\tilde{v} = 0$. Show that $\tilde{Q}$ has the form

$$
\tilde{Q} = \left[
\begin{array}{c|c}
\begin{matrix} & & \\ & \check{Q} & \\ & & \end{matrix} & \begin{matrix} 0 \\ \vdots \\ 0 \end{matrix} \\
\hline
0 \quad \cdots \quad 0 & 1
\end{array}
\right],
$$

where $\check{Q}$ is $(n-1) \times (n-1)$, and deduce that $\tilde{v}_n = v_n \neq 0$.

(c) Now focus on the lower-right-hand 2×3 submatrix of $A_{n-3} - \rho I$. This is

$$
M = \begin{bmatrix} a_{n-1,n-2}^{(n-3)} & a_{n-1,n-1}^{(n-3)} - \rho & a_{n-1,n}^{(n-3)} \\ a_{n,n-2}^{(n-3)} & a_{n,n-1}^{(n-3)} & a_{n,n}^{(n-3)} - \rho \end{bmatrix}.
$$

Using the equation $(A_{n-3} - \rho I)\tilde{v} = 0$ and the fact that $\tilde{v}_n \neq 0$, show that the third column of M is a linear combination of the first two columns. It turns out that M has rank one; all three of its columns are proportional to each other. To get this stronger result, we must make a stronger argument.

(d) Recall that the first transforming matrix Q_0 is a rotator (or other unitary transformation) such that

$$
Q_0^* \begin{bmatrix} a_{11} - \rho \\ a_{21} \\ 0 \\ \vdots \\ 0 \end{bmatrix} = \begin{bmatrix} * \\ 0 \\ 0 \\ \vdots \\ 0 \end{bmatrix}.
$$

Deduce that in the matrix $Q_0^*(A - \rho I)$, the $(2, 1)$ entry is zero. The only nonzero entry in the first column is the $(1, 1)$ entry. Show that Q_0 is a nontrivial rotator, i.e. it's active part has the form $\begin{bmatrix} c & -s \\ s & c \end{bmatrix}$, $s \neq 0$.

(e) Now consider the transformation from $Q_0^*(A - \rho I)$ to $Q_0^*(A - \rho I)Q_0 = A_0 - \rho I$, which creates a bulge in position $(3, 1)$. Consider the submatrix

$$
B_0 = \begin{bmatrix} a_{21}^{(0)} & a_{22}^{(0)} - \rho \\ a_{31}^{(0)} & a_{32}^{(0)} \end{bmatrix},
$$

which contains the bulge element $a_{31}^{(0)}$. Show that B_0 has rank one. Using the fact that Q_0 is a nontrivial rotator, show that $a_{31}^{(0)} \neq 0$. (Here I don't mean that it *might* be nonzero; I mean that it *must* be nonzero.)

(f) Show that Q_1 is a nontrivial rotator. Using what we have deduced about the submatrix B_0, show that the transformation from $A_0 - \rho I$ to $Q_1^*(A_0 - \rho I)$ transforms the entry $a_{32}^{(0)}$ to zero.

(g) Now consider the transformation from $Q_1^*(A_0 - \rho I)$ to $Q_1^*(A_0 - \rho I)Q_1 = A_1 - \rho I$, which creates a bulge in position $(4, 2)$. Show that the submatrix

$$B_1 = \begin{bmatrix} a_{32}^{(1)} & a_{33}^{(1)} - \rho \\ a_{31}^{(1)} & a_{32}^{(1)} \end{bmatrix}$$

has rank one, and the bulge entry $a_{31}^{(1)}$ *must* be nonzero.

(h) Prove by induction on k that the submatrix

$$B_k = \begin{bmatrix} a_{k+2,k+1}^{(k)} & a_{k+2,k+2}^{(k)} - \rho \\ a_{k+3,k+1}^{(k)} & a_{k+3,k+2}^{(k)} \end{bmatrix}$$

(a submatrix of $A_k - \rho I$) has rank one, and the bulge entry $a_{k+3,k+1}^{(k)}$ *must* be nonzero. The submatrices B_k are called *bulge pencils*. For a more general discussion of bulge pencils see [101, Ch. 7] and the primary references listed there.

(i) Now refer back to the matrix M in part (c). Notice that the first two columns of M are exactly the bulge pencil B_{n-3}. Deduce that the first two columns of M are linearly dependent. Deduce further that M has rank one.

(j) Show that the transformation from $A_{n-3} - \rho I$ to $Q_{n-2}^*(A_{n-3} - \rho I)$ transforms the submatrix M to a matrix of the form

$$\begin{bmatrix} b & c & d \\ 0 & e & f \end{bmatrix}$$

that still has rank one, and in which b must be nonzero. Deduce that $e = f = 0$.

(k) Now consider the transformation from $Q_{n-2}^*(A_{n-3} - \rho I)$ to $Q_{n-2}^*(A_{n-3} - \rho I)Q_{n-2} = \hat{A} - \rho I$. Show that the bottom row of $\hat{A} - \rho I$ consists entirely of zeros. Deduce that the bottom row of $\hat{A}$ is $\begin{bmatrix} 0 & \cdots & 0 & \rho \end{bmatrix}$. Thus the eigenvalue ρ can be deflated.

□

rcise 5.6.34 Let $x \in \mathbb{C}^{n \times n}$ be a nonzero vector, and let $e_1 \in \mathbb{C}^{n \times n}$ be the first standard unit vector. In many contexts we need to generate nonsingular transforming matrices G that "introduce zeros" into the vector x, in the sense that $G^{-1}x = \beta e_1$, where β is a nonzero constant. Transformations of this type are the heart of LU and QR decompositions, for example, and have found extensive use in this section. Show that $G^{-1}x = \beta e_1$ for some β if and only if the first column of G is proportional to x.

□

5.7 USE OF FRANCIS'S ALGORITHM TO CALCULATE EIGENVECTORS

In Section 5.3 we observed that inverse iteration can be used to find eigenvectors associated with known eigenvalues. This is a powerful and important technique. In this section we consider a different approach; Francis's algorithm can be used to calculate eigenvalues and eigenvectors simultaneously. In Section 6.3 it will be shown that this use of Francis's algorithm can be viewed as a form of inverse iteration. We will begin by discussing the symmetric case, since it is relatively uncomplicated.

Symmetric Matrices

Let $A \in \mathbb{R}^{n \times n}$ be symmetric and tridiagonal. Francis's algorithm can be used to find all of the eigenvalues and eigenvectors of A at once. After some finite number of iterations, A will have been reduced essentially to diagonal form

$$D = Q^T A Q, \tag{5.7.1}$$

where the main-diagonal entries of D are the eigenvalues of A. If a total of m Francis iterations are taken, then $Q = Q_1 \cdots Q_m$, where Q_i is the transforming matrix for the ith iteration. Each Q_i is a product of $n - 1$ rotators. Thus Q is the product of a large number of rotators. The importance of Q is that its columns are a complete orthonormal set of eigenvectors of A, by Spectral Theorem 5.4.12. It is easy to accumulate Q in the course of performing the Francis algorithm, thereby obtaining the eigenvectors along with the eigenvalues. An additional array is needed for the accumulation of Q. Calling this array Q also, we set $Q = I$ initially. Then for each rotator Q_{ij} that is applied to A (that is, $A \leftarrow Q_{ij}^T A Q_{ij}$), we multiply Q by Q_{ij} on the right ($Q \leftarrow Q Q_{ij}$). The end result is clearly the transforming matrix Q of (5.7.1).

How much does this transformation procedure cost? Each transformation $Q \leftarrow Q Q_{ij}$ alters two columns of Q. Thus $2n$ numbers are updated. The exact flop count depends upon the details of the implementation, but in any event it is $O(n)$. Since each QR step uses $n - 1$ rotators, the cost of updating Q for each complete step is $O(n^2)$ flops. Recalling that the basic cost of a symmetric Francis iteration is $O(n)$ flops, we conclude that the accumulation of Q increases the cost by an order of magnitude. Making the (reasonable) assumption that the total number of Francis iterations is $O(n)$, the total cost of accumulating Q is $O(n^3)$ flops, whereas the total cost of the Francis iterations without accumulating Q is $O(n^2)$.

Exercise 5.7.2 Outline the steps that would be taken to calculate a complete set of eigenvalues and eigenvectors of a symmetric (non-tridiagonal) matrix using Francis's algorithm. Estimate the flop count for each step and the total flop count. □

Exercise 5.7.3 Write a MATLAB, Fortran, or C program that calculates the eigenvectors of a symmetric tridiagonal matrix by a reduction to tridiagonal form followed by Francis's algorithm. □

Unsymmetric Matrices

Let $A \in \mathbb{C}^{n \times n}$ be an upper Hessenberg matrix. If Francis's algorithm is used to calculate the eigenvalues of A, then after some finite number of steps we have essentially

$$T = Q^* A Q, \tag{5.7.4}$$

where T is upper triangular. This is the Schur form (Theorem 5.4.11). As in the symmetric case, we can accumulate the transforming matrix Q.

There is an important difference in the way Francis's algorithm is handled, depending upon whether just eigenvalues are computed or eigenvectors as well. If a zero appears on the subdiagonal of some iterate, then the matrix has the block-triangular form

$$\begin{bmatrix} A_{11} & A_{12} \\ 0 & A_{22} \end{bmatrix}.$$

If only eigenvalues are needed, then A_{11} and A_{22} can be treated separately, and A_{12} can be ignored. However, if eigenvectors are wanted, then A_{12} cannot be ignored. We must continue to update it because it eventually forms part of the matrix T, which is needed for the computation of the eigenvectors. Thus if rows i and j of A_{11} are altered by some rotator, then the same rotator must be applied to rows i and j of A_{12}. Similarly, if columns i and j of A_{22} are altered by some rotator, then the corresponding columns of A_{12} must also be updated.

Once we have obtained the form (5.7.4), we are still not finished. Only the first column of Q is an eigenvector of A. To obtain the other eigenvectors, we must do a bit more work. It suffices to find the eigenvectors of T, since for each eigenvector v of T, Qv is an eigenvector of A. Let us therefore examine the problem of calculating the eigenvectors of an upper-triangular matrix. The eigenvalues of T are t_{11}, t_{22}, $\ldots$, t_{nn}, which we will assume to be distinct. To find an eigenvector associated with the eigenvalue t_{kk}, we must solve the homogeneous equation $(T - t_{kk}I)v = 0$. It is convenient to make the partition

$$T - t_{kk}I = \begin{bmatrix} S_{11} & S_{12} \\ 0 & S_{22} \end{bmatrix}, \qquad v = \begin{bmatrix} v_1 \\ v_2 \end{bmatrix},$$

where $S_{11} \in \mathbb{C}^{k \times k}$ and $v_1 \in \mathbb{C}^k$. Then the equation $(T - t_{kk}I)v = 0$ becomes

$$\begin{aligned} S_{11}v_1 + S_{12}v_2 &= 0 \\ S_{22}v_2 &= 0. \end{aligned}$$

S_{11} and S_{22} are upper triangular. S_{22} is nonsingular because its main-diagonal entries are $t_{jj} - t_{kk}$, $j = k + 1, \ldots, n$, all of which are nonzero. Therefore v_2 must equal zero, and the equations reduce to $S_{11}v_1 = 0$. S_{11} is singular because its (k, k) entry is zero. Making another partition

$$S_{11} = \begin{bmatrix} \hat{S} & r \\ 0^T & 0 \end{bmatrix}, \qquad v_1 = \begin{bmatrix} \hat{v} \\ w \end{bmatrix},$$

where $\hat{S} \in \mathbb{C}^{(k-1)\times(k-1)}$, $r \in \mathbb{C}^{k-1}$, and $\hat{v} \in \mathbb{C}^{k-1}$, the equation $S_{11}v_1 = 0$ becomes

$$\hat{S}\hat{v} + rw = 0.$$

The matrix $\hat{S}$ is upper triangular and nonsingular. Taking w to be any nonzero number, we can solve $\hat{S}\hat{v} = -rw$ for $\hat{v}$ by back substitution to obtain an eigenvector. For example, if we take $w = 1$, we get the eigenvector

$$v = \begin{bmatrix} -\hat{S}^{-1}r \\ 1 \\ 0 \\ \vdots \\ 0 \end{bmatrix}.$$

Exercise 5.7.5 Calculate three linearly independent eigenvectors of the matrix

$$T = \begin{bmatrix} 1 & 1 & 1 \\ 0 & 2 & 1 \\ 0 & 0 & 3 \end{bmatrix}.$$

$\square$

Exercise 5.7.6 How many flops are needed to calculate n linearly independent eigenvectors of the upper-triangular matrix $T \in \mathbb{C}^{n\times n}$ with distinct eigenvalues? $\square$

The case in which T has repeated eigenvalues is more complicated, since T may fail to have n linearly independent eigenvectors; that is, T may be defective. The case of repeated eigenvalues is worked out, in part, in Exercise 5.7.7. Even when the eigenvalues are distinct, the algorithm we have just outlined can sometimes give inaccurate results because the eigenvectors can be ill conditioned. Section 7.1 discusses conditioning of eigenvalues and eigenvectors.

Exercise 5.7.7 Let T be an upper-triangular matrix with a double eigenvalue $\lambda = t_{jj} = t_{kk}$. Give necessary and sufficient conditions under which T has two linearly independent eigenvectors associated with λ, and outline a procedure for calculating the eigenvectors. $\square$

If A is a real matrix, we prefer to work within the real number system as much as possible. Therefore we use the Francis iterations of degree two to reduce A to the form

$$T = Q^T A Q,$$

where $Q \in \mathbb{R}^{n\times n}$ is orthogonal, and $T \in \mathbb{R}^{n\times n}$ is quasi-triangular. This is the Wintner-Murnaghan form (Theorem 5.4.23).

Exercise 5.7.8 Sketch a procedure for calculating the eigenvectors of a quasi-triangular matrix $T \in \mathbb{R}^{n\times n}$ that has distinct eigenvalues. $\square$

Suppose $A \in \mathbb{R}^{n \times n}$ has complex eigenvalues λ and $\overline{\lambda}$. Since the associated eigenvectors are complex, we expect that we will have to deal with complex vectors at some point. In the name of efficiency we would like to postpone the use of complex arithmetic for as long as possible. It turns out that we can avoid complex arithmetic right up to the very end. Suppose $A = QTQ^T$, where T is the quasi-triangular Wintner-Murnaghan form. Using the procedure you sketched in Exercise 5.7.8, we can find an eigenvector v of T associated with λ: $Tv = \lambda v$. Then $w = Qv$ is an eigenvector of A. Both v and w can be broken into real and imaginary parts: $v = v_1 + iv_2$ and $w = w_1 + iw_2$, where v_1, v_2, w_1, and $w_2 \in \mathbb{R}^n$. Since Q is real, it follows easily that $w_1 = Qv_1$ and $w_2 = Qv_2$. Thus the computation of w does not require complex arithmetic; w_1 and w_2 can be calculated individually, entirely in real arithmetic. Notice also that $\overline{w} = w_1 - iw_2$ is the eigenvector of A associated with $\overline{\lambda}$. Thus the two real vectors w_1 and w_2 yield two complex eigenvectors.

·cise 5.7.9 Rework Exercise 5.7.8 so that all calculations are done in real arithmetic. □

5.8 THE SVD REVISITED

Throughout this section, A will denote a nonzero, real $n \times m$ matrix. In Chapter 4 we introduced the singular value decomposition (SVD) and proved that A has an SVD. The proof given there (Exercise 4.1.16) did not use the concepts of eigenvalue and eigenvector. Later on (Exercises 5.2.17, 5.4.46, and 5.4.48) we noticed that singular values and vectors are closely related to eigenvalues and eigenvectors. In the first part of this section we will present a second proof of the SVD theorem that makes use of the eigenvector connection. In the second part of the section we will show how to compute the SVD. The two parts are not heavily interdependent; the reader who is interested only in the computational aspects can confidently skip ahead.

A Second Proof of the SVD Theorem

Recall that $A \in \mathbb{R}^{n \times m}$ has two important spaces associated with it—the *null space* and the *range*, given by

$$\mathcal{N}(A) = \{x \in \mathbb{R}^m \mid Ax = 0\},$$

$$\mathcal{R}(A) = \{Ax \mid x \in \mathbb{R}^m\}.$$

The null space is a subspace of $\mathbb{R}^m$, and the range is a subspace of $\mathbb{R}^n$. Recall that the range is also called the column space of A (Exercise 3.5.13), and its dimension is called the *rank* of A. Finally, recall that $m = \dim(\mathcal{N}(A)) + \dim(\mathcal{R}(A))$. This is Corollary 4.1.9, which can also be proved by elementary means.

The matrices $A^T A \in \mathbb{R}^{m \times m}$ and $AA^T \in \mathbb{R}^{n \times n}$ will play an important role in what follows. Let us therefore explore the properties of these matrices and their relationships to A and A^T.

·cise 5.8.1 **(Review)** Prove that $A^T A$ and AA^T are (a) symmetric and (b) positive semidefinite. □

Theorem 5.8.2 $\mathcal{N}(A^T A) = \mathcal{N}(A)$.

Proof. It is obvious that $\mathcal{N}(A) \subseteq \mathcal{N}(A^T A)$. To prove that $\mathcal{N}(A^T A) \subseteq \mathcal{N}(A)$, suppose $x \in \mathcal{N}(A^T A)$. Then $A^T A x = 0$. An easy computation (Lemma 3.5.9) shows that $\langle A^T A x, x \rangle = \langle Ax, Ax \rangle$. Thus $0 = \langle A^T A x, x \rangle = \langle Ax, Ax \rangle = \|Ax\|_2^2$. Therefore $Ax = 0$; that is, $x \in \mathcal{N}(A)$. □

Corollary 5.8.3 $\mathcal{N}(AA^T) = \mathcal{N}(A^T)$.

Corollary 5.8.4 $\operatorname{rank}(A^T A) = \operatorname{rank}(A) = \operatorname{rank}(A^T) = \operatorname{rank}(AA^T)$.

Proof. By Corollary 4.1.9, $\operatorname{rank}(A) = m - \dim(\mathcal{N}(A))$ and, similarly, $\operatorname{rank}(A^T A) = m - \dim(\mathcal{N}(A^T A))$. Thus $\operatorname{rank}(A^T A) = \operatorname{rank}(A)$. The second equation is a basic result of linear algebra. The third equation is the same as the first, except that the roles of A and A^T have been reversed. □

Proposition 5.8.5 *If v is an eigenvector of $A^T A$ associated with a nonzero eigenvalue λ, then Av is an eigenvector of AA^T associated with the same eigenvalue.*

Proof. Since $\lambda \neq 0$, we have $Av \neq 0$. Furthermore $AA^T(Av) = A(A^T Av) = A(\lambda v) = \lambda(Av)$. □

Corollary 5.8.6 *$A^T A$ and AA^T have the same nonzero eigenvalues, counting multiplicity.*

Proposition 5.8.7 *Let v_1 and v_2 be eigenvectors of $A^T A$. If v_1 and v_2 are orthogonal, then Av_1 and Av_2 are also orthogonal.*

Exercise 5.8.8 Prove Proposition 5.8.7. □

Exercise 5.8.9 Let $B \in \mathbb{C}^{m \times m}$ be a semisimple matrix with linearly independent eigenvectors $v_1, \ldots, v_m \in \mathbb{C}^m$, associated with eigenvalues $\lambda_1, \ldots, \lambda_m \in \mathbb{C}$. Suppose $\lambda_1, \ldots, \lambda_r$ are nonzero and $\lambda_{r+1}, \ldots, \lambda_m$ are zero. Show that $\mathcal{N}(B) = \operatorname{span}\{v_{r+1}, \ldots, v_m\}$ and $\mathcal{R}(B) = \operatorname{span}\{v_1, \ldots, v_r\}$. Therefore $\operatorname{rank}(B) = r$, the number of nonzero eigenvalues. □

The matrices $A^T A$ and AA^T are both symmetric and hence semisimple. Thus Corollary 5.8.6 and Exercise 5.8.9 yield a second proof that they have the same rank, which equals the number of nonzero eigenvalues. Since $A^T A$ and AA^T are generally not of the same size, they cannot have exactly the same eigenvalues. The difference is made up by a zero eigenvalue of the appropriate multiplicity. If $\operatorname{rank}(A^T A) = \operatorname{rank}(AA^T) = r$, and $r < m$, then $A^T A$ has a zero eigenvalue of multiplicity $m - r$. If $r < n$, then AA^T has a zero eigenvalue of multiplicity $n - r$.

Exercise 5.8.10   (a) Give an example of a matrix $A \in \mathbb{R}^{1 \times 2}$ such that $A^T A$ has a zero eigenvalue and AA^T does not. (b) Where does the proof of Proposition 5.8.5 break down in the case $\lambda = 0$? □

The following theorem is the same as Theorem 4.1.3, except that one sentence has been added.

Theorem 5.8.11 *(Geometric SVD Theorem) Let $A \in \mathbb{R}^{n \times m}$ be a nonzero matrix with rank r. Then $\mathbb{R}^m$ has an orthonormal basis $v_1, \ldots, v_m$, $\mathbb{R}^n$ has an orthonormal basis $u_1, \ldots, u_n$, and there exist $\sigma_1 \geq \sigma_2 \geq \ldots \geq \sigma_r > 0$ such that*

$$Av_i = \begin{cases} \sigma_i u_i & i = 1, \ldots, r \\ 0 & i = r+1, \ldots, m \end{cases} \qquad A^T u_i = \begin{cases} \sigma_i v_i & i = 1, \ldots, r \\ 0 & i = r+1, \ldots, n \end{cases}$$
$$(5.8.12)$$

Equations (5.8.12) imply that $v_1, \ldots, v_m$ are eigenvectors of $A^T A$, $u_1, \ldots, u_n$ are eigenvectors of AA^T, and $\sigma_1^2, \ldots, \sigma_r^2$ are the nonzero eigenvalues of $A^T A$ and AA^T.

Proof. The final sentence is essentially a rerun of Exercise 5.2.17. You can easily verify that its assertions are true. This determines how $v_1, \ldots, v_m$ must be chosen. Let $v_1, \ldots, v_m$ be an orthonormal basis of $\mathbb{R}^m$ consisting of eigenvectors of $A^T A$. Corollary 5.4.22 guarantees the existence of this basis. Let $\lambda_1, \ldots, \lambda_m$ be the associated eigenvalues. Since $A^T A$ is positive semidefinite, all of its eigenvalues are real and nonnegative. Assume $v_1, \ldots, v_m$ are ordered so that $\lambda_1 \geq \lambda_2 \geq \cdots \geq \lambda_m$. Since $r = \operatorname{rank}(A^T A)$, it must be that $\lambda_r > 0$ and $\lambda_{r+1} = \cdots = \lambda_m = 0$. For $i = 1, \ldots, r$, define $\sigma_i \in \mathbb{R}$ and $u_i \in \mathbb{R}^n$ by

$$\sigma_i = \|Av_i\|_2 \qquad \text{and} \qquad u_i = \frac{1}{\sigma_i} Av_i.$$

These definitions imply that $Av_i = \sigma_i u_i$ and $\|u_i\|_2 = 1$, $i = 1, \ldots, r$. Proposition 5.8.7 implies that $u_1, \ldots, u_r$ are orthogonal, hence orthonormal.

It is easy to show that $\sigma_i^2 = \lambda_i$ for $i = 1, \ldots, r$. Indeed $\sigma_i^2 = \|Av_i\|_2^2 = \langle Av_i, Av_i \rangle = \langle A^T Av_i, v_i \rangle = \langle \lambda_i v_i, v_i \rangle = \lambda_i$. It now follows easily that $A^T u_i = \sigma_i v_i$, for $A^T u_i = (1/\sigma_i) A^T Av_i = (\lambda_i/\sigma_i) v_i = \sigma_i v_i$.

The proof is now complete, except that we have not defined $u_{r+1}, \ldots, u_n$, assuming $r < n$. By Proposition 5.8.5, the vectors $u_1, \ldots, u_r$ are eigenvectors of AA^T associated with nonzero eigenvalues. Since $AA^T \in \mathbb{R}^{n \times n}$ and $\operatorname{rank}(AA^T) = r$, AA^T must have a null space of dimension $n - r$ (Corollary 4.1.9). Let $u_{r+1}, \ldots, u_n$ be any orthonormal basis of $\mathcal{N}(AA^T)$. Noting that $u_{r+1}, \ldots, u_n$ are all eigenvectors of AA^T associated with the eigenvalue zero, we see that each of the vectors $u_{r+1}, \ldots, u_n$ is orthogonal to each of the vectors $u_1, \ldots, u_r$ (Theorem 5.4.15). Thus $u_1, \ldots, u_n$ is an orthonormal basis of $\mathbb{R}^n$ consisting of eigenvectors of AA^T. Since $\mathcal{N}(AA^T) = \mathcal{N}(A^T)$, we have $A^T u_i = 0$ for $i = r+1, \ldots, n$. This completes the proof. $\square$

Recall from Chapter 4 that the numbers $\sigma_1, \ldots, \sigma_r$ are called the (nonzero) *singular values* of A. These numbers are uniquely determined, since they are the positive square roots of the nonzero eigenvalues of $A^T A$. The vectors $v_1, \ldots, v_m$ are *right singular vectors* of A, and $u_1, \ldots, u_n$ are *left singular vectors* of A. Singular vectors are not uniquely determined; they are no more uniquely determined than any eigenvectors of length 1. Any singular vector can be replaced by its opposite, and if $A^T A$ or AA^T happens to have some repeated eigenvalues, an even more serious lack of uniqueness results.

Computing the SVD

One way to compute the SVD of A is simply to calculate the eigenvalues and eigenvectors of $A^T A$ and AA^T. This approach is illustrated in the following example and exercises. After that we will discuss other, more accurate, approaches, in which the SVD is computed without forming $A^T A$ or AA^T explicitly.

Example 5.8.13 Find the singular values and right and left singular vectors of the matrix

$$A = \begin{bmatrix} 1 & 2 & 0 \\ 2 & 0 & 2 \end{bmatrix}.$$

Since $A^T A$ is 3×3 and AA^T is 2×2, it seems reasonable to work with the latter. We easily compute

$$AA^T = \begin{bmatrix} 5 & 2 \\ 2 & 8 \end{bmatrix},$$

so the characteristic polynomial is $(\lambda-5)(\lambda-8)-4 = \lambda^2-13\lambda+36 = (\lambda-9)(\lambda-4)$, and the eigenvalues of AA^T are $\lambda_1 = 9$ and $\lambda_2 = 4$. The singular values of A are therefore

$$\sigma_1 = 3 \qquad \text{and} \qquad \sigma_2 = 2.$$

The left singular vectors of A are eigenvectors of AA^T. Solving $(\lambda_1 I - AA^T)u = 0$, we find that multiples of $[1, 2]^T$ are eigenvectors of AA^T associated with λ_1. Then solving $(\lambda_2 I - AA^T)u = 0$, we find that the eigenvectors of AA^T corresponding to λ_2 are multiples of $[2, -1]^T$. Since we want representatives with unit Euclidean norm, we take

$$u_1 = \frac{1}{\sqrt{5}} \begin{bmatrix} 1 \\ 2 \end{bmatrix} \qquad \text{and} \qquad u_2 = \frac{1}{\sqrt{5}} \begin{bmatrix} 2 \\ -1 \end{bmatrix}.$$

(What other choices could have been made?) These are the left singular vectors of A. Notice that they are orthogonal, as they must be.

We can find the right singular vectors v_1, v_2, and v_3 by calculating the eigenvectors of $A^T A$. However v_1 and v_2 are more easily found by the formula $v_i = \sigma_i^{-1} A^T u_i$, $i = 1, 2$. Thus

$$v_1 = \frac{1}{3\sqrt{5}} \begin{bmatrix} 5 \\ 2 \\ 4 \end{bmatrix} \qquad \text{and} \qquad v_2 = \frac{1}{\sqrt{5}} \begin{bmatrix} 0 \\ 2 \\ -1 \end{bmatrix}.$$

Notice that these vectors are orthonormal. The third vector must satisfy $Av_3 = 0$. Solving the equation $Av = 0$ and normalizing the solution, we get

$$v_3 = \frac{1}{3} \begin{bmatrix} -2 \\ 1 \\ 2 \end{bmatrix}.$$

In this case we could have found v_3 without reference to A by applying the Gram-Schmidt process to find a vector orthogonal to both v_1 and v_2. Normalizing the vector, we would get $\pm v_3$.

Now that we have the singular values and singular vectors of A, we can easily construct the SVD $A = U\Sigma V^T$ with $U \in \mathbb{R}^{2\times 2}$ and $V \in \mathbb{R}^{3\times 3}$ orthogonal and $\Sigma \in \mathbb{R}^{2\times 3}$ diagonal. We have

$$U = \begin{bmatrix} u_1 & u_2 \end{bmatrix} = \frac{1}{\sqrt{5}} \begin{bmatrix} 1 & 2 \\ 2 & -1 \end{bmatrix},$$

$$\Sigma = \begin{bmatrix} \sigma_1 & 0 & 0 \\ 0 & \sigma_2 & 0 \end{bmatrix} = \begin{bmatrix} 3 & 0 & 0 \\ 0 & 2 & 0 \end{bmatrix},$$

$$V = \begin{bmatrix} v_1 & v_2 & v_3 \end{bmatrix} = \frac{1}{3\sqrt{5}} \begin{bmatrix} 5 & 0 & -2\sqrt{5} \\ 2 & 6 & \sqrt{5} \\ 4 & -3 & 2\sqrt{5} \end{bmatrix}.$$

You can easily check that $A = U\Sigma V^T$.   □

rcise 5.8.14 Write down the condensed SVD $A = \hat{U}\hat{\Sigma}\hat{V}^T$ (Theorem 4.1.10) of the matrix A in Example 5.8.13.   □

rcise 5.8.15 Let A be the matrix of Example 5.8.13. Calculate the eigenvalues and eigenvectors of $A^T A$. Compare them with the quantities calculated in Example 5.8.13.   □

rcise 5.8.16 Compute the SVD of the matrix $A = \begin{bmatrix} 3 & 4 \end{bmatrix}$.   □

rcise 5.8.17 Compute the SVD of the matrix

$$A = \begin{bmatrix} 3 & 2 \\ 2 & 3 \\ 2 & -2 \end{bmatrix}.$$

□

rcise 5.8.18 (a) Compute the SVD of the matrix

$$A = \begin{bmatrix} 3 & 1 \\ 6 & 2 \end{bmatrix}.$$

(b) Compute the condensed SVD $A = \hat{U}\hat{\Sigma}\hat{V}^T$.   □

Loss of Information through Squaring

Computation of the SVD by solving the eigenvalue problem for $A^T A$ or AA^T is also a viable option for larger, more realistic problems, since we have efficient algorithms for solving the symmetric eigensystem problem. However, this approach has a serious disadvantage: if A has small but nonzero singular values, they will be calculated inaccurately. This is a consequence of the "loss of information through squaring" phenomenon (see Example 3.5.25), which occurs when we compute $A^T A$.

We can get some idea why this information loss occurs by considering an example. Suppose the entries of A are known to be correct to about six decimal places. If A has, say, $\sigma_1 \approx 1$ and $\sigma_{17} \approx 10^{-3}$, then σ_{17} is small compared with σ_1, but it is still well above the error level $\epsilon \approx 10^{-5}$ or 10^{-6}. We ought to be able to calculate σ_{17} with some precision, at least two or three decimal places. The entries of $A^T A$ also have about six digits accuracy. Associated with the singular values σ_1 and σ_{17}, $A^T A$ has eigenvalues $\lambda_1 = \sigma_1^2 \approx 1$ and $\lambda_{17} = \sigma_{17}^2 \approx 10^{-6}$. Since λ_{17} is of about the same magnitude as the errors in $A^T A$, we cannot expect to compute it with any accuracy at all.

Direct Computation of the SVD

We now turn our attention to methods that calculate the SVD by operating directly on A. As we noticed earlier in this chapter, the eigenvalue problem can be made easier if we first reduce the matrix to Hessenberg or tridiagonal form. A similar reduction greatly simplifies the SVD problem. Here we will develop a method that reduces the matrix to bidiagonal form and then computes the SVD by a variant of Francis's algorithm.

Other approaches are possible. In Section 7.2 we outline a number of alternative methods for the symmetric eigenvalue problem. Some, perhaps all, of those methods can be adapted to the SVD problem. In particular, Jacobi's method is able to compute the SVD (especially the smallest singular values) more accurately than any algorithm that begins by bidiagonalizing the matrix. Traditionally Jacobi methods have been several times slower than methods based on Francis's algorithm. However, Drmač and Veselić [24, 25] have shown recently that through certain preprocessing steps Jacobi's method can be made to run faster than ever before. They have developed a Jacobi-based SVD code that is comparable in speed to state-of-the-art codes.

Reduction to Bidiagonal Form

When we solve the eigenvalue problem, the reduction has to be done via similarity transformations. For the singular value decomposition $A = U\Sigma V^T$, it is clear that similarity transformations are not required, but the transforming matrices should be orthogonal. Two matrices $A, B \in \mathbb{R}^{n \times m}$ are said to be *orthogonally equivalent* if there exist orthogonal matrices $P \in \mathbb{R}^{n \times n}$ and $Q \in \mathbb{R}^{m \times m}$ such that $B = PAQ$. We will see that we can reduce any matrix to bidiagonal form by an orthogonal equivalence transformation, in which each of the transforming matrices is a product of m or fewer reflectors. The algorithms that we are about to discuss are appropriate for dense matrices. We will not discuss the sparse SVD problem.

cise 5.8.19   Show that if two matrices are orthogonally equivalent, then they have the same singular values, and there are simple relationships between their singular vectors.   □

We continue to assume that we are dealing with a matrix $A \in \mathbb{R}^{n \times m}$, and we will now make the additional assumption that $n \geq m$. This does not imply any loss of generality, for if $n < m$, we can operate on A^T instead of A. If the SVD of A^T is $A^T = U\Sigma V^T$, then the SVD of A is $A = V\Sigma^T U^T$.

A matrix $B \in \mathbb{R}^{n \times m}$ is said to be *bidiagonal* if $b_{ij} = 0$ whenever $i > j$ or $i < j - 1$. This means that B has the form

$$
\begin{bmatrix}
* & * & & & & \\
 & * & * & & & \\
 & & * & & & \\
 & & & \ddots & & * \\
 & & & & & * \\
 & & & & & \\
\end{bmatrix}
$$

with nonzero entries appearing only on two diagonals.

Theorem 5.8.20 *Let $A \in \mathbb{R}^{n \times m}$ with $n \geq m$. Then there exist orthogonal $\hat{U} \in \mathbb{R}^{n \times n}$ and $\hat{V} \in \mathbb{R}^{m \times m}$, both products of a finite number of reflectors, and a bidiagonal $B \in \mathbb{R}^{n \times m}$, such that*

$$A = \hat{U} B \hat{V}^T.$$

There is a finite algorithm to calculate $\hat{U}$, $\hat{V}$, and B.

Proof. We will prove Theorem 5.8.20 by describing the construction. This is the *Golub-Kahan* algorithm [41]. It is quite similar to both the QR decomposition by reflectors (Section 3.2) and the reduction to upper Hessenberg form by reflectors (Section 5.5), so we will just sketch the procedure. The first step creates zeros in the first column and row of A. Let $\hat{U}_1 \in \mathbb{R}^{n \times n}$ be a reflector such that

$$
\hat{U}_1 \begin{bmatrix} a_{11} \\ a_{21} \\ \vdots \\ a_{n1} \end{bmatrix} = \begin{bmatrix} * \\ 0 \\ \vdots \\ 0 \end{bmatrix}.
$$

Then the first column of $\hat{U}_1 A$ consists of zeros, except for the $(1,1)$ entry. Now let $\begin{bmatrix} \hat{a}_{11} & \hat{a}_{12} & \cdots & \hat{a}_{1m} \end{bmatrix}$ denote the first row of $\hat{U}_1 A$, and let $\hat{V}_1$ be a reflector of the form

$$
\hat{V}_1 = \left[\begin{array}{c|ccc} 1 & 0 & \cdots & 0 \\ \hline 0 & & & \\ \vdots & & \check{V}_1 & \\ 0 & & & \end{array} \right],
$$

such that

$$\begin{bmatrix} \hat{a}_{12} & \cdots & \hat{a}_{1m} \end{bmatrix} \check{V}_1 = \begin{bmatrix} * & 0 & \cdots & 0 \end{bmatrix}.$$

Then the first row of $\hat{U}_1 A \hat{V}_1$ consists of zeros, except for the first two entries. Because the first column of $\hat{V}_1$ is e_1, the first column of $\hat{U}_1 A$ is unaltered by right multiplication by $\hat{V}_1$. Thus $\hat{U}_1 A \hat{V}_1$ has the form

$$\hat{U}_1 A \hat{V}_1 = \left[\begin{array}{c|cccc} * & * & 0 & \cdots & 0 \\ \hline 0 & & & & \\ \vdots & & \hat{A} & & \\ 0 & & & & \end{array} \right].$$

The second step of the construction is identical to the first, except that it acts on the submatrix $\hat{A}$. It is easy to show that the reflectors used on the second step do not destroy the zeros created on the first step. After two steps we have

$$\hat{U}_2 \hat{U}_1 A \hat{V}_1 \hat{V}_2 = \left[\begin{array}{c|c|cccc} * & * & 0 & 0 & \cdots & 0 \\ \hline 0 & * & * & 0 & \cdots & 0 \\ \hline 0 & 0 & & & & \\ \vdots & \vdots & & & \tilde{A} & \\ 0 & 0 & & & & \end{array} \right].$$

The third step acts on the submatrix $\tilde{A}$, and so on. After m steps we have

$$\hat{U}_m \cdots \hat{U}_2 \hat{U}_1 A \hat{V}_1 \hat{V}_2 \cdots \hat{V}_{m-2} = \begin{bmatrix} * & * & & & & \\ & * & * & & & \\ & & * & & & \\ & & & \ddots & * & \\ & & & & * & \end{bmatrix} = B.$$

Notice that steps $m - 1$ and m require multiplications on the left only. Let $\hat{U} = \hat{U}_1 \hat{U}_2 \cdots \hat{U}_m$ and $\hat{V} = \hat{V}_1 \hat{V}_2 \cdots \hat{V}_{m-2}$. Then $\hat{U}^T A \hat{V} = B$; that is, $A = \hat{U} B \hat{V}^T$. □

Exercise 5.8.21 Carry out a flop count for this algorithm. Assume that $\hat{U}$ and $\hat{V}$ are not to be assembled explicitly. Recall from Section 3.2 that if $x \in \mathbb{R}^k$ and $U \in \mathbb{R}^{k \times k}$ is a reflector, then the operations $x \to Ux$ and $x^T \to x^T U$ each cost about $4k$ flops. Show that the cost of the right multiplications is slightly less than that of the left multiplications, but for large n and m, the difference is negligible. Notice that the total flop count is about twice that of a QR decomposition by reflectors. □

Exercise 5.8.22 Explain how the reflectors can be stored efficiently. How much storage space is required, in addition to the array that contains A initially? Assume that A is to be overwritten. □

In many applications (e.g. least squares problems) n is much larger than m. In this case it is sometimes more efficient to perform the reduction to bidiagonal form in two stages. In the first stage a QR decomposition of A is performed:

$$A = QR = \begin{bmatrix} Q_1 & Q_2 \end{bmatrix} \begin{bmatrix} \hat{R} \\ 0 \end{bmatrix},$$

where $\hat{R} \in \mathbb{R}^{m \times m}$ is upper triangular. This requires multiplications by reflectors on the left hand side of A only. In the second stage the relatively small matrix $\hat{R}$ is reduced to bidiagonal form $\hat{R} = \tilde{U}\tilde{B}\tilde{V}^T$. All of these matrices are $m \times m$. Then

$$A = \begin{bmatrix} Q_1 & Q_2 \end{bmatrix} \begin{bmatrix} \tilde{U} & 0 \\ 0 & I \end{bmatrix} \begin{bmatrix} \tilde{B} \\ 0 \end{bmatrix} \tilde{V}^T.$$

Letting

$$\hat{U} = \begin{bmatrix} Q_1 & Q_2 \end{bmatrix} \begin{bmatrix} \tilde{U} & 0 \\ 0 & I \end{bmatrix} = \begin{bmatrix} Q_1\tilde{U} & Q_2 \end{bmatrix} \in \mathbb{R}^{n \times n},$$

$$B = \begin{bmatrix} \tilde{B} \\ 0 \end{bmatrix} \in \mathbb{R}^{n \times m},$$

and $\hat{V} = \tilde{V} \in \mathbb{R}^{m \times m}$, we have $A = \hat{U}B\hat{V}^T$.

The advantage of this arrangement is that the right multiplications are applied to the small matrix $\hat{R}$ instead of the large matrix A. They therefore cost a lot less. The disadvantage is that the right multiplications destroy the upper-triangular form of $\hat{R}$. Thus most of the left multiplications must be repeated on the small matrix $\hat{R}$. If the ratio n/m is sufficiently large, the added cost of the left multiplications will be more than offset by the savings in the right multiplications. The break-even point is around $n/m = 5/3$. If $n/m \gg 1$, the flop count is cut nearly in half.

The various applications of the SVD have different requirements. Some require only the singular values, while others require the right or left singular vectors, or both. If any of the singular vectors are needed, then the matrices $\hat{U}$ and/or $\hat{V}$ have to be computed explicitly. Usually it is possible to avoid calculating $\hat{U}$, but there are numerous applications for which $\hat{V}$ is needed. The flop count of Exercise 5.8.21 does not include the cost of computing $\hat{U}$ or $\hat{V}$. If $\hat{U}$ is needed, then there is no point in doing the preliminary QR decomposition.

Computing the SVD of B

Since B is bidiagonal, it has the form

$$B = \begin{bmatrix} \tilde{B} \\ 0 \end{bmatrix},$$

where $\tilde{B} \in \mathbb{R}^{m \times m}$ is bidiagonal. The problem of finding the SVD of A is thus reduced to that of computing the SVD of the small bidiagonal matrix $\tilde{B}$.

For notational convenience we now drop the tilde from $\tilde{B}$ and let $B \in \mathbb{R}^{m \times m}$ be a bidiagonal matrix, say

$$
B = \begin{bmatrix}
\beta_1 & \gamma_1 & & & \\
 & \beta_2 & \gamma_2 & & \\
 & & \beta_3 & \ddots & \\
 & & & \ddots & \gamma_{m-1} \\
 & & & & \beta_m
\end{bmatrix}. \tag{5.8.23}
$$

We will call B a *properly* bidiagonal matrix if (in the notation of (5.8.23)) $\beta_i \neq 0$ and $\gamma_i \neq 0$ for all i.

If B is not properly bidiagonal, the problem of finding its SVD can be reduced to two or more smaller subproblems. First of all, if some γ_k is zero, then

$$
B = \begin{bmatrix} B_1 & 0 \\ 0 & B_2 \end{bmatrix}, \tag{5.8.24}
$$

where $B_1 \in \mathbb{R}^{k \times k}$ and $B_2 \in \mathbb{R}^{(m-k) \times (m-k)}$ are both bidiagonal. The SVDs of B_1 and B_2 can be found separately and then combined to yield the SVD of B. If some β_k is zero, a small amount of work transforms B to a form that can be reduced. The procedure is discussed in Exercise 5.8.47.

We can now assume, without loss of generality, that B is a properly bidiagonal matrix. We can even assume (Exercise.5.8.48) that all of the β_k and γ_k are positive.

Exercise 5.8.25 Let B be a bidiagonal matrix.

(a) Show that BB^T and $B^T B$ are tridiagonal.

(b) Show that if B is properly bidiagonal, then BB^T and $B^T B$ are properly tridiagonal.

□

The SVD problem for B is equivalent to the eigenvalue/vector problem for the properly tridiagonal matrices BB^T and $B^T B$. In the rest of this section we will develop the *Golub-Reinsch* algorithm [42], [9, pp. 160–180], a variant of Francis's algorithm that does the task without explicitly forming either of the products BB^T and $B^T B$. There are several other excellent methods for solving the symmetric eigenvalue problem; an overview is given in Section 7.2. Each of these methods can be adapted to solving the SVD problem. We do not claim that the Golub-Reinsch algorithm is necessarily better than these others.

Our point of departure will not be the pair of matrices BB^T and $B^T B$ but instead the symmetric matrix

$$
C = \begin{bmatrix} 0 & B^T \\ B & 0 \end{bmatrix}. \tag{5.8.26}
$$

Exercise 5.8.27 Let $B \in \mathbb{R}^{m \times m}$ be properly bidiagonal. Consider the $2m \times 2m$ symmetric matrix C given by (5.8.26).

(a) Show that C is a nonsingular matrix. Thus all of the eigenvalues of C are nonzero.

(b) Show that if $\begin{bmatrix} v \\ u \end{bmatrix}$ is an eigenvector of B associated with the eigenvalue σ, then $\begin{bmatrix} v \\ -u \end{bmatrix}$ is an eigenvector of C associated with the eigenvalue $-\sigma$. Thus the eigenvalues of C occur in $\pm\sigma$ pairs.

(c) Compute the matrix C^2 and reflect on the result.

$\square$

Part (b) of Exercise 5.8.27 is a review of Exercise 5.4.48, in which we also showed that $\|u\|_2 = \|v\|_2$. If we assume, without loss of generality, that $\sigma > 0$ and $\|u\|_2 = \|v\|_2 = 1$, then v and u are right and left singular vectors of B, respectively, corresponding to the singular value σ. This is so because of the evident relationships $Bv = \sigma u$ and $B^T u = \sigma v$. Conversely, given singular vectors v and u corresponding to a singular value σ, we can use the singular vectors to build eigenvectors of C associated with the eigenvalues σ and $-\sigma$. Thus the SVD problem for B is equivalent to the eigenvalue problem for C.

Our development makes use of the perfect shuffle permutation, which we introduce in the following exercise. We will use it again for a similar purpose in Section 7.4.

rcise 5.8.28 Let $e_1, e_2, \ldots, e_{2m}$ denote the standard basis vectors in $\mathbb{R}^{2m}$. Define a matrix $P \in \mathbb{R}^{2m \times 2m}$ by $P = \begin{bmatrix} e_1 & e_{m+1} & e_2 & e_{m+2} & \cdots & e_m & e_{2m} \end{bmatrix}$. This is the *perfect shuffle* matrix. Note that P is a permutation matrix; thus it is orthogonal.

(a) Show that if M is any matrix with $2m$ columns, then the transformation $M \to MP$ does a perfect shuffle of the columns of M. In other words, columns $1, 2, \ldots, m$ of M become columns $1, 3, \ldots 2m - 1$ of MP, while columns $m + 1, m + 2, \ldots 2m$ of M become columns $2, 4, \ldots, 2m$ of MP.

(b) Show that if M is any matrix with $2m$ rows, then the transformation $M \to P^T M$ does a perfect shuffle of the rows of M.

From From parts (a) and (b) we deduce that if M is $2m \times 2m$, the transformation $M \to P^T MP$ does a perfect shuffle of the rows and columns. This is an orthogonal similarity transformation. Obviously it can be reversed: If $K = P^T MP$, then the transformation $K \to PKP^T = M$ undoes the shuffle. $\square$

rcise 5.8.29 Let

$$M = \begin{bmatrix} A & C \\ B & D \end{bmatrix},$$

where A, B, C, and D are 4×4.

(a) Write down the 8×8 matrix $P^T MP$, where P is the perfect shuffle matrix.

(b) Write down $P^T M P$ in the special case where $A = D = 0$, $C = B^T$, and B is bidiagonal, the 4×4 version of (5.8.23). This matrix is tridiagonal.

□

We will work with

$$C = \begin{bmatrix} 0 & B^T \\ B & 0 \end{bmatrix}$$

and its shuffled variant

$$\tilde{C} = P^T C P = \left[\begin{array}{cc|cc|cc|c} 0 & \beta_1 & & & & & \\ \beta_1 & 0 & \gamma_1 & & & & \\ \hline & \gamma_1 & 0 & \beta_2 & & & \\ & & \beta_2 & 0 & \gamma_2 & & \\ \hline & & & \gamma_2 & 0 & \beta_3 & \\ & & & & \beta_3 & 0 & \gamma_3 \\ \hline & & & & & \gamma_3 & \ddots \end{array} \right]. \tag{5.8.30}$$

When we work with C we will say that we are in the *unshuffled coordinate system* or we are looking at the *unshuffled (or deshuffled) picture*. When we work with $\tilde{C}$, we are in the *shuffled coordinate system* or we are looking at the *shuffled picture*. Both pictures are useful. Since $\tilde{C}$ is symmetric and tridiagonal, we can compute its eigensystem by Francis's algorithm, which preserves symmetry and tridiagonality. Since the eigenvalues of $\tilde{C}$ (which are the same as the those of C) occur in $\pm\sigma$ pairs, it makes sense to do Francis iterations of degree two with shifts $\pm\rho$ in order to preserve as much structure as possible. With this strategy, the eigenvalues σ and $-\sigma$ will be extracted simultaneously.

We start the iteration by computing the first column of $(\tilde{C} - \rho I)(\tilde{C} + \rho I) = \tilde{C}^2 - \rho^2 I$, which turns out to be

$$\begin{bmatrix} \beta_1^2 - \rho^2 \\ 0 \\ \gamma_1 \beta_1 \\ 0 \\ \vdots \\ 0 \end{bmatrix}. \tag{5.8.31}$$

Exercise 5.8.32 Show that the first column of $\tilde{C}^2 - \rho^2 I$ is given by (5.8.31). You might find it simplest to use the fact that $\tilde{C}^2 - \rho^2 I = P^T (C^2 - \rho^2 I) P$, do the computation in the unshuffled coordinate system, then shuffle.   □

The double Francis iteration begins by applying an orthogonal matrix Q_0^T that transforms (5.8.31) to a multiple of e_1. Clearly this can be done by a rotator (or other orthogonal transformation) that acts on rows 1 and 3 only. When this transformation acts on $\tilde{C}$, it alters only four entries of the matrix, creating a bulge element in position $(1, 4)$. These entries are in odd rows and even columns. When the similarity transformation is completed by multiplying on the right by Q_0, another four entries are changed, and a bulge element is created in position $(4, 1)$. These entries are in even rows and odd columns. Symmetry must be preserved, which implies that the operations that are done during the right multiplication are exactly the same as those done during the left multiplication. Using the case $m = 3$ for illustration, the transformed matrix has the form

$$\begin{bmatrix} & * & \Big| & + & \Big| & \\ * & & \Big| & * & \Big| & \\ \hline & * & \Big| & * & \Big| & \\ + & & \Big| & * & \Big| & * \\ \hline & & \Big| & * & \Big| & * \\ & & \Big| & & \Big| & * \end{bmatrix}. \tag{5.8.33}$$

This matrix has a bulge of degree two of a special form. The bulge occupies positions $(3, 1)$, $(4, 1)$, and $(4, 2)$ (cf. (5.6.8)) but only the tip entry $(4, 1)$ is actually nonzero. Clearly it can be chased forward by a transformation Q_1^T that acts only on rows 2 and 4. This transformation alters only six entries of the matrix, including the annihilated $(4, 1)$ entry. A new bulge is created in position $(2, 5)$. All of these entries are in even rows and odd columns. The similarity transformation is completed by a right multiplication by Q_1 that acts on columns 2 and 4. This annihilates the bulge in position $(1, 4)$ and creates a new bulge in position $(5, 2)$. All of the six altered entries lie in odd rows and even columns. By symmetry the multiplication on the right by Q_1 effects exactly the same operations as the multiplication on the left by Q_1^T. The transformed matrix now has the form

$$\begin{bmatrix} & * & \Big| & & \Big| & \\ * & & \Big| & * & \Big| & + \\ \hline & * & \Big| & * & \Big| & \\ & & \Big| & * & \Big| & * \\ \hline & + & \Big| & * & \Big| & * \\ & & \Big| & & \Big| & * \end{bmatrix}. \tag{5.8.34}$$

The pattern is now clear. The next step will annihilate the bulges in positions $(5, 2)$ and $(2, 5)$ and create new bulges in positions $(6, 3)$ and $(3, 6)$. The resulting matrix has the form

$$\begin{bmatrix} & * & \Big| & & \Big| & \\ * & & \Big| & * & \Big| & \\ \hline & * & \Big| & * & \Big| & + \\ & & \Big| & * & \Big| & * \\ \hline & & \Big| & * & \Big| & * \\ & & \Big| & + & \Big| & * \end{bmatrix}. \tag{5.8.35}$$

We continue the process until the bulge gets chased off of the bottom of the matrix, completing the double Francis iteration.

Notice that the zero entries on the main diagonal of $\tilde{C}$ are preserved by all the operations, so the very special tridiagonal form of $\tilde{C}$ is preserved.

If we perform these double Francis iterations repeatedly with some reasonable shifting strategy, we can expect rapid convergence to a pair of eigenvalues. The bottom of the tridiagonal matrix $\tilde{C}$ has the form

$$
\begin{bmatrix}
\ddots & & & & \\
& \gamma_{m-2} & & & \\
\hline
\gamma_{m-2} & 0 & \beta_{m-1} & & \\
& \beta_{m-1} & 0 & \gamma_{m-1} & \\
\hline
& & \gamma_{m-1} & 0 & \beta_m \\
& & & \beta_m & 0
\end{bmatrix}.
\tag{5.8.36}
$$

The simplest reasonable strategy is the generalized Rayleigh quotient shift strategy, which takes the shifts to be the two eigenvalues of the bottom right 2×2 submatrix, which are easily seen to be $\pm\beta_m$. This strategy yields a cubic local convergence rate, .i.e. $\gamma_{m-1} \to 0$ cubically [101, Ch. 5], allowing a quick deflation of a $\pm$ pair of eigenvalues. However, global convergence is not guaranteed. A better strategy is to find the four eigenvalues of the lower-right-hand 4×4 matrix (two $\pm$ pairs) and take as shifts that pair that is closer to $\pm\beta_m$. In Chapter 6 (Exercise 6.3.22) we will show that this is equivalent to the Wilkinson shift strategy applied to the tridiagonal matrix BB^T. Thus global convergence is guaranteed.

Exercise 5.8.37

(a) Show that the eigenvalues of the lower-right-hand 2×2 submatrix of (5.8.36) are $\pm\beta_m$.

(b) Show that the characteristic polynomial of the lower-right-hand 4×4 submatrix of (5.8.36) is

$$
\lambda^4 - (\beta_{m-1}^2 + \beta_m^2 + \gamma_{m-1}^2)\lambda^2 + \beta_{m-1}^2\beta_m^2.
$$

Thus the eigenvalues can be found by (a careful application of) the quadratic formula.

$\square$

Now let us take a look at this procedure in the unshuffled coordinate system. The deshuffled version of (5.8.33) is

$$
\left[
\begin{array}{ccc|ccc}
 & & & * & + & \\
 & & & * & * & \\
 & & & & * & * \\
\hline
* & * & & & & \\
+ & * & * & & & \\
 & & * & & &
\end{array}
\right]. \tag{5.8.38}
$$

The entries in the $(1, 2)$ block are exactly the numbers that were in the odd rows and even columns of (5.8.33). These were altered by the left multiplication by Q_0^T. In the deshuffled picture (5.8.38) this corresponds to an operation on rows 1 and 2 that creates a bulge in position $(1, 5)$. The similarity transformation is completed by a right multiplication that acts on columns 1 and 2, creating a bulge in position $(5, 1)$. This operations acts only entries of the $(2, 1)$ block, which correspond to entries in even rows and odd columns of the shuffled matrix.

The deshuffled version of (5.8.34) is

$$
\left[
\begin{array}{ccc|ccc}
 & & & * & & \\
 & & & * & * & \\
 & & & + & * & * \\
\hline
* & * & + & & & \\
 & * & * & & & \\
 & & * & & &
\end{array}
\right]. \tag{5.8.39}
$$

In the deshuffled picture we get from (5.8.38) to (5.8.39) by two transformations. One is a left transformation on rows 4 and 5 that annihilates the bulge in position $(5, 1)$ and creates a new one in position $(4, 3)$. The second is a right transformation on columns 4 and 5 that annihilates the bulge in position $(1, 5)$ and creates a new one in position $(3, 4)$. The second operation is exactly the transpose of the first. The first operation corresponds to the left multiplication by Q_1^T in the shuffled picture, acting on entries in even rows and odd columns. The second corresponds to right multiplication by Q_1, which acts on entries in odd rows and even columns.

The deshuffled version of (5.8.35) is

$$
\left[
\begin{array}{ccc|ccc}
 & & & * & & \\
 & & & * & * & + \\
 & & & & * & * \\
\hline
* & * & & & & \\
 & * & * & & & \\
 & + & * & & &
\end{array}
\right]. \tag{5.8.40}
$$

In the unshuffled picture, we get from (5.8.39) to (5.8.40) by a transformation of rows 2 and 3 that annihilates the bulge in position $(3, 4)$ and creates a new bulge at

$(2, 6)$, and a transformation of columns 2 and 3 that annihilates the $(4, 3)$ bulge and creates a new bulge at $(6, 2)$. Of course these two operations are transposes of one another. As an exercise the reader can check what these operations look like in the shuffled picture.

The next step would operate on rows and columns 5 and 6 to annihilate the bulges at $(6, 2)$ and $(2, 6)$. In the case $m = 3$, this completes the iteration, as the bulges have now been chased away. In the case $m > 3$, the bulges would need to be chased further.

In the unshuffled picture it is especially clear that the operations are redundant. Everything that happens in the $(2, 1)$ block happens in transpose in the $(1, 2)$ block. Therefore we can just work on one block and throw away the other. We elect to retain the $(2, 1)$ block, which contains the matrix B initially. Let us now review the iteration as it acts on the single block B.

Referring back to (5.8.31) and its deshuffled variant, let $V_0 \in \mathbb{C}^{m \times m}$ be a rotator (or other orthogonal matrix) acting in the 1–2 plane such that

$$
V_0^T \begin{bmatrix} \beta_1^2 - \rho^2 \\ \gamma_1 \beta_1 \\ 0 \\ \vdots \\ 0 \end{bmatrix} = \begin{bmatrix} * \\ 0 \\ 0 \\ \vdots \\ 0 \end{bmatrix}.
$$

The first transformation is $B \to BV_0$. This recombines the first two columns of B and creates a bulge in the $(3, 1)$ position. In the case $m = 5$, the transformed matrix has the form

$$
\begin{bmatrix} * & * & & & \\ + & * & * & & \\ & & * & * & \\ & & & * & * \\ & & & & * \end{bmatrix},
$$

which can be compared to the $(2, 1)$ block of (5.8.38). For the next step we pick a rotator U_0 acting in the 1–2 plane such that the transformation $B_0 V_0 \to U_0^T B V_0$ annihilates the bulge and creates a new bulge in position $(1, 3)$. Now the transformed matrix has the form.

$$
\begin{bmatrix} * & * & + & & \\ & * & * & & \\ & & * & * & \\ & & & * & * \\ & & & & * \end{bmatrix},
$$

which can be compared to the $(2, 1)$ block of (5.8.39). Next a rotator V_1 acting in the 2–3 plane is applied on the right. The transformation $U_0^T B V_0 \to U_0^T B V_0 V_1$ annihilates the bulge at $(1, 3)$ and creates a new bulge at $(3, 2)$, so that the transformed matrix has the form

$$
\begin{bmatrix}
* & * & & & \\
  & * & * & & \\
  & + & * & * & \\
  &   & * & * & \\
  &   &   & * &
\end{bmatrix}.
$$

The pattern should be clear by now. Subsequent alternating left and right transformations send the bulge to positions $(2, 4)$, $(4, 3)$, $(3, 5)$, $(5, 4)$, and so on, until the bulge is pushed out of the matrix. For general m a total of $m - 1$ transformations on the right and $m - 1$ on the left are needed for the task: The matrix

$$
\hat{B} = U_{m-2}^T \cdots U_0^T B V_0 \cdots V_{m-2} \tag{5.8.41}
$$

is again bidiagonal, and the iteration is complete.

Because this procedure is equivalent to a double Francis iteration on (5.8.30), we know that if we repeat the iterations with a good shift strategy, the entry γ_{m-1} will converge to 0 cubically. Once it is small enough, we can set it to zero and continue to look for the next eigen/singular value on an $(m - 1) \times (m - 1)$ submatrix. After $m - 1$ deflations, the matrix B will have been transformed to a diagonal matrix $\Sigma = \text{diag}\{\sigma_1, \ldots, \sigma_m\}$. We have

$$
\Sigma = U^T B V, \tag{5.8.42}
$$

where U and V are orthogonal matrices, each of which is the product of a large number of rotators. We can arrange the algorithm so that all of the σ_i come out positive, but even if we do not do that, it is easy to make them positive afterwards: If $\sigma_i < 0$, we can reverse its sign by reversing the sign of the ith column of V. This does not effect the orthogonality of the matrix. (Equally well, we could reverse the ith column of U.) We can also assume without loss of generality that $\sigma_1 \geq \sigma_2 \geq \cdots \geq \sigma_m > 0$. This is so because we can easily interchange any two of them: If we interchange columns i and j of both U and V, we interchange σ_i and σ_j. Now if we assume that we have normalized Σ so that $\sigma_1 \geq \sigma_2 \geq \cdots \geq \sigma_m > 0$ and rewrite (5.8.42) as

$$
B = U \Sigma V^T,
$$

we see that this is an SVD of B.

The algorithm that we have just derived is the Golub-Reinsch algorithm [42] for computing the SVD. The derivation given here is a special case of the development of Francis's (implicitly shifted QR) algorithm for the product eigenvalue problem [57], [58], [100], [101, Ch. 8]. Other special cases will be presented in Sections 7.3 and 7.4 in connection with the generalized eigenvalue problem.

The following theorem summarizes compactly the action of the Golub-Reinsch algorithm.

Theorem 5.8.43 *Let $B \in \mathbb{R}^{m \times m}$ be properly bidiagonal, and let $\hat{B}$ be the result of one iteration of the Golub-Reinsch algorithm with shift ρ. Then*

$$\hat{B} = U^T B V,$$

where U and V are orthogonal and $V e_1 = \alpha(B^T B - \rho^2 I)e_1$ for some nonzero α. In words, the first column of V is proportional to the first column of $B^T B - \rho^2 I$

Exercise 5.8.44 Prove Theorem 5.8.43 as follows. Define U and V with reference to (5.8.41), show that $V_0 e_1 = \alpha(B^T B - \rho^2 I)e_1$, then show that $V e_1 = \alpha(B^T B - \rho^2 I)e_1$. $\square$

A Few Final Words on Shifts

Some good shifting strategies were suggested in the paragraph leading up to Exercise 5.8.37. Demmel and Kahan [17] found that it is worthwhile to start with a few iterations with shifts $\pm \rho = 0$. If B has any tiny singular values, these will emerge promptly at the bottom, and B can be deflated. This procedure results in greater accuracy for the tiny singular values.

Once the tiny singular values have been removed, we should switch to a shifting strategy that causes the remaining singular values to emerge quickly. For example, we can compute the two $\pm$ pairs of eigenvalues of the lower-right-hand 4×4 submatrix of $\tilde{C}$ (5.8.36), then take the pair that is closer to $\pm \beta_m$ as the shifts. In case of a tie, take the pair that is smaller in magnitude. Alternatively, one can always take the pair that is smaller in magnitude, in an effort to deflate out the smallest singular values first.

Exercise 5.8.45 Show that the two shifts given by the strategy just described are the square roots of the Wilkinson shift for the tridiagonal matrix BB^T. (Recall that the Wilkinson shift for a symmetric, tridiagonal $A \in \mathbb{R}^{m \times m}$ is the eigenvalue of the lower-right-hand 2×2 submatrix that is closer to a_{nn}. In case of a tie, we take the one that is smaller in magnitude.) $\square$

Additional Exercises

Exercise 5.8.46 Write a Fortran (or MATLAB or C or ...) program that calculates the singular values of a bidiagonal matrix B. Test it on the matrix $B \in \mathbb{R}^{m \times m}$ given by $\beta_i = 1$ and $\gamma_i = 1$ for all i. Try various values of m. The singular values of B are

$$\sigma_j = 2 \cos \frac{j\pi}{2m+1} \qquad j = 1, \ldots, m.$$

$\square$

Exercise 5.8.47 Let $B \in \mathbb{R}^{m \times m}$ have the form (5.8.23). Since B is $m \times m$, we can say that B has m singular values $\sigma_1 \geq \sigma_2 \geq \cdots \geq \sigma_m \geq 0$. If $\text{rank}(B) = r < m$, the last $m - r$ singular values are zero.

(a) Show that $\beta_k = 0$ for some k if and only if zero is a singular value of B.

(b) Suppose $\beta_k = 0$. Show that we can make $\gamma_{k-1} = 0$ by a sequence of $k - 1$ rotators applied to B from the right. The first acts on columns $k - 1$ and k and transforms γ_{k-1} to zero. At the same time it produces a new nonzero entry in position $(k - 2, k)$. The second rotator acts on columns $k - 2$ and k, annihilates the $(k - 2, k)$ entry and creates a new nonzero entry in position $(k - 3, k)$. Subsequent rotators push the extra nonzero entry up the kth column until it disappears off the top of the matrix. The resulting bidiagonal matrix has both $\beta_k = 0$ and $\gamma_{k-1} = 0$.

(c) Supposing still that $\beta_k = 0$, show that we can make $\gamma_k = 0$ by a sequence of $m - k$ rotators applied to B from the left. The procedure is analogous to the procedure of part (b), except that it pushes the unwanted nonzero entry out along the kth row rather than up the kth column.

Once the procedures of parts (b) and (c) have been applied, we have $\beta_k = 0$, $\gamma_{k-1} = 0$, and $\gamma_k = 0$. The bidiagonal matrix now has the form

$$
\begin{bmatrix} B_1 & & \\ & 0 & \\ & & B_2 \end{bmatrix},
$$

where the zero that is shown explicitly is β_k. This is a zero singular value of B. We can move it to the (m, m) position of the matrix by a row and column interchange, if we wish. The rest of the SVD can be found by computing the SVDs of B_1 and B_2 separately and then combining them.   □

cise 5.8.48 Show that if B is properly bidiagonal, then there exist orthogonal, diagonal matrices D_1 and D_2 (these have entries ± 1 on the main diagonal), such that all of the nonzero entries of $D_1 B D_2$ are positive. Thus we can assume, without loss of generality, that all β_k and γ_k are positive.   □

cise 5.8.49 Show that if $A \in \mathbb{R}^{n \times n}$ is a symmetric, positive definite matrix, then its singular values are the same as its eigenvalues, and its right and left singular vectors and (normalized) eigenvectors are all the same. How is the picture changed if A is symmetric but indefinite?   □

cise 5.8.50 Let $A \in \mathbb{R}^{n \times m}$ and $C \in \mathbb{R}^{m \times n}$, and define

$$
F = \begin{bmatrix} 0 & C \\ A & 0 \end{bmatrix} \in \mathbb{R}^{(n+m) \times (n+m)}.
$$

Show that a nonzero scalar μ is an eigenvalue of F if and only if μ^2 is an eigenvalue of both AC and CA. Determine the relationship between eigenvectors of M and eigenvectors of AC and CA.   □

CHAPTER 6

EIGENVALUES AND EIGENVECTORS II

This chapter has two distinct objectives. One is to explain why Francis's algorithm works, and the other is to introduce some methods for solving large, sparse eigenvalue problems. These two projects, which might at first seem to be completely different, have several common elements.

The first section introduces the important notions of *eigenspace* and *invariant subspace*. The latter concept is particularly important for a deeper understanding of eigenvalue problems and algorithms. At the end of Section 6.1 we discuss the computation of invariant subspaces by eigenvalue swapping.

In Section 6.2 we introduce *subspace iteration*, a natural multidimensional generalization of the power method, and *simultaneous iteration*, a practical way of implementing nested subspace iterations that was once an important tool for solving large, sparse eigenvalue problems. We then introduce simultaneous iteration with a change of coordinate system at each step, which brings us close to an explanation of Francis's algorithm.

The one additional concept that is needed for an understanding of Francis's algorithm is that of *Krylov subspaces*. Once we have this, we are able to show in Section 6.3 that Francis's algorithm is indeed simultaneous iteration with changes of coordinate system. This essentially concludes our explanation of Francis's algorithm.

Fundamentals of Matrix Computations, Third Edition. By David S. Watkins
Copyright © 2010 John Wiley & Sons, Inc.

However, we also prove a duality theorem that links Francis's algorithm to inverse iteration and Rayleigh quotient iteration. At the end of the section we make the connection between Francis's iteration and the QR decomposition, which explains why the Francis algorithm sometimes (well, almost always) goes by the name of QR algorithm.

In Section 6.4 we begin our discussion of large, sparse eigenvalue problems, beginning with simultaneous iteration and the shift-and-invert strategy. We then move on to the Arnoldi process, a variant of the Gram-Schmidt process that generates orthogonal bases of Krylov subspaces. The algorithm then seeks approximations to eigenvectors in the Krylov subspaces that are generated. Thus we see that the concept of Krylov subspace is important to both objectives of this chapter.

Section 6.5 builds on Section 6.4 by showing how to do implicit restarts of the Arnoldi process. The implicitly-restarted Arnoldi (IRA) process and its variant, the Krylov-Schur algorithm, are currently the most successful methods for solving large, sparse eigenvalue problems. They are often used in conjunction with the shift-and-invert strategy.

In situations where methods of Section 6.5 are inadequate, for example when the matrix is too large for the shift-and-invert strategy, an alternative to consider is the class of *Jacobi-Davidson* methods. These are discussed in the final section of the chapter.

6.1 EIGENSPACES AND INVARIANT SUBSPACES

Let $A \in \mathbb{C}^{n \times n}$, let λ be any complex number, and consider the set of vectors

$$S_\lambda = \{ v \in \mathbb{C}^n \mid Av = \lambda v \}.$$

You can easily check that S_λ is a subspace of $\mathbb{C}^n$.

Exercise 6.1.1 (a) Show that if v_1, $v_2 \in S_\lambda$, then $v_1 + v_2 \in S_\lambda$. (b) Show that if $v \in S_\lambda$ and $\alpha \in \mathbb{C}$, then $\alpha v \in S_\lambda$. Thus S_λ is a subspace of $\mathbb{C}^n$.   □

Since the equation $Av = \lambda v$ can be rewritten as $(\lambda I - A)v = 0$, we see that $S_\lambda = \mathcal{N}(\lambda I - A)$, the null space of $\lambda I - A$. The space S_λ always contains the zero vector. It contains other vectors if and only if λ is an eigenvalue of A, in which case we call S_λ the *eigenspace* of A associated with λ.

The notion of an invariant subspace is a useful generalization of eigenspace. Since we will want to talk about both real and complex invariant subspaces, we will begin by introducing some notation to make the presentation more flexible. Let $\mathbb{F}$ denote either the real or complex number field. Thus $\mathbb{F}^n$ and $\mathbb{F}^{n \times n}$ will denote either $\mathbb{R}^n$ and $\mathbb{R}^{n \times n}$ or $\mathbb{C}^n$ and $\mathbb{C}^{n \times n}$, depending on the choice of $\mathbb{F}$.

Let $A \in \mathbb{F}^{n \times n}$. A subspace S of $\mathbb{F}^n$ is said to be *invariant* under A if $Ax \in S$ whenever $x \in S$, in other words, $AS \subseteq S$. We will also sometimes bend the language slightly and say that S is an *invariant subspace* of A. The spaces $\{0\}$ and $\mathbb{F}^n$ are both invariant subspaces of A. The next exercise gives some nontrivial examples of invariant subspaces.

First we recall some notation from Chapter 3. Given x_1, x_2, ..., $x_k \in \mathbb{F}^m$, span$\{x_1, \ldots, x_k\}$ denotes the set of all linear combinations of $x_1, \ldots, x_k$. That is,

$$\text{span}\{x_1, \ldots, x_k\} = \left\{ \sum_{i=1}^{k} c_i x_i \mid c_i \in \mathbb{F}, i = 1, \ldots, k \right\}.$$

Notice that there is now some ambiguity in the meaning of span$\{x_1, \ldots, x_k\}$. The coefficients c_i are taken from $\mathbb{F}$. They can be either real or complex, depending on the context. There was no such ambiguity in Chapter 3, because there we always had $\mathbb{F} = \mathbb{R}$.

rcise 6.1.2

(a) Let $v \in \mathbb{F}^n$ be any eigenvector of A. Show that span$\{v\}$ is invariant under A.

(b) Let $\mathcal{S}_\lambda$ be any eigenspace of A (of dimension 1 or greater). Show that $\mathcal{S}_\lambda$ is invariant under A.

(c) Let $v_1, \ldots, v_k \in \mathbb{F}^n$ be any k eigenvectors of A associated with eigenvalues $\lambda_1, \ldots, \lambda_k$. Show that the space $\mathcal{S} = \text{span}\{v_1, \ldots, v_k\}$ is invariant under A. It can be shown that if A is semisimple, every invariant subspace has this form.

(d) Let

$$A = \begin{bmatrix} 1 & 0 & 0 \\ 0 & 2 & 1 \\ 0 & 0 & 2 \end{bmatrix}.$$

Find the eigenspaces of A, and find a space $\mathcal{S}$ that is invariant under A and is not spanned by eigenvectors of A.

$\square$

If we think of A as a linear operator mapping $\mathbb{F}^n$ into $\mathbb{F}^n$, then an invariant subspace can be used to reduce the operator in the following sense: suppose $\mathcal{S}$ is an invariant subspace under A that is neither $\{0\}$ nor $\mathbb{F}^n$, and let $\hat{A}$ denote the restricted operator that acts on the subspace $\mathcal{S}$, that is, $\hat{A} = A \mid_\mathcal{S}$. Then $\hat{A}$ maps $\mathcal{S}$ into $\mathcal{S}$. It is clear that every eigenvector of $\hat{A}$ is an eigenvector of A, and so is every eigenvalue. Thus we can obtain information about the eigensystem of A by studying $\hat{A}$, an operator that is simpler than A in the sense that it acts on a lower-dimensional space. The following sequence of theorems and exercises shows how this reduction can be carried out in practice. We begin by characterizing invariant subspaces in the language of matrices.

Theorem 6.1.3 *Let $\mathcal{S}$ be a subspace of $\mathbb{F}^n$ with a basis $x_1, \ldots, x_k$. Thus $\mathcal{S} = \text{span}\{x_1, \ldots, x_k\}$. Let $\hat{X} = \begin{bmatrix} x_1 & \cdots & x_k \end{bmatrix} \in \mathbb{F}^{n \times k}$. Then $\mathcal{S}$ is invariant under $A \in \mathbb{F}^{n \times n}$ if and only if there exists $\hat{B} \in \mathbb{F}^{k \times k}$ such that*

$$A\hat{X} = \hat{X}\hat{B}.$$

Proof. Suppose there is a $\hat{B}$ such that $A\hat{X} = \hat{X}\hat{B}$. Let b_{ij} denote the (i,j) entry of $\hat{B}$. Equating the jth column of $A\hat{X}$ with the jth column of $\hat{X}\hat{B}$, we see that $Ax_j = \sum_{i=1}^{k} x_i b_{ij} \in \text{span}\{x_1, \ldots, x_k\} = S$. Since $Ax_j \in S$ for $j = 1, \ldots, k$, and $x_1, \ldots, x_k$ span S, it follows that $Ax \in S$ for all $x \in S$. Thus S is invariant under A. Conversely, assume S is invariant. We must construct $\hat{B}$ such that $A\hat{X} = \hat{X}\hat{B}$. For $j = 1, \ldots, k$, $Ax_j \in S$, so there exist constants $b_{1j}, \ldots, b_{kj} \in \mathbb{F}$ such that $Ax_j = x_1 b_{1j} + x_2 b_{2j} + \cdots + x_k b_{kj}$. Define $\hat{B} \in \mathbb{F}^{k \times k}$ to be the matrix whose (i,j) entry is b_{ij}. Then $A[x_1 \cdots x_k] = [x_1 \cdots x_k]\hat{B}$, that is, $A\hat{X} = \hat{X}\hat{B}$. $\quad\square$

Exercise 6.1.4 Let S be an invariant subspace of A with basis $x_1, \ldots, x_k$, and let $\hat{X} = [x_1 \cdots x_k]$. By Theorem 6.1.3 there exists $\hat{B} \in \mathbb{F}^{k \times k}$ such that $A\hat{X} = \hat{X}\hat{B}$.

(a) Show that if $\hat{v} \in \mathbb{F}^k$ is an eigenvector of $\hat{B}$ with eigenvalue λ, then $v = \hat{X}\hat{v}$ is an eigenvector of A with eigenvalue λ. In particular, every eigenvalue of $\hat{B}$ is an eigenvalue of A. What role does the linear independence of $x_1, \ldots, x_k$ play?

The eigenvalues of $\hat{B}$ are called the eigenvalues of A *associated with* the invariant subspace S.

(b) Show that $v \in S$. Thus v is an eigenvector of $A|_S$.

$\square$

Exercise 6.1.5   With A, $\hat{B}$, and S as in the previous exercise, show that $\hat{B}$ is a matrix representation of the linear operator $A|_S$. $\quad\square$

The next theorem shows that whenever we have (a basis for) a nontrivial invariant subspace, we can make a similarity transformation that breaks the eigenvalue problem into two smaller problems.

Theorem 6.1.6 *Let S be invariant under $A \in \mathbb{F}^{n \times n}$. Suppose S has dimension k, with $1 \le k < n$, and let $x_1, \ldots, x_k$ be a basis of S. Let $x_{k+1}, \ldots, x_n$ be any $n - k$ vectors such that the vectors $x_1, \ldots, x_n$ together form a basis for $\mathbb{F}^n$. Let $X_1 = [x_1 \cdots x_k]$, $X_2 = [x_{k+1} \cdots x_n]$, and $X = [x_1 \cdots x_n] \in \mathbb{F}^{n \times n}$. Define $B = X^{-1}AX$. Then B is block upper triangular:*

$$B = \begin{bmatrix} B_{11} & B_{12} \\ 0 & B_{22} \end{bmatrix},$$

where $B_{11} \in \mathbb{F}^{k \times k}$. Furthermore $AX_1 = X_1 B_{11}$.

B is similar to A, so it has the same spectrum. This can be found by computing the spectra of B_{11} and B_{22} separately. Thus the eigenvalue problem has been broken into two smaller problems.

Notice that B_{11} is the same as the matrix $\hat{B}$ that was introduced in Theorem 6.1.3.

If we wish, we can always take $x_1, \ldots, x_k$ to be an orthonormal basis of S and choose the additional vectors $x_{k+1}, \ldots, x_n$ so that $x_1, \ldots, x_n$ together form an orthonormal basis of $\mathbb{F}^n$. This can be achieved by the Gram-Schmidt process, for example. Then X is unitary, and B is unitarily similar to A.

Proof. The equation $B = X^{-1}AX$ is equivalent to $AX = XB$. The jth column of this equation is $Ax_j = \sum_{i=1}^n x_i b_{ij}$. Since $x_1, \ldots, x_n$ are linearly independent, this sum is the unique representation of Ax_j as a linear combination of $x_1, \ldots, x_n$. On the other hand, for $j = 1, \ldots, k$, x_j lies in the invariant subspace $\mathcal{S} = \operatorname{span}\{x_1, \ldots, x_k\}$, so Ax_j lies in $\mathcal{S}$ as well. Thus $Ax_j = c_{1j}x_1 + \cdots + c_{kj}x_k$ for some $c_{1j}, \ldots, c_{kj} \in \mathbb{F}$. By uniqueness of the representation of Ax_j, we have $b_{ij} = c_{ij}$ for $i = 1, \ldots, k$, and more importantly $b_{ij} = 0$ for $i = k + 1, \ldots, n$. This holds for $j = 1, \ldots, k$. In the partition $B = \begin{bmatrix} B_{11} & B_{12} \\ B_{21} & B_{22} \end{bmatrix}$, the block B_{21} consists precisely of those b_{ij} for which $k + 1 \le i \le n$ and $1 \le j \le k$. Thus $B_{21} = 0$. This proves that B is block triangular. The equation $AX_1 = X_1 B_{11}$ now follows immediately from the obvious partition of the equation $AX = XB$. $\qquad\square$

See Exercise 6.1.16 for a second proof of this theorem.

·cise 6.1.7 Prove the converse of Theorem 6.1.6: if $B_{21} = 0$, then $\mathcal{S}$ is invariant under A. $\qquad\square$

Corollary 6.1.8 *Let* $B = X^{-1}AX$*, where* $X = [x_1 \ \ldots \ x_n] \in \mathbb{F}^{n \times n}$ *is nonsingular. Then B is upper triangular if and only if all of the spaces*

$$\operatorname{span}\{x_1, \ldots, x_k\}, \qquad k = 1, \ldots, n - 1$$

are invariant under A.

·cise 6.1.9 Prove Corollary 6.1.8. $\qquad\square$

Theorem 6.1.6 shows that whenever we find a nontrivial invariant subspace, we can make a similarity transformation that breaks the eigenvalue problem into two smaller problems. If we can then find subspaces that are invariant under B_{11} and B_{22}, we can break each of them into smaller subproblems. If we can continue this process indefinitely, we can eventually break the problem down to many small submatrices of size 1×1 and 2×2, from which we can read off the eigenvalues. It is thus reasonable to say that finding invariant subspaces is the fundamental task of eigensystem computations.

Schur's Theorem (Theorem 5.4.11) and its real variant, the Wintner-Murnaghan Theorem (Theorem 5.4.23) show that the breakdown process described in the previous paragraph is always possible in principle. We now see (cf. Corollary 6.1.8) that these are theorems about invariant subspaces.

Francis's algorithm achieves the practical task of breaking the matrix down. In a typical reduction the submatrix B_{22} is tiny and B_{11} is huge. Thus (nearly) all of the remaining work consists of breaking down B_{11}. It would be nice to develop an algorithm that breaks a matrix into submatrices of roughly equal size. This requires the identification of an invariant subspace of dimension approximately $n/2$, which has proved to be a difficult task.[33]

Computing Invariant Subspaces by Eigenvalue Swapping

Suppose a matrix A has been been reduced to the Schur form $A = UTU^*$ by Francis's algorithm or some other means. U is unitary and T is upper triangular. Let $\lambda_i = t_{ii}$, $i = 1, \ldots, n$. These are the eigenvalues of A. From Theorem 6.1.6 we know that the first k columns of U span the invariant subspace associated with the eigenvalues λ_1, $\ldots, \lambda_k$ (Exercise 6.1.4). Now suppose we are looking for some other k-dimensional invariant subspace corresponding, say, to eigenvalues $\mu_1, \ldots, \mu_k$. Then we would like a different Schur decomposition, one with $\mu_1, \ldots, \mu_k$ (in any order) in the first k main-diagonal positions of the upper-triangular matrix. It is therefore natural to ask whether it is possible to modify the given Schur decomposition inexpensively in such a way that the desired eigenvalues are moved into the top positions. This is the question of eigenvalue swapping.

The problem is trivial if A is normal, since then T is diagonal and the columns of U are eigenvectors. We can swap eigenvalues λ_j and λ_k simply by swapping the jth and kth columns of U. In fact swapping is actually unnecessary in this case. If we want the invariant subspace corresponding to eigenvalues $\lambda_{i_1}, \ldots, \lambda_{i_k}$, we can just pick out columns $i_1, i_2, \ldots, i_k$ of U. These k eigenvectors form an orthonormal basis for the desired invariant subspace.

If A is not normal, then T is not diagonal, and the problem is not so trivial. However, it is clear that if we can swap any two adjacent eigenvalues, then we can move the eigenvalues wherever we please on the main diagonal. Therefore it is enough to consider the 2×2 case. Given

$$\begin{bmatrix} \lambda & a \\ 0 & \mu \end{bmatrix}$$

with $\mu \neq \lambda$, we seek a complex rotator (Exercise 3.2.53)

$$\begin{bmatrix} c & -\overline{s} \\ s & \overline{c} \end{bmatrix}$$

with $|c|^2 + |s|^2 = 1$ such that

$$\begin{bmatrix} \overline{c} & \overline{s} \\ -s & c \end{bmatrix} \begin{bmatrix} \lambda & a \\ 0 & \mu \end{bmatrix} \begin{bmatrix} c & -\overline{s} \\ s & \overline{c} \end{bmatrix} = \begin{bmatrix} \mu & \hat{a} \\ 0 & \lambda \end{bmatrix}. \tag{6.1.10}$$

Proposition 6.1.11 *If $\lambda \neq \mu$, the transformation (6.1.10) can be achieved by taking $c = a/r$ and $s = (\mu - \lambda)/r$, where $r = \sqrt{|a|^2 + |\mu - \lambda|^2}$.*

We have restricted our attention to the case $\lambda \neq \mu$. When λ and μ are the same, one normally has no reason to swap them. If λ and μ are distinct but very close, one might not wish to swap them either. We will say more about this below.

---

[33] However, some headway has been made in this area. See [5].

rcise 6.1.12 Prove Proposition 6.1.11 as follows. Let $\begin{bmatrix} b_{11} & b_{12} \\ b_{21} & b_{22} \end{bmatrix}$ denote the product of three matrices on the left-hand side of (6.1.10).

(a) Show that $b_{21} = s(c(\mu - \lambda) - sa)$, which is zero if and only if either $s = 0$ or $c(\mu - \lambda) = sa$. The first choice leads to a trivial transformation that does not swap λ and μ. Show that the second condition is satisfied if c and s are defined as in the statement of Proposition 6.1.11. (This is the only nontrivial solution up to a phase shift $e^{i\theta}$.)

(b) Show that under this choice of c and s we have $b_{11} = \mu$ and $b_{22} = \lambda$.

□

So far we have been assuming that the matrix has been reduced all the way to the triangular Schur form. If the matrix is real, we normally prefer to stay in real arithmetic by using the Wintner-Murnaghan form (Theorem 5.4.23). Then T is quasitriangular, having a 2×2 block for each complex conjugate pair of eigenvalues. Thus we need to be able to swap eigenvalues with 2×2 blocks and 2×2 blocks with each other.

Let us consider the latter case. Given

$$B = \begin{bmatrix} B_{11} & B_{12} \\ 0 & B_{22} \end{bmatrix},$$

where B_{11} and B_{22} are both 2×2 and have no eigenvalues in common, we seek a real orthogonal matrix Q such that

$$Q^T B Q = \begin{bmatrix} C_{11} & C_{12} \\ 0 & C_{22} \end{bmatrix}, \tag{6.1.13}$$

where the eigenvalues of C_{11} (resp. C_{22}) are the same as those of B_{22} (resp. B_{11}). It turns out that this can always be done in principle. A procedure is sketched in Exercise 6.1.22. For more details see [3].

In exact arithmetic the swap of two 2×2 blocks produces $C_{21} = 0$ exactly, as shown in (6.1.13). However, in floating point arithmetic we will obtain a C_{21} that is not exactly zero. If it happens that $\| C_{21} \|$ is not small enough that C_{21} can be safely set to zero, then we must reject the swap. This can happen if the eigenvalues of B_{11} are too close to those of B_{22}. This possibility also has to be watched for during swaps of two 1×1 blocks (single eigenvalues) and swaps of 1×1 blocks with 2×2 blocks. In general it can be dangerous to try to swap eigenvalues that are too close together.

It is also generally undesirable to "separate" eigenvalues that are close together, in the following sense. Suppose the spectrum of A includes a tight cluster, and we want to compute the invariant subspace S associated with a set of eigenvalues that contains some but not all of the eigenvalues in the cluster. Then S will certainly be ill conditioned. That is, a tiny perturbation in A can cause a huge perturbation of S. This can be understood partly by studying the condition numbers of eigenvectors, as in done in Section 7.1. There it is shown that eigenvectors corresponding to eigenvalues in clusters are necessarily ill conditioned. For a discussion of the conditioning of invariant subspaces see, for example, [83] or [101, § 2.7].

If one wishes to obtain an invariant subspace that is well conditioned (and therefore physically meaningful), one must include either all of the eigenvalues in a cluster or none of them. If one makes the commitment never to split a cluster, then it will be possible to do the swaps needed to separate the wanted from the unwanted eigenvalues without ever having to swap eigenvalues that are close to each other.

Additional Exercises

Exercise 6.1.14 If μ is an eigenvalue of A, then the eigenspace $S_\mu(A) = \{v \in \mathbb{C}^n \mid Av = \mu v\}$ has dimension at least 1. Often the dimension is exactly 1, but sometimes it is greater. The *geometric multiplicity* of the eigenvalue μ is defined to be the dimension of the eigenspace $S_\mu(A)$. The geometric multiplicity may be different from the *algebraic multiplicity*, which is defined to be the number of times μ appears as a root of the characteristic equation. By this we mean the number of times the factor $\lambda - \mu$ appears as a factor of the characteristic polynomial $\det(\lambda I - A)$.

(a) Find the algebraic and geometric multiplicities of the eigenvalue $\lambda = 2$ in each of the following matrices.

$$A = \begin{bmatrix} 1 & 2 & 3 \\ 0 & 2 & 3 \\ 0 & 0 & 3 \end{bmatrix} \quad B = \begin{bmatrix} 1 & 0 & 0 \\ 0 & 2 & 1 \\ 0 & 0 & 2 \end{bmatrix} \quad C = \begin{bmatrix} 1 & 0 & 0 \\ 0 & 2 & 0 \\ 0 & 0 & 2 \end{bmatrix}$$

(b) Schur's Theorem (Theorem 5.4.11) guarantees that every matrix $A \in \mathbb{C}^{n \times n}$ is unitarily similar to an upper triangular matrix T: $A = UTU^{-1}$. Show that the algebraic multiplicity of μ as an eigenvalue of A is equal to the number of times μ appears on the main diagonal of T.

(c) Show that if A and B are similar matrices, then $\dim S_\mu(A) = \dim S_\mu(B)$. Thus the geometric multiplicity of an eigenvalue is preserved under similarity transformations.

(d) Show that the geometric multiplicity of μ is less than or equal to the algebraic multiplicity. (Work with the upper triangular matrix T, and suppose the algebraic multiplicity of μ is k. We know that in Schur's Theorem we can arrange the eigenvalues in any order on the main diagonal of T (Exercise 5.4.31). Suppose T is constructed so that its first k main diagonal entries are μ. What does an eigenvector look like? Show that the dimension of the eigenspace cannot exceed k.)

$\square$

Exercise 6.1.15   (Proof of Wintner-Murnaghan Theorem 5.4.23)

(a) Prove that if $A \in \mathbb{R}^{n \times n}$ with $n \geq 3$, then there is a subspace $S \in \mathbb{R}^n$ of dimension 1 or 2 that is invariant under A. (Consider an eigenvector of A.

Its real and imaginary parts span a subspace of $\mathbb{R}^n$. Show that this space is invariant and has dimension 1 or 2.)

(b) Using part (a) and Theorem 6.1.6, prove Theorem 5.4.23 by induction on n.

□

rcise 6.1.16 This exercise derives a second proof of Theorem 6.1.6. Assuming the same notation as in Theorem 6.1.6, define $Y \in \mathbb{F}^{n \times n}$ by $Y^T = X^{-1}$. Let $y_1, \ldots, y_n$ denote the columns of Y, and let $Y_1 = [y_1 \cdots y_k]$ and $Y_2 = [y_{k+1} \cdots y_n]$.

(a) Show that $Y_1^T X_1 = I \in \mathbb{F}^{k \times k}$, $Y_2^T X_2 = I \in \mathbb{F}^{(n-k) \times (n-k)}$, $Y_2^T X_1 = 0$, and $Y_1^T X_2 = 0$.

(b) Show that $Y_1^T A X_1 = B_{11}$, $Y_2^T A X_2 = B_{22}$, $Y_2^T A X_1 = B_{21}$, and $Y_1^T A X_2 = B_{12}$.

(c) Use Theorem 6.1.3 and selected equations from parts (a) and (b) to show that $B_{21} = 0$ if the columns of X_1 span an invariant subspace.

□

rcise 6.1.17 In the notation of Theorem 6.1.6, let $\mathcal{U} = \text{span}\{x_{k+1}, \ldots, x_n\}$.

(a) Show that $\mathcal{U}$ is invariant under A if and only if $B_{12} = 0$. Your solution may resemble the proof of Theorem 6.1.6, or you can use ideas from Exercise 6.1.16.

(b) Suppose $\mathcal{U}$ is invariant under A. Show that if $\hat{v}$ is an eigenvector of B_{22} associated with the eigenvalue λ, then $v = X_2 \hat{v}$ is an eigenvector of A associated with eigenvalue λ.

□

rcise 6.1.18 Suppose the hypotheses of Theorem 6.1.6 are satisfied. Define $Y \in \mathbb{F}^{n \times n}$ by $Y^T = X^{-1}$, let $y_1, \ldots, y_n$ denote the columns of Y, let $Y_1 = [y_1 \cdots y_k]$, $Y_2 = [y_{k+1} \cdots y_n]$, and define $\mathcal{U} = \text{span}\{y_{k+1}, \ldots, y_n\}$.

(a) Show that $\mathcal{U}$ is invariant under A^T.

(b) Show that if $\hat{w}$ is an eigenvector of B_{22}^T associated with the eigenvalue λ, then $w = Y_2 \hat{w}$ is an eigenvector of A^T associated with λ.

□

rcise 6.1.19 We repeat Exercise 6.1.18 with one minor alteration. Take $\mathbb{F} = \mathbb{C}$ and define $Z \in \mathbb{C}^{n \times n}$ by $Z^* = X^{-1}$. Let $z_1, \ldots, z_n$ denote the columns of Z, and define $Z_2 = [z_{k+1} \cdots z_n]$ and $\mathcal{V} = \text{span}\{z_{k+1}, \ldots, z_n\}$.

(a) Show that $\mathcal{V}$ is invariant under A^*.

(b) Show that if $\hat{w}$ is an eigenvector of B_{22}^* associated with the eigenvalue $\overline{\lambda}$, then $w = Z_2 \hat{w}$ is an eigenvector of A^* associated with $\overline{\lambda}$.

(c) It is always possible to choose $x_1, \ldots, x_n$ so that X is unitary, in which case $Z = X$. Use these facts to show that if $\mathcal{S}$ is invariant under A, then $\mathcal{S}^\perp$ is invariant under A^*.

□

Exercise 6.1.20 In this exercise we prove directly (without using bases) that $\mathcal{S}^\perp$ is invariant under A^* if $\mathcal{S}$ is invariant under A. Recall that the inner product in $\mathbb{C}^n$ is defined by $\langle x, y \rangle = \sum_{j=1}^{n} x_j \bar{y}_j$. Let $A \in \mathbb{C}^{n \times n}$.

(a) Show that for all $x, y \in \mathbb{C}^n$, $\langle Ax, y \rangle = \langle x, A^*y \rangle$. This is the complex analogue of Lemma 3.5.9.

(b) Use the result of part (a) to prove directly that if $\mathcal{S}$ is invariant under A, then $\mathcal{S}^\perp$ is invariant under A^*.

(c) Conclude from part (b) that if A is Hermitian ($A = A^*$), then $\mathcal{S}$ is invariant under A if and only if $\mathcal{S}^\perp$ is. Relate this result to Exercise 6.1.17 and Theorem 6.1.6.

□

Exercise 6.1.21 Let A, B, $X \in \mathbb{F}^{n \times n}$ with X nonsingular and $B = X^{-1}AX$. Let $x_1, \ldots, x_n$ denote the columns of X. Prove that B is lower triangular if and only if the spaces $\mathrm{span}\{x_{k+1}, \ldots, x_n\}$, $k = 1, \ldots, n-1$, are all invariant under A. □

Exercise 6.1.22 We consider the general problem of swapping eigenvalues. Suppose

$$B = \begin{bmatrix} B_{11} & B_{12} \\ 0 & B_{22} \end{bmatrix},$$

where B_{11} is $k \times k$ and B_{22} is $m \times m$. Suppose further that B_{11} and B_{22} have no eigenvalues in common. We seek a unitary Q such that

$$Q^*BQ = C = \begin{bmatrix} C_{11} & C_{12} \\ 0 & C_{22} \end{bmatrix},$$

where C_{11} is $m \times m$ and has the same eigenvalues as B_{22}. Then C_{22} has perforce the same eigenvalues as B_{11}. If B is real, then we would like Q and C to be real as well.

(a) Given an X of dimension $k \times m$, let $\mathcal{S}_X$ denote the space spanned by the columns of $\begin{bmatrix} X \\ I \end{bmatrix}$. Show that $\mathcal{S}_X$ is invariant under B if and only if

$$\begin{bmatrix} B_{11} & B_{12} \\ 0 & B_{22} \end{bmatrix} \begin{bmatrix} X \\ I \end{bmatrix} = \begin{bmatrix} X \\ I \end{bmatrix} B_{22}.$$

Moreover, if it is invariant, it must be the invariant subspace associated with the eigenvalues of B_{22}. We need to build a similarity transformation that moves these eigenvalues "to the top".

(b) Deduce that S_X is invariant under B if and only if

$$XB_{22} - B_{11}X = B_{12}. \qquad (6.1.23)$$

This *Sylvester equation* is a system of km linear equations in km unknowns. In Exercise 6.1.25 it is shown that the Sylvester equation has a unique solution if and only if B_{11} and B_{22} have no common eigenvalues. If k and m are small, then this system is not too big. For example, when $k = m = 2$, it is a system of 4 equations in 4 unknowns. It can be solved by conventional means.

(c) Suppose we have solved the Sylvester equation (6.1.23) for X. Consider a QR decomposition

$$\begin{bmatrix} X \\ I \end{bmatrix} = QR = \begin{bmatrix} Q_1 & Q_2 \end{bmatrix} \begin{bmatrix} \hat{R} \\ 0 \end{bmatrix},$$

where Q_1 has m columns and $\hat{R}$ is $m \times m$. Show that the columns of Q_1 span S_X, the desired invariant subspace.

(d) Let

$$Q^*BQ = C = \begin{bmatrix} C_{11} & C_{12} \\ C_{21} & C_{22} \end{bmatrix},$$

where C_{11} is $m \times m$. Show that $C_{21} = 0$, and the eigenvalues of C_{11} are the same as the eigenvalues of B_{22}. Thus this is the desired transformation.

(e) Show that if B is real, then Q and C are real.

□

cise 6.1.24 This study of some basic properties of Kronecker products of matrices will prepare you for Exercise 6.1.25. Let X and Y be matrices of dimensions $\alpha \times \beta$ and $\gamma \times \delta$, respectively. Then the *Kronecker product* or *tensor product* of X and Y is the $\alpha\gamma \times \beta\delta$ matrix $X \otimes Y$ defined by

$$X \otimes Y = \begin{bmatrix} x_{11}Y & x_{12}Y & \cdots & x_{1\beta}Y \\ x_{21}Y & x_{22}Y & \cdots & x_{2\beta}Y \\ \vdots & \vdots & \ddots & \vdots \\ x_{\alpha 1}Y & x_{\alpha 2}Y & \cdots & x_{\alpha\beta}Y \end{bmatrix}.$$

(a) Let X, Y, W, and Z be matrices whose dimensions are such that the products XW and YZ are defined. Show that the product $(X \otimes Y)(W \otimes Z)$ is defined, and $(X \otimes Y)(W \otimes Z) = (XW) \otimes (YZ)$.

(b) Show that if U and V are unitary, then $(U \otimes V)^* = U^* \otimes V^*$, and $U \otimes V$ is unitary.

(c) Show that if S and T are upper triangular, then $S \otimes T$ is upper triangular. Supposing S is $k \times k$ and T is $m \times m$, show that

$$S \otimes I_m - I_k \otimes T$$

is also upper triangular, where I_k and I_m are identity matrices of the appropriate dimensions.

(d) Suppose E is $k \times k$ and C is $m \times m$. Let $E = USU^*$ and $C = VTV^*$ be Schur decompositions. Show that $E \otimes I_m - I_k \otimes C$ has the Schur decomposition

$$(U \otimes V)(S \otimes I_m - I_k \otimes T)(U \otimes V)^*.$$

(e) Using (d), show that if the eigenvalues of E and C are $\lambda_1, \ldots, \lambda_k$ and $\mu_1, \ldots, \mu_m$, respectively, then the eigenvalues of $E \otimes I_m - I_k \otimes C$ are $\lambda_i - \mu_j$, $i = 1, \ldots, k$ and $j = 1, \ldots, m$. Deduce that $E \otimes I_m - I_k \otimes C$ is nonsingular if and only if E and C have no eigenvalues in common.

$\square$

Exercise 6.1.25   Exercise 6.1.24 is a prerequisite for this exercise. Consider a Sylvester equation

$$XF - CX = G, \tag{6.1.26}$$

where F is $m \times m$, C is $k \times k$, G is $k \times m$, and we are to solve for an unknown $k \times m$ matrix X. Given a $k \times m$ matrix $Z = \begin{bmatrix} z_1 & \cdots & z_m \end{bmatrix}$, we define

$$\text{vec}(Z) = \begin{bmatrix} z_1 \\ \vdots \\ z_m \end{bmatrix}.$$

Thus $\text{vec}(Z)$ is just a long column vector that contains the same information as Z.

(a) Show that $\text{vec}(CX) = (I_m \otimes C)\text{vec}(X)$.

(b) Show that $\text{vec}(XF) = (F^T \otimes I_k)\text{vec}(X)$.

(c) Deduce that the system (6.1.26) is equivalent to

$$(F^T \otimes I_k - I_m \otimes C)\text{vec}(X) = \text{vec}(G).$$

Deduce further that (6.1.26) has a unique solution if and only if F and C have no eigenvalues in common. Thus (6.1.23) has a unique solution if and only if B_{11} and B_{22} have no eigenvalues in common.

$\square$

6.2   SUBSPACE ITERATION AND SIMULTANEOUS ITERATION

The main objective of this section is to bring the reader closer to an understanding of Francis's algorithm. We begin by introducing subspace iteration, a straightforward generalization of the power method. We then turn to simultaneous iteration, a practical way of implementing subspace iteration that was in widespread use for solving large, sparse eigenvalue problems around 1970. Simultaneous iteration has the virtue of doing subspace iterations on subspaces of many different dimensions at once. Although it no longer finds much use as a practical algorithm, it is still a useful concept that helps us understand Francis's algorithm. Next we introduce an interesting variant of simultaneous iteration that does a change of coordinate system at each step. In Section 6.3 we will develop one more tool that we need in order to identify Francis's algorithm as simultaneous iteration with changes of coordinate system.

Subspace Iteration

We begin by returning to the power method and recasting it in terms of subspaces. Given a matrix $A \in \mathbb{C}^{n \times n}$ that has a dominant eigenvalue, we can use the power method to calculate a dominant eigenvector v_1 or a multiple thereof. It does not matter which multiple we obtain, since each nonzero multiple is as good an eigenvector as any other. In fact each multiple of v_1 is just a representative of the eigenspace span$\{v_1\}$, which is the real object of interest. Likewise, in the power sequence q, Aq, A^2q, A^3q, ... each of the iterates A^jq can be viewed as a representative of the space span$\{A^jq\}$, which it spans. The operation of rescaling a vector amounts to replacing one representative by another representative of the same one-dimensional space. Thus the power method can be viewed as a process of iterating on subspaces: First a one-dimensional subspace $\mathcal{S} = $ span$\{q\}$ is chosen. Then the iterates

$$\mathcal{S}, \ A\mathcal{S}, \ A^2\mathcal{S}, \ A^3\mathcal{S}, \ \ldots \tag{6.2.1}$$

are formed. This sequence converges linearly to the eigenspace $\mathcal{U} = $ span$\{v_1\}$ in the sense that the angle between $A^j\mathcal{S}$ and $\mathcal{U}$ converges to zero as $j \to \infty$.

It is quite natural to generalize this process to subspaces of dimension greater than one. Thus we can choose a subspace $\mathcal{S}$ of any dimension and form the sequence (6.2.1). It is perhaps not surprising that this sequence will generally converge to a space that is invariant under A. Before proceeding, you should work the following exercise, which reviews a number of concepts and covers the most basic properties of subspace iteration.

rcise 6.2.2 Let $A \in \mathbb{F}^{n \times n}$, where $\mathbb{F} = \mathbb{R}$ or $\mathbb{C}$, and let $\mathcal{S}$ be a subspace of $\mathbb{F}^n$. Then $A\mathcal{S}$ is defined by $A\mathcal{S} = \{Ax \mid x \in \mathcal{S}\}$.

(a) Recall that a nonempty subset $\mathcal{U}$ of $\mathbb{F}^n$ is a *subspace* if and only if (i) x_1, $x_2 \in \mathcal{U} \Rightarrow x_1 + x_2 \in \mathcal{U}$ and (ii) $\alpha \in \mathbb{F}$ and $x \in \mathcal{U} \Rightarrow \alpha x \in \mathcal{U}$. Show that if $\mathcal{S}$ is a subspace of $\mathbb{F}^n$, then $A\mathcal{S}$ is also a subspace of $\mathbb{F}^n$.

(b) By definition, $A^j\mathcal{S} = \{A^jx \mid x \in \mathcal{S}\}$. Show that $A^j\mathcal{S} = A(A^{j-1}\mathcal{S})$.

(c) Show that if $\mathcal{S} = $ span$\{x_1, \ldots, x_k\}$, then $A\mathcal{S} = $ span$\{Ax_1, \ldots, Ax_k\}$.

(d) Recall that the *null space* of A is $\mathcal{N}(A) = \{x \in \mathbb{F}^n \mid Ax = 0\}$. Let S be a subspace of $\mathbb{F}^n$ for which $S \cap \mathcal{N}(A) = \{0\}$. Show that if $x_1, \ldots, x_k$ is a basis of S, then $Ax_1, \ldots, Ax_k$ is a basis of AS. Consequently $\dim(AS) = \dim(S)$.

$\square$

In order to talk about convergence of subspaces, we need to say what we mean by the distance between two subspaces. A detailed discussion would take us too far afield, so we will just say that the *distance* between two subspaces S_1 and S_2, denoted $d(S_1, S_2)$, is defined to be the sine of the largest principal angle between them. Principal angles are defined in [43] and [101, § 2.6], for example. Given a sequence of subspaces (S_j) and a subspace $\mathcal{U}$, all of the same dimension, we will say that (S_j) *converges* to $\mathcal{U}$ (denoted symbolically by $S_j \to \mathcal{U}$) provided that $d(S_j, \mathcal{U}) \to 0$ as $j \to \infty$. We are now ready to state the main theorem on the convergence of subspace iteration.

Theorem 6.2.3 *Let $A \in \mathbb{F}^{n \times n}$ be semisimple with linearly independent eigenvectors $v_1, \ldots, v_n \in \mathbb{F}^n$ and associated eigenvalues $\lambda_1, \ldots, \lambda_n \in \mathbb{F}$, satisfying $|\lambda_1| \geq |\lambda_2| \geq \cdots \geq |\lambda_n|$. Suppose $|\lambda_k| > |\lambda_{k+1}|$ for some k. Let $\mathcal{U}_k = \mathrm{span}\{v_1, \ldots, v_k\}$ and $\mathcal{V}_k = \mathrm{span}\{v_{k+1}, \ldots, v_n\}$. Let S be any k-dimensional subspace of $\mathbb{F}^n$ such that $S \cap \mathcal{V}_k = \{0\}$. Then there exists a constant C such that*

$$d(A^j S, \mathcal{U}_k) \leq C |\lambda_{k+1}/\lambda_k|^j, \quad j = 0, 1, 2, \ldots$$

Thus $A^j S \to \mathcal{U}_k$ linearly with convergence ratio $|\lambda_{k+1}/\lambda_k|$.

Since $\mathcal{U}_k$ is spanned by eigenvectors of A, it is invariant under A. $\mathcal{U}_k$ is called the *dominant* invariant subspace of A of dimension k. The semisimplicity assumption is not crucial; a slightly revised version of Theorem 6.2.3 holds for defective matrices. For a proof see [102] or [101, § 5.1].

Exercise 6.2.4

(a) Show that the condition $|\lambda_k| > |\lambda_{k+1}|$ implies that $\mathcal{N}(A) \subseteq \mathcal{V}_k$.

(b) More generally, show that $\mathcal{N}(A^j) \subseteq \mathcal{V}_k$ for all $j > 0$.

(c) Conclude that $A^j S$ has dimension k for all $j > 0$.

$\square$

It is easy to argue the plausibility of Theorem 6.2.3. Let q be any nonzero vector in S. We can easily show that the iterates $A^j q$ lie (relatively) closer and closer to $\mathcal{U}_k$ as j increases. Indeed q may be expressed uniquely in the form

$$q = c_1 v_1 + c_2 v_2 + \cdots + c_k v_k \qquad \text{(component in } \mathcal{U}_k)$$

$$+ c_{k+1} v_{k+1} + \cdots + c_n v_n. \qquad \text{(component in } \mathcal{V}_k)$$

Since $q \notin \mathcal{V}_k$, at least one of the coefficients $c_1, \ldots, c_k$ must be nonzero. Now

$$A^j q / (\lambda_k)^m = c_1 (\lambda_1 / \lambda_k)^j v_1 + \cdots + c_{k-1} (\lambda_{k-1} / \lambda_k)^j v_{k-1} + c_k v_k$$

$$+ c_{k+1} (\lambda_{k+1} / \lambda_k)^j v_{k+1} + \cdots + c_n (\lambda_n / \lambda_k)^j v_n.$$

The coefficients of the component in $\mathcal{U}_k$ increase, or at least do not decrease, as j increases. In contrast, the coefficients of the component in $\mathcal{V}_k$ tend to zero with rate $|\lambda_{k+1} / \lambda_k|$ or better. Thus each sequence $(A^j q)$ converges to $\mathcal{U}_k$ at the stated rate or better, and consequently the limit of $(A^j \mathcal{S})$ lies in $\mathcal{U}_k$. The limit cannot be a proper subspace of $\mathcal{U}_k$ because it has dimension k.

The hypothesis $\mathcal{S} \cap \mathcal{V}_k = \{0\}$ merits some comment. Is this a stringent condition or is it not? First let us see what it amounts to in the case $k = 1$. In this case we have $\mathcal{S} = \mathrm{span}\{q\}$, and $\mathcal{V}_1 = \mathrm{span}\{v_2, \ldots, v_n\}$. Clearly $\mathcal{S} \cap \mathcal{V}_1 = \{0\}$ if and only if $q \notin \mathcal{V}_1$. This just means that $c_1 \neq 0$ in the unique expansion $q = c_1 v_1 + c_2 v_2 + \cdots + c_n v_n$. You will recall from Section 5.3 that this is the condition for convergence of the basic power method to $\mathrm{span}\{v_1\}$. It is satisfied by almost every $v \in \mathbb{F}^n$; you can easily form a mental image of the case $n = 3$. In the language of subspaces, a one-dimensional subspace $\mathcal{S}$ and an $(n-1)$-dimensional subspace $\mathcal{V}_1$ are almost certain to satisfy $\mathcal{S} \cap \mathcal{V}_1 = \{0\}$. Fortunately the same is true in general, as long as the sum of the dimensions of $\mathcal{S}$ and $\mathcal{V}_k$ does not exceed n.

To get a feel for why this might be so, note first that if the sum of the dimensions *does* exceed n, then $\mathcal{S}$ and $\mathcal{V}_k$ must intersect nontrivially. This is a consequence of the fundamental relationship

$$\dim(\mathcal{S} \cap \mathcal{V}_k) + \dim(\mathcal{S} + \mathcal{V}_k) = \dim(\mathcal{S}) + \dim(\mathcal{V}_k), \tag{6.2.5}$$

since $\dim(\mathcal{S} + \mathcal{V}_k) \leq \dim(\mathbb{F}^n) = n$.

In our present situation, $\dim(\mathcal{S}) = k$ and $\dim(\mathcal{V}_k) = n - k$. The dimensions add to n exactly, so there is enough room in $\mathbb{F}^n$ that $\mathcal{S}$ and $\mathcal{V}_k$ are not forced by (6.2.5) to intersect nontrivially. If two subspaces are not forced to intersect nontrivially, then they almost certainly will not. Consider the situation in $\mathbb{R}^3$. There any two two-dimensional subspaces (planes through the origin) are required to intersect nontrivially because the sum of their dimensions exceeds three. In contrast, a plane and a line are not required to intersect nontrivially, and it is obvious that they almost certainly will not. See also Exercise 6.2.10. We conclude that the condition $\mathcal{S} \cap \mathcal{V}_k = \{0\}$ of Theorem 6.2.3 is not stringent; it is satisfied by practically any $\mathcal{S}$ chosen at random.

Shifts and Multiple Steps

The rate of convergence of the subspaces in Theorem 6.2.3 is governed by the ratio $|\lambda_{k+1} / \lambda_k|$, so it is natural to think about what might be done to decrease it. If we replace A by a shifted matrix $A - \rho I$, the eigenvalues are shifted from λ_j to $\lambda_j - \rho$, so the ratio will change to $|\lambda_{k+1} - \rho| / |\lambda_k - \rho|$. You might think at first that you could improve the ratio by a lot if you could find a ρ that is really close to λ_{k+1}, but this turns out to be overly optimistic. For a given ρ, we must relabel the eigenvalues so that $|\lambda_1 - \rho| \geq |\lambda_2 - \rho| \geq \cdots \geq |\lambda_n - \rho|$ before taking the ratio. If ρ is extremely close to λ_{k+1}, then λ_{k+1} will become λ_n in the new numbering. Nevertheless, shifting does have the potential for improving the convergence rate, and there is the possibility that for some choices of k and ρ the convergence could be made really fast.

Another useful idea is to lump several steps together into a single step. Taking m steps of subspace iteration with shifts $\rho_1, \ldots, \rho_m$ is the same as taking a single step with the matrix

$$p(A) = (A - \rho_1 I)(A - \rho_2 I) \cdots (A - \rho_m I).$$

Since the eigenvalues of $p(A)$ are $p(\lambda_1), \ldots, p(\lambda_n)$, the convergence rate is determined by $|p(\lambda_{k+1})|/|p(\lambda_k)|$. This assumes, of course, that we have renumbered the eigenvalues so that $|p(\lambda_1)| \geq |p(\lambda_2)| \geq \cdots \geq |p(\lambda_n)|$. If we can find a p and a k such that the ratio $|p(\lambda_{k+1})|/|p(\lambda_k)|$ is really small, we can obtain rapid convergence of subspace iterations of dimension k.

In the interest of flexibility we will study subspace iterations driven by $p(A)$ from this point on. The case $p(A) = A$ is included as a special case.

Exercise 6.2.6   Suppose A has eigenvalues $\lambda_1, \ldots, \lambda_n$, and p is a polynomial.

(a)   According to Schur's Theorem 5.4.11, A has a Schur decomposition $A = UTU^*$, where U is unitary and T is upper triangular. Show that $p(A) = Up(T)U^*$.

(b)   Determine the main-diagonal entries of $p(T)$.

(c)   Deduce that the eigenvalues of $p(A)$ are $p(\lambda_1), \ldots, p(\lambda_n)$.

$\square$

Simultaneous Iteration

Our main motivation for studying subspace iteration is to use it as a vehicle for understanding Francis's algorithm. Nevertheless it is interesting to think about how subspace iteration could be implemented as a practical algorithm. In order to carry out subspace iteration in practice, we must choose a basis for S and iterate on the basis vectors simultaneously. Let $q_1^{(0)}, q_2^{(0)}, \ldots, q_k^{(0)}$ be a basis for S. From Exercises 6.2.2 and 6.2.4 we know that if $S \cap V_k = \{0\}$, then $A^j q_1^{(0)}, A^j q_2^{(0)}, \ldots, A^j q_k^{(0)}$ form a basis for $A^j S$. More generally, if we are iterating with $p(A)$ instead of A, we can say that $p(A)^j q_1^{(0)}, p(A)^j q_2^{(0)}, \ldots, p(A)^j q_k^{(0)}$ form a basis for $p(A)^j S$. Thus, in theory, we can simply iterate on a basis of S to obtain bases for $p(A)S$, $p(A)^2 S$, $p(A)^3 S$ and so on. There are two reasons why it is not advisable to do this in practice: (i) The vectors will have to be rescaled regularly in order to avoid underflow or overflow.

(ii) Each of the sequences $q_i^{(0)}$, $p(A)q_i^{(0)}$, $p(A)^2 q_i^{(0)}$, ... independently converges to the dominant eigenspace span$\{v_1\}$ (assuming $|p(\lambda_1)| > |p(\lambda_2)|$). It follows that for large j the vectors $p(A)^j q_1^{(0)}$, ..., $p(A)^j q_k^{(0)}$ all point in nearly the same direction. Thus they form an ill-conditioned basis of $A^j S$. In numerical practice, an ill-conditioned basis determines the space poorly; a small perturbation in one of the basis vectors can make a big change in the space.

Ill-conditioned bases can be avoided by replacing the basis obtained at each step by a well-conditioned basis for the same subspace. This replacement operation can also incorporate the necessary rescaling. There are numerous ways to do this, but the most reliable is to orthonormalize. Thus the following *simultaneous iteration* procedure is recommended.

> Start with an orthonormal basis $q_1^{(0)}, \ldots q_k^{(0)}$ of S
> for $j = 0, 1, 2, \ldots$
>
> $\left[\begin{array}{l} \text{Calculate } p(A)q_1^{(j)}, \ldots, p(A)q_k^{(j)}, \text{ a basis for } p(A)^{j+1} S \\ \text{Orthonormalize } p(A)q_1^{(j)}, \ldots, p(A)q_k^{(j)} \\ \qquad \text{to obtain } q_1^{(j+1)}, \ldots, q_k^{(j+1)} \end{array} \right.$ (6.2.7)

In practice one should use a robust process to perform the orthonormalization. For example, one could use Gram-Schmidt with reorthogonalization (Section 3.4).

The simultaneous iteration procedure (6.2.7) has the agreeable property of iterating on lower-dimensional subspaces at no extra cost. For $i = 1, \ldots, k$, let S_i denote the i-dimensional subspace spanned by $q_1^{(0)}, \ldots, q_i^{(0)}$. Then

$$p(A)S_i = \text{span}\left\{ p(A)q_1^{(0)}, \ldots, p(A)q_i^{(0)} \right\} = \text{span}\left\{ q_1^{(1)}, \ldots, q_i^{(1)} \right\},$$

by the subspace-preserving property (3.4.12) of the Gram-Schmidt procedure. In general

$$p(A)^j S_i = \text{span}\left\{ q_1^{(j)}, \ldots, q_i^{(j)} \right\},$$

so span$\left\{ q_1^{(j)}, \ldots, q_i^{(j)} \right\}$ converges to the invariant subspace span$\{v_1, \ldots, v_i\}$ as $j \to \infty$, provided that appropriate hypotheses are satisfied. Thus simultaneous iteration seeks not only an invariant subspace of dimension k, but subspaces of dimensions $1, 2, \ldots, k - 1$ as well.

Now consider what happens when simultaneous iteration is applied to a complete set of orthonormal vectors $q_1^{(0)}, \ldots, q_n^{(0)} \in \mathbb{F}^n$. We continue to assume that A is semisimple with linearly independent eigenvectors $v_1, \ldots, v_n$. For $k = 1, \ldots, n - 1$, let $S_k = \text{span}\left\{ q_1^{(0)}, \ldots, q_k^{(0)} \right\}$, $\mathcal{U}_k = \text{span}\{v_1, \ldots, v_k\}$, and $\mathcal{V}_k = \text{span}\{v_{k+1}, \ldots, v_n\}$, and assume $S_k \cap \mathcal{V}_k = \{0\}$ and $|p(\lambda_k)| > |p(\lambda_{k+1})|$. Then (for $k = 1, \ldots, n - 1$) $p(A)^j S_k = \text{span}\left\{ q_1^{(j)}, \ldots, q_k^{(j)} \right\} \to \mathcal{U}_k$ linearly with convergence ratio $|p(\lambda_{k+1})/p(\lambda_k)|$ as $j \to \infty$. The main point here is that, since we have a sequence of nested subspace iterations taking place at once, all of the ratios $|p(\lambda_{k+1})/p(\lambda_k)|$, $k = 1, \ldots, n - 1$, are important. If any one of them is small, we make rapid progress. This is what happens in Francis's algorithm.

Simultaneous Iteration with Changes of Coordinate System

Now let us consider a step of simultaneous iteration starting from the standard basis vectors e_1, e_2, ..., e_n. We get $p(A)e_1$, $p(A)e_2$, ..., $p(A)e_n$, which are just the columns of $p(A)$, then we orthonormalize to get, say, q_1, q_2, ..., q_n. Now we could do another iteration, but first let's do a similarity transformation

$$\hat{A} = Q^*AQ,$$

where Q is the unitary matrix whose columns are q_1, q_2, ..., q_n.

A similarity transformation is a change of coordinate system. The linear transformation that was represented by the matrix A in the old coordinate system is represented by $\hat{A}$ in the new system. A vector that was represented by the coordinate vector x in the old coordinate system is represented by $Q^{-1}x = Q^*x$ in the new system. For example, if v is an eigenvector of A associated with the the eigenvalue λ, then Q^*v is an eigenvector of $\hat{A}$ associated with λ.

We took a step of simultaneous iteration, starting with e_1, ..., e_n and ending with q_1, ..., q_n. Then we did a change of coordinate system, so we have to check how this change affects these vectors. Since the columns of Q are q_1, ..., q_n, and these vectors are orthonormal, we easily deduce that $Q^*q_1 = e_1$, ..., $Q^*q_n = e_n$. Thus the change of coordinate system maps q_1, ..., q_n right back to e_1, ..., e_n.

Exercise 6.2.8 Show that if q_1, ..., q_n are orthonormal and $Q = \begin{bmatrix} q_1 & \cdots & q_n \end{bmatrix}$, then $Q^*q_i = e_i$, $i = 1, ..., n$. □

Now if we want to do another step of simultaneous iteration, we have the option of doing it in the new coordinate system. This would just be a matter of repeating what we did before. Thus we would start with e_1, ..., e_n, compute $p(\hat{A})e_1$, ..., $p(\hat{A})e_n$, orthonormalize these vectors to get a new orthonormal basis, then do another change of coordinate system to bring us back to e_1, ..., e_n.

Imagine repeating this process over and over again. Let us pick a k for which $|p(\lambda_{k+1})|/|p(\lambda_k)| < 1$ and focus on this k. Assuming the subspace condition span$\{e_1, ..., e_k\} \cap \mathcal{V}_k = \{0\}$ is satisfied, the subspace iterations will converge to an invariant subspace. But what does that mean in this case? We start with span$\{e_1, ..., e_k\}$ and apply $p(A)$ to get $p(A)$span$\{e_1, ..., e_k\}$ = span$\{q_1, ..., q_k\}$. Then we do a change of coordinate system that maps span$\{q_1, ..., q_k\}$ back to span$\{e_1, ..., e_k\}$. At the same time we get a new matrix $\hat{A}$, which we will now also refer to as A_1. If we do another iteration, we do another change of coordinate system and get yet another matrix A_2. As we repeat the process, we generate a whole sequence of matrices (A_j). Thus we see that this is a version of subspace iteration in which the subspace stays fixed (span$\{e_1, ..., e_k\}$) and the matrix changes. Contrast this with the standard version of subspace iteration in which the matrix stays fixed and the subspace changes.

In this context convergence means that the fixed subspace span$\{e_1, ..., e_k\}$ gets closer and closer to being invariant under A_j as j increases.

rcise 6.2.9 Partition A_j as

$$A_j = \begin{bmatrix} A_{11}^{(j)} & A_{12}^{(j)} \\ A_{21}^{(j)} & A_{22}^{(j)} \end{bmatrix},$$

where $A_{11}^{(j)}$ is $k \times k$.

(a) Prove that span$\{e_1, \ldots, e_k\}$ is invariant under A_j if and only if $A_{21}^{(j)} = 0$.

(b) Show that if span$\{e_1, \ldots, e_k\}$ is invariant under A_j, then the eigenvalues of $A_{11}^{(j)}$ are the eigenvalues of A_j associated with the subspace span$\{e_1, \ldots, e_k\}$.

(c) Show that these results are a special case of Theorem 6.1.6.

□

Exercise 6.2.9 shows that span$\{e_1, \ldots, e_k\}$ is invariant under A_j if and only if A_j is block triangular:

$$A_j = \begin{bmatrix} A_{11}^{(j)} & A_{12}^{(j)} \\ 0 & A_{22}^{(j)} \end{bmatrix}.$$

Of course it is never the case that we have exact invariance; what we actually have is convergence to an invariant subspace. It is reasonable to expect then that we will have $A_{21}^{(j)} \to 0$ linearly with rate $|p(\lambda_{k+1})/p(\lambda_k)|$ as $j \to \infty$. Indeed this is not hard to prove [102], [101, Ch. 5]. It is also true that the eigenvalues of $A_{11}^{(j)}$ converge to the dominant eigenvalues $\lambda_1, \ldots, \lambda_k$ (sorted according to the magnitude of $p(\lambda_i)$). Notice that if the matrices A_j happen to be upper Hessenberg, then $A_{21}^{(j)}$ has only one nonzero entry, namely $a_{k+1,k}^{(j)}$. In this case the statement $A_{21}^{(j)} \to 0$ just means that $a_{k+1,k}^{(j)} \to 0$ as $j \to \infty$.

In typical situations this convergence will be happening not just for one particular choice of k but for many k at once. If it happens for all k, then A_j will converge to upper-triangular form. Typically it will not hold for all k, but it will hold for enough k that (A_j) will converge to block-triangular form with lots of small blocks on the main diagonal.

The attentive reader will have complained by now that this process is too expensive to be practical on several grounds. For one thing, we get a new matrix A_j on each iteration and we seem to have to compute $p(A_j)$ to move the algorithm forward. This would be excessively expensive. If we wish to build a practical algorithm along these lines, we need to find a way to avoid this expense.

As you will recall from Section 5.6, Francis's algorithm computes only the first column of $p(A)$. From Theorem 5.6.14 we know that the Francis iteration does a change of coordinate system $\hat{A} = Q^*AQ$, where q_1, the first column of Q, is proportional to the first columns of $p(A)$. Thus $q_1 = \alpha p(A)e_1$. This is one step of the power method. Then the change of coordinate system maps q_1 back to e_1, and we are ready for the next iteration. Thus the Francis iteration does subspace iterations of dimension $k = 1$ with changes of coordinate system. Actually it does much more:

it does simultaneous iterations of all dimensions k. To prove this we need one more idea, and that will come in the following section.

Additional Exercise

Exercise 6.2.10 Let S and V be subspaces of $\mathbb{F}^n$ of dimension k and $n - k$, with bases s_1, ..., s_k and $u_1, \ldots, u_{n-k}$, respectively. Let $B = [s_1 \cdots s_k \, u_1 \cdots u_{n-k}] \in \mathbb{F}^{n \times n}$. Show that $S \cap V = \{0\}$ if and only if B is nonsingular. (B fails to be nonsingular when and only when $\det(B) = 0$. This is a very special relationship. If one chooses n vectors at random from $\mathbb{F}^n$ and builds a matrix whose columns are those vectors, its determinant is almost certain to be nonzero.)   □

6.3 KRYLOV SUBSPACES AND FRANCIS'S ALGORITHM

We will define Krylov subspaces and present some of their basic properties. Then we will be ready to explain Francis's algorithm Krylov subspaces also play a big role in the solution of large, sparse eigenvalue problems, as will be shown in Sections 6.4 and 6.5. In Chapter 8 we will see that some of the most important iterative methods for solving large, sparse linear systems $Ax = b$ make use of Krylov subspaces.

Krylov Subspaces

We consider a matrix $A \in \mathbb{C}^{n \times n}$, which we think of as a fixed object of study. Given A, each nonzero $x \in \mathbb{C}^n$ generates a nested sequence of *Krylov subspaces* defined as follows:

$$
\begin{aligned}
\mathcal{K}_1(A, x) &= \operatorname{span}\{x\}, \\
\mathcal{K}_2(A, x) &= \operatorname{span}\{x, Ax\}, \\
\mathcal{K}_3(A, x) &= \operatorname{span}\{x, Ax, A^2x\}, \\
\mathcal{K}_4(A, x) &= \operatorname{span}\{x, Ax, A^2x, A^3x\},
\end{aligned}
$$

and so on. In general the jth *Krylov subspace* associated with A and x is $\mathcal{K}_j(A, x) = \operatorname{span}\{x, Ax, A^2x, \ldots, A^{j-1}x\}$. Since $\mathcal{K}_j(A, x)$ is generated by j vectors, its dimension is at most j. If we take A for granted, we can think of the vector x as containing all the information needed to generate the entire sequence of Krylov subspaces $\mathcal{K}_j(A, x)$.

The following theorem establishes an important connection between Krylov subspaces and properly Hessenberg matrices.

Theorem 6.3.1 *Suppose A, H, $Q \in \mathbb{C}^{n \times n}$, and Q is nonsingular. Let $q_1, \ldots, q_n$ denote the columns of Q. If*

$$
H = Q^{-1}AQ
$$

and H is properly upper Hessenberg, then

$$\text{span}\{q_1, \ldots, q_j\} = \mathcal{K}_j(A, q_1), \quad j = 1, 2 \ldots, n.$$

The theorem says that if we make a change of coordinate system and the resulting matrix is properly upper Hessenberg, then the columns of the transforming matrix span the Krylov subspaces generated by the first column. In our application, Q will be unitary, but that hypothesis is not needed for the theorem.

Proof. The proof is by induction on j. The case $j = 1$ asserts that $\text{span}\{q_1\} = \mathcal{K}_1(A, q_1)$, which is trivially true. We will complete the induction proof by showing that for any $j < n$, if $\text{span}\{q_1, \ldots, q_j\} = \mathcal{K}_j(A, q_1)$, then $\text{span}\{q_1, \ldots, q_{j+1}\} = \mathcal{K}_{j+1}(A, q_1)$. The equation $H = Q^{-1}AQ$ can be rewritten as $AQ = QH$. Equating jth columns of this matrix equation, we obtain

$$Aq_j = Qh_j = \sum_{i=1}^{j+1} q_i h_{ij} = \sum_{i=1}^{j} q_i h_{ij} + q_{j+1} h_{j+1,j}. \tag{6.3.2}$$

The sum stops after the $(j + 1)$st term because H is upper Hessenberg: $h_{ij} = 0$ if $i > j + 1$. Since $h_{j+1,j} \neq 0$, we can solve (6.3.2) for q_{j+1} to get

$$q_{j+1} = h_{j+1,j}^{-1} \left(Aq_j - \sum_{i=1}^{j} q_i h_{ij} \right). \tag{6.3.3}$$

By the induction hypothesis, $q_j \in \mathcal{K}_j(A, q_1) = \text{span}\{q_1, \ldots, A^{j-1}q_1\}$, so $Aq_j \in \text{span}\{Aq_1, \ldots, A^j q_1\} \subseteq \mathcal{K}_{j+1}(A, q_1)$. Moreover, again by the induction hypothesis, $q_i \in \mathcal{K}_j(A, q_1) \subseteq \mathcal{K}_{j+1}(A, q_1)$ for $i = 1, \ldots, j$. Therefore, by (6.3.3), $q_{j+1} \in \mathcal{K}_{j+1}(A, q_1)$, and consequently $\text{span}\{q_1, \ldots, q_{j+1}\} \subseteq \mathcal{K}_{j+1}(A, q_1)$. The details are left as an exercise for the reader. Since $\text{span}\{q_1, \ldots, q_{j+1}\}$ has dimension $j + 1$ and $\mathcal{K}_{j+1}(A, q_1)$ can have dimension at most $j + 1$, the spaces must be equal: $\text{span}\{q_1, \ldots, q_{j+1}\} = \mathcal{K}_{j+1}(A, q_1)$. □

The following simpler result is also important.

Proposition 6.3.4 *If A is properly upper Hessenberg, then*

$$\mathcal{K}_j(A, e_1) = \text{span}\{e_1, \ldots, e_j\}, \quad j = 1, 2, \ldots, n.$$

rcise 6.3.5 Prove Proposition 6.3.4 two different ways.

(a) Prove the result directly.

(b) Derive it as a special case of Theorem 6.3.1.

□

Subspace Iteration with Krylov Subspaces

Proposition 6.3.6 *Let $A \in \mathbb{C}^{n \times n}$, $x \in \mathbb{C}^n$, and let p be any polynomial. Then*

$$p(A)\mathcal{K}_k(A, x) = \mathcal{K}_k(A, p(A)x).$$

This proposition is an easy consequence of the fact that $p(A)A = Ap(A)$.

Exercise 6.3.7 Prove Proposition 6.3.6. □

Proposition 6.3.6 says that if we apply subspace iteration to a Krylov subspace, we get another Krylov subspace. Iterating this result, we have

$$p(A)^j \mathcal{K}_k(A, x) = \mathcal{K}_k(A, p(A)^j x), \quad j = 1, 2, 3 \ldots$$

In the power sequence $(p(A)^j x)$, each of the iterates $p(A)^j x$ generates a nested sequence of Krylov subspaces $\mathcal{K}_k(A, p(A)^j x)$, $k = 1, 2, 3, \ldots$, and each of these is equal to $p(A)^j \mathcal{K}_k(A, x)$, the result of j steps of subspace iteration on $\mathcal{K}_k(A, x)$. We draw the following conclusion: whenever we apply the power method to a vector x, we are also implicitly doing subspace iterations on the nested sequence of associated Krylov subspaces $\mathcal{K}_k(A, x)$. $k = 1, 2, 3, \ldots$.

Application to Francis's Algorithm

Consider a single iteration of degree m of Francis's algorithm with shifts $\rho_1, \ldots, \rho_m$. By Theorem 5.6.14, this is a unitary similarity transformation (change of coordinate system)

$$\hat{A} = Q^* A Q,$$

where the first column of Q is proportional to the first column of $p(A) = (A - \rho_1 I) \cdots (A - \rho_m I)$. Letting $q_1, \ldots, q_n$ denote the columns of Q, we have $q_1 = \alpha p(A)e_1$ for some α. This is a step of the power method driven by $p(A)$, starting from e_1. The change of coordinate system maps q_1 back to $Q^* q_1 = e_1$. Thus a Francis iteration amounts to a step of the power method with a change of coordinate system. We already made this observation in Section 6.2, but now we can say more.

The matrix $\hat{A}$ is upper Hessenberg. It could have some zeros on the subdiagonal, in which case we can break the eigenvalue problem for $\hat{A}$ into smaller problems. This is the lucky case; it occurs only when some of the shifts are equal to eigenvalues of A. (See Exercise 6.3.21, Exercise 5.6.33, and [101, Theorem 4.6.1].)

Let us suppose that we have not been so lucky, in which case $\hat{A}$ is properly upper Hessenberg. Then by Theorem 6.3.1, the columns of Q span Krylov subspaces generated by q_1. We noted above that $q_1 = \alpha p(A)e_1$, so q_1 is the result of a step of the power method starting from e_1. Since we are taking a step of the power method, we must also automatically be doing subspace iteration on the nested sequence of Krylov subspaces $\mathcal{K}_k(A, e_1)$ to obtain the nested sequence of Krylov subspaces $\mathcal{K}_k(A, q_1) = \text{span}\{q_1, \ldots, q_k\}$, $k = 1, 2, 3, \ldots$. This fact is made explicit by the following sequence of subspace equations, which is based on Theorem 6.3.1 and Propositions 6.3.4 and 6.3.6:

$$
\begin{aligned}
\text{span}\{q_1, \ldots, q_k\} &= \mathcal{K}_k(A, q_1) \\
&= \mathcal{K}_k(A, p(A)e_1) \\
&= p(A)\mathcal{K}_k(A, e_1) \\
&= p(A)\text{span}\{e_1, \ldots, e_k\}.
\end{aligned}
$$

Thus $\text{span}\{q_1, \ldots, q_k\}$ is the result of one step of subspace iteration starting from $\text{span}\{e_1, \ldots, e_k\}$. The change of coordinate system $\hat{A} = Q^*AQ$ maps the subspace $\text{span}\{q_1, \ldots, q_k\}$ back to $Q^*\text{span}\{q_1, \ldots, q_k\} = \text{span}\{e_1, \ldots, e_k\}$, ready for the next iteration. This holds not just one value of k but for $k = 1, \ldots, n$.

In conclusion we can say that if we do repeated Francis iterations with the same shifts, then we are doing simultaneous iteration with changes of coordinate system, as discussed in Section 6.2.

Because of the convergence of the underlying subspace iterations (Theorem 6.2.3), we can draw the following conclusions. As always, we number the eigenvalues so that $|p(\lambda_1)| \geq |p(\lambda_2)| \geq \cdots \geq |p(\lambda_n)|$. Let (A_j) denote the sequence of iterates generated by Francis's algorithm, and let $a_{k+1,k}^{(j)}$ denote the $(k+1, k)$ entry of A_j. For every k, if $|p(\lambda_k)| > |p(\lambda_{k+1})|$, the subdiagonal entry $a_{k+1,k}^{(j)} \to 0$ linearly with ratio $|p(\lambda_{k+1})/p(\lambda_k)|$. This assumes (Theorem 6.2.3) that a certain subspace condition $\text{span}\{e_1, \ldots, e_n\} \cap \mathcal{V}_k = \{0\}$ holds, but in Exercise 6.3.19 we show that it does in this case.

In this discussion we have assumed for convenience that the same shifts are used on every iteration. As we already mentioned in Section 5.6, the practical Francis algorithm actually chooses new shifts for each iteration. This makes the algorithm harder to analyze but also improves the convergence rate greatly. Typically the convergence is quadratic. We refer the reader back to the discussion of convergence of Francis's algorithm in Section 5.6. See also [102] and [101, Ch. 5]. Exercises 6.3.23 and 6.3.24 touch on convergence theory.

Duality in Subspace Iteration

The following duality theorem provides a link between Francis's algorithm and inverse iteration. It shows that whenever direct (subspace) iteration takes place, inverse (subspace) iteration also takes place automatically.

Theorem 6.3.8 *Let $A \in \mathbb{F}^{n \times n}$ be nonsingular, and let $B = (A^*)^{-1}$. Let $\mathcal{S}$ be any subspace of $\mathbb{F}^n$. Then the two sequences of subspaces*

$$
\mathcal{S}\ ,\ A\mathcal{S}\ ,\ A^2\mathcal{S}\ ,\ldots
$$

$$
\mathcal{S}^\perp,\ B\mathcal{S}^\perp,\ B^2\mathcal{S}^\perp,\ldots
$$

are equivalent in the sense that they yield orthogonal complements. That is,

$$
(A^j\mathcal{S})^\perp = B^j(\mathcal{S}^\perp) \quad \text{for} \quad j = 0, 1, 2, \ldots.
$$

Proof. For every $x, y \in \mathbb{F}^n$, $\langle A^j x, B^j y \rangle = y^*(B^*)^j A^j x = y^*(A^{-1})^j A^j x = y^* x = \langle x, y \rangle$. Thus $A^j x$ and $B^j y$ are orthogonal if and only if x and y are orthogonal. The theorem follows directly from this observation. $\quad\square$

The second subspace sequence in Theorem 6.3.8 is a sequence of inverse iterates, since the iteration matrix is the inverse of A^*.

Now consider a Francis iteration of degree one with shift ρ. Thus $p(A) = A - \rho I$, and we have

$$(A - \rho I)\mathrm{span}\{e_1, \ldots, e_k\} = \mathrm{span}\{q_1, \ldots, q_k\}, \qquad k = 1, \ldots, n.$$

Clearly $\mathrm{span}\{e_1, \ldots, e_k\}^{\perp} = \mathrm{span}\{e_{k+1}, \ldots, e_n\}$ and, since Q is unitary,

$$\mathrm{span}\{q_1, \ldots, q_k\}^{\perp} = \mathrm{span}\{q_{k+1}, \ldots, q_n\},$$

so by Theorem 6.3.8,

$$(A^* - \overline{\rho} I)^{-1}\mathrm{span}\{e_{k+1}, \ldots, e_n\} = \mathrm{span}\{q_{k+1}, \ldots, q_n\}.$$

Of particular interest is the case $k = n - 1$, which yields

$$(A^* - \overline{\rho} I)^{-1}\mathrm{span}\{e_n\} = \mathrm{span}\{q_n\}.$$

This shows that the map $e_n \to q_n$ is a shift-and-invert step with shift $\overline{\rho}$ applied to A^*. The subsequent change of coordinate system maps q_n back to e_n for another step of shifted inverse iteration. Notice that if we take the Rayleigh quotient of e_n with respect to A^*, we have $e_n^T A^* e_n = \overline{a}_{nn}$, so if we take $\overline{\rho} = \overline{a}_{nn}$, i.e. $\rho = a_{nn}$, we are doing Rayleigh quotient iteration. This explains why the shift $\rho = a_{nn}$ is called the *Rayleigh quotient shift* for Francis's iteration.

Francis's Algorithm and the QR Decomposition

Francis's algorithm is also known as the implicitly shifted QR algorithm. This name is a bit mysterious so far, as we have seen no connection between Francis iterations and the QR decomposition. At this point we will take the time to make the connection.

Given $A \in \mathbb{C}^{n \times n}$ and $x \in \mathbb{C}^n$, the *Krylov matrix* associated with A and x is the $n \times n$ matrix

$$K(A, x) = \begin{bmatrix} x & Ax & \cdots & A^{n-1}x \end{bmatrix}.$$

Proposition 6.3.9 *Let* $A \in \mathbb{C}^{n \times n}$.

(a) *If* A *is upper Hessenberg, then* $K(A, e_1)$ *is upper triangular.*

(b) *If* A *is properly upper Hessenberg, then* $K(A, e_1)$ *is nonsingular.*

rcise 6.3.10

(a) Prove Proposition 6.3.9

(b) Relate Proposition 6.3.9 to Proposition 6.3.4.

$\square$

rcise 6.3.11 A Francis iteration is a unitary similarity transformation $\hat{A} = Q^*AQ$, where $Qe_1 = \alpha p(A)e_1$ (Theorem 5.6.14).

(a) Show that $K(A, Qe_1) = QK(\hat{A}, e_1)$.

(b) Show that $p(A)K(A, e_1) = QK(\hat{A}, e_1)$.

$\square$

Theorem 6.3.12 *Consider a Francis iteration* $\hat{A} = Q^*AQ$ *of degree* m *with shifts* $\rho_1, \ldots, \rho_m$. *Let* $p(A) = (A - \rho_1 I) \cdots (A - \rho_m I)$. *Then there is an upper triangular* R *such that*

$$p(A) = QR. \qquad (6.3.13)$$

Proof. From Exercise 6.3.11, $p(A)K(A, e_1) = QK(\hat{A}, e_1)$. A is properly upper Hessenberg and $\hat{A}$ is upper Hessenberg, so by Proposition 6.3.9, $K(A, e_1)$ and $K(\hat{A}, e_1)$ are both upper triangular, and $K(A, e_1)$ is nonsingular. By Exercise 1.7.46, $K(A, e_1)^{-1}$ is also upper triangular. Let $R = K(\hat{A}, e_1) K(A, e_1)^{-1}$. Then R is upper triangular, and clearly $p(A) = QR$. $\qquad\square$

Since Q is unitary, equation (6.3.13) is a QR decomposition of $p(A)$. Clearly $p(A)$ is nonsingular except in those lucky special cases when one or more of the shifts is exactly an eigenvalue of A. For the rest of this discussion, let us suppose that we have not been so lucky; $p(A)$ is nonsingular. Then the QR decomposition is essentially unique (Exercise 3.2.59), so the equations

$$p(A) = QR, \qquad \hat{A} = Q^*AQ \qquad (6.3.14)$$

essentially characterize the Francis iteration.

rcise 6.3.15

(a) Consider a Francis iteration of degree one with zero shift, i.e. $p(A) = A$. Show that A and $\hat{A}$ are related by the equations

$$A = QR, \qquad RQ = \hat{A}.$$

This shows that $\hat{A}$ can be obtained from A by taking a QR decomposition of A, reversing the order of the factors, and multiplying them back together. This is the original unshifted QR algorithm. Its ancestors are the LR and q-d algorithms of H. Rutishauser. (See, for example, [101].)

(b) Now consider a single Francis iteration with a shift, i.e. $p(A) = A - \rho I$. Show that A and $\hat{A}$ are related by the equations

$$A - \rho I = QR, \qquad RQ + \rho I = \hat{A}.$$

This basic QR algorithm with shifts was published independently by Francis [31] and Kublanovskaya [59].

$\square$

We need to get one more result, which we will use in connection with the implicitly-restarted Arnoldi process in Section 6.5.

Theorem 6.3.16 *Let A_m be the result of m Francis iterations of degree one, starting from $A_0 = A$, using shift ρ_j on the jth iteration. Let $p(A) = (A - \rho_1 I) \cdots (A - \rho_m I)$. Then there exist unitary $\hat{Q}_m$ and upper-triangular $\hat{R}_m$ such that*

$$A_m = \hat{Q}_m^* A \hat{Q}_m, \qquad p(A) = \hat{Q}_m \hat{R}_m.$$

This theorem shows that m Francis iterations of degree one are equivalent to a single Francis iteration of degree m with the same shifts. Clearly we can make other similar statements. For example, if m is even, $m/2$ Francis iterations of degree two are equivalent to a single iteration of degree m with the same shifts. A general result along these lines is worked out in Exercise 6.3.23.

Proof. Let $A = A_0, A_1, \ldots, A_m$ be the successive Francis iterates of degree one. By Theorem 6.3.12, the jth iterate satisfies

$$A_{j-1} - \rho_j I = Q_j R_j, \qquad A_j = Q_j^* A_{j-1} Q_j.$$

For $j = 1, 2, 3, \ldots, m$, let

$$\hat{Q}_j = Q_1 Q_2 \cdots Q_j, \quad \hat{R}_j = R_j \cdots R_2 R_1, \quad \text{and} \quad \hat{p}_j(A) = (A - \rho_j I) \cdots (A - \rho_1 I).$$

Clearly $\hat{Q}_j$ is unitary and $\hat{R}_j$ is upper triangular. You can easily check that $A_j = \hat{Q}_j^* A \hat{Q}_j$ for $j = 1, \ldots, m$, and in particular $A_m = \hat{Q}_m^* A \hat{Q}_m$. We will prove by induction on j that

$$\hat{p}_j(A) = \hat{Q}_j \hat{R}_j, \quad j = 1, 2, \ldots, m. \tag{6.3.17}$$

Since $p(A) = \hat{p}_m(A)$, the case $j = m$ gives $p(A) = \hat{Q}_m \hat{R}_m$, which is the desired result.

To prove (6.3.17) by induction on j, you can begin by checking that the case $j = 1$ is trivially true. For the induction step, we can show that $\hat{p}_j(A) = \hat{Q}_j \hat{R}_j$ holds if $\hat{p}_{j-1}(A) = \hat{Q}_{j-1} \hat{R}_{j-1}$ does. We have $\hat{p}_j(A) = (A - \rho_j I)\hat{p}_{j-1}(A)$. Note that $A - \rho_j I = \hat{Q}_{j-1}(A_{j-1} - \rho_j I)\hat{Q}_{j-1}^*$, then put the pieces together to get the desired result. $\square$

Exercise 6.3.18 Work out the details of the proof of Theorem 6.3.16. $\square$

Additional Exercises

Exercise 6.3.19 Francis's algorithm is essentially simultaneous (subspace) iteration with starting subspaces $S_k = \text{span}\{e_1, \ldots, e_k\}$, $k = 1, \ldots, n - 1$. Theorem 6.2.3 states that subspace iteration converges if $|\lambda_k| > |\lambda_{k+1}|$ and the subspace condition $S_k \cap V_k = \{0\}$ holds. Recall that $V_k = \text{span}\{v_{k+1}, \ldots, v_n\}$, which is a subspace of $\mathbb{F}^n$ of dimension $n - k$ that is invariant under A. In this exercise you will show that the subspace condition $S_k \cap V_k = \{0\}$ is automatically satisfied if A is properly upper Hessenberg.

(a) Let V be a subspace of $\mathbb{F}^n$ that is invariant under A. Show that if $v \in V$, then $A^m v \in V$ for $m = 1, 2, 3, \ldots$.

(b) Suppose A is properly upper Hessenberg. Let $v \in \text{span}\{e_1, \ldots, e_k\} = S_k$ be nonzero. (What does v look like? What can you say about Av? $A^2 v$?) Show that the vectors $v, Av, A^2 v, \ldots, A^{n-k} v$ are linearly independent.

(c) Show that v cannot lie in V_k. Hence $S_k \cap V_k = \{0\}$.

$\square$

Exercise 6.3.20 The result from this simple exercise will be used in Exercise 6.3.21. Suppose B is upper triangular and C is upper Hessenberg.

(a) Show that both BC and CB are upper Hessenberg.

(b) Show that if B is also nonsingular and C is properly upper Hessenberg, then both BC and CB are properly upper Hessenberg.

$\square$

Exercise 6.3.21 Consider a Francis iteration $\hat{A} = Q^* A Q$ of degree m with shifts $\rho_1, \ldots, \rho_m$. In this exercise you will show that if none of the shifts is an eigenvalue of A, then $\hat{A}$ is properly upper Hessenberg. As usual, let $p(A) = (A - \rho_1 I) \cdots (A - \rho_m I)$, and assume none of $\rho_1, \ldots, \rho_m$ is an eigenvalue of A. By Theorem 6.3.12, the Francis iteration is essentially characterized by the equations

$$p(A) = QR, \qquad \hat{A} = Q^* A Q.$$

(a) Show that $p(A)$ is nonsingular. Deduce that R is also nonsingular.

(b) Show that $Q = p(A) R^{-1}$ and $Q^{-1} = R p(A)^{-1}$, then deduce that $\hat{A} = R A R^{-1}$.

(c) Using results from Exercise 6.3.20, deduce that $\hat{A}$ is properly upper Hessenberg.

$\square$

Exercise 6.3.22 In Section 5.8 we developed the Golub-Reinsch algorithm, a variant of Francis's algorithm for computing the SVD of an $m \times m$ properly bidiagonal matrix

B. Consider an iteration of the Golub-Reinsch algorithm $\hat{B} = U^T B V$ with shift ρ. Assume ρ is not exactly a singular value of B Thus ρ^2 is not an eigenvalue of $B^T B$ and $B B^T$.

(a) Using Theorem 5.8.43, show that

$$\hat{B}^T \hat{B} = V^T B^T B V, \quad \text{where} \quad V e_1 = \alpha (B^T B - \rho^2 I) e_1.$$

(b) Proceeding as in Exercise 6.3.11, show that there is an upper-triangular matrix R such that

$$B^T B - \rho^2 I = V R, \qquad \hat{B}^T \hat{B} = V^T (B^T B) V.$$

Deduce that the Golub-Reinsch iteration implicitly executes a Francis iteration of degree one with shift ρ^2 on $B^T B$ to produce $\hat{B}^T \hat{B}$.

(c) Show that $U = B V \hat{B}^{-1}$ and $B(B^T B - \rho^2 I) = (B B^T - \rho^2 I) B$. Using these equations and the fact that B and $\hat{B}^{-1}$ are upper triangular, deduce that there is a nonzero γ such that

$$U e_1 = \gamma (B B^T - \rho^2 I) e_1.$$

(d) Show that there is an upper-triangular S such that

$$B B^T - \rho^2 I = U S, \qquad \hat{B} \hat{B}^T = U^T (B B^T) U.$$

Deduce that the Golub-Reinsch iteration implicitly executes a Francis iteration of degree one with shift ρ^2 on $B B^T$ to produce $\hat{B} \hat{B}^T$.

$\square$

Exercise 6.3.23 Consider a sequence of iterations of Francis's algorithm, starting from $A_0 = A$ and producing a sequence (A_j) of unitarily similar matrices. From Exercise 6.3.11 we know that the jth iteration is essentially characterized by equations

$$p_j(A_{j-1}) = Q_j R_j, \qquad A_j = Q_j^* A_{j-1} Q_j.$$

Here we are allowing the possibility of a different polynomial p_j (different shifts) on each iteration, and we make no restriction on the degree of p_j. For $j = 1, 2, 3, \ldots$, let

$$\hat{Q}_j = Q_1 Q_2 \cdots Q_j, \quad \hat{R}_j = R_j \cdots R_2 R_1, \quad \text{and} \quad \hat{p}_j(A) = p_j(A) \cdots p_2(A) p_1(A).$$

Clearly $\hat{Q}_j$ is unitary and $\hat{R}_j$ is upper triangular. In this exercise you will show that the combined effect of j iterations of Francis's algorithm is characterized by the equations

$$\hat{p}_j(A) = \hat{Q}_j \hat{R}_j, \qquad A_j = \hat{Q}_j^* A \hat{Q}_j.$$

Relationships of this type can be used to prove convergence theorems. See [102] or [101, Ch. 5].

(a) Show that $A_j = \hat{Q}_j^* A \hat{Q}_j$.

(b) Prove $\hat{p}_j(A) = \hat{Q}_j \hat{R}_j$ by induction on j. [For the induction step, assume $\hat{p}_{j-1}(A) = \hat{Q}_{j-1}\hat{R}_{j-1}$, and write $\hat{p}_j(A) = p_j(A)\hat{p}_{j-1}(A)$. Show that $p_j(A) = \hat{Q}_{j-1}p_j(A_{j-1})\hat{Q}_{j-1}^*$, and put the pieces together.]

$\square$

cise 6.3.24 We continue to use the notation established in Exercise 6.3.23. For a given k, define spaces $\mathcal{S}_j = \hat{p}_j(A)\mathrm{span}\{e_1, \ldots, e_k\}$. A good shifting strategy will cause rapid convergence of $\mathcal{S}_j$ to a k-dimensional invariant subspace $\mathcal{U}_k$ for at least some choices of k (notably $k = n - m$, if we are doing Francis iterations of degree m). Assuming we are in the generic case (all $\hat{p}_j(A)$ are nonsingular), use the equation $\hat{p}_j(A) = \hat{Q}_j \hat{R}_j$ to show that the space spanned by the first k columns of $\hat{Q}_j$ is exactly $\mathcal{S}_j$. Thus the leading k columns of $\hat{Q}_j$ are converging rapidly to an invariant subspace. This causes rapid convergence of $A_j = \hat{Q}_j^* A \hat{Q}_j$ to block triangular form. See [102] or [101, Ch. 5]. $\square$

6.4 LARGE SPARSE EIGENVALUE PROBLEMS

Most large matrices that occur in applications are sparse. That is, the vast majority of their entries are zeros. If a matrix is sparse enough, it may be worthwhile to store it in a sparse data structure, which stores only the nonzero entries, together with information about where each entry belongs in the matrix. If the matrix is extremely large, there may be no alternative to the sparse data structure, since there may not be enough storage space to store the matrix in the conventional way. If we want to find some eigenvalues of such a matrix, we need to use a method that can work with the sparse data structure. Francis's algorithm and other algorithms that use similarity transformations are inappropriate here, as they cause a great deal of fill-in. See Exercise 6.4.24.

Needed are methods that do not alter the matrix. One algorithm that immediately comes to mind is simultaneous iteration (6.2.7). Here we refer to the basic simultaneous iteration algorithm, which does not change the coordinate system at each step. Looking at (6.2.7) in the case $p(A) = A$, we see that the only way the algorithm uses A is to multiply it by the vectors $q_1^{(m)}, \ldots, q_k^{(m)}$ at each step. The entries of A are never altered in any way. Even if A is stored in a sparse format, it is a simple matter to calculate a matrix-vector product Aq (see Exercise 6.4.25). Thus it is a simple matter to perform simultaneous iteration on a large, sparse matrix. The number of vectors, k, that we use in the simultaneous iteration is typically limited by storage space and computational requirements, so we normally have $k \ll n$, which means that we can compute at most the few (at most k) eigenvalues of largest modulus and the associated eigenvectors. Some of the practical details of simultaneous iteration are studied in the following exercise.

Exercise 6.4.1   Consider the simultaneous iteration algorithm (6.2.7) with $p(A) = A$ and $k \ll n$.

(a) Let $q_1^{(m)}$, ..., $q_k^{(m)}$ be the k orthonormal vectors at the mth iteration of simultaneous iteration (6.2.7), and let $\hat{Q}_m = [q_1^{(m)} \cdots q_k^{(m)}] \in \mathbb{R}^{n \times k}$. This is a tall, skinny matrix. Show that

$$A\hat{Q}_m = \hat{Q}_{m+1}R_{m+1},$$

where R_{m+1} is a $k \times k$ upper-triangular matrix. This is a condensed QR decomposition.

(b) Let $B_m = A\hat{Q}_m \in \mathbb{C}^{n \times k}$ and $C_m = \hat{Q}_m^* B_m \in \mathbb{C}^{k \times k}$. Show that the columns of $\hat{Q}_m$ span an invariant subspace under A if and only if $B_m - \hat{Q}_m C_m = 0$, in which case the eigenvalues of the small matrix C_m are k of the eigenvalues of A. (In practice we can accept $\hat{Q}_m$ as invariant as soon as $\| B_m - \hat{Q}_m C_m \|_1$ is sufficiently small.)

(c) Show how invariant subspaces of dimension $j = 1, 2, \ldots, k - 1$ can be detected by considering leading submatrices of $\hat{Q}_m$, B_m, and C_m. Once an invariant subspace of dimension j has been detected, the first j columns of $\hat{Q}_m$ can be "locked in," and subsequent iterations can operate on the subset $q_{j+1}^{(m)}$, $\ldots, q_k^{(m)}$.

(d) A step of simultaneous iteration has three stages: (i) computation of B_m, (ii) condensed QR decomposition of B_m, and (iii) tests for convergence of invariant subspaces. Estimate the number of flops that each of these stages requires (See Exercise 6.4.25). About how much storage space is needed in all? In typical large applications k is small, and the first stage dominates the computational cost. Show, however, that if k is made large enough, the other stages will dominate the flop count.

$\square$

Up until about 1980, simultaneous iteration was the most popular method for computing a few eigenvalues of a large, sparse matrix. A number of refinements and acceleration techniques were introduced, but ultimately simultaneous iteration was largely supplanted by more sophisticated methods, some of which we will describe in this section.

The Shift-and-Invert Strategy

Before we leave simultaneous iteration, we will describe the *shift-and-invert* strategy, which can be combined with simultaneous iteration to compute interior eigenvalues of

A. As we shall later see, the shift-and-invert strategy can also be used in conjunction with other algorithms.

Simultaneous iteration computes the eigenvalues of A of largest modulus. What if that's not what we want? Suppose we would like to compute the eigenvalues in a certain region near some target value τ. If we shift by τ, and then take the inverse, we get a new matrix $(A - \tau I)^{-1}$ that has the same eigenvectors and invariant subspaces as A but different eigenvalues. Each eigenvalue λ of A corresponds to an eigenvalue $(\lambda - \tau)^{-1}$ of $(A - \tau I)^{-1}$. The largest eigenvalues of $(A - \tau I)^{-1}$ correspond to the eigenvalues of A that are closest to τ. We can find these eigenvalues (and associated invariant subspaces) by applying simultaneous iteration to $(A - \tau I)^{-1}$. This is the shift-and-invert strategy. Of course we don't actually form the inverse matrix; the inverse of a sparse matrix is not sparse. Instead we compute (once!) a sparse LU decomposition of $A - \tau I$. The computation $(A - \tau I)^{-1}q = p$ is then effected by solving for p the equation $(A - \tau I)p = q$ by forward and back substitution, using the factors of the LU decomposition. Once we have the LU decomposition in hand, we can use it over and over again to perform as many computations of type $(A - \tau I)^{-1}q$ as we please. A new LU decomposition is not needed unless we change the target shift τ.

The shift-and-invert technique is quite powerful and has found widespread use over the years. However, there is an important limitation on its applicability. If A is really large, its sparse LU decomposition, which is typically much less sparse than A itself, will require a great deal of storage space. If the computer's memory is not big enough to store the LU decomposition, then we cannot use this technique.

The Arnoldi Process

The Arnoldi process [2] has been around since about 1950, but it did not come into vogue as a method for computing eigenvalues until the 1970's. Variants of the Arnoldi process have now supplanted simultaneous iteration as the method of choice for a wide variety of applications.

Recall that the power method starts with an initial vector q and then computes (multiples of) the vectors $q, Aq, A^2q, \ldots, A^kq, \ldots$. As the algorithm proceeds, it discards the old iterates. At step k, it has only a multiple of A^kq; none of the past information has been saved. Simultaneous iteration operates on subspaces, but it works the same way; the information from the past is discarded.

The idea of the Arnoldi process is to retain and use all of the past information. After k steps we have the $k + 1$ vectors $q, Aq, \ldots, A^kq$, all of which we have saved. We then search for good eigenvector approximations in the $(k+1)$-dimensional space spanned by these vectors. This is exactly the Krylov subspace $\mathcal{K}_{k+1}(A, q)$ defined in Section 6.3. Now we consider some practical matters.

In practice the vectors $q, Aq, \ldots, A^kq$ usually are an ill-conditioned basis for the Krylov subspace $\mathcal{K}_{k+1}(A, q)$. The vector A^jq points more and more in the direction of a dominant eigenvector for A as j is increased. Thus the vectors toward the end of the sequence $q, \ldots, A^kq$ may be pointing in nearly the same direction. To counter this we can replace these vectors by an orthonormal set $q_1, \ldots, q_{k+1}$ for $\mathcal{K}_{k+1}(A, q)$ This can be achieved by the Gram-Schmidt process with one slight modification. Suppose we have generated the first k vectors and now we want to obtain the $(k + 1)$st. If we were working with the original sequence $q, Aq, \ldots, A^{k-1}q$, we would simply multiply $A^{k-1}q$ by A to obtain $A^k q$. However, we do not have the original sequence; we have the orthonormal set $q_1, \ldots, q_k$, and we want q_{k+1}. In principle we can obtain q_{k+1} by orthogonalizing $A^k q$ against $q_1, \ldots, q_k$ using Gram-Schmidt. In practice this is not possible, because we do not have the vector $A^k q$. Thus we must do something else. Instead of multiplying the (unavailable) vector $A^{k-1}q$ by A to obtain $A^k q$, we multiply the (available) vector q_k by A to obtain Aq_k. We then orthogonalize Aq_k against $q_1, \ldots, q_k$ to obtain q_{k+1}. This is the Arnoldi process.

Now let's add a bit more detail. The first step of the Arnoldi process consists of the normalization

$$q_1 = q/\| q \|_2. \tag{6.4.2}$$

On subsequent steps we take

$$\tilde{q}_{k+1} = Aq_k - \sum_{j=1}^{k} q_j h_{jk}, \tag{6.4.3}$$

where h_{jk} is the Gram-Schmidt coefficient

$$h_{jk} = \langle Aq_k, q_j \rangle = q_k^* Aq_j. \tag{6.4.4}$$

This is just the right coefficient to make $\tilde{q}_{k+1}$ orthogonal to q_j. Thus $\tilde{q}_{k+1}$ is orthogonal to $q_1, \ldots, q_k$ but does not have norm 1. We complete the step by normalizing $\tilde{q}_{k+1}$. That is, we take

$$q_{k+1} = \tilde{q}_{k+1}/h_{k+1,k}, \quad \text{where} \quad h_{k+1,k} = \| \tilde{q}_{k+1} \|_2. \tag{6.4.5}$$

It is not hard to show that (in exact arithmetic) this process produces exactly the same sequence of vectors as the Gram-Schmidt process applied to $q, Aq, A^2 q, \ldots, A^k q$. This is partly demonstrated in Exercise 6.4.30.

Just as in the case of the Gram-Schmidt algorithm (3.4.19) and (3.4.25), we can express some of the operations more succinctly in terms of matrix-vector multiplication operations. In the following algorithm we use the MATLAB-like notation $h_{1:k,k}$ for $\begin{bmatrix} h_{1k} & \cdots & h_{kk} \end{bmatrix}^T$ and $q_{1:k}$ for $\begin{bmatrix} q_1 & \cdots & q_k \end{bmatrix}$. We also do the orthogonalization twice, as recommended in Section 3.4, in order to get vectors that are orthogonal to working precision.

> **Arnoldi Process.** Given a nonzero starting vector q, this algorithm produces orthonormal $q_1, q_2, \ldots, q_{m+1}$ such that
>
> $$\text{span}\{q_1, \ldots, q_k\} = \mathcal{K}_k(A, q) \quad \text{for} \quad k = 1, \ldots, m+1.$$

$$q_1 \leftarrow q/\|q\|_2$$
for $k = 1, \ldots, m$

$$
\left[
\begin{array}{ll}
q_{k+1} \leftarrow Aq_k & \\
h_{1:k,k} \leftarrow q_{1:k}^* q_{k+1} & \\
q_{k+1} \leftarrow q_{k+1} - q_{1:k} h_{1:k,k} & \text{(orthogonalize)} \\
s \leftarrow q_{1:k}^* q_{k+1} & \\
q_{k+1} \leftarrow q_{k+1} - q_{1:k} s & \text{(reorthogonalize)} \\
h_{1:k,k} \leftarrow h_{1:k,k} + s & \\
h_{k+1,k} \leftarrow \|q_{k+1}\|_2 & \\
\text{if } h_{k+1,k} = 0 & \\
\quad \left[
\begin{array}{l}
\text{set flag (span}\{q_1, \ldots, q_k\} \text{ is invariant under } A) \\
\text{exit}
\end{array}
\right. & \\
q_{k+1} \leftarrow q_{k+1}/h_{k+1,k} &
\end{array}
\right.
\tag{6.4.6}
$$

In this algorithm the only reference to the matrix is in the line $q_{k+1} \leftarrow Aq_k$. Thus we can apply the algorithm to A as long as we can multiply A by an arbitrary vector.

Before we can show how to use the Arnoldi process to compute eigenvalues, we need to establish a few fundamental relationships.

Proposition 6.4.7 *Suppose q, Aq, $\ldots$, $A^{m-1}q$ are linearly independent. Then $\mathcal{K}_m(A, q)$ is invariant under A if and only if q, Aq, $\ldots$, $A^{m-1}q$, $A^m q$ are linearly dependent.*

Exercise 6.4.8 Prove Proposition 6.4.7. □

Theorem 6.4.9 *Suppose q, Aq, $\ldots$, $A^{m-1}q$ are linearly independent, and $q_1, \ldots, q_m$ are generated by the Arnoldi process (6.4.2), (6.4.3), (6.4.4), (6.4.5). Then*

(a) $\text{span}\{q_1, \ldots, q_k\} = \mathcal{K}_k(A, q)$ *for $k = 1, \ldots, m$.*

(b) *For $k = 1, \ldots, m - 1$, $h_{k+1,k} > 0$.*

(c) $h_{m+1,m} = 0$ *if and only if q, Aq, $\ldots$, $A^m q$ are linearly dependent, which holds in turn if and only if the Krylov subspace $\mathcal{K}_m(A, q)$ is invariant under A. This justifies the flag in Algorithm 6.4.6.*

Theorem 6.4.9 can be proved by induction on k. See Exercise 6.4.30. This theorem has much in common with Theorem 6.3.1. Notice that equation (6.4.3), which produces a new vector, is the same as (6.3.3). But in Theorem 6.4.9 we are proving a bit more. If you are not planning to work through the proof right away, you can at least work the following special case.

Exercise 6.4.10 Prove Theorem 6.4.9 in the case $k = 2$. (The case $k = 1$ is trivial.)

(a) Specifically, show that if q and Aq are linearly independent, then $\tilde{q}_2$ in (6.4.3) is nonzero. Thus $h_{21} > 0$, and the division in (6.4.5) can be performed to obtain q_2. Show that there exist constants c_0 and c_1, with $c_1 \neq 0$, such that $q_2 = c_0 q + c_1 Aq$. Show further that $q_1 \in \mathcal{K}_2(A, q)$, $q_2 \in \mathcal{K}_2(A, q)$, and hence $\text{span}\{q_1, q_2\} \subseteq \mathcal{K}_2(A, q)$. Conversely, show that $q \in \text{span}\{q_1, q_2\}$, $Aq \in \text{span}\{q_1, q_2\}$, and hence $\mathcal{K}_2(A, q) \subseteq \text{span}\{q_1, q_2\}$.

(b) Show that if q and Aq are linearly dependent, then $\tilde{q}_2 = 0$, $h_{21} = 0$, and span$\{q\}$ is invariant under A. In this case q is an eigenvector of A.

$\square$

Matrix Representations of the Arnoldi Process

From (6.4.3) and (6.4.5) we easily deduce that

$$Aq_k = \sum_{j=1}^{k+1} q_j h_{jk}, \qquad k = 1, 2, 3, \dots \qquad (6.4.11)$$

From Theorem 6.4.9 we know that these relationships hold for $k = 1, \dots, m$ if q, Aq, $\dots$, $A^m q$ are linearly independent. These $m + 1$ vector equations can be combined into a single matrix equation as follows. Define

$$Q_m = \begin{bmatrix} q_1 & \cdots & q_m \end{bmatrix} \in \mathbb{C}^{n \times m}$$

and

$$H_{m+1,m} = \begin{bmatrix} h_{11} & h_{12} & \cdots & h_{1,m-1} & h_{1m} \\ h_{21} & h_{22} & \cdots & h_{2,m-1} & h_{2m} \\ 0 & h_{32} & \cdots & h_{3,m-1} & h_{3m} \\ \vdots & & \ddots & & \vdots \\ 0 & & & h_{m,m-1} & h_{mm} \\ 0 & 0 & \cdots & 0 & h_{m+1,m} \end{bmatrix} \in \mathbb{C}^{m+1 \times m}.$$

Since Q_m has orthonormal columns, it is an isometry. $H_{m+1,m}$ is a non-square upper-Hessenberg matrix with strictly positive entries on the subdiagonal. From (6.4.11) we immediately have

$$AQ_m = Q_{m+1} H_{m+1,m}, \qquad (6.4.12)$$

for the kth column of (6.4.12) is exactly (6.4.11). This matrix equation summarizes neatly the relationships between A and the quantities that are generated by the Arnoldi process. Another useful expression is obtained by rewriting the right-hand side of (6.4.12). Let H_m denote the square upper Hessenberg matrix obtained by deleting the bottom row of $H_{m+1,m}$. If we separate the last column of Q_{m+1} and the bottom row of $H_{m+1,m}$ from the matrix product, we obtain $Q_{m+1}H_{m+1,m} = Q_m H_m + q_{m+1} \begin{bmatrix} 0 & \cdots & 0 & h_{m+1,m} \end{bmatrix}$. Thus

$$AQ_m = Q_m H_m + q_{m+1} h_{m+1,m} e_m^T, \qquad (6.4.13)$$

where e_m denotes the mth standard basis vector in $\mathbb{R}^m$.

Exercise 6.4.14 Verify equations (6.4.11), (6.4.12), and (6.4.13). $\square$

The following simple result, which is proved in Exercise 6.4.33, will be used in the next section. It is a variant of the implicit-Q theorem.

Proposition 6.4.15 *Suppose $q_1, \ldots, q_{m+1}$ are orthonormal vectors,*

$$Q_m = \begin{bmatrix} q_1 & \cdots & q_m \end{bmatrix},$$

and H_m is an upper Hessenberg matrix with $h_{j+1,j} > 0$ for $j = 1, \ldots, m$. Although these may have been obtained by any means whatsoever, suppose they satisfy (6.4.13). Then $q_1, \ldots, q_{m+1}$ must be exactly the vectors produced by the Arnoldi process with starting vector q_1. In other words, given a matrix A, the objects in (6.4.13) are uniquely determined by the first column of Q_m.

If q, Aq, $\ldots$, $A^m q$ are linearly independent, then $h_{m+1,m} \neq 0$. However, if they are dependent, we have $h_{m+1,m} = 0$, and (6.4.13) becomes $AQ_m = Q_m H_m$. We then conclude from Theorem 6.1.3 that the space spanned by the columns of Q_m is invariant under A. (This is consistent with Theorem 6.4.9, since $\mathcal{R}(Q_m) = \mathrm{span}\{q_1, \ldots, q_m\} = \mathcal{K}_m(A, q)$.) Furthermore, by Exercise 6.1.4 the eigenvalues of H_m are eigenvalues of A. If m is not too big, we can easily compute these m eigenvalues using, say, Francis's algorithm. This begins to show how the Arnoldi process can deliver eigenvalues.

Our plan is to apply the Arnoldi process to find a few eigenvalues of extremely large matrices. Each step of the process adds another basis vector q_k, which takes up significant storage space. We also need to store the entries h_{jk} of the Hessenberg matrix. These storage requirements severely limit the number of steps we will be able to take. Thus we will typically have to stop after m steps, where $m \ll n$.

Nevertheless, there is nothing to stop us from imagining what happens if we take more steps. If q, Aq, $\ldots A^{n-1}q$ are independent, we can take n steps and end up with $q_1, \ldots, q_n$, an orthonormal basis of $\mathbb{C}^n$. If we then try to take one more step, we will get $h_{n+1,n} = 0$, since the $n + 1$ vectors q, Aq, $\ldots A^n q$ in the n-dimensional space $\mathbb{C}^n$ must be linearly dependent. Thus (6.4.13) becomes $AQ_n = Q_n H_n$. The square matrix Q_n is unitary, H_n is properly upper Hessenberg, and we have $H_n = Q_n^{-1} A Q_n$. Thus the Arnoldi process, if carried to completion, computes a unitary similarity transformation to upper Hessenberg form. In Section 5.5 you learned how to accomplish this same task using reflectors. The process developed in Section 5.5 is generally preferred over the Arnoldi process for this particular task. See Exercise 6.4.34.

Now let us return to our main task, which is to obtain information about the eigenvalues of A by taking only a few steps of the Arnoldi process. This can now be viewed as a partial similarity transformation. A is reduced part way to upper Hessenberg form, in the sense that the first few columns of the transforming matrix and the upper left hand corner of the Hessenberg matrix are produced. We have already seen that if we get $h_{m+1,m} = 0$ at some point, then $\mathrm{span}\{q_1, \ldots, q_m\}$ is invariant under A, and the m eigenvalues of H_m are eigenvalues of A. However, we will seldom be so lucky as to encounter a small invariant subspace, unless we choose q very carefully. We'll find out how to do this by the implicitly restarted

Arnoldi process in the next section, but for now let us suppose we do not have such a special q. Suppose we have built $q_1, \ldots, q_m$, and we do not have $h_{m+1,m} = 0$. If $h_{m+1,m}$ is small, we might reasonably hope that we are close to an invariant subspace and the eigenvalues of H_m are close to eigenvalues of A. It turns out that even if $h_{m+1,m}$ is not small, some of the eigenvalues of H_m may be good approximations to eigenvalues of A.

Theorem 6.4.16 Let Q_m, H_m, and $h_{m+1,m}$ be generated by the Arnoldi process, so that (6.4.13) holds. Let μ be an eigenvalue of H_m with associated eigenvector x, normalized so that $\| x \|_2 = 1$. Let $v = Q_m x \in \mathbb{C}^n$ (also with $\| v \|_2 = 1$). Then

$$\| Av - \mu v \|_2 = |h_{m+1,m}| \, |x_m|, \tag{6.4.17}$$

where x_m denotes the mth (and last) component of x.

Exercise 6.4.18 Use (6.4.13) to prove Theorem 6.4.16.   □

The vector v introduced in Theorem 6.4.16 is called a *Ritz vector* of A associated with the subspace $\mathcal{K}_m(A, q) = \operatorname{span}\{q_1, \ldots, q_m\}$, so called because it is a Rayleigh-Ritz-Galerkin approximation to an eigenvector of A (See Exercise 6.4.35). The scalar μ is called the *Ritz value* associated with v. The pair (μ, v) is called a *Ritz pair*.

If (μ, v) is an eigenpair of A, then the residual $Av - \mu v$ is zero. If (μ, v) is not an eigenpair, the residual norm $\| Av - \mu v \|_2$ will not be zero, but it will be close to zero if (μ, v) is a good approximation to an eigenpair. Conversely, if $\| Av - \mu v \|_2$ is small, we have reason to expect that (μ, v) is close to an eigenpair of A. It turns out that the relationship is not so simple: if the eigenpair that is being approximated is ill conditioned (See Section 7.1), a small residual does not guarantee a good approximation. Nevertheless, the residual norm does give some indication of the quality of the approximation: a small residual guarantees that (μ, v) is an exact eigenpair of a matrix that is close to A, as Exercise 6.4.36 shows. Finally, as we shall see in Section 7.1, all eigenvalues of normal matrices are well conditioned. Thus if A is normal, a small residual guarantees that μ is near an eigenvalue of A. Recall that the class of normal matrices includes all real, symmetric matrices and other important classes of matrices.

Theorem 6.4.16 shows that the residual norm of the Ritz pair (μ, v) may be small even if $h_{m+1,m}$ is not, if the last entry of the eigenvector x is small. If m is not large, it is an easy matter to compute the eigenvalues and eigenvectors of H_m and check the quantities $|h_{m+1,m}| \, |x_m|$ for all eigenvectors x. If any of the residuals is small, we have reason to hope that we have a good approximation.

Experience has shown that some of the Ritz values will be good approximations to eigenvalues of A long before m is large enough that $h_{m+1,m}$ becomes small. Typically the eigenvalues on the periphery of the spectrum are approximated first, as the following example illustrates. See also Exercise 6.4.26.

Example 6.4.19 Using the following sequence of MATLAB instructions we built a random complex sparse matrix of dimension 144.

```
G = numgrid('N',14);
```

```
B = delsq(G);
A = sprandn(B) + i*sprandn(B);
```

Since this example matrix is only of modest size, we can easily compute its eigen-values by `eigvals = eig(full(A))`. We implemented Algorithm 6.4.6 in MATLAB and ran it for 40 steps, using the vector of all ones as the starting vector q, to get 40 orthonormal vectors and a 40×40 upper Hessenberg matrix H_{40}. We computed the eigenvalues and eigenvectors of H_{40} using `eig`. In Figure 6.1 we have plotted the true eigenvalues as pluses and the eigenvalues of H_{40} (Ritz values) as circles. The left-hand plot shows all of the Ritz values. We observe that many

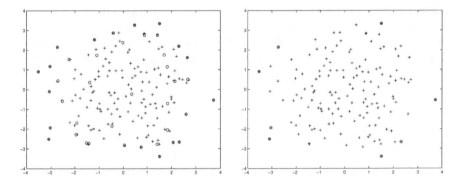

Figure 6.1   Eigenvalues and Ritz values of a random sparse matrix

of the eigenvalues on the periphery of the spectrum are well approximated by Ritz values, while none of the interior eigenvalues are approximated well.

For each of the 40 Ritz pairs we computed the residual norm using the formula $|h_{41,40}|\,|x_{40}|$ from Theorem 6.4.16. The eight smallest residuals are listed in Table 6.1. In this example $h_{41,40} \approx 2.4$, so the smallness of each of the residuals is

residual ($|h_{41,40}|\,|x_{40}|$)	true error
1.1×10^{-04}	1.3×10^{-06}
1.5×10^{-04}	6.0×10^{-06}
1.4×10^{-03}	2.5×10^{-04}
1.4×10^{-03}	1.2×10^{-04}
1.7×10^{-03}	7.1×10^{-05}
2.3×10^{-03}	1.8×10^{-04}
6.0×10^{-03}	5.9×10^{-04}
6.4×10^{-03}	1.0×10^{-03}

Table 6.1   Best eight Ritz pair residuals and corresponding Ritz value errors

attributable entirely to smallness of x_{40}, the last component of the eigenvector. Since we know the true eigenvalues in this case, we can compute the actual errors in the Ritz values. These are also shown in Table 6.1. In this benign case the errors are even smaller than the residuals. The eight best Ritz values are plotted, along with all 144 eigenvalues, in the right hand plot in Figure 6.1. $\qquad\Box$

The reasons for the outer eigenvalues being approximated best are explored in Exercises 6.4.37 and 6.4.41.

Shift-and-Invert Arnoldi

Since the peripheral eigenvalues are found first, the Arnoldi process is not directly useful for computing interior eigenvalues. However, if the matrix A is not prohibitively large, the shift-and-invert strategy, which we introduced earlier in this section, can be combined with Arnoldi for this purpose. If we wish to find interior eigenvalues near the target τ, we can apply the Arnoldi process to $(A - \tau I)^{-1}$, whose peripheral eigenvalues $\mu_i = (\lambda_i - \tau)^{-1}$ correspond to the eigenvalues λ_i that are closest to τ. For each μ_i that we compute, we easily obtain $\lambda_i = \tau + 1/\mu_i$. The eigenvectors of $(A - \tau I)^{-1}$ are the same as the eigenvectors of A (Exercise 6.4.29).

The Symmetric Lanczos Process

When A is Hermitian, that is, $A = A^*$, the Arnoldi process takes a much simpler form. In most Hermitian applications the matrix is in fact real, so let's just assume that $A \in \mathbb{R}^{n \times n}$ and $A = A^T$. Then all of the eigenvalues are real, and so are the eigenvectors. More precisely, $\mathbb{R}^n$ has an orthonormal basis consisting of eigenvectors of A. Given a starting vector $q \in \mathbb{R}^{n \times n}$, after m steps of the Arnoldi process we have

$$AQ_m = Q_m T_m + q_{m+1} h_{m+1,m} e_m^T, \qquad (6.4.20)$$

where Q_m is a real matrix with orthonormal columns, and T_m is a real upper Hessenberg matrix. The algorithm is simplified because T_m inherits the symmetry of A.

Exercise 6.4.21 Show that $T_m = Q_m^T A Q_m$, and T_m is symmetric if A is. Deduce that T_m is tridiagonal. $\qquad\Box$

Since T_m is tridiagonal, most of the t_{jk} in the Arnoldi formula

$$\tilde{q}_{k+1} = A q_k - \sum_{j=1}^{k} q_j t_{jk}$$

are zero; there are only two nonzero terms in the sum. If we introduce the notation

$$T_m = \begin{bmatrix} \alpha_1 & \beta_1 & & \\ \beta_1 & \alpha_2 & \ddots & \\ & \ddots & \ddots & \beta_{m-1} \\ & & \beta_{m-1} & \alpha_m \end{bmatrix},$$

we obtain the three-term recurrence

$$\tilde{q}_{k+1} = Aq_k - q_k\alpha_k - q_{k-1}\beta_{k-1},$$

or, normalizing $\tilde{q}_{k+1}$,

$$q_{k+1}\beta_k = Aq_k - q_k\alpha_k - q_{k-1}\beta_{k-1}. \tag{6.4.22}$$

This special case of the Arnoldi process is called the *symmetric Lanczos process* [61].

Symmetric Lanczos Process without Reorthogonalization. Given $A = A^T \in \mathbb{R}^{n \times n}$ and a nonzero starting vector $q \in \mathbb{R}^n$, this algorithm produces orthonormal $q_1, q_2, \ldots, q_m$ such that $\text{span}\{q_1, \ldots, q_k\} = \text{span}\{q, Aq, \ldots, A^{k-1}q\}$ for $k = 1, \ldots, m$.

$$q_1 = q/\|q\|_2$$
for $k = 1, \ldots, m - 1$

$$\begin{array}{l} q_{k+1} \leftarrow Aq_k \\ \alpha_k \leftarrow q_k^T q_{k+1} \\ q_{k+1} \leftarrow q_{k+1} - q_k\alpha_k \\ \text{if } k > 1, \text{ then } q_{k+1} \leftarrow q_{k+1} - q_{k-1}\beta_{k-1} \\ \beta_k \leftarrow \|q_{k+1}\|_2 \\ \text{if } \beta_k = 0 \\ \quad \begin{array}{l} \text{set flag (span}\{q_1, \ldots, q_k\} \text{ is invariant under } A) \\ \text{exit} \end{array} \\ q_{k+1} \leftarrow q_{k+1}/\beta_k \end{array} \tag{6.4.23}$$

The coefficient α_k is computed as an inner product, just as in the Arnoldi process, but there is no need to compute β_{k-1} in this way because it was already computed as the scaling factor β_k on the previous step. This is a consequence of symmetry of T_m.

More significant consequences of symmetry are that the algorithm is much cheaper and has much lower storage requirements, if it is executed without reorthogonalization, as shown in (6.4.23). If only eigenvalues are wanted, only the coefficients α_k and β_k need to be saved, as these are nonzero entries of T_m. At each step only the two most recently computed vectors are needed; the others can be discarded. Thus the storage requirement is much less. At the kth step, only one inner product is computed instead of k, so the work is much less. For more information about this way of using the Lanczos process, see the books by Cullum and Willoughby [12].

The principle difficulty of using the symmetric Lanczos process without reorthogonalization is that the orthonormality of the q_k gradually deteriorates due to roundoff errors (Exercise 6.4.28.) If, as is usually the case, we are interested in maintaining orthonormality, then we must keep all the q_k and orthogonalize each new vector against them (twice). Thus the storage and computation savings are nullified. Notice that we also need to save all the q_k if we want to compute eigen*vectors* of A.

Additional Exercises

Exercise 6.4.24 Use MATLAB's test matrix west0479 to demonstrate that similarity transformations can cause a great deal of fill in.

    (a) Show that the reduction to upper Hessenberg form fills in the entire upper part of the matrix. For example,

```
load west0479
A = west0479;
nnza = nnz(A)
spy(A)
H = hess(full(A));    % hess operates on full matrices.
nnzh = nnz(H);
spy(H)
```

    (b) Use MATLAB's qr command to perform a QR decomposition of A (not H). Then compute $\hat{A} = RQ$. This is an iteration of the basic QR algorithm (Exercise 6.3.15 (a)). Make a spy plot of the result.

□

Exercise 6.4.25 The sparse data structure used by MATLAB lists the nonzero entries of A in a list s of length nz. In addition there are two lists of integers row and col, also of length nz, that specify where each entry belongs in A. For example, if the seventh entry in the list is $s(7) = 1.234 \times 10^5$, and $row(7) = 45$ and $col(7) = 89$, then $a_{45,89} = 1.234 \times 10^5$. (A more detailed description was given in Section 1.6.) Suppose A is stored in this way. Write a pseudocode algorithm that computes $y = Ax$, assuming the vector x (and also y) is stored in the conventional way. Your algorithm should make just one pass through the list s and perform $2nz$ flops in all.

□

Exercise 6.4.26 Write a MATLAB script that implements the Arnoldi algorithm with reorthogonalization. You may use Algorithm 6.4.6 as a pattern if you wish. Use a random vector (q = randn(n,1)) or the vector of all ones (q = ones(n,1)) as a starting vector. Use eig to compute the eigenvalues of the Hessenberg matrix H_m that is generated. These are the Ritz values, some of which may be good estimates of eigenvalues. To test your code, generate a sparse matrix by

```
num = 20;   % or other appropriate number
G = numgrid('N',num);
B = delsq(G);
A = sprandn(B) + i*sprandn(B);
```

as in Example 6.4.19. (Use help numgrid, help delsq, and so on for information about these functions.) You can get a larger or smaller test matrix by making num larger or smaller. (You will get a matrix of dimension (num − 2)².) Check the nonzero pattern of A by using spy. Using a random starting vector or other starting vector of your choice, run the Arnoldi process for 15, 30, 45, and 60 steps to get Hessenberg matrices of sizes 15, 30, 45, and 60. (Actually you can just do one run, because the smaller Hessenberg matrices are submatrices that can be extracted from H_{60}.) Calculate the Ritz values for each of the four cases (for example, `ritz15 = eig(h(1:15,1:15))`), and in each case plot the Ritz values and the true eigenvalues together. For example, you could use commands like these:

```
hold off
eigval = eig(full(A));
plot(real(eigval),imag(eigval),'r+')
hold on
plot(real(ritz15),imag(ritz15),'bo')
```

Notice that the quality of approximation improves as the number of steps is increased, and the outer eigenvalues are approximated best. □

cise 6.4.27 Repeat Exercise 6.4.26 taking A to be the test matrix `west0479`.

```
load west0479
A = west0479;
```

□

cise 6.4.28 This exercise explores the loss of orthogonality in the Arnoldi process without reorthogonalization. If you have not already done so, write a MATLAB script that implements the Arnoldi algorithm with reorthogonalization. Modify it so that the reorthogonalization can be turned on or off, depending on the value of some switching variable. For each of the following matrices, take 100 Arnoldi steps without reorthogonalization. Then compute the inner products $q_1^* q_k$ for $k = 2, \ldots, 100$. Notice how the orthogonality deteriorates as k increases. Also compute $\| I_m - Q_m^* Q_m \|$ (with $m = 100$), which should be zero in theory. Now repeat the Arnoldi process with reorthogonalization and compute $\| I_m - Q_m^* Q_m \|$.

(a) Use the matrix `west0479`.

```
load west0479
A = west0479;
spy(A)
n = size(A,1)
```

(b) Now try a discrete Laplacian matrix of dimension 324 (from the interior of a 20×20 square grid). This sparse matrix is symmetric.

```
A = delsq(numgrid('N',20));
n = size(A,1)
```

(c) Now, to show that loss of orthogonality does not always occur, try a matrix like the one from Exercise 6.4.26.

```
B = delsq(numgrid('N',20));
A = sprandn(B) + i*sprandn(B);
n = size(A,1)
```

□

Exercise 6.4.29 Modify your code from Exercise 6.4.26 to do the shift-and-invert Arnoldi process. This requires only a few modifications. An LU decomposition of $(A - \tau I)$, where τ is the target shift, needs to be done once at the beginning:

```
[L,U] = lu(A-tau*speye(size(A)));
```

`speye` is a sparse identity matrix. Then whenever an operation of the form

$$q \leftarrow (A - \tau I)^{-1}q$$

needs to be done, it can be effected by the operations `q = L\q; q = U\q;`. The eigenvalues that are computed are eigenvalues of $(A - \tau I)^{-1}$, so your code will need to transform them to eigenvalues of A. From $\mu = (\lambda - \tau)^{-1}$, we obtain $\lambda = \tau + 1/\mu$. Test your program by using it to calculate the ten eigenvalues of `west0479` that are closest to $\tau = 6 + 0i$. Take as many Arnoldi steps as you need to get the residuals down around 10^{-16}. Compute the residuals by the inexpensive method of Theorem 6.4.16. Compare your computed eigenvalues with the "true" eigenvalues computed by Francis's algorithm (`lam = eig(full(west0479));`). Find the eigenvalues that are closest to 6 by using MATLAB's `sort` command:

```
[dummy,order] = sort(lam-tau);
lam = lam(order);
```

□

Exercise 6.4.30 Prove Theorem 6.4.9 by induction on k. You might like to proceed as follows. Assuming that $q, Aq, A^2q, \ldots, A^{m-1}q$ are linearly independent, we must show that $\mathrm{span}\{q_1, \ldots, q_k\} = \mathcal{K}_k(A, q)$ for $k = 1, \ldots, m$.

(a) Establish that the case $k = 1$ holds, i.e. $\mathrm{span}\{q_1\} = \mathcal{K}_1(A, q)$.

(b) To establish the general result, let $j < m$. You must show that if

$$\mathrm{span}\{q_1, \ldots, q_k\} = \mathcal{K}_k(A, q) \quad \text{for } k = 1, \ldots, j,$$

then $\mathrm{span}\{q_1, \ldots, q_{j+1}\} = \mathcal{K}_{j+1}(A, q)$. As part of the induction hypothesis we have $\mathrm{span}\{q_1, \ldots, q_j\} = \mathcal{K}_j(A, q)$, so $q_j \in \mathcal{K}_j(A, q)$. Thus there exist constants $c_0, \ldots, c_{j-1}$ such that

$$q_j = c_0 q + c_1 Aq + c_2 A^2 q + \cdots + c_{j-1} A^{j-1} q. \tag{6.4.31}$$

Show that $c_{j-1} \neq 0$. (You can use $\mathrm{span}\{q_1, \ldots, q_{j-1}\} = \mathcal{K}_{j-1}(A, q)$ here.) Now multiply (6.4.31) by A to obtain an expression for Aq_j, which you can use in (6.4.3) to show that $\tilde{q}_{j+1} \neq 0$. Thus $h_{j+1,j} > 0$, and we can do the division in (6.4.5) to obtain q_{j+1}. Show that there exist constants $d_0, \ldots, d_j$ such that

$$q_{j+1} = d_0 q + d_1 Aq + d_2 A^2 q + \cdots + d_j A^j q. \tag{6.4.32}$$

Thus $q_{j+1} \in \mathcal{K}_{j+1}(A, q)$. Show that $\mathrm{span}\{q_1, \ldots, q_{j+1}\} \subseteq \mathcal{K}_{j+1}(A, q)$.

(c) Continuing from part (b), show that $d_j \neq 0$ in (6.4.32). Using (6.4.32) and the induction hypothesis $\mathrm{span}\{q_1, \ldots, q_j\} = \mathcal{K}_j(A, q)$, show that $A^j q \in$

span$\{q_1, \ldots, q_{j+1}\}$ and hence $\mathcal{K}_{j+1}(A, q) \subseteq$ span$\{q_1, \ldots, q_{j+1}\}$. Conclude that span$\{q_1, \ldots, q_{j+1}\} = \mathcal{K}_{j+1}(A, q)$. This completes the induction proof of part (a) of Theorem 6.4.9. Along the way we have proved part (b) as well.

(d) Using relationships developed in the previous parts of this exercise, prove that if $q, Aq, A^2q, \ldots, A^mq$ are linearly dependent, then $\tilde{q}_{m+1} = 0$ and $h_{m+1,m} = 0$. Conversely, if $q, Aq, A^2q, \ldots, A^mq$ are linearly independent, then $\tilde{q}_{m+1} \neq 0$ and $h_{m+1,m} \neq 0$. This was already proved in part (b). Now complete the proof of Theorem 6.4.9, part (c).

(e) This is an alternate version of part (c). Use the result span$\{q_1, \ldots, q_{j+1}\} \subseteq \mathcal{K}_{j+1}(A, q)$ from part (b) and a dimension argument to prove that

$$\text{span}\{q_1, \ldots, q_{j+1}\} = \mathcal{K}_{j+1}(A, q).$$

$\square$

rcise 6.4.33 Prove Proposition 6.4.15 by two methods.

(a) Show that if (6.4.13) is satisfied, then so is (6.4.11), which is equivalent to (6.4.3) combined with (6.4.5). Show that the coefficients h_{jk} in (6.4.3) are uniquely determined by the requirement that $\tilde{q}_{k+1}$ be orthogonal to $q_1, \ldots, q_k$. Show that $h_{k+1,k}$ is uniquely determined by the requirements that it be positive and that $\| q_{k+1} \|_2 = 1$. Prove by induction that all q_j are uniquely determined by q_1.

(b) The second approach uses Krylov matrices. In Exercise 6.3.10 we introduced square Krylov matrices; here we use rectangular Krylov matrices as well. Given $A \in \mathbb{C}^{n \times n}$, $x \in \mathbb{C}^n$, and $j \leq n$, the *Krylov matrix* $K_j(A, x) \in \mathbb{C}^{n \times j}$ is the matrix with columns $x, Ax, A^2x, \ldots, A^{j-1}x$.

   (i) Let H_m be as in (6.4.13); that is, H_m is upper Hessenberg with positive subdiagonal entries. Show that the Krylov matrix $K_m(H_m, e_1)$ is upper triangular and has all of its main-diagonal entries positive.

   (ii) Show that if (6.4.13) is satisfied, then $AQ_my = Q_mH_my$ for all $y \in \mathbb{C}^m$ satisfying $e_m^T y = 0$.

   (iii) Still assuming (6.4.13) and using the result from part (ii), prove by induction on j that $A^j Q_m e_1 = Q_m H_m^j e_1$ for $j = 0, 1, \ldots, m-1$. Deduce that $K_m(A, q_1) = Q_m K_m(H_m, e_1)$.

   (iv) Using the uniqueness of QR decompositions (Theorem 3.4.8), prove that Q_m is uniquely determined by q_1.

$\square$

rcise 6.4.34 Count the flops required to transform a matrix to upper Hessenberg form by running the Arnoldi process to completion (n steps of Algorithm 6.4.6). Count the cost of the orthogonalization separately from the matrix-vector multiplications Aq_k.

If the vectors q_k are to be orthonormal to working precision, we must reorthogonalize, as shown in Algorithm 6.4.6. Show that if we do this, the cost of the orthogonalization alone is more than the cost of the reduction to upper Hessenberg form using reflectors (Algorithm 5.5.2). Assuming A is a dense matrix, show that the total cost of the Arnoldi algorithm is about $6n^3$ flops.   □

Exercise 6.4.35 Let $A \in \mathbb{C}^{n \times n}$, and let S be a k-dimensional subspace of $\mathbb{C}^n$. Then a vector $v \in S$ is called a *Ritz vector* of A from S if and only if there is a $\mu \in \mathbb{C}$ such that the *Rayleigh-Ritz-Galerkin* condition

$$Av - \mu v \perp S$$

holds, that is, $\langle Av - \mu v, s \rangle = 0$ for all $s \in S$. The scalar μ is called the *Ritz value* of A associated with v. Let $q_1, \ldots, q_k$ be an orthonormal basis of S, let $Q = [\, q_1 \ \cdots \ q_k \,]$, and let $B = Q^* A Q \in \mathbb{C}^{k \times k}$.

(a) Since $v \in S$, there is a unique $x \in \mathbb{C}^k$ such that $v = Qx$. Show that v is a Ritz vector of A with associated Ritz value μ if and only if x is an eigenvector of B with associated eigenvalue μ. In particular, there are k Ritz values of A associated with S, namely the k eigenvalues of B.

(b) Show that if v is a Ritz vector with Ritz value μ, then $\mu = v^* A v / v^* v$. That is, μ is the Rayleigh quotient of v. (More generally, we can think of $B = Q^* A Q$ as a $k \times k$ *matrix Rayleigh quotient*.)

(c) Suppose Q_m and H_m are generated by applying the Arnoldi process to A. Show that the eigenvalues of H_m are Ritz values of A associated with the subspace spanned by the columns of Q_m.

□

Exercise 6.4.36 Let (μ, v) be an approximate eigenpair of A with $\|v\|_2 = 1$, and let $r = Av - \mu v$ (the residual). Let $\epsilon = \|r\|_2$ and $E = -rv^*$. Show that (μ, v) is an eigenpair of $A + E$, and $\|E\|_2 = \epsilon$. This shows that if the residual norm $\|r\|_2$ is small, then (μ, v) is an exact eigenpair of a matrix that is close to A. Thus (μ, v) is a good approximate eigenpair of A in the sense of backward error.   □

Exercise 6.4.37

(a) Show that a vector v is in the Krylov subspace $\mathcal{K}_m(A, q)$ if and only if v can be expressed as $p(A)q$, where p is a polynomial of degree $m - 1$ or less. Thus $\mathcal{K}_m(A, q) = \{p(A)q \mid p \in \mathcal{P}_{m-1}\}$, where $\mathcal{P}_{m-1}$ denotes the set of polynomials of degree less than or equal to $m - 1$.

(b) Show that if v is an eigenvector of A with eigenvalue λ, then v is an eigenvector of $p(A)$ with eigenvalue $p(\lambda)$.

(c) Suppose A is a semisimple matrix with linearly independent eigenvectors v_1, $\ldots, v_n$ and associated eigenvalues $\lambda_1, \ldots, \lambda_n$. Then

$$q = c_1 v_1 + c_2 v_2 + \cdots + c_n v_n$$

for some scalars $c_1, \ldots, c_n$, which are not known in advance. If q was chosen "at random," then (typically) none of the c_i will be extremely small, and q will have a significant contribution from each of the eigenvectors. Let us assume that this is the case, and that $\| v_j \| = 1$ for all j. Show that if $v = p(A)q$, then

$$v = c_1 p(\lambda_1) v_1 + c_2 p(\lambda_2) v_2 + \ldots + c_n p(\lambda_n) v_n.$$

Show that if there is a $p \in \mathcal{P}_{m-1}$ such that, for some j, $|p(\lambda_j)|$ is much larger $|p(\lambda_i)|$ for all $i \neq j$, then the Krylov subspace $\mathcal{K}_m(A, q)$ will contain a vector that is close to the eigenvector v_j. Under these circumstances we normally would expect that one of the Ritz pairs from $\mathcal{K}_m(A, q)$ will approximate the eigenpair (λ_j, v_j) well.

(d) Suppose λ_j is well separated from the other eigenvalues of A. How might one build a polynomial $p(z) = (z - z_1)(z - z_2) \cdots (z - z_{m-1})$ such that $|p(\lambda_j)| \gg |p(\lambda_i)|$ for all $i \neq j$? (Think about the placement of the zeros z_i.) How would the task be more difficult if λ_j were not well separated from the other eigenvalues of A? For these reasons the Arnoldi process tends to find eigenvalues on the periphery of the spectrum first, since these are better isolated from the rest of the eigenvalues. These questions are studied more closely for symmetric matrices in Exercise 6.4.41.

□

cise 6.4.38 In this exercise you will develop the basic properties of Chebyshev polynomials, which are perhaps the most useful polynomials in approximation theory. In Exercise 6.4.41 you will use the Chebyshev polynomials to study the rate of convergence of the symmetric Lanczos process. Later on (Exercise 8.9.13), you will use them in the analysis of the conjugate-gradient algorithm for solving positive-definite linear systems $Ax = b$.

For $x \in [-1, 1]$, define an auxiliary variable $\theta \in [0, \pi]$ by $x = \cos \theta$. The value of θ is uniquely determined by x; in fact $\theta = \arccos x$. Now for $m = 0, 1, 2, \ldots$ define a function T_m on $[-1, 1]$ by

$$T_m(x) = \cos m\theta.$$

(a) Show that $|T_m(x)| \leq 1$ for all $x \in [-1, 1]$, $T_m(1) = 1$, $T_m(-1) = (-1)^m$, $|T_m(x)| = 1$ at $m + 1$ distinct points in $[-1, 1]$, and $T_m(x) = 0$ at m distinct points in $(-1, 1)$.

(b) Use the trigonometric identity

$$\cos(\alpha + \beta) + \cos(\alpha - \beta) = 2 \cos \alpha \cos \beta$$

to show that

$$T_{m+1}(x) + T_{m-1}(x) = 2x T_m(x) \qquad m = 1, 2, 3, \ldots.$$

This yields the recursion

$$T_{m+1}(x) = 2xT_m(x) - T_{m-1}(x) \qquad m = 1, 2, 3, \ldots. \qquad (6.4.39)$$

(c) Determine $T_0(x)$ and $T_1(x)$ directly from the definition, then use (6.4.39) to calculate $T_2(x)$, $T_3(x)$, and $T_4(x)$. Note that each is a polynomial in x and can therefore be extended in a natural way beyond the interval $[-1, 1]$. Graph the polynomials T_2, T_3, and T_4, focusing on the interval $[-1, 1]$, but notice that they grow rapidly once x leaves the interval.

(d) Using the recursion (6.4.39), prove by induction on m that for all m, T_m is a polynomial of degree m. (From part (a) we know that T_m has all m of its zeros in $(-1, 1)$. Therefore T_m must grow rapidly once x leaves that interval.)

(e) Not only does T_m grow rapidly once x leaves $[-1, 1]$. For a given x not in $[-1, 1]$, $T_m(x)$ grows rapidly as a function of m. To show this, we need another characterization of the Chebyshev polynomials. For $x \in [1, \infty)$ define an auxiliary variable $t \in [0, \infty)$ by $x = \cosh t$. (Recall that $\cosh t$, the *hyperbolic cosine* of t, is defined by $\cosh t = \frac{1}{2}(e^t + e^{-t})$.) For $m = 0, 1, 2, \ldots$ define a function $\tilde{T}_m$ on $[1, \infty)$ by

$$\tilde{T}_m(x) = \cosh mt.$$

Use the hyperbolic identity

$$\cosh(\alpha + \beta) + \cosh(\alpha - \beta) = 2 \cosh \alpha \cosh \beta$$

to show that

$$\tilde{T}_{m+1}(x) + \tilde{T}_{m-1}(x) = 2x\tilde{T}_m(x) \qquad m = 1, 2, 3, \ldots,$$

hence

$$\tilde{T}_{m+1}(x) = 2x\tilde{T}_m(x) - \tilde{T}_{m-1}(x) \qquad m = 1, 2, 3, \ldots,$$

which is exactly the recursion (6.4.39). Compute $\tilde{T}_0$ and $\tilde{T}_1$ directly from the definition, and prove by induction on m that $\tilde{T}_m = T_m$ for all m. Thus

$$T_m(x) = \cosh mt, \quad \text{where} \quad x = \cosh t. \qquad (6.4.40)$$

(f) If $x > 1$, then $x = \cosh t$, where $t > 0$. Let $\rho = e^t > 1$. Use (6.4.40) to show that $T_m(x) \geq \frac{1}{2}\rho^m$. Thus $T_m(x)$ grows exponentially as a function of m.

(g) Use MATLAB (which has built-in functions `cosh` and `acosh`) to compute $t = \cosh^{-1} x$ and $\rho = e^t$ when $x = 1.1$. compute $T_{10}(1.1)$, $T_{20}(1.1)$, and $T_{30}(1.1)$. Compare the ratio $T_{30}(1.1)/T_{20}(1.1)$ with e^{10t}. To what can you attribute the good agreement?

(h) Show that $T_m(-x) = (-1)^m T_m(x)$ for all m and x. Therefore we have the same growth of $|T_m(x)|$ for $x \in (-\infty, -1)$ as for $x \in (1, \infty)$.

□

cise 6.4.41 This exercise makes use of the Chebyshev polynomials, introduced in the previous exercise, to pursue the theme of Exercise 6.4.37 in greater detail. Let $A \in \mathbb{R}^{n \times n}$ be a symmetric matrix with eigenvalues $\lambda_1 > \lambda_2 \geq \cdots \geq \lambda_n$, with $\lambda_2 > \lambda_n$. We can obtain polynomials p_m with Chebyshev behavior on the interval $[\lambda_n, \lambda_2]$ by performing a transformation that maps $[\lambda_n, \lambda_2]$ onto $[-1, 1]$. These polynomials will have the property that $p_m(\lambda_1)$ is large and $p_m(\lambda_j)$ is bounded by 1 for $j \neq 1$.

(a) Show that the transformation

$$w = 1 + 2 \frac{x - \lambda_2}{\lambda_2 - \lambda_n}$$

maps $[\lambda_n, \lambda_2]$ onto $[-1, 1]$.

(b) Define a polynomial $p_m \in \mathcal{P}_m$ by

$$p_m(x) = T_m(w) = T_m \left(1 + 2 \frac{x - \lambda_2}{\lambda_2 - \lambda_n} \right).$$

Show that $|p_m(\lambda_i)| \leq 1$ for $i \geq 2$ and

$$|p_m(\lambda_1)| \geq \frac{1}{2} \rho^m,$$

where

$$\rho \geq 1 + 2 \frac{\lambda_1 - \lambda_2}{\lambda_2 - \lambda_n} > 1. \tag{6.4.42}$$

Use part (f) of Exercise 6.4.38.

(c) Let $v_1, \ldots, v_n$ denote orthonormal eigenvectors of A associated with the eigenvalues $\lambda_1, \ldots, \lambda_n$, respectively. Let q be a starting vector for the symmetric Lanczos algorithm with $\|q\|_2 = 1$, and suppose $q = c_1 v_1 + c_2 v_2 + \cdots + c_n v_n$ with $c_1 \neq 0$. Define $\tilde{p}_m$ by $\tilde{p}_m(x) = p_m(x)/(c_1 p_m(\lambda_1))$, where p_m is as defined in part (b). Let $w_m = \tilde{p}_{m-1}(A)q \in \mathcal{K}_m(A, q)$. Show that there is a constant C such that

$$\|w_m - v_1\|_2 \leq C \rho^{-m}, \qquad m = 1, 2, 3, \ldots,$$

where ρ is as in (6.4.42). Thus, for large m, the Krylov subspace $\mathcal{K}_m(A, q)$ contains a vector that is close to the eigenvector v_1.

(d) Adapt the arguments given above to show that for large m, $\mathcal{K}_m(A, q)$ contains vectors that are close to v_n, the eigenvector associated with λ_n. Assume $\lambda_1 > \lambda_{n-1} > \lambda_n$ and $c_n \neq 0$.

(e) Suppose $\lambda_1 > \lambda_2 > \lambda_3 > \lambda_n$. For $m = 1, 2, 3, \ldots$, define a polynomial $\hat{p}_m \in \mathcal{P}_m$ by

$$\hat{p}_m(x) = (x - \lambda_1) T_{m-1} \left(1 + 2 \frac{x - \lambda_3}{\lambda_3 - \lambda_n} \right).$$

Use $\hat{p}_m$ to show that for sufficiently large m, the Krylov subspace $\mathcal{K}_m(A, q)$ contains vectors that are close to v_2. Assume $c_2 \neq 0$.

(f) Suppose $\lambda_2 > \lambda_3 > \lambda_4 > \lambda_n$. Construct polynomials that can be used to show that for sufficiently large m, $\mathcal{K}_m(A, q)$ contains vectors that are close to v_3, assuming $c_3 \neq 0$.

$\square$

Exercise 6.4.43   This exercise is for those who are familiar with the three-term recurrence (Stieltjes process) for generating orthogonal polynomials. First note that the Lanczos procedure is governed by a three-term recurrence (6.4.22). Then show that the three-term recurrence for generating orthogonal polynomials can be viewed as the Lanczos algorithm applied to a certain linear operator acting on an infinite-dimensional space. (*Note:* The recurrence (6.4.39) for the Chebyshev polynomials is a special case.) $\square$

6.5   IMPLICIT RESTARTS

This section, which builds on the previous one, introduces one of the most effective known methods for computing eigenvalues of large matrices, the implicitly restarted Arnoldi process, and its variant, the Krylov-Schur algorithm. For more information see [4].

The Implicitly Restarted Arnoldi Process

The rate at which the Arnoldi process begins to give good approximations to eigenpairs depends very much upon the choice of starting vector q. In most cases we cannot hope to find a good q at the outset; commonly q is chosen at random, due to a lack of information. However, once we have taken a few Arnoldi steps, we have new information that we might be able to use to find a new vector $\hat{q}$ that would have been a better starting vector. Supposing we can find such a $\hat{q}$, might it be worthwhile to start a new Arnoldi process with $\hat{q}$ instead of continuing the old one? If so, might it be worthwhile to restart repeatedly, with a better starting vector each time?

Indeed, repeated restarts are worthwhile, and that is exactly what the implicitly restarted Arnoldi process (IRA) does. After a short Arnoldi run, it restarts with a new vector, does another short Arnoldi run, restarts again, and so on. Because of the nature of the process by which new starting vectors are chosen, it turns out not to be necessary to start each new run from scratch, rather it can be picked up in midstream, so to speak. For this reason the restarts are called *implicit*.

After a number of implicit restarts, the process terminates with a low-dimensional invariant subspace containing the desired eigenvectors. It is then a simple matter to extract those eigenvectors and the associated eigenvalues.

MATLAB has a function called `eigs`, which computes a few eigenvalues of a sparse matrix by IRA. See Exercise 6.5.27. A Fortran implementation of IRA called ARPACK [64] is available for free.

A major advantage of restarts is that less storage space is needed. If we restrict ourselves to short Arnoldi runs of length m, say, then we only have to store about m vectors. Each time we restart, we free up memory. Furthermore, since only a few vectors are kept, it is not too expensive to reorthogonalize them as necessary to ensure that they remain truly orthonormal to working precision.

Now that we have the basic idea of the process and have noted some potential advantages, an obvious question arises: What constitutes a good starting vector? To begin with an extreme example, suppose we are interested in a certain eigenvector. Then it would be nice to start the Arnoldi process with that eigenvector. If we are so lucky as to start with such a q, then the process will terminate in one step with the eigenvector q_1 and an associated eigenvalue. In many cases we are interested in more than one eigenpair. Suppose we are able to start with a q that is a linear combination of a small number of eigenvectors, say $q = c_1 v_1 + \cdots + c_k v_k$ with $k \ll n$. Then the Arnoldi process will terminate in k or fewer steps with (typically) span$\{q_1, \ldots, q_k\}$ equal to the invariant subspace span$\{v_1, \ldots, v_k\}$ (see Exercise 6.5.28).

This suggests that we try to build a q that is a linear combination of just a few desired eigenvectors. To keep the discussion simple, let us assume that A is a semisimple matrix with linearly independent eigenvectors $v_1, \ldots, v_n$ and associated eigenvalues $\lambda_1, \ldots, \lambda_n$. For definiteness, let us suppose we are looking for the k eigenvalues of largest modulus and their associated eigenvectors. Let us number the eigenvalues so that $|\lambda_1| \geq |\lambda_2| \geq \cdots \geq |\lambda_n|$ and assume that $|\lambda_k| > |\lambda_{k+1}|$. Then the desired eigenvectors are $v_1, \ldots, v_k$.

Since $v_1, \ldots, v_n$ form a basis for $\mathbb{C}^n$, we can certainly write, for any q,

$$q = c_1 v_1 + c_2 v_2 + \cdots + c_n v_n$$

for some uniquely determined but unknown $c_1, c_2, \ldots, c_n$. If we choose q at random, it is likely to have significant components in the directions of all eigenvectors. That is, none of the c_j will be zero or exceptionally close to zero. This is far from what we want. Since the desired eigenvectors are $v_1, \ldots, v_k$, we would like to have $c_{k+1} = c_{k+2} = \cdots = c_n = 0$. Our task, then, is to find a new vector

$$\hat{q} = \hat{c}_1 v_1 + \hat{c}_2 v_2 + \cdots + \hat{c}_n v_n,$$

in which $\hat{c}_1, \ldots, \hat{c}_k$ have been augmented and $\hat{c}_{k+1}, \ldots, \hat{c}_n$ have been diminished. Suppose we take $\hat{q} = p(A)q$, where p is some polynomial. Then (see Exercise 6.4.37)

$$\hat{q} = c_1 p(\lambda_1) v_1 + c_2 p(\lambda_2) v_2 + \cdots + c_n p(\lambda_n) v_n. \tag{6.5.1}$$

If we can choose p so that $p(\lambda_1), \ldots, p(\lambda_k)$ are large in comparison with $p(\lambda_{k+1})$, $\ldots, p(\lambda_n)$, we will have made progress. This is what the implicitly restarted Arnoldi process seeks to do.

We now describe an iteration of the implicitly restarted Arnoldi process (IRA). Suppose we want to find k eigenvalues and the associated invariant subspace. We pick a j that is comparable to k, and we make Arnoldi runs of length $m = k + j$. Some reasonable choices are $j = k, 2k, 4k$. Starting from the vector q, after m steps we have generated $Q_m = [\, q_1 \quad \cdots \quad q_m \,]$ with orthonormal columns and upper Hessenberg H_m such that

$$AQ_m = Q_m H_m + q_{m+1} h_{m+1,m} e_m^T, \qquad (6.5.2)$$

as in (6.4.13). Since m is small, we easily compute (by Francis's algorithm) the m eigenvalues of H_m. These are the Ritz values of A associated with the subspace $\mathcal{R}(Q_m) = \text{span}\{q_1, \ldots, q_m\}$. Let us call them $\mu_1, \ldots, \mu_m$ and order them so that $|\mu_1| \geq |\mu_2| \geq \cdots \geq |\mu_m|$. The largest ones, $\mu_1, \ldots, \mu_k$, are estimates of the k largest eigenvalues of A (the desired eigenvalues), while $\mu_{k+1}, \ldots, \mu_m$ approximate other (undesired) parts of the spectrum. Although these may be poor approximations, they at least give us a crude picture of the location of the spectrum of A.

IRA then performs j iterations of Francis's algorithm on H_m, using j shifts $\nu_1, \ldots, \nu_j$ in the region of spectrum that we want to suppress. The most popular choice is to take $\nu_1 = \mu_{k+1}, \nu_2 = \mu_{k+1}, \ldots, \nu_j = \mu_m$. If we make this choice, we are doing the *exact shift* variant of IRA. The Francis iterations are inexpensive because m is small. Their combined effect is a unitary similarity transformation

$$\hat{H}_m = V_m^{-1} H_m V_m, \qquad (6.5.3)$$

where

$$p(H_m) = V_m R_m, \qquad (6.5.4)$$

V_m is unitary, R_m is upper triangular, and p is a polynomial of degree j with zeros $\nu_1, \ldots, \nu_j$:

$$p(z) = (z - \nu_1)(z - \nu_2) \cdots (z - \nu_j),$$

as was shown in Theorem 6.3.16. The Francis iterations preserve upper Hessenberg form, so $\hat{H}_m$ is also upper Hessenberg. Let $\hat{Q}_m = Q_m V_m$, and in particular let $\hat{q}_1$ be the first column of $\hat{Q}_m$.

The next iteration of IRA consists of another Arnoldi run of m steps, starting from $\hat{q}_1$. However, as we have already indicated, we need not start the Arnoldi run from scratch. To see why this is so, multiply (6.5.2) by V_m on the right, and use (6.5.3) to obtain

$$A\hat{Q}_m = \hat{Q}_m \hat{H}_m + q_{m+1} h_{m+1,m} e_m^T V_m. \qquad (6.5.5)$$

Exercise 6.5.29 shows that the row vector $e_m^T V_m$ has exactly $m - j - 1$ leading zeros. Thus if we drop the last j entries from this vector, we obtain a vector of the form βe_k^T, where β is some nonzero scalar. Therefore, if we drop the last j columns from equation (6.5.5), we obtain $A\hat{Q}_k = \hat{Q}_{k+1} \hat{H}_{k+1,k} + q_{m+1} h_{m+1,m} \beta e_k^T$ or

$$A\hat{Q}_k = \hat{Q}_k \hat{H}_k + \breve{q}_{k+1} \breve{h}_{k+1,k} e_k^T + q_{m+1} h_{m+1,m} \beta e_k^T, \qquad (6.5.6)$$

where we have used $\check{q}_{k+1}$ to denote the $(k+1)$st column of $\hat{Q}_m$ and $\check{h}_{k+1,k}$ to denote the $(k+1, k)$ entry of $\hat{H}_{k+1,k}$. Define

$$\hat{q}_{k+1} = \gamma(\check{q}_{k+1}\check{h}_{k+1,k} + q_{m+1}h_{m+1,m}\beta), \tag{6.5.7}$$

where γ is a positive scalar chosen so that $\|\hat{q}_{k+1}\|_2 = 1$.

cise 6.5.8 Show that q_{m+1} is orthogonal to $\check{q}_{k+1}$. Use this fact to show that $\check{q}_{k+1}\check{h}_{k+1,k} + q_{m+1}h_{m+1,m}\beta$ is nonzero. Thus there exists a $\gamma > 0$ such that $\|\hat{q}_{k+1}\|_2 = 1$. Show that $\hat{q}_{k+1}$ is orthogonal to $\hat{q}_1, \ldots, \hat{q}_k$.   □

Let $\hat{h}_{k+1,k} = 1/\gamma$. Then (6.5.6) becomes

$$A\hat{Q}_k = \hat{Q}_k\hat{H}_k + \hat{q}_{k+1}\hat{h}_{k+1,k}e_k^T, \tag{6.5.9}$$

which is identical to (6.4.13), except for the hats on the symbols and the m having been replaced by k. Thus Proposition 6.4.15 implies that the columns of $\hat{Q}_k$ are exactly the vectors that would be built by the Arnoldi process, starting from $\hat{q}_1$.

Thus IRA need not start from scratch when building up an Arnoldi sequence from $\hat{q}_1$. Instead it can extract $\hat{Q}_k$ and $\hat{H}_k$ (submatrices of $\hat{Q}_m = Q_m V_m$ and $\hat{H}_m = V_m^{-1}H_m V_m$, respectively) and start with step k. Thus $k - 1$ Arnoldi steps are avoided.

IRA is summarized in the following algorithm.

Implicitly Restarted Arnoldi (IRA) Process. Given $A \in \mathbb{C}^{n \times n}$ and a nonzero starting vector $q \in \mathbb{R}^n$, this algorithm produces orthonormal $q_1, q_2, \ldots, q_k$ that span an invariant subspace. If an invariant subspace is not found in $imax$ iterations, the algorithm terminates with an error flag.

$$
\begin{array}{l}
q_1 \leftarrow q/\|q\|_2 \\
k_r \leftarrow 0 \\
\text{for } ii = 1, 2, \ldots, imax \\
\quad\left[\begin{array}{l}
\text{take } m - k_r \text{ Arnoldi steps (Algorithm 6.4.6)} \\
\quad\text{to produce } Q_m \text{ and } H_m, m = k + j. \\
\text{if } |h_{k+1,k}| < tol, \text{ exit } (\mathcal{R}(Q_k) \text{ is invariant under } A) \\
\text{find } j \text{ shifts } \nu_1, \ldots, \nu_j \\
\text{take } j \text{ steps of Francis's algorithm with shifts } \nu_1, \ldots, \nu_j \\
\quad\text{to obtain transformation matrix } V_m \\
H_m \leftarrow V_m^{-1}H_m V_m \\
H_k \leftarrow H_m(1:k, 1:k) \\
Q_k \leftarrow Q_m V_m(:, 1:k) \\
k_r \leftarrow k
\end{array}\right. \\
\text{set flag (failed to converge in } imax \text{ iterations)}
\end{array} \tag{6.5.10}
$$

In the interest of clarity, some details have been left out of (6.5.10). In particular, the convergence test should be more comprehensive; it should test all of the subdiagonal entries $h_{i+1,i}$ for $i = 1, \ldots, m - 1$. If any one of them is effectively zero, then the first i columns of Q_k span an invariant subspace of dimension i. If $i < k$, then the iterations should continue, but the first i columns of Q_k should be locked in for the remainder of the computation. This saves some work in the update of Q_k.

Moreover, one has the possibility of constructing $\hat{q}_{k+1}$ and $\hat{h}_{k+1,k} = 1/\gamma$ as in (6.5.7), thereby saving one Arnoldi step on the restart.

One other point deserves mention. It is often helpful to work with spaces whose dimension is slightly greater than the number of eigenvalues wanted. For example, if we want s eigenvalues, we might work with spaces of dimension $k = s + 2$.

Nothing has been said in (6.5.10) about how the shifts should be chosen. As we have indicated above, they could be taken to be the j smallest eigenvalues of H_m. This is a good choice if the largest eigenvalues are sought. If, on the other hand, the eigenvalues of A that are furthest to the right in the complex plane are wanted, one can take the shifts to be the j leftmost eigenvalues of H_m, for example. Recall that we always want to choose shifts that are in the region of the spectrum that we are trying to suppress.

Carrying this idea further, if one is looking for eigenvalues that are close to some target τ (not on the periphery of the spectrum), one might take as shifts those eigenvalues of H_m that are furthest from τ. However, this strategy works poorly because it attempts to enhance interior eigenvalues, which the Arnoldi process does not produce well, while trying to suppress peripheral eigenvalues, which the Arnoldi process is best at producing. A much more successful approach is to use the shift-and-invert strategy, that is, to apply IRA to the shifted, inverted operator $(A - \tau I)^{-1}$. This approach is feasible whenever we have the time and space to compute the LU decomposition of $(A - \tau I)$.

Why IRA Works

We stated above that an IRA iteration replaces the starting vector q by a new starting vector $\hat{q} = p(A)q$, where p is a polynomial chosen to suppress the unwanted eigenvectors and enhance the wanted ones. Now we will show that this is so. We begin with the Arnoldi configuration

$$AQ_m = Q_m H_m + q_{m+1}h_{m+1,m}e_m^T.$$

Clearly this equation continues to hold even after insertion of a shift:

$$(A - \nu_1 I)Q_m = Q_m(H_m - \nu_1 I) + E_1, \qquad (6.5.11)$$

where the remainder term $E_1 = q_{m+1}h_{m+1,m}e_m^T$ is identically zero except for the last column. Now consider applying a second shift. We can obtain an expression for $(A - \nu_2 I)(A - \nu_1 I)Q_m$ by multiplying (6.5.11) by $(A - \nu_2 I)$. The resulting equation contains an expression $(A - \nu_2 I)Q_m$, which can be eliminated by applying (6.5.11) with ν_1 replaced by ν_2. The resulting expression has two remainder terms (terms involving E_1), which can be combined to yield

$$(A - \nu_2 I)(A - \nu_1 I)Q_m = Q_m(H_m - \nu_2 I)(H_m - \nu_1 I) + E_2,$$

where $E_2 = (A - \nu_2 I)E_1 + E_1(H_m - \nu_1 I)$ is identically zero except in the last two columns. More generally, we have the following theorem, which is proved in Exercise 6.5.30.

Theorem 6.5.12 *Suppose*

$$AQ_m = Q_m H_m + q_{m+1} h_{m+1,m} e_m^T,$$

and let p be a polynomial of degree $j < m$. Then

$$p(A)Q_m = Q_m p(H_m) + E_j,$$

where where all but the last j columns of $E_j \in \mathbb{C}^{n \times m}$ are identically zero.

Now let us recall what IRA does with Q_m and H_m. It chooses shifts $\nu_1, \ldots, \nu_j$ from the region of the spectrum that is to be suppressed and does j steps of Francis's algorithm on H_m using $\nu_1, \ldots, \nu_j$ as shifts. This results in $\hat{H}_m = V_m^{-1} H_m V_m$, where V_m is the unitary factor in the QR decomposition $p(H_m) = V_m R_m$, as we recall from (6.5.3) and (6.5.4). Here $p(z) = (z - \nu_1) \cdots (z - \nu_j)$. The transforming matrix V_m is used to update Q_m to $\hat{Q}_m = Q_m V_m$. If we now apply Theorem 6.5.12, we obtain

$$p(A)Q_m = Q_m V_m R_m + E_j = \hat{Q}_m R_m + E_j. \tag{6.5.13}$$

Since we are interested in the first column, we multiply this equation by e_1 to get

$$p(A)Q_m e_1 = \hat{Q}_m R_m e_1 + E_j e_1. \tag{6.5.14}$$

The left-hand side of this equation is just $p(A)q_1$. On the right-hand side we have $R_m e_1 = \alpha e_1$, where $\alpha = r_{11} \neq 0$, because R_m is upper triangular. Therefore the first term on the right-hand side is just $\alpha \hat{q}_1$. The second term is zero because the first column of E_j is zero. Thus we have

$$\hat{q}_1 = \beta p(A) q_1, \tag{6.5.15}$$

where $\beta = \alpha^{-1}$. This is exactly what we desired. Since $p(z)$ is zero at $\nu_1, \ldots, \nu_j$, it follows that p will take on small values near these points and large values away from these points. Referring back to (6.5.1), we see that eigenvectors corresponding to eigenvalues near $\nu_1, \ldots, \nu_j$ will be suppressed, while those corresponding to eigenvalues away from $\nu_1, \ldots, \nu_j$ will be enhanced.

IRA chooses different shifts on each iteration, but imagine for a moment a version that does not change the shifts. After i iterations the starting vector will be proportional to $p(A)^i q_1$, as we see by applying (6.5.15) i times. Thus IRA is just effecting the power method driven by $p(A)$. The eigenvectors that correspond to the largest eigenvalues $p(\lambda)$ of $p(A)$ will be favored.

The power method is subspace iteration with subspaces of dimension 1. Theorem 6.5.16 (below) shows that IRA also effects subspace iterations on higher dimensional spaces. Thus the convergence theory of subspace iteration can be brought to bear in the analysis of the convergence of IRA. In practice the subspace iterations are nonstationary because different shifts are chosen at each step. This improves the algorithm by making it adaptive, hence flexible, but it makes the convergence analysis more difficult.

Theorem 6.5.16 *Suppose one iteration of IRA transforms $q_1, \ldots, q_k$ to $\hat{q}_1, \ldots, \hat{q}_k$. Let $p(z) = (z - \nu_1) \cdots (z - \nu_j)$, where $\nu_1, \ldots, \nu_j$ are the shifts used by IRA. Then*

$$\text{span}\{\hat{q}_1, \ldots, \hat{q}_i\} = p(A)\text{span}\{q_1, \ldots, q_i\}, \quad \text{for } i = 1, \ldots, k.$$

Proof. In the discussion following Proposition 6.3.6 we pointed out that any time we are doing power iterations, we are automatically doing subspace iterations on the associated Krylov subspaces. That is what is happening here. By Theorem 6.4.9, $\text{span}\{q_1, \ldots, q_i\} = \mathcal{K}_i(A, q_1)$ and $\text{span}\{\hat{q}_1, \ldots, \hat{q}_i\} = \mathcal{K}_i(A, \hat{q}_1)$. Furthermore (6.5.15) implies that $\mathcal{K}_i(A, \hat{q}_1) = p(A)\mathcal{K}_i(A, q_1)$ by Proposition 6.3.6. $\quad\square$

Exercise 6.5.17 Provide a direct proof of Theorem 6.5.16 that uses (6.5.14), exploiting the special form of the error term E_j. $\quad\square$

The Krylov-Schur Algorithm

The Krylov-Schur algorithm [86] and its symmetric variant, the thick restart Lanczos process [107], are methods similar to IRA that were proposed during the decade after its introduction. It turns out that these methods are equivalent to the exact-shift variant of IRA, as is shown in [101, § 9.3], for example. So, in a sense, Krylov-Schur is nothing new. However, it does have certain organizational advantages.

The Krylov-Schur algorithm begins with m steps of the Arnoldi process to build vectors satisfying an Arnoldi configuration

$$AQ_m = Q_m H_m + q_{m+1}h_{m+1,m}e_m^T. \tag{6.5.18}$$

Then, just as in IRA, the most promising k-dimensional subspace is kept, j vectors are discarded ($k + j = m$), and the Arnold process resumes at step k. However, the details are different.

Using Francis's algorithm a Schur or real Schur (Wintner-Murnaghan) decomposition

$$H_m = VTV^*$$

can be computed, such that

$$T = \begin{bmatrix} T_{11} & T_{12} \\ 0 & T_{22} \end{bmatrix},$$

T_{11} is $k \times k$ and contains the Ritz values (eigenvalues of T) that we wish to retain, and T_{22} is $j \times j$ and contains the Ritz values that we wish to discard. For example, if we want the k eigenvalues of greatest magnitude, the k Ritz values with greatest magnitude should be retained in T_{11}. Getting the desired separation of the eigenvalues may require some swapping, as discussed in the second part of Section 6.1.

Let $\tilde{Q}_m = Q_m V$ and $\tilde{z} = h_{m+1,m}e_m^T V$. Multiply (6.5.18) by V on the right to obtain

$$A\tilde{Q}_m = \tilde{Q}_m T + q_{m+1}\tilde{z}^T. \tag{6.5.19}$$

We will keep the first k columns of $\tilde{Q}_m$ and discard the rest. If we just keep the first k columns of (6.5.19) and take into account the block-triangular structure of T, we obtain

$$A\tilde{Q}_k = \tilde{Q}_k T_{11} + \hat{q}_{k+1}z^T, \tag{6.5.20}$$

where $\tilde{Q}_k$ denotes $n \times k$ matrix consisting of the first k columns of $\tilde{Q}_m$, $\hat{q}_{k+1} = q_{m+1}$, and z^T is the row vector consisting of the first k entries of $\tilde{z}^T$. We have kept T_{11} and discarded the rest.

At this point there is an opportunity to check for convergence and lock in any invariant subspaces that have converged, but we will postpone that discussion for now. Supposing that we have not locked in an invariant subspace, we must transform (6.5.20) to an Arnoldi configuration like (6.5.9) so that we can proceed with a new run of Arnoldi steps.[34] We begin by building a reflector W that creates zeros in z^T: $z^T W = \alpha e_k^T$ for some α. Applying W to (6.5.20) on the right, we obtain

$$A(\tilde{Q}_k W) = (\tilde{Q}_k W)(W^* T_{11} W) + \hat{q}_{k+1}\alpha e_k^T. \tag{6.5.21}$$

We then transform $W^* T_{11} W$ to upper Hessenberg form by a process similar to the one described in Section 5.5, except that it creates zeros row by row from bottom to top. (Recall that the algorithm of Section 5.5 creates zeros column by column from left to right). Let X be the unitary transforming matrix that effects this reduction. That is $X^*(W^* T_{11} W)X = \hat{H}_k$, where $\hat{H}_k$ is upper Hessenberg. Let $\hat{Q}_k = \tilde{Q}_k W X$. Then, multiplying 6.5.21 on the right by X, we obtain

$$A\hat{Q}_k = \hat{Q}_k \hat{H}_k + \hat{q}_{k+1}\hat{h}_{k+1,k}e_k^T, \tag{6.5.22}$$

where $\hat{h}_{k+1,k} = \alpha$. The remainder term is unaffected by this transformation because $e_k^T X = e_k^T$.

rcise 6.5.23 Sketch an algorithm that transforms a matrix $M \in \mathbb{C}^{k \times k}$ to upper Hessenberg form row by row from bottom to top. Verify that if $X \in \mathbb{C}^{k \times k}$ is the transforming matrix constructed by this algorithm, then $e_k^T X = e_k^T$. $\quad\square$

Equation (6.5.22) is an Arnoldi configuration. From here we can do j Arnoldi steps to build up to a new Arnoldi configuration of dimension $m = k + j$, like (6.5.18), then do another restart, and so on.

Checking for Convergence

One of the strengths of the Krylov-Schur algorithm is the ease with which converged invariant subspaces can be locked in. As we mentioned above, the convergence

---

[34]Actually it is possible to defer this transformation, as is done in both [86] and [107], but there seems to be no advantage in putting it off.

check takes place at (6.5.20). The matrix T_{11} is either upper triangular or quasi-triangular. To keep the discussion simple, let us suppose at first that it is triangular. Suppose the first i entries of z are zero (or small enough to be set to zero), so that $z^T = \begin{bmatrix} 0^T & z_2^T \end{bmatrix}$, where $z_2 \in \mathbb{C}^{k-i}$. Partition T_{11} and $\tilde{Q}_k$ in conformity with this partition of z^T, i.e.

$$T_{11} = \begin{bmatrix} S_{11} & S_{12} \\ 0 & S_{22} \end{bmatrix}, \tag{6.5.24}$$

where $S_{11} \in \mathbb{C}^{i \times i}$, and

$$\tilde{Q}_k = \begin{bmatrix} \tilde{Q}_i & \check{Q}_i \end{bmatrix},$$

where $\tilde{Q}_i \in \mathbb{C}^{n \times i}$. Substituting these forms into (6.5.20) and retaining only the first i columns, we have

$$A\tilde{Q}_i = \tilde{Q}_i S_{11}.$$

This shows that the columns of $\tilde{Q}_i$ span an invariant subspace of A, and the eigenvalues of S_{11} are eigenvalues of A associated with this subspace. These can now be locked in; that is, they can be held fixed for the remainder of the computation.[35]

In the case where T_{11} is not triangular but quasi-triangular, we can do the same thing. We just have to respect the block structure; in particular our choice of i must be one that does not split a 2×2 block. In other words, the partition (6.5.24) has to be valid.

Once we are done checking for convergence, we continue the algorithm by transforming (6.5.20) back to an Arnoldi configuration as described above, but now the transformation is simplified. Writing (6.5.20) in partitioned form, we have

$$A \begin{bmatrix} \tilde{Q}_i & \check{Q}_i \end{bmatrix} = \begin{bmatrix} \tilde{Q}_i & \check{Q}_i \end{bmatrix} \begin{bmatrix} S_{11} & S_{12} \\ 0 & S_{22} \end{bmatrix} + \hat{q}_{k+1} \begin{bmatrix} 0 & z_2^T \end{bmatrix}. \tag{6.5.25}$$

The transformation of z^T to αe_k^T only needs to act on the last $k - i$ columns. Let $W \in \mathbb{C}^{(k-i) \times (k-i)}$ be a reflector such that $z_2^T W = \alpha e_{k-i}^T$. Multiplying (6.5.25) on the right by $\operatorname{diag}\{I_i, W\}$, we obtain

$$A \begin{bmatrix} \tilde{Q}_i & \check{Q}_i W \end{bmatrix} = \begin{bmatrix} \tilde{Q}_i & \check{Q}_i W \end{bmatrix} \begin{bmatrix} S_{11} & S_{12}W \\ 0 & W^* S_{22} W \end{bmatrix} + \hat{q}_{k+1} \alpha e_k^T. \tag{6.5.26}$$

S_{11} and $\tilde{Q}_i$ are untouched by this transformation. Only $W^* S_{22} W$ needs to be transformed to upper Hessenberg form (row by row from bottom to top). This also leaves S_{11} and $\tilde{Q}_i$ untouched. These remain unchanged (locked in) from this point on.

---

[35]The convergence check can be enhanced by eigenvalue swapping, as described in the second half of Section 6.1. Suppose we have $z^T = \begin{bmatrix} 0 & \cdots & 0 & z_{i+1} & \cdots & z_n \end{bmatrix}$, where z_{i+1} is not small enough to be set to zero. Then we can swap the eigenvalue $t_{i+1,i+1}$ to the bottom of T_{11}. This gives a new z with a new z_{i+1}, which might be small enough to be set to zero. If it is, we set it to zero and go to work on z_{i+2}. If not, we swap the new $t_{i+1,i+1}$ (almost) to the bottom and get a new z_{i+1} to examine. Continuing in this way, we can examine all of the Ritz values in T_{11} for possible convergence.

Additional Exercises

rcise 6.5.27 MATLAB's built-in `eigs` function computes a few eigenvalues of a sparse (or full) matrix by the implicitly-restarted Arnoldi (IRA) method. In MATLAB type `help eigs` to learn how the `eigs` command works. Use `eigs` to calculate the ten eigenvalues of `west0479` closest to $\tau = 6 + 0i$. Use a residual tolerance of 10^{-16}. How many iterations did `eigs` take? Did this method take more or less work than straight shift-and-invert Arnoldi? □

rcise 6.5.28 Let $v_1, \ldots, v_k$ be eigenvectors of A, and suppose $q = c_1 v_1 + \cdots + c_k v_k$.

(a) Show that for $j = 1, 2, 3, \ldots$, we have $A^j q \in \text{span}\{v_1, \ldots, v_k\}$. Deduce that the Krylov subspaces $\mathcal{K}_m(A, q)$ lie in $\text{span}\{v_1, \ldots, v_k\}$ for all m.

(b) Show that $\mathcal{K}_k(A, q)$ is invariant under A. Deduce that the Arnoldi process starting with q will terminate in k steps or fewer with $\text{span}\{q_1, \ldots, q_m\}$ ($m \leq k$) an invariant subspace contained in $\text{span}\{v_1, \ldots, v_k\}$.

(c) Let $\lambda_1, \ldots, \lambda_k$ be the eigenvalues of A associated with $v_1, \ldots, v_k$, respectively. Show that if the λ_j are all distinct and $c_j \neq 0$ for all j, then q, Aq, A^2q, ..., $A^{k-1}q$ are linearly independent, and the Arnoldi process terminates in exactly k steps with $\text{span}\{q_1, \ldots, q_k\} = \mathcal{K}_k(A, q) = \text{span}\{v_1, \ldots, v_k\}$.

(d) Suppose some of the eigenvalues $\lambda_1, \ldots, \lambda_k$ are equal. Show that if there are only s distinct eigenvalues, $s < k$, among $\lambda_1, \ldots, \lambda_k$, then q can be written as a linear combination of s or fewer eigenvectors, and the Arnoldi process will terminate in s steps or fewer.

□

rcise 6.5.29 Let j be a non-negative integer. A matrix B is called j-*Hessenberg* if $b_{ik} = 0$ whenever $i - k > j$. A j-Hessenberg matrix is said to be *properly* j-Hessenberg if $b_{ij} \neq 0$ whenever $i - k = j$.

(a) What are the common names for 0-Hessenberg and 1-Hessenberg matrices? Give two examples of 2-Hessenberg matrices of dimension 5, one properly 2-Hessenberg and one not.

(b) Suppose $B \in \mathbb{C}^{n \times n}$ is properly j-Hessenberg, $C \in \mathbb{C}^{n \times n}$ is properly k-Hessenberg, and $j + k < m$. Show that BC is properly $(j + k)$-Hessenberg.

(c) Show that if $B \in \mathbb{C}^{m \times m}$ is properly j-Hessenberg ($j < m$), then the first $m - j$ columns of B are linearly independent.

(d) Show that if H_m is properly upper Hessenberg and p is a polynomial of degree j, say $p(z) = (z - \nu_1)(z - \nu_2) \cdots (z - \nu_j)$, with $j < m$, then $p(H_m)$ is properly j-Hessenberg.

(e) Partition the QR decomposition $p(H_m) = V_m R_m$ of (6.5.4) as follows:

$$\begin{bmatrix} B_1 & B_2 \end{bmatrix} = \begin{bmatrix} V_1 & V_2 \end{bmatrix} \begin{bmatrix} R_{11} & R_{12} \\ 0 & R_{22} \end{bmatrix},$$

where $B_1 \in \mathbb{C}^{m \times (m-j)}$, $V_1 \in \mathbb{C}^{m \times (m-j)}$, and $R_{11} \in \mathbb{C}^{(m-j) \times (m-j)}$. From parts (c) and (d), deduce that the columns of B_1 are linearly independent. Show that this implies that R_{11} is nonsingular, and $V_1 = B_1 R_{11}^{-1}$.

(f) Deduce from part (e) that V_m is a properly j-Hessenberg matrix. Conclude that the first $m - j - 1$ entries of the row vector $e_m^T V_m$ are zero, and the $m - j$th entry (which we called β in (6.5.6)) is nonzero.

$\square$

Exercise 6.5.30 Prove Theorem 6.5.12 by induction on j:

(a) Show that the theorem holds when $j = 1$. In this case $p(z) = \alpha_1(z - \nu_1)$, where α_1 is a nonzero constant.

(b) Show that the theorem holds when $j = 2$. In this case $p(z) = \alpha_2(z - \nu_1)(z - \nu_2)$. This step is just for practice; it is not crucial to the proof of the Theorem.

(c) Show that if the theorem holds for polynomials of degree $j - 1$, then it holds for polynomials of degree j.

$\square$

6.6 THE JACOBI-DAVIDSON AND RELATED ALGORITHMS

The Arnoldi process builds up subspaces of increasing dimension by adding a vector at each step. This is done in such a way that at each step the space is a Krylov subspace: $\operatorname{span}\{q_1, \ldots, q_k\} = \mathcal{K}_k(A, q_1)$. Now we will consider methods that increase the subspace dimension by one at each step but do not use Krylov subspaces.

Suppose we have orthonormal vectors $q_1, \ldots, q_k$, and we wish to add q_{k+1}. We want to pick this vector so that the expanded space is an improvement on the current space, in the sense that it contains better approximations to eigenvectors. For guidance we look to the space that we already have in hand. Let $Q_k = \begin{bmatrix} q_1 & \cdots & q_k \end{bmatrix}$, as before, and let $B_k = Q_k^* A Q_k$. In the Arnoldi process this is just the Hessenberg matrix H_k, but now B_k need not be upper Hessenberg. Since k is not too big, we can compute its eigenvalues and eigenvectors easily. Let (μ, x) be some selected eigenpair (for example, take the largest μ), and let $q = Q_k x$. Then (μ, q) is a Ritz pair of A associated with the subspace $\operatorname{span}\{q_1, \ldots, q_k\}$ (see Exercise 6.4.35), and it can be viewed as an approximation to an eigenpair of A. The norm of the residual $r = Aq - \mu q$ gives an indication of the quality of (μ, q) as an approximate eigenpair.

Several important methods make use of the residual to determine the new subspace vector q_{k+1}. In each case r is used to determine a second vector s such that

$s \notin \text{span}\{q_1, \ldots, q_k\}$. Then q_{k+1} is obtained by orthonormalizing s against q_1, $\ldots, q_k$ by the Gram-Schmidt process. The methods differ in how they obtain s from r.

The simplest possibility is to take $s = r$. This leads to a method that is equivalent to the Arnoldi process. See Exercise 6.6.6.

A second possibility is to take $s = (D - \mu I)^{-1} r$, where D is the diagonal matrix that has the same main-diagonal entries as A. This leads to *Davidson's method*, which has been used extensively in quantum chemistry calculations. In these applications the matrices are symmetric and extremely large. They are also strongly diagonally dominant, which means that the main-diagonal entries are much larger than the entries off the main diagonal. This property is crucial to the success of Davidson's method. Notice that the computation of s is quite inexpensive, as $(D - \mu I)^{-1}$ is a diagonal matrix.

A third way of choosing s leads to the *Jacobi-Davidson* method, which we motivate as follows. If q is close to an eigenvector of A, then a small correction $\tilde{s}$ can make $q + \tilde{s}$ an exact eigenvector. Thus

$$A(q + \tilde{s}) = (\mu + \tilde{\nu})(q + \tilde{s}), \qquad (6.6.1)$$

where $\tilde{\nu}$ is a small correction to the Ritz value μ. Furthermore, we may take the correction to be orthogonal to q, i.e. $q^* \tilde{s} = 0$. The Jacobi-Davidson method chooses s to be an approximation to $\tilde{s}$. Suppose $\|\tilde{s}\| = O(\epsilon)$ and $|\tilde{\nu}| = O(\epsilon)$, where $\epsilon \ll 1$. Expanding (6.6.1) we obtain $Aq + A\tilde{s} = \mu q + \tilde{\nu} q + \mu \tilde{s} + \tilde{\nu} \tilde{s}$, or

$$(A - \mu I)\tilde{s} - q\tilde{\nu} = -r + \tilde{\nu} \tilde{s}.$$

We expect the term $\tilde{\nu} \tilde{s}$ to have a negligible effect on this equation, because $\|\tilde{\nu} \tilde{s}\| = O(\epsilon^2)$. The Jacobi-Davidson method takes s and ν to be the solution to the equation obtained by ignoring this term while enforcing the orthogonality condition $q^* s = 0$. Thus

$$(A - \mu I)s - q\nu = -r, \qquad q^* s = 0,$$

or

$$\begin{bmatrix} A - \mu I & q \\ q^* & 0 \end{bmatrix} \begin{bmatrix} s \\ -\nu \end{bmatrix} = \begin{bmatrix} -r \\ 0 \end{bmatrix}. \qquad (6.6.2)$$

The reader might recognize this as Newton's method.[36] It is also a variant of Rayleigh quotient iteration (Exercise 6.6.7).

To summarize, the Jacobi-Davidson method obtains a Ritz pair (μ, q) from $\text{span}\{q_1, \ldots, q_k\}$, computes the residual $r = Aq - \mu q$, then solves (6.6.2) to obtain s (and ν). The new vector q_{k+1} is obtained by orthogonalizing s against $q_1, \ldots, q_k$ by some version of Gram Schmidt. The new space $\text{span}\{q_1, \ldots, q_{k+1}\}$ contains both q and s, so it also contains $q + s$, an excellent approximation to an eigenvector.

What we have just described is known as the "exact" variant of Jacobi-Davidson, since the equation (6.6.2) is solved exactly. In practice, if A is extremely large, it will

---

[36] Newton's method solves a linear system obtained by ignoring the (hopefully) small nonlinear terms.

be impractical to solve (6.6.2). Therefore the Jacobi-Davidson process is normally used in an "inexact" mode, which is to say that (6.6.2) is solved approximately by an iterative method such as those discussed in Section 8.10. To this end (6.6.2) is usually reformulated as follows. Assuming without loss of generality that $\| q \|_2 = 1$, let $P = I - qq^*$. Then P is the orthoprojector onto the space $\mathcal{S}$ consisting of all vectors orthogonal to q.

Exercise 6.6.3 Let $P = I - qq^*$, where $\| q \|_2 = 1$. Let $\mathcal{S}$ be the set of vectors in $\mathbb{C}^n$ that are orthogonal to q.

(a) Verify that $P^2 = P$ and $P = P^*$. Thus P is an orthoprojector.

(b) Show that $Ps = s$ if and only if $s \in \mathcal{S}$. In particular, we know from Exercise 6.4.35 that $r \in \mathcal{S}$, where $r = Aq - \mu q$ and (μ, q) is a Ritz pair of A. Thus $Pr = r$.

(c) Show that $Pv = 0$ if and only if v is a multiple of q.

$\square$

A solution of (6.6.2) must satisfy $(A - \mu I)s - qv = -r$ and, since $Pq = 0$ and $Pr = r$,

$$P(A - \mu I)s = -r. \tag{6.6.4}$$

We must also enforce the condition $s \perp q$. which is equivalent to the condition $Ps = s$. Because of this, equation (6.6.4) is often written as

$$P(A - \mu I)Ps = -r.$$

This is the form of the Jacobi-Davidson equation that is usually seen in the literature. It is this equation that is normally solved approximately, while strictly enforcing the condition $s \perp q$.

An accurate solution is not needed; a very crude approximation is often good enough. The only penalty for using a poor approximation is that the vector that is added to the space is of reduced quality. This means that more steps will need to be taken before a sufficiently accurate approximate eigenpair is obtained. We do not mind taking many steps if the cost of each step is not too great. For more information see [4] and [79].

At each step of the Jacobi-Davidson procedure, we have k Ritz pairs to choose from, where k is the current dimension of the space. Which pair we actually choose for our next step depends upon our objective. If we wish to find the largest eigenvalues of A, we should choose the largest Ritz value. If, on the other hand, we wish to find interior eigenvalues near the target value τ, it makes sense to choose the Ritz value that is closest to τ. However, in this context it is usually better to work with *harmonic* Ritz values instead of the standard Ritz values. See Exercise 6.6.8.

Each step of the Davidson or Jacobi-Davidson algorithm adds a new vector. After many steps we may wish to dispose of the less promising vectors and keep only a small subspace that contains the best estimates of eigenvectors, as is done in IRA.

This is easier in the current context than it is in IRA; the subspaces that we are building are not Krylov subspaces, so we do not need to worry about preserving that property. We can proceed as follows, for example. Suppose we have $m = k + j$ orthonormal vectors, the columns of the matrix $Q \in \mathbb{C}^{n \times m}$, and we want to discard j columns and keep a k-dimensional subspace. Let $B = Q^*AQ$. Compute the Schur decomposition $B = UTU^*$. U is unitary, and T is upper triangular. The main diagonal entries of T are the eigenvalues of B, which are Ritz values of A with respect to the current subspace. These can be made to appear in any order in T by eigenvalue swapping. Suppose we order them so that

$$T = \begin{bmatrix} T_{11} & T_{12} \\ 0 & T_{22} \end{bmatrix},$$

where $T_{11} \in \mathbb{C}^{k \times k}$ contains the k "most promising" Ritz values, the ones that we want to keep. Let $\hat{Q} = QU$, and partition $\hat{Q}$ conformably with T, i.e. $\hat{Q} = \begin{bmatrix} \hat{Q}_1 & \hat{Q}_2 \end{bmatrix}$, where $\hat{Q}_1 \in \mathbb{C}^{n \times k}$. Then $T_{11} = \hat{Q}_1^* A \hat{Q}_1$, which implies that the eigenvalues of T_{11} are the Ritz values of A with respect to the space $\mathcal{R}(\hat{Q}_1)$. If we now keep $\hat{Q}_1$ and discard $\hat{Q}_2$, we will have retained the desired Ritz values and a space that contains their associated Ritz vectors. This process is known as *purging*.

rcise 6.6.5 Check the claims of the previous paragraph. In particular, show that $T_{11} = \hat{Q}_1^* A \hat{Q}_1$. □

In situations where more than one eigenvalue is sought, it will often happen that some eigenvalues converge before the others do. Then we have an invariant subspace that we wish to preserve while looking for a larger invariant subspace. It is a simple matter to lock in the converged space. We simply move the converged vectors to the front of our basis (if they are not already there) and keep them there.

Additional Exercises

rcise 6.6.6 Let v be a Ritz vector from span$\{q_1, \ldots, q_k\}$. Then $v = c_1 q_1 + \cdots + c_k q_k$ for some constants $c_1, \ldots, c_k$. Typically $c_k \neq 0$. Consider a method that expands the subspace at each step by picking a Ritz pair (μ, v) at each step, computing the residual $r = Av - \mu v$, and orthonormalizing the residual against the previously determined vectors. Using induction on k, show that if $c_k \neq 0$ at each step, then span$\{q_1, \ldots, q_k\} = \mathcal{K}_k(A, q_1)$ for all k. Thus this method is equivalent to the Arnoldi process. What happens if $c_k = 0$ at some step? □

rcise 6.6.7 Let (μ, q) be a Ritz pair of A from some subspace. Show that $\mu = q^* Aq / q^* q$. Thus μ is the Rayleigh quotient of q. Show that if s is determined by the Jacobi-Davidson equation (6.6.2), then $q + s = \nu(A - \mu I)^{-1} q$. Therefore $q + s$ is the result of one step of Rayleigh quotient iteration with starting vector q. □

rcise 6.6.8 In this exercise we introduce *harmonic* Ritz pairs. We use the same notation as in Exercise 6.4.35. Let τ be a complex (target) value that is not an eigenvalue of

A. Then μ is a *harmonic Ritz value* of A with target τ with respect to the space $\mathcal{S}$ if $(\mu - \tau)^{-1}$ is an ordinary Ritz value of $(A - \tau I)^{-1}$. From Exercise 6.4.35 we see that μ satisfies this property if and only if $(\mu - \tau)^{-1}$ is an eigenvalue of $Q^*(A - \tau I)^{-1}Q$. Although this is a small ($k \times k$) matrix, it can be difficult to evaluate if A is large. It turns out to be easier to obtain Harmonic Ritz values with respect to a different space. Let $\mathcal{U} = (A - \tau I)\mathcal{S}$.

(a) Let $u \in \mathbb{C}^n$. Show that $u \in \mathcal{U}$ if and only if there is an $x \in \mathbb{C}^k$ such that $u = (A - \tau I)Qx$. Show that the vector x corresponding to a fixed $u \in \mathcal{U}$ is uniquely determined.

(b) Show that μ is a harmonic Ritz value of A with target τ with respect to the space $\mathcal{U}$ if and only if there is a $u \in \mathcal{U}$ such that

$$(A - \tau I)^{-1}u - (\mu - \tau)^{-1}u \perp \mathcal{U},$$

and that this holds in turn if and only if there is an $x \in \mathbb{C}^k$ such that

$$(A - \mu I)Qx \perp \mathcal{U}.$$

Let $v = Qx$. We call v a *harmonic Ritz vector* associated with the harmonic Ritz value μ. Notice that v belongs to $\mathcal{S}$, not $\mathcal{U}$.

(c) Let $Y = (A - \tau I)Q \in \mathbb{C}^{n \times k}$. Show that μ is a harmonic Ritz value (with τ and $\mathcal{U}$), with associated harmonic Ritz vector $v = Qx$ if and only if

$$Y^*Yx = (\mu - \tau)Y^*Qx.$$

This equation is an example of a *generalized* eigenvalue problem (see Section 7.4). Since the matrices Y^*Y and Y^*Q are small ($k \times k$), this problem can be solved inexpensively.

(d) Show that μ is a harmonic Ritz value (with τ and $\mathcal{U}$) if and only if $\mu - \tau$ is an eigenvalue of $(Y^*Y)(Y^*Q)^{-1}$ and $(Y^*Q)^{-1}(Y^*Y)$, assuming that $(Y^*Q)^{-1}$ exists.

$\square$

CHAPTER 7

EIGENVALUES AND EIGENVECTORS III

This chapter discusses four important topics that could have been included in Chapter 5, except that that chapter is already long enough. Section 7.1 discusses the sensitivity of eigenvalues and eigenvectors, and introduces condition numbers for these objects. Section 7.2 surveys a variety of methods that can be used as alternatives to Francis's algorithm on symmetric matrices. Some of these methods are superior to Francis's algorithm for symmetric, tridiagonal eigenvalue problems. Section 7.3 discusses the question of computing the eigenvalues of a product AC, where A and C are given separately. Finally, Section 7.4 provides a brief introduction to the *generalized eigenvalue problem* $Av = \lambda Bv$ and algorithms for solving it.

The sections are mostly independent and can be read in any order, except that Section 7.3 is a prerequisite for the last part of Section 7.4.

7.1 SENSITIVITY OF EIGENVALUES AND EIGENVECTORS

Since the matrix A whose eigensystem we wish to calculate is never known exactly, it is important to study how the eigenvalues and eigenvectors are affected by perturbations of A. Thus in this section we will ask the question of how close the eigenvalues and eigenvectors of $A + \delta A$ are to those of A if $\| \delta A \| / \| A \|$ is small. This would

Fundamentals of Matrix Computations, Third Edition. By David S. Watkins
Copyright © 2010 John Wiley & Sons, Inc.

be an important question even if the uncertainty in A were our only concern, but of course there is also another reason for asking it. We noted in Chapter 3 that any algorithm that transforms a matrix by rotators or reflectors constructed as prescribed in Section 3.2 is backward stable. Francis's algorithm is of this type. This means that Francis's algorithm determines the exact eigenvalues of a matrix $A + \delta A$ where $\| \delta A \| / \| A \|$ is tiny. If we can show that the eigenvalues of $A + \delta A$ are close to those of A, we will know that our answers are accurate. Of course it will turn out that we cannot always guarantee accurate eigenvalues; the accuracy depends upon certain condition numbers.

A related question is that of residuals. Suppose we have calculated an approximate eigenvalue λ and associated eigenvector v, and we wish to know whether they are accurate. It is natural to calculate the residual $r = Av - \lambda v$ and check whether it is small. Suppose $\| r \|$ *is* small. Does this guarantee that λ and v are accurate? As the following theorem shows, this question also reduces to that of the sensitivity of A.

Theorem 7.1.1 *Let $A \in \mathbb{C}^{n \times n}$, let v be an approximate eigenvector of A with $\| v \|_2 = 1$, λ an associated approximate eigenvalue, and r the residual: $r = Av - \lambda v$. Then λ and v are an exact eigenpair of some perturbed matrix $A + \delta A$, where $\| \delta A \|_2 = \| r \|_2$.*

Proof. If you have worked Exercise 6.4.36, then you have already done this. Let $\delta A = -rv^*$. Then $\| \delta A \|_2 = \| r \|_2 \| v \|_2 = \| r \|_2$, and $(A + \delta A)v = Av - rv^*v = Av - r = \lambda v$. $\qquad\qquad\square$

Exercise 7.1.2   Verify that $\| rv^* \|_2 = \| r \|_2 \| v \|_2$ for all r, $v \in \mathbb{C}^n$. (On the left-hand side of this equation we have the matrix 2-norm; on the right-hand side we have the vector 2-norm, the Euclidean norm.) This is the complex analogue of Exercise 2.1.31. $\square$

Sensitivity of Eigenvalues

First of all, the eigenvalues of a matrix depend continuously on the entries of the matrix. This is so because the coefficients of the characteristic polynomial are continuous functions of the matrix entries, and the zeros of the characteristic polynomial, that is, the eigenvalues, depend continuously on the coefficients. Various approaches to proving this fact are discussed in [15], [93], [101], [103] and the references cited therein. This means that we can put as small a bound as we please on how far the eigenvalues can wander, simply by making the perturbation of the matrix sufficiently small. But this information is too vague. It would be more useful to have a number κ such that if we perturb the matrix by ϵ, then the eigenvalues are perturbed by at most $\kappa \epsilon$, at least for sufficiently small ϵ. Then κ would give an upper bound for how badly the eigenvalues can behave and would serve as a sort of overall condition number for the spectrum. It turns out that we can get such a κ if the matrix is semisimple. This is the content of our next theorem.

Theorem 7.1.3 *(Bauer-Fike) Let $A \in \mathbb{C}^{n \times n}$ be a semisimple matrix, and suppose $V^{-1}AV = D$, where V is nonsingular and D is diagonal. Let $\delta A \in \mathbb{C}^{n \times n}$ be some* *perturbation of A, and let μ be an eigenvalue of $A + \delta A$. Then A has an eigenvalue λ such that*

$$|\mu - \lambda| \leq \kappa_p(V)\|\delta A\|_p \qquad (7.1.4)$$

for $1 \leq p \leq \infty$.

This theorem shows that $\kappa_p(V)$ (which was defined in Section 2.2) is an overall condition number for the spectrum of A.[37] We could nondimensionalize (7.1.4) by dividing both sides by $\|A\|_p$, but that would be an unnecessary complication. **Proof.** Let $\delta D = V^{-1}(\delta A)V$. Then

$$\|\delta D\|_p \leq \|V^{-1}\|_p\|\delta A\|_p\|V\|_p = \kappa_p(V)\|\delta A\|_p. \qquad (7.1.5)$$

Since $D + \delta D$ is similar to $A + \delta A$, μ is an eigenvalue of $D + \delta D$. Let x be an associated eigenvector. If μ happens to be an eigenvalue of A, we are done, so suppose it is not. Then $\mu I - D$ is nonsingular, and the equation $(D + \delta D)x = \mu x$ can be rewritten as $x = (\mu I - D)^{-1}(\delta D)x$. Thus $\|x\|_p \leq \|(\mu I - D)^{-1}\|_p\|\delta D\|_p\|x\|_p$. Cancelling out the positive factor $\|x\|_p$ and rearranging, we find that

$$\|(\mu I - D)^{-1}\|_p^{-1} \leq \|\delta D\|_p. \qquad (7.1.6)$$

The matrix $(\mu I - D)^{-1}$ is diagonal with main diagonal entries $(\mu - \lambda_i)^{-1}$, where $\lambda_1, \ldots, \lambda_n$ are the eigenvalues of A. It follows very easily that $\|(\mu I - D)^{-1}\|_2 = |\mu - \lambda|^{-1}$, where λ is the eigenvalue of A that is closest to μ. Thus (7.1.6) can be rewritten as

$$|\mu - \lambda| \leq \|\delta D\|_p.$$

Combining this inequality with (7.1.5), we get our result. $\qquad\qquad\square$

cise 7.1.7 Verify the following facts, which were used in the proof of Theorem 7.1.3

(a) Let Δ be a diagonal matrix with main diagonal entries $\delta_1, \ldots, \delta_n$. Verify that for $1 \leq p \leq \infty$, $\|\Delta\|_p = \max_{1 \leq i \leq n} |\delta_i|$.

(b) Verify that $\|(\mu I - D)^{-1}\|_p = |\mu - \lambda|^{-1}$, where λ is the eigenvalue of the diagonal matrix D that is closest to μ.

$\qquad\qquad\square$

The columns of the transforming matrix V are eigenvectors of A, so the condition number $\kappa_p(V)$ is a measure of how close to being linearly dependent the eigenvectors are: the larger the condition number, the closer they are to being dependent. From this viewpoint it would seem reasonable to assign an overall condition number of infinity to the eigenvalues of a defective matrix, since such matrices do not even have n linearly independent eigenvectors. Support for this viewpoint is given in Exercise 7.1.32.

---

[37]To be precise, it is an *upper bound* for a condition number.

If A is Hermitian or even normal, V can be taken to be unitary (cf. Theorems 5.4.12 and 5.4.16). Unitary V satisfy $\kappa_2(V) = 1$, so the following corollary holds.

Corollary 7.1.8 *Let $A \in \mathbb{C}^{n \times n}$ be normal, let δA be any perturbation of A, and let μ be any eigenvalue of $A + \delta A$. Then A has an eigenvalue λ such that*

$$|\mu - \lambda| \leq \|\delta A\|_2.$$

Corollary 7.1.8 can be summarized by saying that the eigenvalues of a normal matrix are perfectly conditioned. If a normal matrix is perturbed slightly, the resulting perturbation of the eigenvalues is no greater than the perturbation of the matrix elements.

A weakness of Theorem 7.1.3 is that it gives a single overall condition number for the eigenvalues. In fact it can happen that some of the eigenvalues are well conditioned while others are ill conditioned. This is true for both semisimple and defective matrices. It is therefore important to develop individual condition numbers for the eigenvalues. Again we will restrict our attention to the semisimple case; in fact, we will assume distinct eigenvalues.

Our discussion of individual condition numbers will depend on the notion of *left eigenvectors*. Let λ be an eigenvalue of A. Then $A - \lambda I$ is singular, so there exists a nonzero $x \in \mathbb{C}^n$ such that $(A - \lambda I)x = 0$, that is,

$$Ax = \lambda x. \tag{7.1.9}$$

It is equally true that there is a nonzero $y \in \mathbb{C}^n$ such that $y^*(A - \lambda I) = 0$, that is,

$$y^* A = \lambda y^*. \tag{7.1.10}$$

Any nonzero vector y^* that satisfies (7.1.10) is called a *left eigenvector* of A associated with the eigenvalue λ. Actually we will be somewhat casual with the nomenclature and refer to y itself as a left eigenvector of A, as if y and y^* were the same thing. Furthermore, to avoid confusion we will refer to any nonzero x satisfying (7.1.9) as a *right eigenvector* of A. A left eigenvector of A is just (the conjugate transpose of) a right eigenvector of A^*.

Theorem 7.1.11 *Let $A \in \mathbb{C}^{n \times n}$ have distinct eigenvalues $\lambda_1, \lambda_2, \ldots, \lambda_n$ with associated linearly independent right eigenvectors $x_1, \ldots, x_n$ and left eigenvectors $y_1, \ldots, y_n$. Then*

$$y_j^* x_i \quad \begin{cases} = 0 & \text{if } i \neq j \\ \neq 0 & \text{if } i = j \end{cases}$$

(Two sequences of vectors that satisfy these relationships are said to be biorthogonal.*)*

Proof. Suppose $i \neq j$. From the equation $Ax_i = \lambda_i x_i$ it follows that $y_j^* A x_i = \lambda_i y_j^* x_i$. On the other hand, $y_j^* A = \lambda_j y_j^*$, so $y_j^* A x_i = \lambda_j y_j^* x_i$. Thus $\lambda_i y_j^* x_i = \lambda_j y_j^* x_i$. Since $\lambda_i \neq \lambda_j$, it must be true that $y_j^* x_i = 0$.

It remains to be shown that $y_i^* x_i \neq 0$. Let's assume that $y_i^* x_i = 0$ and get a contradiction. From our assumption and the first part of the proof we see that

$y_i^* x_k = 0$ for $k = 1, \ldots, n$. The vectors $x_1, \ldots, x_n$ are linearly independent, so they form a basis for $\mathbb{C}^n$. Notice that $y_i^* x_k$ is just the complex inner product $\langle x_k, y_i \rangle$, so y_i is orthogonal to $x_1, \ldots, x_n$, hence to every vector in $\mathbb{C}^n$. In particular it is orthogonal to itself, which implies $y_i = 0$, a contradiction. $\square$

The next theorem establishes the promised condition numbers for individual eigenvalues. Let's take a moment to set the scene. Suppose $A \in \mathbb{C}^{n \times n}$ has n distinct eigenvalues, and let λ be one of them. Let δA be a small perturbation satisfying $\| \delta A \|_2 = \epsilon$. Since the eigenvalues of A are distinct, and they depend continuously on the entries of A, we can assert that if ϵ is sufficiently small, $A + \delta A$ will have exactly one eigenvalue $\lambda + \delta\lambda$ that is close to λ. In Theorem 7.1.12 we will assume that all of these conditions hold.

Theorem 7.1.12 *Let $A \in \mathbb{C}^{n \times n}$ have n distinct eigenvalues. Let λ be an eigenvalue with associated right and left eigenvectors x and y, respectively, normalized so that $\| x \|_2 = \| y \|_2 = 1$. Let $s = y^* x$. (Then by Theorem 7.1.11 $s \neq 0$.) Define*

$$\kappa = \frac{1}{|s|} = \frac{1}{|y^* x|}.$$

Let δA be a small perturbation satisfying $\| \delta A \|_2 = \epsilon$, and let $\lambda + \delta\lambda$ be the eigenvalue of $A + \delta A$ that approximates λ. Then

$$|\delta\lambda| \leq \kappa\epsilon + O(\epsilon^2).$$

Thus κ is a condition number for the eigenvalue λ.[38]

Theorem 7.1.12 is actually valid for any simple eigenvalue, regardless of whether or not the matrix is semisimple.

Proof. From Theorem 7.1.3 we know that $|\delta\lambda| \leq \kappa_p(V)\epsilon$. This says that λ is not merely continuous in A, it is even Lipschitz continuous. This condition can be expressed briefly by the statement $|\delta\lambda| = O(\epsilon)$. It turns out that the same is true of the eigenvector as well: $A + \delta A$ has an eigenvector $x + \delta x$ associated with the eigenvalue $\lambda + \delta\lambda$, such that $\delta x = O(\epsilon)$. This depends on the fact that λ is a simple eigenvalue. For a proof see [101, § 2.7] or [103, p. 67]. Expanding the equation $(A + \delta A)(x + \delta x) = (\lambda + \delta\lambda)(x + \delta x)$ and using the fact that $Ax = \lambda x$, we find that

$$(\delta A)x + A(\delta x) + O(\epsilon^2) = (\delta\lambda)x + \lambda(\delta x) + O(\epsilon^2).$$

Left multiplying by y^* and using the equation $y^* A = \lambda y^*$, we obtain

$$y^*(\delta A)x + O(\epsilon^2) = (\delta\lambda)y^* x + O(\epsilon^2),$$

---

[38] λ is a *simple* eigenvalue; that is, its algebraic multiplicity is 1. Therefore x and y are chosen from one-dimensional eigenspaces. This and the fact that they have norm 1 guarantees that they are uniquely determined up to complex scalars of modulus 1. Hence s is determined up to a scalar of modulus 1, and κ is uniquely determined; that is, it is well defined.

Eigenvalue (λ_i)	Condition number (κ_i)
10, 1	4.5×10^3
9, 2	3.6×10^4
8, 3	1.3×10^5
7, 4	2.9×10^5
6, 5	4.3×10^5

Table 7.1 Condition numbers of eigenvalues of matrix A of Example 7.1.15

and therefore

$$\delta\lambda = \frac{y^*(\delta A)x}{y^*x} + O(\epsilon^2).$$

Taking absolute values and noting that $|y^*(\delta A)x| \leq \|y\|_2 \|\delta A\|_2 \|x\|_2 = \epsilon$, we are done.   □

The relationship between κ and the overall condition number given by Theorem 7.1.3 is investigated in Exercise 7.1.29.

Exercise 7.1.13 Show that the condition number κ always satisfies $\kappa \geq 1$.   □

Exercise 7.1.14 Let $A \in \mathbb{C}^{n \times n}$ be Hermitian ($A^* = A$), and let x be a right eigenvector associated with the eigenvalue λ.

(a) Show that x is also a left eigenvector of A.

(b) Show that the condition number of the eigenvalue λ is $\kappa = 1$.

This gives a second confirmation that the eigenvalues of a Hermitian matrix are perfectly conditioned. The results of this exercise also hold for normal matrices. See Exercise 7.1.29.   □

Example 7.1.15 The 10×10 matrix

$$A = \begin{bmatrix} 10 & 10 & & & & \\ & 9 & 10 & & & \\ & & 8 & 10 & & \\ & & & \ddots & \ddots & \\ & & & & 2 & 10 \\ & & & & & 1 \end{bmatrix}$$

is a scaled-down version of an example from [103, p. 90]. The entries below the main diagonal and those above the superdiagonal are all zero. The eigenvalues are obviously $10, 9, 8, \ldots, 2, 1$. We calculated the right and left eigenvectors, and thereby obtained the condition numbers, which are shown in Table 7.1. The eigenvalues are listed in pairs because they possess a certain symmetry. Notice that the condition numbers are fairly large, but the extreme eigenvalues are not as ill conditioned as the

Eigenvalues of A	10, 1	9, 2	8, 3	7, 4	6, 5
Eigenvalues of A_ϵ	10.0027	8.9740	8.0909	6.6614	6.4192
	.9973	2.0260	2.9091	4.3386	4.5808
Perturbations	.0027	.0260	.0909	.3386	.4192
$\kappa_i \epsilon$	.0045	.0361	.1326	.2931	.4281

Table 7.2   Comparison of eigenvalues of A and A_ϵ for $\epsilon = 10^{-6}$

ones in the middle of the spectrum. Let A_ϵ be the matrix that is the same as A, except that the (10,1) entry is ϵ instead of 0. This perturbation of norm ϵ should cause a perturbation in λ_i for which $\kappa_i \epsilon$ is a rough bound. Table 7.2 gives the eigenvalues of A_ϵ for $\epsilon = 10^{-6}$, as calculated by Francis's algorithm. It also shows how much the eigenvalues deviate from those of A and gives the numbers $\kappa_i \epsilon$ for comparison. As you can see, the numbers $\kappa_i \epsilon$ give good order-of-magnitude estimates of the actual perturbations. Notice also that the extreme eigenvalues are still quite close to the original values, while those in the middle have wandered quite far. We can expect that if ϵ is made much larger, some of the eigenvalues will lose their identities completely. Indeed, for $\epsilon = 10^{-5}$ the eigenvalues of A_ϵ, as computed by Francis's algorithm, are

$$10.026 \qquad 8.68 \pm .29i \qquad 6.64 \pm .98i$$
$$.974 \qquad 2.32 \pm .29i \qquad 4.36 \pm .98i$$

Only the two most extreme eigenvalues are recognizable. The others have collided in pairs to form complex conjugate eigenvalues. We conclude this example with two remarks. 1.) We also calculated the overall condition number $\kappa_2(V)$ given by Theorem 7.1.3, and found that it is about 2.5×10^6. This is clearly a gross overestimate of all of the individual condition numbers. 2.) The ill conditioning of the eigenvalues of A can be explained in terms of A's *departure from normality*. From Corollary 7.1.8 we know that the eigenvalues of a normal matrix are perfectly conditioned, and it is reasonable to expect that a matrix that is in some sense nearly normal would have well-conditioned eigenvalues. In Exercise 5.4.42 we found that a matrix that is upper triangular is normal if and only if it is a diagonal matrix. Our matrix A appears to be far from normal since it is upper triangular yet far from diagonal. Such matrices are said to have a high *departure from normality*.   □

cise 7.1.16   MATLAB's command `condeig` is a variant of `eig` that computes the condition numbers given by Theorem 7.1.12. Type `help condeig` for information on how to use this command. Use `condeig` to check the numbers in Table 7.1.   □

Sensitivity of Eigenvectors

In the proof of Theorem 7.1.12 we used the fact that if A has distinct eigenvalues, then the eigenvectors are Lipschitz continuous functions of the entries of A. This means that if x is an eigenvector of A, then any slightly perturbed matrix $A + \delta A$

will have an eigenvector $x + \delta x$ such that $\| \delta x \|_2 \approx \kappa_x \epsilon$. Here $\epsilon = \| \delta A \|_2$, and κ_x is a positive constant independent of δA. This is actually not hard to prove using notions from classical matrix theory and the theory of analytic functions; see Chapter 2 of [103]. Here we will take the existence of κ_x for granted and study its value, for κ_x is a condition number for the eigenvector x.

As we shall see, the computation of condition numbers for eigenvectors is more difficult than for eigenvalues, both in theory and in practice. We will proceed by degrees, and we will not attempt to cover every detail. We are guided by [83]. An important strength of this approach is that it can be generalized to give condition numbers for invariant subspaces. See [83] and [101, § 2.7].

To begin with, let us assume that our matrix, now called T, is a block upper-triangular matrix

$$T = \begin{bmatrix} \lambda & w^T \\ 0 & \hat{T} \end{bmatrix}, \qquad (7.1.17)$$

where the eigenvalues of $\hat{T}$ are all distinct from λ. Thus λ is a simple eigenvalue of T and has a one-dimensional eigenspace spanned by $e_1 = [\, 1, \, 0, \, \cdots, \, 0\,]^T$. Suppose we perturb T by changing only the zero part. Thus we take

$$T + \delta T = \begin{bmatrix} \lambda & w^T \\ y & \hat{T} \end{bmatrix}, \qquad (7.1.18)$$

where $\| y \|_2$ is small. Let us suppose that $\| \delta T \|_2 / \| T \|_2 = \| y \|_2 / \| T \|_2 = \epsilon \ll 1$. Then, if ϵ is small enough, $T + \delta T$ has an eigenvalue $\lambda + \delta \lambda$ near λ and an associated eigenvector $\begin{bmatrix} 1 \\ z \end{bmatrix}$ near e_1, which means that $\| z \|_2$ is small. Our task is to figure out just how small $\| z \|_2$ is. More precisely, we wish to find a condition number κ such that $\| z \|_2 \leq \kappa \epsilon$.

To this end, we write down the equation that must be satisfied by this eigenpair of $T + \delta T$. We have

$$\begin{bmatrix} \lambda & w^T \\ y & \hat{T} \end{bmatrix} \begin{bmatrix} 1 \\ z \end{bmatrix} = \begin{bmatrix} 1 \\ z \end{bmatrix} (\lambda + \delta \lambda). \qquad (7.1.19)$$

This can be written as two separate equations, the first of which is a scalar equation that tells us immediately that

$$\delta \lambda = w^T z. \qquad (7.1.20)$$

If we use this in the second equation of (7.1.19), we obtain $y + \hat{T} z = z(\lambda + w^T z)$, which we rewrite as

$$(\hat{T} - \lambda I) z = -y + z(w^T z). \qquad (7.1.21)$$

This is a nonlinear equation that we could solve for z in principle. It is nonlinear because components of z are multiplied by components of z in the term $z(w^T z)$. The nonlinearity would make the equation more difficult to handle, but fortunately we can ignore it. Since $\| z \| = O(\epsilon)$, we have $\| z(w^T z) \| = O(\epsilon^2)$, so this term is insignificant. Therefore, since $\hat{T} - \lambda I$ is nonsingular, we can multiply (7.1.21) by $(\hat{T} - \lambda I)^{-1}$ to obtain

$$z = -(\hat{T} - \lambda I)^{-1}y + O(\epsilon^2)$$

and hence

$$\| z \|_2 \leq \| (\hat{T} - \lambda I)^{-1} \|_2 \| y \|_2 + O(\epsilon^2). \qquad (7.1.22)$$

This result yields immediately the following lemma, which gives a condition number for the eigenvector e_1.

Lemma 7.1.23 *Let T be a block-triangular matrix of the form (7.1.17), where λ is distinct from the eigenvalues of $\hat{T}$. Let $T + \delta T$ be a perturbation of T of the special form (7.1.18), where $\| \delta T \|_2 / \| T \|_2 = \epsilon$. Then, if ϵ is sufficiently small, $T + \delta T$ has an eigenvector $e_1 + \delta v$ such that*

$$\| \delta v \|_2 \leq \left(\| (\hat{T} - \lambda I)^{-1} \|_2 \| T \|_2 \right) \epsilon + O(\epsilon^2).$$

Thus

$$\| (\hat{T} - \lambda I)^{-1} \|_2 \| T \|_2$$

is a condition number for the eigenvector e_1 with respect to perturbations of T of the special form given by (7.1.18).

Proof. Since $\| \delta v \|_2 = \| z \|_2$ and $\| \delta T \|_2 = \| y \|_2$, this result is an immediate consequence of (7.1.22). □

A couple of remarks are in order. 1.) This analysis also yields an expression for the perturbation of the eigenvalue λ, namely (7.1.20). The relationship between this expression and the condition number given by Theorem 7.1.12 is explored in Exercise 7.1.34. 2.) Equation (7.1.20) yields immediately $|\delta \lambda| \leq \| w \|_2 \| z \|_2$, which suggests (correctly) that the norm of w affects the sensitivity of the eigenvalue λ. Although w is not mentioned in Lemma 7.1.23, it also affects the sensitivity of the eigenvector: it determines, in part, how small ϵ must be in order for Lemma 7.1.23 to be valid. For details see [83] or [101, § 2.7]. In our analysis we obliterated w by replacing $z(w^T z)$ by the expression $O(\epsilon^2)$.

Now consider a general matrix A with a simple eigenvalue λ and corresponding eigenvector x, normalized so that $\| x \|_2 = 1$. Let δA be a perturbation such that $\| \delta A \|_2 / \| A \|_2 = \epsilon \ll 1$. Then $A + \delta A$ has an eigenvector $x + \delta x$ near x. We aim to derive a bound on $\| \delta x \|_2 / \| x \|_2 = \| \delta x \|_2$.

Let V be any unitary matrix whose first column is x, and let $T = V^{-1}AV$. (For example, we could take the Schur decomposition of A. See Theorem 5.4.11.) Then T has the form

$$T = \begin{bmatrix} \lambda & w^T \\ 0 & \hat{T} \end{bmatrix},$$

where $\hat{T}$ has eigenvalues different from λ. Let $\delta T = V^{-1}\delta AV$. Then $A + \delta A = V(T + \delta T)V^{-1}$, and $\| \delta T \|_2 / \| T \|_2 = \| \delta A \|_2 / \| A \|_2 = \epsilon$. Writing

$$T + \delta T = \begin{bmatrix} \lambda + \delta t_{11} & w^T + \delta w^T \\ y & \hat{T} + \delta \hat{T} \end{bmatrix},$$

we have $|\delta t_{11}| \leq \|\delta T\|_2$, $\|\delta \hat{T}\|_2 \leq \|\delta T\|_2$, and $\|y\|_2 \leq \|\delta T\|_2$. If ϵ is small enough, then $\lambda + \delta t_{11}$ is distinct from all of the eigenvalues of $\hat{T} + \delta \hat{T}$, and we can apply Lemma 7.1.23 to the matrix

$$\tilde{T} = \begin{bmatrix} \lambda + \delta t_{11} & w^T + \delta w^T \\ 0 & \hat{T} + \delta \hat{T} \end{bmatrix}.$$

The perturbation $\delta \tilde{T} = \begin{bmatrix} 0 & 0 \\ y & 0 \end{bmatrix}$ gives $\tilde{T} + \delta \tilde{T} = T + \delta T$. We conclude that $T + \delta T$ has an eigenvector $e_1 + \delta v$ such that

$$\|\delta v\|_2 \leq \left(\|(\hat{T} + \delta \hat{T} - (\lambda + \delta t_{11})I)^{-1}\|_2 \|T\|_2 \right) \frac{\|y\|_2}{\|T\|_2} + O(\epsilon^2).$$

Now let $\delta x = V\delta v$. Since also $x = Ve_1$ (first column of V), we have $x + \delta x = V(e_1 + \delta v)$. Since $e_1 + \delta v$ is an eigenvector of $T + \delta T$, $x + \delta x$ is an eigenvector of $A + \delta A$. The unitary transformation V preserves Euclidean norms, and $\|y\|_2 \leq \|\delta A\|_2$, so

$$\frac{\|\delta x\|_2}{\|x\|_2} \leq \left(\|(\hat{T} + \delta \hat{T} - (\lambda + \delta t_{11})I)^{-1}\|_2 \|A\|_2 \right) \frac{\|\delta A\|_2}{\|A\|_2} + O(\epsilon^2)$$

if ϵ is sufficiently small. This result seems to say that we can use

$$\|(\hat{T} + \delta \hat{T} - (\lambda + \delta t_{11})I)^{-1}\|_2 \|A\|_2$$

as a condition number for the eigenvector x. Its obvious defect is that it depends upon δt_{11} and $\delta \hat{T}$, quantities that are unknown and unknowable. Since they are small quantities, we simply ignore them. (This move is justified in [101, § 2.7].) Thus we take

$$\|(\hat{T} - \lambda I)^{-1}\|_2 \|A\|_2 \tag{7.1.24}$$

as the condition number of x.

Computation of the condition number (7.1.24) is no simple matter. Not only is a Schur-like decomposition $A = VTV^{-1}$ required, but we must also compute $\|(\hat{T} - \lambda I)^{-1}\|_2$. In practice we can estimate this quantity just as one estimates $\|A^{-1}\|$, as discussed at the end of Section 2.2, and thereby obtain an estimate of the condition number. An estimate is usually good enough, since we are only interested in the magnitude of the condition number, not its exact value. A condition number estimator for this problem is included in LAPACK [1].

What general conclusions can we draw from the condition number (7.1.24)? If we use the Schur decomposition $A = VTV^{-1}$ (Theorem 5.4.11), $\hat{T}$ is upper triangular and has the eigenvalues of A, other than λ, on its main diagonal. Then $(\hat{T} - \lambda I)^{-1}$ is also upper triangular and has main diagonal entries of the form $(\lambda_i - \lambda)^{-1}$, where λ_i are the other eigenvalues of A. It follows easily that

$$\|(\hat{T} - \lambda I)^{-1}\|_2 \geq \max_i |\lambda_i - \lambda|^{-1} = \frac{1}{\min_i |\lambda_i - \lambda|}. \tag{7.1.25}$$

This number will be large if A has other eigenvalues that are extremely close to λ. We conclude that if λ is one of a cluster of (at least two) tightly packed eigenvalues, then its associated eigenvector will be ill conditioned.

rcise 7.1.26 Show that if U is upper triangular, then $\| U \|_2 \geq \max_i |u_{ii}|$. Show that equality holds if U is diagonal. □

If A is normal, then the Schur matrix T is diagonal, and equality holds in (7.1.25). In this case we also have $\| A \|_2 = \max_i |\lambda_i|$, so the condition number of the eigenvector associated with simple eigenvalue λ_j is

$$\frac{\max_i |\lambda_i|}{\min_{i \neq j} |\lambda_i - \lambda_j|}. \tag{7.1.27}$$

We emphasize that this is for normal matrices only. In words, if A is normal, the condition number of the eigenvector associated with λ_j is inversely proportional to the distance from λ_j to the next nearest eigenvalue. Exercises 7.1.35 and 7.1.36 study the sensitivity of eigenvectors of normal matrices.

If A is not normal, then (7.1.27) is merely a lower bound on the condition number. It is often a good estimate; however, it can sometimes be a severe underestimate, because the gap in the inequality (7.1.25) can be very large when $\hat{T}$ is not normal. We conclude that it is possible for an eigenvector to be ill conditioned even though its associated eigenvalue is well separated from the other eigenvalues.

Additional Exercises

rcise 7.1.28 In Exercise 5.3.19 we ascertained that the MATLAB test matrix west0479 has an eigenvalue $\lambda \approx 17.548546093831 + 34.237822957500i$, and in Exercise 5.3.21 we computed a right eigenvector by one step of inverse iteration. By the same method we can get a left eigenvector. Therefore we can compute the condition number of λ given by Theorem 7.1.12. Here is some sample code, adapted from Exercise 5.3.19.

```
load west0479;
A = west0479;
n = 479;
lam = 17.548546093831 +34.237822957500i;
[L,U,P] = lu(A - lam*speye(n));
vright = ones(n,1);
vright = P*vright;
vright = L\vright;
vright = U\vright;
vright = vright/norm(vright);
vleft  = ones(n,1);
vleft = U'\vleft;
vleft = L'\vleft;
vleft = P'*vleft;
vleft = vleft/norm(vleft);
```

(a) This code segment computes a decomposition $A - \lambda I = P^T LU$, where P is a permutation matrix. (Type `help lu` for more information.) Show that the instructions involving `vleft` compute a left eigenvector of A (i.e. a right eigenvector of A^*) by one step of inverse iteration.

(b) Add more code to compute $\|r\|_2$, where $r = Av - \lambda v$ is the residual for the right eigenvector, and compute an analogous residual for the left eigenvector. Both of the residuals should be tiny.

(c) Add more code to compute the condition number of λ given by Theorem 7.1.12. You should get $\kappa \approx 5 \times 10^5$. Thus λ is not very well conditioned.

(d) Using your computed values of κ and $\|r\|_2$, together with Theorems 7.1.1 and 7.1.12, decide how many of the digits in the expression $\lambda \approx 17.548546093831 + 34.237822957500i$ can be trusted.

□

Exercise 7.1.29 Let $A \in \mathbb{C}^{n \times n}$ have distinct eigenvalues $\lambda_1, \ldots, \lambda_n$ with associated right and left eigenvectors $v_1, \ldots, v_n$ and $w_1, \ldots, w_n$, respectively. By Theorem 7.1.11 we know that $w_i^* v_i \neq 0$ for all i, so we can assume without loss of generality that the vectors have been scaled so that $w_i^* v_i = 1$, $i = 1, \ldots, n$. This scaling does not determine the eigenvectors uniquely, but once v_i is chosen, then w_i^* is uniquely determined and vice versa. Let $V \in \mathbb{C}^{n \times n}$ be the matrix whose ith column is v_i, and let $W^* \in \mathbb{C}^{n \times n}$ be the matrix whose ith row is w_i^*.

(a) Show that $W^* = V^{-1}$.

(b) Show that the condition number κ_i associated with the ith eigenvalue is given by $\kappa_i = \|v_i\|_2 \|w_i\|_2$.

(c) Show that $\kappa_i \leq \kappa_2(V)$ for all i. Thus the overall condition number from Theorem 7.1.3 (with $p = 2$) always overestimates the individual condition numbers.

(d) Show that if A is normal, $\kappa_i = 1$ for all i.

□

Exercise 7.1.30 The Gerschgorin disk theorem is an interesting classical result that has been used extensively in the study of perturbations of eigenvalues. In this exercise you will prove the most basic version of the Gerschgorin disk theorem, and in Exercise 7.1.31 you will use the Gerschgorin theorem to derive a second proof of Theorem 7.1.3. We begin by defining the n Gerschgorin disks associated with a matrix $A \in \mathbb{C}^{n \times n}$. Associated with the ith row of A we have the ith off-diagonal row sum

$$s_i = \sum_{j \neq i} |a_{ij}|,$$

the sum of the absolute values of the entries of the ith row, omitting the main-diagonal entry. The ith *Gerschgorin disk* is the set of complex numbers z such that $|z - a_{ii}| \le s_i$. This is a closed circular disk in the complex plane, the set of complex numbers whose distance from a_{ii} does not exceed s_i. The matrix A has n Gerschgorin disks associated with it, one for each row. The Gerschgorin Disk Theorem states simply that the eigenvalues of A lie within the union of its n Gerschgorin disks. Prove this theorem by showing that each eigenvalue λ must lie within one of the Gerschgorin disks: Let λ be an eigenvalue of A, and let $x \ne 0$ be an associated eigenvector. Let the largest entry of x be the kth entry: $|x_k| = \max_j |x_j| > 0$. Show that

$$(\lambda - a_{ii})x_i = \sum_{j \ne i} a_{ij}x_j$$

for all i. Using this equation in the case $i = k$, taking absolute values, and the using the triangle inequality and the maximum property of $|x_k|$, deduce that λ lies within the kth Gerschgorin disk of A.   □

rcise 7.1.31 Apply the Gerschgorin disk theorem to the matrix $D + \delta D$ that appears in the proof of Theorem 7.1.3 to obtain a second proof of Theorem 7.1.3 (Bauer-Fike) for the case $p = 1$. For more delicate applications of this useful theorem see [103].   □

rcise 7.1.32 For small $\epsilon \ge 0$ and $n \ge 2$ consider the $n \times n$ matrix

$$J(\epsilon) = \begin{bmatrix} 0 & 1 & & & & \\ & 0 & 1 & & & \\ & & 0 & 1 & & \\ & & & \ddots & \ddots & \\ & & & & 0 & 1 \\ \epsilon & & & & & 0 \end{bmatrix}.$$

The entries that have been left blank are zeros. Notice that $J(0)$ is a Jordan block, a severely defective matrix. It has only the eigenvalue 0, repeated n times, with a one-dimensional eigenspace (cf. Exercise 5.2.14).

(a) Show that the characteristic polynomial of $J(\epsilon)$ is $\lambda^n - \epsilon$. Show that every eigenvalue of $J(\epsilon)$ satisfies $|\lambda| = \epsilon^{1/n}$. Thus they all lie on the circle of radius $\epsilon^{1/n}$ centered on 0. (In fact the eigenvalues are the nth roots of ϵ: $\lambda_k = \epsilon^{1/n}e^{2\pi ik/n}$, $k = 1, \ldots, n$.)

(b) Sketch the graph of the function $f_n(\epsilon) = \epsilon^{1/n}$ for small $\epsilon \ge 0$ for a few values of n, e.g. $n = 2, 3, 4$. Compare this with the graph of $g_\kappa(\epsilon) = \kappa\epsilon$ for a few values of κ. Show that $f'_n(\epsilon) \to \infty$ as $\epsilon \to 0$.

(c) Let $A = J(0)$, whose only eigenvalue is 0, and let $\delta A_\epsilon = J(\epsilon) - J(0)$. Let μ_ϵ be any eigenvalue of $A + \delta A_\epsilon$. Show that there is no real number κ such that $|\mu_\epsilon - 0| \le \kappa\|\delta A\|_p$ for all $\epsilon > 0$. Thus there is no finite condition number for the eigenvalues of A. (Remark: The eigenvalues of $J(\epsilon)$ are continuous in ϵ but not Lipschitz continuous.)

(d) Consider the special case $n = 13$ and $\epsilon = 10^{-13}$. Observe that $\|\delta A\|_p = 10^{-13}$ but the eigenvalues of $A + \delta A$ all satisfy $|\lambda| = 1/10$. The eigenvalues are perturbed by an amount that is a trillion times larger than the perturbation in A.

□

Exercise 7.1.33 Let A be the defective matrix

$$A = \begin{bmatrix} 0 & 1 \\ 0 & 0 \end{bmatrix}.$$

Find left and right eigenvectors of A associated with the eigenvalue 0, and show that they satisfy $y^* x = 0$. Does this contradict Theorem 7.1.11?   □

Exercise 7.1.34 Let T have the form (7.1.17).

(a) Show that T has a left eigenvector of the form $\begin{bmatrix} 1 & u^T \end{bmatrix}$, where $u^T(T - \lambda I) = -w^T$.

(b) Show that $\|u\|_2 < \kappa$, the condition number for λ given by Theorem 7.1.12.

(c) Using (7.1.20) and (7.1.21), show that $\delta\lambda = u^T y + O(\epsilon^2)$, and deduce that

$$|\delta\lambda| \leq \kappa\|\delta T\|_2 + O(\epsilon^2).$$

This gives a second proof of Theorem 7.1.12 for perturbations of this special kind.

□

Exercise 7.1.35 The matrix

$$A = \begin{bmatrix} \epsilon & 0 & 0 \\ 0 & -\epsilon & 0 \\ 0 & 0 & 1 \end{bmatrix}$$

is Hermitian, hence normal, and its eigenvalues are obviously ϵ, $-\epsilon$, and 1. Suppose $0 < \epsilon \ll 1$, so that the eigenvalues $\pm\epsilon$ are poorly separated. Let

$$\delta A = \begin{bmatrix} -\epsilon & \epsilon & 0 \\ \epsilon & \epsilon & 0 \\ 0 & 0 & 0 \end{bmatrix}.$$

Obviously $\|\delta A\| = O(\epsilon)$. Do the following using pencil and paper (or in your head).

(a) Find the eigenvectors of A associated with the eigenvalues $\pm\epsilon$.

(b) Find the eigenvalues of $A + \delta A$ and note that two of them are of order ϵ. Find the eigenvectors of $A + \delta A$ associated with these two eigenvalues, and notice that they are far from eigenvectors of A.

(c) Show that the two eigenvectors of part (a) span the same two-dimensional subspace as the eigenvectors of part (b). This is an invariant subspace of both A and $A + \delta A$. Although the individual eigenspaces are ill conditioned, the two-dimensional invariant subspace is well conditioned.

□

cise 7.1.36 Use MATLAB to investigate the sensitivity of the eigenvectors of the normal matrix

$$A = \begin{bmatrix} 1 + \gamma & 0 \\ 0 & 1 - \gamma \end{bmatrix}$$

for small γ. Build a perturbation δA by

```
deltaA = [ 0 eps ; eps 0 ]*randn
```

MATLAB's constant `eps` is the "machine epsilon," which is twice the unit roundoff u. Use MATLAB's `eig` command to compute the eigenvectors of $A + \delta A$ for the three values $\gamma = 10^{-9}$, 10^{-12}, and 10^{-15} (e.g. `gamma = 1e-9`). In each case compute the distance from the eigenvectors of $A + \delta A$ to the eigenvectors of A. Are your results generally consistent with (7.1.27)? □

7.2 METHODS FOR THE SYMMETRIC EIGENVALUE PROBLEM

For dense, nonsymmetric eigenvalue problems, Francis's algorithm, following reduction to Hessenberg form, remains supreme. However, in the symmetric case there is serious competition. A rich variety of approaches has been developed and, toward the end of the 20th century, new faster and more accurate algorithms were discovered and fine tuned. These include Cuppen's divide-and-conquer algorithm, the differential quotient-difference algorithm, and the MRRR algorithm for computing eigenvectors.

Accuracy

Results in Section 7.1 imply that the eigenvalues of a symmetric matrix are well conditioned in the following sense: If $A + E$ is a slight perturbation of a symmetric matrix A, say $\| E \|_2 / \| A \|_2 = \epsilon$, then each eigenvalue μ of $A + E$ is close to an eigenvalue λ of A, in the sense that

$$|\mu - \lambda| \leq \epsilon \| A \|_2. \tag{7.2.1}$$

Since Francis's algorithm is normwise backward stable, it computes the exact eigenvalues of a matrix $A + E$ that is very near to A. The approximate eigenvalues computed by Francis's algorithm satisfy (7.2.1) with ϵ equal to a modest multiple of the computer's unit roundoff u. This is a good result, but it has been found that some algorithms (applied to certain classes of matrices) can get the better result

$$|\mu - \lambda| \le \epsilon|\lambda| \tag{7.2.2}$$

for all λ. This is a more stringent bound, which says that all eigenvalues, whether large or tiny, are computed to full precision. In contrast, the bound (7.2.1) implies only that the eigenvalues that are of about the same magnitude as $\|A\|_2$ are computed to full precision; tiny eigenvalues are not necessarily computed to high relative accuracy. For example, an eigenvalue with magnitude $10^{-6}\|A\|_2$ could have six digits less precision than the larger eigenvalues. This is in fact what happens when Francis's algorithm is used. Some of the algorithms to be discussed in this section do better than that; they achieve the stronger bound (7.2.2).

These are nice results, but our enthusiasm must be tempered by the following observation. Suppose we have computed the eigenvalues of a matrix A with high relative accuracy. In most applications we do not know our matrix to perfect precision, and even if we do, we cannot store it to perfect precision in floating-point arithmetic. Thus in general we really want the eigenvalues of $A + E$, where E is some small error term. We therefore consider the following question. Suppose all of the entries of A are correct to high relative accuracy. Does it follow that the eigenvalues are determined to high relative accuracy? In other words, suppose the entries of the error term E satisfy $|e_{ij}| \le \epsilon|a_{ij}|$ for all i and j. Let λ_i and $\lambda_i + \nu_i$, $i = 1, \dots, n$, denote the eigenvalues of A and $A + E$, respectively. Can we conclude that $|\nu_i| \le \epsilon|\lambda_i|$ for all i? The answer turns out to be no, as is shown in the following example.

Example 7.2.3 Consider the symmetric matrix

$$A = \begin{bmatrix} 1 + \epsilon & 1 \\ 1 & 1 + \epsilon \end{bmatrix} \quad \text{and} \quad E = \begin{bmatrix} 0 & \epsilon \\ \epsilon & 0 \end{bmatrix}$$

where ϵ is tiny. The eigenvalues of A are easily seen to be $2 + \epsilon$ and ϵ, and those of $A + E$ are $2 + 2\epsilon$ and 0. Thus the small eigenvalue has been perturbed by 100%, even though the conditions $|e_{ij}| \le \epsilon|a_{ij}|$ are satisfied. $\qquad\square$

Conclusion: If we have computed the eigenvalues of A to high relative accuracy, it does not follow that we have computed the eigenvalues of $A + E$ to high relative accuracy.

The Jacobi Method

The Jacobi method is one of the oldest numerical methods for the eigenvalue problem. It is older than matrix theory itself, dating back to Jacobi's 1846 paper [55]. Like most numerical methods, it was little used in the precomputer era. It enjoyed a brief renaissance in the 1950s, but in the 1960s it was supplanted as the method of choice by Francis's algorithm. Later it attracted renewed interest because of its inherent parallelism. We will outline the method briefly. The first edition of this book [98] contained a much more detailed discussion of Jacobi's method, including parallel Jacobi schemes.

Let us begin by considering a 2×2 real symmetric matrix

$$A = \begin{bmatrix} a & b \\ b & d \end{bmatrix}.$$

It is not hard to show that there is a rotator

$$Q = \begin{bmatrix} c & -s \\ s & c \end{bmatrix}$$

such that $Q^T A Q$ is diagonal:

$$\begin{bmatrix} c & s \\ -s & c \end{bmatrix} \begin{bmatrix} a & b \\ b & d \end{bmatrix} \begin{bmatrix} c & -s \\ s & c \end{bmatrix} = \begin{bmatrix} \lambda_1 & 0 \\ 0 & \lambda_2 \end{bmatrix}.$$

This solves the 2×2 symmetric eigenvalue problem. The details are worked out in Exercise 7.2.47.

Finding the eigenvalues of a 2×2 matrix is not an impressive feat. Now let us see what we can do with an $n \times n$ matrix. Of the algorithms that we will discuss in this section, Jacobi's method is the only one that does not begin by reducing the matrix to tridiagonal form. Instead it moves the matrix directly toward diagonal form by setting off-diagonal entries to zero, one after the other. It is now clear that we can set any off diagonal entry a_{ij} to zero by an appropriate plane rotator: Just apply in the (i, j) plane the rotator that diagonalizes

$$\begin{bmatrix} a_{ii} & a_{ij} \\ a_{ij} & a_{jj} \end{bmatrix}.$$

Rotators that accomplish this task are called *Jacobi rotators*.

The *classical Jacobi method* searches the matrix for the largest off-diagonal entry and sets it to zero. Then it searches again for the largest remaining off-diagonal entry and sets that to zero, and so on. Since an entry that has been set to zero can be made nonzero again by a later rotator, we cannot expect to reach diagonal form in finitely many steps (and thereby violate Abel's Theorem). At best we can hope that the infinite sequence of Jacobi iterates will converge to diagonal form. Exercise 7.2.53 shows that this hope is realized. Indeed the convergence becomes quite swift, once the matrix is sufficiently close to diagonal form. This method was employed by Jacobi [55] in the solution of an eigenvalue problem that arose in a study of the perturbations of planetary orbits. The system was of order seven because there were then seven known planets. In the paper Jacobi stressed that the computations (by hand!) are quite easy. That was easy for him to say, for the actual computations were done by a student.

The classical Jacobi procedure is quite appropriate for hand calculation but inefficient for computer implementation. For a human being working on a small matrix it is a simple matter to identify the largest off-diagonal entry. the hard part is the arithmetic. However, for a computer working on a larger matrix, the arithmetic is easy;

the search for the largest entry is the expensive part. See Exercise 7.2.52. Because the search is so time consuming, a class of variants of Jacobi's procedure, called *cyclic Jacobi methods*, was introduced. A cyclic Jacobi method sweeps through the matrix, setting entries to zero in some prespecified order and paying no attention to the magnitudes of the entries. In each complete *sweep*, each off-diagonal entry is set to zero once. For example, we could perform a *sweep by columns*, which would create zeros in the order

$$(2,1), (3,1), \ldots, (n,1), (3,2), (4,2), \ldots, (n,2), \ldots, (n,n-1). \qquad (7.2.4)$$

Alternatively one could sweep by rows or by diagonals, for example. We will refer to the method defined by the ordering (7.2.4) as the *special cyclic Jacobi method*. One can show that repeated special cyclic Jacobi sweeps result in convergence to diagonal form [28]. The analysis is more difficult than it is for classical Jacobi.

We remarked earlier that the classical Jacobi method converges quite swiftly, once the off-diagonal entries become fairly small. The same is true of cyclic Jacobi methods. In fact the convergence is quadratic, in the sense that if the off-diagonal entries are all $O(\epsilon)$ after a given sweep, they will be $O(\epsilon^2)$ after the next sweep. Exercise 7.2.54 shows clearly, if not rigorously, why this is so. Once all the off-diagonal entries are small enough, we are done.

The accuracy of Jacobi's method depends upon the stopping criterion. Demmel and Veselic [18] have shown that if the algorithm is run until each a_{ij} is tiny relative to both a_{ii} and a_{jj}, then all eigenvalues are computed to high relative accuracy. That is, (7.2.2) is achieved. Thus the Jacobi method is more accurate than Francis's algorithm. (But see Example 7.2.3.)

If eigenvectors are wanted, they can be obtained by accumulating the product of the Jacobi rotators. A complete set of eigenvectors, orthonormal to working precision, is produced.

For either task, just computing eigenvalues or computing both eigenvalues and eigenvectors, the Jacobi method is several times slower than a reduction to tridiagonal form, followed by Francis's algorithm.

Algorithms for Tridiagonal Matrices

In Section 5.5 we saw how to reduce any real, symmetric matrix to tridiagonal form

$$A = \begin{bmatrix} \alpha_1 & \beta_1 & & & & \\ \beta_1 & \alpha_2 & \beta_2 & & & \\ & \beta_2 & \alpha_3 & \ddots & & \\ & & \ddots & \ddots & \beta_{n-2} & \\ & & & \beta_{n-2} & \alpha_{n-1} & \beta_{n-1} \\ & & & & \beta_{n-1} & \alpha_n \end{bmatrix} \qquad (7.2.5)$$

by an orthogonal similarity transformation. Without loss of generality, we will assume that all of the β_j are positive. From there we can reduce the tridiagonal matrix to diagonal form by Francis's algorithm, for example. However, there are several alternatives for this second step.

Factored Forms of Tridiagonal Matrices

If all of the leading principal submatrices of A are nonsingular (for example, if A is positive definite), then by Theorem 1.7.30, A has a decomposition $A = LDL^T$, where L is unit lower triangular and D is diagonal. It is a routine matter to compute this decomposition by Gaussian elimination. When A is tridiagonal, the computation is especially simple and inexpensive. We will derive it here in a few lines, even though it is a special case of an algorithm that we already discussed in Chapter 1. Since A is tridiagonal, L must be bidiagonal, so the equation $A = LDL^T$ can be written more explicitly as

$$
\begin{bmatrix}
\alpha_1 & \beta_1 & & \\
\beta_1 & \alpha_2 & \ddots & \\
& \ddots & \ddots & \beta_{n-1} \\
& & \beta_{n-1} & \alpha_n
\end{bmatrix}
$$

$$
=
\begin{bmatrix}
1 & & & \\
l_1 & 1 & & \\
& \ddots & \ddots & \\
& & l_{n-1} & 1
\end{bmatrix}
\begin{bmatrix}
d_1 & & & \\
& d_2 & & \\
& & \ddots & \\
& & & d_n
\end{bmatrix}
\begin{bmatrix}
1 & l_1 & & \\
& 1 & \ddots & \\
& & \ddots & l_{n-1} \\
& & & 1
\end{bmatrix}.
$$

$$(7.2.6)$$

rcise 7.2.7

(a) Multiply the matrices together on the right hand side of (7.2.6) and equate the entries on the right and left sides to obtain a system of $2n - 1$ equations.

(b) Verify that the entries of L and D are obtained from those of A by the following algorithm.

$$
\begin{aligned}
&d_1 \leftarrow \alpha_1 \\
&\text{for } j = 1, \dots, n-1 \\
&\quad \begin{bmatrix} l_j \leftarrow \beta_j / d_j \\ d_{j+1} \leftarrow \alpha_{j+1} - d_j l_j^2 \end{bmatrix}
\end{aligned}
\qquad (7.2.8)
$$

(c) About how many flops does this algorithm take?

$\square$

The matrix A is determined by the $2n-1$ parameters $\alpha_1, \dots, \alpha_n$ and $\beta_1, \dots, \beta_{n-1}$. If we perform the decomposition $A = LDL^T$, we obtain an equal number of parameters $l_1, \dots, l_{n-1}$ and $d_1, \dots, d_n$, which encode the same information. Exercise 7.2.7

shows that the second parametrization can be obtained from the first in about $4n$ flops. Conversely, we can retrieve $\alpha_1, \ldots, \alpha_n$ and $\beta_1, \ldots, \beta_{n-1}$ from $l_1, \ldots, l_{n-1}$ and $d_1, \ldots, d_n$ with comparable speed. In principle either parametrization is an equally good representation of A.

In practice, however, there is a difference. The parameters $\alpha_1, \ldots, \alpha_n$ and $\beta_1, \ldots, \beta_{n-1}$ determine all eigenvalues of A to high absolute accuracy, but not to high relative accuracy. That is, if A has some tiny eigenvalues, then small relative perturbations of the parameters can cause large relative changes in those eigenvalues (Example 7.2.3). The parameters $l_1, \ldots, l_{n-1}$ and $d_1, \ldots, d_n$ are better behaved in that regard. In the positive definite case, they determine all of the eigenvalues of LDL^T, even the tiniest ones, to high relative accuracy [17, 56, 71]. This property nearly always holds in the non-positive case as well, although it is not guaranteed. Any factorization LDL^T whose parameters determine the eigenvalues to high relative accuracy is called a *relatively robust representation* (RRR) of the matrix.

These observations have the following consequence. If we want to produce highly accurate algorithms, we should develop algorithms that operate on the factored form of the matrix. The algorithms presented below do just that. Once a factored form is obtained, the tridiagonal matrix is never again explicitly formed.

As we have seen, a common operation in eigenvalue computations is to shift the origin, that is, to replace A by $A - \rho I$, where ρ is a shift. As a first task, let us see how we can effect a shift on a factored form of A. Say we have A in the factored form LDL^T. Our task is to compute unit lower-bidiagonal $\hat{L}$ and diagonal $\hat{D}$ such that $LDL^T - \rho I = \hat{L}\hat{D}\hat{L}^T$. By comparing entries in this equation, one easily checks that $\hat{L}$ and $\hat{D}$ can be computed by the following algorithm.

$$
\begin{aligned}
&\hat{d}_1 \leftarrow d_1 - \rho \\
&\text{for } j = 1, \ldots, n-1 \\
&\quad \left[
\begin{aligned}
&\hat{l}_j \leftarrow l_j d_j / \hat{d}_j \\
&\hat{d}_{j+1} \leftarrow d_{j+1} + d_j l_j^2 - \hat{d}_j \hat{l}_j^2 - \rho
\end{aligned}
\right.
\end{aligned}
\tag{7.2.9}
$$

Exercise 7.2.10 Show that if $LDL^T - \rho I = \hat{L}\hat{D}\hat{L}^T$, then

(a) $d_j + d_{j-1} l_{j-1}^2 - \rho = \hat{d}_j + \hat{d}_{j-1} \hat{l}_{j-1}^2$ for $j = 1, \ldots, n$, where $l_0 = \hat{l}_0 = d_0 = \hat{d}_0 = 0$.

(b) $l_j d_j = \hat{l}_j \hat{d}_j$ for $j = 1, \ldots, n-1$.

(c) Verify that (7.2.9) computes $\hat{L}$ and $\hat{D}$.

$\square$

A stabler version of the algorithm is obtained by introducing the auxiliary quantities

$$
s_j = d_{j-1} l_{j-1}^2 - \hat{d}_{j-1} \hat{l}_{j-1}^2 - \rho = \hat{d}_j - d_j,
\tag{7.2.11}
$$

with $s_1 = -\rho$. One easily checks that

$$
s_{j+1} = \hat{l}_j l_j s_j - \rho.
\tag{7.2.12}
$$

This gives the so-called *differential* form of the algorithm.

Given L, D, and ρ, this algorithm produces $\hat{L}$ and $\hat{D}$ such that $\hat{L}\hat{D}\hat{L}^T = LDL^T - \rho$.

$$
\begin{aligned}
& s_1 \leftarrow -\rho \\
& \text{for } j = 1, \dots, n - 1 \\
& \qquad \begin{bmatrix} \hat{d}_j \leftarrow d_j + s_j \\ \hat{l}_j \leftarrow l_j d_j / \hat{d}_j \\ s_{j+1} \leftarrow \hat{l}_j l_j s_j - \rho \end{bmatrix} \\
& \hat{d}_n \leftarrow d_n + s_n
\end{aligned}
\tag{7.2.13}
$$

Obviously the flop count for (7.2.13) is $O(n)$.

cise 7.2.14

(a) Show that if s_j are defined by (7.2.11) with $s_1 = -\rho$, then the recursion (7.2.12) is valid. (For example, start with $s_{j+1} = d_j l_j l_j - \hat{d}_j \hat{l}_j \hat{l}_j - \rho$, and make a couple of substitutions using the equation $l_j d_j = \hat{l}_j \hat{d}_j$.)

(b) Verify that (7.2.13) performs the same function as (7.2.9).

$\square$

A useful alternative to the LDL^T decomposition is obtained by disturbing the symmetry of A in a constructive way. It is easy to show that the symmetric, tridiagonal matrix A of (7.2.5) can be transformed to the form

$$
\tilde{A} = \Delta^{-1} A \Delta = \begin{bmatrix}
\alpha_1 & 1 & & & & \\
\beta_1^2 & \alpha_2 & 1 & & & \\
& \beta_2^2 & \alpha_3 & \ddots & & \\
& & \ddots & \ddots & 1 & \\
& & & \beta_{n-2}^2 & \alpha_{n-1} & 1 \\
& & & & \beta_{n-1}^2 & \alpha_n
\end{bmatrix},
\tag{7.2.15}
$$

where $\Delta = \mathrm{diag}\{\delta_1, \dots, \delta_n\}$ is a diagonal transforming matrix (Exercise 7.2.55). Conversely, any matrix of the form

$$
\tilde{A} = \begin{bmatrix}
\alpha_1 & 1 & & & & \\
\gamma_1 & \alpha_2 & 1 & & & \\
& \gamma_2 & \alpha_3 & \ddots & & \\
& & \ddots & \ddots & 1 & \\
& & & \gamma_{n-2} & \alpha_{n-1} & 1 \\
& & & & \gamma_{n-1} & \alpha_n
\end{bmatrix}
\tag{7.2.16}
$$

with all $\gamma_j > 0$ is diagonally similar to a symmetric matrix of the form (7.2.5) with $\beta_j = \sqrt{\gamma_j}$, $j = 1, \dots, n-1$. It might seem like heresy to destroy the symmetry of a matrix, but it turns out that the form (7.2.16) is often useful.

If the matrix $\tilde{A}$ in (7.2.16) has all leading principal submatrices nonsingular, then it has an LU decomposition. In the context of eigenvalue computations, the symbol R is often used instead of U, so we will speak of an LR decomposition and write $\tilde{A} = \tilde{L}\tilde{R}$. The factors have the particularly simple form

$$
\begin{bmatrix}
1 & & & & & \\
\tilde{l}_1 & 1 & & & & \\
& \tilde{l}_2 & 1 & & & \\
& & \ddots & \ddots & & \\
& & & \tilde{l}_{n-2} & 1 & \\
& & & & \tilde{l}_{n-1} & 1
\end{bmatrix}
\begin{bmatrix}
\tilde{r}_1 & 1 & & & & \\
& \tilde{r}_2 & 1 & & & \\
& & \tilde{r}_3 & \ddots & & \\
& & & \ddots & 1 & \\
& & & & \tilde{r}_{n-1} & 1 \\
& & & & & \tilde{r}_n
\end{bmatrix}. \tag{7.2.17}
$$

The $2n - 1$ parameters $\tilde{l}_1, \ldots, \tilde{l}_{n-1}$ and $\tilde{r}_1, \ldots, \tilde{r}_n$ encode the same information as $\alpha_1, \ldots, \alpha_n$ and $\beta_1, \ldots, \beta_{n-1}$, in principle. However, just as in the case of the LDL^T decomposition, the entries of $\tilde{L}$ and $\tilde{R}$ usually determine the eigenvalues to high relative accuracy, whereas those of $\tilde{A}$ do not. Thus it makes sense to develop algorithms that operate directly on the parameters $\tilde{l}_1, \ldots, \tilde{l}_{n-1}$ and $\tilde{r}_1, \ldots, \tilde{r}_n$, never forming the tridiagonal matrix $\tilde{A}$ explicitly.

Obviously the parameters $\tilde{l}_1, \ldots, \tilde{l}_{n-1}$ and $\tilde{r}_1, \ldots, \tilde{r}_n$ must be closely related to the parameters $l_1, \ldots, l_{n-1}$ and $d_1, \ldots, d_n$ of the decomposition $A = LDL^T$. Indeed, it turns out that $\tilde{r}_j = d_j$ and $\tilde{l}_j = l_j\delta_j/\delta_{j+1}$, where the δ_j are the entries of the diagonal matrix Δ of (7.2.15).

Exercise 7.2.18 Let $A = LDL^T$ as in (7.2.6), $\tilde{A} = \Delta^{-1}A\Delta$ as in (7.2.15), and $\tilde{A} = \tilde{L}\tilde{R}$ as in (7.2.17).

(a) Show that $\tilde{L} = \Delta^{-1}L\Delta$ and $\tilde{R} = \Delta^{-1}DL^T\Delta$.

(b) Deduce that $\tilde{l}_j = l_j\delta_j/\delta_{j+1}$ for $j = 1, \ldots, n-1$ and $\tilde{r}_j = d_j$ for $j = 1, \ldots, n$.

$\square$

We noted above how to effect a shift of origin on an LDL^T decomposition without explicitly forming the product A ((7.2.13) and Exercise 7.2.14). Naturally we can do the same thing with an LR decomposition. Suppose we have $\tilde{A} = \tilde{L}\tilde{R}$ in factored form, and we wish to compute $\hat{L}\hat{R}$ such that $\tilde{A} - \rho I = \hat{L}\hat{R}$ without explicitly forming $\tilde{A}$. One easily shows (Exercise 7.2.56) that the following algorithm produces $\hat{L}$ and $\hat{R}$ from $\tilde{L}$, $\tilde{R}$, and ρ.

Given $\tilde{L}$, $\tilde{R}$, and ρ, this algorithm produces $\hat{L}$ and $\hat{R}$ such that $\hat{L}\hat{R} = \tilde{L}\tilde{R} - \rho$.

$$
\begin{aligned}
&t_1 \leftarrow -\rho \\
&\text{for } j = 1, \ldots, n-1 \\
&\quad \begin{bmatrix} \hat{r}_j \leftarrow \tilde{r}_j + t_j \\ q \leftarrow \tilde{l}_j/\hat{r}_j \\ \hat{l}_j \leftarrow \tilde{r}_j q \\ t_{j+1} \leftarrow t_j q - \rho \end{bmatrix} \\
&\hat{r}_n = \tilde{r}_n + t_n
\end{aligned}
\tag{7.2.19}
$$

The Slicing or Bisection Method

The slicing method, also known as the bisection or Sturm sequence method, is based on the notions of congruence and inertia. We begin by noting that for any symmetric matrix $A \in \mathbb{R}^{n \times n}$ and any $S \in \mathbb{R}^{n \times n}$, SAS^T is also symmetric. Now let A and B be any two symmetric matrices in $\mathbb{R}^{n \times n}$. We say that B is *congruent* to A if there exists a nonsingular matrix $S \in \mathbb{R}^{n \times n}$ such that $B = SAS^T$. It is important that S is nonsingular.

Exercise 7.2.20 Show that (a) A is congruent to A, (b) if B is congruent to A, then A is congruent to B, (c) if A is congruent to B and B is congruent to C, then A is congruent to C. In other words, congruence is an equivalence relation. ☐

The eigenvalues of a symmetric $A \in \mathbb{R}^{n \times n}$ are all real. Let $\nu(A)$, $\zeta(A)$, and $\pi(A)$ denote the number of negative, zero, and positive eigenvalues of A, respectively. The ordered triple $(\nu(A), \zeta(A), \pi(A))$ is called the *inertia* of A. The key result of our development is Sylvester's law of inertia, which states that congruent matrices have the same inertia.

Theorem 7.2.21 *(Sylvester's Law of Inertia) Let A, $B \in \mathbb{R}^{n \times n}$ be symmetric, and suppose B is congruent to A. Then $\nu(A) = \nu(B)$, $\zeta(A) = \zeta(B)$, and $\pi(A) = \pi(B)$.*

For a proof see Exercise 7.2.57. We will use Sylvester's law to help us slice the spectrum into subsets. Let $\lambda_1, \ldots, \lambda_n$ be the eigenvalues of A, ordered so that $\lambda_1 \geq \lambda_2 \geq \cdots \geq \lambda_n$. Suppose that for any given $\rho \in \mathbb{R}$ we can determine the inertia of $A - \rho I$. The number $\pi(A - \rho I)$ equals the number of eigenvalues of A that are greater than ρ. If $\pi(A - \rho I) = i$, with $0 < i < n$, then

$$\lambda_n \leq \cdots \leq \lambda_{i+1} \leq \rho < \lambda_i \leq \cdots \leq \lambda_1.$$

This splits the spectrum into two subsets. We will see that by repeatedly slicing the spectrum with systematically chosen values of ρ, we can determine all the eigenvalues of A with great precision.

Exercise 7.2.22 Suppose you have a subroutine that can calculate $\pi(A - \rho I)$ for any value of ρ quickly. Devise an algorithm that uses this subroutine repeatedly to calculate all of the eigenvalues of A in some specified interval with error less than some specified tolerance $\epsilon > 0$. ☐

We now observe that if A is tridiagonal, we do in fact have inexpensive ways to calculate $\pi(A - \rho I)$ for any ρ. We need only calculate the decomposition

$$A - \rho I = L_\rho D_\rho L_\rho^T, \tag{7.2.23}$$

with L_ρ unit lower triangular and D_ρ diagonal, in $O(n)$ flops (Exercise 7.2.7). This equation is a congruence between $A - \rho I$ and D_ρ, so $\pi(A - \rho I) = \pi(D_\rho)$. Since D_ρ is diagonal, we find $\pi(D_\rho)$ by counting the number of positive elements on its main diagonal.

If we wish, we can factor A once into LDL^T and then use algorithm (7.2.13) to obtain the desired shifted factorizations without ever assembling any of the matrices $A - \rho I$ explicitly. Alternatively, we can transform A to the unsymmetric form $\tilde{A}$ and compute the decomposition $\tilde{L}\tilde{R}$. By Exercise 7.2.18, the main diagonal entries of $\tilde{R}$ are the same as those of D, so this decomposition also reveals the inertia of A. To obtain decompositions of shifted matrices, we use algorithm (7.2.19).

In either (7.2.13) or (7.2.19) it can happen that $\hat{d}_j = 0$ (resp $\hat{r}_j = 0$) for some j, which will lead to a division by zero. This poses no problem; for example, one can simply replace the zero by an extremely tiny number $\hat{\epsilon}$ (e.g. 10^{-307}) and continue. This has the same effect as replacing $\hat{d}_j$ by $\hat{d}_j + \hat{\epsilon}$, which perturbs the spectrum by a negligible amount.

Now that we know how to calculate $\pi(A - \rho I)$ for any value of ρ, let us see how we can systematically determine the eigenvalues of A. A straightforward bisection approach works very well. Suppose we wish to find all eigenvalues in the interval $(a, b]$. We begin by calculating $\pi(A - aI)$ and $\pi(A - bI)$ to find out how many eigenvalues lie in the interval. If $\pi(A - aI) = i$ and $\pi(A - bI) = j$, then

$$a < \lambda_i \leq \cdots \leq \lambda_{j+1} \leq b,$$

so there are $i - j$ eigenvalues in the interval. Now let $\rho = (a + b)/2$, the midpoint of the interval, and calculate $\pi(A - \rho I)$. From this we can deduce how many eigenvalues lie in each of the intervals $(a, \rho]$ and $(\rho, b]$. More precise information can be obtained by bisecting each of these subintervals, and so on. Any interval that is found to contain no eigenvalues can be removed from further consideration. An interval that contains a single eigenvalue can be bisected repeatedly until the eigenvalue has been located with sufficient precision. If we know that $\lambda_k \in (\rho_1, \rho_2]$, where $\rho_2 - \rho_1 < 2\epsilon$, then the approximation $\lambda_k \approx (\rho_1 + \rho_2)/2$ is in error by less than ϵ.

The LR and Quotient-Difference Algorithms

Another approach to computing the eigenvalues of a symmetric, tridiagonal matrix is the *quotient-difference* (qd) algorithm. We begin by introducing the LR algorithm, which will serve as a stepping stone to the qd algorithm. To this end we transform A to the nonsymmetric form $\tilde{A} = \Delta^{-1}A\Delta$ (7.2.16). For notational simplicity, we will now leave off the tildes and work with the nonsymmetric matrix

$$A = \begin{bmatrix} \alpha_1 & 1 & & & & \\ \gamma_1 & \alpha_2 & 1 & & & \\ & \gamma_2 & \alpha_3 & \ddots & & \\ & & \ddots & \ddots & 1 & \\ & & & \gamma_{n-2} & \alpha_{n-1} & 1 \\ & & & & \gamma_{n-1} & \alpha_n \end{bmatrix}. \tag{7.2.24}$$

If A has all leading principal submatrices nonsingular, then it has an LR decomposition.

$$
\begin{bmatrix}
1 & & & & & \\
l_1 & 1 & & & & \\
& l_2 & 1 & & & \\
& & \ddots & \ddots & & \\
& & & l_{n-2} & 1 & \\
& & & & l_{n-1} & 1
\end{bmatrix}
\begin{bmatrix}
r_1 & 1 & & & & \\
& r_2 & 1 & & & \\
& & r_3 & \ddots & & \\
& & & \ddots & 1 & \\
& & & & r_{n-1} & 1 \\
& & & & & r_n
\end{bmatrix}. \quad (7.2.25)
$$

If we reverse the order of these factors and multiply them back together, we obtain

$$
\hat{A} =
\begin{bmatrix}
r_1 + l_1 & 1 & & & & \\
r_2 l_1 & r_2 + l_2 & 1 & & & \\
& r_3 l_2 & r_3 + l_3 & \ddots & & \\
& & \ddots & \ddots & 1 & \\
& & & r_{n-1} l_{n-2} & r_{n-1} + l_{n-1} & 1 \\
& & & & r_n l_{n-1} & r_n
\end{bmatrix}, \quad (7.2.26)
$$

a matrix of the same form as A. The transformation from A to $\hat{A}$, which is easily seen to be a similarity transformation, is one step of the *basic LR algorithm*, which is analogous to the basic QR algorithm (Exercise 6.3.15), the historical antecedent of Francis's algorithm. If we iterate the process, the subdiagonal entries will converge gradually to zero, revealing the eigenvalues on the main diagonal. Just as for the QR algorithm, we can accelerate the convergence by using shifts of origin.[39]

A sequence of LR iterations produces a sequence of similar matrices $A = A_0, A_1,$ $A_2, A_3, \ldots$. Each matrix is obtained from the previous one via an LR decomposition. We arrive at the quotient-difference (qd) algorithm by shifting our attention from the sequence of matrices (A_j) to the sequence of intervening LR decompositions. The qd algorithm jumps from one LR decomposition to the next, bypassing the intermediate tridiagonal matrix. The advantage of this change of viewpoint is that it leads to more accurate algorithms since, as we have already mentioned, the spectrum of the matrix is more accurately encoded by the factored form.

It is not hard to figure out how to jump from one LR decomposition to the next. Suppose we have a decomposition LR of some iterate A. To complete an LR iteration in the conventional way, we would form $\hat{A} = RL$. To start the next iteration, we would subtract a shift ρ from $\hat{A}$ and perform a new decomposition $\hat{A} - \rho I = \hat{L}\hat{R}$. Our task now is to obtain the $\hat{L}$ and $\hat{R}$ directly from L and R (and ρ) without ever forming the intermediate matrix $\hat{A}$. Equating the matrix entries in the equation $RL - \rho I = \hat{L}\hat{R}$, we easily determine (Exercise 7.2.59) that the parameters $\hat{l}_j$ and $\hat{r}_j$

---

[39]The general theory of GR algorithms, which includes LR, QR, Francis's algorithm, and more general bulge chasing algorithms, is presented in [101].

can be obtained from the l_j and r_j by the simple algorithm

$$
\begin{aligned}
&\hat{l}_0 \leftarrow 0 \\
&\text{for } j = 1 \ldots, n-1 \\
&\quad \left[\begin{aligned}
\hat{r}_j &\leftarrow r_j + l_j - \rho - \hat{l}_{j-1} \\
\hat{l}_j &\leftarrow l_j r_{j+1}/\hat{r}_j
\end{aligned} \right. \\
&\hat{r}_n \leftarrow r_n - \rho - \hat{l}_{n-1}
\end{aligned}
\tag{7.2.27}
$$

This is one iteration of the *shifted quotient-difference* (qds) algorithm. If it is applied iteratively, it effects a sequence of iterations of the LR algorithm without ever forming the "intermediate" tridiagonal matrices. As $l_{n-1} \to 0$, r_n converges to an eigenvalue.

There is a slight difference in the way the shifts are handled. At each iteration we subtract a shift ρ, and we do not bother to restore the shift at the end of the iteration. Instead, we keep track of the accumulated shift. At each iteration we add the new shift ρ to the accumulated shift, which we can call $\hat{\rho}$. As each eigenvalue emerges, we must realize that it is the eigenvalue of a shifted matrix. To get the correct eigenvalue of the original matrix, we have to add on the accumulated shift $\hat{\rho}$.

To obtain a numerically superior algorithm, we have to rearrange (7.2.27). We introduce auxiliary quantities $s_j = r_j - \rho - \hat{l}_{j-1}, j = 1, \ldots, n$, where $\hat{l}_0 = 0$. Then we can write (7.2.27) as

Shifted differential quotient-difference (dqds) iteration

$$
\begin{aligned}
&s_1 \leftarrow r_1 - \rho \\
&\text{for } j = 1, \ldots, n-1 \\
&\quad \left[\begin{aligned}
\hat{r}_j &\leftarrow s_j + l_j \\
q &\leftarrow r_{j+1}/\hat{r}_j \\
\hat{l}_j &\leftarrow l_j q \\
s_{j+1} &\leftarrow s_j q - \rho
\end{aligned} \right. \\
&\hat{r}_n \leftarrow s_n
\end{aligned}
\tag{7.2.28}
$$

The details are worked out in Exercise 7.2.60.

In the positive-definite case, the algorithm dqds (7.2.28) is extremely stable. One easily shows that in that case all of the quantities r_j and l_j are positive (Exercise 7.2.61). Looking at (7.2.28) in the case $\rho = 0$, we see that the s_j are also necessarily positive. It follows that there are no subtractions in (7.2.28). In the one addition operation, two positive quantities are added. Therefore there is no cancellation in (7.2.28), every arithmetic operation is done with high relative accuracy (see Section 2.5), and all quantities are computed to high relative precision. If the process is repeated to convergence, the resulting eigenvalue will be obtained to high relative accuracy, regardless of how tiny it is. For numerical confirmation try Exercise 7.2.62.

This all assumes that $\rho = 0$ at every step. Even if ρ is nonzero, we get the same outcome as long as ρ is small enough that $A - \rho I$ is positive definite. In that case it can be shown [70] that the s_j remain positive, so the addition in (7.2.28) is still that of two positive numbers. Some cancellation is inevitable when the shift is subtracted, but whatever is subtracted is no greater than what will be restored when the accumulated shift is added to a computed eigenvalue in the end. It follows (that is, it can be shown) that every eigenvalue is computed with high relative accuracy.

Thus we achieve high relative accuracy by using a shifting strategy that always underestimates the smallest eigenvalue among those that have not already been deflated out. The eigenvalues are found in order, from smallest to largest.

However, we must temper these conclusions by remembering Example 7.2.3 and the fact that the matrix whose eigenvalues we have just computed with high relative accuracy may not be exactly the matrix whose eigenvalues we intended to compute. Moreover, if the tridiagonal matrix was obtained by reduction of some full symmetric matrix T to tridiagonal form, there is no guarantee that we have computed the eigenvalues of T to high relative accuracy. The tridiagonal reduction algorithm (5.5.4) is normwise backward stable, but it does not preserve tiny eigenvalues of T to high relative accuracy.[40]

Historical Remark: Our order of development of this subject has been almost exactly the opposite of the historical order. Rutishauser introduced the qd algorithm (but not the differential form) in 1954 [74, 75] as a method for computing poles of meromorphic functions. A few years later he placed his parameters in matrices and introduced the more general LR algorithm [76]. This provided, in turn, the impetus for the development of the QR algorithm around 1960 [31, 59] and it's implicitly shifted version, which we call Francis's algorithm [32]. For many years Francis's algorithm was seen as the pinnacle of these developments, and in some senses it is. However, by the early nineties Fernando and Parlett [27, 70] realized that differential versions of the qd algorithm could be used to solve the symmetric tridiagonal eigenvalue problem (and the related singular value problem) with greater accuracy, thus giving new life to the qd algorithm.

Computing Eigenvectors by Inverse Iteration

Once we have calculated some eigenvalues by bisection or the dqds algorithm, if we want the corresponding eigenvectors, we can find them by inverse iteration, using the computed eigenvalues as shifts, as described in Section 5.3. For each eigenvalue λ_k this requires an LDL^T (or related) decomposition of $A - \lambda_k I$, which costs $O(n)$ flops, and one or two steps of inverse iteration, also costing $O(n)$ flops. Thus the procedure is quite economical.

In exact arithmetic, the eigenvectors of a symmetric matrix are orthogonal. A weakness of the inverse iteration approach is that it produces (approximate) eigenvectors that are not truly orthogonal. In particular, the computed eigenvectors associated with a tight cluster of eigenvalues can be far from orthogonal. One remedy is to apply the Gram-Schmidt process to these vectors to orthonormalize them. This costs $O(nm^2)$ work for a cluster of m eigenvalues, so it is satisfactory if m is not too large. However, for a matrix with huge clusters of eigenvalues, the added cost is unacceptable. A better remedy is to use *twisted factorizations*.

---

[40]Of course, all eigenvalues *are* computed accurately relative to $\|T\|$, by which we mean that each approximation μ to an eigenvalue λ satisfies $|\mu - \lambda|/\|T\| \approx Cu$ for some C that isn't much bigger than one.

Computing Accurate Eigenvectors Using Twisted Factorizations

The super-accurate eigenvalues obtained by dqds can help us to compute super-accurate, numerically orthogonal eigenvectors, provided we perform inverse iteration in an appropriate way. Again it is important to work with factored forms (relatively robust representations) of the matrix rather than the matrix itself. Another key is to use special factorizations called *twisted factorizations*.

The *MRRR* (Multiple Relatively Robust Representation) algorithm, an algorithm built using these factorizations, takes one step of inverse iteration, at a cost of $O(n)$ flops, for each eigenvector. Thus a complete set of eigenvalues and eigenvectors of a symmetric, tridiagonal matrix can be obtained in $O(n^2)$ flops. More generally, k eigenvalues and corresponding eigenvectors can be found in $O(nk)$ flops. Since each eigenvector is determined with very high accuracy, the computed eigenvectors are automatically orthogonal to working precision. Thus the potentially expensive Gram-Schmidt step mentioned above is avoided. This is one situation where the high relative accuracy really pays off. The flop count $O(nk)$ is optimal in the following sense. Since k eigenvectors are determined by nk numbers in all, and at least one flop (not to mention fetching, storing, etc.) must be expended in the computation of each number, there is no way to avoid $O(nk)$ work.

An implementation of the MRRR algorithm is included in LAPACK [1]. The details of the algorithm are complicated. Here we will just indicate some of the basic ideas.

We assume once again that we have a symmetric, tridiagonal matrix stored in the factored form LDL^T. We have already seen how to obtain that factored form of a shifted matrix

$$LDL^T - \rho I = \hat{L}\hat{D}\hat{L}^T \qquad (7.2.29)$$

by (7.2.13). For our further development we need a companion factorization

$$LDL^T - \rho I = \check{U}\check{D}\check{U}^T, \qquad (7.2.30)$$

where $\check{U}$ is unit upper bidiagonal:

$$\check{U} = \begin{bmatrix} 1 & \check{u}_1 & & & & \\ & 1 & \check{u}_2 & & & \\ & & 1 & \ddots & & \\ & & & \ddots & \check{u}_{n-2} & \\ & & & & 1 & \check{u}_{n-1} \\ & & & & & 1 \end{bmatrix}.$$

From Section 1.7 we know that LU decompositions, and hence also LDL^T decompositions such as (7.2.29), are natural byproducts of Gaussian elimination, performed in the usual top-to-bottom manner. Our preference for going from top to bottom and from left to right is simply a prejudice. If one prefers, one can start from the bottom, using the (n, n) entry as a pivot for the elimination of all of the entries above it in the nth column, and then move upward and to the left. If one performs Gaussian elimination in this way, one obtains as a byproduct a decomposition of the form UL, of which the UDU^T decomposition in (7.2.30) is a variant. Thus we expect to be able to compute the parameters $\check{d}_j$ and $\check{u}_j$ in (7.2.30) by a bottom-to-top process. Indeed, one easily checks (Exercise 7.2.63) that the following algorithm does the job.

$$
\begin{aligned}
&p_n \leftarrow d_n - \rho \\
&\text{for } j = n, \ldots, 2 \\
&\left[
\begin{array}{l}
\check{d}_j \leftarrow p_j + d_{j-1} l_{j-1}^2 \\
q_{j-1} \leftarrow d_{j-1}/\check{d}_j \\
\check{u}_{j-1} \leftarrow l_{j-1} q_{j-1} \\
p_{j-1} \leftarrow p_j q_{j-1} - \rho
\end{array}
\right. \\
&\check{d}_1 \leftarrow p_1
\end{aligned}
\tag{7.2.31}
$$

This is very similar to (7.2.28), the biggest difference being that it works from bottom to top.

Related to both (7.2.29) and (7.2.30) are the *twisted factorizations*. The kth twisted factorization has the form

$$
LDL^T - \rho I = N_k D_k N_k^T,
\tag{7.2.32}
$$

where D_k is diagonal and N_k is "twisted," partly lower and partly upper triangular:

$$
N_k =
\begin{bmatrix}
1 & & & & & & \\
\hat{l}_1 & 1 & & & & & \\
 & \ddots & \ddots & & & & \\
 & & \hat{l}_{k-1} & 1 & \check{u}_k & & \\
 & & & & \ddots & \ddots & \\
 & & & & & 1 & \check{u}_{n-1} \\
 & & & & & & 1
\end{bmatrix}.
\tag{7.2.33}
$$

There are n twisted factorizations of $LDL^T - \rho I$, corresponding to $k = 1, \ldots, n$. The twisted factorizations for the cases $k = 1$ and $k = n$ are (7.2.30) and (7.2.29), respectively. It is easy to compute a twisted factorization. The entries $\hat{l}_1, \ldots, \hat{l}_{k-1}$ are easily seen to be the same as the $\hat{l}_j$ in (7.2.29). Likewise the entries $\check{u}_n, \ldots, \check{u}_k$ are the same as in (7.2.30). The diagonal matrix D_k has the form

$$
\text{diag}\left\{ \hat{d}_1, \ldots, \hat{d}_{k-1}, \delta_k, \check{d}_{k+1}, \ldots, \check{d}_n \right\},
$$

where $\hat{d}_1, \ldots, \hat{d}_{k-1}$ are from (7.2.29), and $\check{d}_{k+1}, \ldots, \check{d}_n$ are from (7.2.30). The only entry that cannot be grabbed directly from either (7.2.29) or (7.2.30) is δ_k, the "middle" entry of D_k. Checking the (k, k) entry of the equation (7.2.32), we find that

$$
d_k + d_{k-1} l_{k-1}^2 - \rho = \hat{d}_{k-1} \hat{l}_{k-1}^2 + \delta_k + \check{d}_{k+1} \check{u}_k^2,
$$

and therefore

$$
\delta_k = d_k + d_{k-1} l_{k-1}^2 - \rho - \hat{d}_{k-1} \hat{l}_{k-1}^2 - \check{d}_{k+1} \check{u}_k^2.
$$

Referring back to (7.2.13), (7.2.31), and Exercise 7.2.63, we find that δ_k can also be expressed as

$$\delta_k = s_k + p_{k+1}q_k. \tag{7.2.34}$$

This is a more robust formula.

Exercise 7.2.35 Check the assertions of the previous paragraph.   ☐

We now see that we can compute all n twisted factorizations at once. We just need to compute (7.2.29) and (7.2.30) by algorithms (7.2.13) and (7.2.31), respectively, saving the auxiliary quantities s_j, p_j, and q_j. We use these to compute the δ_k in (7.2.34). This gives us all of the ingredients for all n twisted factorizations for $O(n)$ flops.

The MRRR algorithm uses the twisted factorizations to compute eigenvectors associated with well-separated eigenvalues. Using a computed eigenvalue as the shift ρ, we compute the twisted factorizations $N_k D_k N_k^T$ simultaneously, as just described. In theory the D_k matrix in each factorization should have a zero entry on the main diagonal, since $LDL^T - \rho I$ should be exactly singular if ρ is an eigenvalue. Further inspection shows that precisely the entry δ_k should be zero. In practice, due to roundoff errors, none of the δ_k will be exactly zero. Although any of the twisted factorizations can be used to perform a step of inverse iteration, from a numerical standpoint it is best to use the one for which $|\delta_k| = \min_j |\delta_j|$. Specifically, we solve

$N_k D_k N_k^T x = e_k \delta_k$, where e_k is the kth standard basis vector. The vector x, or $x/\|x\|_2$ if one wishes, is a very accurate eigenvector. The factor δ_k is included just for convenience. In fact, the computation is even easier than it looks. Exercise 7.2.64 shows that x satisfies

$$N_k^T x = e_k,$$

which can be solved by "outward substitution" as follows.

$$
\begin{aligned}
&x_k \leftarrow 1 \\
&\text{for } j = k-1, ..., 1 \\
&\quad [\ x_j \leftarrow -l_j x_{j+1} \\
&\text{for } j = k, ..., n-1 \\
&\quad [\ x_{j+1} \leftarrow -\breve{u}_j x_j
\end{aligned}
\tag{7.2.36}
$$

This procedure requires $n-1$ multiplications and no additions or subtractions. Thus there are no cancellations, and x is computed to high relative accuracy. For more details and an error analysis see [20].

For clustered eigenvalues a more complex strategy is needed. Shift the matrix by ρ, where ρ is very near the cluster. Compute a new representation by Algorithm (7.2.13). This shifted matrix has many eigenvalues near zero. Although the absolute separation of these eigenvalues is no different than it was before the shift, their relative separation is now much greater, as they are all now much smaller in magnitude. Of course their relative accuracy is also much smaller than it was, so they need to be recomputed or refined. Once they have been recomputed to high relative accuracy, their eigenvectors can be computed using twisted factorizations, as described above. These highly accurate eigenvectors are numerically orthogonal to each other as well as to all of the other computed eigenvectors. This process must be repeated for each cluster. For details consult [19], [20], and [71].

Cuppen's Divide-and-Conquer Algorithm

We include here a brief description of Cuppen's divide-and-conquer algorithm. For more details see [13], [23]. The algorithm is implemented in LAPACK [1].

A symmetric, tridiagonal matrix A can be rewritten as $A = \tilde{A} + H$, where

$$\tilde{A} = \begin{bmatrix} \tilde{A}_1 & 0 \\ 0 & \tilde{A}_2 \end{bmatrix}$$

$$= \left[\begin{array}{ccccc|ccccc} \alpha_1 & \beta_1 & & & & & & & & \\ \beta_1 & \alpha_2 & \ddots & & & & & & & \\ & \ddots & \ddots & \beta_{i-1} & & & & & & \\ & & \beta_{i-1} & (\alpha_i - \beta_i) & & 0 & & & & \\ \hline & & & 0 & & (\alpha_i - \beta_i) & \beta_{i+1} & & & \\ & & & & & \beta_{i+1} & \alpha_{i+2} & \ddots & & \\ & & & & & & \ddots & \ddots & \beta_{n-1} \\ & & & & & & & \beta_{n-1} & \alpha_n \end{array}\right]$$

and H is a very simple matrix of rank one whose only nonzero entries consist of the submatrix

$$\begin{bmatrix} \beta_i & \beta_i \\ \beta_i & \beta_i \end{bmatrix}$$

in the "middle" of H. This can be done for any choice of i, but our interest is in the case $i \approx n/2$, for which the submatrices $\tilde{A}_1$ and $\tilde{A}_2$ are of about the same size.

The eigenproblem for $\tilde{A}$ is easier than that for A, since it consists of two separate problems for $\tilde{A}_1$ and $\tilde{A}_2$, which can be solved independently, perhaps in parallel. Suppose we have the eigenvalues and eigenvectors of $\tilde{A}$. Can we deduce the eigensystem of $A = \tilde{A} + H$ by a simple updating procedure? As we shall see, the answer is yes. Cuppen's algorithm simply applies this idea recursively. Here is a thumbnail sketch of the algorithm: If A is 1×1, its eigenvalue and eigenvector are returned. Otherwise, A is decomposed to $\tilde{A} + H$ as shown above, and the algorithm computes the eigensystems of $\tilde{A}_1$ and $\tilde{A}_2$ by calling itself. This gives the eigensystem of $\tilde{A}$. Then the eigensystem of $A = \tilde{A} + H$ is obtained by the updating procedure (described below). This simple divide-and-conquer idea works very well. In practice we might prefer not to carry the subdivisions all the way down to the 1×1 level. Instead we can set some threshold size (e.g. 10×10) and compute eigensystems of matrices that are below that size by Francis's algorithm or some other method.

Rank-One Updates

It remains to show how to obtain the eigenvalues and eigenvectors of A from those of $\tilde{A}$, where $A = \tilde{A} + H$ and H has rank one.

Exercise 7.2.37   Let $H \in \mathbb{R}^{n \times n}$ be a matrix of rank one.

(a) Show that there exist nonzero vectors $u, v \in \mathbb{R}^n$ such that $H = uv^T$.

(b) Show that if H is symmetric, then there exist nonzero $\rho \in \mathbb{R}$ and $w \in \mathbb{R}^n$ such that $H = \rho w w^T$. Show that w can be chosen so that $\| w \|_2 = 1$, in which case ρ is uniquely determined. To what extent is w nonunique?

(c) Let $H = \rho w w^T$, where $\| w \|_2 = 1$. Determine the complete eigensystem of H.

$\square$

Now we can write $A = \tilde{A} + \rho w w^T$, where $\| w \|_2 = 1$. Since we have the entire eigensystem of $\tilde{A}$, we have $\tilde{A} = \tilde{Q} \tilde{D} \tilde{Q}^T$, where $\tilde{Q}$ is an orthogonal matrix whose columns are eigenvectors, and $\tilde{D}$ is a diagonal matrix whose main-diagonal entries are eigenvalues of $\tilde{A}$. Thus

$$A = \tilde{Q}(\tilde{D} + \rho z z^T)\tilde{Q}^T,$$

where $z = \tilde{Q}^T w$. We seek an orthogonal Q and diagonal D such that $A = QDQ^T$. If we can find an orthogonal $\hat{Q}$ and a diagonal D such that $\tilde{D} + \rho z z^T = \hat{Q} D \hat{Q}^T$, then $A = QDQ^T$, where $Q = \tilde{Q}\hat{Q}$. Thus it suffices to consider the matrix $\tilde{D} + \rho z z^T$.

The eigenvalues of $\tilde{D}$ are its main-diagonal entries, $\tilde{d}_1, \ldots, \tilde{d}_n$, and the associated eigenvectors are $e_1, \ldots, e_n$, the standard basis vectors of $\mathbb{R}^n$. The next exercise shows that it can happen that some of these are also eigenpairs of $\tilde{D} + \rho z z^T$.

Exercise 7.2.38   Let z_i denote the ith component of z. Suppose $z_i = 0$ for some i.

(a) Show that the ith column of $\tilde{D} + \rho z z^T$ is $\tilde{d}_i e_i$ and the ith row is $\tilde{d}_i e_i^T$.

(b) Show that $\tilde{d}_i$ is an eigenvalue of $\tilde{D} + \rho z z^T$ with associated eigenvector e_i.

(c) Show that if $q \in \mathbb{R}^n$ is an eigenvector of $\tilde{D} + \rho z z^T$ with associated eigenvalue $\lambda \neq \tilde{d}_i$, then q_i, the ith component of q, is zero.

$\square$

From this exercise we see that if $z_i = 0$, we get an eigenpair for free. Furthermore the ith row and column of $\tilde{D} + \rho z z^T$ can be ignored during the computation of the other eigenpairs. thus we effectively work with a submatrix; that is, we deflate the problem.

Another deflation opportunity arises if two or more of the $\tilde{d}_i$ happen to be equal.

cise 7.2.39 Suppose $\tilde{d}_1 = \tilde{d}_2 = \cdots = \tilde{d}_k$. Construct a reflector U such that $U(\tilde{D} + \rho z z^T)U^T = \tilde{D} + \rho \tilde{z} \tilde{z}^T$, and $\tilde{z}$ has $k - 1$ entries equal to zero. At what point are you using the fact that $\tilde{d}_1 = \cdots = \tilde{d}_k$? $\qquad\square$

Exercise 7.2.39 shows that whenever we have k equal $\tilde{d}_i$, we can deflate $k - 1$ eigenpairs.

Thanks to these deflation procedures, we can now assume without loss of generality that all of the z_i are nonzero and the $\tilde{d}_i$ are all distinct. We may also assume, by reordering if necessary, that $\tilde{d}_1 < \tilde{d}_2 < \cdots < \tilde{d}_n$. It is not hard to see what the eigenpairs of $\tilde{D} + \rho z z^T$ must look like. Let λ be an eigenvalue with associated eigenvector q. Then $(\tilde{D} + \rho z z^T)q = \lambda q$; that is,

$$(\tilde{D} - \lambda I)q + \rho z(z^T q) = 0. \tag{7.2.40}$$

It is not hard to show that the scalar $z^T q$ is nonzero and that λ is distinct from $\tilde{d}_1, \tilde{d}_2, \ldots, \tilde{d}_n$.

cise 7.2.41

(a) Use (7.2.40) to show that if $z^T q = 0$, then $\lambda = \tilde{d}_i$ and q is a multiple of e_i for some i. Conclude that $z_i = 0$, in contradiction to one of our assumptions.

(b) Use the ith row of (7.2.40) to show that if $\lambda = \tilde{d}_i$, then either $z^T q = 0$ or $z_i = 0$. Thus $\lambda \neq \tilde{d}_i$.

$\qquad\qquad\qquad\square$

Since $\lambda \neq \tilde{d}_i$ for all i, $(\tilde{D} - \lambda I)^{-1}$ exists. Multiplying equation (7.2.40) by $(\tilde{D} - \lambda I)^{-1}$, we obtain

$$q + \rho(\tilde{D} - \lambda I)^{-1}z(z^T q) = 0. \tag{7.2.42}$$

Multiplying this equation by z^T on the left and then dividing by the nonzero scalar $z^T q$, we see that

$$1 + \rho z^T(\tilde{D} - \lambda I)^{-1}z = 0. \tag{7.2.43}$$

Since $(\tilde{D} - \lambda I)^{-1}$ is a diagonal matrix, (7.2.43) can also be written as

$$1 + \rho \sum_{i=1}^{n} \frac{z_i^2}{\tilde{d}_i - \lambda} = 0. \tag{7.2.44}$$

Every eigenvalue of $\tilde{D} + \rho z z^T$ must satisfy (7.2.44), which is known as the *secular equation*. A great deal can be learned about the solutions of (7.2.44) by studying the function

$$f(\lambda) = 1 + \rho \sum_{i=1}^{n} \frac{z_i^2}{\tilde{d}_i - \lambda}. \tag{7.2.45}$$

This is a rational function with the n distinct poles $\tilde{d}_1, \tilde{d}_2, \ldots, \tilde{d}_n$.

Exercise 7.2.46 Let f be as in (7.2.45)

(a) Compute the derivative of f. Show that if $\rho > 0$ ($\rho < 0$), then f is increasing (resp. decreasing) on each subinterval on which it is defined.

(b) Sketch the graph of f for each of the cases $\rho > 0$ and $\rho < 0$. Show that the secular equation $f(\lambda) = 0$ has exactly one solution between each pair of poles of f.

(c) Let $d_1 < d_2 < \cdots < d_n$ denote the solutions of the secular equation. Show that if $\rho > 0$, then $\tilde{d}_i < d_i < \tilde{d}_{i+1}$ for $i = 1, \ldots, n-1$.

(d) Recall from Exercise 5.4.55 that the trace of a matrix is equal to the sum of its eigenvalues. Applying this fact to the matrix $\tilde{D} + \rho z z^T$, show that $d_n < \tilde{d}_n + \rho$ if $\rho > 0$. (For a more general result see Exercise 7.2.65.)

(e) Show that if $\rho < 0$, then $\tilde{d}_{i-1} < d_i < \tilde{d}_i$ for $i = 2, \ldots, n$.

(f) Using a trace argument as in part (d), show that if $\rho < 0$, then $\tilde{d}_1 + \rho < d_1$. (This is extended in Exercise 7.2.65)

□

Thanks to Exercise 7.2.46, we know that each of the $n-1$ intervals $(\tilde{d}_i, \tilde{d}_{i+1})$ contains exactly one eigenvalue of $\tilde{D} + \rho z z^T$, and the nth eigenvalue is located in either $(\tilde{d}_n, \tilde{d}_n + \rho)$ or $(\tilde{d}_1 + \rho, \tilde{d}_1)$, depending upon the sign of ρ. Using this information, we can solve the secular equation (7.2.44) numerically and thereby determine the eigenvalues of $\tilde{D} + \rho z z^T$ with as much accuracy as we please. For this we could employ the bisection method, which was introduced earlier in this section, for example. However, much faster methods are available. A great deal of effort went into the development of the solver that is used in the LAPACK [1] implementation of this algorithm.

Once the eigenvalues of $\tilde{D} + \rho z z^T$ have been found, the eigenvectors can be obtained using (7.2.42). For each eigenvalue λ, an associated eigenvector q is given by

$$q = c(\tilde{D} - \lambda I)^{-1} z,$$

where c is any nonzero scalar. Thus the components of q are given by

$$q_i = \frac{c z_i}{\tilde{d}_i - \lambda} \qquad i = 1, \ldots, n.$$

In this brief sketch we have ignored numerous details.

Additional Exercises

Exercise 7.2.47 In this exercise you will show that the matrix

$$A = \begin{bmatrix} a & b \\ b & d \end{bmatrix}$$

can always be diagonalized by a rotator

$$Q = \begin{bmatrix} c & -s \\ s & c \end{bmatrix},$$

a *Jacobi rotator*.

(a) Show that

$$Q^T A Q = \begin{bmatrix} c^2 a + s^2 d + 2csb & (c^2 - s^2)b + cs(d - a) \\ (c^2 - s^2)b + cs(d - a) & c^2 d + s^2 a - 2csb \end{bmatrix}.$$

(b) Show that if $a = d$, then $Q^T A Q$ can be made diagonal by taking $c = s = 1/\sqrt{2}$. Otherwise, letting

$$\hat{t} = \frac{2b}{a - d}, \tag{7.2.48}$$

show that $Q^T A Q$ is diagonal if and only if

$$\hat{t} = \frac{2cs}{c^2 - s^2} = \tan 2\theta,$$

where $c = \cos\theta$ and $s = \sin\theta$. There is a unique $\theta \in (-\pi/4, \pi/4)$ for which $\hat{t} = \tan 2\theta$.

(c) Show that for any $\theta \in (-\pi/4, \pi/4)$,

$$\tan\theta = \frac{\sin 2\theta}{1 + \cos 2\theta},$$

$$\cos 2\theta = \frac{1}{\sqrt{1 + \tan^2 2\theta}}, \qquad \sin 2\theta = \cos 2\theta \tan 2\theta,$$

and

$$\tan\theta = \frac{\tan 2\theta}{1 + \sqrt{1 + \tan^2 2\theta}}.$$

Conclude that the tangent $t = \tan\theta$ of the angle that makes $Q^T A Q$ diagonal is given by

$$t = \frac{\hat{t}}{1 + \sqrt{1 + \hat{t}^2}}, \tag{7.2.49}$$

where $\hat{t}$ is as in (7.2.48). Show that $c = \cos\theta$ and $s = \sin\theta$ can be obtained from t by

$$c = \frac{1}{\sqrt{1 + t^2}} \qquad \text{and} \qquad s = ct.$$

(d) There is a (very slight) chance of overflow in (7.2.48) if $a - b$ is tiny. This can be eliminated by working with the reciprocal whenever $|a - b| < |2b|$. Let

$$\hat{k} = \frac{a - d}{2b}.$$

Then $\hat{k} = \cot 2\theta$, where θ is the angle of the desired rotator. Show that $t = \tan \theta$ is given by

$$t = \frac{\text{sign}(\hat{k})}{|\hat{k}| + \sqrt{1 + \hat{k}^2}}. \qquad (7.2.50)$$

(Rewrite (7.2.49) in terms of $\hat{k} = 1/\hat{t}$.)

(e) Show that

$$Q^T A Q = \begin{bmatrix} a + tb & 0 \\ 0 & d - tb \end{bmatrix}.$$

$\square$

Exercise 7.2.51 Use the formulas developed in Exercise 7.2.47 to diagonalize the matrices

$$\begin{bmatrix} 2 & 1 \\ 1 & 3 \end{bmatrix} \quad \text{and} \quad \begin{bmatrix} 5 & 6 \\ 6 & 1 \end{bmatrix}.$$

Do each computation two ways, once using (7.2.49), and once using (7.2.50). Check your answers by computing the eigenvalues of the matrices by some other means. $\square$

Exercise 7.2.52 Show that the search for the largest off-diagonal entry of an $n \times n$ matrix costs $O(n^2)$ work, while the cost of applying a Jacobi rotator is $O(n)$. Thus, for large n, the classical Jacobi method expends the vast majority of its effort on the searches. $\square$

Exercise 7.2.53 Let $\hat{A} = Q^T A Q$, where Q is the Jacobi rotator that sets a_{ij} to zero. Let D and $\hat{D}$ be diagonal matrices, and let E and $\hat{E}$ be symmetric matrices with zeros on the main diagonal, uniquely determined by the equations $A = D + E$ and $\hat{A} = \hat{D} + \hat{E}$.

(a) Prove that

$$\|\hat{E}\|_F^2 = \|E\|_F^2 - 2a_{ij}^2 \quad \text{and} \quad \|\hat{D}\|_F^2 = \|D\|_F^2 + 2a_{ij}^2.$$

Thus the Jacobi rotator transfers weight $2a_{ij}^2$ from off-diagonal onto main diagonal entries.

(b) In the classical Jacobi method we always annihilate the off-diagonal entry of greatest magnitude. Prove that the largest off-diagonal entry always satisfies $a_{ij}^2 \geq \|E\|_F^2/N$, where $N = n(n-1)$. Infer that $\|\hat{E}\|_F^2 \leq (1 - 1/N)\|E\|_F^2$. Thus after m steps of the classical Jacobi method, $\|E\|_F^2$ will have been reduced at least by a factor of $(1 - 2/N)^m$. Thus the classical Jacobi method is guaranteed to converge, and the convergence is no worse than linear. Exercise 7.2.54 will show that the convergence is in fact quadratic.

$\square$

Exercise 7.2.54 Let $A \in \mathbb{R}^{n \times n}$ be a symmetric matrix with distinct eigenvalues $\lambda_1, \ldots, \lambda_n$, and let $\delta = \min\{|\lambda_i - \lambda_j| \mid i \neq j\}$. We will continue to use the notation $A = D + E$

established in Exercise 7.2.53. Suppose that at the beginning of a cyclic Jacobi sweep, $\| E \|_F = \epsilon$, where ϵ is small compared with δ. then $|a_{ij}| \le \epsilon$ for all $i \ne j$. The elements must stay this small since $\| E \|_F$ in nonincreasing. Suppose further that $\| E \|_F$ is small enough that the main-diagonal entries of A are fairly close to the eigenvalues.

(a) Show that each rotator generated during the sweep must satisfy $|s| \le O(\epsilon)$ and $c \approx 1$.

(b) Using the result of part (a), show that once each a_{ik} is set to zero, subsequent rotations can make it no bigger than $O(\epsilon^2)$. Thus, at the end of a complete sweep, every off-diagonal entry is $O(\epsilon^2)$. This means that the convergence is quadratic.

A closer analysis reveals that quadratic convergence occurs even when the eigenvalues are not distinct. □

cise 7.2.55

(a) Compute the matrix $\Delta^{-1} A \Delta$, where A is as in (7.2.5), and

$$\Delta = \mathrm{diag}\{\delta_1, \ldots, \delta_n\}$$

with all $\delta_j \ne 0$. Show how to choose the δ_j so that $\Delta^{-1} A \Delta$ has the form (7.2.15). Notice that δ_1 (or any one of the δ_j) can be chosen arbitrarily, after which all of the other δ_j are uniquely determined.

(b) Show that any matrix of the form (7.2.16) with all γ_j positive is diagonally similar to a matrix of the form (7.2.5) with $\beta_j = \sqrt{\gamma_j}$, $j = 1, \ldots, n-1$.

(c) Check that the decomposition $\tilde{A} = \tilde{L}\tilde{R}$ given by (7.2.17) holds with $\tilde{r}_j = \alpha_j - \tilde{l}_{j-1}$, $j = 1, \ldots, n$ (if we define $\tilde{l}_0 = 0$), and $\tilde{l}_j = \gamma_j / \tilde{r}_j$, $j = 1, \ldots, n-1$.

□

cise 7.2.56 This exercise shows how to apply a shift to a tridiagonal matrix $\tilde{A}$ in factored form without explicitly forming $\tilde{A}$. Suppose we have a decomposition $\tilde{A} = \tilde{L}\tilde{R}$ of the form (7.2.17), and we want to obtain the decomposition of a shifted matrix: $\tilde{A} - \rho I = \hat{L}\hat{R}$.

(a) Show that the entries of $\hat{L}$ and $\hat{R}$ can be obtained directly from those of $\tilde{L}$ and $\tilde{R}$ by the algorithm

$$
\begin{aligned}
&\hat{r}_1 \leftarrow \tilde{r}_1 - \rho \\
&\text{for } j = 1, \ldots, n-1 \\
&\quad \left[
\begin{aligned}
&\hat{l}_j \leftarrow \tilde{l}_j \tilde{r}_j / \hat{r}_j \\
&\hat{r}_{j+1} \leftarrow \tilde{r}_{j+1} + \tilde{l}_j - \hat{l}_j - \rho
\end{aligned}
\right.
\end{aligned}
$$

This is the *obvious* form of the algorithm.

(b) Introduce auxiliary quantities t_j by $t_1 = -\rho$ and $t_{j+1} = \tilde{l}_j - \hat{l}_j - \rho = \hat{r}_{j+1} - \tilde{r}_{j+1}$, j= $1, \ldots, n-1$. Show that the algorithm from part (a) can be rewritten as (7.2.19), which is the *differential* form of the algorithm. It is more accurate than the version from part (a).

$\square$

Exercise 7.2.57 In this exercise you will prove Sylvester's law of inertia. Let $A, B \in \mathbb{R}^{n \times n}$ be symmetric matrices.

(a) Show that the dimension of the null space of A is equal to the number of eigenvalues of A that are zero.

(b) Show that if B is congruent to A, then $\mathcal{N}(A)$ and $\mathcal{N}(B)$ have the same dimension. Deduce that $\zeta(A) = \zeta(B)$ and $\nu(A) + \pi(A) = \nu(B) + \pi(B)$.

(c) Show that if B is congruent to A, then there exist diagonal matrices D and E and a nonsingular matrix C such that $E = C^T D C$, the eigenvalues of A are the main-diagonal entries of D, and the eigenvalues of B are the main-diagonal entries of E.

(d) Continuing from part (c), the number of positive entries on the main diagonal of D and E are equal to $\pi(A)$ and $\pi(B)$, respectively, and we would like to show that these are the same. Let us assume that they are different and show that this leads to a contradiction. Without loss of generality, assume E has more positive entries than D does. Show that there is a nonzero vector x such that
$$\begin{aligned} x_i = 0 &\qquad \text{if } e_{ii} \leq 0 \\ (Cx)_i = 0 &\qquad \text{if } d_{ii} > 0. \end{aligned}$$
(*Hint:* These are homogeneous linear equations in n unknowns. How many equations are there?)

(e) Let x be the nonzero vector from part (d), and let $z = Cx$. Show that $x^T E x > 0$, $z^T D z \leq 0$, and $x^T E x = z^T D z$, which yields a contradiction.

(f) Show that $(\nu(A), \zeta(A), \pi(A)) = (\nu(B), \zeta(B), \pi(B))$.

$\square$

Exercise 7.2.58 Show that if $A - \rho I = LR$ and $RL + \rho I = \hat{A}$, then $\hat{A} = L^{-1}AL$. Thus the LR iteration is a similarity transformation. $\square$

Exercise 7.2.59 Let L and R be as in (7.2.17), let $\hat{A} = RL$ and $\hat{A} - \rho I = \hat{L}\hat{R}$, where $\hat{L}$ and $\hat{R}$ have the same general form as in (7.2.17).

(a) Compute $\hat{A}$, $\hat{L}$, and $\hat{R}$, and verify that the entries of the factors $\hat{L}$ and $\hat{R}$ can be obtained directly from those of L and R by the algorithm (7.2.27).

(b) Show that if $l_{n-1} = 0$, then r_n is an eigenvalue of A and $\hat{A}$.

$\square$

cise 7.2.60 Let $s_j = r_j - \rho - \hat{l}_{j-1}$, $j = 1, \ldots, n$, where $\hat{l}_0 = 0$. Show that in algorithm (7.2.27)

(a) $s_j = \hat{r}_j - l_j$, $j = 1, \ldots, n - 1$.

(b) $s_{j+1} = r_{j+1} - r_{j+1}l_j/\hat{r}_j - \rho = r_{j+1}s_j/\hat{r}_j - \rho$, $j = 1, \ldots, n - 1$.

(c) Confirm that (7.2.27) and (7.2.28) are equivalent in exact arithmetic.

$\square$

cise 7.2.61 Let A be a tridiagonal matrix of the form (7.2.16) with all $\gamma_j > 0$, and suppose A is diagonally similar to a positive definite matrix. Then A has a decomposition $A = LR$ of the form (7.2.17). Show that the quantities $r_1, \ldots, r_n$ in R are all positive (cf. Exercise 7.2.18). Then show that the entries $l_1, \ldots, l_{n-1}$ of L are positive as well. $\square$

cise 7.2.62 Consider the 5×5 matrix

$$A = LR = \begin{bmatrix} 1 & & & & \\ m & 1 & & & \\ & m & 1 & & \\ & & m & 1 & \\ & & & m & 1 \end{bmatrix} \begin{bmatrix} 1 & 1 & & & \\ & 1 & 1 & & \\ & & 1 & 1 & \\ & & & 1 & 1 \\ & & & & 1 \end{bmatrix},$$

where $m = 123456$.

(a) Using MATLAB, multiply L by R to get A explicitly, then use the `eig` command to compute the eigenvalues of A. Notice that there are four large eigenvalues around 123456 and one tiny one. The normwise backward stability of Francis's algorithm (which `eig` uses) guarantees that the four large eigenvalues are accurate. The fifth eigenvalue is surely near zero, but there is no guarantee that `eig` has computed it accurately in a relative sense.

(b) Prove that the determinant of A is 1. Recall from Exercise 5.4.56 that the determinant of a matrix is the product of its eigenvalues. Thus $\det(A) = \lambda_1\lambda_2\lambda_3\lambda_4\lambda_5$. Compute $\lambda_1\lambda_2\lambda_3\lambda_4\lambda_5$ using the eigenvalues you computed in part (a), and notice that this product is far from 1. This proves that λ_5, the tiny eigenvalue, has not been computed accurately; it is off by a factor of about a million. To get the correct value, use the equation $\lambda_1\lambda_2\lambda_3\lambda_4\lambda_5 = 1$ to solve for λ_5: $\lambda_5 = 1/(\lambda_1\lambda_2\lambda_3\lambda_4)$. Since the large eigenvalues are certainly accurate, and this formula for λ_5 involves only multiplications and divisions (no cancellations), it gives the correct value of λ_5 to 15 decimal places or so. Use `format long e` to view all digits of λ_5.

(c) Write a MATLAB program that implements the dqds algorithm (7.2.28). Apply the dqds algorithm with zero shift to the factors of A. If your program is correct, it will give the correct value of the tiny eigenvalue to about five decimal places after only one iteration. Perform a second iteration and get the correct tiny eigenvalue to sixteen decimal places accuracy.

$\square$

Exercise 7.2.63 Show that the equation (7.2.30) implies the following relationships.

(a) $d_j + d_{j-1}l_{j-1}^2 - \rho = \breve{d}_j + \breve{d}_{j+1}\breve{u}_j^2$ for $j = 1, \ldots, n$, where we set $d_0 = l_0 = \breve{d}_{n+1} = \breve{u}_n = 0$.

(b) $l_j d_j = \breve{d}_{j+1}\breve{u}_j$ for $j = 1, \ldots, n - 1$.

(c) The relationships from parts (a) and (b) can be used to construct an algorithm for computing $\breve{U}$ and $\breve{D}$. However, a better *differential* form of the algorithm is obtained by introducing the auxiliary quantities

$$p_j = d_j - \breve{d}_{j+1}\breve{u}_j^2 - \rho,$$

with $p_n = d_n - \rho$. Show that $p_j = \breve{d}_j - d_{j-1}l_{j-1}^2$ for $j = 2, \ldots, n$, and $p_j = p_{j+1}(d_j/\breve{d}_{j+1}) - \rho$ for $j = 2, \ldots, n - 1$.

(d) Verify that algorithm (7.2.31) produces the parameters that define $\breve{U}$ and $\breve{D}$.

$\square$

Exercise 7.2.64 Consider a twisted factorization $N_k D_k N_k^T$, as in (7.2.32), with N_k as in (7.2.33).

(a) Devise an "inward substitution" algorithm to solve systems of the form $N_k z = w$ for z, where w is a given vector. Show that the algorithm is simplified drastically when $w = e_k$, the kth standard unit vector; in fact, $N_k e_k = e_k$, so $z = e_k$.

(b) Show that $N_k D_k N_k^T x = e_k \delta_k$ if and only if $N_k^T x = e_k$.

(c) Devise an algorithm to solve systems of the form $N_k^T x = y$ by "outward substitution."

(d) Show that when $y = e_k$, the algorithm from part (c) reduces to (7.2.36).

$\square$

Exercise 7.2.65 This exercise extends the eigenvalue bounds of Exercise 7.2.46. Prove the following results using the equation $\text{tr}(\tilde{D} + \rho z z^T) = \text{tr}(\tilde{D}) + \rho \text{tr}(z z^T)$ and the inequalities proved in Exercise 7.2.45.

(a) If $\rho > 0$, then $d_i < \tilde{d}_i + \rho$ for $i = 1, \ldots, n$.

(b) If $\rho < 0$, then $d_i > \tilde{d}_i + \rho$ for $i = 1, \ldots, n$.

(c) Regardless of the sign of ρ, there exist positive constants $c_1, \ldots, c_n$ such that $c_1 + \cdots + c_n = 1$ and $d_i = \tilde{d}_i + c_i \rho$ for $i = 1, \ldots, n$.

$\square$

rcise 7.2.66 Let $A = \begin{bmatrix} 2 & 2 \\ 2 & 5 \end{bmatrix}$.

(a) Write A in the form $\tilde{D} + \rho z z^T$, where $\tilde{D}$ is diagonal.

(b) Graph the function $f(\lambda)$ of (7.2.45) given by this choice of $\tilde{D}$, ρ, and z.

(c) Calculate the eigenvalues and eigenvectors of A by performing a rank-one update.

$\square$

7.3 PRODUCT EIGENVALUE PROBLEMS

Consider the problem of computing eigenvalues and eigenvectors of a product AC, where A and C are two given matrices. We already studied a special case of this problem in Section 5.8. In the course of computing the SVD of a matrix A, we obtained the complete eigensystems of the products $A^T A$ and $A A^T$ without ever actually computing these products. Now we want to do the same thing with A and C. We want eigensystem information about the product AC, but we want to get it by working with the factors A and C, never computing the product AC or the related product CA.

A second application arises in Section 7.4, where we consider the generalized eigenvalue problem $Av = \lambda Bv$. If B is nonsingular, the generalized eigenvalue problem is equivalent to the standard eigenvalue problem for the product AB^{-1} or $B^{-1}A$. In Section 7.4 we will see how to solve these eigenvalue problems working directly with A and B, never forming the products. An added twist there is that the data we have to work with are A and B, not A and B^{-1}.

One can also consider products $A_1 A_2 \cdots A_k$ of more than two factors. The methodology developed in this section can be extended to those cases. See [100] or [101, Ch. 8] and works of Kressner [57, 58]. The inspiration for the approach taken in this section comes from [58].

Now consider a product of two matrices, AC, and the related product CA. An easy way to study these products is to consider the larger matrix F defined by (7.3.2).

Theorem 7.3.1 *Let $A \in \mathbb{C}^{n \times m}$ and $C \in \mathbb{C}^{m \times n}$. Then the square matrices AC and CA have the same nonzero eigenvalues. The nonzero eigenvalues of*

$$F = \begin{bmatrix} 0 & A \\ C & 0 \end{bmatrix} \tag{7.3.2}$$

are the square roots of the nonzero eigenvalues of AC and CA.

Exercise 7.3.3 Prove Theorem 7.3.1. In particular, ...

(a) Let λ be a nonzero eigenvalue of AC, and let v be an associated eigenvector. Show that Cv is an eigenvector of CA associated with the eigenvalue λ. Thus every nonzero eigenvalue of AC is an eigenvalue of CA. Note: As part of the proof you must check that $Cv \neq 0$.

(b) conversely, show that every nonzero eigenvalue of CA is an eigenvalue of AC.

(c) What is the situation with zero eigenvalues?

(d) Suppose $\begin{bmatrix} v \\ w \end{bmatrix}$ is an eigenvector of F associated with the nonzero eigenvalue μ. Show that $\begin{bmatrix} v \\ -w \end{bmatrix}$ is an eigenvector of F associated with $-\mu$. Thus the nonzero eigenvalues of F occur in $\pm\mu$ pairs.

(e) Continuing from part (d), show that v and w are eigenvectors of AC and CA, respectively, associated with the eigenvalue $\lambda = \mu^2$. Do not neglect to check that neither v nor w can be the zero vector.

(f) Conversely, let λ be a nonzero eigenvalue of AC and CA, and let μ be either square root of λ. Starting from an eigenvector of either AC or CA, show how to construct an eigenvector of F associated with the eigenvalue μ.

(g) Deduce that the nonzero eigenvalues of F are exactly the square roots of the nonzero eigenvalues of AC and CA.

(h) Compute the matrix F^2. Comment on the result.

$\square$

From Theorem 7.3.1 and its proof in Exercise 7.3.3, we see that we can get all of the spectral information about AC and CA by studying the matrix

$$F = \begin{bmatrix} 0 & A \\ C & 0 \end{bmatrix},$$

whose formation does not require the computation of any products. From this point on, we will focus on the matrix F.

Also from this point on we will assume that A and C are square, both $n \times n$. We do this simply for convenience; it is not at all essential. The interested reader can work out the details for the non-square case. Let $P \in \mathbb{R}^{2n \times 2n}$ be the perfect shuffle permutation matrix (Exercises 5.8.28 and 5.8.29), and let $F_s = P^T F P$. The unitary similarity transformation $F \to F_s$ does a perfect shuffle of the rows and columns of

F. In the case $n = 4$, for example, we have

$$
F =
\begin{bmatrix}
 & & & & a_{11} & a_{12} & a_{13} & a_{14} \\
 & & & & a_{21} & a_{22} & a_{23} & a_{24} \\
 & & & & a_{31} & a_{32} & a_{33} & a_{34} \\
 & & & & a_{41} & a_{42} & a_{43} & a_{44} \\
\hline
c_{11} & c_{12} & c_{13} & c_{14} & & & & \\
c_{21} & c_{22} & c_{23} & c_{24} & & & & \\
c_{31} & c_{32} & c_{33} & c_{34} & & & & \\
c_{41} & c_{42} & c_{43} & c_{44} & & & &
\end{bmatrix}
$$

and

$$
F_s =
\begin{bmatrix}
 & a_{11} & & a_{12} & & a_{13} & & a_{14} \\
c_{11} & & c_{12} & & c_{13} & & c_{14} & \\
\hline
 & a_{21} & & a_{22} & & a_{23} & & a_{24} \\
c_{21} & & c_{22} & & c_{23} & & c_{24} & \\
\hline
 & a_{31} & & a_{32} & & a_{33} & & a_{34} \\
c_{31} & & c_{32} & & c_{33} & & c_{34} & \\
\hline
 & a_{41} & & a_{42} & & a_{43} & & a_{44} \\
c_{41} & & c_{42} & & c_{43} & & c_{44} &
\end{bmatrix}
. \qquad (7.3.4)
$$

When we work with F_s we say we are in the *shuffled coordinate system* or we are looking at the *shuffled picture*. When we work with F we are in the *unshuffled* or *deshuffled coordinate system* and we are looking at the *deshuffled picture*. Both pictures are useful.

Reduction to Hessenberg Form

We start with the shuffled picture. Consider the reduction of F_s to upper Hessenberg form by the standard algorithm, which was described in Section 5.5. The first step of the algorithm operates on rows 2 through $2n$ and creates zeros in the first column from the third position on down. Looking at the first column of F_s in (7.3.4), we see that the entries in odd rows are already zero, so the transformation that creates the zeros can act on the even rows only. This left-multiplication transformation acts only on entries of C. When the similarity transformation is completed by a right multiplication, only the even columns, which contain entries of A, are affected.

The second step creates zeros in the second column below the third row. Noticing that the entries in even positions are already zero, we deduce that the desired zeros can be created by a left-multiplication transformation that acts only on odd rows 3, $5, \ldots, 2n - 1$. This affects only entries of A and does not touch the first row. The similarity transformation is completed by a right multiplication that acts only on odd columns $3, 5, \ldots, 2n - 1$. This affects only entries of C, and it does not touch the first column. Thus the zeros that were created on the first step are preserved. Moreover, all of the many zeros that were present in F_s to begin with have been preserved all along.

The third step will create the desired zeros in the third column. The left transformation acts only on elements of C and does not touch the first row of C. The right transformation acts only on elements of A and does not touch the first column of A. Continuing in this manner for $2n - 2$ steps, we transform F_s to the upper Hessenberg form

$$
\left[
\begin{array}{c|c|c|c}
\begin{matrix} h_{11} \\ t_{11} \end{matrix} & \begin{matrix} h_{12} \\ t_{12} \end{matrix} & \begin{matrix} h_{13} \\ t_{13} \end{matrix} & \begin{matrix} h_{14} \\ t_{14} \end{matrix} \\
\hline
\begin{matrix} h_{21} \\ \, \end{matrix} & \begin{matrix} h_{22} \\ t_{22} \end{matrix} & \begin{matrix} h_{23} \\ t_{23} \end{matrix} & \begin{matrix} h_{24} \\ t_{24} \end{matrix} \\
\hline
 & \begin{matrix} h_{32} \\ \, \end{matrix} & \begin{matrix} h_{33} \\ t_{33} \end{matrix} & \begin{matrix} h_{34} \\ t_{34} \end{matrix} \\
\hline
 & & \begin{matrix} h_{43} \\ \, \end{matrix} & \begin{matrix} h_{44} \\ t_{44} \end{matrix}
\end{array}
\right].
$$

If we now deshuffle this matrix, we obtain

$$
\left[
\begin{array}{cccc|cccc}
 & & & & h_{11} & h_{12} & h_{13} & h_{14} \\
 & & & & h_{21} & h_{22} & h_{23} & h_{24} \\
 & & & & & h_{32} & h_{33} & h_{34} \\
 & & & & & & h_{43} & h_{44} \\
\hline
t_{11} & t_{12} & t_{13} & t_{14} & & & & \\
 & t_{22} & t_{23} & t_{24} & & & & \\
 & & t_{33} & t_{34} & & & & \\
 & & & t_{44} & & & &
\end{array}
\right].
$$

Thus we see that A and C have been transformed to an upper Hessenberg matrix H and an upper triangular matrix T, respectively. These developments have led us to the following theorem.

Theorem 7.3.5 *Let $A, C \in \mathbb{C}^{n \times n}$. Then there exist unitary $Q, U \in \mathbb{C}^{n \times n}$ such that the matrix $H = Q^* A U$ is upper Hessenberg and the matrix $T = U^* C Q$ is upper triangular. These unitary equivalence transformations can also be combined to form the single unitary similarity transformation*

$$
\begin{bmatrix} 0 & H \\ T & 0 \end{bmatrix} = \begin{bmatrix} Q^* & 0 \\ 0 & U^* \end{bmatrix} \begin{bmatrix} 0 & A \\ C & 0 \end{bmatrix} \begin{bmatrix} Q & 0 \\ 0 & U \end{bmatrix}. \tag{7.3.6}
$$

The matrices H, T, Q, and U can be computed by a direct method that takes $O(n^3)$ flops.

We also note that the products $HT = Q^*(AC)Q$ and $TH = U^*(CA)U$ are upper Hessenberg matrices that are unitarily similar to the products we began with.

Proof. It is a simple matter to check that the unitary similarity transformation (7.3.6) is equivalent to the two equations $H = Q^* A U$ and $T = U^* C Q$. We will focus on obtaining (7.3.6). The proof is contained in the above construction of the Hessenberg form of the shuffled matrix. In what follows, we are simply restating that construction in the unshuffled coordinate system. Thus we start with the unshuffled matrix

$$F = \begin{bmatrix} 0 & A \\ C & 0 \end{bmatrix}.$$

The first step of the reduction starts by building a unitary matrix U_1 such that U_1^*C contains zeros in the first column from the second entry on. In other words, the first column of U_1^*C is in upper triangular form. Then the unitary similarity transformation

$$\begin{bmatrix} I & 0 \\ 0 & U_1^* \end{bmatrix} \begin{bmatrix} 0 & A \\ C & 0 \end{bmatrix} \begin{bmatrix} I & 0 \\ 0 & U_1 \end{bmatrix} = \begin{bmatrix} 0 & AU_1 \\ U_1^*C & 0 \end{bmatrix}$$

is performed. The left multiplication acts only on C and the right multiplication acts only on A. This creates the desired zeros in the first column of C and performs exactly the same operations as in the first step of the reduction of the shuffled matrix to Hessenberg form.

The second step starts by building a unitary matrix Q_1 such that the "A matrix" $Q_1^*AU_1$ has zeros in the first column from the third entry on. In other words, the first column of $Q_1^*AU_1$ is in upper Hessenberg form. Q_1 has the form

$$Q_1 = \begin{bmatrix} 1 & | & 0 & \cdots & 0 \\ \hline 0 & | & & & \\ \vdots & | & & \tilde{Q}_1 & \\ 0 & | & & & \end{bmatrix},$$

so the transformation $AU_1 \rightarrow Q_1^*AU_1$ does not alter the first row of AU_1. Next the similarity transformation

$$\begin{bmatrix} Q_1^* & 0 \\ 0 & I \end{bmatrix} \begin{bmatrix} 0 & AU_1 \\ U_1^*C & 0 \end{bmatrix} \begin{bmatrix} Q_1 & 0 \\ 0 & I \end{bmatrix} = \begin{bmatrix} 0 & Q_1^*AU_1 \\ U_1^*CQ_1 & 0 \end{bmatrix}$$

is carried out. This creates the desired zeros in the first column of the "A matrix" and does not destroy the zeros in the "C matrix" U_1^*C that were created on the first step, as the transformation $U_1^*C \rightarrow U_1^*CQ_1$ does not touch the first column. The operations performed here are exactly the same as in the second step of the reduction of the shuffled matrix to Hessenberg form.

The third step begins by generating a unitary matrix U_2 such that the "C matrix" $U_2^*U_1^*CQ_1$ has the desired zeros in the second column. This transformation does not alter the first row of the matrix. Then the similarity transformation

$$\begin{bmatrix} I & 0 \\ 0 & U_2^* \end{bmatrix} \begin{bmatrix} 0 & Q_1^*AU_1 \\ U_1^*CQ_1 & 0 \end{bmatrix} \begin{bmatrix} I & 0 \\ 0 & U_2 \end{bmatrix} = \begin{bmatrix} 0 & Q_1^*AU_1U_2 \\ U_2^*U_1^*CQ_1 & 0 \end{bmatrix}$$

creates the desired zeros in the "C matrix" without destroying the zeros that were created previously in either of the matrices. The operations performed here are exactly the same as in the third step of the reduction of the shuffled matrix to upper Hessenberg form.

The fourth step generates a unitary Q_2 such that Q_2^* creates the desired zeros in the second column of the "A matrix" without disturbing any of the zeros that had been created previously, and so on. After $2n - 3$ steps the matrix F will have been transformed to the form

$$\begin{bmatrix} 0 & H \\ T & 0 \end{bmatrix},$$

where H is upper Hessenberg and T is upper triangular. Defining $U = U_1 U_2 \cdots U_{n-1}$ and $Q = Q_1 Q_2 \cdots Q_{n-2}$, we have (7.3.6). $\qquad\square$

Exercise 7.3.7

(a) Write a pseudocode algorithm that implements the procedure outlined in the proof above. Assume each U_i and Q_i is built from an elementary reflector of the appropriate size, and indicate clearly which rows or columns of each matrix are affected by each transformation. Overwrite A and C by H and T, respectively. The information about the reflectors can be stored mainly in the areas where zeros are produced in the two matrices, or U and Q can be generated explicitly.

(b) Count the flops for the algorithm in part (a), and show that the flop count is $O(n^3)$, regardless of whether or not the unitary matrices U and Q are constructed explicitly.

$\qquad\square$

Exercise 7.3.8   Check that the operations in the construction in the proof of Theorem 7.3.5 are exactly the same as the operations in the reduction of the shuffled matrix to upper Hessenberg form. $\qquad\square$

In Section 5.5 we demonstrated that every square matrix can be reduced to upper Hessenberg form. Theorem 7.3.5 looks like a nice extension of that result to products of matrices. However, the perfect-shuffle viewpoint demonstrates that it is really just a special case.

Computation of Eigenvalues by Francis's Algorithm

Suppose we have reduced our matrices A and C to the form given by Theorem 7.3.5. Recycling notation, we now have a matrix

$$F = \begin{bmatrix} 0 & A \\ C & 0 \end{bmatrix},$$

where A is upper Hessenberg and C is upper triangular. If A is not properly upper Hessenberg, then we can reduce our problem to two or more smaller subproblems, so let us assume that A is properly upper Hessenberg. It is also possible to do a reduction if C is singular (Exercise 7.3.22), so we will assume that C is nonsingular; all of its main-diagonal entries are nonzero.

We return to the shuffled picture. The shuffled version of F has the form exemplified in the case $n = 4$ by

$$
F_s =
\begin{bmatrix}
a_{11} & a_{12} & a_{13} & a_{14} \\
c_{11} & c_{12} & c_{13} & c_{14} \\
\hline
a_{21} & a_{22} & a_{23} & a_{24} \\
& c_{22} & c_{23} & c_{24} \\
\hline
& a_{32} & a_{33} & a_{34} \\
& & c_{33} & c_{34} \\
\hline
& & a_{43} & a_{44} \\
& & & c_{44}
\end{bmatrix}.
$$

This is a properly upper Hessenberg matrix, so we can apply Francis iterations to it and preserve the Hessenberg structure. We know in addition from Theorem 7.3.1 that matrices of this form (shuffled or not) have eigenvalues occurring in $\pm\mu$ pairs. It therefore makes sense to exploit this structure by doing Francis iterations of degree two with shifts $\pm\rho$. Then eigenvalues will be found and extracted in $\pm\mu$ pairs. More generally we could do Francis iterations of degree $2m$ with shifts $\pm\rho_1, \ldots, \pm\rho_m$, but let us stick with the simplest case for now.

To initiate a Francis iteration of degree two with shifts $\pm\rho$, we must compute the vector

$$
p = p(F_s)e_1 = (F_s - \rho I)(F_s + \rho I)e_1 = (F_s^2 - \rho^2 I)e_1,
$$

which turns out to be

$$
p =
\begin{bmatrix}
a_{11}c_{11} - \rho^2 \\
0 \\
a_{21}c_{11} \\
\vdots \\
0
\end{bmatrix}. \tag{7.3.9}
$$

cise 7.3.10 Verify that the first column of $F_s^2 - \rho^2 I$ has the form shown in (7.3.9). You might find it easier to work in the unshuffled coordinate system: Compute the first column of the unshuffled $F^2 - \rho^2 I$, then shuffle the result. $\qquad\square$

Let Q_0 be a unitary transformation acting in the 1–3 plane such that

$$
Q_0^* p =
\begin{bmatrix}
* \\
0 \\
0 \\
\vdots \\
0
\end{bmatrix}.
$$

The Francis iteration on F_s is initiated by a transformation to $Q_0^* F_s Q_0$. The transformation $F_s \to Q_0^* F_s$ recombines rows 1 and 3. It acts only on entries from A

and does not in any way disturb the pattern of zeros in F_s. The transformation $Q_0^* F_s \rightarrow Q_0^* F_s Q_0$ recombines columns 1 and 3. It acts only on entries from C and does disturb the pattern of zeros by introducing a nonzero entry in position $(4, 1)$, shown by the cross:

$$
\begin{bmatrix}
 & a & & a & & a & & a \\
c & & c & & c & & c & \\
 & a & & a & & a & & a \\
+ & & c & & c & & c & \\
 & & & a & & a & & a \\
 & & & & c & & c & \\
 & & & & & a & & a \\
 & & & & & & c &
\end{bmatrix}.
$$

This matrix has a bulge of degree 2 of a very special form: The two bulge entries in the $(3, 1)$ and $(4, 2)$ positions are zero. We need to chase this bulge. Let Q_1 be a unitary transformation acting in the 2–4 plane such that left multiplication by Q_1^* sets the $(4, 1)$ entry to zero. This transformation acts only on "c" entries in rows 2 and 4 and does not introduce any new nonzero entries. When the similarity transform is completed by a right multiplication by Q_1, only "a" entries in columns 2 and 4 are affected, and a new bulge entry is produced in position $(5, 2)$:

$$
\begin{bmatrix}
 & a & & a & & a & & a \\
c & & c & & c & & c & \\
 & a & & a & & a & & a \\
 & & c & & c & & c & \\
 & + & & a & & a & & a \\
 & & & c & & c & & \\
 & & & & & a & & a \\
 & & & & & & c &
\end{bmatrix}.
$$

The pattern should now be clear. The next transformation will operate on rows 3 and 5, annihilating the bulge. When the similarity transformation is completed by acting on columns 3 and 5, a new bulge is created in the $(6, 3)$ position. Continuing in this manner, we chase the bulge until it is pushed off the bottom of the matrix, at which point the Francis iteration is complete.

At the end of the Francis iteration, the matrix is again in Hessenberg form. Moreover, the many zeros that are interspersed among the "a" and "c" entries have also been preserved. In other words, when the matrix is deshuffled, it still has the form

$$
\begin{bmatrix}
0 & A \\
C & 0
\end{bmatrix}.
$$

If Francis iterations of this type are repeated with reasonable shift choices, we can expect the entry $a_{n,n-1}$ to converge to zero quadratically. Once it is small enough to be neglected, the submatrix

$$\begin{bmatrix} 0 & a_{nn} \\ c_{nn} & 0 \end{bmatrix} \tag{7.3.11}$$

will house a pair of eigenvalues $\pm\sqrt{a_{nn}c_{nn}}$.

We should say a few words about shift choice. The simplest is to take the shifts $\pm\rho$ to be the eigenvalues of (7.3.11), namely $\pm\sqrt{a_{nn}c_{nn}}$. This will normally result in quadratic convergence. A better choice is to consider the 4×4 trailing submatrix

$$\left[\begin{array}{cc|cc} & a_{n-1,n-1} & & a_{n-1,n} \\ c_{n-1,n-1} & & c_{n-1,n} & \\ \hline & a_{n,n-1} & & a_{n,n} \\ & & c_{n,n} & \end{array}\right], \tag{7.3.12}$$

which has two pairs of eigenvalues $\pm\rho_1$, $\pm\rho_2$. The pair that is closer to $\pm\sqrt{a_{nn}c_{nn}}$ could be taken as the shift.

Now let's see how the Francis iteration looks in the unshuffled coordinate system. Initially we have

$$F = \left[\begin{array}{cccc|cccc} & & & & a_{11} & a_{12} & a_{13} & a_{14} \\ & & & & a_{21} & a_{22} & a_{23} & a_{24} \\ & & & & & a_{32} & a_{33} & a_{34} \\ & & & & & & a_{43} & a_{44} \\ \hline c_{11} & c_{12} & c_{13} & c_{14} & & & & \\ & c_{22} & c_{23} & c_{24} & & & & \\ & & c_{33} & c_{34} & & & & \\ & & & c_{44} & & & & \end{array}\right].$$

Then

$$p = (F^2 - \rho^2 I)e_1 = \begin{bmatrix} a_{11}c_{11} - \rho^2 \\ a_{21}c_{11} \\ 0 \\ \vdots \\ 0 \end{bmatrix}. \tag{7.3.13}$$

The Francis iteration is initiated by a unitary transformation Q_0 acting in the 1–2 plane such that $Q_0^* p = \begin{bmatrix} * & 0 & \cdots & 0 \end{bmatrix}^T$. Applying Q_0^* to F on the left, we recombine the first two rows of A. When we apply Q_0 on the right, we recombine the first two columns of C, and in the process we create a bulge entry:

$$
\left[
\begin{array}{cccc|cccc}
 & & & & a & a & a & a \\
 & & & & a & a & a & a \\
 & & & & & a & a & a \\
 & & & & & & a & a \\
\hline
c & c & c & c & & & & \\
+ & c & c & c & & & & \\
 & & c & c & & & & \\
 & & & c & & & &
\end{array}
\right].
$$

Next we create a unitary matrix U_1 acting in the $(n + 1)$–$(n + 2)$ plane such that when U_1^* is applied on the left, it recombines the first two rows of the "c" matrix and annihilates the bulge. (In the shuffled picture, we called this matrix Q_1. For the unshuffled picture it is better to use one symbol "Q" for transformations that act on the top half and another symbol "U" for transformations that act on the bottom half. See Exercise 7.3.15 below.) When U_1 is applied on the right, it recombines the first two columns of the "a" matrix and creates a new bulge entry:

$$
\left[
\begin{array}{cccc|cccc}
 & & & & a & a & a & a \\
 & & & & a & a & a & a \\
 & & & & + & a & a & a \\
 & & & & & a & a \\
\hline
c & c & c & c & & & & \\
 & c & c & c & & & & \\
 & & c & c & & & & \\
 & & & c & & & &
\end{array}
\right].
$$

The next transformation (which we will now call Q_1) acts on rows 2 and 3 of the "a" matrix to remove the bulge. It also acts on columns 2 and 3 of the "c" matrix to create a new bulge:

$$
\left[
\begin{array}{cccc|cccc}
 & & & & a & a & a & a \\
 & & & & a & a & a & a \\
 & & & & & a & a & a \\
 & & & & & & a & a \\
\hline
c & c & c & c & & & & \\
 & c & c & c & & & & \\
 & + & c & c & & & & \\
 & & & c & & & &
\end{array}
\right].
$$

Continuing in this way, the bulge is passed back and forth between the "a" and "c" matrices, moving downward all the while, until it disappears off the bottom.

This appears to be a neat extension of Francis's algorithm, but the shuffled viewpoint shows that it is really just a special case. We also remark that the Golub-Reinsch algorithm for the SVD, which was presented in Section 5.8, is just the symmetric case of this algorithm.

Proposition 7.3.14 *The Francis iteration of degree two with shifts* $\pm\rho$ *on*

$$F = \begin{bmatrix} 0 & A \\ C & 0 \end{bmatrix}$$

automatically effects Francis iterations of degree one with shift ρ^2 *on the products* AC *and* CA.

cise 7.3.15 Prove Proposition 7.3.14.

(a) Show that the double Francis iteration on F with shifts $\pm\rho$ produces a new matrix $\hat{F}$ satisfying

$$\hat{F} = \begin{bmatrix} 0 & \hat{A} \\ \hat{C} & 0 \end{bmatrix} = \begin{bmatrix} Q^* & 0 \\ 0 & U^* \end{bmatrix} \begin{bmatrix} 0 & A \\ C & 0 \end{bmatrix} \begin{bmatrix} Q & 0 \\ 0 & U \end{bmatrix},$$

where Q and U are unitary.

(b) Using Theorem 6.3.12, deduce that there exist upper-triangular matrices R and S such that

$$\begin{bmatrix} AC - \rho^2 I & 0 \\ 0 & CA - \rho^2 I \end{bmatrix} = \begin{bmatrix} Q & 0 \\ 0 & U \end{bmatrix} \begin{bmatrix} R & 0 \\ 0 & S \end{bmatrix}.$$

(c) Deduce further that

$$\hat{A}\hat{C} = Q^*(AC)Q, \quad \text{where} \quad AC - \rho^2 I = QR.$$

This is a Francis iteration of degree one on AC with shift ρ^2.

(d) Finally, deduce that

$$\hat{C}\hat{A} = U^*(CA)U, \quad \text{where} \quad CA - \rho^2 I = US.$$

This is a Francis iteration of degree one on CA with shift ρ^2.

For parts (c) and (d), see the discussion following the proof of Theorem 6.3.12. Technically we have proved Proposition 7.3.14 only for the (generic) case when ρ^2 is not an eigenvalue of AC and CA, since only then are the QR decompositions essentially unique. However, a more detailed argument would show that the proposition remains true when ρ^2 is an eigenvalue. □

The Real Case

If the matrices are real, we would like to be able to stay within real arithmetic while using complex conjugate shifts. This can be achieved by Francis iterations of degree 4 on the shuffled matrix. To get appropriate shifts we can look (for example) at the lower-right hand 4×4 submatrix (7.3.12), whose eigenvalues are two pairs $\pm\rho_1$ and $\pm\rho_2$. These can be real or purely imaginary, or the four of them together can form a complex conjugate quadruple ρ, $-\rho$, $\bar{\rho}$, $-\bar{\rho}$. In all of these cases, the polynomial $p(z) = (z - \rho_1)(z + \rho_1)(z - \rho_2)(z + \rho_2)$, which is used to start the quadruple Francis step, has real coefficients. In fact, it is just the characteristic polynomial of the real matrix (7.3.12). Thus the vector $p = p(F)e_1$ is real, and the entire quadruple Francis iteration can be done in real arithmetic. We continue with the notation

$$F = \begin{bmatrix} 0 & A \\ C & 0 \end{bmatrix}$$

and the assumptions that A is properly upper Hessenberg and C is upper triangular and nonsingular.

Exercise 7.3.16 Deshuffle the matrix (7.3.12) and deduce that its eigenvalues are the square roots of the eigenvalues of the product

$$\begin{bmatrix} a_{n-1,n-1} & a_{n-1,n} \\ a_{n,n-1} & a_{n,n} \end{bmatrix} \begin{bmatrix} c_{n-1,n-1} & c_{n-1,n} \\ 0 & c_{n,n} \end{bmatrix}.$$

Deduce that the eigenvalues are two $\pm$ pairs, either of which can be real or purely imaginary, or the four of them together can form a quadruple of the form ρ, $-\rho$, $\bar{\rho}$, $-\bar{\rho}$. □

Exercise 7.3.17 Suppose we have four shifts $\pm\rho_1$, $\pm\rho_2$, obtained by any method, such that each pair is real or purely imaginary, or the four shifts together constitute a complex quadruple ρ, $-\rho$, $\bar{\rho}$, $-\bar{\rho}$. Let $\tau_1 = \rho_1^2$ and $\tau_2 = \rho_2^2$.

(a) Show that either τ_1 and τ_2 are both real or $\tau_2 = \bar{\tau}_1$.

(b) Let $p(z) = (z - \rho_1)(z + \rho_1)(z - \rho_2)(z + \rho_2)$, the polynomial that is used to start the Francis iteration of degree four. Show that

$$p(z) = z^4 - (\tau_1 + \tau_2)z^2 + \tau_1\tau_2.$$

Deduce that this polynomial always has real coefficients. Thus the vector $p = p(F)e_1$ is real.

(c) Show that only the first three entries of $p = p(F)e_1$ are nonzero and that p is the extension by zeros of the vector

$$\tilde{p} = [(AC)^2 - (\tau_1 + \tau_2)(AC) + \tau_1\tau_2 I]e_1.$$

(d) Show that the three nonzero entries of p can be obtained by the computation

$$\left(\begin{bmatrix} a_{11} & a_{12} \\ a_{21} & a_{22} \\ 0 & a_{32} \end{bmatrix} \begin{bmatrix} c_{11} & c_{12} \\ 0 & c_{22} \end{bmatrix} - \begin{bmatrix} t & 0 \\ 0 & t \\ 0 & 0 \end{bmatrix} \right) \begin{bmatrix} a_{11}c_{11} \\ a_{21}c_{11} \end{bmatrix} + \begin{bmatrix} d \\ 0 \\ 0 \end{bmatrix},$$

where $t = \tau_1 + \tau_2$ and $d = \tau_1\tau_2$. This costs $O(1)$ flops.

(e) Suppose the shifts are chosen as in Exercise 7.3.16. Show that p can be obtained without ever computing the eigenvalues of (7.3.12) or their squares.

☐

The shuffled picture is important because it shows that the algorithm is just a special case of Francis's algorithm on a single matrix, but for an actual implementation it is easiest to think in terms of the unshuffled picture. From Exercise 7.3.17 we know that the vector $p = p(F)e_1$ is nonzero only in its first three entries., and these are easily computed. The Francis iteration is initiated by a unitary transformation Q_0 acting in 1–2–3 space such that $Q_0^* p = \begin{bmatrix} * & 0 & \cdots & 0 \end{bmatrix}^T$. Applying Q_0^* to F on the left, we recombine the first three rows of A, creating a small bulge. When we apply Q_0 on the right, we recombine the first three columns of C, creating a bigger bulge:

$$
\left[
\begin{array}{ccccc|ccccc}
 & & & & & a & a & a & a & a \\
 & & & & & a & a & a & a & a \\
 & & & & & + & a & a & a & a \\
 & & & & & & a & a & a \\
 & & & & & & & a & a \\
\hline
c & c & c & c & c & & & & & \\
+ & c & c & c & c & & & & & \\
+ & + & c & c & c & & & & & \\
 & & c & c & & & & & & \\
 & & & c & & & & & & \\
\end{array}
\right].
$$

Next we create a unitary matrix U_1 acting in $(n+1)$–$(n+2)$–$(n+3)$ space such that when U_1^* is applied on the left, it recombines the first three rows of the "c" matrix and annihilates the bulge entries in the first column. When U_1 is applied on the right, it recombines the first three columns of the "a" matrix and enlarges the bulge:

$$
\left[
\begin{array}{ccccc|ccccc}
 & & & & & a & a & a & a & a \\
 & & & & & a & a & a & a & a \\
 & & & & & + & a & a & a & a \\
 & & & & & + & + & a & a & a \\
 & & & & & & & a & a \\
\hline
c & c & c & c & c & & & & & \\
 & c & c & c & c & & & & & \\
 & + & c & c & c & & & & & \\
 & & c & c & & & & & & \\
 & & & c & & & & & & \\
\end{array}
\right].
$$

The next transformation acts on rows 2, 3, and 4 of the "a" matrix to remove the bulge entries from the first column. It also acts on columns 2, 3, and 4 of the "c"

matrix to enlarge the bulge there:

$$
\left[
\begin{array}{ccccc|ccccc}
 & & & & & a & a & a & a & a \\
 & & & & & a & a & a & a & a \\
 & & & & & & a & a & a & a \\
 & & & & & + & a & a & a \\
 & & & & & & & a & a \\
\hline
c & c & c & c & c & & & & & \\
 & c & c & c & c & & & & & \\
 & + & c & c & c & & & & & \\
 & + & + & c & c & & & & & \\
 & & & & c & & & & &
\end{array}
\right].
$$

Continuing in this way, we gradually move the bulges downward until they are chased off the bottom of the matrix.

Exercise 7.3.18 Sketch the quadruple Francis iteration in the shuffled coordinate system. □

Proposition 7.3.19 *The Francis iteration of degree four with shifts* $\pm\rho_1$, $\pm\rho_2$ *on*

$$
F = \begin{bmatrix} 0 & A \\ C & 0 \end{bmatrix}
$$

automatically effects Francis iterations of degree two with shifts ρ_1^2 *and* ρ_2^2 *on the products* AC *and* CA.

Exercise 7.3.20 Prove Proposition 7.3.19.

(a) Show that the quadruple Francis iteration on F with shifts $\pm\rho_1$, $\pm\rho_2$ produces a new matrix $\hat{F}$ satisfying

$$
\hat{F} = \begin{bmatrix} 0 & \hat{A} \\ \hat{C} & 0 \end{bmatrix} = \begin{bmatrix} Q^* & 0 \\ 0 & U^* \end{bmatrix} \begin{bmatrix} 0 & A \\ C & 0 \end{bmatrix} \begin{bmatrix} Q & 0 \\ 0 & U \end{bmatrix},
$$

where Q and U are unitary.

(b) Using Theorem 6.3.12, deduce that there exist upper-triangular matrices R and S such that

$$
\begin{bmatrix} p(AC) & 0 \\ 0 & p(CA) \end{bmatrix} = \begin{bmatrix} Q & 0 \\ 0 & U \end{bmatrix} \begin{bmatrix} R & 0 \\ 0 & S \end{bmatrix},
$$

where $p(z) = (z - \rho_1^2)(z - \rho_2^2)$.

(c) Deduce further that

$$
\hat{A}\hat{C} = Q^*(AC)Q, \quad \text{where} \quad (AC - \rho_1^2 I)(AC - \rho_2^2 I) = QR.
$$

This is a Francis iteration of degree two on AC with shifts ρ_1^2 and ρ_2^2.

(d) Finally, deduce that

$$\hat{C}\hat{A} = U^*(CA)U, \quad \text{where} \quad (CA - \rho_1^2 I)(CA - \rho_2^2 I) = US.$$

This is a Francis iteration of degree two on CA with shift ρ_1^2 and ρ_2^2

As in Exercise 7.3.15, we are relying here on the discussion following the proof of Theorem 6.3.12. This proves Proposition 7.3.19 for the (generic) case when the shifts are not eigenvalues of AC and CA. A more detailed argument would show that the proposition remains true when one or both of the shifts are eigenvalues. □

Additional Exercises

·cise 7.3.21 Outline a general Francis iteration of degree $2m$ with shifts $\pm\rho_1, \ldots, \pm\rho_m$ on the matrix F. Work with the shuffled picture or the unshuffled picture (or both), whichever you prefer. □

·cise 7.3.22 Suppose A is upper Hessenberg, C is upper triangular, and $c_{kk} = 0$ for some k. Then AC must have a zero eigenvalue. This exercise shows how to deflate out the zero eigenvalue. In the course of the procedure we will transform the matrices A and C. For simplicity we will not give the transformed matrices new names; we will continue to refer to them as A and C.

(a) The procedure begins with a (unitary) transformation acting on rows 1 and 2 of A that sets the entry a_{21} to zero. In order to complete a similarity transformation on

$$\begin{bmatrix} 0 & A \\ C & 0 \end{bmatrix}, \tag{7.3.23}$$

we must apply the complex conjugate transformation to columns 1 and 2 of C. Show that this makes a bulge in position c_{21}.

(b) To eliminate the bulge in c_{21} we apply a transformation to the first two rows of C. Explain what needs to be done to complete a similarity transformation, and show that this causes a bulge to appear in position a_{31}.

(c) Show further that the submatrix

$$\begin{bmatrix} a_{21} & a_{22} \\ a_{31} & a_{32} \end{bmatrix}$$

has rank one.

(d) The bulge at a_{31} is eliminated by a transformation on rows 2 and 3. Show that this transformation sets both a_{31} and a_{32} to zero. Show further that when the similarity transformation is completed, a bulge is created at c_{32}.

(e) Outline the continuing bulge chase. Show that each time a bulge is created in position $a_{i,i-2}$, the 2×2 submatrix

$$\begin{bmatrix} a_{i-1,i-2} & a_{i-1,i-2} \\ a_{i,i-2} & a_{i,i-1} \end{bmatrix}$$

has rank one. Show that on the subsequent transformation both $a_{i,i-2}$ and $a_{i,i-1}$ are set to zero.

(f) Show that when the bulge chase reaches columns $k-1$ and k of C, the bulge disappears because $c_{kk} = 0$. Show that at this point A is upper Hessenberg with $a_{k,k-1} = 0$. This concludes the the first portion of the deflation.

(g) Devise an analogous procedure for transforming $a_{k+1,k}$ to zero. This procedure begins by acting on columns $n-1$ and n of A to set $a_{n-1,n}$ to zero. This starts an upward bulge chase which continues until the bulge vanishes at $c_{k+1,k}$ because $c_{kk} = 0$. Show that at that moment A is upper Hessenberg and $a_{k+1,k} = 0$. Of course $a_{k,k-1}$ is also still zero.

(h) Explain how the problem can now be broken apart, exposing one zero eigenvalue of AC and breaking the remaining problem into two smaller subproblems.

(i) Show that the downward bulge chase is just a Francis iteration of degree two on (7.3.23) with shifts ± 0. (The upward bulge chase is also a Francis iteration. It is an *upward* Francis iteration of degree two with shifts ± 0.)

$\square$

Exercise 7.3.24 (continuation of previous problem) Figure out how to proceed when C has more than one zero entry on the main diagonal. $\square$

7.4   THE GENERALIZED EIGENVALUE PROBLEM

Numerous applications give rise to a more general form of eigenvalue problem

$$Av = \lambda Bv,$$

where $A, B \in \mathbb{C}^{n \times n}$. This problem has a rich theory, and numerous algorithms for solving it have been devised. This section provides only a brief overview.

Eigenvalue problems of this type can arise from systems of linear differential equations of the form

$$B\dot{x} = Ax - b. \tag{7.4.1}$$

These differ from those studied in Section 5.1 by the introduction of the coefficient matrix B in front of the derivative term. We can solve such problems by essentially the same methodology as in we used in Section 5.1. First we consider the homogeneous system

$$B\dot{x} = Ax. \tag{7.4.2}$$

As in Section 5.1, we seek solutions of the homogeneous problem with the simple form $x(t) = g(t)v$, where $g(t)$ is a nonzero scalar-valued function of t, and v is a nonzero vector that is independent of time. Substituting this form into (7.4.2), we find that

$$\frac{\dot{g}}{g}Bv = Av,$$

which implies that $\dot{g}/g$ must be constant. Call the constant λ. Then $\dot{g} = \lambda g$ (which implies $g(t) = ce^{\lambda t}$), and

$$Av = \lambda Bv.$$

This is the generalized eigenvalue problem. Each solution (λ, v) yields a solution $e^{\lambda t}v$ of (7.4.2). Thus we must solve the generalized eigenvalue problem in order to solve the homogeneous problem (7.4.2). Apart from that, the solution procedure is the same as for the problems we discussed in Section 5.1.

Example 7.4.3 The electrical circuit in Figure 7.1 differs from the ones we looked at in Section 5.1 in that it has an inductor in the interior wire, which is shared by two loop currents. As before, we can obtain a system of differential equations for

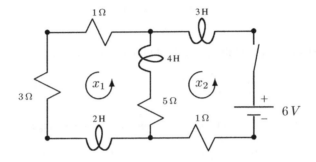

Figure 7.1 Circuit with inductor shared by two loop currents

this circuit by applying Kirchhoff's voltage law: the sum of the voltage drops around each loop is zero. For the first loop, the voltage drops across the 1Ω and 3Ω resistors are $1x_1$ and $3x_1$ volts, respectively. The voltage drop across the 2 H inductor is $2\dot{x}_1$ volts. The drop across the 5Ω resistor is $5(x_1 - x_2)$ volts, since the net upward current in the middle wire is $x_1 - x_2$ amps. Similarly, the drop across the 4 H inductor is $4(\dot{x}_1 - \dot{x}_2)$ volts. Thus the equation for the first loop is

$$x_1 + 3x_1 + 2\dot{x}_1 + 5(x_1 - x_2) + 4(\dot{x}_1 - \dot{x}_2) = 0$$

or

$$6\dot{x}_1 - 4\dot{x}_2 = -9x_1 + 5x_2.$$

Similarly, the equation for the second loop is

$$-4\dot{x}_1 + 7\dot{x}_2 = 5x_1 - 6x_2 + 6.$$

Combining the two equations into a single matrix equation, we obtain

$$\begin{bmatrix} 6 & -4 \\ -4 & 7 \end{bmatrix} \begin{bmatrix} \dot{x}_1 \\ \dot{x}_2 \end{bmatrix} = \begin{bmatrix} -9 & 5 \\ 5 & -6 \end{bmatrix} \begin{bmatrix} x_1 \\ x_2 \end{bmatrix} - \begin{bmatrix} 0 \\ -6 \end{bmatrix}. \qquad (7.4.4)$$

which is a system of differential equations of the form $B\dot{x} = Ax - b$, as in (7.4.1).[41]

As a first step toward solving this system, we find a single solution z of (7.4.4). The simplest solution is a steady-state solution, which can be found by setting the derivative terms in (7.4.4) to zero. Doing so, we obtain the linear system

$$\begin{bmatrix} -9 & 5 \\ 5 & -6 \end{bmatrix} \begin{bmatrix} z_1 \\ z_2 \end{bmatrix} = \begin{bmatrix} 0 \\ -6 \end{bmatrix}.$$

Since the coefficient matrix is nonsingular, there is a unique steady-state solution, which we can determine by solving the system, either by pencil and paper or using MATLAB. Using MATLAB, we find that

$$z = \begin{bmatrix} 1.0345 \\ 1.8621 \end{bmatrix} \tag{7.4.5}$$

amperes.

The next step is to find the general solution of the homogeneous problem

$$\begin{bmatrix} 6 & -4 \\ -4 & 7 \end{bmatrix} \begin{bmatrix} \dot{x}_1 \\ \dot{x}_2 \end{bmatrix} = \begin{bmatrix} -9 & 5 \\ 5 & -6 \end{bmatrix} \begin{bmatrix} x_1 \\ x_2 \end{bmatrix}.$$

As we have seen above, this requires solving the generalized eigenvalue problem

$$\begin{bmatrix} 6 & -4 \\ -4 & 7 \end{bmatrix} \begin{bmatrix} v_1 \\ v_2 \end{bmatrix} = \lambda \begin{bmatrix} -9 & 5 \\ 5 & -6 \end{bmatrix} \begin{bmatrix} v_1 \\ v_2 \end{bmatrix}.$$

This is small enough that we could solve it by pencil and paper, but we will use MATLAB instead. To get MATLAB's `eig` function to solve the generalized eigenvalue problem $Av = \lambda Bv$, we simply type `[V,D] = eig(A,B)`. When invoked in this way, `eig` uses one of two algorithms, depending on the properties of A and B. One is a variant of Francis's algorithm called the Moler-Stewart or QZ algorithm. The output from `[V,D] = eig(A,B)` is a matrix V whose columns are (if possible) linearly independent eigenvectors and a diagonal matrix D whose main diagonal entries are the associated eigenvalues. Together they satisfy $AV = BVD$. As you can easily check, this means that $Av_j = \lambda_j Bv_j$, $j = 1, \ldots, n$, where v_j denotes the jth column of V, and λ_j is the (j, j) entry of D. Using `eig` in this way to solve our generalized eigenvalue problem, we obtain the eigenvalues

$$\lambda_1 = -1.5493 \quad \text{and} \quad \lambda_2 = -0.7199,$$

and associated eigenvectors

$$v_1 = \begin{bmatrix} 0.9708 \\ 0.2399 \end{bmatrix} \quad \text{and} \quad v_2 = \begin{bmatrix} 0.4126 \\ 0.9109 \end{bmatrix}.$$

---

[41]The matrices have additional interesting structure. Both A and B are symmetric, B is positive definite and A is negative definite. Alternatively, we can flip some signs and write the system as $B\dot{x} + Ax = b$, in which both B and A are positive definite.

Since these vectors are linearly independent, the general solution of the homogeneous problem is

$$c_1 e^{-1.5493t} \begin{bmatrix} 0.9708 \\ 0.2399 \end{bmatrix} + c_2 e^{-0.7199t} \begin{bmatrix} 0.4126 \\ 0.9109 \end{bmatrix},$$

where c_1 and c_2 are arbitrary constants. We now obtain the general solution to the nonhomogeneous problem by adding in the steady-state solution (7.4.5):

$$x(t) = \begin{bmatrix} 1.0345 \\ 1.8621 \end{bmatrix} + c_1 e^{-1.5493t} \begin{bmatrix} 0.9708 \\ 0.2399 \end{bmatrix} + c_2 e^{-0.7199t} \begin{bmatrix} 0.4126 \\ 0.9109 \end{bmatrix}. \quad (7.4.6)$$

Notice that both of the exponential functions decay as t increases. This is so because both eigenvalues are negative. Therefore, whatever the loop currents may initially be they will (quickly) settle down to their steady-state values. The system is stable.

Now suppose the switch in the circuit is closed at time zero. Then the subsequent values of the loop currents are given by the unique solution of the system that satisfy the initial conditions

$$x(0) = \begin{bmatrix} 0 \\ 0 \end{bmatrix}.$$

Setting $t = 0$ in (7.4.6) and inserting these initial values, we obtain the linear system

$$\begin{bmatrix} 0.9708 & 0.4126 \\ 0.2399 & 0.9109 \end{bmatrix} \begin{bmatrix} c_1 \\ c_2 \end{bmatrix} = \begin{bmatrix} -1.0345 \\ -1.8621 \end{bmatrix}.$$

Solving this, we obtain $c_1 = -0.2215$ and $c_2 = -1.9859$.

A similar example is considered in Exercise 7.4.30. □

Basic Properties of the Generalized Eigenvalue Problem

Consider the generalized eigenvalue problem.

$$Av = \lambda Bv. \quad (7.4.7)$$

This is a generalization of the standard eigenvalue problem, reducing to standard problem in the case $B = I$. A nonzero $v \in \mathbb{C}^n$ that satisfies (7.4.7) for some value of λ is called an *eigenvector* of the ordered pair (A, B), and λ is called the associated *eigenvalue*. Clearly (7.4.7) holds if and only if $(\lambda B - A)v = 0$, so λ is an eigenvalue of (A, B) if and only if the *characteristic equation*

$$\det(\lambda B - A) = 0$$

is satisfied. One easily checks that $\det(\lambda B - A)$ is a polynomial of degree n or less. It is called the *characteristic polynomial* of the pair (A, B). The expression $\lambda B - A$ is often called a *matrix pencil*, and one speaks of the eigenvalues, eigenvectors, characteristic equation, and so forth, of the matrix pencil.

rcise 7.4.8 Verify that $\det(\lambda B - A)$ is a polynomial, and its degree is at most n. □

Most generalized eigenvalue problems can be reduced to standard eigenvalue problems. For example, if B is nonsingular, the eigenvalues of the pair (A, B) are just the eigenvalues of $B^{-1}A$.

Exercise 7.4.9 Suppose B is nonsingular.

(a) Show that v is an eigenvector of (A, B) with associated eigenvalue λ if and only if v is an eigenvector of $B^{-1}A$ with eigenvalue λ.

(b) Show that v is an eigenvector of (A, B) with associated eigenvalue λ if and only if Bv is an eigenvector of AB^{-1} with eigenvalue λ.

(c) Show that $B^{-1}A$ and AB^{-1} are similar matrices.

(d) Show that the characteristic equation of (A, B) is essentially the same as that of $B^{-1}A$ and AB^{-1}.

$\square$

From Exercise 7.4.9 we see that if B is nonsingular, then the generalized eigenvalue problem is a product eigenvalue problem of the type we studied in Section 7.3, with $C = B^{-1}$. If we are willing to compute B^{-1} explicitly, then we can use the reduction to Hessenberg form and the variant of Francis's algorithm developed in Section 7.3 to solve the generalized eigenvalue problem. Unfortunately the computation of B^{-1} can cause a loss of accuracy. Moreover, there are applications in which B is not invertible. Therefore we prefer to develop algorithms that work with A and B, not A and B^{-1}.

From Exercise 7.4.9 we also see that if B is nonsingular, the pair (A, B) must have exactly n eigenvalues, counting multiplicity. If B is singular, the situation is different. Consider the following example.

Example 7.4.10 Let

$$A = \begin{bmatrix} 1 & 0 \\ 0 & -1 \end{bmatrix} \quad \text{and} \quad B = \begin{bmatrix} 1 & 0 \\ 0 & 0 \end{bmatrix},$$

and notice that B is singular. You can easily verify that the characteristic equation of (A, B) is $\lambda - 1 = 0$, whose degree is less than the order of the matrices. Evidently the pair (A, B) has only one eigenvalue, namely $\lambda = 1$. As we will see in a moment, it is reasonable to attribute to (A, B) a second eigenvalue $\lambda = \infty$. $\square$

Suppose the pair (A, B) has a nonzero eigenvalue λ. Then, making the substitution $\mu = 1/\lambda$ in (7.4.7) and multiplying through by μ, we obtain

$$Bv = \mu Av,$$

which is the generalized eigenvalue problem for the ordered pair (B, A). We conclude that the nonzero eigenvalues of (B, A) are the reciprocals of the nonzero eigenvalues of (A, B). Now suppose B is singular. Then zero is an eigenvalue of (B, A). In light of the reciprocal relationship we have just noted, it is reasonable to regard ∞ as an eigenvalue of (A, B). We will do just that. With this convention, most generalized eigenvalue problems have n eigenvalues. There is, however, a class of problems that is not so well behaved. Let's look at an example.

Example 7.4.11 Let

$$A = \begin{bmatrix} 1 & 0 \\ 0 & 0 \end{bmatrix} \quad \text{and} \quad B = \begin{bmatrix} 1 & 0 \\ 0 & 0 \end{bmatrix}.$$

Then $\det(\lambda B - A) = 0$ for all λ, so every complex number is an eigenvalue.   □

cise 7.4.12

(a) Let (A, B) be as in Example 7.4.11. Find an eigenvector associated with each eigenvalue. Find the nullspaces $\mathcal{N}(A)$ and $\mathcal{N}(B)$.

(b) Let A and B be any two matrices for which $\mathcal{N}(A) \cap \mathcal{N}(B) \neq \{0\}$. Show that every complex number is an eigenvalue of the pair (A, B), and every nonzero $v \in \mathcal{N}(A) \cap \mathcal{N}(B)$ is an eigenvector of (A, B) associated with every λ.

□

Any pair (A, B) for which the characteristic polynomial $\det(\lambda B - A)$ is identically zero is said to be *singular*. The term *singular pencil* is frequently used. If (A, B) is not singular, it is said to be *regular*. Certainly (A, B) is regular whenever A or B is a nonsingular matrix. As we have just seen, (A, B) is singular whenever $\mathcal{N}(A)$ and $\mathcal{N}(B)$ have a nontrivial intersection. It can also happen that a pair is singular even if $\mathcal{N}(A) \cap \mathcal{N}(B) = \{0\}$.

Example 7.4.13 Let

$$A = \begin{bmatrix} 1 & 0 & 0 \\ 0 & 0 & 0 \\ 0 & 0 & 1 \end{bmatrix} \quad \text{and} \quad B = \begin{bmatrix} 0 & 1 & 0 \\ 0 & 0 & 1 \\ 0 & 0 & 0 \end{bmatrix}.$$

Then the pair (A, B) is singular, as $\det(\lambda B - A) = 0$ for all λ. Both A and B are singular matrices; however, $\mathcal{N}(A) \cap \mathcal{N}(B) = \{0\}$.   □

cise 7.4.14 Verify the assertions of Example 7.4.13.   □

It can also happen that a pair (A, B) is regular, even though both A and B are singular matrices.

Example 7.4.15 Let

$$A = \begin{bmatrix} 1 & 0 \\ 0 & 0 \end{bmatrix} \quad \text{and} \quad B = \begin{bmatrix} 0 & 0 \\ 0 & 1 \end{bmatrix}.$$

Then A and B are singular, but (A, B) is a regular pair.   □

Exercise 7.4.16 Let A and B be as in Example 7.4.15.

(a) Show that both A and B are singular matrices.

(b) Calculate the characteristic polynomial of the pair (A, B), and conclude that (A, B) is regular. What are the eigenvalues of (A, B)?

□

For the rest of this section we will restrict our attention to regular pencils. For more information on singular pencils see [16, 95, 104] and Volume II of [36].

In most applications at least one of A and B is nonsingular. Whenever this is the case, it is possible to reduce the generalized eigenvalue problem to a standard eigenvalue problem for one of the matrices AB^{-1}, $B^{-1}A$, BA^{-1}, or $A^{-1}B$. Although this is a possibility that should not be ruled out, there are a number of reasons why it might not be the best course of action. Suppose we compute AB^{-1}. If B is ill conditioned (in the sense of Chapter 2), the eigenvalues of the computed AB^{-1} can be poor approximations of the eigenvalues of (A, B), even though some or all of the eigenvalues of (A, B) are well conditioned (Exercise 7.4.31). A second reason not to compute AB^{-1} is that A and B might be symmetric. In many applications it is desirable to preserve the symmetry, and AB^{-1} will typically not be symmetric. Finally, A and B might be sparse, in which case we would like to exploit the sparseness. But AB^{-1} will not be sparse, for the inverse of a sparse matrix is typically not sparse. For these reasons it is useful to develop algorithms that solve the generalized eigenvalue problem directly.

Equivalence Transformations

We know that two matrices have the same eigenvalues if they are similar, a fact which is exploited in numerous algorithms. If we wish to develop algorithms in the same spirit for the generalized problem, we need to know what sorts of transformations we can perform on pairs of matrices without altering the eigenvalues. It turns out that an even larger class than similarity transformations works. Two pairs of matrices (A, B) and $(\tilde{A}, \tilde{B})$ are said to be *equivalent* if there exist nonsingular matrices U and V such that $\tilde{A} = UAV$ and $\tilde{B} = UBV$. We can write the relationship more briefly as $\tilde{A} - \lambda\tilde{B} = U(A - \lambda B)V$. In the next exercise you will show that equivalent pairs have the same eigenvalues.

Exercise 7.4.17 Suppose (A, B) and $(\tilde{A}, \tilde{B})$ are equivalent pairs: $\tilde{A} - \lambda\tilde{B} = U(A - \lambda B)V$.

(a) Show that λ is an eigenvalue of (A, B) with associated eigenvector v if and only if λ is an eigenvalue of $(\tilde{A}, \tilde{B})$ with associated eigenvector $V^{-1}v$.

(b) Show that (A, B) and $(\tilde{A}, \tilde{B})$ have essentially the same characteristic equation.

□

Exercise 7.4.18 Investigate the equivalence transformations given by each of the following pairs of nonsingular matrices, assuming that the indicated inverses exist:

(a) $U = B^{-1}$, $V = I$, (b) $U = I$, $V = B^{-1}$, (c) $U = A^{-1}$, $V = I$, (d) $U = I$, $V = A^{-1}$. $\quad\square$

Computing Generalized Eigensystems: The Symmetric Case

If $A, B \in \mathbb{R}^{n \times n}$ are symmetric matrices, and B is positive definite, the pair (A, B) is called a *symmetric* pair. Given a symmetric pair (A, B), B has a Cholesky decomposition $B = R^T R$, where R is upper triangular. The equivalence transformation given by $U = R^{-T}$ and $V = R^{-1}$ transforms the generalized eigenvalue problem $Av = \lambda Bv$ to the standard eigenvalue problem $R^{-T} A R^{-1} y = \lambda y$, where $y = Rv$. Notice that the coefficient matrix $\tilde{A} = R^{-T} A R^{-1}$ is symmetric. This proves that a symmetric pair has real eigenvalues, and it has n linearly independent eigenvectors. The latter are not orthogonal in the conventional sense, but they are orthogonal with respect to an appropriately chosen inner product.

cise 7.4.19   The symmetric matrix $\tilde{A} = R^{-T} A R^{-1}$ has orthonormal eigenvectors $v_1, \ldots, v_n$, so $R^{-1} v_1, \ldots, R^{-1} v_n$ are eigenvectors of (A, B). Prove that these are orthonormal with respect to the *energy* inner product $\langle x, y \rangle_B = y^T B x$. $\quad\square$

It is interesting that the requirement that B be positive definite is essential; it is not enough simply to specify that A and B be symmetric. It can be shown (cf. [69, p. 304]) that every $C \in \mathbb{R}^{n \times n}$ can be expressed as a product $C = AB^{-1}$, where A and B are symmetric. This means that the eigenvalue problem for any C can be reformulated as a generalized eigenvalue problem $Av = \lambda Bv$, where A and B are symmetric. Thus one easily finds examples of pairs (A, B) for which A and B are symmetric but the eigenvalues are complex.[42]

One way to solve the eigenvalue problem for the symmetric pair (A, B) is to perform the transformation to $R^{-T} A R^{-1}$ and use one of the many available techniques for solving the symmetric eigenvalue problem. This has two of the same drawbacks as the transformation to AB^{-1}: 1.) If B is ill conditioned, the eigenvalues will not be determined accurately, and 2.) $R^{-T} A R^{-1}$ does not inherit any sparseness that A and B might possess. The second drawback is not insurmountable. If B is sparse, its Cholesky factor R will usually also be somewhat sparse. For example, if B is banded, R is also banded. Because of this it is possible to apply sparse eigenvalue methods such as the Lanczos algorithm (6.4.23) to $\tilde{A} = R^{-T} A R^{-1}$ at a reasonable cost. The only way in which this method uses $\tilde{A}$ is to compute vectors $\tilde{A}x$ for a sequence of choices of x. For $\tilde{A}$ of the given form we can compute $\tilde{A}x$ by first solving $Ry = x$ for y, then computing $z = Ay$, then solving $R^T w = z$ for w. Clearly $w = \tilde{A}x$. If R is sparse, the two systems $Ry = x$ and $R^T w = z$ can be solved cheaply.

In many applications both A and B are positive definite. If A is much better conditioned than B, it might make sense to interchange the roles of A and B and take a Cholesky decomposition of A.

---

[42]To be honest, the condition of positive definiteness of B is a bit stronger than we really need. If $\alpha A + \beta B$ is positive definite for some real α and β, then the pair has real eigenvalues.

There are numerous other approaches to solving the symmetric generalized eigenvalue problem. A number of methods for the symmetric standard eigenvalue problem can be adapted to the generalized problem. For example, both the Jacobi method and the bisection method, described in Section 7.2, have been adapted to the generalized problem. See [69] for details and references.

Exercise 7.4.20 If B is symmetric and positive definite, then it has a spectral decomposition $B = VDV^T$ where V is orthogonal and D is diagonal and positive definite. Show how this decomposition can be used to reduce the symmetric generalized eigenvalue problem to a standard symmetric problem.   ☐

Exercise 7.4.21   Two pairs (A, B) and $(\tilde{A}, \tilde{B})$ are said to be *congruent* if there exists a nonsingular $X \in \mathbb{R}^{n \times n}$ such that $\tilde{A} = XAX^T$ and $\tilde{B} = XBX^T$. Clearly any two congruent pairs are equivalent, but not conversely. Show that if (A, B) is a symmetric pair (and in particular B is positive definite), then $(\tilde{A}, \tilde{B})$ is also a symmetric pair. Thus we can retain symmetry by using congruence transformations. The transformation to a standard eigenvalue problem using the Cholesky factor and the transformation suggested in Exercise 7.4.20 are both examples of congruence transformations.   ☐

Computing Generalized Eigensystems: The General Problem

We now consider the problem of finding the eigenvalues of a pair (A, B) that is not symmetric. We will develop an analogue of the reduction to upper Hessenberg form and a variant of Francis's algorithm. Unfortunately these algorithms do not preserve symmetry.

Schur's Theorem 5.4.11 guarantees that every matrix in $\mathbb{C}^{n \times n}$ is unitarily similar to an upper triangular matrix. The analogous result for the generalized eigenvalue problem is

Theorem 7.4.22   *(Generalized Schur Theorem) Let $A, B \in \mathbb{C}^{n \times n}$. Then there exist unitary $Q, Z \in \mathbb{C}^{n \times n}$ and upper triangular $T, S \in \mathbb{C}^{n \times n}$ such that $Q^* AZ = T$ and $Q^* BZ = S$. Thus*

$$Q^*(A - \lambda B)Z = T - \lambda S.$$

A proof is worked out in Exercise 7.4.32.

The pair (T, S) is equivalent to (A, B), so it has the same eigenvalues. But the eigenvalues of (T, S) are evident because

$$\det(\lambda S - T) = (\lambda s_{11} - t_{11})(\lambda s_{22} - t_{22}) \cdots (\lambda s_{nn} - t_{nn}).$$

If there is a k for which both s_{kk} and t_{kk} are zero, the characteristic polynomial is identically zero; that is, the pair is singular. Otherwise the eigenvalues are t_{kk}/s_{kk}, $i = 1, \ldots, n$. Some of these quotients may be ∞.

The numbers s_{kk} and t_{kk} give more information than the ratio s_{kk}/t_{kk}. If both s_{kk} and t_{kk} are near zero, the pair (A, B) is nearly singular in the sense that it is close to a singular pair.

·cise 7.4.23 Use Theorem 7.4.22 to verify the following facts. Suppose (A, B) is a regular pair. Then the degree of the characteristic polynomial is less than or equal to n. It is exactly n if and only if B is nonsingular. If B is singular, then the pair (A, B) has at least one infinite eigenvalue. The degree of the characteristic polynomial is equal to n minus the number of infinite eigenvalues. □

For real (A, B) there is a real decomposition analogous to the Wintner-Murnaghan theorem (Theorem 5.4.23). In this variant, Q and Z are real, orthogonal matrices, S is upper triangular, and T is quasitriangular.

Reduction to Hessenberg-Triangular Form

Theorem 7.3.5 shows that any A and C can be transformed to $H = Q^*AU$ and $T = U^*CQ$, where H is upper Hessenberg and T is upper triangular. Q and U are unitary. This gives unitary similarity transforms of the products: $HT = Q^*(AC)Q$ and $TH = U^*(CA)U$. The generalized eigenvalue problem is a product eigenvalue problem for the products AB^{-1} and $B^{-1}A$. If we apply Theorem 7.3.5 with C taken to be B^{-1}, we get $H = Q^*AU$ and $T = U^*B^{-1}Q$. Inverting the second equation and taking $S = T^{-1}$, we have $S = Q^*BU$. Combining this with the first equation, we have a unitary equivalence transformation $Q^*(A - \lambda B)U = H - \lambda S$, where H is upper Hessenberg and S is upper triangular.

Theorem 7.4.24 *Let A, $B \in \mathbb{C}^{n \times n}$. Then there exist unitary Q, $U \in \mathbb{C}^{n \times n}$, upper Hessenberg $H \in \mathbb{C}^{n \times n}$ and upper triangular $S \in \mathbb{C}^{n \times n}$ such that*

$$Q^*(A - \lambda B)U = H - \lambda S.$$

There is a stable, direct algorithm to calculate Q, U, H, and S in $O(n^3)$ flops. If A and B are real, Q, U, H, and S can be taken to be real.

Proof. The paragraph preceding the statement of the theorem constitutes a proof (except for the stability claim) in the case that B has an inverse. We will now outline a second construction, due to Moler and Stewart [65], which works directly with B, not B^{-1}, is guaranteed to be backward stable, and works even if B is singular.

The transformation will be effected by a sequence of reflectors and rotators applied on the left and right of A and B. Each transformation applied on the left (right) makes a contribution to Q^* (resp. U). Every transformation applied to A must also be applied to B and vice versa. We know from Chapter 3 that B can be reduced to upper triangular form by a sequence of reflectors applied on the left. Applying this sequence of reflectors to both B and A, we obtain a new (A, B), for which B is upper triangular. The rest of the algorithm consists of reducing A to upper Hessenberg form without destroying the upper triangular form of B. The first major step is to introduce zeros in the first column of A. This is done one element at a time, from bottom to top. First the $(n, 1)$ entry is set to zero by a rotator (or reflector) acting in the $(n, n - 1)$ plane, applied to A on the left. This alters rows $n - 1$ and n of A. The same transformation must also be applied to B. You can easily check that this operation will introduce one nonzero entry in the lower triangle of B, in the $(n, n-1)$ position. This blemish can be removed by the applying the appropriate rotator to columns $n-1$ and n of B. That is, we apply a rotation in the $(n, n-1)$ plane to B on the right. Applying the same rotation to A, we do not disturb the zero that we previously introduced in column 1. We next apply a rotation to rows $n-2$ and $n-1$ of A in such a way that a zero is produced in position $(n-1, 1)$. Applying the same rotation to B, we introduce a nonzero in position $(n-1, n-2)$. This can be removed by applying a rotator to columns $n-1$ and $n-2$. This operation does not disturb the zeros in the first column of A. Continuing in this manner we can produce zeros in positions $(n-2, 1)$, $(n-3, 1)$, ..., $(3, 1)$. This scheme cannot be used to produce a zero in position $(2, 1)$. In order to do that we would have to apply a rotator to rows 1 and 2 of A. The same rotator applied to B would produce a nonzero in position $(2, 1)$. We could eliminate that entry by applying a rotator to columns 1 and 2 of B. Applying the same rotator to columns 1 and 2 of A, we would destroy all of the zeros that we had so painstakingly created in column 1. Thus we must leave the $(2, 1)$ entry of A as it is and move on to column 2. The second major step is to introduce zeros in column 2 in positions $(n, 2)$, $(n-1, 2)$, ..., $(4, 2)$ by the same scheme as we used to create zeros in column 1. You can easily check that none of the rotators that are applied on the right operate on either column 1 or 2, so the zeros that have been created so far are maintained. We then proceed to column 3, and so on. After $n-2$ major steps, we are done. $\square$

Exercise 7.4.25 Count the flops required to execute the algorithm outlined in the proof of Theorem 7.4.24, (a) assuming Q and U are not to be assembled, (b) assuming Q and U are to be assembled. In either case the count is $O(n^3)$. $\square$

Francis's Algorithm for the Generalized Eigenvalue Problem

We can now restrict our attention to pairs (A, B) for which A is upper Hessenberg and B is upper triangular. We can even assume that A is properly upper Hessenberg, since otherwise the eigenvalue problem can be reduced to subproblems in the obvious way. We will also assume that B is nonsingular. If it is not, then the pair (A, B) has an infinite eigenvalue (or is a singular pair), and this can be removed by a method that is outlined in Exercise 7.4.34.

In Section 7.3 we developed a variant (a special case) of Francis's algorithm that computes the eigenvalues of the products AC and CA while working with the factors directly. A is assumed to be upper Hessenberg, and C is assumed upper triangular. In that algorithm each transformation $A \to Q^*A$ has to be matched by a transformation $C \to CQ$, and each transformation $C \to U^*C$ has to be matched by a transformation $A \to AU$.

We now adapt the algorithm to the product eigenvalue problem for AB^{-1} and $B^{-1}A$. Substituting B^{-1} for C, we see that we need to make transformations of the types $B^{-1} \to B^{-1}Q$ and $B^{-1} \to U^*B^{-1}$. Since we prefer to work directly with B rather than B^{-1}, we invert these transformations. Instead of $B^{-1} \to B^{-1}Q$, we will do the equivalent transformation $B \to Q^*B$. Instead of $B^{-1} \to U^*B^{-1}$ we will do the equivalent transformation $B \to BU$. In this way we will avoid ever having to work with B^{-1}. Each transformation $A \to Q^*A$ is accompanied by a transformation $B \to Q^*B$, and each transformation $A \to AU$ is accompanied by a transformation $B \to BU$. These are unitary equivalence transformations on the pair (A, B).

Now let us look at the details of the algorithm. Referring back to (7.3.13) and multiplying through by $b_{11} = c_{11}^{-1}$, we see that we need to initiate the Francis iteration by computing a vector

$$p = \begin{bmatrix} a_{11} - \tau b_{11} \\ a_{21} \\ 0 \\ \vdots \\ 0 \end{bmatrix},$$

where $\tau = \rho^2$ is a shift chosen to approximate an eigenvalue of AB^{-1}. The shift strategies suggested in Section 7.3 can be adapted to the current situation. A unitary transformation Q_0 acting in the 1–2 plane such that $Q_0^*p = \begin{bmatrix} * & 0 & \cdots & 0 \end{bmatrix}^T$ is built and applied to A on the left. The same transformation is also applied to B on the left, resulting in a bulge in B. In the case $n = 4$, the situation is

$$\begin{bmatrix} a & a & a & a \\ a & a & a & a \\ & a & a & a \\ & & a & a \end{bmatrix} \quad \begin{bmatrix} b & b & b & b \\ + & b & b & b \\ & & b & b \\ & & & b \end{bmatrix}.$$

Now a transformation U_1 acting in the 1-2 plane is built and applied on the right to both matrices. It acts on columns 1 and 2, annihilating the bulge in one matrix and creating a new bulge in the other:

$$\begin{bmatrix} a & a & a & a \\ a & a & a & a \\ + & a & a & a \\ & & a & a \end{bmatrix} \quad \begin{bmatrix} b & b & b & b \\ & b & b & b \\ & & b & b \\ & & & b \end{bmatrix}.$$

The next transformation Q_1^* acts in the 2–3 plane, multiplying both matrices on the left, annihilating the bulge in one matrix and creating a yet another bulge in the other:

$$\begin{bmatrix} a & a & a & a \\ a & a & a & a \\ & a & a & a \\ & & a & a \end{bmatrix} \quad \begin{bmatrix} b & b & b & b \\ & b & b & b \\ & + & b & b \\ & & & b \end{bmatrix}.$$

The Francis iteration continues in this manner until the bulge has been chased off the bottom, at which point the iteration is complete.

Repeated iterations will normally cause the entry $a_{n,n-1}$ to converge to zero quadratically, isolating an eigenvalue a_{nn}/b_{nn}. The problem can then be deflated and the next eigenvalue sought.

The Real Case

In the real case we would like to be able to work entirely in real arithmetic while utilizing complex conjugate shifts. This can be achieved by applying Francis iterations of degree four to the matrix

$$F = \begin{bmatrix} 0 & A \\ B^{-1} & 0 \end{bmatrix},$$

as described in Section 7.3. By Proposition 7.3.19, these are equivalent to Francis iterations of degree two on AB^{-1} and $B^{-1}A$. Here we will show how to do this while working directly with B, not B^{-1}. This is the Moler-Stewart algorithm [65], which is commonly known as the QZ algorithm.

Referring back to Exercise 7.3.17, we see that to start the iteration we need to compute the first three entries (the only nonzero entries) of

$$p = (AB^{-1} - \tau_1 I)(AB^{-1} - \tau_2 I)e_1,$$

where τ_1 and τ_2 are shifts that approximate eigenvalues of AB^{-1}. These are either real or a complex conjugate pair.

Exercise 7.4.26

    (a) Show that if the shifts are chosen as in Exercise 7.3.16, the lower-right-hand 2×2 submatrix of B^{-1} is needed, and this is

$$\begin{bmatrix} b_{n-1,n-1}^{-1} & -b_{n-1,n-1}^{-1}b_{n-1,n}b_{n,n}^{-1} \\ 0 & b_{n,n}^{-1} \end{bmatrix}.$$

    (b) Referring back to Exercise 7.3.17, show that the only other part of B^{-1} that is needed for the computation of p is the upper-left-hand 2×2 submatrix, which is

$$\begin{bmatrix} b_{11}^{-1} & -b_{11}^{-1}b_{12}b_{22}^{-1} \\ 0 & b_{22}^{-1} \end{bmatrix}.$$

    (c) Show that this matrix can also be conveniently expressed as

$$\begin{bmatrix} b_{11}^{-1} & 0 \\ 0 & 1 \end{bmatrix} \begin{bmatrix} 1 & -b_{12} \\ 0 & 1 \end{bmatrix} \begin{bmatrix} 1 & 0 \\ 0 & b_{22}^{-1} \end{bmatrix}.$$

    (d) Write a pseudocode algorithm that computes the three nonzero entries of p. A minor simplification is achieved by multiplying through by b_{11}.

                                                                      □

Once we have the vector p, we can perform the quadruple Francis iteration on

$$\begin{bmatrix} 0 & A \\ B^{-1} & 0 \end{bmatrix}$$

exactly as described in Section 7.3, except that each transformation of the form $B^{-1} \to B^{-1}Q$ or $B^{-1} \to U^*B^{-1}$ is replaced by $B \to Q^*B$ or $B \to BU$, respectively. The first step is to produce a unitary Q_0 such that $Q_0^*p = \begin{bmatrix} * & 0 & \cdots & 0 \end{bmatrix}^T$. Then Q_0^* is applied on the left to both A and B. This operation recombines rows 1 through 3 of A and B, creating new A and B matrices, each of which has a bulge:

$$\begin{bmatrix} a & a & a & a & a \\ a & a & a & a & a \\ + & a & a & a & a \\ & & a & a & a \\ & & & a & a \end{bmatrix} \quad \begin{bmatrix} b & b & b & b & b \\ + & b & b & b & b \\ + & + & b & b & b \\ & & & b & b \\ & & & & b \end{bmatrix}.$$

The next step is to build a unitary transformation U_0 acting on 1–2–3 space such that the transformation $B \to BU_0$ annihilates the bulge entries in positions $(2,1)$ and $(3,1)$. We have not yet shown how to do this, and this is the one novel aspect of the algorithm. The procedure is worked out in Exercise 7.4.27. We must also apply U_0 to A on the right. This recombines columns 1 through 3 of A and adds a row to the bulge. At this point we have

$$\begin{bmatrix} a & a & a & a & a \\ a & a & a & a & a \\ + & a & a & a & a \\ + & + & a & a & a \\ & & & a & a \end{bmatrix} \quad \begin{bmatrix} b & b & b & b & b \\ & b & b & b & b \\ + & & b & b & b \\ & & & b & b \\ & & & & b \end{bmatrix}.$$

The next step is to produce a Q_1 such that when Q_1^* is applied to A on the left, it recombines rows 2 through 4 and eliminates the bulge entries from positions $(3,1)$ and $(4,1)$. When the same transformation is applied to B, it adds a row to the bulge in B. We now have

$$\begin{bmatrix} a & a & a & a & a \\ a & a & a & a & a \\ & a & a & a & a \\ & + & a & a & a \\ & & & a & a \end{bmatrix} \quad \begin{bmatrix} b & b & b & b & b \\ & b & b & b & b \\ & + & b & b & b \\ & + & + & b & b \\ & & & & b \end{bmatrix}.$$

The next step is to produce a U_1 which, when applied to B on the right, eliminates the bulge entries in positions $(3,2)$ and $(4,2)$ by the process worked out in Exercise 7.4.27. Once this transformation has been applied to both B and A, we have

$$\begin{bmatrix} a & a & a & a & a \\ a & a & a & a & a \\ & a & a & a & a \\ & + & a & a & a \\ & + & + & a & a \end{bmatrix} \quad \begin{bmatrix} b & b & b & b & b \\ & b & b & b & b \\ & & b & b & b \\ & & + & b & b \\ & & & & b \end{bmatrix}.$$

We continue this process until the bulges have been pushed off the bottom of the matrix, at which point the Francis iteration is complete.

If these iterations are repeated with a good shifting strategy, the entry $a_{n-1,n-2}$ will normally tend to zero quadratically. Once it becomes small enough to be considered zero, we can break off the small subpencil

$$
\begin{bmatrix} a_{n-1,n-1} & a_{n-1,n} \\ a_{n,n-1} & a_{n,n} \end{bmatrix} - \lambda \begin{bmatrix} b_{n-1,n-1} & b_{n-1,n} \\ b_{n,n-1} & b_{n,n} \end{bmatrix},
$$

which has either two real eigenvalues or a conjugate pair of complex eigenvalues. Sometimes it will happen that $a_{n,n-1}$ goes to zero, isolating a single real eigenvalue a_{nn}/b_{nn}. In either case we can deflate and continue the iterations on a smaller problem.

We developed the algorithm under the assumption that B is nonsingular. If B is singular, it must have some zeros on the main diagonal, and these can be removed by the method outlined in Exercise 7.4.34. It is recommended that this be done. However, the algorithm works perfectly well in the case of singular B [65, 97, 99]. All that is needed is that the main-diagonal entries at the very top and bottom: b_{11}, b_{22}, $b_{n-1,n-1}$, b_{nn}, be nonzero.

Exercise 7.4.27 Suppose B is nonsingular and upper triangular, except that it contains a bulge of degree two. Thus

$$
B = \begin{bmatrix} B_{11} & B_{12} & B_{13} \\ & B_{22} & B_{23} \\ & & B_{33} \end{bmatrix},
$$

where B_{11} and B_{33} are upper triangular and B_{22} is the 3×3 matrix that contains the bulge.

(a) Show that B^{-1} is also upper triangular, except for a bulge in the same position. Specifically

$$
B^{-1} = \begin{bmatrix} B_{11}^{-1} & C_{12} & C_{13} \\ & B_{22}^{-1} & C_{23} \\ & & B_{33}^{-1} \end{bmatrix},
$$

where B_{11}^{-1} and B_{33}^{-1} are upper triangular.

If we want to chase the bulge in B^{-1} further by a transformation on the left, we need to produce a 3×3 unitary transformation U_0 such that

$$
U_0^* B_{22}^{-1} = \begin{bmatrix} * & * & * \\ 0 & * & * \\ 0 & * & * \end{bmatrix}.
$$

For this task we can take U_0 to be an elementary reflector, as discussed in Section 3.2 and used routinely in numerous algorithms since then. If we want to make the equivalent transformation on B rather than B^{-1}, we need to build U_0 so that

$$
B_{22} U_0 = \begin{bmatrix} * & * & * \\ 0 & * & * \\ 0 & * & * \end{bmatrix}. \tag{7.4.28}
$$

We need to develop a way of doing this.

(b) Show that U_0 satisfies (7.4.28) if and only if the first column of U_0 is orthogonal to the second and third columns of B_{22}^*

(c) Consider a decomposition $B_{22}^* = QL$, where Q is unitary and L is lower triangular. Let $Q = \begin{bmatrix} q_1 & q_2 & q_3 \end{bmatrix}$. Show that If x_2 and x_3 denote the second and third columns of B_{22}^*, then span$\{x_2, x_3\} \subseteq$ span$\{q_2, q_3\}$. Deduce that q_1 is orthogonal to the second and third columns of B_{22}^*. Thus we could take $U_0 = Q$. Part (d) shows how to build Q as a product of two reflectors. A more efficient procedure is to compute Q, then take U_0 to be a reflector such that $U_0^* q_1 = \begin{bmatrix} * & 0 & 0 \end{bmatrix}^T$. This reflector will satisfy $U_0 e_1 = \alpha q_1$, so it will also fill the bill.

(d) Sketch an algorithm to compute the QL decomposition of an arbitrary $n \times n$ matrix A. This is just like the QR decomposition algorithm, except that it moves from right to left and creates zeros in the top half of the matrix. Q is a product of $n - 1$ reflectors.

□

Additional Exercises

rcise 7.4.29 Using MATLAB, check that the calculations in Example 7.4.3 are correct. Plot the solutions for $0 \leq t \leq 5$. (You may find it helpful to review Exercises 5.1.19 and 5.1.20.) □

rcise 7.4.30 In working this exercise, do not overlook the advice in Exercises 5.1.19 and 5.1.20. In Figure 7.2 all of the resistances are 1 Ω and all of the inductances are 1 H, except for the three that are marked otherwise. Initially the loop currents are all zero, because the switch is open. Suppose the switch is closed at time 0.

(a) Write down a system of four differential equations for the four unknown loop currents. Write your system in the matrix form $B\dot{x} = Ax - b$.

(b) Solve the system $Az = b$ to obtain a steady-state solution of the differential equation.

(c) Get the eigenvalues and eigenvectors of A, and use them to construct the general solution of the homogeneous equation $B\dot{z} = Az$.

(d) Deduce the solution of the initial value problem $B\dot{x} = Ax - b$, $x(0) = 0$.

(e) Plot the loop currents $x_1(t), \ldots, x_4(t)$ for $0 \leq t \leq 6$ seconds. (Checkpoint: Do these look realistic?)

□

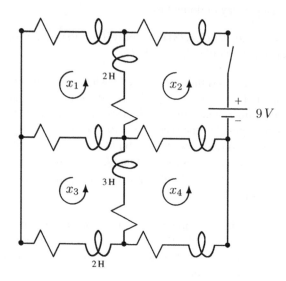

Figure 7.2   Solve for the time-varying loop currents.

Exercise 7.4.31 Perform the following experiment using MATLAB. Build two matrices A and B as follows:

```
n = 4;
M = randn(n);
A = M'*M
cond(A)
N = randn(n);
B = N'*N
lam = eig(B);
shift = min(lam) - 10*eps;
B = B - shift*eye(n);
cond(B)
```

These are random positive definite matrices. On average A should be reasonably well conditioned. However, B has been modified to make it extremely ill conditioned.

(a) Explain why B should be ill conditioned.

(b) Compute the eigenvalues of the pair (A, B) two different ways:

```
format long e
lg = eig(A,B);
lb = eig(A*inv(B));
```

Either way you get one huge eigenvalue and $n - 1$ smaller eigenvalues. Compare the two results. Probably they will not agree very well.

(c) Decide which eigenvalues are better by computing some eigenvectors (by [V,D] = eig(A,B) ,e.g.) and computing each residual norm $\| Av - \lambda Bv \|$ twice, once using the lg "eigenvalue" and once using the lb "eigenvalue."

□

·cise 7.4.32 In this exercise you will prove the Generalized Schur theorem, which states that every pair (A, B) is unitarily equivalent to an upper-triangular pair (T, S): $Q^*(A - \lambda B)Z = T - \lambda S$. The proof is by induction on n, the dimension of the matrices.

(a) Confirm that Theorem 7.4.22 is true when $n = 1$.

(b) Now suppose we have a problem of size $n > 1$. We need to reduce it to one of size $n - 1$ so that we can use induction. The key is to produce a pair of vectors v and $w \in \mathbb{C}^n$ such that $\| v \|_2 = 1$, $\| w \|_2 = 1$,

$$Av = \alpha w \quad \text{and} \quad Bv = \beta w, \tag{7.4.33}$$

where α and β are complex scalars. Suppose first that B is nonsingular. Then the pair (A, B) has n eigenvalues. Let μ be any one of the eigenvalues, and let v be a corresponding eigenvector such that $\| v \|_2 = 1$. Define w to be an appropriate multiple of Av, and show that (7.4.33) is satisfied for appropriate choices of α and β satisfying $\mu = \alpha/\beta$.

(c) Now suppose B is singular. Show that if we choose v so that $Bv = 0$, then (7.4.33) is satisfied for appropriate choices of w, α, and β. (Consider two cases: $Av \neq 0$ (infinite eigenvalue) and $Av = 0$ (singular pencil).)

(d) We have now established that in all cases we can find v and w with unit norm, such that (7.4.33) holds. Let V and W be unitary matrices whose first columns are v and w, respectively. Show that W^*AV and W^*BV are both block triangular; specifically,

$$W^*(\lambda B - A)V = \lambda \begin{bmatrix} \beta & * & \cdots & * \\ 0 & & & \\ \vdots & & \hat{B} & \\ 0 & & & \end{bmatrix} - \begin{bmatrix} \alpha & * & \cdots & * \\ 0 & & & \\ \vdots & & \hat{A} & \\ 0 & & & \end{bmatrix}.$$

(e) Prove Theorem 7.4.22 by induction on n.

□

·cise 7.4.34 Consider a pair (A, B), where A is upper Hessenberg, B is upper triangular, and $b_{jj} = 0$ for some j. Show that an infinite eigenvalue can be deflated from the pencil after a finite sequence of rotations.

(a) First outline an algorithm that deflates the infinite eigenvalue at the top of the pencil: The first rotator acts on B from the right and sets $b_{j-1,j-1}$ to zero.

When this rotator is applied to A, it creates a bulge, which can then be removed by a rotator applied on the left. This rotator must also be applied to B. Next another rotator is applied to B on the right to set $b_{j-2,j-2}$ to zero. This creates another bulge in A, which can be annihilated by a rotator acting on the left. Show that we can continue this way until the entry b_{11} has been set to zero, the modified B is upper triangular (and $b_{22} = 0$ also), and A is upper Hessenberg. Apply one more rotator on the left to A to set a_{21} to zero. Now show that an infinite eigenvalue can be deflated at the top.

(b) Outline an algorithm similar to the one in part (a) that chases the infinite eigenvalue to the bottom.

If $j \leq n/2$, the algorithm of part (a) is cheaper; otherwise the algorithm of part (b) is the more economical. $\quad\square$

CHAPTER 8

ITERATIVE METHODS FOR LINEAR SYSTEMS

In this chapter we return to the problem of solving a linear system $Ax = b$, where A is $n \times n$ and nonsingular. This problem can be solved without difficulty, even for fairly large values of n, by Gaussian elimination on today's computers. However, once n becomes very large (e.g. several thousand) and the matrix A becomes very sparse (e.g. 99.9% of its entries are zeros), iterative methods become more efficient. This chapter begins with a section that shows how such large, sparse problems can arise. Then the classical iterative methods are introduced and analyzed. From there we move on to a discussion of descent methods, including the powerful conjugate gradient method for solving positive definite systems. The important idea of preconditioning is introduced along the way. The conjugate gradient method is just one of a large family of Krylov subspace methods. The chapter concludes with a brief discussion of Krylov subspace methods for indefinite and nonsymmetric problems.

We restrict our attention to real systems throughout the chapter. However, virtually everything said here can be extended to the complex case.

Fundamentals of Matrix Computations, Third Edition. By David S. Watkins
Copyright © 2010 John Wiley & Sons, Inc.

8.1   A MODEL PROBLEM

Large sparse matrices arise routinely in the numerical solution of partial differential equations (PDE). We will proceed by stages, beginning with a simple ordinary differential equation (ODE). This is a *one-dimensional* problem, in the sense that there is one independent variable, x. Suppose a function $f(x)$ is given for $0 \leq x \leq 1$, and we wish to find a function $u(x)$ satisfying the ODE

$$-u''(x) = f(x) \qquad 0 < x < 1 \tag{8.1.1}$$

and the boundary conditions

$$u(0) = T_0 \qquad \text{and} \qquad u(1) = T_1. \tag{8.1.2}$$

Here T_0 and T_1 are two given numbers. Boundary value problems of this type arise in numerous settings. For example, if we wish to determine the temperature distribution $u(x)$ inside a uniform wall, we must solve such a problem. The numbers T_0 and T_1 represent the ambient temperatures on the two sides of the wall, and the function $f(x)$ represents a heat source within the wall.

This is a very simple problem. Depending on the nature of f, you may be able to solve it easily by integrating f a couple of times and using the boundary conditions to solve for the constants of integration. Let us not pursue this approach; our plan is to use this problem as a stepping stone to a harder boundary value problem involving a PDE. Instead, let us pursue an approximate solution by the finite difference method, as we did in Example 1.2.12.

Let $h = 1/m$ be a small increment, where m is some large integer (e.g. 100 or 1000), and mark off equally spaced points x_i on the interval $[0, 1]$. Thus $x_0 = 0$, $x_1 = h$, $x_2 = 2h, \ldots, x_m = mh = 1$, or briefly $x_i = ih$ for $i = 0, 1, 2, \ldots, m$. We will approximate the solution of (8.1.1) on $x_0, x_1, \ldots, x_m$. Since (8.1.1) holds at each of these points, we have

$$-u''(x_i) = f(x_i) \qquad i = 1, \ldots, m - 1.$$

As we noted in Example 1.2.12 (and Exercise 1.2.21), a good approximation for the second derivative is

$$u''(x_i) \approx \frac{u(x_{i-1}) - 2u(x_i) + u(x_{i+1})}{h^2}. \tag{8.1.3}$$

Substituting this approximation into the ODE, we obtain, for $i = 1, \ldots, m - 1$,

$$\frac{-u(x_{i-1}) + 2u(x_i) - u(x_{i+1})}{h^2} \approx f(x_i).$$

These are approximations, not equations, but it is not unreasonable to expect that if we treat them as equations and solve them exactly, we will get good approximations to the true solution at the mesh points x_i. Thus we let $u_1, \ldots, u_{m-1}$ denote approximations to $u(x_1), \ldots, u(x_{m-1})$ obtained by solving the equations exactly:

$$-u_{i-1} + 2u_i - u_{i+1} = h^2 f(x_i), \qquad i = 1, \ldots, m - 1. \tag{8.1.4}$$

We have multiplied through by h^2 for convenience. Since $x_0 = 0$ and $x_m = 1$, it makes sense to define $u_0 = T_0$ and $u_m = T_1$, in agreement with the boundary conditions (8.1.2). The boundary value u_0 is used in (8.1.4) when $i = 1$, and u_m is used when $i = m - 1$. Equation (8.1.4) is actually a system of $m - 1$ linear equations for the $m - 1$ unknowns $u_1, \ldots, u_{m-1}$, so it can be written as a matrix equation

$$Au = b,$$

where u is now a column vector containing the unknowns,

$$A = \begin{bmatrix} 2 & -1 & & & & \\ -1 & 2 & -1 & & & \\ & -1 & 2 & \ddots & & \\ & & \ddots & \ddots & -1 & \\ & & & -1 & 2 & -1 \\ & & & & -1 & 2 \end{bmatrix}, \tag{8.1.5}$$

and

$$b = \begin{bmatrix} h^2 f(x_1) + T_0 \\ h^2 f(x_2) \\ h^2 f(x_3) \\ \vdots \\ h^2 f(x_{m-2}) \\ h^2 f(x_{m-1}) + T_1 \end{bmatrix}.$$

Notice that the boundary values end up in b, since they are known quantities. The coefficient matrix A is nonsingular (Exercise 8.1.14), so the system has a unique solution. Once we have solved it, we have approximations to $u(x)$ at the grid points x_i.

The matrix A has numerous useful properties. It is obviously symmetric. It is even positive definite (Exercise 8.1.15 or Exercise 8.3.37). It is also extremely sparse, tridiagonal, in fact. The reason for this is clear: each unknown is directly connected only to its nearest neighbors.

Such a simple system does not require iterative methods for its solution. This is the simplest case of a banded matrix. The band is preserved by the Cholesky decomposition, and the system can be solved in $O(m)$ flops. If Cholesky's method works, then so does Gaussian elimination without pivoting, which also preserves the tridiagonal structure. This variant is actually preferable, as it avoids calculating $m - 1$ square roots. But now let us get to our main task, which is to generate examples of large sparse systems for which the use of iterative methods is advantageous.

A Model PDE (Two-Dimensional)

Let Ω denote a bounded region in the plane with boundary $\partial\Omega$, let $f(x, y)$ be a given function defined on Ω, and let $g(x, y)$ be defined on $\partial\Omega$. Consider the problem of finding a function $u(x, y)$ satisfying the partial differential equation

$$-\frac{\partial^2 u}{\partial x^2} - \frac{\partial^2 u}{\partial y^2} = f \quad \text{in } \Omega \qquad (8.1.6)$$

subject to the boundary condition

$$u = g \quad \text{on } \partial\Omega. \qquad (8.1.7)$$

The PDE (8.1.6) is called Poisson's equation. It can be used to model a broad range of phenomena, including heat flow, chemical diffusion, fluid flow, elasticity, and electrostatic potential. In a typical heat flow problem $u(x, y)$ represents the temperature at point (x, y) in a homogeneous post or pillar whose cross-section is Ω. The function g represents the specified temperature on the surface of the post, and $f(x, y)$ represents a heat source within the post. Since there are two independent variables, x and y, we call this a two-dimensional problem.

For certain special choices of Ω, f and g, the exact solution can be found, but usually we have to settle for an approximation. Let us consider how we might solve (8.1.6) numerically by the finite difference method. To minimize complications, we restrict our attention to the case where $\Omega = [0, 1] \times [0, 1]$, the unit square. Let $h = 1/m$ be an increment, and lay out a computational grid of points in Ω with spacing h. The (i, j)th grid point is $(x_i, y_j) = (ih, jh)$, with $i, j = 0, \ldots, m$. A grid with $h = 1/5$ is shown in Figure 8.1.

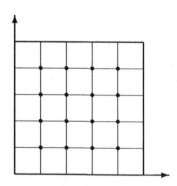

Figure 8.1   Computational grid with $h = 1/5$ on $\Omega = [0, 1] \times [0, 1]$

We will approximate the PDE (8.1.6) by a system of difference equations defined on the grid. The approximation that we used for the ordinary derivative $u''(x)$ can also be used for the partial derivatives, since a partial derivative is just an ordinary derivative taken with respect to one variable while the other variable is held fixed. Thus the approximation (8.1.3) yields the approximation

$$\frac{\partial^2 u}{\partial x^2}(x_i, y_j) \approx \frac{u(x_{i-1}, y_j) - 2u(x_i, y_j) + u(x_{i+1}, y_j)}{h^2}.$$

The y variable is held fixed. Similarly we have for the y-derivative

$$\frac{\partial^2 u}{\partial y^2}(x_i, y_j) \approx \frac{u(x_i, y_{j-1}) - 2u(x_i, y_j) + u(x_i, y_{j+1})}{h^2}.$$

Substituting these approximations into (8.1.6), we find that a solution of (8.1.6) satisfies the approximation

$$\frac{-u(x_{i-1}, y_j) + 2u(x_i, y_j) - u(x_{i+1}, y_j)}{h^2} + \frac{-u(x_i, y_{j-1}) + 2u(x_i, y_j) - u(x_i, y_{j+1})}{h^2} \approx f(x_i, y_j)$$

at the interior grid points $i, j = 1, \ldots, m - 1$. These are approximations, not equations, but, again, if we treat them as equations and solve them exactly, we should get a good approximation of the true solution $u(x, y)$. Consider, therefore, the system of equations

$$\frac{-u_{i-1,j} + 2u_{i,j} - u_{i+1,j}}{h^2} + \frac{-u_{i,j-1} + 2u_{i,j} - u_{i,j+1}}{h^2} = f_{i,j},$$

which becomes, after minor rearrangement,

$$-u_{i,j-1} - u_{i-1,j} + 4u_{i,j} - u_{i+1,j} - u_{i,j+1} = h^2 f_{i,j}, \qquad i, j = 1, \ldots, m - 1. \tag{8.1.8}$$

The shorthand $f_{i,j} = f(x_i, y_j)$ has been introduced.

Each equation involves five of the unknown values, whose relative location in the grid is shown in the left-hand diagram of Figure 8.2. The weights with which the five

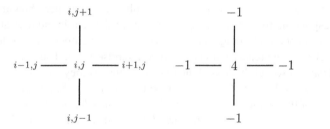

Figure 8.2 Five-point stencil

unknowns are combined are shown in the right-hand diagram of Figure 8.2. This is called the *five-point stencil* for approximating Poisson's equation.

Boundary values of $u_{i,j}$, which occur in the equations for mesh points that are adjacent to the boundary, can be determined by the boundary conditions. For example, equations for the mesh points $(i, m - 1)$ (adjacent to the top boundary) contain the "unknown" $u_{i,m}$, which can be specified by the boundary condition $u_{i,m} = g(x_i, 1)$. With this understanding, (8.1.8) can be seen to be a system of $(m - 1)^2$ equations in the $(m - 1)^2$ unknowns $u_{i,j}$, $i, j = 1, \ldots, m - 1$. If we can solve these equations, we will have approximations $u_{i,j} \approx u(x_i, y_j)$ to the solution of (8.1.6) with boundary conditions (8.1.7) at the grid points.

The equations are linear, so they can be written as a matrix equation $Au = b$. Before we can do this, we need to decide on the order in which the unknowns $u_{i,j}$ should be packed into a single vector u. (The ordering was obvious in the ODE case.) There is one unknown for each grid point (Figure 8.1). We can order the $u_{i,j}$ by sweeping through the grid by rows, by columns, or by diagonals, for example. Let us do it by rows. The first row contains $u_{1,1}, \ldots, u_{m-1,1}$, the second row contains $u_{1,2}, \ldots, u_{m-1,2}$, and so on. Thus u will be the column vector defined by

$$u^T = [u_{1,1}, \ldots, u_{m-1,1}, u_{1,2}, \ldots, u_{m-1,2}, \ldots, u_{m-1,m-1}].$$

We use the same ordering to pack the $h^2 f_{i,j}$ (plus boundary terms, where appropriate) into the vector b. Having decided on an ordering, we can now write the equations (8.1.8) as a matrix equation $Au = b$.

Exercise 8.1.9 Determine the form of the matrix A corresponding to the system (8.1.8) and the ordering of equations and unknowns specified in the previous paragraph. □

Just as in the one-dimensional (ODE) case, the matrix A is very sparse. Since each equation in (8.1.8) involves only five of the unknowns, each row of A has at most five nonzero entries. This is true regardless of the ordering. Again the reason for the sparseness is obvious: each unknown interacts directly only with its four nearest neighbors. In the ordering we have chosen (and in every ordering), A has 4's on its main diagonal. These are the 4's that multiply $u_{i,j}$ in the equation corresponding to the (i, j)th grid point. Each 4 is immediately preceded by and followed by a -1, corresponding to terms $u_{i,j-1}$ and $u_{i,j+1}$, with exceptions near boundaries. The -1's corresponding to $u_{i-1,j}$ and $u_{i+1,j}$ appear further out in the row, at a distance of $m - 1$ from the main diagonal. This is all rather cumbersome to describe but should be clear to the reader who has worked Exercise 8.1.9. The nonzero entries in A are not packed so tightly around the main diagonal as they were in the ODE case; the semi-band width is m. One might hope to achieve a tight packing by changing the ordering of the unknowns. This would require an ordering that places every grid point directly after two of its nearest neighbors and directly before the other two, which is clearly impossible. The ordering we have chosen is about as good as we are going to get, if minimization of bandwidth is our objective. This is a major difference between the one-dimensional and two-dimensional cases.

The description of A is simplified if we use block matrix notation. The dimension of A is $n \times n$, where $n = (m - 1)^2$. We can describe it as an $(m - 1) \times (m - 1)$ block matrix, where each of the blocks is $(m - 1) \times (m - 1)$, as follows:

$$A = \begin{bmatrix} T & -I & & & & \\ -I & T & -I & & & \\ & -I & T & \ddots & & \\ & & \ddots & \ddots & -I & \\ & & & -I & T & -I \\ & & & & -I & T \end{bmatrix}, \tag{8.1.10}$$

where I denotes the $(m-1) \times (m-1)$ identity matrix, and T is the $(m-1) \times (m-1)$ tridiagonal matrix

$$
T = \begin{bmatrix}
4 & -1 & & & & \\
-1 & 4 & -1 & & & \\
& -1 & 4 & \ddots & & \\
& & \ddots & \ddots & -1 & \\
& & & -1 & 4 & -1 \\
& & & & -1 & 4
\end{bmatrix}.
$$

This way of writing A shows clearly that it is symmetric. It is also nonsingular (Exercise 8.1.16) and even positive definite (Exercise 8.1.17). Thus the system $Au = b$ has a unique solution, which approximates the solution of the PDE.

Now let us consider solving the system $Au = b$. The most obvious difference between the one and two-dimensional cases is that now the matrix dimension is much higher. Now we have $n = (m-1)^2$, compared with $n = m-1$ in the one-dimensional case. Thus sheer size can be a problem. For example, if we choose an interval $h = 1/101$, corresponding to $m = 101$, we would have to solve a system of 100 equations in the one-dimensional case or 10,000 equations in the two-dimensional case. Our ultimate objective is to solve a PDE. If our approximate solution is not good enough, we can get a more accurate one by taking a smaller h, that is, a finer grid. This, of course, increases the size of the system $Au = b$. It poses no serious problem in the one-dimensional case. The size of the matrix is $O(m)$, and so is the solution time by banded Gaussian elimination. If we cut h in half, we double the work. The situation is much worse in the two-dimensional case. The size of the matrix is about m^2, so cutting h in half quadruples the size of the matrix. For example, if we take $h = 1/201$, we have to solve a system of order 40,000.

The size of the system is only part of the problem. There is also the question of bandwidth, which we have already mentioned. In the one-dimensional case, the semi-bandwidth is always two, regardless of how big the system is. In the two-dimensional case, the semi-bandwidth is m, so it increases as we decrease h. Let us consider the cost of solving $Au = b$ by a banded Cholesky decomposition, as described in Section 1.5. There we noted that the flop count for solving a system of dimension n and semi-bandwidth s is about $ns^2/2$. Here we have $n = (m-1)^2$ and $s = m$, so $ns^2/2 \approx \frac{1}{2}m^4$. Thus we see that each time we halve h (doubling m), we multiply the amount of computational work by a factor of sixteen.

Storage space is also a consideration. The matrix A has fewer than $5n$ nonzero entries (≤ 5 per row), so it can be stored very compactly. (Indeed there is no real reason to store A at all; it consists of the numbers 4 and -1, placed in a regular pattern. If we know the pattern, we know the matrix.) If we want to use Cholesky's method, we have to compute and store the Cholesky factor, which inherits the band structure of A but is not nearly so sparse. Whereas the huge majority of the entries within A's band are zeros, R's band is filled in with nonzeros, which have to be stored. The required storage space $ns \approx m^3$. Thus our storage requirement is multiplied by eight each time we halve h.

Are these costs unavoidable, or is there some way around them? It is not hard to derive some basic lower bounds on storage and computational requirements. If the matrix has a very simple form, we may not have to store it at all. However, it seems inevitable that we will have to store the solution u, which consists of n numbers. Thus n is a lower bound on the storage requirement. For our two-dimensional model PDE, $n \approx m^2$. Compared with this figure, the m^3 storage required by Cholesky's method looks like it might be excessive. This turns out to be true. Almost all of the iterative methods that we will discuss in this chapter require at most a modest multiple of n storage locations.[43]

Now let us consider computational costs. Each of the n entries of u has to be computed, and each of these will presumably require at least one flop, so n is also a lower bound on the computational cost. If we solve our model PDE problem by banded Cholesky, the flop count is $\frac{1}{2}m^4 \approx \frac{1}{2}n^2$, so here too it looks like there might be room for improvement. Although it may be too much to hope for a method that needs only some n flops to solve the problem, we might still hope to find one that does the job in Cn flops, where C is some constant. It turns out that the best iterative methods are able to achieve this, at least for highly regular problems like our model problem. The constant C is typically large, so $Cn > \frac{1}{2}n^2$ unless n is fairly large. Once n is big enough, the iterative method will win.[44]

A Three-Dimensional Problem

We live in a three-dimensional world, or so it appears. Hence many applications have three independent variables. The three-dimensional analogue of (8.1.6) is the 3-D Poisson equation

$$-\frac{\partial^2 u}{\partial x^2} - \frac{\partial^2 u}{\partial y^2} - \frac{\partial^2 u}{\partial z^2} = f \quad \text{in } \Omega, \tag{8.1.11}$$

where Ω is a region in $\mathbb{R}^3$. To keep the discussion simple, let us take $\Omega = [0,1]^3$, the unit cube. Assume boundary conditions of the form (8.1.7).

Discretization of three-dimensional PDE's like this one leads to very large systems of equations.

Exercise 8.1.12 Write down a system of equations analogous to (8.1.8) that approximates the solution of (8.1.11). How many unknowns does your system have? (Assume $h = 1/m$.) How many unknowns appear in each equation? □

Exercise 8.1.13 Consider the system of equations derived in the previous exercise. Suppose the unknowns $u_{i,j,k}$ and the $f_{i,j,k}$ are ordered in a systematic way, and the system is written as a matrix equation $Au = b$. Just as in the one- and two-dimensional cases,

---

[43] One notable exception is the popular GMRES method for nonsymmetric systems. See Section 8.10.
[44] This comparison is not quite fair to the direct methods. If we reorder the matrix by the approximate minimum degree algorithm, as discussed in Section 1.6, or by nested dissection, the flop count can be reduced. With the nested-dissection ordering, the flop count is $O(n^{3/2})$ [37]. However, it remains true that iterative methods win if n is large enough.

A is positive definite and banded. What is the semi-bandwidth of A? If we wish to solve this system by a banded Cholesky decomposition, how much storage space will be needed? How many flops will it take? □

The previous two exercises demonstrate the difficulties of trying to solve three-dimensional problems by Gaussian elimination. Even with the fairly modest mesh size $h = 1/100$, we get a matrix with dimension $n \approx 10^6$ and semi-bandwidth $s \approx 10^4$. The solution by Cholesky's method takes $\frac{1}{2} \times 10^{14}$ flops and requires the storage of 10^{10} numbers. Each time h is halved, the flop count goes up by a factor of 128, and the storage requirement is multiplied by 32. These numbers show that we very quickly move into the range where iterative methods have the advantage. Indeed, iterative methods are indispensable to the solution of these problems.

The tendency for computational problems to become much more difficult as the dimension increases is commonly known as *the curse of dimensionality*.

Some Final Remarks

It is not the task of this book to study discretizations of PDE; here we have considered only the most basic technique. For more detailed coverage consult the vast literature, for example [7, 89].

Our model problem is just the simplest of a large class of elliptic boundary value problems. More complicated problems arise when one considers irregularly shaped regions, differential operators with variable coefficients, and other types of boundary conditions. The iterative methods discussed in this chapter can be brought to bear to solve a wide variety of these problems.

One class of techniques for solving the model problem (8.1.8) that will not be discussed in this book is the so-called *fast Poisson solvers*. These are direct methods based on the fast Fourier transform, cyclic reduction, or both. They are nearly optimal for the model problem and other simple problems, but their range of application is limited. Fast Poisson solvers are sometimes incorporated into preconditioners for more complicated problems. An overview of fast Poisson solvers is given in [7].

Additional Exercises

rcise 8.1.14 In this exercise you will show that the matrix A in (8.1.5) is nonsingular. One characterization of nonsingularity is that if $Aw = 0$, then $w = 0$ as well. Suppose, therefore, that $Aw = 0$.

(a) Show that the equation $Aw = 0$ is equivalent to the homogeneous system of difference equations

$$-w_{i-1} + 2w_i - w_{i+1} = 0, \qquad i = 1, \ldots, m - 1,$$

with boundary conditions $w_0 = 0$ and $w_m = 0$. Show further that

$$w_i = \frac{w_{i-1} + w_{i+1}}{2}, \qquad i = 1, \ldots, m - 1.$$

In words, each interior value w_i is the average of its neighbors.

(b) let $\mu = \min\{w_0, \ldots, w_m\}$. Prove that if $w_i = \mu$ at some interior point, then $w_{i-1} = w_{i+1} = \mu$ as well. (It is perhaps easier to show the contrapositive: if $w_{i-1} > \mu$ or $w_{i+1} > \mu$, then $w_i > \mu$.) Conclude that $\mu = 0$.

(c) Let $\mu = \max\{w_0, \ldots, w_m\}$. Prove that $\mu = 0$.

(d) Prove that A is nonsingular.

In working this problem you have developed and used a discrete version of the *maximum principle* for harmonic functions. □

Exercise 8.1.15 In Exercise 5.4.38 you showed that a symmetric matrix is positive definite if and only if all of its eigenvalues are positive. Use this fact, Gerschgorin's Theorem (Exercise 7.1.30), and the result of the previous exercise to prove that the matrix A in (8.1.5) is positive definite. □

Exercise 8.1.16 Use the methodology of Exercise 8.1.14 to prove that the matrix A in (8.1.10) (and (8.1.8)) is nonsingular. □

Exercise 8.1.17 Use the methodology of Exercise 8.1.15 to prove that the matrix A in (8.1.10) (and (8.1.8)) is positive definite. □

8.2 THE CLASSICAL ITERATIVE METHODS

We are interested primarily in sparse matrices, but for the description of the algorithms we can let $A \in \mathbb{R}^{n \times n}$ be (almost) any nonsingular matrix. The minimal requirement is that all of the main diagonal entries a_{ii} be nonzero. Let $b \in \mathbb{R}^n$. Our objective is to solve the linear system $Ax = b$ or $Au = b$. In the early chapters we used x to denote the unknown vector, but in the examples in the previous section we used u. (This change was made because we had another use for the symbol x.) Now we will revert to the use of x for the unknown vector, but we will not hesitate to switch back to u when it seems appropriate.

Iterative methods require an initial guess $x^{(0)}$, a vector in $\mathbb{R}^n$ that approximates the true solution. Once we have $x^{(0)}$, we use it to generate a new guess $x^{(1)}$, which is then used to generate yet another guess $x^{(2)}$, and so on. In this manner we generate a sequence of iterates $(x^{(k)})$ which (we hope) converges to the true solution x.

In practice we will not iterate forever. Once $x^{(k)}$ is sufficiently close to the solution (as indicated, e.g., by the magnitude of $\| b - Ax^{(k)} \|$), we stop and accept $x^{(k)}$ as an adequate approximation to the solution. How soon we stop will depend on how accurate an approximation we need.

The iterative methods that we are going to study do not require a good initial guess. If no good approximation to x is known, we can take $x^{(0)} = 0$. Of course, we should take advantage of a good initial guess if we have one, for then we can get to the solution in fewer iterations than we otherwise would.

The ability to exploit a good initial guess and the possibility of stopping early if only a crude approximant is needed are two important advantages of iterative methods over direct methods like Gaussian elimination. The latter has no way of exploiting a good initial guess. It simply executes a predetermined sequence of operations and delivers the solution at the end. If you stop it early, it gives you nothing.

Jacobi's Method

Each of the methods of this section can be described completely by specifying how a given iterate $x^{(k)}$ is used to generate the next iterate $x^{(k+1)}$. Suppose, therefore, that we have $x^{(k)}$, and consider the following simple idea for improving on it: Use the ith equation to correct the ith unknown. The ith equation in the system $Ax = b$ is

$$\sum_{j=1}^{n} a_{ij}x_j = b_i,$$

which can be rewritten (solved for x_i) as

$$x_i = \frac{1}{a_{ii}} \left(b_i - \sum_{j \neq i} a_{ij}x_j \right), \tag{8.2.1}$$

since, as we have assumed, $a_{ii} \neq 0$. If we replace x by $x^{(k)}$ in (8.2.1), equality will no longer hold in general. Let us define $x_i^{(k+1)}$ to be the adjusted value that would make the ith equation true:

$$x_i^{(k+1)} = \frac{1}{a_{ii}} \left(b_i - \sum_{j \neq i} a_{ij}x_j^{(k)} \right). \tag{8.2.2}$$

Making this adjustment for $i = 1, \ldots, n$, we obtain our next iterate $x^{(k+1)}$. This is Jacobi's method.

Of course, $x^{(k+1)}$ is not the exact solution to $Ax = b$, because the correction $x_i^{(k)} \to x_i^{(k+1)}$, which "fixes" the ith equation, also affects each of the other equations in which the unknown x_i appears. The hope is that repeated application of (8.2.2) for $k = 0, 1, 2, \ldots$ will result in convergence to the true solution x.

Example 8.2.3 If we apply Jacobi's method (8.2.2) to the system

$$\begin{bmatrix} 5 & 2 & 1 & 1 \\ 2 & 6 & 2 & 1 \\ 1 & 2 & 7 & 2 \\ 1 & 1 & 2 & 8 \end{bmatrix} \begin{bmatrix} x_1 \\ x_2 \\ x_3 \\ x_4 \end{bmatrix} = \begin{bmatrix} 29 \\ 31 \\ 26 \\ 19 \end{bmatrix}$$

with initial guess $x^{(0)} = 0$, we obtain

$$x^{(10)} = \begin{bmatrix} 3.902 \\ 2.899 \\ 1.914 \\ 0.935 \end{bmatrix}, \quad x^{(20)} = \begin{bmatrix} 3.9965 \\ 2.9964 \\ 1.9970 \\ 0.9977 \end{bmatrix}, \quad x^{(30)} = \begin{bmatrix} 3.99988 \\ 2.99987 \\ 1.99989 \\ 0.99992 \end{bmatrix},$$

which steadily approach the exact solution $x = [4, \ 3, \ 2, \ 1]^T$. Any desired accuracy (limited only by roundoff errors) can be obtained by taking enough steps. For example, $x^{(80)}$ agrees with the true solution to twelve decimal places.

Notice that the main-diagonal entries of the coefficient matrix are large relative to the off-diagonal entries. This aids convergence, as the analysis in the next section will show. □

Exercise 8.2.4 Let D denote the diagonal matrix whose main diagonal entries are the same as those of A. Show that equation (8.2.2), which defines Jacobi's method, can be written as a matrix equation

$$x^{(k+1)} = D^{-1}\left[(D - A)x^{(k)} + b\right]. \qquad (8.2.5)$$

Show further that

$$x^{(k+1)} = x^{(k)} + D^{-1}r^{(k)},$$

where $r^{(k)} = b - Ax^{(k)}$ is the residual after k iterations. □

Exercise 8.2.6 Write a simple computer program that performs Jacobi iterations (8.2.2), and use it to verify the results shown in Example 8.2.3. Notice how easy this is to program in Fortran, C, or whatever language. A crude MATLAB code based on either of the equations derived in Exercise 8.2.4 is particularly easy. □

Exercise 8.2.7 Use the code you wrote for Exercise 8.2.6 to solve the system

$$\begin{bmatrix} 5 & 1 & 1 \\ 1 & 4 & 1 \\ 1 & 1 & 3 \end{bmatrix} \begin{bmatrix} x_1 \\ x_2 \\ x_3 \end{bmatrix} = \begin{bmatrix} 10 \\ 12 \\ 12 \end{bmatrix}$$

by Jacobi's method, starting with $x^{(0)} = 0$. Observe the iterates $x^{(k)}$ and the residuals $r^{(k)} = b - Ax^{(k)}$. How many iterations does it take to make $\|r^{(k)}\|_2 < 10^{-5}$? For this value of k, how well does $x^{(k)}$ approximate the true solution? □

It is easy to apply Jacobi's method to the model problem (8.1.8). The equation associated with the (i, j)th grid point is

$$-u_{i-1,j} - u_{i,j-1} + 4u_{i,j} - u_{i,j+1} - u_{i+1,j} = h^2 f_{i,j}.$$

We use this equation to correct the corresponding unknown, $u_{i,j}$. Solving the equation for $u_{i,j}$, we have

$$u_{i,j} = \frac{1}{4}\left[u_{i-1,j} + u_{i,j-1} + u_{i,j+1} + u_{i+1,j} + h^2 f_{i,j}\right].$$

Given a current iterate $u^{(k)}$, we obtain a new iterate $u^{(k+1)}$ by

$$u_{i,j}^{(k+1)} = \frac{1}{4}\left[u_{i-1,j}^{(k)} + u_{i,j-1}^{(k)} + u_{i,j+1}^{(k)} + u_{i+1,j}^{(k)} + h^2 f_{i,j}\right].$$

As usual, the boundary values of $u^{(k)}$, which are needed in the computation for those grid points that are adjacent to the boundary, are determined by the boundary conditions.

Notice that execution of Jacobi's method does not require writing the system in the form $Au = b$. There is no need to assemble or store the matrix (8.1.10) nor to pack $u_{i,j}$ and $f_{i,j}$ into long vectors.

Example 8.2.8 Jacobi's method was applied to the model problem with $f = 0$ and g given by

$$g(x,y) = \begin{cases} 0 & \text{if } x = 0 \\ y & \text{if } x = 1 \\ (x - 1)\sin x & \text{if } y = 0 \\ x(2 - x) & \text{if } y = 1. \end{cases}$$

The initial guess was $u^{(0)} = 0$. Four different grid sizes, $h = \frac{1}{20}, \frac{1}{40}, \frac{1}{80}$, and $\frac{1}{160}$, were used. The iterations were continued until the change in u from one iteration to the next did not exceed 10^{-8}. More precisely, the iterations were stopped as soon as

$$\frac{\| u^{(k+1)} - u^{(k)} \|_2}{\| u^{(k+1)} \|_2} < 10^{-8}.$$

The results are shown in Table 8.1 We observe that convergence is slow, and it

h	Matrix dimension	Iterations to convergence
1/20	361	1090
1/40	1521	3908
1/80	6241	13817
1/160	25281	48033

Table 8.1 Jacobi's method applied to model problem

becomes slower as the grid is refined. On the positive side, the program was easy to write and debug.   □

Gauss-Seidel Method

Now consider the following simple modification of Jacobi's method. The system to be solved is $Ax = b$, whose ith equation is

$$\sum_{i=1}^{n} a_{i,j}x_j = b_i.$$

As before, we use the ith equation to modify the ith unknown, but now we consider doing the process sequentially. First we use the first equation to compute $x_1^{(k+1)}$, then we use the second equation to compute $x_2^{(k+1)}$, and so on. By the time we get to the ith equation, we have already computed $x_1^{(k+1)}, \ldots, x_{i-1}^{(k+1)}$. In computing $x_i^{(k+1)}$, we could use these newly calculated values, or we could use the old values $x_1^{(k)}, \ldots, x_{i-1}^{(k)}$. The Jacobi method uses the old values; Gauss-Seidel uses the new. That is the only difference. Thus, instead of (8.2.2), Gauss-Seidel performs

$$x_i^{(k+1)} = \frac{1}{a_{ii}} \left(b_i - \sum_{j \neq i} a_{ij} x_j^{(*)} \right),$$

where $x_j^{(*)}$ denotes the most up-to-date value for the unknown x_j. More precisely, we can write a Gauss-Seidel iteration as follows:

for $i = 1, \ldots, n$
$$\left[\; x_i^{(k+1)} = \frac{1}{a_{ii}} \left(b_i - \sum_{j=1}^{i-1} a_{ij} x_j^{(k+1)} - \sum_{j=i+1}^{n} a_{ij} x_j^{(k)} \right) \right. \tag{8.2.9}$$

There is no need to maintain separate storage locations for $x_i^{(k)}$ and $x_i^{(k+1)}$; everything is done in a single x array. As soon as $x_\alpha^{(k+1)}$ has been computed, it is stored in place of $x_\alpha^{(k)}$ (which will never be needed again) in location x_α in the array. Thus the iteration (8.2.9) takes the form

for $i = 1, \ldots, n$
$$\left[\; x_i \leftarrow \frac{1}{a_{ii}} \left(b_i - \sum_{j \neq i} a_{ij} x_j \right) \right. \tag{8.2.10}$$

in practice.

The order in which the corrections are made is important. If they were made in, say, the reverse order $i = n, \ldots, 1$, the iteration would have a different outcome. We will always assume that a Gauss-Seidel iteration will be performed in the standard order $i = 1, \ldots, n$, as indicated in (8.2.10), unless otherwise stated. The question of orderings is important; more will be said about it in connection with the model problem (8.1.8).

The fact that we can store each new x_i value immediately in place of the old one is an advantage of the Gauss-Seidel method over Jacobi. For one thing, it makes the programming easier. It also saves storage space; Jacobi's method needs to store two copies of x, since $x^{(k)}$ needs to be kept until the computation of $x^{(k+1)}$ is complete. If the system we are solving has millions of unknowns, each copy of x will occupy several megabytes of storage space.

On the other hand, Jacobi's method has the advantage that all of the corrections (8.2.2) can be performed simultaneously; the method is inherently parallel. Gauss-Seidel, in contrast, is inherently sequential, or so it seems. As we shall see, parallelism can be recovered by reordering the equations.

We now investigate the performance of the Gauss-Seidel method by means of some simple examples.

Example 8.2.11 If we apply the Gauss-Seidel method to the same small system as in Example 8.2.3, we obtain, for example,

$$x^{(5)} = \begin{bmatrix} 4.00131 \\ 2.99929 \\ 1.99990 \\ 0.99995 \end{bmatrix} \qquad x^{(10)} = \begin{bmatrix} 4.000000142 \\ 2.999999878 \\ 2.000000018 \\ 0.999999993 \end{bmatrix},$$

which are considerably better than the iterates produced by Jacobi's method. Whereas it takes Jacobi's method 80 iterations to obtain an approximation that agrees with the true solution to twelve decimal places, Gauss-Seidel achieves the same accuracy after only 18 iterations. $\qquad\square$

cise 8.2.12 Given $A \in \mathbb{R}^{n \times n}$, define matrices D, E, and F as follows: $A = D - E - F$, D is diagonal, E is strictly lower triangular, and F is strictly upper triangular. This specifies the matrices completely. For example, $-E$ is the strictly lower triangular part of A.

(a) Show that the Gauss-Seidel iteration (8.2.9) can be written as a matrix equation

$$x^{(k+1)} = D^{-1} \left(b + E x^{(k+1)} + F x^{(k)} \right).$$

(b) The vector $x^{(k+1)}$ appears in two places in the equation that you derived in part (a). Solve this equation for $x^{(k+1)}$ to obtain

$$x^{(k+1)} = (D - E)^{-1} \left(b + F x^{(k)} \right). \tag{8.2.13}$$

(c) Let $M = D - E$. Thus M is the lower triangular part of A, including the main diagonal. Show that $F = M - A$, then verify the formula

$$x^{(k+1)} = x^{(k)} + M^{-1} r^{(k)},$$

where $r^{(k)} = b - A x^{(k)}$ is the residual.

$\qquad\square$

cise 8.2.14 Write a simple computer program that performs Gauss-Seidel iterations, and use it to verify the results shown in Example 8.2.11. For example, a simple MATLAB code can be built around either of the formulas derived in Exercise 8.2.12, or a Fortran or C code can be built around (8.2.10). $\qquad\square$

cise 8.2.15 Use the code you wrote for Exercise 8.2.14 to solve the system

$$\begin{bmatrix} 5 & 1 & 1 \\ 1 & 4 & 1 \\ 1 & 1 & 3 \end{bmatrix} \begin{bmatrix} x_1 \\ x_2 \\ x_3 \end{bmatrix} = \begin{bmatrix} 10 \\ 12 \\ 12 \end{bmatrix}$$

by the Gauss-Seidel method, starting with $x^{(0)} = 0$. Observe the iterates $x^{(k)}$ and the residuals $r^{(k)}$. How many iterations does it take to make $\| r^{(k)} \|_2 < 10^{-5}$? For this value of k, how well does $x^{(k)}$ approximate the true solution? How do your results compare with those of Exercise 8.2.7? $\qquad\square$

Now let us see how the Gauss-Seidel method can be applied to the model problem (8.1.8). The equation associated with the (i, j)th grid point is

$$-u_{i-1,j} - u_{i,j-1} + 4u_{i,j} - u_{i,j+1} - u_{i+1,j} = h^2 f_{i,j}.$$

We use it to correct $u_{i,j}$. In the spirit of (8.2.10), we write

$$u_{i,j} \leftarrow \frac{1}{4} \left[u_{i-1,j} + u_{i,j-1} + u_{i,j+1} + u_{i+1,j} + h^2 f_{i,j} \right]. \qquad (8.2.16)$$

As we have already noted, the outcome of the Gauss-Seidel iteration depends on the order in which the corrections are made. For example, one can sweep through the computational grid by rows (see Figure 8.1), performing the corrections in the order $(1, 1), (2, 1), \ldots, (m - 1, 1), (1, 2), (2, 2), \ldots, (m - 1, 2), \ldots, (m - 1, m - 1)$. We will call this the *standard row ordering*. There is an analogous *standard column ordering*. There are also *reverse* row and column orderings, obtained by reversing these. Each ordering gives a slightly different outcome. There are many other orderings that one might consider using.

Example 8.2.17 The Gauss-Seidel method, using the standard row ordering, was applied to the model problem (8.1.8) under exactly the same conditions as in Example 8.2.8. The results are shown in Table 8.2. Comparing with Example 8.2.8,

h	Matrix dimension	Iterations to convergence
1/20	361	581
1/40	1521	2082
1/80	6241	7389
1/160	25281	25877

Table 8.2   Standard Gauss-Seidel applied to model problem

we see that the Gauss-Seidel method needs only slightly more than half as many iterations as the Jacobi method to obtain the same accuracy. It is still fair to say that the convergence is slow, and there is still the problem that the number of iterations increases as the grid is made finer. The largest problem considered here is of modest size. Extrapolating from these results, we can expect Gauss-Seidel will converge very slowly on really big problems. $\qquad\square$

Red-Black and Multicolor Gauss-Seidel

Before moving on to a better method, let us spend a little more time on orderings. Red-black orderings are illustrated by Figure 8.3. Grid points are alternately labelled

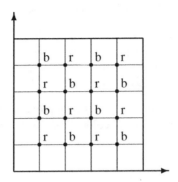

Figure 8.3 Computational grid of red and black points

red and *black* in a checkerboard pattern. A *red-black ordering* is one in which all of the red points are updated first, then the black points are updated. Notice that every red (resp. black) point has as nearest neighbors only black (resp. red) points (and, in some cases, boundary points). Thus the Gauss-Seidel correction of each red point depends only on the values of u at black points and vice versa. This implies that all of the red points can be updated simultaneously. Once the red points have been done, all of black points can be updated simultaneously. Red-black orderings restore parallelism. Red-black Gauss-Seidel converges at the same rate as standard row-ordered Gauss-Seidel on the model problem (8.1.8). Numerical examples will be given later.

Now consider orderings in the context of the general system $Ax = b$. The standard ordering is $i = 1, \ldots, n$. Any other ordering can be obtained by applying the standard ordering to a system $\hat{A}\hat{x} = \hat{b}$ obtained from $Ax = b$ by rearranging the equations (and rearranging the unknowns accordingly). Thus the question of orderings can be translated to a question of permuting the rows and columns of A.

A red-black ordering is one that corresponds to a coefficient matrix with the block structure

$$\hat{A} = \begin{bmatrix} D_1 & B_{12} \\ B_{21} & D_2, \end{bmatrix},$$

where D_1 and D_2 are (square) diagonal matrices of roughly the same size. The first block of rows consists of red equations (with their corresponding red unknowns), and the second block consists of black equations[45]. All of the unknowns of a given color can be updated simultaneously.

---

[45] This reordering is exactly the deshuffling operation used in conjunction with product eigenvalue problems in Sections 5.8 and 7.3

Obviously most systems do not admit a red-black ordering. Our model problem (8.1.8) does, and so does its three-dimensional analogue, but many large, sparse problems that arise in applications do not. In these cases a *multicolor* ordering involving more than two colors is often possible. For example, a four-color ordering corresponds to a coefficient matrix with the block structure

$$\hat{A} = \begin{bmatrix} D_1 & B_{12} & B_{13} & B_{14} \\ B_{21} & D_2 & B_{23} & B_{24} \\ B_{31} & B_{32} & D_3 & B_{34} \\ B_{41} & B_{42} & B_{43} & D_4 \end{bmatrix},$$

where $D_1, \ldots, D_4$ are all diagonal matrices of approximately the same size. All of the unknowns within a given block or "color class" can be updated simultaneously.

Symmetric Gauss-Seidel

One iteration of symmetric Gauss-Seidel consists of two standard Gauss-Seidel iterations, one in the forward direction followed by one in the reverse direction. Thus the first half of the iteration is (8.2.10), and the second half is (8.2.10) with "$i = n, \ldots, 1$" in place of the standard ordering.

Exercise 8.2.18 In Exercise 8.2.12 you showed that the Gauss-Seidel iteration can be expressed in matrix form as (8.2.13).

(a) Derive the analogous formula for the reverse Gauss-Seidel method.

(b) The symmetric Gauss-Seidel iteration consists of a forward Gauss-Seidel step, transforming $x^{(k)}$ to $x^{(k+\frac{1}{2})}$, followed by a reverse Gauss-Seidel step, transforming $x^{(k+\frac{1}{2})}$ to $x^{(k+1)}$. Show that the symmetric Gauss-Seidel iteration satisfies

$$(D - F)x^{(k+1)} = E(D - E)^{-1}Fx^{(k)} + \left[I + E(D - E)^{-1}\right]b.$$

(c) Show that $I + E(D - E)^{-1} = D(D - E)^{-1}$. Then use this fact together with the equation from part (b) to show that the symmetric Gauss-Seidel iteration satisfies

$$Mx^{(k+1)} = Nx^{(k)} + b,$$

where $M = (D - E)D^{-1}(D - F)$, and $N = (D - E)D^{-1}E(D - E)^{-1}F = ED^{-1}F$. Show that M is symmetric if A is.

(d) Show that the M and N determined in part (c) satisfy $A = M - N$. Then show that the symmetric Gauss-Seidel iteration satisfies

$$x^{(k+1)} = x^{(k)} + M^{-1}r^{(k)}$$

with this choice of M, where $r^{(k)}$ denotes, once again, the residual $b - Ax^{(k)}$.

□

Successive Overrelaxation (SOR)

The process of correcting an equation by modifying one unknown is sometimes called *relaxation*. Before the correction, the equation is not quite true; like an assemblage of parts that does not fit together quite right, it is in a state of tension. The correction of one variable relaxes the tension. The Gauss-Seidel method performs *successive relaxation*. That is, it moves from equation to equation, relaxing one after the other.[46]

In many cases convergence can be accelerated substantially by overrelaxing. This means that rather than making a correction for which the equation is satisfied exactly, we make a somewhat bigger correction. In the simplest case one chooses a *relaxation factor* $\omega > 1$ and overcorrects by that factor at each step. This is *successive overrelaxation* (SOR):

$$\text{for } i = 1, \ldots, n$$
$$\begin{bmatrix} \hat{x} \leftarrow \dfrac{1}{a_{ii}} \left(b_i - \sum_{j \neq i} a_{ij} x_j \right) \\ \delta \leftarrow \hat{x} - x_i \\ x_i \leftarrow x_i + \omega\delta \end{bmatrix} \qquad (8.2.19)$$

Comparing (8.2.19) with (8.2.10), we see that SOR collapses to Gauss-Seidel when $\omega = 1$. One could also use a factor ω that is less than one, which amounts to underrelaxation, but this normally slows convergence.

Since SOR is "successive" (like Gauss-Seidel) rather than "simultaneous" (like Jacobi), it needs to keep only one copy of the vector x.

Not much is gained by applying overrelaxation to the small problem featured in Examples 8.2.3 and 8.2.11, so let us move directly to the model problem (8.1.8). Referring to the Gauss-Seidel step (8.2.16), we see that the corresponding SOR step is

$$\hat{u} \leftarrow \tfrac{1}{4} \left[u_{i-1,j} + u_{i,j-1} + u_{i,j+1} + u_{i+1,j} + h^2 f_{i,j} \right]$$
$$\delta \leftarrow \hat{u} - u_{i,j} \qquad (8.2.20)$$
$$u_{i,j} \leftarrow u_{i,j} + \omega\delta.$$

Just as for Gauss-Seidel, the outcome of an SOR iteration depends on the order in which the updates are made. Commonly used orderings include the standard row and column orderings and the red-black ordering illustrated by (8.3). One can also do *symmetric* SOR (abbreviated SSOR), which consists of alternating forward and backward standard row (or column) sweeps.

Example 8.2.21 The SOR method, using the standard row ordering, was applied to the model problem (8.1.8) under exactly the same conditions as in Examples 8.2.8 and 8.2.17. Several values of the relaxation factor ω were tried. The results are shown in Table 8.3. Iterations were halted as soon as

$$\frac{\left\| u^{(k+1)} - u^{(k)} \right\|_2}{\left\| u^{(k+1)} \right\|_2} < 10^{-8}.$$

---

[46]In contrast, Jacobi's method performs *simultaneous relaxation*.

The choice $\omega = 1$ corresponds to Gauss-Seidel; the numbers in the second row are

	1/20	1/40	1/80	1/160
$\omega = 0.8$	845	3018	10676	27208
1.0	581	2082	7389	25877
1.5	204	756	2714	9611
1.8	89	252	981	3541
1.9	180	179	450	1769
1.95	354	361	355	832
1.97	589	585	602	643
2.0	∞	∞	∞	∞

Table 8.3   Iteration counts for SOR applied to model problem

identical to those in Table 8.2. We note that underrelaxation ($\omega = 0.8$) is detrimental and overrelaxation can be very beneficial, particularly if an ω that is close to optimal is used. For the finest mesh, the iteration count is cut by a factor of 40 when $\omega = 1.97$. Since SOR steps take only slightly more work than Gauss-Seidel steps, the savings in iterations translate almost directly into savings in computer time.

From the numbers in Table 8.3 we see that the optimal value of ω depends on the mesh size (and in general it depends on the coefficient matrix A). For $h = 1/20$ the best ω is somewhere around 1.8, for $h = 1/40$ it is around 1.9, and for $h = 1/80$ it is around 1.95. For most problems the optimal ω is not known, but the numbers in Table 8.3 suggest that even a crude guess can yield substantial savings. However, SOR diverges when $\omega \geq 2$.

For our model problem the optimal ω is known to be

$$\omega_b = \frac{2}{1 + \sin \pi h}.$$

When $h = 1/160$ it is approximately 1.9615. Using this value of ω, SOR converges in 518 iterations.   □

As an exercise, you can explore the convergence of the red-black SOR and SSOR methods (Exercise 8.2.23). On the model problem (and many similar ones) red-black and standard row ordered SOR converge at the same rate. SSOR is somewhat slower but less sensitive to the choice of the relaxation parameter ω.

The computations reported in Example 8.2.21 were performed by a simple Fortran program based on (8.2.20). An iteration with the standard row ordering is effected by doubly nested loops:

$$
\begin{array}{l}
\text{for } j = 1, \ldots, m - 1 \\
\left\lceil \begin{array}{l}
\text{for } i = 1, \ldots, m - 1 \\
\quad \left\lceil \begin{array}{l}
\hat{u} \leftarrow \frac{1}{4} \left(u_{i-1,j} + u_{i,j-1} + u_{i,j+1} + u_{i+1,j} + h^2 f_{i,j} \right) \\
\delta \leftarrow \hat{u} - u_{i,j} \\
u_{i,j} \leftarrow u_{i,j} + \omega \delta.
\end{array} \right.
\end{array} \right.
\end{array}
$$

The implementations of red-black SOR and SSOR are only slightly more complicated. In particular, it was not necessary to form or manipulate the coefficient matrix A or any related matrix.

In Exercises 8.2.4 and 8.2.12 you derived the matrix formulas (8.2.5) and (8.2.13) for the Jacobi and Gauss-Seidel iterations, respectively. A similar expression can be derived for the SOR method (Exercise 8.2.24). These formulas are useful for analyzing the convergence of the methods, as will be illustrated in Section 8.3. They are also handy for writing simple MATLAB programs to apply to small examples. However, it is important to realize that these formulas should not be used for large problems, as the resulting code would be horribly inefficient. Instead one should write code as described in the previous paragraph.

Block Iterative Methods

All of the iterative methods that we have derived in this section have block variants. Suppose the system $Ax = b$ has been subdivided into blocks:

$$A = \begin{bmatrix} A_{11} & A_{12} & \cdots & A_{1q} \\ A_{21} & A_{22} & \cdots & A_{2q} \\ \vdots & \vdots & \ddots & \vdots \\ A_{q1} & A_{q2} & \cdots & A_{qq} \end{bmatrix} \begin{bmatrix} x_1 \\ x_2 \\ \vdots \\ x_q \end{bmatrix} = \begin{bmatrix} b_1 \\ b_2 \\ \vdots \\ b_q \end{bmatrix},$$

where the main diagonal blocks A_{ii} are all square and nonsingular. The entries x_i and b_i now refer to subvectors of x and b of the appropriate lengths. A block Jacobi iteration is just like an ordinary Jacobi iteration, except that it acts on blocks instead of the matrix entries. Instead of (8.2.2), we have

$$x_i^{(k+1)} = A_{ii}^{-1} \left(b_i - \sum_{j \neq i} A_{ij} x_j^{(k)} \right). \tag{8.2.22}$$

Block Gauss-Seidel and SOR iterations are derived in similar fashion. In order that such a block algorithm be efficient, it is necessary that the main diagonal blocks A_{ii} be simple enough that their inverses (or LU decompositions) can be obtained easily.

Consider, for example, the matrix of the model problem (8.1.8), written in the block form (8.1.10). To apply a block algorithm using this blocking, we need to be able to LU decompose the tridiagonal matrix T. As we know, this is simple. Each block corresponds to one row (or *line*) of unknowns in the computational grid, so block algorithms based on this partition are known as *line* iterations, for example, line Gauss-Seidel or line SOR.

Red-Black (Gauss-Seidel or SOR) iterations can be viewed as block iterations based on a 2×2 blocking

$$\begin{bmatrix} D_1 & A_{12} \\ A_{21} & D_2 \end{bmatrix}.$$

More generally, a multicolor iteration with q colors is a block iteration based on a $q \times q$ blocking. Here the matrices on the main diagonal are trivial to invert, as they are diagonal matrices.

Additional Exercises

Exercise 8.2.23 Write programs that solve the model problem (8.1.8) by standard SOR, red-black SOR, and SSOR. Use each of them to solve the model problem with $f = 0$ and g given by

$$g(x,y) = \begin{cases} 0 & \text{if } x = 0 \\ y & \text{if } x = 1 \\ (x - 1)\sin x & \text{if } y = 0 \\ x(2 - x) & \text{if } y = 1, \end{cases}$$

as in Example 8.2.8. Use mesh sizes $h = 1/80$ and $h = 1/160$. Starting from $u^{(0)} = 0$, record how many iterations are needed until

$$\frac{\| u^{(k+1)} - u^{(k)} \|_2}{\| u^{(k+1)} \|_2} < 10^{-8}.$$

Since SSOR iterations take twice as much work as the other iterations, count each SSOR iteration as two iterations. For each of the three methods try $\omega = 1.0, 1.5, 1.8,$ 1.9, 1.95, 1.97, 1.98. Compare the performance of the three methods.   □

Exercise 8.2.24 (Matrix representation of SOR)

(a) In the SOR iteration (8.2.19), show that successive iterates $x^{(k)}$ and $x^{(k+1)}$ are related by

$$a_{ii}x_i^{(k+1)} = a_{ii}x_i^{(k)} + \omega \left[b_i - \sum_{j=1}^{i-1} a_{ij}x_j^{(k+1)} - \sum_{j=i}^{n} a_{ij}x_j^{(k)} \right].$$

(b) Write $A = D - E - F$, as in Exercise 8.2.12. Derive the following expressions for the relationship between successive SOR iterates $x^{(k)}$ and $x^{(k+1)}$.

$$Dx^{(k+1)} = Dx^{(k)} + \omega \left[b + Ex^{(k+1)} + (F - D)x^{(k)} \right].$$

$$\left(\frac{1}{\omega}D - E \right) x^{(k+1)} = \left(\frac{1-\omega}{\omega}D + F \right) x^{(k)} + b. \qquad (8.2.25)$$

Thus $Mx^{(k+1)} = Nx^{(k)} + b$, where $M = \frac{1}{\omega}D - E$, $N = \frac{1-\omega}{\omega}D + F$, and $A = M - N$.

(c) Show that

$$x^{(k+1)} = x^{(k)} + M^{-1}r^{(k)},$$

where $r^{(k)} = b - Ax^{(k)}$.

□

cise 8.2.26 (Matrix representation of SSOR) This exercise generalizes Exercise 8.2.18

(a) Find the formula analogous to (8.2.25) for an SOR iteration in the reverse direction.

(b) The symmetric SOR (SSOR) iteration consists of a forward SOR step, transforming $x^{(k)}$ to $x^{(k+\frac{1}{2})}$, followed by a reverse SOR step, transforming $x^{(k+\frac{1}{2})}$ to $x^{(k+1)}$. Show that the SSOR iteration satisfies

$$\left(\tfrac{1}{\omega}D - F\right)x^{(k+1)} = \left(\tfrac{1-\omega}{\omega}D + E\right)\left(\tfrac{1}{\omega}D - E\right)^{-1}\left(\tfrac{1-\omega}{\omega}D + F\right)x^{(k)}$$
$$+ \left[I + \left(\tfrac{1-\omega}{\omega}D + E\right)\left(\tfrac{1}{\omega}D - E\right)^{-1}\right]b.$$

(c) Show that $I + \left(\tfrac{1-\omega}{\omega}D + E\right)\left(\tfrac{1}{\omega}D - E\right)^{-1} = \tfrac{2-\omega}{\omega}D\left(\tfrac{1}{\omega}D - E\right)^{-1}$. Then use this fact together with the equation from part (b) to show that the SSOR iteration satisfies

$$Mx^{(k+1)} = Nx^{(k)} + b,$$

where

$$M = \tfrac{\omega}{2-\omega}\left(\tfrac{1}{\omega}D - E\right)D^{-1}\left(\tfrac{1}{\omega}D - F\right),$$

and

$$N = \tfrac{\omega}{2-\omega}\left(\tfrac{1-\omega}{\omega}D + E\right)D^{-1}\left(\tfrac{1-\omega}{\omega}D + F\right).$$

Show that M is symmetric if A is.

(d) Show that the M and N determined in part (c) satisfy $A = M - N$. Then show that the SSOR iteration satisfies

$$x^{(k+1)} = x^{(k)} + M^{-1}r^{(k)}$$

with this choice of M, where $r^{(k)} = b - Ax^{(k)}$.

$\square$

cise 8.2.27 (Smoothing property of Gauss-Seidel iterations) Consider the linear system $Ax = b$, where A is the 10×10 version of the tridiagonal matrix (8.1.5) and

$$b = \begin{bmatrix} 2 & -2 & 2 & -1 & 0 & 0 & 1 & -2 & 2 & -2 \end{bmatrix}^T.$$

You can easily check that the exact solution to this system is

$$x = \begin{bmatrix} 1 & 0 & 1 & 0 & 0 & 0 & 0 & -1 & 0 & -1 \end{bmatrix}^T.$$

Starting with $x^{(0)} = 0$, perform at least four Gauss-Seidel iterations. Since the exact solution of the system is known, you can calculate the errors $e^{(i)} = x - x^{(i)}$. Make plots of the errors; that is, given an error vector e, plot the points (i, e_i) and connect them with line segments (plot(e) or plot(1:10,e) in MATLAB). Notice that the plot of $e^{(0)}$ is quite jagged, but each successive plot is less so. The plot of $e^{(4)}$ is quite smooth.

Multigrid methods are fast iterative methods that solve problems like (8.1.8) rapidly by applying smoothing operations and coarse-grid corrections recursively. Gauss-Seidel is a popular smoothing operator for multigrid algorithms.   □

Exercise 8.2.28

(a) Repeat Exercise 8.2.27 using Jacobi iterations in place of Gauss-Seidel. Notice that the Jacobi method does not smooth the error. Thus the basic Jacobi method is useless as a multigrid smoothing operator.

(b) The *damped Jacobi method* with *damping factor* $\omega < 1$ is the same as the ordinary Jacobi method, except that the correction at each step is damped by the factor ω. Thus, instead of $x^{(k+1)} = x^{(k)} + D^{-1}r^{(k)}$ (cf. Exercise 8.2.4), damped Jacobi makes the correction

$$x^{(k+1)} = x^{(k)} + \omega D^{-1}r^{(k)}.$$

Repeat Exercise 8.2.27 using the Damped Jacobi method with $\omega = 1/2$ and $\omega = 2/3$. The damped Jacobi iteration with $1/2 \leq \omega \leq 2/3$ is an effective smoother.

□

8.3 CONVERGENCE OF ITERATIVE METHODS

Splittings

We unify the convergence theory of iterative methods by introducing the notion of a splitting. Given an $n \times n$ nonsingular matrix A, a *splitting* of A is nothing but an additive decomposition $A = M - N$, where M is nonsingular. The matrix M is called the *splitting matrix*. As we shall later see, it can also be called a *preconditioner*.

Every splitting gives rise to an iterative method as follows: Use the splitting to rewrite the system $Ax = b$ in the form $Mx = Nx + b$ or $x = M^{-1}Nx + M^{-1}b$. Then define an iteration by either of the equivalent equations

$$x^{(k+1)} = M^{-1}Nx^{(k)} + M^{-1}b$$

or

$$Mx^{(k+1)} = Nx^{(k)} + b. \tag{8.3.1}$$

Clearly this scheme requires that M be nonsingular. From a practical standpoint it is also important that M be "easy to solve," meaning that systems with M as the coefficient matrix are easy to solve. If the iterates converge to some x, then that x satisfies $Mx = Nx + b$, hence $Ax = b$. For the sake of rapid convergence, we would like $M \approx A$ and $N \approx 0$. From this viewpoint the best choice is $M = A$, which gives convergence in one iteration, but this M violates the "easy to solve" requirement. What is needed is a good compromise: M should approximate A as well as possible without being too complicated.

All of the iterative methods that we have discussed so far can be induced by splittings of A. Table 8.4 lists the splitting matrices M associated with the classical iterations, as determined in Exercises 8.2.4, 8.2.12, 8.2.24, and 8.2.26. The matrices D, E, and F refer to the decomposition $A = D - E - F$, where D, $-E$, and $-F$ are the diagonal, lower triangular, and upper triangular parts of A, respectively. In

Method	Splitting Matrix (M)	Exercise
Jacobi	D	8.2.4
Gauss-Seidel	$D - E$	8.2.12
SOR	$\frac{1}{\omega}D - E$	8.2.24
SSOR	$\frac{\omega}{2-\omega}\left(\frac{1}{\omega}D - E\right)D^{-1}\left(\frac{1}{\omega}D - F\right)$	8.2.26

Table 8.4 Splitting matrices for classical iterations

each case only M has been listed because N can be inferred from the relationship $N = M - A$.

Example 8.3.2 A very simple method that plays a surprisingly big role in convergence theory is *Richardson's method*, which is induced by the splitting $M = \frac{1}{\omega}I$, $N = \frac{1}{\omega}I - A$, where ω is a *damping factor* chosen to make M approximate A as well as possible. Thus Richardson's iteration is

$$x^{(k+1)} = (I - \omega A)x^{(k)} + \omega b.$$

The properties of Richardson's method are explored in Exercises 8.3.19, 8.3.20, 8.3.22, and 8.3.37. □

cise 8.3.3 Suppose we apply Richardson's method to the model problem (8.1.8). What would be the obvious choice of ω to try, given that we want M to approximate A as well as possible? For this choice of ω, with which classical iteration does Richardson's method coincide? □

Convergence

We consider an iteration of the form (8.3.1). At each step there is an *error* $e^{(k)} = x - x^{(k)}$, which is just the difference between the true solution and its approximant after k iterations. As we are in the process of solving a problem, we will not know what $e^{(k)}$ is (because x is not yet known), but that does not stop us from talking about it. For a given splitting $A = M - N$, we would like to prove that $e^{(k)} \to 0$ as $k \to \infty$. Moreover, since we cannot wait forever in practice, we would like to show, if possible, that the $e^{(k)}$ become small quickly. The true solution of $Ax = b$ also satisfies $Mx = Nx + b$. Subtracting the equation (8.3.1) from this, we find that

$Me^{(k+1)} = Ne^{(k)}$. Thus

$$e^{(k+1)} = Ge^{(k)},$$

where $G = M^{-1}N = I - M^{-1}A$. As this equation holds for all k, we have $e^{(1)} = Ge^{(0)}$, $e^{(2)} = Ge^{(1)} = G^2e^{(0)}$, and, in general,

$$e^{(k)} = G^k e^{(0)}. \tag{8.3.4}$$

The vector $e^{(0)}$ is our initial error; its size depends on the initial guess. Equation (8.3.4) shows that regardless of what the initial guess was, $e^{(k)} \to 0$ if $G^k \to 0$. Since powers of a matrix are what matters, it should not be surprising that the convergence theory of iterative methods resembles the analysis of the power method for computing a dominant eigenvector. In particular, the eigenvalues of the iteration matrix play a crucial role. But now the scenario is somewhat different; we want the iterates to converge to zero, not to an eigenvector.

Let us assume that the iteration matrix G is semisimple. (For the non-semisimple case, see Exercise 8.3.27.) This means that G has n linearly independent eigenvectors $v_1, v_2, \ldots, v_n$, which form a basis of $\mathbb{C}^n$. Let $\lambda_1, \lambda_2, \ldots, \lambda_n$ denote the corresponding eigenvalues of G: $Gv_i = \lambda_i v_i$, $i = 1, \ldots, n$. We will focus on the errors $e^{(k)}$ directly rather than the matrices G^k (but see Exercise 8.3.26). Since the eigenvectors $v_1, \ldots, v_n$ form a basis of $\mathbb{C}^n$, we can express the initial error $e^{(0)}$ as a linear combination of them:

$$e^{(0)} = c_1 v_1 + c_2 v_2 + \cdots + c_n v_n. \tag{8.3.5}$$

The scalars $c_1, \ldots, c_n$ are unknown, since $e^{(0)}$ is, and they might be complex numbers, even though $e^{(0)}$ is real. None of this causes any problems for the analysis. Applying G to (8.3.5), we find that

$$e^{(1)} = Ge^{(0)} = c_1 \lambda_1 v_1 + c_2 \lambda_2 v_2 + \cdots + c_n \lambda_n v_n,$$

and in general

$$e^{(k)} = G^k e^{(0)} = c_1 \lambda_1^k v_1 + c_2 \lambda_2^k v_2 + \cdots + c_n \lambda_n^k v_n.$$

Consequently

$$\| e^{(k)} \| \le |c_1| \, |\lambda_1|^k \| v_1 \| + |c_2| \, |\lambda_2|^k \| v_2 \| + \cdots + |c_n| \, |\lambda_n|^k \| v_n \|,$$

where the norm can be any vector norm. Since $|\lambda_i|^k \to 0$ if and only if $|\lambda_i| < 1$, we conclude that $\| e^{(k)} \| \to 0$ for every initial guess $x^{(0)}$ if and only if $\max_i |\lambda_i| < 1$.

Let $\sigma(G)$ denote the spectrum of G, the set of eigenvalues. The *spectral radius*, denoted $\rho(G)$, is the maximum distance of an eigenvalue from the origin. Thus

$$\rho(G) = \max_{\lambda \in \sigma(G)} |\lambda|.$$

Using this new terminology we can restate our conclusion from the previous paragraph: The iterations converge (for any starting vector) if and only if $\rho(G) < 1$.

The spectral radius also gives information about the rate of convergence. If, say, λ_1 is the eigenvalue of greatest modulus ($|\lambda_1| = \rho(G)$), we eventually have

$$e^{(k)} \approx c_1 \lambda_1^k v_1,$$

since the other λ_i^k tend to zero more quickly than λ_1^k does, so

$$\frac{\| e^{(k+1)} \|}{\| e^{(k)} \|} \approx |\lambda_1| = \rho(G) \qquad (8.3.6)$$

for sufficiently large k. Thus the convergence is linear with convergence ratio $\rho(G)$. The smaller $\rho(G)$ is, the faster the iterations converge.

This argument depends on the assumption $c_1 \neq 0$, which holds for almost all choices of $x^{(0)}$. For those (exceedingly rare) vectors that yield $c_1 = 0$, the convergence is typically faster in theory. (Why?)

The approximation (8.3.6) is certainly valid for large k. It does not necessarily hold for small k.

We have now nearly (cf. Exercise 8.3.27) proved the following Theorem.

Theorem 8.3.7 *The iteration (8.3.1) converges to the true solution of $Ax = b$ for every initial guess $x^{(0)}$ if and only if the spectral radius of the iteration matrix $G = I - M^{-1}A$ is less than one. The convergence is linear. The average convergence ratio never exceeds $\rho(G)$; it equals $\rho(G)$ for almost all choices of $x^{(0)}$.*

In general it is difficult to calculate $\rho(G)$, but $\rho(G)$ is known in some special cases. In particular, for the model problem (8.1.8), $\rho(G)$ is known for the Richardson, Jacobi, Gauss-Seidel, and SOR (standard and red-black) iterations. There is not room here to derive all of these results; fortunately there are some excellent references, for example, [47], [49], [96], [108]. We will just scratch the surface.

The Jacobi iteration matrix is $G = D^{-1}(E + F)$. For the model problem (8.1.8), we have $G = \frac{1}{4}(E + E^T)$. In Exercise 8.3.35 it is shown that in this case $\rho(G) = \cos(\pi h)$, where h, as always, is the mesh size. It is not surprising that $\rho(G)$ would depend on h, particularly in light of the results of Example 8.2.8. We have $\rho(G) < 1$, so Jacobi's method is convergent for any h. However, as $h \to 0$, $\cos(\pi h) \to 1$, which implies that the convergence becomes very slow as h is made small. To get a more quantitative view of the situation, recall the Taylor expansion $\cos z = 1 - z^2/2! + O(z^4)$, which implies

$$\rho(G) \approx 1 - \frac{\pi^2}{2} h^2. \qquad (8.3.8)$$

Each time h is cut by a factor of 2, the distance between $\rho(G)$ and 1 is cut by a factor of about 4. For example, when $h = 1/20, 1/40, 1/80$, and $1/160$ we have $\rho(G) = .9877, .9969, .9992$, and $.9998$, respectively.

It is clear that with a spectral radius like .9998, convergence will be very slow. Let us estimate how many iterations are needed to reduce the error by a factor of 10. By (8.3.6) we have

$$\frac{\| e_{k+j} \|}{\| e_k \|} \approx \rho(G)^j.$$

Thus, to shrink the error by a factor of 10, we will need about j iterations, where $\rho(G)^j \approx 10^{-1}$. Applying logarithms and solving for j, we find that

$$j \approx \frac{\log 10}{-\log \rho(G)}.$$

For example, for $\rho = .9998$, we get $j \approx 12000$. That is, about 12000 iterations are needed for each reduction of the error by a factor of 10. If we wish to reduce the error by a factor of, say, 10^{-8}, we will need about $8 \times 12000 \approx 96000$ iterations.

In the computation of j, any base of logarithms can be used. Base 10 looks inviting, but for theoretical work, base e is preferred. The *asymptotic convergence rate* of an iterative method, denoted $R_\infty(G)$, is defined by

$$R_\infty(G) = -\log_e \rho(G). \tag{8.3.9}$$

The subscript ∞ is meant to convey the idea that this number is useful when the number of iterations is large (i.e. as $k \to \infty$) and not necessarily for small k.

Exercise 8.3.10 Show that $1/R_\infty(G)$ is approximately the number of iterations needed to reduce the error by a factor of e (once k is sufficiently large). □

When $\rho(G) = 1$ we have $R_\infty(G) = 0$, indicating no convergence. When $\rho(G) < 1$, we have $R_\infty(G) > 0$. The smaller $\rho(G)$ is, the larger $R_\infty(G)$ will be. As $\rho(G) \to 0$, we have $R_\infty(G) \to \infty$.

A doubling of $R_\infty(G)$ implies a doubling of the convergence rate, in the sense that only about half as many iterations will be required to reduce the error by a given factor.

When Jacobi's method is applied to the model problem, the spectral radius is given by (8.3.8). Using this and the Taylor expansion $-\log(1-z) = z + \frac{1}{2}z^2 + \frac{1}{3}z^3 + \cdots$, we find that

$$R_\infty(G) = \frac{\pi^2}{2}h^2 + O(h^4).$$

Stating this result more succinctly, we have $R_\infty(G) = O(h^2)$. This means that each time we cut h in half, we cut $R_\infty(G)$ by a factor of 4, which means, in turn, that about four times as many iterations will be needed to attain a given accuracy. This theoretical conclusion is borne out by the numbers in Example 8.2.8.

Exercise 8.3.11 Calculate $\rho(G)$ and $R_\infty(G)$ for Jacobi's method applied to the model problem (8.1.8) with $h = .02$ and $h = .01$. For each of these choices of h, about how many iterations are required to reduce the error by a factor of 10^{-3}? □

The numbers in Example 8.2.17 suggest that the Gauss-Seidel method converges somewhat faster than the Jacobi method does. Indeed this is true for the model problem and any problem that admits a red-black ordering. Let G_J and G_{GS} denote the iteration matrices for Jacobi and red-black Gauss-Seidel, respectively. In Exercise 8.3.38 you will show that $\rho(G_{GS}) = \rho(G_J)^2$. The standard Gauss-Seidel iteration has the same spectral radius as red-black Gauss-Seidel because the standard ordering is a so-called *consistent ordering* [49], [96], [108]. This means that one Gauss-Seidel iteration is about as good as two Jacobi iterations; Gauss-Seidel converges twice as fast as Jacobi. The numbers in Examples 8.2.8 and 8.2.17 bear this out.

cise 8.3.12 Show that if $\rho(G_{GS}) = \rho(G_J)^2$, then $R_\infty(G_{GS}) = 2R_\infty(G_J)$. $\square$

For the model problem (8.1.8) we have $\rho(G_{GS}) = \rho(G_J)^2 = \cos^2 \pi h$ and $R_\infty(G_{GS}) = \pi^2 h^2 + O(h^4)$. Gauss-Seidel is just as bad as Jacobi in the sense that $R_\infty(G_{GS}) = O(h^2)$. Each time h is halved, the number of iterations needed is multiplied by four. Clearly something better is needed.

Example 8.2.21 indicates that the SOR method with a good relaxation factor ω is substantially better than Gauss-Seidel. Indeed it can be shown that SOR with the optimal ω satisfies $R_\infty(G) \approx 2\pi h$. This and other results are summarized in

Method	$\rho(G)$	$R_\infty(G)$
Jacobi	$\cos \pi h$	$\frac{1}{2}\pi^2 h^2 + O(h^4)$
Gauss-Seidel	$\cos^2 \pi h$	$\pi^2 h^2 + O(h^4)$
SOR (optimal ω)	$1 - 2\pi h + O(h^2)$	$2\pi h + O(h^2)$
SSOR (good ω)	$\leq 1 - \pi h + O(h^2)$	$\geq \pi h + O(h^2)$

Table 8.5 Convergence rates of classical iterations on model problem (8.1.8)

Table 8.5. Since $R_\infty(G) = O(h)$, each halving of h only doubles the number of iterations required for SOR to achieve a given reduction in error. Thus SOR promises to perform significantly better than Gauss-Seidel when h is small.

cise 8.3.13 For $h = .02$ and $h = .01$, estimate $R_\infty(G)$ for SOR with optimal ω on the model problem (8.1.8). For both of these h, about how many iterations are required to reduce the error by a factor of 10^{-3}? Compare your results with those of Exercise 8.3.11 $\square$

Although SOR is a huge improvement over Jacobi and Gauss-Seidel, it still is not optimal. Ideally one would like an iterative method for which $R_\infty(G)$ is independent of h. Then the number of iterations required to reach a given accuracy would also be independent of h. If the cost of an iteration is $O(n)$, that is, proportional to the number of equations, then the cost of solving the system to a given accuracy will also be $O(n)$, since the number of iterations is independent of n. Multigrid and domain decomposition techniques [10], [47], [80] achieve this ideal for the model problem (8.1.8) and a wide variety of other systems obtained from discretization of partial differential equations. A discussion of these important techniques is beyond the scope of this book.

The convergence theory of the classical iterative methods is laid out in [47], [96], and [108], for example. For the model Poisson problem the theory is complete, and most of the theory carries over to other positive definite systems that admit a red-black ordering: the convergence rates of all of the classical iterative methods are determined by the spectral radius of the Jacobi iteration matrix.

So far we have restricted our attention to simple iterative methods with no memory. These methods construct $x^{(k+1)}$ from $x^{(k)}$; they make no use of $x^{(k-1)}$ or earlier iterates. An advantage of these methods is that they economize on memory: old iterates need not be saved. One might wonder, nevertheless, whether there might possibly be some gain from saving a few previous iterates and somehow making use of the information that they contain. They would give an idea of the general trend of the iterations and might be used to extrapolate to a much better estimate of the solution. This train of ideas was pursued successfully by Golub and Varga, who (independently) invented the Chebyshev semi-iterative method, also known as Chebyshev acceleration. A related technique is conjugate gradient acceleration. Both of these acceleration techniques generate $x^{(k+1)}$ from just $x^{(k)}$ and $x^{(k-1)}$. A good reference is [49].

Instead of developing acceleration techniques, we will take a different approach. The conjugate gradient algorithm will be derived as an example of a descent method. We will see that the convergence of descent algorithms can be enhanced significantly by use of preconditioners. It turns out that applying the conjugate gradient method with preconditioner M (as described later in this chapter) is equivalent to applying conjugate gradient acceleration (as described in [49]) to a basic iteration (8.3.1) with splitting matrix M.

Additional Exercises

Exercise 8.3.14 This and the next few exercises explore the computational costs of the Gauss-Seidel and SOR methods. The coefficient matrix of the model problem (8.1.8) with $h = 1/m$ has $n \approx m^2$ equations in as many unknowns. We cannot hope to solve $Au = b$ in fewer than $O(m^2)$ flops. In Section 8.1 we observed that banded Gaussian elimination requires $\frac{1}{2}m^4$ flops.

    (a) Show that the number of flops required to do one iteration of Gauss-Seidel or SOR is a modest multiple of m^2.

    (b) Show that the number of Gauss-Seidel iterations needed to decrease the error by a fixed factor ϵ is approximately Cm^2, where $C = -(\log_e \epsilon)/\pi^2$.

    (c) Estimate the overall flop count for solving $Au = b$ by Gauss-Seidel. How does this compare to the flop count for banded Gaussian elimination?

    (d) Estimate the number of SOR iterations (with optimal ω) needed to decrease the error by a fixed factor ϵ.

    (e) Show that the overall flop count for SOR with optimal ω is much less than for banded Gaussian elimination, if h is sufficiently small.

    (f) How does SOR compare with a method that takes $O(m^2)$ flops per iteration and converges in some number of iterations that is independent of h?

$\square$

cise 8.3.15 Repeat Exercise 8.3.14 for the three-dimensional model problem (see (8.1.11) and Exercises 8.1.12 and 8.1.13), bearing in mind that some of the conclusions may be different. In particular, how do the iterative methods compare to banded Gaussian elimination? (The spectral radii for the three-dimensional problem are the same as for the two-dimensional problem.)   □

cise 8.3.16 The solution of the model problem (8.1.8) is an approximation to the solution of a boundary value problem. The smaller h is, the better the approximation will be, and the more worthwhile it will be to solve $Au = b$ accurately. A more careful flop count will take this into account. If the exact solution of the boundary value problem is smooth enough, the solution of (8.1.8) will differ from it by approximately Ch^2, where C is a constant. It therefore makes sense to solve $Au = b$ with accuracy ch^2, where $c < C$.

(a) Rework parts (b) through (e) of Exercise 8.3.14 under the assumption that we need to do enough iterations to reduce the error by a factor ϵh^2. Show that this just changes the results by a factor $K \log m$ (when m is large).

(b) Calculate $\log m$ for $m = 100, 10000, 1000000$. This is meant to remind you of how slowly the logarithm grows.

(c) Is the exponent of h important? Suppose we need to reduce the error by a factor ϵh^s, where s is some number other than 2. Does this change the general conclusion of the analysis?

□

cise 8.3.17 Consider solving the model problem (8.1.8) with $f = 0$ and g given by

$$g(x,y) = \begin{cases} 0 & \text{if } x = 0 \\ y & \text{if } x = 1 \\ (x-1)\sin x & \text{if } y = 0 \\ x(2-x) & \text{if } y = 1, \end{cases}$$

as in Example 8.2.8, using mesh size $h = 1/10$.

(a) Referring to Table 8.5, calculate $\rho(G)$ and $R_\infty(G)$ for Jacobi's method applied to this problem. Based on these quantities, about how many iterations should it take to reduce the error by a factor of 100?

(b) Write a computer program that tests the theory. A crude MATLAB code will be good enough. Have the program begin by computing an accurate solution (error $< 10^{-10}$) of $Au = b$ by SOR or any other method. Use this as an "exact solution" against which to compare your Jacobi iterates. Do Jacobi iterates, starting with $u^{(0)} = 0$. At each step calculate the norm of the error: $\|e^{(k+1)}\|_2 = \|u - u^{(k+1)}\|_2$ and the ratio $\|e^{(k+1)}\|_2 / \|e^{(k)}\|_2$. Observe the ratios for 100 steps or so. Are they close to what you would expect?

(c) Observe how many iterations it takes to reduce $\|e^{(k)}\|_2$ from 10^{-1} to 10^{-3}. Does this agree well with theory?

(d) The first iterations make more progress toward convergence than subsequent iterations do. How might this effect be explained?

□

Exercise 8.3.18   Repeat the previous exercise using Gauss-Seidel (standard row ordering) instead of Jacobi.   □

Exercise 8.3.19   (Convergence of Richardson's method) Consider a linear system $Ax = b$, where the eigenvalues of A are real and positive. For example, A could be positive definite. Let λ_1 and λ_n denote the smallest and largest eigenvalues, respectively, of A. Let $G_\omega = I - \omega A$ denote the iteration matrix of Richardson's method applied to A.

(a) Show that all of the eigenvalues of G_ω are less than one.

(b) Prove that Richardson's method converges if and only if $\omega < 2/\lambda_n$.

(c) Prove that the optimal value of ω (minimizing $\rho(G_\omega)$) is $\omega_b = \frac{2}{\lambda_n + \lambda_1}$. More precisely, show that

$$\rho(G_\omega) = \begin{cases} 1 - \omega\lambda_1 & \text{if } \omega \le \omega_b \\ \frac{\lambda_n - \lambda_1}{\lambda_n + \lambda_1} & \text{if } \omega = \omega_b \\ \omega\lambda_n - 1 & \text{if } \omega \ge \omega_b. \end{cases}$$

(d) Show that if A is symmetric and positive definite, then

$$\rho(G_{\omega_b}) = \frac{\kappa_2(A) - 1}{\kappa_2(A) + 1},$$

where $\kappa_2(A)$ denotes the spectral condition number of A (cf. Theorem 4.2.4 and Exercise 5.4.47). Thus (optimal) Richardson's method converges rapidly when A is well conditioned and slowly when A is ill conditioned.

(e) Show that the convergence rate of (optimal) Richardson's method is

$$R_\infty(G_{\omega_b}) = -\log_e \left(\frac{1 - \kappa^{-1}}{1 + \kappa^{-1}} \right),$$

where $\kappa^{-1} = 1/\kappa_2(A)$. Use Taylor expansions to show that $R_\infty(G_{\omega_b}) = 2\kappa^{-1} + O(\kappa^{-2})$.

□

Exercise 8.3.20   Show that any iteration that is generated by a splitting $A = M - N$ can be expressed as a correction based on the residual $r^{(k)} = b - Ax^{(k)}$:

$$x^{(k+1)} = x^{(k)} + M^{-1}r^{(k)}. \tag{8.3.21}$$

In particular, Richardson's iteration can be written as $x^{(k+1)} = x^{(k)} + \omega r^{(k)}$.   □

·cise 8.3.22 (Preconditioners) If we wish to solve a system $Ax = b$, we can equally well solve an equivalent system $M^{-1}Ax = M^{-1}b$, where M is any nonsingular matrix. When we use a matrix M in this way, we call it a *preconditioner*. Preconditioners will figure prominently in Sections 8.6 and 8.7. Good splitting matrices make good preconditioners. Verify the following fact: Every iterative method defined by a splitting $A = M - N$ can be viewed as Richardson's method (with $\omega = 1$) applied to the preconditioned system $M^{-1}Ax = M^{-1}b$. This accounts for the importance of Richardson's method. □

·cise 8.3.23 Damping is a way of taming a nonconvergent iteration to get it to converge. Given a splitting matrix M, which gives the iteration (8.3.21), the corresponding *damped iteration* with *damping factor* $\omega < 1$ is defined by

$$x^{(k+1)} = x^{(k)} + \omega M^{-1} r^{(k)}. \tag{8.3.24}$$

Clearly the splitting matrix of the damped iteration is $\frac{1}{\omega}M$. We have already seen two examples of damped iterations: (i) The damped Jacobi method (Exercise 8.2.28), (ii) Richardson's method.

Show that if $M^{-1}A$ has real, positive eigenvalues, then the damped iteration associated with M converges if ω is sufficiently small. □

·cise 8.3.25 In principle one can perform iterations (8.3.24) with $\omega > 1$, which would be *extrapolated iterations*. This looks like the same idea as SOR, but it is not. Show that extrapolated Gauss-Seidel is different from SOR. (*Note:* SOR is a good idea; extrapolated Gauss-Seidel is not.) □

·cise 8.3.26 In this and the next exercise, you will analyze the convergence of iterative methods by examining the matrix powers G^k. In this exercise we assume G is semisimple. Then G is similar to a diagonal matrix: $G = V\Lambda V^{-1}$, where $\Lambda = \text{diag}\{\lambda_1, \ldots, \lambda_n\}$. The main diagonal entries of Λ are the eigenvalues of G (and the columns of V are the corresponding eigenvectors). Show that for all k,

$$\| G^k \|_2 \leq \kappa_2(V) \left[\rho(G) \right]^k.$$

Thus $G^k \to 0$ linearly if $\rho(G) < 1$. □

·cise 8.3.27 This exercise studies convergence in the case when the iteration matrix G is defective. Every G has the form $G = XJX^{-1}$, where $J = \text{diag}\{J_1, \ldots, J_p\}$ is in Jordan canonical form (see, e.g., [53], [60], or [101]). The *Jordan blocks* J_i are square matrices of various sizes, each having the form

$$J_i = \begin{bmatrix} \lambda & 1 & & & \\ & \lambda & 1 & & \\ & & \lambda & \ddots & \\ & & & \ddots & 1 \\ & & & & \lambda \end{bmatrix},$$

where λ is an eigenvalue of G. Each is upper triangular with λ's on the main diagonal, 1's on the superdiagonal, and zeros elsewhere. A 1×1 Jordan block has the form $\begin{bmatrix} \lambda \end{bmatrix}$. A semisimple matrix is one whose Jordan blocks are all 1×1.

(a) Prove that the powers of a Jordan block have the form

$$J_i^k = \begin{bmatrix} \lambda^k & k\lambda^{k-1} & \frac{k(k-1)}{2}\lambda^{k-2} & \cdots & \\ & \lambda^k & k\lambda^{k-1} & \ddots & \\ & & \lambda^k & \ddots & \frac{k(k-1)}{2}\lambda^{k-2} \\ & & & \ddots & k\lambda^{k-1} \\ & & & & \lambda^k \end{bmatrix}.$$

Specifically, J_i^k has $k+1$ nonzero diagonals (or less, if the dimension of J_i is less), and the jth diagonal from the main diagonal has entries $\binom{k}{j}\lambda^{k-j}$.

(b) Given a non-negative integer j, define a function p_j by $p_j(k) = \binom{k}{j}$. Show that p_j is a polynomial of degree j.

(c) Suppose the Jordan block J_i is $m \times m$, and $\lambda \neq 0$. Show that for sufficiently large k, $\| J_i^k \|_\infty = p(k)|\lambda|^k$, where p is a polynomial of degree $m-1$:

$$p(k) = \sum_{j=0}^{m-1} |\lambda|^{-j} \binom{k}{j}.$$

(Recall that $\| B \|_\infty = \max_i \sum_j |b_{ij}|$.)

(d) Suppose the Jordan block J_i is $m \times m$, and $\lambda = 0$. Show that $J_i^k = 0$ for $k \geq m$.

(e) Let m be the size of the largest Jordan block of J associated with an eigenvalue λ satisfying $|\lambda| = \rho(G)$. Show that for sufficiently large k, the powers of J satisfy $\| J^k \|_\infty = p(k)\rho(G)^k$, where p is a polynomial of degree $m-1$.

(f) Show that for large k

$$\frac{\| J^{k+1} \|_\infty}{\| J^k \|_\infty} \approx \rho(G).$$

$\square$

cise 8.3.28

(a) The coefficient matrix for the one-dimensional model problem is the tridiagonal matrix (8.1.5). Show that the associated Jacobi iteration matrix is

$$H = \frac{1}{2} \begin{bmatrix} 0 & 1 & & & \\ 1 & 0 & 1 & & \\ & 1 & 0 & \ddots & \\ & & \ddots & \ddots & 1 \\ & & & 1 & 0 \end{bmatrix}.$$

(b) We will calculate the spectral radius of H by finding all of its eigenvalues. The dimension of H is $m - 1$. Show that v is an eigenvector of H with eigenvalue λ if and only if the difference equation

$$v_{j+1} - 2\lambda v_j + v_{j-1} = 0, \quad j = 1, \ldots, m - 1, \tag{8.3.29}$$

holds with boundary conditions

$$v_0 = 0 = v_m. \tag{8.3.30}$$

(c) Since the linear, homogeneous difference equation (8.3.29) is of second order, it has two linearly independent solutions (i.e., it has a two-dimensional solution space). Show that a geometric progression $v_j = z^j$ (with $z \neq 0$) is a solution of (8.3.29) if and only if z is a solution of the *characteristic equation*

$$z^2 - 2\lambda z + 1 = 0. \tag{8.3.31}$$

(d) This quadratic equation has two solutions z_1 and z_2. Show that

$$z_1 z_2 = 1 \tag{8.3.32}$$

and

$$\lambda = \frac{1}{2} (z_1 + z_2). \tag{8.3.33}$$

(e) If the solutions of (8.3.31) satisfy $z_1 \neq z_2$, the general solution of (8.3.29) is $v_j = c_1 z_1^j + c_2 z_2^j$, where c_1 and c_2 are arbitrary constants. Show that in this case, any nonzero solution that also satisfies the boundary conditions (8.3.30) must have (i) $c_2 = -c_1$, (ii) $z_1^{2m} = 1$, (iii) $z_1 = e^{k\pi i/m}$, where k is not a multiple of m (remember the assumption $z_1 \neq z_2$). Each choice of k yields an eigenvector of H. We can normalize the eigenvector by choosing c_1. Show that if we take $c_1 = \frac{1}{2i}$ (and hence $c_2 = -\frac{1}{2i}$), then the eigenvector v is given by (iv) $v_j = \sin(jk\pi/m)$.

(f) Associated with each eigenvector is an eigenvalue. Use (8.3.33) to show that the eigenvalue associated with $z_1 = e^{k\pi i/m}$ is $\lambda_k = \cos(\pi hk)$, where $h = \frac{1}{m}$

is the mesh size. Show that letting $k = 1, \ldots, m - 1$, we get $m - 1$ distinct eigenvalues satisfying $1 > \lambda_1 > \lambda_2 > \cdots > \lambda_{m-1} > -1$. Since H has at most $m - 1$ eigenvalues, these are all of them. (You can easily check that values of k outside the range $1, \ldots, m - 1$ and not multiples of m just give these same eigenvalues over and over again.)

(g) Show that $\lambda_{m-k} = -\lambda_k$ for $k = 1, \ldots, m - 1$. Conclude that the spectral radius of H is $\cos(\pi h)$.

$\square$

Exercise 8.3.34 This exercise, which overlaps some with Exercise 6.1.24, serves as preparation for Exercise 8.3.35. We recall from Exercise 6.1.24 the definition of the Kronecker product of two matrices. Let X and Y be matrices of dimensions $\alpha \times \beta$ and $\gamma \times \delta$, respectively. Then the *Kronecker product* or *tensor product* of X and Y is the $\alpha\gamma \times \beta\delta$ matrix $X \otimes Y$ defined by

$$X \otimes Y = \begin{bmatrix} x_{11}Y & x_{12}Y & \cdots & x_{1\beta}Y \\ x_{21}Y & x_{22}Y & \cdots & x_{2\beta}Y \\ \vdots & \vdots & \ddots & \vdots \\ x_{\alpha 1}Y & x_{\alpha 2}Y & \cdots & x_{\alpha\beta}Y \end{bmatrix}.$$

(a) Let X, Y, W, and Z be matrices whose dimensions are such that the products XW and YZ are defined. Show that the product $(X \otimes Y)(W \otimes Z)$ is defined, and $(X \otimes Y)(W \otimes Z) = (XW) \otimes (YZ)$. (Notice the following important special case. If u and v are column vectors such that Xu and Yv are defined, then $u \otimes v$ is a (long) column vector such that $(X \otimes Y)(u \otimes v) = Xu \otimes Yv$.)

(b) Suppose X is $m \times m$ and Y is $n \times n$. Suppose λ is an eigenvalue of X with associated eigenvector u, and μ is an eigenvalue of Y with associated eigenvector v.

   (i) Show that $\lambda\mu$ is an eigenvalue of $X \otimes Y$ with associated eigenvector $u \otimes v$.

   (ii) Let I_k denote the $k \times k$ identity matrix. Show that $\lambda + \mu$ is an eigenvalue of $X \otimes I_n + I_m \otimes Y$ with associated eigenvector $u \otimes v$.

(c) Show that if $u_1, \ldots, u_\alpha$ is a set of α orthonormal vectors (e.g. eigenvectors), and $v_1, \ldots, v_\beta$ is a set of β orthonormal vectors, then $u_i \otimes v_j$, $i = 1, \ldots, \alpha$, $j = 1, \ldots, \beta$, is a set of $\alpha\beta$ orthonormal vectors.

$\square$

Exercise 8.3.35 Work Exercise 8.3.34 before you do this one.

(a) The coefficient matrix A of the two-dimensional model problem (8.1.8) is given by (8.1.10). Show that

$$A = \hat{A} \otimes I_{m-1} + I_{m-1} \otimes \hat{A},$$

where $\hat{A}$ is the one-dimensional model matrix given by (8.1.5).

(b) Show that the Jacobi iteration matrix associated with (8.1.8) is given by

$$G = \frac{1}{2} (H \otimes I_{m-1} + I_{m-1} \otimes H),$$

where H is as in Exercise 8.3.28.

(c) Show that the $(m-1)^2$ eigenvalues of G are

$$\lambda_{jk} = \frac{1}{2} (\cos(jh\pi) + \cos(kh\pi)), \quad j, k = 1, \dots, m-1.$$

(d) Show that $\rho(G) = \cos \pi h$.

□

cise 8.3.36 Work out the analogue of Exercise 8.3.35 for the three-dimensional model problem. Conclude that the iteration matrix for Jacobi's method has spectral radius $\cos \pi h$ in this case as well.

□

cise 8.3.37

(a) Show that the eigenvalues of the tridiagonal matrix $\hat{A}$ (8.1.5) from the one-dimensional model problem are

$$\lambda_k = 2 - 2 \cos \left(\frac{\pi k}{m} \right) = 4 \sin^2 \left(\frac{\pi k}{2m} \right), \quad k = 1, \dots, m-1.$$

(*Hint:* This is an easy consequence of results in Exercise 8.3.28. Notice that $\hat{A} = 2(I - H)$.)

(b) Show that the eigenvalues of the coefficient matrix of the two-dimensional model problem (8.1.8) are

$$\lambda_{jk} = 4 \left[\sin^2 \left(\frac{\pi j}{2m} \right) + \sin^2 \left(\frac{\pi k}{2m} \right) \right] \quad j, k = 1, \dots, m-1.$$

(See Exercise 8.3.35).

(c) Show that if Richardson's method is applied to the model problem (8.1.8), the optimal ω is 1/4 (Exercise 8.3.19). Show that with this value of ω, Richardson's method is the same as Jacobi's method (on this particular problem).

(d) Show that $\kappa_2(A) = \cot^2 (\pi h/2)$ for the model problem (8.1.8). Use this figure to calculate the convergence rate for Richardson's method with optimal ω. Show that it agrees with the convergence rate for Jacobi's method calculated in Exercise 8.3.35.

□

Exercise 8.3.38 Consider a system $Ax = b$ that admits a red-black ordering. Thus (after reordering, if necessary)

$$A = \begin{bmatrix} D_1 & B_{12} \\ B_{21} & D_2 \end{bmatrix},$$

where D_1 and D_2 are diagonal matrices.

(a) Show that the iteration matrix for Jacobi's method applied to this system is

$$G_J = \begin{bmatrix} 0 & C_1 \\ C_2 & 0 \end{bmatrix}, \tag{8.3.39}$$

where $C_1 = -D_1^{-1}B_{12}$ and $C_2 = -D_2^{-1}B_{21}$.

(b) Show that the red-black Gauss-Seidel iteration matrix for A is

$$G_{GS} = \begin{bmatrix} 0 & C_1 \\ 0 & C_2C_1 \end{bmatrix}.$$

(c) Using Theorem 7.3.1, show that $\rho(G_{GS}) = \rho(G_J)^2$.

□

Exercise 8.3.40 If the coefficient matrix A is positive definite, it is often advisable to use splittings $A = M - N$ for which M is also positive definite. Show that the following methods have (symmetric and) positive definite splitting matrices, assuming A is positive definite: (a) Richardson, (b) Jacobi, (c) symmetric Gauss-Seidel, (d) SSOR ($0 < \omega < 2$). □

Exercise 8.3.41 Let A be a symmetric matrix, and consider a splitting $A = M - N$. If M is not symmetric, then $A = M^T - N^T$ is another splitting of A. This exercise shows how to use these two splittings together to build one that is symmetric. Consider an iteration that consists of two half steps, one with splitting matrix M, the other with splitting matrix M^T:

$$Mx^{(k+1/2)} = Nx^{(k)} + b,$$
$$M^T x^{(k+1)} = N^T x^{(k+1/2)} + b.$$

(a) Show that $M^T x^{(k+1)} = N^T M^{-1} N x^{(k)} + (I + N^T M^{-1})b$.

(b) Show that $I + N^T M^{-1} = HM^{-1}$, where $H = M + N^T = M + M^T - A = A + N + N^T$. Show that $N^T M^{-1} N = M^T - HM^{-1}A$.

(c) Show that if H is nonsingular, then

$$\hat{M}x^{(k+1)} = \hat{N}x^{(k)} + b,$$

where $\hat{M} = MH^{-1}M^T$, and $A = \hat{M} - \hat{N}$. This is the splitting for the combined iteration.

(d) Show that the splitting matrix $\hat{M}$ is symmetric. Show that $\hat{M}$ is positive definite if H is.

(e) Show that the SSOR iteration is of this form (assuming A is symmetric). What is H in the SSOR case? Show that if A is positive definite, then H is positive definite if and only if the relaxation factor ω satisfies $0 < \omega < 2$.

$\square$

8.4 DESCENT METHODS; STEEPEST DESCENT

We continue to study the linear system $Ax = b$. From this point on, unless otherwise stated, we shall assume A is symmetric and positive definite. We begin by showing that the problem of solving $Ax = b$ can be reformulated as a minimization problem. We then proceed to a study of methods for solving the minimization problem. Define a function $J : \mathbb{R}^n \to \mathbb{R}$ by

$$J(y) = \tfrac{1}{2}y^T Ay - y^T b. \qquad (8.4.1)$$

Then the vector that minimizes J is exactly the solution of $Ax = b$.

Theorem 8.4.2 *Let $A \in \mathbb{R}^{n \times n}$ be positive definite, let $b \in \mathbb{R}^n$, and define J as in (8.4.1). Then there is exactly one $x \in \mathbb{R}^n$ for which*

$$J(x) = \min_y J(y),$$

and this x is the solution of $Ax = b$.

Proof. The function J is quadratic in the variables $y_1, \ldots, y_n$. We shall discover its minimum by the simple technique of completing the square. Let x denote the solution of $Ax = b$. Then

$$
\begin{aligned}
J(y) &= \tfrac{1}{2}y^T Ay - y^T Ax \\
&= \tfrac{1}{2}y^T Ay - y^T Ax + \tfrac{1}{2}x^T Ax - \tfrac{1}{2}x^T Ax \\
&= \tfrac{1}{2}(y - x)^T A(y - x) - \tfrac{1}{2}x^T Ax. \qquad (8.4.3)
\end{aligned}
$$

The term $\tfrac{1}{2}x^T Ax$ is independent of y, so $J(y)$ is minimized exactly when $\tfrac{1}{2}(y - x)^T A(y - x)$ is minimized. Since A is positive definite, we know that this term is positive unless $y - x = 0$. Thus it takes its minimum value when and only when $y = x$. $\square$

In many problems the function J has physical significance.

Example 8.4.4 In elasticity problems, $J(y)$ denotes the potential energy of the system in configuration y. The configuration that the system actually assumes at equilibrium is the one for which the potential energy is minimized. An example is the simple mass-spring system that was discussed in Example 1.2.10. The matrix A is called the *stiffness matrix*. The term $\frac{1}{2}y^T Ay$ is the *strain energy*, the energy stored in the springs due to stretching or compression. The term $-y^T b$ is the work done by the system against the external forces, which are represented by the vector b. Some details are worked out in Exercise 8.4.18. □

Additional insight is obtained by calculating ∇J, the gradient of J. Recalling that

$$\nabla J = \left[\frac{\partial J}{\partial y_1}, \ldots, \frac{\partial J}{\partial y_n}\right]^T \text{ and performing the routine computation, we find that}$$

$$\nabla J = Ay - b.$$

This is just the negative of the residual of y as an approximation to the solution of $Ax = b$. Clearly the only point at which the gradient is zero is the solution of $Ax = b$. Thus we see (again) that the only vector that can minimize J is the solution of $Ax = b$.

Descent methods solve $Ax = b$ by minimizing J. These are iterative methods. Each descent method begins with an initial guess $x^{(0)}$ and generates a sequence of iterates $x^{(0)}, x^{(1)}, x^{(2)}, x^{(3)}, \ldots$ such that at each step $J(x^{(k+1)}) \leq J(x^{(k)})$, and preferably $J(x^{(k+1)}) < J(x^{(k)})$. In this sense we get closer to the minimum at each step. If at some point we have $Ax^{(k)} = b$ or nearly so, we stop and accept $x^{(k)}$ as the solution. Otherwise we take another step. The step from $x^{(k)}$ to $x^{(k+1)}$ has two ingredients: (i) choice of a search direction, and (ii) a line search in the chosen direction. Choosing a search direction amounts to choosing a vector $p^{(k)}$ that indicates the direction in which we will travel to get from $x^{(k)}$ to $x^{(k+1)}$. Several strategies for choosing $p^{(k)}$ will be discussed below. Once a search direction has been chosen, $x^{(k+1)}$ will be chosen to be a point on the line $\{x^{(k)} + \alpha p^{(k)} \mid \alpha \in \mathbb{R}\}$. Thus we will have

$$x^{(k+1)} = x^{(k)} + \alpha_k p^{(k)}$$

for some real α_k. The process of choosing α_k from among all $\alpha \in \mathbb{R}$ is the line search. We want to choose α_k in such a way that $J(x^{(k+1)}) \leq J(x^{(k)})$. One way to ensure this is to choose α_k so that

$$J(x^{(k+1)}) = \min_{\alpha \in \mathbb{R}} J(x^{(k)} + \alpha p^{(k)}).$$

If α_k is chosen in this way, we say that the line search is *exact*. Otherwise, we say that it is *inexact*.

For some types of functions an exact line search can be a formidable task, but for quadratic functions like (8.4.1), line searches are trivial. The following theorem shows that the correct value of α can be obtained from a formula.

Theorem 8.4.5 *Let $x^{(k+1)} = x^{(k)} + \alpha_k p^{(k)}$ be obtained by an exact line search. Then*

$$\alpha_k = \frac{p^{(k)T} r^{(k)}}{p^{(k)T} A p^{(k)}},$$

where $r^{(k)} = b - Ax^{(k)}$.

Proof. Let $g(\alpha) = J(x^{(k)} + \alpha p^{(k)})$. The minimizer of g is α_k. A routine computation shows that $g(\alpha) = J(x^{(k)}) - \alpha p^{(k)T} r^{(k)} + \frac{1}{2}\alpha^2 p^{(k)T} A p^{(k)}$. This is a quadratic polynomial in α whose unique minimum can be found by solving the equation $g'(\alpha) = 0$. Since $g'(\alpha) = -p^{(k)T} r^{(k)} + \alpha p^{(k)T} A p^{(k)}$, we have $\alpha^{(k)} = p^{(k)T} r^{(k)} / p^{(k)T} A p^{(k)}$.

$\square$

Remark 8.4.6 Notice that $\alpha_k = 0$ if and only if $p^{(k)T} r^{(k)} = 0$. We want to avoid this situation because we would rather have $x^{(k+1)} \neq x^{(k)}$. The equation $p^{(k)T} r^{(k)} = 0$ just says that the search direction is orthogonal to the residual. Thus we would normally want to choose $p^{(k)}$ so that it is not orthogonal to $r^{(k)}$. This is always possible, unless $r^{(k)} = 0$ (in which case $x^{(k)}$ is the solution). If we always choose $p^{(k)}$ so that $p^{(k)T} r^{(k)} \neq 0$, we will always have the strict inequality $J(x^{(k+1)}) < J(x^{(k)})$.

$\square$

Let us now consider some examples of descent methods.

cise 8.4.7 In this exercise the initial guess will be denoted $x^{(1)}$ for notational simplicity. Similarly the first search direction will be denoted $p^{(1)}$. Consider a method in which the first n search directions $p^{(1)}, \dots, p^{(n)}$ are taken to be the standard unit vectors $e_1, \dots, e_n$, the next n search directions $p^{(n+1)}, \dots, p^{(2n)}$ are $e_1, \dots, e_n$ again, so are $p^{(2n+1)}, \dots, p^{(3n)}$, and so on. Suppose an exact line search is performed at each step.

(a) Show that each group of n steps is one iteration of the Gauss-Seidel method.

(b) Show that it can happen that $J(x^{(k+1)}) = J(x^{(k)})$ even if $x^{(k)} \neq x$.

(c) Prove that the situation described in part (b) cannot persist for n consecutive steps. In other words, show that if $x^{(k)} \neq x$, then $J(x^{(k+n)}) < J(x^{(k)})$.

Thus each Gauss-Seidel iteration lowers the potential energy. This idea leads to a proof that the Gauss-Seidel method applied to a positive definite matrix always converges.

$\square$

cise 8.4.8 Let ω be a fixed number. Consider a method that chooses the search directions as in the previous exercise but takes $x^{(k+1)} = x^{(k)} + \omega \alpha_k p^{(k)}$, where α_k is the increment determined by an exact line search. This method reduces to the previous method if $\omega = 1$. If $\omega \neq 1$, the line search is inexact.

(a) Show that each group of n steps is an iteration of the SOR method with relaxation factor ω.

(b) Show that if $p^{(k+1)T} r^{(k)} \neq 0$, then (i) $J(x^{(k+1)}) < J(x^{(k)})$ if $0 < \omega < 2$, (ii) $J(x^{(k+1)}) = J(x^{(k)})$ if $\omega = 0$ or $\omega = 2$, and (iii) $J(x^{(k+1)}) > J(x^{(k)})$ if $\omega < 0$ or $\omega > 2$.

(c) Show that if $x^{(k)} \neq x$, then (i) $J(x^{(k+n)}) < J(x^{(k)})$ if $0 < \omega < 2$, (ii) $J(x^{(k+n)}) = J(x^{(k)})$ if $\omega = 0$ or $\omega = 2$, and (iii) $J(x^{(k+n)}) > J(x^{(k)})$ if $\omega < 0$ or $\omega > 2$.

This leads to a proof that SOR applied to a positive definite matrix always converges if and only if $0 < \omega < 2$.

$\square$

As we have observed previously, SOR can beat Gauss-Seidel significantly if ω is chosen well. This shows that exact line searches are not necessarily better than inexact searches in the long run.

Steepest Descent

The method of *steepest descent* takes $p^{(k)} = r^{(k)}$ and performs exact line searches. Since $r^{(k)} = -\nabla J(x^{(k)})$, the search direction is the direction of steepest descent of J from the point $x^{(k)}$.

To search in the direction of steepest descent is a perfectly natural idea. Unfortunately it doesn't work particularly well. Steepest descent is worth studying, nevertheless, for at least two reasons: (i) It is a good vehicle for introducing the idea of preconditioning. (ii) Minor changes transform the steepest descent algorithm into the powerful conjugate-gradient method.

Example 8.4.9 If we apply the steepest descent method to the same small system as in Example 8.2.3, we obtain

$$x^{(5)} = \begin{bmatrix} 3.9525 \\ 3.0508 \\ 1.9973 \\ 1.0147 \end{bmatrix}, \quad x^{(10)} = \begin{bmatrix} 3.9980 \\ 3.0011 \\ 1.9989 \\ 0.9997 \end{bmatrix}, \quad x^{(15)} = \begin{bmatrix} 3.99993 \\ 3.00008 \\ 1.99999 \\ 1.00002 \end{bmatrix},$$

which is not much better than we got using Jacobi's method. After 42 iterations, our approximation agrees with the true solution to twelve decimal places. $\square$

It is a simple matter to program the steepest descent algorithm. Let us consider some of the implementation issues. It will prove worthwhile to begin by writing down a generic descent algorithm. At each step our approximate solution is updated by

$$x^{(k+1)} = x^{(k)} + \alpha_k p^{(k)}. \tag{8.4.10}$$

If we are doing exact line searches, we will have calculated α_k using the formula given in Theorem 8.4.5. This requires, among other things, multiplying the matrix A by the vector $p^{(k)}$. The cost of this operation depends on how sparse A is. In many applications the matrix-vector product is the most expensive step of the algorithm, so we should try not to do too many of them. We also need the residual $r^{(k)} = b - Ax^{(k)}$, which seems to require an additional matrix-vector product $Ax^{(k)}$. We can avoid this by using the simple recursion

$$r^{(k+1)} = r^{(k)} - \alpha_k Ap^{(k)}, \tag{8.4.11}$$

which is an easy consequence of (8.4.10), to update the residual from one iteration to the next. Now the matrix-vector product is $Ap^{(k)}$, which we will have already calculated as part of the computation of α_k.

Exercise 8.4.12   Use (8.4.10) to derive (8.4.11). $\square$

Introducing the auxiliary vector $q^{(k)} = Ap^{(k)}$, we get a generic descent algorithm with exact line search.

Prototype Generic Descent Algorithm (exact line search)

$$r^{(0)} \leftarrow b - Ax^{(0)}$$
$$p^{(0)} \leftarrow \quad ?$$
for $k = 0, 1, 2, \ldots$

$$\left[\begin{array}{l} q^{(k)} \leftarrow Ap^{(k)} \\ \alpha_k \leftarrow p^{(k)T} r^{(k)} / p^{(k)T} q^{(k)} \\ x^{(k+1)} \leftarrow x^{(k)} + \alpha_k p^{(k)} \\ r^{(k+1)} \leftarrow r^{(k)} - \alpha_k q^{(k)} \\ p^{(k+1)} \leftarrow \quad ? \end{array} \right. \qquad (8.4.13)$$

All that is needed is a rule for specifying the search direction. We get steepest descent by setting $p^{(j)} = r^{(j)}$.

In practice we should stop once the iterates have converged to our satisfaction. In Section 8.2 we used a criterion involving the difference between two successive iterates. That criterion can also be used here, but there are other possibilities. For example, the steepest descent algorithm calculates $p^T r = r^T r = \| r \|_2^2$ on each iteration, which makes it easy to use a stopping criterion based on the norm of the residual. Whatever criterion we use, we accept $x^{(k+1)}$ as an adequate approximation of the solution as soon as the criterion is satisfied. We must also realize that if we set a tolerance that is too strict, roundoff errors may prevent termination entirely. It is therefore essential to place a limit l on the number of iterations.

We do not need separate storage locations for $x^{(0)}$, $x^{(1)}$, $x^{(2)}$, and so on; we can have a single vector variable x, which starts out containing the initial guess, carries each of the iterates in turn, and ends up containing the final solution x. Similarly, a single vector r can be used to store all of the residuals in turn. Initially r can be used to store the right-hand-side vector b, which appears only in the calculation of the initial residual $r^{(0)}$. Similar remarks apply to $p^{(k)}$, $q^{(k)}$, and α_k.

Taking these remarks into account we obtain a more refined statement of the algorithm.

Generic Descent Algorithm (with exact line search) for solving $Ax = b$. On entry the initial guess is stored in x, and the vector b is stored in r. The algorithm returns in x its best estimate of the solution and a flag that indicates whether or not the specified tolerance was achieved.

$$r \leftarrow r - Ax$$
$$p \leftarrow \quad ?$$
$$k \leftarrow 0$$
do until satisfied or $k = l$

$$\left[\begin{array}{l} q \leftarrow Ap \\ \alpha \leftarrow p^T r / p^T q \\ x \leftarrow x + \alpha p \\ r \leftarrow r - \alpha q \\ p \leftarrow \quad ? \\ k \leftarrow k + 1 \end{array} \right. \qquad (8.4.14)$$

if not satisfied, set flag

Set $p \leftarrow r$ to get steepest descent.

The matrix A is used to compute the matrix-vector product Ax in the initial residual. Aside from that, it is needed to calculate the matrix-vector product Ap once per iteration. The matrix-vector products are normally performed by a user-supplied subroutine, which accepts a vector p as input and returns the vector Ap. The details of the computation depend on the form of A. If A is a dense matrix stored in the conventional way, one can write a straightforward matrix-vector product algorithm that takes about $2n^2$ flops to do the computation. However, if A is sparse, one should exploit the sparseness to produce Ap in far fewer than $2n^2$ flops.

Suppose, for example, A is the matrix of the model problem (8.1.8). Then each row of A will have at most five nonzero entries, so each element of the vector Ap can be calculated in roughly ten flops. Thus the entire vector Ap can be calculated in some $10n$ flops; the cost is $O(n)$ instead of $O(n^2)$.

Let us consider some of the details of applying steepest descent to the model problem (8.1.8). Throughout our discussion of descent algorithms, we have treated the vectors x, r, p, etc. as ordinary column vectors in $\mathbb{R}^n$. For example, $p^T = [p_1, p_2, \ldots, p_n]$, and the inner product $p^T r$ in Algorithm 8.4.14 means

$$\sum_{i=1}^{n} p_i r_i.$$

In the model problem, it is more natural to store the unknown (which we now call u rather than x) in a doubly subscripted array $u_{i,j}, i,j = 1, \ldots, m-1$. Each entry $u_{i,j}$ represents an approximation to $u(x_i, y_j)$, the value of the solution of a PDE at the grid point (x_i, y_j). Fortunately there is no need to depart from this notation when we apply Algorithm 8.4.14 to this problem; we need only interpret the operations in the algorithm appropriately. All of the vectors are stored as doubly subscripted arrays. The update $u \leftarrow u + \alpha p$ means $u_{ij} \leftarrow u_{ij} + \alpha p_{ij}, i,j = 1, \ldots, m-1$. The inner product $p^T r$ means

$$\sum_{j=1}^{m-1} \sum_{i=1}^{m-1} p_{ij} r_{ij}.$$

The matrix-vector product $q \leftarrow Ap$ is

$$q_{ij} \leftarrow -p_{i-1,j} - p_{i,j-1} + 4p_{i,j} - p_{i,j+1} - p_{i+1,j}, \qquad i,j = 1, \ldots, m-1. \quad (8.4.15)$$

As usual, the matrix A is not stored in any way, shape, or form. More implementation details are given in Exercise 8.4.19.

Example 8.4.16 We applied steepest descent to the model problem under the same conditions as in Example 8.2.8. The results are summarized in Table 8.6. Comparing with the table in Example 8.2.8, we observe that the steepest descent algorithm is just as slow as Jacobi's method. Indeed, the similarity in their performance is striking.

From (8.4.15) we see that cost of the operation $q \leftarrow Ap$ is comparable to that of one Jacobi or Gauss-Seidel iteration. However, steepest descent has the additional inner products $p^T r$ and $p^T q$ and the updates $u \leftarrow u + \alpha p$ and $r \leftarrow r - \alpha q$, each of which takes about $2n \approx 2m^2$ flops. Thus the work to do one steepest descent iteration is somewhat higher than that for a Jacobi or Gauss-Seidel iteration.   □

h	Matrix dimension	Iterations to convergence
$1/20$	361	1114
$1/40$	1521	4010
$1/80$	6241	14252
$1/160$	25281	49755

Table 8.6   Steepest descent method applied to model problem

Geometric Interpretation of Steepest Descent

The objective of a descent method is to minimize the function $J(y)$. From (8.4.3) we know that J has the form

$$J(y) = \tfrac{1}{2}(y - x)^T A(y - x) - \gamma,$$

where x is the solution of $Ax = b$, and γ is constant. Since A is symmetric, there exists an orthogonal matrix U such that $U^T A U$ is a diagonal matrix Λ, by Theorem 5.4.20. The main diagonal entries of Λ are the eigenvalues of A, which are positive. Introducing new coordinates $z = U^T(y - x)$ and dropping the inessential constant γ and the factor $\tfrac{1}{2}$, we see that minimizing $J(y)$ is equivalent to minimizing

$$\tilde{J}(z) = z^T \Lambda z = \sum_{i=1}^{n} \lambda_i z_i^2. \tag{8.4.17}$$

To get a picture of the function $\tilde{J}$, consider the 2×2 case. Now $\tilde{J}$ is a function of two variables, so its contours or level surfaces $\tilde{J}(z_1, z_2) = c$ are curves in the plane. From (8.4.17), the contours have the form

$$\lambda_1 z_1^2 + \lambda_2 z_2^2 = c,$$

which are concentric ellipses centered on the origin. They are ellipses, not hyperbolas, because the eigenvalues λ_1 and λ_2 have the same sign.

The orthogonal coordinate transformation $z = U^T(y - x)$ preserves lengths and angles, so the contours of J are also ellipses of the same shape. For example, the contours of the function $J(y)$ associated with the matrix

$$A = \begin{bmatrix} 6 & 2 \\ 2 & 3 \end{bmatrix}$$

are shown in Figure 8.4. The solution $x = \begin{bmatrix} x_1 & x_2 \end{bmatrix}^T$ lies at the center of the ellipses.

The semiaxes of the ellipses are $\sqrt{c/\lambda_1}$ and $\sqrt{c/\lambda_2}$, whose ratio is $\sqrt{\lambda_2/\lambda_1}$. The eigenvalues of a positive definite matrix are the same as its singular values, so the spectral condition number is equal to the ratio of largest to smallest eigenvalue (see

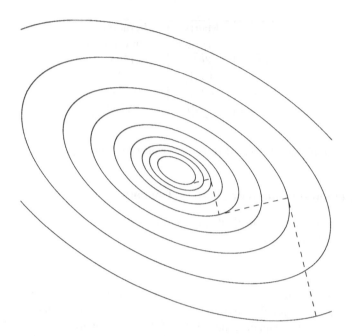

Figure 8.4 Steepest descent in the 2×2 case

Theorem 4.2.4 and Exercise 5.4.47). Thus the ratio of the semiaxes is $\sqrt{\kappa_2(A)}$. We conclude that the shape of the contours depends on the condition number of A. The greater the condition number, the more eccentric the ellipses are. Figure 8.4 shows the well-conditioned case.

The dotted lines in Figure 8.4 represent four steps of the steepest descent algorithm. From a given point, the search proceeds in the direction of steepest descent, which is orthogonal to the contour line (the direction of no descent). The exact line search follows the search line to the point at which J is minimized. J decreases as long as the search line cuts through the contours. The minimum occurs at the point at which the search line is tangent to a contour. (After that, J begins to increase.) Since the next search direction will be orthogonal to the contour at that point, we see that each search direction is orthogonal to the previous one. Thus the search bounces back and forth in the canyon formed by the function $J(y)$ and proceeds steadily toward the minimum.

The minimum is reached quickly if A is well conditioned. In the best case, $\lambda_1 = \lambda_2$, the contours are circles, the direction of steepest descent (from any starting point) points directly to the center, and the exact minimum is reached in one iteration. If A is well conditioned, the contours will be nearly circular, the direction of steepest descent will point close to the center, and the method will converge rapidly. If, on the other hand, A is somewhat ill conditioned, the contours will be highly eccentric ellipses. From a given point, the direction of steepest descent is likely to point

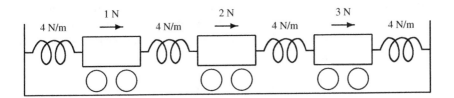

Figure 8.5   System of three carts

nowhere near the center; it is not a good search direction. In this case the function J forms a steep, narrow canyon. The steepest descent algorithm bounces back and forth in this canyon, taking very short steps, and approaching the minimum with agonizing slowness. This phenomenon does not require extreme ill conditioning. Even if the system is only modestly ill conditioned, and well worth solving from the standpoint of numerical accuracy, the convergence can be very slow.

So far we have been discussing the 2×2 case, but the same difficulties arise in general. In the 3×3 case the level sets are ellipsoids in $\mathbb{R}^3$, and in general they are hyperellipsoids in $\mathbb{R}^n$, whose roundness or lack thereof depends upon $\kappa_2(A)$. Each steepest descent step proceeds in the direction orthogonal to the level hypersurface and continues to a point of tangency with a hypersurface. If the hyperellipsoids are highly eccentric, progress will be slow.

The Jacobi, Gauss-Seidel, and Richardson algorithms all can be viewed as descent algorithms (Exercises 8.4.7 and 8.4.21), and all have difficulty negotiating the steep, narrow canyon. See, for example, Exercise 8.4.23. You could also refer back to Exercise 8.3.19, in which you showed by other means that the convergence of Richardson's method is adversely affected by ill conditioning. Even mild ill conditioning is enough to cause very slow convergence. See Exercise 8.4.25.

Additional Exercises

·cise 8.4.18 Suppose a linear spring has stiffness k N/m. Then if we stretch (or compress) it s meters from its equilibrium position, it exerts a restoring force of ks Newtons. To compute the work needed to stretch the spring x meters from its equilibrium position, we integrate the force: $\int_0^x ks \, ds = \frac{1}{2}kx^2$ Joules. The work is equal to the strain energy, the energy stored in the spring due to the deformation. Now consider a system of three masses and four springs, as shown in Figure 8.5. This is the same system as we considered in Example 1.2.10. If external forces are applied to the carts as shown, they will move from their original equilibrium position to a new equilibrium position.

(a) Each of the springs has stiffness 4 N/m. Show that if the carts 1, 2, and 3 are displaced by y_1, y_2, and y_3 meters from their original equilibrium positions (positive displacements being to the right, for example), the total strain energy of the four springs is

$$\frac{4}{2}\left(y_1^2 + (y_2 - y_1)^2 + (y_3 - y_2)^2 + (-y_3)^2\right)$$

Joules. Assume that the springs were completely relaxed, neither stretched nor compressed, initially.[47]

(b) Show that

$$(y_2 - y_1)^2 = \begin{bmatrix} y_1 & y_2 \end{bmatrix} \begin{bmatrix} 1 & -1 \\ -1 & 1 \end{bmatrix} \begin{bmatrix} y_1 \\ y_2 \end{bmatrix}.$$

Then show that the total strain energy of the four springs is $\frac{1}{2}y^T A y$, where A is the same matrix as in Example 1.2.10, and $y = [y_1, y_2, y_3]^T$.

(c) Show that if external forces are 1, 2 and 3 N to the right, as shown, the potential energy gained due to work against those forces is $-y^T b$ Joules, where $b = [1, 2, 3]^T$.

Thus the total potential energy of the system in configuration y is $J(y) = \frac{1}{2}y^T A y - y^T b$. The actual configuration x that the carts assume in the end is the one for which J is minimized, and that is the one for which $Ax = b$. □

Exercise 8.4.19 Write a program that applies the method of steepest descent to the model problem (8.1.8). Use it to solve the model problem with $f = 0$ and g given by

$$g(x, y) = \begin{cases} 0 & \text{if } x = 0 \\ y & \text{if } x = 1 \\ (x - 1)\sin x & \text{if } y = 0 \\ x(2 - x) & \text{if } y = 1, \end{cases}$$

as in Example 8.2.8. Use mesh sizes $h = 1/20, 1/40, 1/80$, and $1/160$. Starting from $u^{(0)} = 0$, record how many iterations are needed until

$$\|r^{(k+1)}\|_2 < 10^{-8}\left(\|b\|_2 + \|A\|_2 \|x^{(k+1)}\|_2\right). \tag{8.4.20}$$

You can use the approximation $\|A\|_2 \approx 8$, which is justified by the result of Exercise 8.3.37, part (b).

Compare your results with those of Example 8.4.16. The stopping criterion (8.4.20) is less stringent than the one used in Example 8.4.16, so your iterations numbers should be a bit lower. The criterion (8.4.20) will be justified in Section 8.5. Some implementation details follow.

---

[47]This assumption simplifies the computation, but it is not essential.

When (8.1.8) is written as a matrix equation $Au = b$, all of the terms associated with the boundary conditions are included in the right-hand-side vector b. Therefore, your program needs to incorporate all of the boundary conditions into b at the start. Consider, for example, the equation associated with the grid point (x_1, y_3). We have

$$-u_{1,2} - u_{0,3} + 4u_{1,3} - u_{2,3} - u_{1,4} = h^2 f(x_1, y_3).$$

Since $u_{0,3}$ belongs to a boundary grid point, its value is given by the boundary conditions: $u_{0,3} = g(x_0, y_3)$. Since this term is not an unknown, it should be moved to the right-hand side of the equation. Thus the equation takes the form

$$-u_{1,2} + 4u_{1,3} - u_{2,3} - u_{1,4} = h^2 f(x_1, y_3) + g(x_0, y_3),$$

and

$$b_{1,3} = h^2 f(x_1, y_3) + g(x_0, y_3).$$

In this way, each $b_{i,j}$ that is adjacent to a boundary point will include a boundary term. Each corner point (e.g. $b_{1,1}$) will include two boundary terms. (Since our test problem has $f = 0$, *all* of the nonzero entries in b will be due to boundary terms.) Once the boundary terms have been loaded into b, they can be forgotten; they are not used anywhere else in the program.

The matrix-vector multiply should be effected by a separate subroutine or procedure, the heart of which is a doubly-nested loop like

$$\text{for } j = 1, \ldots, m-1$$
$$\left[\begin{array}{l} \text{for } i = 1, \ldots, m-1 \\ \quad \left[\; q_{ij} \leftarrow -p_{i-1,j} - p_{i,j-1} + 4p_{i,j} - p_{i,j+1} - p_{i+1,j}. \right. \end{array} \right.$$

If this procedure is to give correct results, the boundary terms must be set to zero. (Since the boundary conditions have been pushed into the vector b, they do not play a role in the matrix A.) If your code sets the boundary values $p_{i,0}$, $p_{i,m}$, $p_{0,j}$, $p_{m,j}$ to zero at the start (and similarly for u) and leaves them zero, the procedure will function correctly.

You might also find it useful to write a separate inner product procedure to calculate the inner products needed for the computation of α.   □

cise 8.4.21 Show that the steepest descent algorithm is just Richardson's method with variable damping, and Richardson's method is steepest descent with an inexact line search.   □

cise 8.4.22 Show that the level surfaces of the function $J(y) = \frac{1}{2} y^T A y - b^T y$ are hyperellipsoids in $\mathbb{R}^n$ with semiaxes $\sqrt{c/\lambda_i}$, $i = 1, \ldots, n$, where $\lambda_1, \ldots, \lambda_n$ are the eigenvalues of A.   □

cise 8.4.23 Draw a picture like Figure 8.4 that shows the progress of the Gauss-Seidel method. Draw another picture that shows why Gauss-Seidel converges slowly if A is ill-conditioned.   □

Exercise 8.4.24 Draw a picture that suggests that SOR will outperform Gauss-Seidel if ω is chosen well. $\qquad\qquad\qquad\qquad\qquad\qquad\qquad\qquad\qquad\qquad\qquad\qquad\qquad\qquad\quad$ □

Exercise 8.4.25 Let A be the matrix of the model problem (8.1.8).

(a) Show that
$$\kappa_2(A) = \cot^2(\pi h/2) \approx (2/\pi h)^2.$$
(See Exercise 8.3.37.)

(b) What relationship do you observe between $\kappa_2(A)$ and known facts about the convergence rates of the Jacobi and Gauss-Seidel methods?

(c) Calculate $\kappa_2(A)$ when $h = 1/40$, $1/80$, and $1/160$. A is ill conditioned but not extremely so in these cases.

$\qquad\qquad\qquad\qquad\qquad\qquad\qquad\qquad\qquad\qquad\qquad\qquad\qquad\qquad\qquad\qquad\qquad\qquad\qquad\quad$ □

8.5 ON STOPPING CRITERIA

This brief section was added at the urging of C. C. Paige. Whenever we use an iterative method, we need a criterion for deciding when to stop. In most cases it doesn't much matter which stopping criterion we use, but there are some situations where it really does matter.

So far in this chapter we have used a couple of criteria. In the examples in Section 8.2 on the classical iterative methods, we stopped when two successive iterates were sufficiently close together, i.e.

$$\left\| x^{(k+1)} - x^{(k)} \right\|_2 < \epsilon \| x^{(k+1)} \|_2,$$

where ϵ is some small number like 10^{-8}. We used this criterion because it is simple, traditional, and convenient. However, we provided no theoretical justification for it.

Another criterion that is in widespread use is to stop when the residual is small enough:

$$\left\| r^{(k+1)} \right\|_2 < \epsilon \| b \|_2. \tag{8.5.1}$$

This is a convenient choice for descent methods because the residual is computed at each step anyway. It also has the following theoretical justification.

Recall from Section 2.5 that a method is called *backward stable* if it produces a computed solution $\hat{x}$ that is the exact solution of a nearby problem $(A+\delta A)\hat{x} = b+\delta b$, where *nearby* means that $\| \delta A \|/\| A \|$ and $\| \delta b \|/\| b \|$ are tiny. Here we can use whatever vector norm we please and a compatible matrix norm, but we will stick to 2-norms for simplicity. Let us now refine this definition by saying that for any $\epsilon > 0$, a computed solution $\hat{x}$ is *ϵ-backward-stable*, or is *backward stable at level* ϵ, if it satisfies an equation $(A + \delta A)\hat{x} = b + \delta b$ for which $\| \delta A \|_2/\| A \|_2 < \epsilon$ and $\| \delta b \|_2/\| b \|_2 < \epsilon$.

Now suppose that our iterative method has produced a current iterate $x^{(k+1)}$ and corresponding residual $r^{(k+1)} = b - Ax^{(k+1)}$ satisfying (8.5.1). Then the current iterate satisfies the equation $Ax^{(k+1)} = b + \delta b$, where $\delta b = -r^{(k+1)}$. In this equation we have $\delta A = 0$ and $\|\delta b\|_2 / \|b\|_2 < \epsilon$. Thus $x^{(k+1)}$ is an ϵ-backward-stable approximate solution to our problem. Conclusion: Stopping condition (8.5.1) is *sufficient* for backward stability at level ϵ. This serves as justification for (8.5.1) to some extent.

It is natural to ask whether (8.5.1) is also *necessary* for ϵ-backward stability. The answer turns out to be no; the less stringent criterion

$$\| r^{(k+1)} \|_2 < \epsilon \left(\|b\|_2 + \|A\|_2 \|x^{(k+1)}\|_2 \right), \tag{8.5.2}$$

which we introduced in Exercise 8.4.19, is necessary and sufficient. This fact, which is proved in Exercise 8.5.3, is a very strong recommendation for its use. Once the iterations terminate, we know we have an ϵ-backward-stable solution. Moreover, the iterations stop *as soon as* we have an ϵ-backward stable solution. No iterations are wasted. Notice that if the relative uncertainty of the data A and b is on the order of ϵ, we might as well stop as soon as we reach an ϵ-backward-stable solution. There is no point in iterating further. This is not a new idea [73].

Since condition (8.5.2) is less stringent than (8.5.1), it always results in earlier termination. Sometimes the difference is extreme; it can happen that an iteration terminates in a timely fashion if (8.5.2) is used but never terminates if (8.5.1) is used. For example, it has been shown [67] that (under certain conditions) a method known as MGS-GMRES is backward stable. This means that it will eventually satisfy (8.5.2) for fairly small values of ϵ (e.g. 10^{-14} in double precision). On the other hand, it is easy to build examples where (8.5.1) never comes close to being satisfied for such small ϵ.

·cise 8.5.3 Let A be a nonsingular matrix, let $x^{(k+1)}$ be an approximation to the solution of $Ax = b$, and let $r^{(k+1)} = b - Ax^{(k+1)}$.

(a) Show that if $x^{(k+1)}$ is an ϵ-backward-stable approximate solution of $Ax = b$, then

$$r^{(k+1)} = -\delta b + \delta A x^{(k+1)},$$

and consequently

$$\| r^{(k+1)} \|_2 \leq \|\delta b\|_2 + \|\delta A\|_2 \|x^{(k+1)}\|_2$$

for some δA and δb satisfying $\|\delta A\|_2 / \|A\|_2 < \epsilon$ and $\|\delta b\|_2 / \|b\|_2 < \epsilon$.

(b) Continuing from part (a), deduce that $x^{(k+1)}$ and $r^{(k+1)}$ satisfy (8.5.2).

(c) Conversely we can show that $x^{(k+1)}$ is ϵ-backward stable if (8.5.2) is satisfied. This requires some preparation; we won't use (8.5.2) until part (e). Let

$$\beta_k = \frac{\|b\|_2}{\|b\|_2 + \|A\|_2 \|x^{(k+1)}\|_2} \quad \text{and} \quad \alpha_k = \frac{\|A\|_2 \|x^{(k+1)}\|_2}{\|b\|_2 + \|A\|_2 \|x^{(k+1)}\|_2},$$

so that $\beta_k + \alpha_k = 1$. Define

$$\delta b = -\beta_k r^{(k+1)} \quad \text{and} \quad \delta A = \alpha_k \frac{r^{(k+1)} x^{(k+1)T}}{\left\| x^{(k+1)} \right\|_2^2}.$$

Show that

$$(A + \delta A)x^{(k+1)} = b + \delta b.$$

(d) Show that for any rank-1 matrix $M = uv^T$, $\| M \|_2 = \| u \|_2 \| v \|_2$. Deduce that $\| \delta A \|_2 = \alpha_k \| r^{(k+1)} \|_2 / \| x^{(k+1)} \|_2$.

(e) Using (8.5.2) and the definition of α_k, show that

$$\frac{\| \delta A \|_2}{\| A \|_2} < \epsilon.$$

(f) Using (8.5.2) and the definition of β_k, show that

$$\frac{\| \delta b \|_2}{\| b \|_2} < \epsilon.$$

(g) From the results of parts (c), (e), and (f), deduce that $x^{(k+1)}$ is an ϵ-backward-stable approximate solution to $Ax = b$, if (8.5.2) is satisfied.

$\square$

8.6 PRECONDITIONERS

In Section 8.4 we observed that the surface that represents the function

$$J(y) = \frac{1}{2} y^T A y - y^T b$$

is a steep, narrow canyon if A is not well conditioned. This causes the steepest descent method (among others) to converge very slowly. A remedy for this problem is to transform the system $Ax = b$ to an equivalent one $\tilde{A}\tilde{x} = \tilde{b}$, for which $\tilde{A}$ is better conditioned than A is. This procedure is called *preconditioning*. We can then apply the descent method to the transformed or *preconditioned* system.

There are many ways to make such a transformation. For example, let M be any simple matrix that approximates A, that is, any splitting matrix, but now we call it a *preconditioner*. Multiply the equation $Ax = b$ by M^{-1} on the left to obtain a transformed system $\tilde{A}\tilde{x} = \tilde{b}$, where $\tilde{A} = M^{-1}A$, $\tilde{x} = x$, and $\tilde{b} = M^{-1}b$. Unfortunately, the resulting $\tilde{A}$ is not symmetric. If M is positive definite, we can make a transformation that preserves the symmetry by making use of the Cholesky decomposition $M = R^T R$. Instead of using $M^{-1}A$, we can use the related matrix $R^{-T}AR^{-1}$. Specifically, we can multiply the equation $Ax = b$ on the left by R^{-T}, and insert the identity matrix $R^{-1}R$ between A and x, to get $(R^{-T}AR^{-1})(Rx) = R^{-T}b$. This is $\tilde{A}\tilde{x} = \tilde{b}$, where $\tilde{A} = R^{-T}AR^{-1}$, $\tilde{x} = Rx$, and $\tilde{b} = R^{-T}b$.

cise 8.6.1 Show that $\tilde{A} = R^{-T}AR^{-1}$ is a positive definite matrix, given that A is. In particular, $\tilde{A}$ is symmetric. □

The straightforward approach to preconditioning is simply to apply Algorithm 8.4.14 to the transformed system. Every instance of A, r, x, etc. is replaced by a transformed quantity $\tilde{A}$, $\tilde{r}$, $\tilde{x}$, etc. The algorithm yields $\tilde{x}$ in relatively few iterations, because $\tilde{A}$ is better conditioned than A is. Then we get the solution to our original system by $x = R^{-1}\tilde{x}$.

A more efficient approach is to carry out equivalent operations in the original coordinate system. If this is done, the entire algorithm can be performed without ever generating the transformed quantities. As we shall soon see, the big bonus of this approach is that it makes the computation of R unnecessary. Execution of the descent method will require only M^{-1}.

It is an easy exercise to write down Algorithm 8.4.14 for the transformed system, then translate line by line back into the untransformed coordinate system. We start out with an algorithm in which all of the vectors have tildes on them. Most importantly, $\tilde{A}$ is used in place of A. Then for each line we figure out an equivalent operation in the original coordinate system. For example, the line $\tilde{x} \leftarrow \tilde{x} + \alpha\tilde{p}$ can be transformed to $x \leftarrow x + \alpha p$ by the transformations $x = R^{-1}\tilde{x}$ and $p = R^{-1}\tilde{p}$. (We do not bother to introduce a symbol $\tilde{\alpha}$, because α is a scalar, not a vector. It does not undergo any coordinate transformation.) The transformation $p = R^{-1}\tilde{p}$ serves as the definition of p. It is reasonable to define p by the same transformation law as is used for x, as this allows us to continue to interpret p as the search direction that is taken to get from one iterate to the next. If we define q by $q = R^T\tilde{q}$, the line $\tilde{q} \leftarrow \tilde{A}\tilde{p}$ transforms to $q \leftarrow Ap$. If we want r to denote the residual as it always has up to now, the correct transformation is $r = R^T\tilde{r}$. Then the line $\tilde{r} \leftarrow \tilde{r} - \alpha\tilde{q}$ transforms to $r \leftarrow r - \alpha q$. Notice also that $\tilde{p}^T\tilde{r} = p^Tr$ and $\tilde{p}^T\tilde{q} = p^Tq$, so the tildes can be dropped from the computation of α.

So far the transformation process appears quite boring. We started with Algorithm 8.4.14 in the transformed coordinate system, involving $\tilde{p}$, $\tilde{q}$, $\tilde{r}$, etc., then defined p, q, and r in such a way that we could rewrite the equations with the tildes removed. The result is the original Algorithm 8.4.14, or so it appears at first.

The one thing we still need to discuss is the transformation of the lines $\tilde{p} \leftarrow ?$. Now let us focus our attention specifically on the steepest descent method. In the original steepest descent method we had $p = r$, so preconditioned steepest descent will have $\tilde{p} = \tilde{r}$. Now here is the important point. When we transform the equation $\tilde{p} = \tilde{r}$ back to the original coordinate system, we do not get $p = r$, because p and r obey different transformation laws.[48] Since $\tilde{p} = Rp$ and $\tilde{r} = R^{-T}r$, we have $p = R^{-1}R^{-T}r$. Recalling that $M = R^TR$, we see that $p = M^{-1}r$. With this transformation, the

---

[48]There are two transformation laws. In the language of duality, the vectors x and p, which satisfy $z = R^{-1}\tilde{z}$, are *primal*, and the vectors b, r, and q, which satisfy $z = R^T\tilde{z}$, are *dual*.

symbol R disappears from the algorithm. Thus the preconditioned steepest descent algorithm with preconditioner M is as follows.

Preconditioned Steepest Descent Algorithm (with preconditioner M) for solving $Ax = b$. On entry the initial guess is stored in x, and the vector b is stored in r. The algorithm returns in x its best estimate of the solution and a flag that indicates whether or not the specified tolerance was achieved.

$$r \leftarrow r - Ax$$
$$p \leftarrow M^{-1}r$$
$$k \leftarrow 0$$
do until converged or $k = l$
$$\begin{bmatrix} q \leftarrow Ap \\ \alpha \leftarrow p^T r / p^T q \\ x \leftarrow x + \alpha p \\ r \leftarrow r - \alpha q \\ p \leftarrow M^{-1}r \\ k \leftarrow k + 1 \end{bmatrix}$$
if not converged, set flag

(8.6.2)

Algorithm 8.6.2 is identical to Algorithm 8.4.14, except for the lines $p \leftarrow M^{-1}r$. Indeed, Algorithm 8.6.2 is an instance of Algorithm 8.4.14. The only effect of applying a preconditioner to the steepest descent algorithm is that the search direction is changed from r to $M^{-1}r$.

Exercise 8.6.3   Check all of the details of the foregoing development. Verify that Algorithm 8.6.2 is indeed the correct translation of the steepest descent algorithm for $\tilde{A}\tilde{x} = \tilde{b}$ where $\tilde{A} = R^{-T}AR^{-1}$.   □

Examples of Preconditioners

Our derivation assumed that M is positive definite. We thus search for preconditioners among those iterative methods that have positive definite splitting matrices.

Example 8.6.4 Jacobi's method has $M = D$, the diagonal part of A. This M is positive definite if A is. The Jacobi preconditioner, which is also known as the *diagonal preconditioner*, is particularly easy to apply. The operation $p \leftarrow M^{-1}r$ amounts to $p_i \leftarrow r_i/a_{ii}$, $i = 1, \ldots n$. Unfortunately it is not a very powerful preconditioner.   □

Exercise 8.6.5   Show that the convergence of the steepest descent algorithm on the model problem (8.1.8) is not accelerated at all by the Jacobi preconditioner.   □

In problems for which the main-diagonal entries of the coefficient matrix vary markedly in magnitude, the Jacobi preconditioner is often effective.

cise 8.6.6 Apply the steepest descent algorithm to the system

$$\begin{bmatrix} 10 & 1 \\ 1 & 1 \end{bmatrix} \begin{bmatrix} x_1 \\ x_2 \end{bmatrix} = \begin{bmatrix} 13 \\ 4 \end{bmatrix}.$$

Starting with $x^{(0)} = 0$, perform three iterations using (a) no preconditioner, (b) the Jacobi preconditioner. You may do this by hand or, better yet, write a simple MATLAB script. Note the extreme effectiveness of the preconditioner in this case. If you are using MATLAB, you might like to compare the condition numbers of A and $\tilde{A} = D^{-1/2}AD^{-1/2}$. □

On small problems like the one in Exercise 8.6.6, there is no harm in forming the matrix M^{-1} explicitly and using it to perform the preconditioning step $p \leftarrow M^{-1}r$. On big problems it is wasteful to do so. Normally the preconditioning operation is handled by a subroutine that takes r as input and returns p, and normally the subroutine does this without forming M or M^{-1}. For example, a Jacobi preconditioning subroutine will just take each entry of r and divide it by the corresponding main diagonal entry of A.

The procedure for applying the Jacobi preconditioner is perfectly obvious. Now let us consider how to apply a more complicated one. Consider any iterative method that is generated by a splitting matrix M. Recall that the iteration is defined by $Mx^{(k+1)} = Nx^{(k)} + b$, where $A = M - N$. If we perform one iteration to get from $x^{(0)}$ to $x^{(1)}$, we have

$$x^{(1)} = x^{(0)} + M^{-1}r^{(0)}, \tag{8.6.7}$$

where $r^{(0)} = b - Ax^{(0)}$. This was shown in general in Exercise 8.3.20 and for special cases in Exercises 8.2.4, 8.2.12, 8.2.18, 8.2.24, and 8.2.26. For example, if we perform one iteration of SOR, getting from $x^{(0)}$ to $x^{(1)}$, the two iterates are related by (8.6.7), where $M = \frac{1}{\omega}D - E$. But the SOR iteration is executed by sweeping through the x vector, correcting the entries one after the other; we never form or even think about the matrix M or its inverse. If we start from $x^{(0)} = 0$, then $r^{(0)} = b$, and (8.6.7) becomes $x^{(1)} = M^{-1}b$. This tells us that when we perform one iteration on the system $Ax = b$, starting with $x^{(0)} = 0$, the result is $M^{-1}b$. (Although we have the SOR example in mind, the conclusion is valid for any iterative method that is generated by a splitting.) For the purpose of preconditioning, we want to compute $M^{-1}r$, not $M^{-1}b$. This can be effected by operating on the system $Ax = r$ instead of $Ax = b$. In summary, to compute $p = M^{-1}r$, apply one iteration, starting from $x^{(0)} = 0$, to the system $Ax = r$. The result is p.

We do not envision using SOR as a preconditioner; it's splitting matrix is not symmetric. A better candidate is SSOR.

Example 8.6.8 In the symmetric case, the splitting matrix for SSOR is

$$M = \frac{\omega}{2 - \omega} \left(\frac{1}{\omega}D - E \right) D^{-1} \left(\frac{1}{\omega}D - E^T \right),$$

where D and $-E$ are the diagonal and lower triangular parts of A, respectively (Exercise 8.2.26). As long as $0 < \omega < 2$, and A is positive definite, M is positive definite too (Exercise 8.3.40). This M looks complicated. Fortunately it is easy to apply M^{-1} without actually forming it. We simply perform one SSOR iteration on the system $Ax = r$, starting from initial guess zero. Recall that an SSOR iteration consists of a forward SOR iteration followed by a backward one.

Table 8.7 lists some results from applying steepest descent with the SSOR preconditioner to the model problem (8.1.8) under the same conditions as in Example 8.2.8. We took $h = 1/160$ and tried several values for the relaxation factor ω. The iteration

Preconditioner	ω	Iterations to convergence
none	N/A	49755
SSOR	1.0	7289
SSOR	1.5	2655
SSOR	1.8	734
SSOR	1.9	356
SSOR	1.95	260

Table 8.7   Steepest descent method with SSOR preconditioner

count 49755 for no preconditioner is the same as in Example 8.4.16. We observe that the SSOR preconditioner delivers a dramatic improvement, regardless of how ω is chosen. If ω is chosen well, the results are excellent. Comparing with Example 8.2.21, we see that the number of iterations needed here with $\omega = 1.95$ is much less than for SOR with the optimal ω. To be fair, we should point out that the a steepest descent iteration with SSOR preconditioner requires several times as much arithmetic as an SOR iteration. Even taking this factor into account, we see that steepest descent with the SSOR preconditioner is competitive with SOR. $\square$

The SSOR preconditioner works quite well on our model problem, and it is easy to use, so it will figure prominently in our examples. For more difficult real-world problems, more powerful preconditioners are preferred.

Some of the most popular preconditioners (and some of the first to be tried) are the ILU preconditioners, which are based on incomplete LU decompositions. Since we are focusing on the positive definite case, we shall restrict our attention to *incomplete Cholesky* preconditioners. If A is sparse, its Cholesky factor is normally much less sparse. A great deal of fill-in occurs during the elimination process. An incomplete Cholesky decomposition is an approximation $A \approx R^T R$, where R is an upper triangular matrix that is much sparser than the true Cholesky factor. For example, R could have exactly the same sparsity pattern as A. One could produce such an R by carrying out the ordinary Cholesky decomposition algorithm but not allowing any fill-in. Numbers that start out zero stay zero. For a better approximation one can allow partial fill-in. Criteria for deciding which entries are allowed to fill in can be based on element size or element location, for example. A great many strategies have been tried. Once an incomplete Cholesky decomposition has been produced, the product $M = R^T R$ can serve as a preconditioner. Application of the preconditioner requires forward and backward substitution on the extremely sparse triangular factors.

Other preconditioners are based on multigrid and domain decomposition techniques [10], [47], [80]. When applied to the model problem (8.1.8) and a wide variety of similar problems, the best of these are optimal, in the sense that they produce a preconditioned matrix $\tilde{A}$ for which the condition number is bounded: $\kappa_2(\tilde{A}) \leq K$, where K is independent of the mesh size h. It follows that the number of iterations needed to achieve a given accuracy is more or less independent of h.

Additional Exercises

cise 8.6.9 Write a program that executes the steepest descent algorithm with the SSOR preconditioner on the model problem (8.1.8). Your preconditioning subroutine should consist of a pair of nested loops, the second of which looks something like this:

$$
\begin{array}{l}
\text{for } j = m - 1, \ldots, 1 \\
\left[\begin{array}{l}
\text{for } i = m - 1, \ldots, 1 \\
\left[\begin{array}{l}
\hat{p} \leftarrow .25(p_{i,j+1} + p_{i,j-1} + p_{i+1,j} + p_{i-1,j} + r_{i,j}) \\
p_{i,j} = p_{i,j} + \omega(\hat{p} - p_{i,j})
\end{array} \right.
\end{array} \right.
\end{array}
$$

Before entering the first loop, p should be zero (including boundary values). To test your program, use the same f, g, and $u^{(0)}$ as in Exercise 8.4.19. Terminate iterations when

$$
\left\| r^{(k+1)} \right\|_2 < \epsilon \left(\| b \|_2 + \| A \|_2 \| x^{(k+1)} \|_2 \right) \tag{8.6.10}
$$

with $\epsilon = 10^{-8}$. Take $h = 1/160$ and try at least $\omega = 1.8, 1.85, 1.9, 1.95,$ and 1.99 (but not 2.0). Observe that the preconditioner performs satisfactorily over this fairly wide range of ω values. $\quad\square$

cise 8.6.11 Our development of Algorithm 8.6.2 assumes that the splitting matrix is positive definite. Nevertheless, there is nothing to stop us from running Algorithm 8.6.2 with a nonsymmetric M. In this case we have no justification for the choice of search direction $p = M^{-1}r$, but we are still performing a descent algorithm with exact line searches.

(a) Try using SOR as a preconditioner for steepest descent on the model problem. How does it compare with SSOR? with no preconditioner?

(b) Try a preconditioner that does m sweeps of SOR instead of one. Try $m = 2$, 3, and 4, at least. Notice that as m is increased, the preconditioner gets better. Of course, it also consumes more time.

$\quad\square$

cise 8.6.12 Given a preconditioner or splitting matrix M, show that the preconditioned steepest descent algorithm is the same as the iteration generated by the splitting matrix $\frac{1}{\omega} M$, except that the damping parameter varies from one iteration to the next. (See Exercises 8.3.20 and 8.3.23.) $\quad\square$

Exercise 8.6.13 Let A and M be positive definite matrices, and let $M = R^T R$. Define $A_1 = M^{-1} A$ and $A_2 = R^{-T} A R^{-1}$.

(a) Show that A_1 and A_2 are similar. That is, $A_2 = S^{-1} A_1 S$ for some nonsingular matrix S.

(b) Show that Richardson's method applied to either A_1 or A_2 will converge if the damping parameter ω is small enough, and the convergence rate is the same for both. (See Exercise 8.3.19.)

$\square$

8.7 THE CONJUGATE-GRADIENT METHOD

All of the iterative methods that we have discussed to this point are limited by their lack of memory. Each uses only information about $x^{(k)}$ to get to $x^{(k+1)}$. All information from earlier iterations is forgotten. The conjugate-gradient (CG) method [50] is a simple variation on steepest descent that performs better because it has a memory.

Our approach will be to introduce the algorithm right away, compare and contrast it with the steepest descent method, and observe how well it performs. Once we have done that, we will derive the algorithm and study its theoretical properties. We begin with the basic CG algorithm with no preconditioner:

> **Conjugate-Gradient Algorithm** for solving $Ax = b$. On entry the initial guess is stored in x, and the vector b is stored in r. The algorithm returns in x its best estimate of the solution and a flag that indicates whether or not the specified tolerance was achieved.

$$
\begin{aligned}
&r \leftarrow r - Ax \\
&p \leftarrow r \\
&\nu \leftarrow r^T r \\
&k \leftarrow 0 \\
&\text{do until converged or } k = l \\
&\quad \left[\begin{aligned}
&q \leftarrow Ap \\
&\mu \leftarrow p^T q \\
&\alpha \leftarrow \nu / \mu \\
&x \leftarrow x + \alpha p \\
&r \leftarrow r - \alpha q \\
&\nu_+ \leftarrow r^T r \\
&\beta \leftarrow \nu_+ / \nu \\
&p \leftarrow r + \beta p \\
&\nu \leftarrow \nu_+ \\
&k \leftarrow k + 1
\end{aligned} \right. \\
&\text{if not converged, set flag}
\end{aligned}
\tag{8.7.1}
$$

In appearance this algorithm differs only slightly from steepest descent. The computation of α is organized a bit differently, but this difference is cosmetic. The line searches are still exact; the CG algorithm is an instance of Algorithm 8.4.14.

Initially $p \leftarrow r$, so the first step is steepest descent. On subsequent steps there is a difference. Instead of $p \leftarrow r$, we have $p \leftarrow r + \beta p$. The residual or steepest descent direction still plays an important role in determining the new search direction, but now the old search direction also matters. This is the one point at which memory of past iterations is used. This slight change makes a huge difference.

Example 8.7.2 We applied the Conjugate-Gradient algorithm to the model problem under the same conditions as in Example 8.2.8. The results are summarized in Table 8.8. Comparing with the table in Example 8.4.16, we see that the CG algorithm

h	Iterations to convergence
1/20	59
1/40	117
1/80	230
1/160	444

Table 8.8   Conjugate-gradient method (with no preconditioner) applied to model problem

is far superior to steepest descent. Indeed its performance is more in line with that of SOR (Example 8.2.21) or steepest descent preconditioned by SSOR (Example 8.6.8). An advantage of CG over these other two methods is that it does not involve any ω whose optimal value must be guessed. □

Why is CG so much better than steepest descent? To keep the discussion simple, let us make the (inessential) assumption that $x^{(0)} = 0$. Then, after j iterations of the form $x^{(k+1)} = x^{(k)} + \alpha_k p^{(k)}$, we have

$$x^{(j)} = \alpha_0 p^{(0)} + \alpha_1 p^{(1)} + \ldots + \alpha_{j-1} p^{(j-1)}.$$

Thus $x^{(j)}$ lies in the space $\mathcal{S}_j$ spanned by the j search directions $p^{(0)}, \ldots, p^{(j-1)}$. This is true for both algorithms; however, they pick out different search directions. Of course, they also compute different coefficients α_k. Interestingly, the two different sets of search directions span the same space. In both cases, $\mathcal{S}_j = \text{span}\{b, Ab, A^2 b, \ldots, A^{j-1} b\}$, a Krylov subspace. Since steepest descent does exact line searches, $x^{(j)}$ minimizes the energy function J over the j lines $x^{(k)} + \alpha p^{(k)}$, $k = 0, \ldots, j - 1$. The union of these lines is a tiny subset of $\mathcal{S}_j$. As we shall see, the CG algorithm does much better. By looking backward just slightly, it manages to pick out the $x^{(j)}$ that minimizes J over the entire space $\mathcal{S}_j$.

Preconditioned Conjugate Gradient Algorithm

The CG algorithm is closely related to steepest descent. Its first step is the same as steepest descent, and each subsequent search direction depends in part on the direction of steepest descent. It is therefore reasonable to hope that CG, like steepest descent, can benefit from preconditioning [11].

The procedure for preconditioning CG is the same as for steepest descent. The following algorithm results.

> **Preconditioned Conjugate-Gradient Algorithm** (with preconditioner M) for solving $Ax = b$. On entry the initial guess is stored in x, and the vector b is stored in r. The algorithm returns in x its best estimate of the solution and a flag that indicates whether or not the specified tolerance was achieved.

$$
\begin{aligned}
&r \leftarrow r - Ax \\
&s \leftarrow M^{-1}r \\
&\nu \leftarrow r^T s \\
&p \leftarrow s \\
&k \leftarrow 0 \\
&\text{do until converged or } k = l \\
&\quad\left[\begin{aligned}
&q \leftarrow Ap \\
&\mu \leftarrow p^T q \\
&\alpha \leftarrow \nu/\mu \\
&x \leftarrow x + \alpha p \\
&r \leftarrow r - \alpha q \\
&s \leftarrow M^{-1}r \\
&\nu_+ \leftarrow r^T s \\
&\beta \leftarrow \nu_+/\nu \\
&p \leftarrow s + \beta p \\
&\nu \leftarrow \nu_+ \\
&k \leftarrow k + 1
\end{aligned}\right. \\
&\text{if not converged, set flag}
\end{aligned}
\tag{8.7.3}
$$

We derived (8.7.3) by applying CG to the transformed system $\tilde{A}\tilde{x} = \tilde{b}$, where $\tilde{A} = R^{-T}AR^{-1}$, $\tilde{x} = Rx$, $\tilde{b} = R^{-T}b$, and $R^T R = M$. We then translated each expression to an equivalent expression in the original coordinate system. We introduced a new vector $s = M^{-1}r$ for convenience.

The big advantage of transforming back to original coordinate system is that it eliminates the need to calculate R. Only M^{-1} appears in Algorithm 8.7.3.

Exercise 8.7.4 Verify that Algorithm 8.7.3 is indeed the correct translation of the conjugate-gradient algorithm for $\tilde{A}\tilde{x} = \tilde{b}$. □

Algorithm 8.7.3 is yet another instance of Algorithm 8.4.14. In particular, the line searches are exact.

The cost of executing CG is only slightly greater than that of steepest descent. Because the computation of the coefficients α and β has been arranged carefully, only two inner products need to be computed on each iteration, which is the same as for steepest descent. All other costs are virtually the same, except that CG has the additional vector update $p \leftarrow s + \beta p$, which costs $2n$ flops. The storage space required by CG is $5n$, for the vectors x, r, s, p, and q, plus whatever is needed (if any) to store A and M^{-1}. This compares with $4n$ for steepest descent and n for SOR.

Example 8.7.5 Table 8.9 shows the results of applying the CG method with the SSOR preconditioner to the model problem (8.1.8) under the same conditions as in Example 8.2.8. As in Example 8.6.8, we took $h = 1/160$ and tried several values for the relaxation factor ω. The iteration counts are quite low, and they are

Preconditioner	ω	Iterations to convergence
none	N/A	444
SSOR	1.00	184
SSOR	1.50	106
SSOR	1.60	93
SSOR	1.70	79
SSOR	1.80	64
SSOR	1.90	50
SSOR	1.95	46

Table 8.9   Conjugate-Gradient method with SSOR preconditioner

fairly insensitive to the choice of ω. Comparing with Example 8.2.21, we see that preconditioned CG is clearly superior to SOR. □

In Exercise 8.7.6 you will see that the advantage of preconditioned CG increases as h is made smaller.

Additional Exercises

cise 8.7.6 Write a program that applies the conjugate-gradient algorithm with the SSOR preconditioner to the model problem (8.1.8). The instructions in Exercises 8.4.19 and 8.6.9 for executing the steps $p \leftarrow Aq$ and $s \leftarrow M^{-1}r$ are also applicable here. To test your program, use the same f, g, and $u^{(0)}$ as in Exercise 8.4.19. Terminate iterations when

$$\|r^{(k+1)}\|_2 < \epsilon \left(\|b\|_2 + \|A\|_2 \|x^{(k+1)}\|_2 \right) \tag{8.7.7}$$

with $\epsilon = 10^{-8}$.

(a) Try out your code under the conditions of Example 8.7.5. You should get similar but lower iteration counts, since you are using a less stringent stopping criterion.

(b) Consider the model problem with $h = 1/501$. What is the dimension of the associated matrix? What is its bandwidth (assuming the natural row ordering)? Approximately what fraction of the matrix entries are nonzero?

(c) With $h = 1/501$, try $\omega = 1.8, 1.85, 1.9, 1.95$ and 1.99. Observe that the preconditioner performs satisfactorily over this range of ω values.

(d) Apply SOR with $h = 1/501$ for comparison against preconditioned CG. (Use your code from Exercise 8.2.23.) Try several values of ω, for example 1.97, 1.98, 1.99, and 1.9875 (optimal for $h = 1/501$). Note the sensitivity to the choice of ω. How does SOR compare with CG preconditioned by SSOR?

(e) Choose a value of ω, say 1.9 or 1.95, and run your preconditioned CG code with that chosen ω and $h = 1/20, 1/40, 1/80, 1/160, 1/320$, and $1/640$. Record the number of iterations i_h in each case. Notice that i_h grows as h is decreased, but the growth is moderate. (Optimally we would like to see no growth. There is still room for improvement.)

□

Exercise 8.7.8 Try replacing SSOR by SOR as a preconditioner for CG. How does it work?

□

Exercise 8.7.9 A conjugate-gradient routine `pcg` is supplied with MATLAB. Read about `pcg` in the MATLAB `help` facilities, and try it out on a discrete Laplacian matrix. Use commands like the following:

```
m = 40;
A = delsq(numgrid('N',m+1));
n = size(A,1)
sol = ones(n,1); % ``all ones'' solution vector.
b = A*sol;       % right-hand side vector.
tol = 1e-12; maxit = 1000;
[x,flag,relres,iter] = pcg(A,b,tol,maxit);
err = norm(sol-x)
```

The matrix A generated here is exactly the matrix of our model problem (8.1.8) with $h = 1/m$. A "nested dissection" ordering is used [37]. The most common way to use `pcg` is to supply the matrix A explicitly in MATLAB's sparse matrix format, and that is what we are doing here. This is far from the most efficient way to use this matrix or to execute the CG algorithm, but it is convenient and works reasonably well on moderate-sized problems. Here we are running `pcg` with a residual tolerance of 10^{-12}, a maximum of 1000 on the number of iterations, and no preconditioner. Try this out with various values of m, e.g. $m = 40, 80, 160$, and check how many iterations are needed in each case.

□

Exercise 8.7.10 MATLAB also provides a routine `cholinc` for computing incomplete Cholesky preconditioners, (see Section 8.6). Read about `cholinc` in the MATLAB `help` facilities. Repeat Exercise 8.7.9 using a an incomplete Cholesky preconditioner $M = R^T R$ based on a drop tolerance of 10^{-2}:

```
droptol = 1e-2;
R = cholinc(A,droptol);
spy(A), spy(R), spy(R'+R),
[x,flag,relres,iter] = pcg(A,b,tol,maxit,R',R);
```

Compare your results with those of Exercise 8.7.9.

□

cise 8.7.11 Repeat Exercise 8.7.10 with fixed m, say $m = 160$, varying the drop tolerance. Try, for example, 10^0, 10^{-1}, 10^{-2}, and 10^{-3}. Check the effect on the number of CG iterations needed and the number of nonzeros in the incomplete Cholesky factor. Compare with the number of nonzeros in the complete Cholesky factor of A, as computed by `chol`. □

cise 8.7.12 Repeat Exercises 8.7.9, 8.7.10, and 8.7.11, using Wathen matrices of various sizes. Type `A = gallery('wathen',20,15)`, for example. For a bigger matrix replace the 20 and 15 by larger numbers. Type `help private/wathen` for information about Wathen matrices. □

8.8 DERIVATION OF THE CG ALGORITHM

The conjugate-gradient algorithm is a descent method that performs exact line searches. To complete the formal definition of the algorithm, we need only specify how the search directions are chosen. After some preparation we will be in a position to derive the search directions in a natural way.

In Chapter 3 we introduced the inner product

$$\langle x, y \rangle = \sum_{i=1}^{n} x_i y_i = y^T x = x^T y \tag{8.8.1}$$

and used it extensively in our discussion of the least-squares problem. It was the only inner product we needed at that time, and we spoke of it as if it were the only inner product in the world. Now it is time to broaden our view of the concept. Given any positive definite matrix H we can define the *inner product induced by H* by

$$\langle x, y \rangle_H = y^T H x = x^T H y.$$

The inner product (8.8.1), which we will now call the *standard inner product*, is the inner product induced by I.

The inner product induced by H has the same basic algebraic properties as the standard inner product. In particular, $\langle x, x \rangle_H > 0$ if $x \neq 0$. Of course the connection with the Euclidean norm, $\| x \|_2 = \sqrt{\langle x, x \rangle}$, does not hold. Instead we can use the H inner product to induce a different norm

$$\| x \|_H = \sqrt{\langle x, x \rangle_H}.$$

A descent method solves the positive definite system $Ax = b$ by minimizing $J(y) = \frac{1}{2} y^T A y - y^T b$. Using our new notation we can rewrite J as

$$J(y) = \frac{1}{2} \langle y, y \rangle_A - \langle b, y \rangle = \frac{1}{2} \| y \|_A^2 - \langle b, y \rangle.$$

In elasticity problems the term $\frac{1}{2} \langle y, y \rangle_A$ represents the strain energy stored in the deformed elastic structure, so the inner product and norm induced by A are commonly called the *energy inner product* and *energy norm*, respectively.

Thinking of y as an approximation to x, the solution of $Ax = b$, we define the error $e = x - y$. Completing the square, as in (8.4.3), we find that

$$J(y) = \tfrac{1}{2}\|e\|_A^2 - \tfrac{1}{2}\|x\|_A^2.$$

The energy norm of x is a fixed constant. Thus minimizing J is the same as minimizing the energy norm of the error.

A method that uses exact line searches minimizes the energy norm of the error along a line at each step. This is a one-dimensional minimization. Our objective now is to develop a method that remembers information from past steps so that it can minimize over higher dimensional subspaces. By the jth step, we hope to minimize over a j-dimensional subspace.

Regardless of how we choose the search directions, the following relationships hold. At each step we have $x^{(k+1)} = x^{(k)} + \alpha_k p^{(k)}$. Starting from $x^{(0)}$, j such steps bring us to

$$x^{(j)} = x^{(0)} + \alpha_0 p^{(0)} + \ldots + \alpha_{j-1} p^{(j-1)}.$$

At step k the error is $e^{(k)} = x - x^{(k)}$. Clearly the errors satisfy the recursion $e^{(k+1)} = e^{(k)} - \alpha_k p^{(k)}$, and after j steps,

$$e^{(j)} = e^{(0)} - \left(\alpha_0 p^{(0)} + \ldots + \alpha_{j-1} p^{(j-1)}\right). \tag{8.8.2}$$

Ideally we would like to have chosen the coefficients $\alpha_0, \ldots, \alpha_{j-1}$ so that the energy norm $\|e^{(j)}\|_A$ is as small as possible. By (8.8.2) this is the same as minimizing

$$\left\|e^{(0)} - \textstyle\sum_{k=0}^{j-1} \alpha_k p^{(k)}\right\|_A,$$

which amounts to finding the best approximation to $e^{(0)}$ from the subspace $\mathcal{S}_j = \operatorname{span}\left\{p^{(0)}, \ldots, p^{(j-1)}\right\}$.

In Chapter 3 we developed the basic characterization theorem (Theorem 3.5.15) for best approximation in Euclidean space. The shortest distance from a point to a line is along the perpendicular. More generally, the best approximation to a vector v from a subspace $\mathcal{S}$ is the unique $s \in \mathcal{S}$ for which the error $v - s$ is orthogonal to every vector in $\mathcal{S}$. In that result, approximations are measured with respect to the Euclidean norm, and orthogonality is reckoned with respect to the standard inner product. In our present scenario we seek a best approximation with respect to the energy norm. Fortunately the basic characterization is valid in this context as well, provided we reckon orthogonality with respect to the energy inner product. Indeed, the theorem is valid for any inner product and its induced norm.

Let H be any positive definite matrix. If $\langle v, w \rangle_H = 0$, we say that the vectors v and w are H-*orthogonal* and write $v \perp_H w$.

Theorem 8.8.3 *Let $H \in \mathbb{R}^{n \times n}$ be a positive definite matrix, let $v \in \mathbb{R}^n$, and let $\mathcal{S}$ be a subspace of $\mathbb{R}^n$. Then there is a unique $s \in \mathcal{S}$ such that*

$$\|v - s\|_H = \min_{w \in \mathcal{S}} \|v - w\|_H.$$

The vector s is characterized by the condition $v - s \perp_H w$ for all $w \in \mathcal{S}$.

This theorem reduces to Theorem 3.5.15 in the case $H = I$. It can be proved by a straightforward generalization of the proof of Theorem 3.5.15.

Applying Theorem 8.8.3 to our current scenario, we see that $\| e^{(j)} \|_A$ is minimized when $p \in \mathcal{S}_j$ is chosen so that the error $e^{(j)} = e^{(0)} - p$ satisfies

$$e^{(j)} \perp_A p^{(i)} \quad \text{for } i = 0, \dots, j - 1. \tag{8.8.4}$$

Two vectors that are orthogonal with respect to the energy inner product are said to be *conjugate*. Our goal now is to develop methods for which the error at each step is conjugate to all of the previous search directions.

The following proposition shows that part of (8.8.4) is achieved by performing exact line searches. As usual $r^{(k+1)}$ denotes the residual $b - Ax^{(k+1)}$.

Proposition 8.8.5 *Let* $x^{(k+1)} = x^{(k)} + \alpha_k p^{(k)}$ *be obtained from an exact line search. Then* $r^{(k+1)} \perp p^{(k)}$ *and* $e^{(k+1)} \perp_A p^{(k)}$.

Proof. Successive residuals are related by the recursion $r^{(k+1)} = r^{(k)} - \alpha_k A p^{(k)}$ (which appears as $r \leftarrow r - \alpha q$ in all of our algorithms). Thus

$$\langle r^{(k+1)}, p^{(k)} \rangle = \langle r^{(k)}, p^{(k)} \rangle - \alpha_k \langle A p^{(k)}, p^{(k)} \rangle = 0,$$

since $\alpha_k = \langle r^{(k)}, p^{(k)} \rangle / \langle A p^{(k)}, p^{(k)} \rangle$, by Theorem 8.4.5.

It is easy to check that the error and the residual are connected by the simple equation $A e^{(k+1)} = r^{(k+1)}$. Therefore $\langle e^{(k+1)}, p^{(k)} \rangle_A = \langle A e^{(k+1)}, p^{(k)} \rangle = \langle r^{(k+1)}, p^{(k)} \rangle = 0$. □

From the proof it is clear that Proposition 8.8.5 is basically a restatement of Theorem 8.4.5. Both are special cases of Theorem 8.8.3, as Exercise 8.8.27 shows.

Proposition 8.8.5 is geometrically obvious. The minimum of J on the line $x^{(k)} + \alpha p^{(k)}$ occurs when the directional derivative of J in the search direction is zero. The directional derivative is just the dot product of the gradient with the direction, so the directional derivative is zero exactly when the gradient (in this case, the residual) is orthogonal to the search direction.

According to Proposition 8.8.5, after the first step we have $e^{(1)} \perp_A p^{(0)}$. This is condition (8.8.4) in the case $j = 1$. It is clear from (8.8.4) that we would like to keep all subsequent errors conjugate to $p^{(0)}$. Since the errors are related by the recursion $e^{(k+1)} = e^{(k)} - \alpha_k p^{(k)}$, we can accomplish this by forcing all subsequent search directions to be conjugate to $p^{(0)}$. If we pick $p^{(1)}$ so that $p^{(1)} \perp_A p^{(0)}$ and perform an exact line search, we get an $x^{(2)}$ for which the error satisfies $e^{(2)} \perp_A p^{(1)}$. We thus have $e^{(2)} \perp_A p^{(i)}$ for $i = 0, 1$, which is (8.8.4) for $j = 2$. We can now keep all subsequent errors conjugate to both $p^{(0)}$ and $p^{(1)}$ by making all subsequent search directions conjugate to $p^{(0)}$ and $p^{(1)}$.

By now it is clear that we can achieve (8.8.4) by choosing our search directions in such a way that $p^{(i)} \perp_A p^{(j)}$ for all $i \neq j$ and doing exact line searches. A method with these characteristics is called a *conjugate direction method*.

Theorem 8.8.6 *If a method employs conjugate search directions and performs exact line searches, then the errors* $e^{(j)}$ *satisfy (8.8.4) for all* j. *Furthermore*

$$\| e^{(j)} \|_A = \min_{p \in \mathcal{S}_j} \| e^{(0)} - p \|_A,$$

where $\mathcal{S}_j = \text{span}\{p^{(0)}, \dots, p^{(j-1)}\}$.

Proof. We prove (8.8.4) by induction on j. For $j = 1$, (8.8.4) holds because the line search is exact. Now let j be arbitrary and assume we have $e^{(j)} \perp_A p^{(i)}$ for $i = 0, \ldots, j - 1$. We have to show that $e^{(j+1)} \perp_A p^{(i)}$ for $i = 0, \ldots, j$. Since $e^{(j+1)} = e^{(j)} - \alpha_j p^{(j)}$, we have $\langle e^{(j+1)}, p^{(i)} \rangle_A = \langle e^{(j)}, p^{(i)} \rangle_A - \alpha_j \langle p^{(j)}, p^{(i)} \rangle_A = 0 - 0 = 0$ for $i = 0, \ldots, j - 1$. Thus we need only establish $e^{(j+1)} \perp_A p^{(j)}$. But this is true because the line search on step $j + 1$ is exact.

Once we have (8.8.4), the optimality property of $e^{(j)}$ follows from Theorem 8.8.3, as we have already seen. $\qquad\square$

Choice of Search Directions

The conjugate gradient method is a conjugate-direction method that chooses search directions by A-orthogonalizing the residual against all previous search directions. The first search direction is $p^{(0)} = r^{(0)}$. Thus the first step of CG is the same as the first step of steepest descent. After k steps with conjugate search directions $p^{(0)}, \ldots, p^{(k)}$, the CG algorithm chooses

$$p^{(k+1)} = r^{(k+1)} - \sum_{i=0}^{k} c_{ki} p^{(i)}, \tag{8.8.7}$$

where the coefficients c_{ki} are chosen so that $p^{(k+1)} \perp_A p^{(i)}$ for $i = 0, \ldots k$. One easily checks that $p^{(k+1)} \perp_A p^{(i)}$ is satisfied if and only if

$$c_{ki} = \frac{\langle r^{(k+1)}, p^{(i)} \rangle_A}{\langle p^{(i)}, p^{(i)} \rangle_A}, \quad i = 0, \ldots, k. \tag{8.8.8}$$

This is just the Gram-Schmidt process applied to the energy inner product.

Exercise 8.8.9 Verify that if $p^{(k+1)}$ is given by (8.8.7), then $p^{(k+1)} \perp_A p^{(i)}$ if and only if c_{ki} is given by (8.8.8). $\qquad\square$

As long as $r^{(k+1)} \neq 0$, meaning the algorithm has not terminated, (8.8.7) yields a nonzero $p^{(k+1)}$. To see this, note that the connection $Ae^{(k+1)} = r^{(k+1)}$ implies that $\langle e^{(k+1)}, p^{(i)} \rangle_A = \langle r^{(k+1)}, p^{(i)} \rangle$. Therefore (8.8.4), with $j = k + 1$, implies

$$r^{(k+1)} \perp p^{(i)} \quad \text{for } i = 0, \ldots, k. \tag{8.8.10}$$

It now follows directly from (8.8.7) that

$$\langle p^{(k+1)}, r^{(k+1)} \rangle = \langle r^{(k+1)}, r^{(k+1)} \rangle = \| r^{(k+1)} \|_2^2. \tag{8.8.11}$$

Thus $\langle p^{(k+1)}, r^{(k+1)} \rangle > 0$ as long as $r^{(k+1)} \neq 0$. This implies that $p^{(k+1)} \neq 0$. It also implies (see Remark 8.4.6) that each CG step results in a strict decrease in energy: $J(x^{(k+1)}) < J(x^{(k)})$.

We will make further use of both (8.8.10) and (8.8.11) later.

Exercise 8.8.12 Verify (8.8.10) and (8.8.11). $\qquad\square$

cise 8.8.13 Prove that the conjugate gradient algorithm arrives at the exact solution of $Ax = b$ in at most n iterations. $\square$

Efficient Implementation

Now that the search directions have been specified, the description of the CG algorithm is formally complete. However, it remains to be shown that the algorithm can be implemented efficiently. Formulas (8.8.7) and (8.8.8) look like expensive computations, and it appears that all of the previous search directions need to be saved. Fortunately it turns out that all of the c_{ki} are zero, except c_{kk}. Thus only one coefficient needs to be computed by (8.8.8), and only the most recent search direction $p^{(k)}$ needs to be kept for (8.8.7), which turns into the elegant statement $p \leftarrow r + \beta p$ in Algorithm 8.7.1.

Proving that the c_{ki} are zero requires some preparation. We begin by recalling some terminology that we introduced in Section 6.3. Given a vector v and a positive integer j, the *Krylov subspace* $\mathcal{K}_j(A, v)$ is defined by

$$\mathcal{K}_j(A, v) = \text{span}\{v, Av, A^2v, \dots, A^{j-1}v\}.$$

Theorem 8.8.14 *After j steps of the conjugate-gradient algorithm (with $r^{(k)} \neq 0$ at each step) we have*

$$\text{span}\{p^{(0)}, \dots, p^{(j)}\} = \text{span}\{r^{(0)}, \dots, r^{(j)}\} = \mathcal{K}_{j+1}(A, r^{(0)}).$$

Proof. The proof is by induction on j. The theorem is trivially true for $j = 0$, since $p^{(0)} = r^{(0)}$. Now assume that the spaces are equal for $j = k$, and we will show that they are also equal for $j = k + 1$. Our first step is to show that

$$\text{span}\{r^{(0)}, \dots, r^{(k+1)}\} \subseteq \mathcal{K}_{k+2}(A, r^{(0)}). \tag{8.8.15}$$

In light of the induction hypothesis, it suffices to show that $r^{(k+1)} \in \mathcal{K}_{k+2}(A, r^{(0)})$. Recalling the recurrence $r^{(k+1)} = r^{(k)} - \alpha_k Ap^{(k)}$, we check the status of $Ap^{(k)}$. By assumption $p^{(k)} \in \mathcal{K}_{k+1}(A, r^{(0)}) = \text{span}\{r^{(0)}, \dots, A^k r^{(0)}\}$, so

$$Ap^{(k)} \in \text{span}\{Ar^{(0)}, \dots, A^{k+1}r^{(0)}\} \subseteq \mathcal{K}_{k+2}(A, r^{(0)}).$$

Furthermore $r^{(k)} \in \mathcal{K}_{k+1}(A, r^{(0)}) \subseteq \mathcal{K}_{k+2}(A, r^{(0)})$, so

$$r^{(k+1)} = r^{(k)} - \alpha_k Ap^{(k)} \in \mathcal{K}_{k+2}(A, r^{(0)}).$$

This establishes (8.8.15).

Our next step is to establish

$$\text{span}\{p^{(0)}, \dots, p^{(k+1)}\} \subseteq \text{span}\{r^{(0)}, \dots, r^{(k+1)}\}. \tag{8.8.16}$$

By the induction hypothesis,

$$p^{(i)} \in \text{span}\{r^{(0)}, \dots, r^{(k)}\} \quad \text{for } i = 0, \dots, k, \tag{8.8.17}$$

so it suffices to show that $p^{(k+1)} \in \text{span}\{r^{(0)}, \dots, r^{(k+1)}\}$. But this follows immediately from (8.8.7), using (8.8.17) once again.

Putting (8.8.16) and (8.8.15) together, we see that the three subspaces of interest are nested. We can show that they are equal by demonstrating that they all have the same dimension. Since $\mathcal{K}_{k+2}(A, r^{(0)})$ is spanned by a set of $k+2$ vectors, its dimension is at most $k+2$. If we can show that the dimension of $\text{span}\{p^{(0)}, \dots, p^{(k+1)}\}$ is exactly $k+2$, it will follow that all three spaces have dimension $k+2$ and are therefore equal. But we already know that $p^{(0)}, \dots, p^{(k+1)}$ are nonzero and orthogonal with respect to the energy inner product. Therefore they are linearly independent; they form a basis for $\text{span}\{p^{(0)}, \dots, p^{(k+1)}\}$, whose dimension is therefore $k+2$. $\square$

The next result is an immediate consequence of Theorems 8.8.6 and 8.8.14.

Corollary 8.8.18 *The error after j steps of the CG algorithm satisfies*

$$e^{(j)} \perp_A \mathcal{K}_j(A, r^{(0)}), \qquad j = 1, 2, 3, \dots.$$

The search directions $p^{(0)}, \dots, p^{(k)}$ form an energy-orthogonal basis of the Krylov subspace $\mathcal{K}_{k+1}(A, r^{(0)})$. The residuals $r^{(0)}, \dots, r^{(k)}$ also form a basis of the Krylov subspace, and they also have an orthogonality property.

Corollary 8.8.19 *The residuals $r^{(0)}$, $r^{(1)}$, $r^{(2)}$, $\dots$ produced by the CG algorithm are orthogonal with respect to the standard inner product. Hence*

$$r^{(j)} \perp \mathcal{K}_j(A, r^{(0)}), \qquad j = 1, 2, 3, \dots.$$

Proof. We need to establish that $r^{(k+1)} \perp r^{(0)}, \dots r^{(k)}$ for all k. This follows immediately from (8.8.10), since $\text{span}\{p^{(0)}, \dots, p^{(k)}\} = \text{span}\{r^{(0)}, \dots, r^{(k)}\}$. $\square$

Now we are in a position to prove that CG can be implemented economically.

Theorem 8.8.20 *For $i = 0, \dots, k-1$, the coefficients c_{ki} defined by (8.8.8) are zero.*

Proof. Referring to (8.8.8), we see that we need to show that $\langle r^{(k+1)}, p^{(i)} \rangle_A = 0$ for $i = 0, \dots, k-1$. Since $\langle r^{(k+1)}, p^{(i)} \rangle_A = \langle r^{(k+1)}, Ap^{(i)} \rangle$, it suffices to show that $r^{(k+1)} \perp Ap^{(i)}$ for $i < k$. We can do this by applying Theorem 8.8.14 twice. We have $p^{(i)} \in \text{span}\{p^{(0)}, \dots, p^{(i)}\} = \text{span}\{r^{(0)}, \dots, A^i r^{(0)}\}$, so $Ap^{(i)} \in \text{span}\{Ar^{(0)}, \dots, A^{i+1}r^{(0)}\} \subseteq \mathcal{K}_{i+2}(A, r^{(0)}) = \text{span}\{r^{(0)}, \dots, r^{(i+1)}\}$. Since $r^{(k+1)}$ is orthogonal to $r^{(0)}, \dots, r^{(i+1)}$ if $i < k$, we conclude that $r^{(k+1)} \perp Ap^{(i)}$ if $i < k$. $\square$

Thanks to Theorem 8.8.20, (8.8.7) reduces to the much simpler expression

$$p^{(k+1)} = r^{(k+1)} + \beta_k p^{(k)}, \tag{8.8.21}$$

where $\beta_k = -c_{kk}$. This shows that $p^{(k+1)}$ can be calculated inexpensively and that the cost does not grow with k. It shows furthermore that there is no need to save the old search directions $p^{(0)}, \ldots, p^{(k-1)}$, so the storage requirement does not grow with k. This is a remarkable fact. We embarked on the derivation of the CG algorithm with the idea that we would improve on the method of steepest descent by making use of information from previous steps. We found that by doing so we could minimize the energy norm of the error over larger and larger subspaces. But now (8.8.21) demonstrates that we can accomplish this without having to save much. All the history we need is encapsulated in $p^{(k)}$.

Theorem 8.8.20 relies heavily on the symmetry property $A = A^T$. The importance of symmetry is easily overlooked, because we have buried A in an inner product. If A were not symmetric, the energy inner product would not be an inner product, since we would not have $\langle v, w \rangle_A = \langle w, v \rangle_A$. If you worked through all the details of this section, you undoubtedly noticed that we used this property from time to time, sometimes in the form $\langle Av, w \rangle = \langle v, Aw \rangle$. The ability to group A with either vector in the inner product is crucial to the development. For example, in the proof of Theorem 8.8.20 we used the identity $\langle r^{(k+1)}, p^{(i)} \rangle_A = \langle r^{(k+1)}, Ap^{(i)} \rangle$.

With one more refinement, our development is complete.

Proposition 8.8.22 *The coefficient β_k in (8.8.21) is given by*

$$\beta_k = \frac{\langle r^{(k+1)}, r^{(k+1)} \rangle}{\langle r^{(k)}, r^{(k)} \rangle}.$$

cise 8.8.23 Recalling that $\beta_k = -c_{kk}$, use (8.8.8), (8.4.11), Theorem 8.4.5, and (8.8.11) to prove Proposition 8.8.22.   □

We summarize our development as a prototype CG algorithm.

Prototype Conjugate-Gradient Algorithm

$x^{(0)} \leftarrow$ initial guess
$r^{(0)} \leftarrow b - Ax^{(0)}$
$p^{(0)} \leftarrow r^{(0)}$
for $k = 0, 1, 2, \ldots$

$$\begin{bmatrix} \alpha_k \leftarrow \langle r^{(k)}, r^{(k)} \rangle / \langle Ap^{(k)}, p^{(k)} \rangle \\ x^{(k+1)} \leftarrow x^{(k)} + \alpha_k p^{(k)} \\ r^{(k+1)} \leftarrow r^{(k)} - \alpha_k Ap^{(k)} \\ \beta_k \leftarrow \langle r^{(k+1)}, r^{(k+1)} \rangle / \langle r^{(k)}, r^{(k)} \rangle \\ p^{(k+1)} \leftarrow r^{(k+1)} + \beta_k p^{(k)} \end{bmatrix} \qquad (8.8.24)$$

Thanks to (8.8.11) and Proposition 8.8.22, $\langle r^{(k)}, r^{(k)} \rangle$ appears three times in the loop. Once it has been computed, it gets used over and over. It can even be used in the convergence criterion.

It is a simple matter to translate Algorithm 8.8.24 into Algorithm 8.7.1.

cise 8.8.25 Derive Algorithm 8.7.1 from Algorithm 8.8.24.   □

Relationship with the Symmetric Lanczos Process

Theorem 8.8.14 shows that the residuals of the CG process form orthogonal bases for Krylov subspaces: $\operatorname{span}\{r^{(0)}, \ldots, r^{(j-1)}\} = \mathcal{K}_j(A, r^{(0)})$, $j = 1, 2, 3, \ldots$. This suggests a connection between the CG algorithm and the Arnoldi process, which also generates orthogonal bases for Krylov subspaces. Recall that when A is real and symmetric, the Arnoldi process is called the *symmetric Lanczos process*. If we start the symmetric Lanczos process with q_1 equal to a multiple of $r^{(0)}$, then the vectors q_1, q_2, q_3, $\ldots$ that it produces will be proportional to $r^{(0)}$, $r^{(1)}$, $r^{(2)}$, $\ldots$, since an orthogonal basis for a sequence of nested spaces is uniquely determined up to scalar multiples. (This is essentially the same as the uniqueness of QR decompositions.) Thus the CG and Lanczos processes are producing essentially the same quantities. The connection between CG and symmetric Lanczos is explored further in Exercise 8.8.28. Because of this connection, it is possible to obtain information about the spectrum of A while running the CG algorithm. Of course, if a preconditioner is used, the information is about the eigenvalues of the transformed matrix $\tilde{A}$, not A.

Additional Exercises

Exercise 8.8.26 Prove Theorem 8.8.3. In other words, show that the proof of Theorem 3.5.15 remains valid in the more general context. $\square$

Exercise 8.8.27 Consider the step $x^{(k+1)} = x^{(k)} + \alpha_k p^{(k)}$ by an exact line search. Show that $\alpha_k p^{(k)}$ is the best approximation to $e^{(k)}$ from the one-dimensional space spanned by $p^{(k)}$. Then apply Theorem 8.8.3 to conclude that $e^{(k+1)} \perp_A p^{(k)}$. Thus Proposition 8.8.5 is a special case of Theorem 8.8.3. $\square$

Exercise 8.8.28 In this exercise you will take a look at the relationships between the CG and symmetric Lanczos algorithms. Referring to the prototype CG algorithm (8.8.24), the update $x^{(k+1)} \leftarrow x^{(k)} + \alpha_k p^{(k)}$ is of great interest to us since it brings us closer and closer to the true solution. Notice, however, that it is not essential to the functioning of the algorithm. Since $x^{(k)}$ is not used in the computation of the coefficients α_k and β_k, we could remove that line from the code, and it would still function perfectly, producing the sequences of $p^{(k)}$ and $r^{(k)}$ vectors. Thus the essential updates in (8.8.24) are $r^{(k+1)} = r^{(k)} + \alpha_k A p^{(k)}$ and $p^{(k+1)} = r^{(k+1)} + \beta_k p^{(k)}$. This is a pair of two-term recurrences. In contrast, the Lanczos algorithm runs a single three-term recurrence. In what follows we will view the vectors $p^{(k)}$ and $r^{(k)}$ as columns of matrices, so let us modify the notation, writing p_{k+1} and r_{k+1} in place of $p^{(k)}$ and $r^{(k)}$. Define $R_m = \begin{bmatrix} r_1 & \cdots & r_m \end{bmatrix}$, and define P_m analogously.

(a) Let q_1, q_2, q_3, $\ldots$ be the orthonormal vectors produced by the symmetric Lanczos process, starting with q_1, a multiple of $r_1 = r^{(0)}$. Let $Q_m = \begin{bmatrix} q_1 & \cdots & q_m \end{bmatrix}$. Show that $Q_m = R_m D_m$, where D_m is a nonsingular diagonal matrix, assuming all r_j are nonzero.

(b) Show that the recurrences $p^{(k+1)} = r^{(k+1)} + \beta_k p^{(k)}$, $k = 0, \ldots, m - 2$, together with $p^{(0)} = r^{(0)}$, can be rewritten as a matrix equation $Q_m = P_m U_m$, where U_m is a bidiagonal, upper-triangular matrix.

(c) Show that the recurrences $r^{(k+1)} = r^{(k)} + \alpha_k A p^{(k)}$, $k = 0, \ldots, m - 1$, can be rewritten as a matrix equation $AP_m = Q_m L_m + q_{m+1}\gamma_m e_m^T$, where L_m is bidiagonal and lower triangular, and γ_m is a scalar.

(c) Multiply the matrix equation from part (c) by U_m on the right to obtain the matrix equation

$$AQ_m = Q_m T_m + q_{m+1}\tilde{\gamma}_m e_m^T, \qquad (8.8.29)$$

where $T_m = L_m U_m$. Show that T_m is tridiagonal. (The equation $T_m = L_m U_m$ is an LU decomposition of T_m.)

(d) Show that T_m is symmetric. (Multiply (8.8.29) by Q_m^T on the left and use orthonormality of the columns of Q_m.)

Equation (8.8.29) is exactly the same as (6.4.20), which is the matrix statement of the three-term recurrence of the symmetric Lanczos process. Thus we have obtained the Lanczos recursion from the CG recursions. Conversely, we can obtain the CG recursions from the Lanczos recursion: starting from (8.8.29), we can perform an LU decomposition of T_m, introduce vectors P_m by $P_m = Q_m U_m^{-1}$, and arrive at a pair of two-term recurrences equivalent to the CG algorithm. $\square$

8.9 CONVERGENCE OF THE CG ALGORITHM

In this section we explore the convergence properties of the CG algorithm. All of the results given here are stated for the unpreconditioned algorithm. To get analogous results for the preconditioned CG method, just replace A by the transformed matrix $\tilde{A} = R^{-T} A R^{-1}$.

Theorem 8.9.1 *The CG algorithm, applied to an $n \times n$ positive definite system $Ax = b$, arrives at the exact solution in n or fewer steps.*

Proof. If we have taken $n - 1$ steps without getting to x, the nonzero residuals $r^{(0)}, \ldots, r^{(n-1)}$ form an orthogonal basis of $\mathbb{R}^n$, by Corollary 8.8.19. After n steps,

$$r^{(n)} \perp r^{(i)} \quad \text{for } i = 0, \ldots, n - 1.$$

Since $r^{(n)}$ is orthogonal to an entire basis of $\mathbb{R}^n$, we must have $r^{(n)} = 0$, $e^{(n)} = 0$, and $x^{(n)} = x$. $\square$

cise 8.9.2   Devise two other proofs of Theorem 8.9.1 based on (a) $e^{(j)} \perp_A p^{(i)}$, (b) minimization of energy norm. $\square$

Theorem 8.9.1 holds for any conjugate-direction method that performs exact line searches. At each step the dimension of the space in which the search is taking place is reduced by one because each new search direction is conjugate to all of the previous ones.

Theorem 8.9.1 is an interesting property, but it is not of immediate practical importance. Consider, for example, model problem (8.1.8) with $h = 1/160$. The dimension of the coefficient matrix is $159^2 = 25877$, so the CG algorithm (ignoring roundoff errors) is guaranteed to get to the exact solution in 25877 steps or fewer. In fact we would like to get close in a lot fewer steps than that, and Example 8.7.5 shows that we can. In that example, the CG algorithm with SSOR preconditioner was able to get a satisfactory approximation in about 50 iterations.

We will ignore the finite-termination property, by and large, and continue to view CG as an iterative method. Above all, we want to know how quickly the iterates get close to x. The following theorem gives some information along those lines.

Theorem 8.9.3 *The conjugate gradient errors* $e^{(j)} = x - x^{(j)}$ *satisfy*

$$\| e^{(j)} \|_A \le 2 \| e^{(0)} \|_A \left(\frac{\sqrt{\kappa_2(A)} - 1}{\sqrt{\kappa_2(A)} + 1} \right)^j.$$

A proof of Theorem 8.9.3 is outlined in Exercise 8.9.13. It is built on material that is yet to come in this section and the properties of Chebyshev polynomials developed in Exercise 6.4.38.

Theorem 8.9.3, like all rigorous error bounds, tends to be pessimistic because it is built by stringing together a number of estimates, some of which may be quite crude. Theorem 8.9.1 shows that for $j \ge n$ it is certainly very pessimistic. Of course, we normally expect to stop long before j approaches n.

Even though it is pessimistic, Theorem 8.9.3 does give us useful information about the performance of CG. The convergence rate is no worse than linear, with convergence ratio

$$\frac{\sqrt{\kappa_2(A)} - 1}{\sqrt{\kappa_2(A)} + 1}.$$

From this ratio it is clear that we can guarantee rapid convergence by making A well conditioned. This reinforces the idea, stated initially for the steepest descent algorithm, that performance can be improved by using a preconditioner that lowers the condition number of the matrix.

Example 8.9.4 Consider the matrix A of the model problem (8.1.8). In Exercise 8.3.37 you showed that $\kappa_2(A) = \cot^2(\pi h/2)$. Thus $\kappa_2(A) \approx \left(\frac{2}{\pi h} \right)^2$ and

$$\frac{\sqrt{\kappa_2(A)} - 1}{\sqrt{\kappa_2(A)} + 1} \approx \frac{1 - \pi h/2}{1 + \pi h/2} \approx 1 - \pi h.$$

To make a comparison with the SOR method, we refer to Table 8.5, which states that SOR with optimal ω has linear convergence with ratio $\rho(G) \approx 1 - 2\pi h$. This suggests that optimal SOR will converge in about half as many iterations as CG with no preconditioner. However, it is not fair to compare the two convergence rates too closely. The rate for SOR is precise, whereas the rate for CG is a crude upper bound. It would be fairer to say that, at worst, unpreconditioned CG converges at roughly the same rate as optimal SOR. This conclusion is consistent with what we observed in Examples 8.2.21 and 8.7.2.

We can expect CG to do much better if we use a preconditioner that effectively replaces A by a transformed matrix $\tilde{A}$ for which $\kappa_2(\tilde{A}) \ll \kappa_2(A)$. $\qquad$ $\square$

cise 8.9.5 Use Taylor expansions to verify the approximations made in Example 8.9.4.
$\qquad$ $\square$

The proof of Theorem 8.9.3 will make use of the fact that CG minimizes the energy norm of the error over subspaces of greater and greater dimension (Theorem 8.8.6). In the remainder of this section, we elaborate on this optimality property and derive some consequences that allow us to prove Theorem 8.9.3 and other useful properties of the CG method.

Theorem 8.9.6 *The iterates of the CG algorithm satisfy*

$$\| x - x^{(j)} \|_A = \min \left\{ \| x - y \|_A \mid y \in x^{(0)} + \mathcal{K}_j(A, r^{(0)}) \right\}.$$

Proof. This is basically a restatement of Theorem 8.8.6. Theorem 8.8.14 shows that the space $\mathcal{S}_j$ of Theorem 8.8.6 is none other than the Krylov subspace $\mathcal{K}_j(A, r^{(0)})$. Given $p \in \mathcal{S}_j$, let $y = x^{(0)} + p \in x^{(0)} + \mathcal{S}_j$. Then, since $x - y = e^{(0)} - p$, we see that minimizing $\| e^{(0)} - p \|_A$ over all p in the subspace $\mathcal{S}_j$ is the same as minimizing $\| x - y \|_A$ over all y in the shifted subspace $x^{(0)} + \mathcal{S}_j$. $\qquad$ $\square$

The Krylov subspace $\mathcal{K}_j(A, r^{(0)})$ is the set of all linear combinations of $r^{(0)}$, $Ar^{(0)}, \ldots, A^{j-1}r^{(0)}$. Thus a generic member of $\mathcal{K}_j(A, r^{(0)})$ has the form

$$c_0 r^{(0)} + c_1 Ar^{(0)} + \cdots + c_{j-1} A^{j-1} r^{(0)} = (c_0 I + c_1 A + \cdots c_{j-1} A^{j-1}) r^{(0)} = q(A) r^{(0)},$$

where $q(z) = c_0 + c_1 z + \cdots + c_{j-1} z^{j-1}$ is a polynomial of degree less than j. Conversely, given any polynomial q of degree less than j, the vector $q(A) r^{(0)}$ belongs to $\mathcal{K}_j(A, r^{(0)})$. Let $\mathcal{P}_{j-1}$ denote the space of all polynomials of degree less than j. Then

$$\mathcal{K}_j(A, r^{(0)}) = \left\{ q(A) r^{(0)} \mid q \in \mathcal{P}_{j-1} \right\}.$$

We have noted in Theorem 8.9.6 that $x^{(j)}$ is the optimal member of $x^{(0)} + \mathcal{K}_j(A, r^{(0)})$ in a certain sense. Let z denote an arbitrary element of $x^{(0)} + \mathcal{K}_j(A, r^{(0)})$. Then $z = x^{(0)} + q(A) r^{(0)}$ for some $q \in \mathcal{P}_{j-1}$. Viewing z as an approximation to x, we consider $e = x - z$. Then, recalling that $r^{(0)} = Ae^{(0)}$, we have

$$e = x - x^{(0)} - q(A) r^{(0)} = e^{(0)} - q(A) Ae^{(0)} = (I - q(A)A) e^{(0)} = p(A) e^{(0)},$$

where $p(z) = 1 - zq(z)$. Note that $p \in \mathcal{P}_j$ and $p(0) = 1$. Conversely, for any $p \in \mathcal{P}_j$ satisfying $p(0) = 1$, there exists $q \in \mathcal{P}_{j-1}$ such that $p(z) = 1 - zq(z)$. Thus $e = x - z$ for some $z \in x^{(0)} + \mathcal{K}_j(A, r^{(0)})$ if and only if $e = p(A) e^{(0)}$ for some $p \in \mathcal{P}_j$ satisfying $p(0) = 1$. This observation allows us to rewrite Theorem 8.9.6 as follows.

Theorem 8.9.7 *Let $x^{(j)}$ be the jth iterate of the CG method, and let $e^{(j)} = x - x^{(j)}$. Then*

$$\| e^{(j)} \|_A = \min_{\substack{p \in \mathcal{P}_j \\ p(0)=1}} \| p(A)e^{(0)} \|_A,$$

where $\mathcal{P}_j$ denotes the space of polynomials of degree less than or equal to j.

This theorem enables us to obtain a useful bound on $\| e^{(j)} \|_A$ in terms of the eigenvalues of A. To this end, let $v_1, \ldots, v_n$ be a complete orthonormal set of eigenvectors of A. Then $v_1, \ldots, v_n$ is a basis of $\mathbb{R}^n$, so

$$e^{(0)} = c_1 v_1 + c_2 v_2 + \cdots + c_n v_n,$$

where $c_1, \ldots, c_n$ are uniquely determined (but unknown) coefficients. Let $\lambda_1, \ldots, \lambda_n$ denote the eigenvalues associated with $v_1, \ldots, v_n$, respectively. Then $p(A)v_k = p(\lambda_k)v_k$ for all k, so

$$
\begin{aligned}
\| p(A)e^{(0)} \|_A^2 &= (p(A)e^{(0)})^T A p(A)e^{(0)} \\
&= \sum_{k=1}^{n} \sum_{l=1}^{n} c_k c_l p(\lambda_l) \lambda_k p(\lambda_k) v_l^T v_k \\
&= \sum_{k=1}^{n} c_k^2 \lambda_k p(\lambda_k)^2.
\end{aligned}
$$

Combining this expansion with Theorem 8.9.7, we obtain

$$\| e^{(j)} \|_A^2 = \min_{\substack{p \in \mathcal{P}_j \\ p(0)=1}} \sum_{k=1}^{n} c_k^2 \lambda_k p(\lambda_k)^2. \tag{8.9.8}$$

To get a result that does not make reference to the coefficients $c_1, \ldots, c_n$, notice that

$$\| e^{(0)} \|_A^2 = e^{(0)T} A e^{(0)} = \sum_{k=1}^{n} \lambda_k c_k^2,$$

and therefore

$$\sum_{k=1}^{n} c_k^2 \lambda_k p(\lambda_k)^2 \leq \max_k p(\lambda_k)^2 \| e^{(0)} \|_A^2.$$

Combining this observation with (8.9.8), we obtain the following theorem.

Theorem 8.9.9 *Let $e^{(j)} = x - x^{(j)}$ be the error after j iterations of the CG algorithm applied to a matrix A with eigenvalues $\lambda_1, \ldots, \lambda_n$. Then*

$$\| e^{(j)} \|_A \leq \left(\min_{\substack{p \in \mathcal{P}_j \\ p(0)=1}} \max_{1 \leq k \leq n} |p(\lambda_k)| \right) \| e^{(0)} \|_A,$$

where $\mathcal{P}_j$ denotes the space of polynomials of degree less than $j + 1$.

Theorem 8.9.3 follows from Theorem 8.9.9 by taking p to be an appropriately normalized Chebyshev polynomial. See Exercise 8.9.13.

If we divide both sides of the inequality in Theorem 8.9.9 by $\| e^{(0)} \|_A$, we obtain

$$\frac{\| e^{(j)} \|_A}{\| e^{(0)} \|_A} \leq \min_{\substack{p \in \mathcal{P}_j \\ p(0)=1}} \max_{k=1,\ldots,n} |p(\lambda_k)|. \tag{8.9.10}$$

The left-hand side is a ratio that measures how much better $x^{(j)}$ approximates x than $x^{(0)}$ does. In problems for which we have no *a priori* information about the solution, the initial guess $x^{(0)} = 0$ is as good as any. In this case we have $e^{(0)} = x$, and the left-hand side of (8.9.10) becomes $\| x - x^{(j)} \|_A / \| x \|_A$, which is the relative error in the approximation of x by $x^{(j)}$. Thus the relative error is bounded above by a quantity that depends only on the eigenvalues of A. This is the "no information" case. In cases where we have some *a priori* information about x, we can usually choose an $x^{(0)}$ for which $\| e^{(0)} \|_A < \| x \|_A$, thereby obtaining a better result.

Theorem 8.9.9 leads to yet another proof of the finite termination property of CG. Taking $j = n$ we see that for any $p \in \mathcal{P}_n$ with $p(0) = 1$ we have

$$\| e^{(n)} \|_A \leq \left(\max_{1 \leq k \leq n} |p(\lambda_k)| \right) \| e^{(0)} \|_A. \tag{8.9.11}$$

Now let $p(z) = c(z - \lambda_1) \cdots (z - \lambda_n) \in \mathcal{P}_n$, where c is chosen that $p(0) = 1$. (This is possible because all of the λ_j are nonzero.) For this p the right-hand side of the (8.9.11) is zero, so $e^{(n)} = 0$, i.e. $x^{(n)} = x$.

The finite termination property can be sharpened in the following way.

Theorem 8.9.12 *Suppose A has only j distinct eigenvalues. Then the CG algorithm terminates with the exact solution in at most j steps.*

Proof. Let $\mu_1, \ldots, \mu_j$ be the distinct eigenvalues of A. Let $p(z) = c(z - \mu_1) \cdots (z - \mu_j)$, where c is chosen so that $p(0) = 1$. Then $p \in \mathcal{P}_j$ and $p(\lambda_k) = 0$ for $k = 1, \ldots, n$, so $e^{(j)} = 0$ by Theorem 8.9.9. $\qquad \square$

Thus termination will take place quite early if the matrix that has only a few distinct eigenvalues of high multiplicity. More generally, consider a matrix whose eigenvalues are located in j tight clusters, where $j \ll n$. Choose $p(z) = c(z - \mu_1)(z - \mu_2) \cdots (z - \mu_j)$, where $\mu_1, \ldots, \mu_j$ are representatives of the j distinct clusters. Since every eigenvalue λ_k is close to one of the μ_m, and every $p(\mu_m)$ is zero, $\max_k |p(\lambda_k)|$ will be small. Therefore, by Theorem 8.9.9, $\| e^{(j)} \|_A$ will be small. We conclude that if A is a matrix whose eigenvalues lie in j tight clusters, then x_j will be close to x.

A preconditioner that makes the condition number small is actually pushing the eigenvalues into a single cluster, as the condition number is just the ratio λ_n/λ_1. Now our discussion shows that a single cluster is not really needed. A preconditioner that can push the eigenvalues into a few clusters will also be effective. In particular, if the eigenvalues are pushed into one cluster with a few outliers, this is no worse than having a few clusters.

Additional Exercises

Exercise 8.9.13 In this exercise you will prove Theorem 8.9.3. Let the eigenvalues of A be numbered in the order $\lambda_1 \leq \cdots \leq \lambda_n$. From Theorem 8.9.9 we see that we can show that $\| e^{(j)} \|_A$ is small by finding a polynomial $p_j \in \mathcal{P}_j$ for which $p_j(0) = 1$, and $p_j(\lambda_1), p_j(\lambda_2), \ldots, p_j(\lambda_n)$ are all small. One way to ensure that all of the $p_j(\lambda_m)$ are small is to make p_j small on the entire interval $[\lambda_1, \lambda_n]$. The best tool for this task is a shifted, scaled Chebyshev polynomial.

(a) Determine the affine transformation $w = \alpha x + \beta$ that maps $[\lambda_1, \lambda_n]$ onto $[-1, 1]$, with $\lambda_n \mapsto -1$ and $\lambda_1 \mapsto 1$. Show that this transformation maps $0 \mapsto (\lambda_n + \lambda_1)/(\lambda_n - \lambda_1)$.

(b) Let
$$
p_j(x) = \frac{T_j\left(\frac{\lambda_n + \lambda_1 - 2x}{\lambda_n - \lambda_1}\right)}{T_j\left(\frac{\lambda_n + \lambda_1}{\lambda_n - \lambda_1}\right)},
$$
where T_j is the jth Chebyshev polynomial (Exercise 6.4.38). Then evidently $p_j \in \mathcal{P}_j$ and $p_j(0) = 1$. Show that
$$
\max_{x \in [\lambda_1, \lambda_n]} |p_j(x)| = \frac{1}{T_j\left(\frac{\lambda_n + \lambda_1}{\lambda_n - \lambda_1}\right)},
$$
and
$$
\frac{\lambda_n + \lambda_1}{\lambda_n - \lambda_1} = \frac{\kappa_2(A) + 1}{\kappa_2(A) - 1}.
$$

(c) Show that if $t > 0$ and $x = \cosh t > 1$, then $e^t = \cosh t + \sinh t = x + \sqrt{x^2 - 1}$. Show further that if $x = (\kappa_2(A) + 1)/(\kappa_2(A) - 1)$, then $e^t = (\sqrt{\kappa_2(A)} + 1)/(\sqrt{\kappa_2(A)} - 1)$.

(d) In part (f) of Exercise 6.4.38 we showed that if $x > 1$, and $t > 0$ is such that $x = \cosh t$, then $T_j(x) \geq \frac{1}{2}(e^t)^j$ for all j. Use this bound with $x = (\kappa_2(A) + 1)/(\kappa_2(A) - 1)$, together with the identity from part (c), to show that (for this particular x)
$$
T_j(x) \geq \frac{1}{2}\left(\frac{\sqrt{\kappa_2(A)} + 1}{\sqrt{\kappa_2(A)} - 1}\right)^j.
$$

(e) Put the pieces together to prove Theorem 8.9.3.

□

Exercise 8.9.14 In this exercise you will show that the convergence rate of the CG method is sometimes much better than Theorem 8.9.3 would indicate. Before working this exercise, you need to have worked Exercise 8.9.13. Suppose the spectrum of A

consists of a tightly packed cluster of eigenvalues plus a few outliers. Let $\lambda_1, \ldots,$ λ_i denote the outliers, and suppose the remaining eigenvalues are ordered so that $\lambda_{i+1} \leq \cdots \leq \lambda_n$. Define a "reduced condition number" $\hat{\kappa} = \lambda_n/\lambda_{i+1}$. This is the condition number of the operator obtained by restricting A to the invariant subspace associated with the non-outlier eigenvalues. If the furthest outlier is far from the tightly packed core, $\hat{\kappa}$ will be significantly smaller than $\kappa_2(A)$. Let $q_i(x) = c(x - \lambda_1)(x - \lambda_2) \cdots (x - \lambda_i)$, where c is chosen so that $q_i(0) = 1$. For each j define

$$p_{i+j}(x) = q_i(x)\frac{T_j\left(\frac{\lambda_n+\lambda_{i+1}-2x}{\lambda_n-\lambda_{i+1}}\right)}{T_j\left(\frac{\lambda_n+\lambda_{i+1}}{\lambda_n-\lambda_{i+1}}\right)}.$$

Then evidently $p_{i+j} \in \mathcal{P}_{i+j}$, $p_{i+j}(0) = 1$, $p_{i+j}(\lambda_k) = 0$ for $k = 1, \ldots, i$, and $|p_{i+j}|$ is small on $[\lambda_{i+1}, \lambda_n]$. Let $C = \max\limits_{x \in [\lambda_{i+1}, \lambda_n]} |q_i(x)|$. Drawing on techniques from Exercise 8.9.13, prove that

$$\left\| e^{(i+j)} \right\|_A \leq 2C \left\| e^{(0)} \right\|_A \left(\frac{\sqrt{\hat{\kappa}} - 1}{\sqrt{\hat{\kappa}} + 1}\right)^j.$$

This shows that after the ith iteration, the convergence rate is governed by $\hat{\kappa}$ instead of $\kappa_2(A)$. Thus we have much faster convergence than we would have expected from looking at $\kappa_2(A)$ alone. $\square$

·cise 8.9.15 Show that the iterates of the preconditioned CG algorithm with preconditioner M satisfy

$$\begin{aligned}
\left\| x - x^{(j)} \right\|_A &= \min\left\{ \left\| x - y \right\|_A \mid y \in x^{(0)} + \mathcal{K}_j(M^{-1}A, s^{(0)}) \right\} \\
&= \min\limits_{\substack{p \in \mathcal{P}_j \\ p(0)=1}} \left\| p(M^{-1}A)e^{(0)} \right\|_A,
\end{aligned}$$

where $s^{(0)} = M^{-1}r^{(0)}$. $\square$

8.10 KRYLOV SUBSPACE METHODS FOR INDEFINITE AND NONSYMMETRIC PROBLEMS

The CG algorithm is one of a large class of algorithms known as *Krylov subspace methods* because they generate sequences of Krylov subspaces. CG was designed for symmetric, positive-definite matrices. For other types of matrices other Krylov subspace methods have been developed. In this section we will give a brief overview of some of the more popular methods. For more information see [6], [33], [45], or [46].

Let us begin with symmetric, indefinite matrices. First of all, even though the CG algorithm was designed for positive-definite matrices, it is possible to run CG on indefinite problems, often successfully. However, the algorithm can occasionally break down, since the computation of α_k in (8.8.24) requires division by $\langle Ap^{(k)}, p^{(k)} \rangle$, which can be zero if A is indefinite.

It is natural to ask whether there are algorithms for symmetric indefinite matrices that are like CG but do not break down. Fortunately the answer is yes. The symmetric Lanczos process (6.4.23), which generates a sequence of Krylov subspaces via the economical 3-term recurrence

$$q_{k+1}\beta_k = Aq_k - q_k\alpha_k - q_{l-1}\beta_{k-1},$$

works for all symmetric matrices, definite or not.[49] Thus we can use the Lanczos process to generate the Krylov subspaces.

But now there is a second question. How do we pick an appropriate approximation from the Krylov subspace? CG minimizes $\| e^{(j)} \|_A$, the energy norm of the error. Since $r^{(j)} = Ae^{(j)}$, we have $\| e^{(j)} \|_A^2 = e^{(j)T} Ae^{(j)} = r^{(j)T} A^{-1} r^{(j)} = \| r^{(j)} \|_{A^{-1}}^2$, so CG is also minimizing the A^{-1}-norm of the residual. In the indefinite case, neither $\| x \|_A$ nor $\| x \|_{A^{-1}}$ is meaningful. The definitions $\| x \|_A = \sqrt{\langle Ax, x \rangle}$ and $\| x \|_{A^{-1}} = \sqrt{\langle A^{-1}x, x \rangle}$ make no sense, since the quantities under the square root signs can be negative. Thus we need another criterion for choosing an approximate solution. The simplest idea is to choose $x^{(j)}$ so that $\| r^{(j)} \|_2$, the Euclidean norm of the residual, is minimized. Just as for CG, the minimization is taken over the translated Krylov subspace $x^{(0)} + \mathcal{K}_j(A, r^{(0)})$. The resulting algorithm is called MINRES (MINimum RESidual) [68].

This procedure can also be carried out for nonsymmetric problems. In this case the Krylov subspaces have to be generated by the Arnoldi process (6.4.6) instead of the symmetric Lanczos process. The algorithm that does this, choosing $x^{(j)}$ to minimize $\| r^{(j)} \|_2$ at each step, is called GMRES (Generalized Minimum RESidual) [77]. The Arnoldi process does not use short recurrences. We see from (6.4.3) that the computation of each new vector requires all of the previous vectors. Thus with each step of the algorithm, storage space for one more vector is needed. The amount of arithmetic per step also increases linearly. These considerations place severe restrictions on the number of GMRES iterations if A is extremely large. Therefore it is a common practice to run GMRES in "restart mode." GMRES(m) is the variant of GMRES that restarts the Arnoldi process every m iterations, using the most recent iterate as the new initial guess. In this mode the storage requirement is held to $O(m)$ vectors. Unfortunately convergence is not guaranteed. It is difficult to know in advance how big m needs to be in order to ensure success. In spite of its shortcomings, GMRES has been very popular.

The algorithms that we have discussed so far choose the new iterate at each step by a minimization criterion. Other algorithms use a criterion based on orthogonalization. Recall that the CG algorithm chooses $x^{(j)}$ so that the error $e^{(j)}$ is orthogonal

---

[49]In Exercise 8.8.28 the connection between the Lanczos and CG recursions was discussed. It was observed that to get from the Lanczos to the CG recursion an LU decomposition of the tridiagonal matrix T is needed. In the positive definite case, the LU decomposition always exists, but in the indefinite case it may not. The CG algorithm breaks down in exactly those situations where the LU decomposition (without pivoting) fails to exist.

to $\mathcal{K}_j(A, r^{(0)})$ in the energy inner product, that is, $e^{(j)} \perp_A \mathcal{K}_j(A, r^{(0)})$ (Corollary 8.8.18). This is equivalent to the residual being orthogonal to $\mathcal{K}_j(A, r^{(0)})$ in the standard inner product: $r^{(j)} \perp \mathcal{K}_j(A, r^{(0)})$ (Corollary 8.8.19).

For CG (with positive definite A) the orthogonality property is linked directly to the minimality of $\| e^{(j)} \|_A$ and $\| r^{(j)} \|_{A^{-1}}$ via Theorem 8.8.3, using the energy inner product. In the indefinite and nonsymmetric cases this connection breaks down, since we no longer have the energy inner product at our disposal. This does not stop us from building an algorithm that chooses $x^{(j)}$ so that the *Galerkin condition*

$$r^{(j)} \perp \mathcal{K}_j(A, r^{(0)}) \tag{8.10.1}$$

is satisfied. The algorithm SYMMLQ [68] operates on symmetric indefinite matrices, generates Krylov subspaces by the symmetric Lanczos process, and chooses $x^{(j)}$ at each step in such a way that (8.10.1) is satisfied. This criterion does not correspond to the minimization of $r^{(j)}$ in some norm, so SYMMLQ is different from MINRES. The algorithm can break down occasionally, because it can happen that (8.10.1) has either no solution or infinitely many solutions. Breakdown at one step does not preclude moving on to the next step, since the underlying symmetric Lanczos process does not break down. However, MINRES is generally preferred over SYMMLQ.

Another question that arises naturally is this. In the nonsymmetric case, are there Krylov subspace methods that use short recurrences, thereby circumventing the storage and execution time difficulties of GMRES? Perhaps we can build such a method based on an orthogonality criterion like (8.10.1) instead of a minimization criterion. It turns out that we can, but we have to give up on orthogonality and settle for the weaker condition known as biorthogonality. The biconjugate-gradient (BiCG) algorithm is a generalization of the CG algorithm that generates sequences of dual vectors $\tilde{r}^{(0)}, \tilde{r}^{(1)}, \tilde{r}^{(2)}, \ldots$ and $\tilde{p}^{(0)}, \tilde{p}^{(1)}, \tilde{p}^{(2)}, \ldots$ along with the primal vectors.

Prototype BiCG Algorithm

$x^{(0)} \leftarrow$ initial guess
$r^{(0)} \leftarrow b - Ax^{(0)}$
$p^{(0)} \leftarrow r^{(0)}$
$\tilde{r}^{(0)} \leftarrow$ user's choice s.t. $\langle r^{(0)}, \tilde{r}^{(0)} \rangle \neq 0$
$\tilde{p}^{(0)} \leftarrow \tilde{r}^{(0)}$
for $k = 0, 1, 2, \ldots$

$$
\begin{aligned}
&\alpha_k \leftarrow \langle r^{(k)}, \tilde{r}^{(k)} \rangle / \langle Ap^{(k)}, \tilde{p}^{(k)} \rangle \qquad\qquad (8.10.2)\\
&x^{(k+1)} \leftarrow x^{(k)} + \alpha_k p^{(k)}\\
&r^{(k+1)} \leftarrow r^{(k)} - \alpha_k Ap^{(k)}\\
&\tilde{r}^{(k+1)} \leftarrow \tilde{r}^{(k)} - \alpha_k A^T \tilde{p}^{(k)}\\
&\beta_k \leftarrow \langle r^{(k+1)}, \tilde{r}^{(k+1)} \rangle / \langle r^{(k)}, \tilde{r}^{(k)} \rangle\\
&p^{(k+1)} \leftarrow r^{(k+1)} + \beta_k p^{(k)}\\
&\tilde{p}^{(k+1)} \leftarrow \tilde{r}^{(k+1)} + \beta_k \tilde{p}^{(k)}
\end{aligned}
$$

Comparing (8.10.2) with (8.8.24), we see that they are almost identical, except that (8.10.2) has extra recurrences to generate the dual vectors. If A is symmetric and $\tilde{r}^{(0)} = r^{(0)}$, then $\tilde{r}^{(k)} = r^{(k)}$ and $\tilde{p}^{(k)} = p^{(k)}$ for all k, and (8.10.2) collapses to (8.8.24).

The vectors generated by (8.10.2) satisfy several useful relationships, which are proved in Exercise 8.10.6. Generalizing Theorem 8.8.14 we have

$$\text{span}\{p^{(0)}, \ldots, p^{(j)}\} = \text{span}\{r^{(0)}, \ldots, r^{(j)}\} = \mathcal{K}_{j+1}(A, r^{(0)})$$

and the dual relationship

$$\text{span}\{\tilde{p}^{(0)}, \ldots, \tilde{p}^{(j)}\} = \text{span}\{\tilde{r}^{(0)}, \ldots, \tilde{r}^{(j)}\} = \mathcal{K}_{j+1}(A^T, \tilde{r}^{(0)})$$

for all j. In addition the $r^{(j)}$ sequence is biorthogonal to the $\tilde{r}^{(j)}$ sequence, that is, $\langle r^{(j)}, \tilde{r}^{(k)} \rangle = 0$ if $j \neq k$. Normally we have $\langle r^{(k)}, \tilde{r}^{(k)} \rangle \neq 0$. If not, the algorithm breaks down, since it then becomes impossible to compute β_k. Restating the biorthogonality conditions in terms of subspaces, we have the *Petrov-Galerkin* conditions

$$r^{(j)} \perp \mathcal{K}_j(A^T, \tilde{r}^{(0)}) \quad \text{and} \quad \tilde{r}^{(j)} \perp \mathcal{K}_j(A, r^{(0)}),$$

generalizing Corollary 8.8.19. The $p^{(j)}$ and $\tilde{p}^j$ sequences also satisfy a sort of biorthogonality condition, namely $\langle Ap^{(j)}, p^{\widetilde{(k)}} \rangle = \langle p^{(j)}, A^T p^{\widetilde{(k)}} \rangle = 0$ if $j \neq k$. Normally $\langle Ap^{(k)}, p^{\widetilde{(k)}} \rangle \neq 0$. If not, the algorithm breaks down, since it is then impossible to compute α_k.

The convergence of BiCG is sometimes erratic. A number of improvements have been proposed. The algorithm QMR (Quasi Minimum Residual) [34, 35] uses essentially the BiCG recurrences to generate the Krylov subspaces but uses a criterion for choosing the approximate solution $x^{(j)}$ that is based on a quasi-minimization of the residual. Because the recurrences can sometimes break down, QMR is also sometimes used with a "look ahead" feature that circumvents breakdowns.

A feature of BiCG and QMR that is sometimes a drawback is that each step requires a matrix-vector multiplication by A^T. In some applications the definition of A is so complex (a long subroutine) that it is not clear how to compute A^T. In such situations one needs an algorithm that does not use A^T. One such algorithm is BiCGSTAB (BiCG STABilized) [94], which replaces the A^T evaluation by a second A evaluation and also offers much smoother convergence. There is also a transpose-free version of QMR. Another popular algorithm of this type is called CGS (CG Squared) [82].

The algorithms BiCG, QMR, BiCGSTAB, and CGS are all based on short recurrences, so the memory and computation requirements per step do not increase as the algorithm proceeds, in contrast to GMRES.

All of the methods mentioned in this section can be enhanced by the use of preconditioners. In the nonsymmetric case there is no need for a symmetric preconditioner. Incomplete LU decompositions are popular, but there is no end to the variety of preconditioners that have been tried.

Of the many iterative methods for solving nonsymmetric systems, there is no clear best choice.

MATLAB provides implementations of GMRES, BiCG, BiCGSTAB, CGS, and QMR (without "look ahead"). These are not the most efficient implementations, but they are convenient.

cise 8.10.3 Using the MATLAB m-file condif.m shown in Exercise 1.9.3, generate a matrix A = condif(n,c). Take $n = 50$ (or larger) and c = [0.2 5]. This 2500×2500 nonsymmetric sparse matrix is a discretized convection-diffusion operator. Try out some of MATLAB's iterative solvers on this matrix. For example, try

```
n = 50; c = [.2 5]; A = condif(n,c);
nn = n^2; sol = randn(nn,1); % random solution vector
b = A*sol;                    % right-hand side vector
tol = 1e-12; maxit = 1000;
x = bicg(A,b,tol,maxit)
error = norm(x-sol)
```

Type help bicg for more information on how to use BiCG. After experimenting with BiCG, try QMR, BiCGSTAB, CGS, and GMRES.   □

cise 8.10.4 MATLAB provides a routine luinc that computes incomplete LU decompositions to use as preconditioners. Repeat Exercise 8.10.3 using a preconditioner. For example, try

```
droptol = 1e-1;
[M1,M2] = luinc(A,droptol);
spy(A)
spy(M1+M2)
x = bicg(A,b,tol,maxit,M1,M2)
error = norm(x-sol)
```

Try the various iterative methods. How does the preconditioner affect your results? Try a variety of drop tolerances. How does varying the drop tolerance affect the sparseness of M1 and M2 and the effectiveness of the preconditioner?   □

cise 8.10.5 Read more about MATLAB's iterative solvers. Figure out how to make them produce (residual) convergence histories. Then print out some convergence histories on semilog plots. For example, try

```
[x,flag,relres,iter,resvec1] = bicg(A,b,tol,maxit,M1,M2);
x1 = 1:size(resvec1,1);
semilogy(x1,resvec1,'b-'); % makes a blue solid line
```

Try comparing the convergence histories of BiCG, QMR, and GMRES, for example, in a single plot.   □

cise 8.10.6 Consider the BiCG algorithm (8.10.2) with the line $p^{(k+1)} \leftarrow r^{(k+1)} + \beta_k p^{(k)}$ replaced by an expression of the form (8.8.7) and the dual line $\tilde{p}^{(k+1)} \leftarrow \tilde{r}^{(k+1)} + \beta_k \tilde{p}^{(k)}$ replaced by an analogous expression with coefficients $\tilde{c}_{ki}$.

(a) Suppose the biorthogonality conditions $\langle Ap^{(i)}, \tilde{p}^{(j)} \rangle = \langle p^{(i)}, A^T \tilde{p}^{(j)} \rangle = 0$ iff $i \neq j$ hold for $i \leq k$ and $j \leq k$. Show that $\langle Ap^{(k+1)}, \tilde{p}^{(j)} \rangle = 0$ and $\langle p^{(j)}, A^T \tilde{p}^{(k+1)} \rangle = 0$ for $j < k+1$ if and only if

$$c_{ki} = \frac{\langle Ar^{(k+1)}, \tilde{p}^{(i)} \rangle}{\langle Ap^{(i)}, \tilde{p}^{(i)} \rangle} \quad \text{and} \quad \tilde{c}_{ki} = \frac{\langle p^{(i)}, A^T \tilde{r}^{(k+1)} \rangle}{\langle p^{(i)}, A^T \tilde{p}^{(i)} \rangle}, \quad i = 0, \ldots, k.$$

$$(8.10.7)$$

Thus we will use these values for c_{ki} and $\tilde{c}_{ki}$ from this point on, to ensure biorthogonality.

(b) Extending Theorem 8.8.14, prove by induction on j that

$$\text{span}\{p^{(0)},\dots,p^{(j)}\} = \text{span}\{r^{(0)},\dots,r^{(j)}\} = \mathcal{K}_{j+1}(A,r^{(0)})$$

and

$$\text{span}\{\tilde{p}^{(0)},\dots,\tilde{p}^{(j)}\} = \text{span}\{\tilde{r}^{(0)},\dots,\tilde{r}^{(j)}\} = \mathcal{K}_{j+1}(A^T,\tilde{r}^{(0)})$$

so long as $\langle Ap^{(j)},\tilde{p}^{(j)}\rangle \neq 0$.

(c) Using the recurrences for $r^{(k+1)}$ and $\tilde{r}^{(k+1)}$ from (8.10.2), the biorthogonality from part (a), the subspace equalities from part (b), and the definition of α_k from (8.10.2), prove by induction that $\langle r^{(i)},\tilde{r}^{(j)}\rangle = 0$ if $i \neq j$.

(d) Using arguments like those in the proof of Theorem 8.8.20, show that $c_{ki} = 0$ and $\tilde{c}_{ki} = 0$ if $i < k$.

(e) Show that $c_{kk} = \tilde{c}_{kk} = -\beta_k$, where β_k is as defined in (8.10.2). This completes the justification of (8.10.2).

$\square$

References

1. E. Anderson et al. *LAPACK Users' Guide.* SIAM, Philadelphia, Third edition, 1999. www.netlib.org/lapack/lug/. (Cited on pp. 11, 103, 132, 206, 213, 480, 498, 501, 504.)

2. W. E. Arnoldi. The principle of minimized iterations in the solution of the matrix eigenvalue problem. *Quart. Appl. Math.,* 9:17–29, 1951. (Cited on p. 439.)

3. Z. Bai and J. W. Demmel. On swapping blocks in real Schur form. *Linear Algebra Appl.,* 186:73–95, 1993. (Cited on p. 415.)

4. Z. Bai, J. W. Demmel, J. J. Dongarra, A. Ruhe, and H. van der Vorst, editors. *Templates for the Solution of Algebraic Eigenvalue Problems: A Practical Guide.* SIAM, Philadelphia, 2000. (Cited on pp. 456, 468.)

5. Z. Bai, J. W. Demmel, and M. Gu. Inverse free parallel spectral divide and conquer algorithms for nonsymmetric eigenproblems. *Numer. Math,* 76:279–308, 1997. (Cited on p. 414.)

6. R. Barrett et al. *Templates for the Solution of Linear Systems: Building Blocks for Iterative Methods.* SIAM, Philadelphia, 1994. (Cited on p. 621.)

7. G. A. Birkhoff and R. E. Lynch. *Numerical Solution of Elliptic Problems.* SIAM Studies in Applied Mathematics. SIAM, Philadelphia, 1984. (Cited on p. 553.)

8. A. Björck. *Numerical Methods for Least Squares Problems.* SIAM, Philadelphia, 1996. (Cited on pp. 232, 278, 279, 284.)

Fundamentals of Matrix Computations, Third Edition. By David S. Watkins
Copyright © 2010 John Wiley & Sons, Inc.

9. R. H. Chan, C. Greif, and D. P. O'Leary, editors. *Milestones in Matrix Computation: Selected Works of Gene H. Golub, with Commentaries.* Oxford University Press, 2007. (Cited on p. 398.)

10. T. F. Chan and T. P. Mathew. Domain decomposition algorithms. *Acta Numerica*, pages 61–143, 1994. (Cited on pp. 573, 601.)

11. P. Concus, G. H. Golub, and D. P. O'Leary. A generalized conjugate gradient method for the numerical solution of elliptic partial differential equations. In J. R. Bunch and D. J. Rose, editors, *Sparse Matrix Computations*, New York, 1976. Academic Press. (Cited on p. 604.)

12. J. K. Cullum and R. A. Willoughby. *Lanczos Algorithms for Large Symmetric Eigenvalue Computations.* Birkhaüser, Boston, 1985. (Cited on p. 447.)

13. J. J. M. Cuppen. A divide and conquer method for the symmetric tridiagonal eigenproblem,. *Numer. Math.*, 36:177–195, 1981. (Cited on p. 501.)

14. T. A. Davis. *Direct Methods for Sparse Linear Systems.* Fundamentals of Algorithms. SIAM, Philadelphia, 2006. (Cited on pp. 55, 65, 65, 71, 108.)

15. J. W. Demmel. *Applied Numerical Linear Algebra.* SIAM, Philadelphia, 1997. (Cited on pp. ix, 141, 141, 472.)

16. J. W. Demmel and B. Kågström. The generalized Schur decomposition of an arbitrary pencil $A - \lambda B$: robust software with error bounds and applications. *ACM Trans. Math. Software*, 19:160–201, 1993. (Cited on p. 532.)

17. J. W. Demmel and W. M. Kahan. Accurate singular values of bidiagonal matrices,. *SIAM J. Sci. Stat. Comput.*, 11:873–912, 1990. (Cited on pp. 406, 490.)

18. J. W. Demmel and K. Veselić. Jacobi's method is more accurate than QR. *SIAM J. Matrix Anal. Appl.*, 13:1204–1246, 1992. (Cited on p. 488.)

19. I. S. Dhillon. *A New $O(n^2)$ Algorithm for the Symmetric Tridiagonal Eigenvalue/Eigenvector Problem.* PhD thesis, University of California, Berkeley, 1997. (Cited on p. 501.)

20. I. S. Dhillon and B. N. Parlett. Orthogonal eigenvectors and relative gaps. LAPACK Working Note 154. Submitted to SIAM J. Matrix Anal. Appl., 2000. (Cited on pp. 500, 501.)

21. J. J. Dongarra, J. R. Bunch, C. B. Moler, and G. W. Stewart. *LINPACK Users' Guide.* SIAM, Philadelphia, 1979. (Cited on p. xv.)

22. J. J. Dongarra, F. G. Gustavson, and A. Karp. Implementing linear algebra algorithms for dense matrices on a vector pipeline machine. *SIAM Review*, 26:91–112, 1984. (Cited on p. 92.)

23. J. J. Dongarra and D. C. Sorensen. A fully parallel algorithm for the symmetric eigenvalue problem. *SIAM J. Sci. Stat. Comput.*, 8:s139–s154, 1987. (Cited on p. 501.)

24. Z. Drmač and K. Veselić. New fast and accurate Jacobi SVD algorithm. I. *SIAM J. Matrix Anal. Appl.*, 29:1322–1342, 2008. (Cited on p. 394.)

25. Z. Drmač and K. Veselić. New fast and accurate Jacobi SVD algorithm. II. *SIAM J. Matrix Anal. Appl.*, 29:1343–1362, 2008. (Cited on p. 394.)

26. I. S. Duff, A. M. Erisman, and J. K. Reid. *Direct Methods for Sparse Matrices.* Oxford University Press, 1986. (Cited on p. 108.)

27. K. V. Fernando and B. N. Parlett. Accurate singular values and differential qd algorithms. *Numer. Math.*, 67:191–229, 1994. (Cited on p. 497.)

28. G. E. Forsythe and P. Henrici. The cyclic Jacobi method for computing the principal values of a complex matrix. *Trans. Amer. Math. Soc.*, 94:1–23, 1960. (Cited on p. 488.)

29. G. E. Forsythe and C. B. Moler. *Computer Solution of Linear Algebraic Systems.* Prentice-Hall, Englewood Cliffs, NJ, 1967. (Cited on p. xv.)

30. L. V. Foster. Gaussian elimination with partial pivoting can fail in practice. *SIAM J. Matrix Anal. Appl.*, 15:1354–1362, 1994. (Cited on p. 164.)

31. J. G. F. Francis. The QR transformation, part I. *Computer J.*, 4:265–272, 1961. (Cited on pp. 434, 497.)

32. J. G. F. Francis. The QR transformation, part II. *Computer J.*, 4:332–345, 1961. (Cited on pp. 358, 497.)

33. R. W. Freund, G. H. Golub, and N. M. Nachtigal. Iterative solution of linear systems. *Acta Numerica*, 1:57–100, 1992. (Cited on p. 621.)

34. R. W. Freund and N. M. Nachtigal. QMR: A quasi-minimal residual method for non-hermitian linear systems. *Numer. Math.*, 60:315–339, 1991. (Cited on p. 624.)

35. R. W. Freund and N. M. Nachtigal. An implementation of the QMR method based on coupled two-term recurrences. *SIAM J. Sci. Comput.*, 15:313–337, 1994. (Cited on p. 624.)

36. F. R. Gantmacher. *The Theory of Matrices.* Chelsea Publishing Co., New York, 1959. (Cited on pp. 335, 532.)

37. A. George and J. W. Liu. *Computer Solution of Large Sparse Positive Definite Systems.* Prentice-Hall, Englewood Cliffs, NJ, 1981. (Cited on pp. xv, 58, 61, 65, 71, 552, 606.)

38. J. R. Gilbert, C. Moler, and R. Schreiber. Sparse matrices in MATLAB: Design and implementation. *SIAM J. Matrix Anal. Appl*, 13:333–356, 1992. (Cited on p. 55.)

39. L. Giraud, J. Langou, M. Rozložník, and J. van den Eshof. Rounding error analysis of the classical gram-schmidt orthogonalization process. *Numer. Math.*, 101:87–100, 2005. (Cited on p. 233.)

40. G. H. Golub. Numerical methods for solving linear least squares problems. *Numer. Math.*, 7:206–216, 1965. (Cited on p. 215.)

41. G. H. Golub and W. Kahan. Calculating the singular values and pseudo-inverse of a matrix. *SIAM J. Numer. Anal.*, 2:205–224, 1965. (Cited on p. 395.)

42. G. H. Golub and C. Reinsch. Singular value decomposition and least squares solutions. *Numer. Math.*, 14:403–420, 1970. Also published as contribution I/10 in [105]. (Cited on pp. 398, 405.)

43. G. H. Golub and C. F. Van Loan. *Matrix Computations.* Johns Hopkins University Press, Baltimore, Third edition, 1996. (Cited on pp. ix, xii, 33, 87, 107, 196, 252, 272, 422.)

44. R. Granat, B. Kågström, and D. Kressner. A novel parallel QR algorithm for hybrid distributed memory HPC systems. Technical Report 2009–15, Seminar for Applied Mathematics, ETH, April 2009. (Cited on pp. 373, 374.)

45. A. Greenbaum. *Iterative Methods for Solving Linear Systems*. SIAM, Philadelphia, 1997. (Cited on p. 621.)

46. M. H. Gutknecht. Lanczos-type solvers for nonsymmetric linear systems of equations. *Acta Numerica*, 6:271–397, 1997. (Cited on p. 621.)

47. W. Hackbusch. *Iterative Solution of Large, Sparse Systems of Equations*, volume 95 of *Applied Mathematical Sciences*. Springer-Verlag, 1994. (Cited on pp. 571, 573, 573, 601.)

48. C. R. Hadlock. *Field Theory and Its Classical Problems*. The Carus Mathematical Monographs. Mathematical Association of America, 1978. (Cited on p. 310.)

49. L. A. Hageman and D. M. Young. *Applied Iterative Methods*. Academic Press, New York, 1981. (Cited on pp. 571, 573, 574, 574.)

50. M. R. Hestenes and E. Stiefel. Methods of conjugate gradients for solving linear systems. *J. Res. Nat. Bur. Standards*, 49:409–436, 1952. (Cited on p. 602.)

51. D. J. Higham and N. J. Higham. *MATLAB Guide*. SIAM, Philadelphia, 2000. (Cited on p. x.)

52. N. J. Higham. *Accuracy and Stability of Numerical Algorithms*. SIAM, Philadelphia, Second edition, 2002. (Cited on pp. 12, 114, 131, 132, 141, 141, 150, 160, 180, 180, 208, 232, 284.)

53. R. A. Horn and C. A. Johnson. *Matrix Analysis*. Cambridge University Press, 1985. (Cited on pp. 332, 335, 577.)

54. A. S. Householder. *The Theory of Matrices in Numerical Analysis*. Blaisdell, New York, 1964. Reprinted by Dover, New York, 1975. (Cited on p. xv.)

55. C. G. J. Jacobi. Über ein leichtes Verfahren die in der Theorie der Säculärstörungen vorkommenden Gleichungen numerisch aufzulösen. *J. Reine Angew. Math.*, 30:51–94, 1846. (Cited on pp. 486, 487.)

56. W. Kahan. Accurate eigenvalues of a symmetric tridiagonal matrix. Technical Report CS 41, Computer Science Department, Stanford University, 1966. (Cited on p. 490.)

57. D. Kressner. *Numerical Methods for General and Structured Eigenproblems*. Springer, New York, 2005. (Cited on pp. 405, 511.)

58. D. Kressner. The periodic QR algorithm is a disguised QR algorithm. *Linear Algebra Appl.*, 417:423–433, 2006. (Cited on pp. 405, 511, 511.)

59. V. N. Kublanovskaya. On some algorithms for the solution of the complete eigenvalue problem. *USSR Comput. Math. and Math. Phys.*, 1:637–657, 1962. (Cited on pp. 434, 497.)

60. P. Lancaster and M. Tismenetsky. *The Theory of Matrices*. Academic Press, Second edition, 1985. (Cited on pp. 332, 335, 577.)

61. C. Lanczos. An iteration method for the solution of the eigenvalue problem of linear differential and integral operators. *J. Res. Nat. Bur. Stand.*, 45:255–282, 1950. (Cited on p. 447.)

62. A. N. Langville and C. D. Meyer. *Google's PageRank and Beyond: The Science of Search Engine Rankings*. Princeton University Press, 2006. (Cited on p. 319.)

63. C. L. Lawson and R. J. Hanson. *Solving Least Squares Problems*. Prentice-Hall, Englewood Cliffs, NJ, 1974. (Cited on pp. xv, 221.)

64. R. B. Lehoucq, D. C. Sorensen, and C. Yang. *ARPACK Users' Guide: Solution of Large-Scale Eigenvalue Problems with Implicitly Restarted Arnoldi Methods*. SIAM, Philadelphia, 1998. http://www.caam.rice.edu/software/ARPACK/UG/. (Cited on p. 457.)

65. C. B. Moler and G. W. Stewart. An algorithm for generalized matrix eigenvalue problems. *SIAM J. Numer. Anal.*, 10:241–256, 1973. (Cited on pp. 535, 538, 540.)

66. M. L. Overton. *Numerical Computing with IEEE Floating Point Arithmetic*. SIAM, Philadelphia, 2001. (Cited on p. 141.)

67. C. C. Paige, M. Rozložnik, and Z. Strakoš. Modified gram-schmidt (mgs), least squares, and backward stability of mgs-gmres. *SIAM J. Matrix Anal. Appl.*, 28:264–284, 2006. (Cited on p. 595.)

68. C. C. Paige and M. Saunders. Solution of sparse indefinite systems of linear equations. *SIAM J. Numer. Anal.*, 12:617–629, 1975. (Cited on pp. 622, 623.)

69. B. N. Parlett. *The Symmetric Eigenvalue Problem*. Prentice-Hall, Englewood Cliffs, NJ, 1980. Reprinted by SIAM, Philadelphia, 1997. (Cited on pp. xv, 329, 365, 378, 533, 534.)

70. B. N. Parlett. The new qd algorithm. *Acta Numerica*, 4:459–491, 1995. (Cited on pp. 496, 497.)

71. B. N. Parlett and I. S. Dhillon. Relatively robust representations of symmetric tridiagonals. *Linear Algebra Appl.*, 309:121–151, 2000. (Cited on pp. 490, 501.)

72. B. N. Parlett and C. Reinsch. Balancing a matrix for calculation of eigenvalues and eigenvectors. *Numer. Math.*, 13:293–304, 1969. also published as contribution II/11 in [105]. (Cited on p. 342.)

73. J.L. Rigal and J. Gaches. On the compatibility of a given solution with the data of a linear system. *J. ACM*, 14:543–548, 1967. (Cited on p. 595.)

74. H. Rutishauser. Der Quotienten-Differenzen-Algorithmus. *Z. angew. Math. Physik*, 5:233–251, 1954. (Cited on p. 497.)

75. H. Rutishauser. *Der Quotienten-Differenzen-Algorithmus*. Number 7 in Mitt. Inst. angew. Math. ETH. Birkhäuser, Basel, 1957. (Cited on p. 497.)

76. H. Rutishauser. Solution of eigenvalue problems with the LR-transformation. *Nat. Bur. Standards Appl. Math. Series*, 49:47–81, 1958. (Cited on p. 497.)

77. Y. Saad and M. Schultz. GMRES: a generalized minimal residual algorithm for solving nonsymmetric linear systems. *SIAM J. Sci. Statist. Comput.*, 7:856–869, 1986. (Cited on p. 622.)

78. R. D. Skeel. Iterative refinement implies numerical stability for Gaussian elimination. *Math. Comp.*, 35:817–832, 1980. (Cited on p. 180.)

79. G. L. G. Sleijpen and H. A. van der Vorst. A Jacobi-Davidson iteration method for linear eigenvalue problems. *SIAM J. Matrix Anal. Appl.*, 17:401–425, 1996. (Cited on p. 468.)

80. B. F. Smith, P. E. Bjørstad, and W. D. Gropp. *Domain Decomposition: Parallel Multilevel Methods for Elliptic Partial Differential Equations*. Cambridge University Press, 1996. (Cited on pp. 573, 601.)

81. B. T. Smith et al. *Matrix Eigensystem Routines — EISPACK Guide*. Springer-Verlag, New York, Second edition, 1976. (Cited on p. xv.)

82. P. Sonneveld. CGS, a fast Lanczos-type solver for nonsymmetric linear systems. *SIAM J. Sci. Statist. Comput.*, 10:36–52, 1989. (Cited on p. 624.)

83. G. W. Stewart. Error and perturbation bounds for subspaces associated with certain eigenvalue problems. *SIAM Review*, 15:727–764, 1973. (Cited on pp. 416, 478, 478, 479.)

84. G. W. Stewart. *Introduction to Matrix Computations*. Academic Press, New York, 1973. (Cited on p. xv.)

85. G. W. Stewart. *Matrix Algorithms, Volume I: Basic Decompositions*. SIAM, Philadelphia, 1998. (Cited on p. ix.)

86. G. W. Stewart. A Krylov-Schur algorithm for large eigenvalue problems. *SIAM J. Matrix Anal. Appl.*, 23:601–614, 2001. (Cited on pp. 462, 463.)

87. G. W. Stewart. *Matrix Algorithms, Volume II: Eigensystems*. SIAM, Philadelphia, 2001. (Cited on p. ix.)

88. V. Strassen. Gaussian elimination is not optimal. *Numer. Math*, 13:354–356, 1969. (Cited on pp. 11, 107.)

89. J. C. Strikwerda. *Finite Difference Schemes and Partial Differential Equations*. Chapman and Hall, New York, 1989. (Cited on p. 553.)

90. A. E. Taylor and D. C. Lay. *Introduction to Functional Analysis*. Krieger, Malabar, FL, Second edition, 1986. (Cited on pp. 246, 311.)

91. L. N. Trefethen and D. Bau, III. *Numerical Linear Algebra*. SIAM, Philadelphia, 1997. (Cited on pp. ix, 164, 170.)

92. A. M. Turing. Rounding-off errors in matrix processes. *Quart. J. Appl. Math.*, 1:287–308, 1948. (Cited on p. 123.)

93. E. E. Tyrtyshnikov. *A Brief Introduction to Numerical Analysis*. Birkhäuser, Boston, 1997. (Cited on p. 472.)

94. H. A. van der Vorst. Bi-CGSTAB: A fast and smoothly converging variant of Bi-CG for the solution of nonsymmetric linear systems. *SIAM J. Sci. Statist. Comput.*, 13:631–644, 1992. (Cited on p. 624.)

95. P. Van Dooren. The computation of Kronecker's canonical form of a singular pencil. *Linear Algebra Appl.*, 27:103–141, 1979. (Cited on p. 532.)

96. R. S. Varga. *Matrix Iterative Analysis*. Prentice-Hall, Englewood Cliffs, NJ, 1962. (Cited on pp. 571, 573, 573.)

97. R. C. Ward. The combination shift QZ algorithm. *SIAM J. Numer. Anal.*, 12:835–853, 1975. (Cited on p. 540.)

98. D. S. Watkins. *Fundamentals of Matrix Computations*. John Wiley and Sons, New York, First edition, 1991. (Cited on p. 486.)

99. D. S. Watkins. Performance of the QZ algorithm in the presence of infinite eigenvalues. *SIAM J. Matrix Anal. Appl.*, 22:364–375, 2000. (Cited on p. 540.)

100. D. S. Watkins. Product eigenvalue problems. *SIAM Rev.*, 47:3–40, 2005. (Cited on pp. 405, 511.)

101. D. S. Watkins. *The Matrix Eigenvalue Problem: GR and Krylov Subspace Methods*. SIAM, Philadelphia, 2007. (Cited on pp. 335, 365, 371, 373, 374, 374, 374, 385,

402, 405, 416, 422, 422, 427, 430, 431, 433, 437, 437, 462, 472, 475, 478, 479, 480, 495, 511, 577.)

102. D. S. Watkins and L. Elsner. Convergence of algorithms of decomposition type for the eigenvalue problem. *Linear Algebra Appl.*, 143:19–47, 1991. (Cited on pp. 365, 422, 427, 431, 437, 437.)

103. J. H. Wilkinson. *The Algebraic Eigenvalue Problem,*. Clarendon Press, Oxford University, 1965. (Cited on pp. xv, 208, 472, 475, 476, 478, 483.)

104. J. H. Wilkinson. Kronecker's canonical form and the QZ algorithm. *Linear Algebra Appl.*, 28:285–303, 1979. (Cited on p. 532.)

105. J. H. Wilkinson and C. Reinsch, editors. *Handbook for Automatic Computation, Volume II, Linear Algebra*. Springer-Verlag, New York, 1971. (Cited on pp. xv, 380, 629, 631.)

106. S. J. Wright. A collection of problems for which Gaussian elimination with partial pivoting is unstable. *SIAM J. Statist. Comput.*, 14:231–238, 1993. (Cited on p. 164.)

107. K. Wu and H. Simon. Thick-restart Lanczos method for large symmetric eigenvalue problems. *SIAM J. Matrix Anal. Appl.*, 22:602–616, 2000. (Cited on pp. 462, 463.)

108. D. M. Young. *Iterative Solution of Large Linear Systems*. Academic Press, New York, 1971. (Cited on pp. 571, 573, 573.)

Index

Index of MATLAB Terms

643

PURE AND APPLIED MATHEMATICS

A Wiley-Interscience Series of Texts, Monographs, and Tracts

Founded by RICHARD COURANT
Editors Emeriti: MYRON B. ALLEN III, DAVID A. COX, PETER HILTON,
HARRY HOCHSTADT, PETER LAX, JOHN TOLAND

*Now available in a lower priced paperback edition in the Wiley Classics Library.
†Now available in paperback.

*Now available in a lower priced paperback edition in the Wiley Classics Library.
†Now available in paperback.

Printed and bound by CPI Group (UK) Ltd, Croydon, CR0 4YY

16/04/2025

14658367-0004